Contemporary Astronomy

Contemporary Astronomy

Third Edition

Jay M. Pasachoff
Field Memorial Professor of Astronomy
Director of the Hopkins Observatory

Williams College
Williamstown, Massachusetts

 SAUNDERS GOLDEN SUNBURST SERIES

SAUNDERS COLLEGE PUBLISHING
Philadelphia   New York   Chicago
San Francisco   Montreal   Toronto
London   Sydney   Tokyo   Mexico City
Rio de Janeiro   Madrid

Address orders to:
383 Madison Avenue
New York, NY 10017

Address editorial correspondence to:
West Washington Square
Philadelphia, PA 19105

Text Typeface: 10/12 Times Roman
Compositor: The Clarinda Company
Acquisitions Editor: John Vondeling
Developmental Editor: Lloyd Black
Project Editors: Patrice L. Smith and Robin C. Bonner
Copyeditor: Janis Moore
Art Director: Carol Bleistine
Art/Design Assistant: Virginia A. Bollard
Text Design: William Boehm
Cover Design: Lawrence R. Didona
Text Artwork: Linda Maugeri and Larry Ward
Production Manager: Tim Frelick
Assistant Production Manager: Maureen Iannuzzi

Cover Photo: Rho Ophiuchi is in the blue reflection nebula at top and Antares is in the
 nebula at bottom, alongside the globular cluster M4. The reddish nebulosity at lower
 right surrounds sigma Oph.
Cover credit: © 1979 Royal Observatory, Edinburgh

**Library of Congress Cataloging in
Publication Data**

Pasachoff, Jay M.
 Contemporary astronomy.

 Includes bibliographies and index.

 1. Astronomy. I. Title.
QB45.P29 1984 520 84–10590
ISBN 0–03–071641–1

ISBN 0–03–71641–1

 6   032   987654

CBS COLLEGE PUBLISHING
Saunders College Publishing
Holt, Rinehart and Winston
The Dryden Press

Preface

*A*stronomy, as a science, combines the best of new and old traditions. Every year we find something new or unexpected: a pulsar sending radio pulses 642 times each second, x-ray and gamma-ray bursts, black holes in the center of our galaxy and in the Large Magellanic Cloud, volcanoes on Io and on Venus, interacting galaxies and quasars, and evidence that the universe may have inflated very rapidly during its first second of existence are but a few examples from recent times. At the same time, astronomy has a rich and fascinating history.

I have attempted, in this book, to describe the state of astronomy as it is now. I have tried to fit the latest discoveries in their places alongside established results in order to give a contemporary picture of the state of our science.

Contemporary Astronomy is written for students with no background in mathematics or physics. It discusses astronomy in non-mathematical terms. In an attempt to make the material easier to understand, I have followed educational theory, taking findings of Piaget to heart and thus paying special attention to developing material from the concrete to the abstract, while also considering the learning theories of Bruner. (I discuss this further in the Teacher's Guide.) The many examples of celestial objects demonstrating different phenomena should also help comprehension.

In my own classes I have found that many students take astronomy because they have heard a bit about black holes or some other fascinating new phenomenon and want to know more. I have thus organized this book to plunge right into real astronomy from the beginning. This differs from the traditional approach of first presenting many chapters of history of astronomy, philosophy of science, and basic physics. These topics are treated where appropriate throughout the text.

The book, after an introductory Part I, begins with the stars (Part II). After all, so much of astronomy deals with them, and the methods we use in studying the stars are generally applicable throughout much of astronomy. Also, students want to get to pulsars and black holes as soon as possible. These objects are included in Part III on stellar evolution, which is a main theme of the book. This edition contains discussions of new models for planetary nebulae and for supernovae. Then we consider our own galaxy (Part IV) and continue outward through other galaxies, quasars, and cosmological consideration of the universe as a whole (Part V). There are recent jumps in our understanding of quasars; the earliest moments of the universe and the "inflationary universe" model are also discussed. The chapters on the solar system, with a chapter on the possibility of finding extra-terrestrial life, close the book (Part VI). (These solar-system chapters, however, can be taken up earlier in the course if desired.)

Contemporary Astronomy includes a wide range of topics that are of current interest to astronomers and in which current research is very active. My point of view is that we astronomers find astronomy fascinating, and that students will find it so too if the reasons for our excitement are explained. I have attempted to cover all fields of astronomy, without major gaps.

As part of my attempt to show students why astronomy is exciting instead of just a collection of facts, I have described the recent space exploration of each planet in somewhat chronological terms while bringing in the point of view of comparative planetology. Thus the students can see not only what we know about the planets at present but also how modern research has changed our views. This should give a base for keeping up with still newer research in years to come, and an appreciation of what has been going on recently.

A major goal of this book is to describe astronomy in clear, thorough, understandable, and colloquial terms. A major theme is the expansion of our senses to all parts of the spectrum. We see time and again throughout the entire book how observations of gamma rays, x-rays, ultraviolet light, infrared light, and radio waves are fit together with observations of visible light and with theory to improve our knowledge of the universe and the objects in it. Such new results appear throughout the entire book, rather than being segregated in a few chapters. The exciting survey of the whole sky in the infrared from the IRAS spacecraft is an example of new material covered in this edition.

This third edition represents a consolidation and reworking of the second edition to streamline the book and to make it easier to use in one-semester courses. I have incorporated almost all the many suggestions sent in by professors and students; I have also considered the results of a survey carried out by the publisher and I thank those who responded. I have continued the many teaching aids that were well received, such as lists of aims, the chapter summaries, and the numbering of all sections and subsections. I have increased the number of questions and problems at the end of the chapters by over 20 percent, have added lists of key words to aid in studying, and have better singled out the mathematical sections so that they can be assigned or not at the teacher's discretion. At the suggestion of users, I have included additional numerical questions and provided answers for selected questions so that students can better check their understanding while studying. The questions are purposely varied to include some that can be answered merely by reading the text and others that require more thought; a test bank including both word and multiple-choice questions is part of the Teacher's Guide.

Of course, the whole book has been brought up to date. I have also added the latest black-and-white and color photographs.

At no time should students find themselves memorizing material instead of understanding concepts. The summaries of each chapter act as an aid in determining just what is most important to remember and to understand. The glossary and detailed index should also help students find references and explanations. The glossary contains definitions of all key words. (Key words for Part openers appear with the following chapter's key words.) I have continually cited the appendices in the text, to help guide students to them.

My stress has continually been on how to understand what is going on, and why we astronomers think or act as we do. My criterion has been to include mostly material that can be remembered for years and that provides a basic understanding of the topic, rather than facts that are merely learned for examinations and soon forgotten. In order to give a feel of what it is like to be an astronomer, I have at several places in the text described just what an astronomer does in different circumstances, such as observing at a large telescope or with a satellite. I myself have been fortunate to be able to observe with a variety of optical, radio, and space telescopes in this country and abroad, and I am glad to have the opportunity to describe the excitement an observer feels.

In view of the importance of governmental support to scientific research, and the need for everyone to be informed about the value of such research, I have described from time to time the benefits that might accrue to society from certain types of astronomical research.

Acknowledgments

The publishers and I have placed a heavy premium on accuracy and have made certain that the manuscript and proof have been read not only by students for clarity and style but also by several astronomers for their professional comments.

Readers of particular sections in the areas of their professional expertise include J. Craig Wheeler (University of Texas), supernovae; Robert Kirshner (University of Michigan), supernovae; Karen Kwitter (Williams College), planetary nebulae; Marek Demianski (Williams College), cosmology; Yervant Terzian (Cornell University), principle of equivalence; Paul Steinhardt (University of Pennsylvania), inflationary universe; Alan R. Guth (MIT), inflationary universe; R. Edward Nather (University of Texas), novae; James Houck (Cornell University), infrared; Clark Chapman (Planetary Science Institute), comets and asteroids. I particularly thank Paul Steinhart (University of Pennsylvania) for his detailed work with me in drafting an explanation of the inflationary universe that is both understandable and accurate. I thank Gustav Tammann (University of Basel), John Huchra (Center for Astrophysics), and Gerard de Vaucouleurs (University of Texas) for their comments on my descriptions of galaxies and the expansion of the universe.

Readers of the entire third edition manuscript include Laurence A. Marschall (Gettysburg College); Anne G. Young (Rochester Institute of Technology); W. N. Hubin (Kent State University); David L. Talent (Abilene Christian University).

The contributions made by readers of the prior edition and its alternative versions, and by scientists who have commented on parts of the manuscript, are lasting ones. I thank Thomas T. Arny, James G. Baker, Laszlo Baksay, Bruce E. Bohannan, Kenneth Brecher, Bernard Burke, James W. Christy, Martin Cohen, Peter Conti, Lawrence Cram, Dale Cruikshank, Morris Davis, Raymond Davis, Jr., Dennis de Cicco, Gerard de Vaucouleurs, Richard B. Dunn, John A. Eddy, James L. Elliot, Farouk El-Baz, David S. Evans, J. Donald Fernie, William R. Forman, Peter V. Foukal, George D. Gatewood, John E. Gaustad, Tom Gehrels, Riccardo Giacconi, Owen Gingerich, Stephen T. Gottesman, Jonathan E. Grindlay, Ian Halliday, U. O. Herrmann, Robert F. Howard, H. W. Ibser, Christine Jones, Bernard J. T. Jones, Agris Kalnajs, David E. Koltenbah, Jerome Kristian, Edwin C. Krupp, Karl F. Kuhn, Marc L. Kutner, Karen B. Kwitter, John Lathrop, Lawrence S. Lerner, Jeffrey L. Linsky, Sarah Lee Lippincott, Bruce Margon, Brian Marsden, Janet Mattei, R. Newton Mayall, Everett Mendelsohn, George K. Miley, Freeman D. Miller, Alan T. Moffet, William R. Moomaw, David D. Morrison, David Park, Carl B. Pilcher, James B. Pollack, B. E. Powell, Edward L. Robinson, Herbert Rood, Maarten Schmidt, David N. Schramm, Leon W. Schroeder, Richard L. Sears, P. Kenneth

Seidelmann, Maurice Shapiro, Joseph I. Silk, Lewis E. Snyder, Theodore Spickler, Hyron Spinrad, Alan Stockton, Robert G. Strom, Jean Pierre Swings, Eugene Tademaru, Joseph H. Taylor, Jr., Joe S. Tenn, David Theison, Laird Thompson, M. Nafi Toksöz, Juri Toomre, Kenneth D. Tucker, Brent Tully, Barry Turner, Jurrie van der Woude, Peter van de Kamp, Joseph Veverka, Gerald J. Wasserburg, Leonid Weliachew, Ray Weymann, John A. Wheeler, J. Craig Wheeler, Ewen A. Whitaker, Reinhard A. Wobus, LeRoy A. Woodward, and Susan Wyckoff.

I am grateful to Nancy Pasachoff Kutner for her excellent work on the index.

I thank many people at Saunders College Publishing for their efforts on my books. John J. Vondeling and Lloyd Black merit special thanks for their continued support. Patrice Smith, Robin Bonner, Carol Bleistine, Tim Frelick, Virginia Bollard, and Margaret Mary Kerrigan worked hard on many aspects of production. New artwork was drawn by Linda Maugeri and Larry Ward. Many of the first edition drawings were executed from my sketches by George Kelvin of Science Graphics.

I remember fondly the influence that Donald H. Menzel of the Harvard College Observatory had on my career over a period of many years.

I appreciate the editorial assistance in Williamstown of Susan Welsch, Karen Kowitz, and Susan Shepard, and in Honolulu of Martine Westermann.

Various members of my family have provided vital and valuable editorial services. My father contributed so much to the book over the years. I also appreciate the work of my mother and my wife. Now that our daughters Eloise and Deborah are nine and seven, respectively, they are writing a lot on their own, but they still have time to help me with this book.

Notes to Teachers

The *Teacher's Guide to Contemporary Astronomy,* 3rd edition, contains possible syllabi for courses of different lengths, labs, tests, and answers to questions, and lists films, tapes, and other audio and visual aids for use as supplementary materials. The publishers and I are making available selected artwork and photographs in the format of overhead transparencies that can be projected in class. For information about the overhead transparencies, please write Textbook Marketing, Saunders College Publishing, 383 Madison Avenue, New York, NY 10017, with a copy to me.

With economics a constant theme in our daily newspapers, it must be recognized that economic pressures directly affect textbook distribution. Publishers try to estimate needs more precisely so they don't have to carry stock, while bookstores try to place their orders as late as possible, often months after professors send in their orders. These two factors together sometimes lead to book shortages just before semesters begin, something that is affecting all publishers. If any professors would write (or telephone) me about any problems with their orders, I would be glad to see that the matter is followed up by humans rather than by computers and that books will be available and delivered on time.

Notes to Readers

A number of sections and boxes have been marked with asterisks (*). This indicates that they are not in the main line of discussion, and can be omitted without loss of understanding of later chapters. I have marked problems with numerical answers by †, and included the answers to some of them (marked with ††) at the bottom of the page.

After you read each chapter a first time, keeping the printed aims in mind, you should go carefully through the list of key words, trying to identify or define each one. If you cannot, look up the words in the glossary, and also find the definition that appears next to the word the first time it is used in the chapter. (The index may help you here.)

Next read through the summary and make certain you understand each point; I have included only the points that I consider most important. Next, read through the chapter again especially carefully, making certain you understand how the chapter fulfills the aims printed at its beginning. Finally, answer the questions at the end of the chapter.

I am extremely grateful to all of the individuals named above for their assistance. Of course, it is I who have put this all together, and I alone am responsible for any errors that have crept through our sieve. I would appreciate hearing from readers, not just about typographical or other errors, but also with suggestions for presentation of topics or even with comments about specific points that need clarification. I invite readers to write me c/o Williams College, Hopkins Observatory, Williamstown, Mass. 01267. I promise a personal response to each writer.

Jay M. Pasachoff

Williamstown, Massachusetts

Contents Overview

Part VI The Solar System

Contents

Appendixes

Studying the Universe

Part I

*L*et us consider that the time between the origin of the universe and the year 2000 is one day. Then it wasn't until 4 P.M. that the earth formed; the first fossils are from 10 P.M. The first humans appeared only 2 seconds ago, and it is only 3/1000 second since Columbus discovered America. The year 2000 will arrive in only 1/10,000 second.

Still, the sun should shine another 8 hours; an astronomical time scale is much greater than the time scale of our daily lives. Astronomers use a wide range of technology and theories to find out about the universe, what is in it, and what its future will be. This book surveys what we have found and how we look.

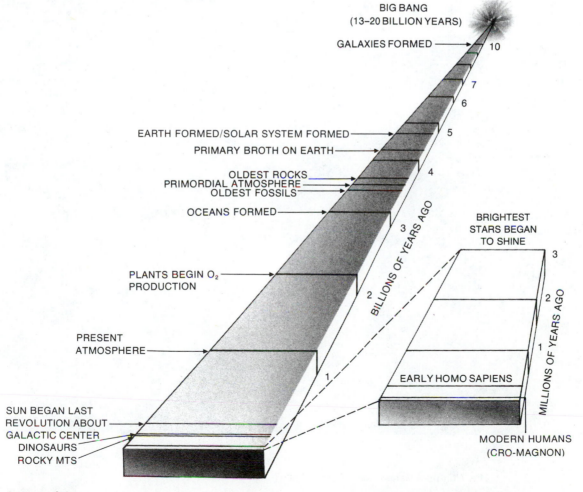

BIG BANG
(13–20 BILLION YEARS)

GALAXIES FORMED

EARTH FORMED/SOLAR SYSTEM FORMED

PRIMARY BROTH ON EARTH

OLDEST ROCKS
PRIMORDIAL ATMOSPHERE
OLDEST FOSSILS

OCEANS FORMED

PLANTS BEGIN O₂ PRODUCTION

PRESENT ATMOSPHERE

SUN BEGAN LAST REVOLUTION ABOUT GALACTIC CENTER
DINOSAURS
ROCKY MTS

BILLIONS OF YEARS AGO

BRIGHTEST STARS BEGAN TO SHINE

MILLIONS OF YEARS AGO

EARLY HOMO SAPIENS

MODERN HUMANS
(CRO-MAGNON)

A sense of time

The Whirlpool Galaxy, M51, in the constellation Canes Venatici

The Universe: An Overview   1

Aims: To get a feeling for the variety of objects in the universe and a sense of scale

The universe is a place of great variety—after all, it has everything in it! Some of the things astronomers study are of a size and scale that we humans can easily comprehend: the planets, for instance. Most astronomical objects, however, are so large and so far away that our minds have trouble grasping their sizes and distances.

Moreover, astronomers study the very small in addition to the very large. The radiation we receive from distant bodies is emitted by atoms, which are much too small to see with the unaided eye. Also, the properties of the large astronomical objects are often determined by changes that take place on a minuscule scale—that of atoms. Thus the astronomer must be an expert in the study of objects the size of atoms as well as in the study of objects the size of galaxies.

Such a variety of objects at very different distances from us or with very different properties often must be studied with widely differing techniques. Clearly, different tests are required to analyze the properties of solid particles like Martian soil (analyzed using equipment in a spacecraft sitting on the Martian surface) than are required to study the light or radio waves from a gaseous body like a quasar deep in space in order to interpret its composition. However, one unifying method does link much of astronomy. This method is *spectroscopy,* a procedure that analyzes components of the light or other radiation that we receive from distant objects and studies these components in detail. Throughout this book, we shall return to spectroscopic methods time and again, to study not only visible light but also other types of radiation. Not all of astronomy is spectroscopy, of course; astronomers also gain important information through other methods. They may make images—pictures—of, for example, planets, the sun, and galaxies, study the variation of the amount of light coming from a star or a galaxy over time, or visit the moon and planets.

The explosion of astronomical research in the last few decades has been fueled by our new ability to study radiation other than light—gamma rays, x-rays, ultraviolet radiation, infrared radiation, and radio waves. Astronomers' use of their new abilities to study such radiation is a major theme of this book. All the kinds of radiation together make up the *electromagnetic spectrum,* which will be discussed in Chapter 2. As we shall see there, we can think of radiation as waves, and all the types of radiation have similar properties except for the length of the waves. Still, although x-rays and visible light may be similar, our normal experiences tell us that very different techniques are necessary to study them.

The earth's atmosphere shields us from most kinds of radiation, though light waves and radio waves do penetrate the atmosphere. Over the the last 50 years, radio astronomy has become a major foundation of our astronomical knowledge. For the last 25 years, we have been able to send satellites into orbit outside the earth's atmosphere, and we are no longer limited to the study of radio and visible (light) radiation. Many of the fascinating discoveries of this decade—the probable observation of a black hole, for example—were made because of our newly extended senses. We will discuss how astronomers use all parts of the spectrum to help us understand the universe. We will see also how we get information from direct sampling of bodies in our solar system (as we shall discuss in Part IV) and from cosmic rays (particles whizzing through space, which we shall discuss in Section 10.3). Perhaps one day we will also be able to observe gravitational waves (as we shall discuss in Section 10.10), whose existence is predicted by Einstein's general theory of relativity (Section 7.11).

1.1 A Sense of Scale

Let us try to get a sense of scale of the universe, starting with sizes that are part of our experience and then expanding toward the infinitely large. In the margin, we can keep track of the size of our field of view as we expand in powers of 100: each diagram will show a square 100 times greater on a side.

We shall use the metric system, which is commonly used by scientists. The basic unit of length is the meter, which is equivalent to 39.37 inches, slightly more than a yard. Prefixes are used (Appendix 1) in conjunction with the word "meter," abbreviated "m," to define new units. The most frequently used prefixes are "milli-," meaning 1/1000, "centi-," meaning 1/100, and "kilo-," meaning 1000 times. Thus 1 millimeter is 1/1000 of a meter, or about .04 inch, and a kilometer is 1000 meters, or about 5/8 mile. We will keep track of the powers of 10 by which we multiply 1 m by writing the number of tens we multiply together as an exponent; 1000 m, for example, is 10^3 m.

We can also keep track of distance in units that are based on the length of time that it takes light to travel. The speed of light is, according to Einstein's 1905 special theory of relativity (Section 7.11), the greatest speed that is physically attainable. Light travels at 300,000 km/s* (186,000 miles/s), fast enough to circle the earth 7 times in a single second. Even at that fantastic speed, we shall see that it would take years for us to reach the stars. Similarly, it has taken years for the light we see from stars to reach us, so we are really seeing the stars as they were years ago. In a sense, we are looking backward in time. The distance that light travels in a year is called a *light year;* note that the light year is a unit of length rather than a unit of time even though the term "year" appears in it.

Literary scholars could say that a "milli-Helen" is a unit of beauty sufficient to launch one ship.

The *distance* travelled by an object is equal to the *rate* at which the object is travelling (its velocity) times the *time* spent travelling ($d = vt$).

Example: How far does light travel in 1 hour? *Answer:* 1 hour = 3,600 seconds. $D = vt =$ 300,000 km/s $\times$ 3600 s = 1,080,000,000 km (just over 1 billion km).

Example: How fast does light travel, expressed in the unit light years per year? *Answer:* Transform the equation to read "$v = d/t$." Then $v = 1$ light year/1 year = 1 light year/year.

---

*SI, Système International, is the international form of the metric system now in use. In SI units, *km* stands for "kilometer" and *s* stands for "second." A minus sign in the exponent stands for "one over," or "divided by." Thus s^{-1} means "1 over seconds" or "per second." A slash is equivalent to an exponent of minus 1. Thus km s^{-1} = km/s, and both are read "kilometers per second."

1.1a Survey of the Universe

Figure 1–1　1 mm = 0.1 cm

Let us begin our journey through space with a view of something 1 mm across. Here we see a velvety tree ant (Fig. 1–1), observed through a scanning electron microscope. Every step we take will show a region 100 times larger in diameter than the previous picture.

A square 100 times larger on each side is 10 centimeters × 10 centimeters. (Since the area of a square is the length of a side squared, the area of a 10 cm square is 10,000 times the area of a 1 mm square.) The area encloses a flower (Fig. 1–2).

Figure 1–2　10 cm = 100 mm

Figure 1–3　10 m = 1000 cm

As we move far enough away to see an area 10 meters on a side, we are seeing an area approximately that taken up by half a tennis court (Fig. 1–3).

A square 100 times larger on each side is now 1 kilometer square, about 250 acres. An aerial view of several square blocks in New York City shows how big an area this is (Fig. 1–4).

Figure 1–4　1 km = 10^3 m

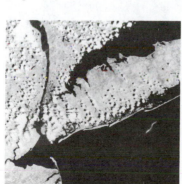

Figure 1–5　100 km = 10^5 m

The next square, 100 km on a side, encloses a major city, New York, and some of its suburbs. Note that though we are still bound to the limited area of the earth, the area we can see is increasing rapidly (Fig. 1–5).

A square 10,000 km on a side covers nearly the entire earth (Fig. 1–6).

Figure 1–6   10,000 km = 10⁷ m

Figure 1–7   1,000,000 km = 10^9 m = 3 lt sec

When we have receded 100 times farther, we see a square 100 times larger in diameter: 1 million kilometers across. It encloses the orbit of the moon around the earth (Fig. 1–7). We can measure with our wristwatches the amount of time that it takes light to travel this distance. If we were carrying on a conversation by radio with someone at this distance, there would be pauses of noticeable length after we finished speaking before we heard an answer. This is because radio waves, even at the speed of light, take that amount of time to travel. Astronauts on the moon have to get used to these pauses when speaking to earth. Laser pulses from earth bounced off the moon to find the distance to the moon (Appendices 2 and 4) take a noticeable time to return. This photograph was taken by the Voyager 1 spacecraft en route to Jupiter and Saturn. Eastern Asia, the western Pacific Ocean, and part of the Arctic are on the illuminated portion of the earth (bottom). Because the moon is many times fainter than the earth, it was artificially brightened in the computer by a factor of three relative to the earth so that it would show up better in this print, though it still appears faint.

When we look on from 100 times farther away still, we see an area 100 million kilometers across, 2/3 the distance from the earth to the sun. We can now see the sun and the two innermost planets in our field of view (Fig. 1–8). Our spacecraft have now visited or passed the moon, Mercury, Venus, Mars, Jupiter, Saturn, and their moons; the results will be discussed at length in Part VI.

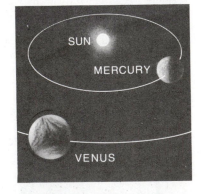

Figure 1–8 10^{11} m = 5 lt min

Figure 1–9 10^{13} m = 8 lt hrs

An area 10 billion kilometers across shows us the entire solar system in good perspective. It takes light 8 hours to travel this distance. The outer planets have become visible and are receding into the distance as our journey outward continues (Fig. 1–9). This artist's conception shows a Voyager spacecraft near Saturn.

From 100 times farther away, we see little that is new. The solar system seems smaller and we see the vastness of the empty space around us. We have not yet reached the scale at which another star besides the sun is in a cube of this size (Fig. 1–10).

ORBIT OF PLUTO
5.9 BILLION KM

Figure 1–10 10^{15} m = 38 lt days

• BARNARD'S
STAR

• SUN

α CENTAURI

PROXIMA
CENTAURI

LALANDE 21185

WOLF 359 •

Figure 1–11 10^{17} m = 10 ly

As we continue to recede from the solar system, the nearest stars finally come into view. We are seeing an area 10 light years across, which contains only a few stars (Fig. 1–11), most of whose names are unfamiliar (Appendix 7). Part II of this book discusses the properties of the stars.

By the time we are 100 times farther away, we can see a fragment of our galaxy, the Milky Way Galaxy (Fig. 1–12). We see not only many individual stars but also many clusters of stars and many areas of glowing, reflecting, or opaque gas or dust called nebulae. There is a lot of material between the stars (most of which is invisible to our eyes) that can be studied with radio telescopes on earth or in infared, ultraviolet, or x-rays with telescopes in space. Part IV of this book is devoted to the study of our galaxy and its contents.

Figure 1–12 10^{19} m = 10^3 ly

Figure 1–13 10^{21} m = 10^5 ly

In a field of view 100 times larger in diameter, we can now see an entire galaxy. The photograph (Fig. 1–13) shows a galaxy called M74, located in the direction of the constellation Pisces, though it is far beyond the stars in that constellation. This galaxy shows arms wound in spiral form. Our galaxy also has spiral arms, though they are wound more tightly.

Next we move sufficiently far away so that we can see an area 10 million light years across (Fig. 1–14). There are 10^{25} centimeters in 10 million light years, about as many centimeters as there are grains of sand in all the beaches of the earth. Our galaxy is in a cluster of galaxies, called the Local Group, that would take up only 1/3 of our angle of vision. In this group are all types of galaxies, which we will discuss in Chapter 14. The photograph shows part of a cluster of galaxies in the constellation Leo.

Figure 1–14 10^{23} m = 10^7 ly

Some of the units used in astronomy and the prefixes used with them are listed in Appendix 1.

$$10^0 = 1$$
$$10^1 = 10$$
$$10^2 = 100$$
$$10^3 = 1000$$

Note that anything to the zeroth power is 1 and anything to the first power is itself.

Box 1.1 Scientific Notation

In astronomy we often find ourselves writing numbers that have strings of zeros attached, so we use what is called either *scientific notation* or *exponential notation*, to simplify our writing chores. Scientific notation helps prevent making mistakes when copying long strings of numbers.

In scientific notation, which we used in Figures 1–4 to 1–14, we merely count the number of zeros, and write the result as a superscript to the number 10. Thus the number 100,000,000, a 1 followed by 8 zeros, is written 10^8. The superscript is called the *exponent*. We also say that "10 is raised to the eighth **power**." When a number is not a power of 10, we divide it into two parts: a number between 1 and 10, and a power of 10. Thus the number 3645 is written as 3.645×10^3. The exponent shows how many places the decimal point was moved to the left.

Scientific notation often simplifies calculation. When a number with an exponent is multiplied by the same number with a different exponent, the result is the same number with an exponent found by adding the two original exponents. Thus $10^5 \times 10^7 = 10^{5+7} = 10^{12}$.

Also, to add $3 \times 10^5 + 4 \times 10^4$, we must change either the 10^5 or the 10^4 in order to relate them to each other. (If we think of the 10^4 as being similar to an apple, we can think of 10^5 as being similar to a box containing 10 apples. The problem is thus similar to adding 3 boxes of apples to 4 apples; we must first convert 3 boxes of apples to 30 apples before we add the 4 apples.) Thus we change the 3×10^5 into 30×10^4 and have $30 \times 10^4 + 4 \times 10^4 = 34 \times 10^4 = 3.4 \times 10^5$.

The examples in the previous paragraph have just used the rule that multiplications and divisions are always done before additions and subtractions, unless some terms are separated by parentheses. So we do not have to explicitly write parentheses around 3×10^5 or around 4×10^4.

Besides multiplying two numbers, astronomers often perform calculations in which they want to raise to a given exponent a number that is already written in terms of an exponent. To do so, we merely multiply the two exponents. Thus, if we want to raise 10^3 to the fifth power, we easily find $(10^3)^5 = 10^{3 \times 5} = 10^{15}$. Such a problem might come up if we want to find, say, 1000^5, since 1000 can also be written as 10^3.

We can represent numbers less than one by using negative exponents. A minus sign in the exponent of a number means that the number is actually one divided by what the quantity would be if the exponent were positive. Thus $10^{-2} = 1/10^2$. One can compute the exponent that follows the minus sign by counting the number of places by which the decimal point has to be moved to the right until it is at the right of the first nonzero digit. Thus, for example, $0.000001435 = 1.435 \times 10^{-6}$, since the decimal point on the left side has to be moved six places to the right to come after the digit 1. Positive exponents, similarly, are the number of places that the decimal point has to be moved to the left to be on the right side of the first digit.

If we could see a field of view 1 billion light years across, our Local Group of galaxies would appear as but one of many clusters. It is difficult to observe on such a large scale.

Before we could enlarge our field of view another 100 times we might see a supercluster—a cluster of clusters of galaxies. We would be seeing almost to the distance of the quasars, which are the topic of Chapter 15. Quasars, the most distant objects known, seem to be explosive events in the cores of galaxies. Light from the most distant quasars observed may have taken 10 billion years to reach us on earth. We are thus looking back to times billions of years ago. Since we think that the universe began 13 to 20 billion years ago, we are looking back almost to the beginning of time.

We even think that we have detected radiation from the universe's earliest years. A combination of radio, ultraviolet, x-ray, and optical studies, together with theoretical work and experiments with giant atom smashers on earth, is allowing us to explore the past and predict the future of the universe. All this is discussed in Chapters 16 and 17.

1.2 The Value of Astronomy

Throughout history, observations of the heavens have led to discoveries that have had major impact on people. Even the dawn of mathematics may have followed ancient observations of the sky, made in order to keep track of seasons and seasonal floods in the fertile areas of the earth. Observations of the motions of the moon and the planets, which are free of such complicating terrestrial forces as friction and which are massive enough so that gravity dominates their motions, led to an understanding of gravity and of the forces that govern all motion.

We can consider the regions of space studied by astronomers as a cosmic labo-

Figure 1–15 The cluster of galaxies in the constellation Hercules.

ratory where we can study matter or radiation, often under conditions that we cannot duplicate on earth.

Many of the discoveries of tomorrow—perhaps the control of nuclear fusion or the discovery of new sources of energy, or perhaps something so revolutionary that it cannot now be predicted—will undoubtedly be based on discoveries made through such basic research as the study of astronomical systems. Considered in this sense, astronomy is an investment in our future.

The impact of astronomy on our conception of the universe has been strong through the years. Discoveries that the earth is not in the center of the universe, or that the universe has been expanding for billions of years, affect our philosophical conceptions of ourselves and our relations to space and time.

Yet most of us study astronomy not for its technological and philosophical benefits but for its grandeur and inherent interest. We must stretch our minds to understand the strange objects and events that take place in the far reaches of space. The effort broadens us and continually fascinates us all. Ultimately, we study astronomy because of its fascination and mystery.

Key Words

spectroscopy, electromagnetic spectrum, light year, scientific notation, exponential notation, exponent

Questions

1. Why do we say that our senses have been expanded in recent years?

†2. The speed of light is 3×10^5 km/s. Express this number in m/s and in cm/s.

†3. During the Apollo explorations of the moon, we had a direct demonstration of the finite speed of light when we heard ground controllers speak to the astronauts. The sound from the astronauts' earpieces was sometimes picked up by the astronauts' microphones and retransmitted to earth as radio signals, which travel at the speed of light. We then heard our words repeated. What is the time delay between the original and the "echo," assuming that no other delays were introduced in the signal? (The distance to the moon and the speed of light are given in Appendix 2.)

†4. The time delay in sending commands to the Voyager 2 spacecraft when it was near Saturn was 45 minutes. What was the distance in km from the earth to Saturn at that time? What was the distance in Astronomical Units (A. U.), where one Astronomical Unit is the average radius of the earth's orbit? (Appendix 2)

†5. The distance to the Andromeda galaxy is 2×10^6 light years. If we could travel at one-tenth the speed of light, how long would a round trip take?

†6. How long would it take to travel to Andromeda, which is 2×10^6 light years away, at 1000 km/hr, the speed of a jet plane?

7. List the following in order of increasing size: (a) light year, (b) distance from earth to sun, (c) size of Local Group, (d) size of football stadium, (e) size of our galaxy, (f) distance to a quasar.

8. Of the examples of scale in this chapter, which would you characterize as part of "everyday" experience? What range of scale does this encompass? How does this range compare with the total range covered in the chapter?

9. What is the largest of the scales discussed in this chapter that could reasonably be explored in person by humans with current technology?

†10. (a) Write the following in scientific notation: 4642; 70,000; 34.7. (b) Write the following in scientific notation: 0.254; 0.0046; 0.10243. (c) Write out the following in an ordinary string of digits: 2.54×10^6; 2.004×10^2.

†11. What is (a) $(2 \times 10^5) + (4.5 \times 10^5)$; (b) $(5 \times 10^7) + (6 \times 10^8)$; (c) $(5 \times 10^3) (2.5 \times 10^7)$; (d) $(7 \times 10^6) (8 \times 10^4)$? Write the answers in scientific notation, with only one digit to the left of the decimal point.

†12. What percentage of the age of the universe has elapsed since the appearance of *Homo sapiens* on the earth?

---

† This indicates a question requiring a numerical solution.

Topic for Discussion

What is the value of astronomy to you? How do you rank National Science Foundation (NSF) and National Aeronautics and Space Administration (NASA) funds for research with respect to other national needs? Reanswer this question when you have completed this course.

Figure 1–16 The loosely wound spiral galaxy NGC45, which contains so little interstellar dust that it is transparent, allowing us to see galaxies through it in this new photograph with the 2.5-m telescope at the Las Campanas Observatory in Chile.

The 4-meter Mayall telescope at the Kitt Peak National Observatory of the National Optical Astronomy Observatories.

Light and Telescopes

2

Aims: To understand the electromagnetic spectrum and the different types of observing techniques and instruments used in astronomy.

If you forced a group of astronomers to choose one and only one instrument to be marooned with on a desert island, they would probably choose—a computer. The notion that astronomers spend most of their time at telescopes is far from the case in modern times. Still, telescopes have been and continue to be very important to the development of astronomy, and in this chapter we will discuss how we study the stars with telescopes and other observational devices.

Although the word "telescope" makes most of us think of objects with lenses and long tubes, that type of telescope is only used to observe "light." Studying ordinary "visible" light, the kind we see with our eyes, is not the only way we can study the universe. We shall see in this chapter that light is only one type of *radiation*—a certain way in which energy moves through space. Other types of radiation are gamma rays, x-rays, ultraviolet light, infrared light, and radio waves. They are all fundamentally identical to ordinary light, though in practice we must usually use different methods to observe them.

First, however, we discuss the properties of radiation that enable scientists to study the universe; after all, we cannot touch a star, and though we have brought bits of the moon back to earth for study, we are not able to do the same for even the nearest planets.

2.1 The Spectrum

It was discovered over 300 years ago that when ordinary light is passed through a prism, a band of color like the rainbow comes out the other side. Thus "white light" is composed of all the colors of the rainbow (Color Plate 5).

These colors are always spread out in a specific order, which has traditionally been remembered by the initials of the friendly fellow, **ROY G. BIV,** which stand for **R**ed, **O**range, **Y**ellow, **G**reen, **B**lue, **I**ndigo, **V**iolet. No matter what rainbow you watch, or what prism you use, the order of the colors never changes.

We can understand why light contains the different colors if we think of light as waves of radiation. These waves are all travelling at the same speed, 3×10^8 m/s (186,000 miles/s). This speed is normally called the *speed of light* (even though light normally travels at a slightly lower speed because the light is not usually in a perfect vacuum; "the speed of light" usually really means "the speed of light in a vacuum").

The distance between one crest of the wave to the next or one trough to the next, or in fact between any point on a wave and the similar point on the next wave,

These days, though, one no longer sees "indigo" as a separate color of the same importance as the others listed.

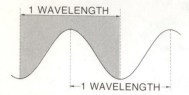

Figure 2–1 The wavelength is the length over which a wave repeats.

Chemists studying atoms join astronomers in using angstroms as a measuring unit, although purists now claim that we should use only SI units, which allow only powers of 1000 or of 1/1000 meters. In this use of the metric system, 1 angstrom = 0.1 nanometer. Astronomers won't give up the use of angstroms without a struggle.

is called the *wavelength* (Fig. 2–1). Light of different wavelengths appears as different colors. It is as simple as that. Red light has approximately 1½ times the wavelength of blue light. Yellow light has a wavelength in between the two. Actually there is a continuous distribution of wavelengths, and one color blends subtly into the next.

The wavelengths of light are very short: just a few ten-millionths of a meter. Astronomers use a unit of length called an angstrom, named after the Swedish physicist A.J. Ångstrom. One angstrom (1 Å) is 10^{-10} m. (Remember that this is 1 divided by 10^{10}, which is .000 000 000 1 m.) The wavelength of violet light is approximately 4000 angstrom units, usually written 4000 angstroms (4000 Å). Yellow light is approximately 6000 Å, and red light is approximately 6500 Å in wavelength.

The human eye is not sensitive to radiation of wavelengths much shorter than 4000 Å or much longer than 6600 Å, but other devices exist that can measure light at shorter and longer wavelengths. At wavelengths shorter than violet, the radiation is called *ultraviolet;* at wavelengths longer than red, the radiation is called *infrared.* (Note that this is not "infared," a common misspelling; "infra" is a Latin prefix that means "below.")

All these types of radiation—visible or invisible—have much in common. Many of us are familiar with the fields of force produced by a magnet—a magnetic field—and the field of force produced by a charged object like a hair comb on a dry day—an electric field. It has been known for over a century that fields that vary rapidly behave in a way that no one would guess from the study of magnets and combs. In fact, light, x-rays, and radio waves are all examples of rapidly varying electric fields and magnetic fields that have become detached from their sources and move rapidly through space. For this reason, these radiations are referred to as *electromagnetic radiation.*

We can draw the entire *electromagnetic spectrum,* often simply called the "spectrum" (plural: *spectra*), ranging from radiation of wavelength shorter than 1 Å to radiation of wavelength many meters long and longer (Fig. 2–2).

Note that from a scientific point of view there is no qualitative difference between types of radiation at different wavelengths. They can all be thought of as electromagnetic waves, waves of varying electric and magnetic fields. Light waves comprise but one limited range of wavelengths. When an electromagnetic wave has a wavelength of 1 Å, we call it an x-ray. When it has a wavelength of 5000 Å, we call it light. When it has a wavelength of 1 cm (which is 10^8 Å), we call it a radio wave. Of course, there are obvious practical differences in the methods by which we detect x-rays, light, and radio waves, but the principles that govern their existence are the same.

Figure 2–2 The electromagnetic spectrum.

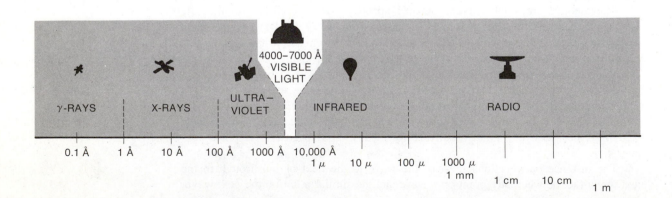

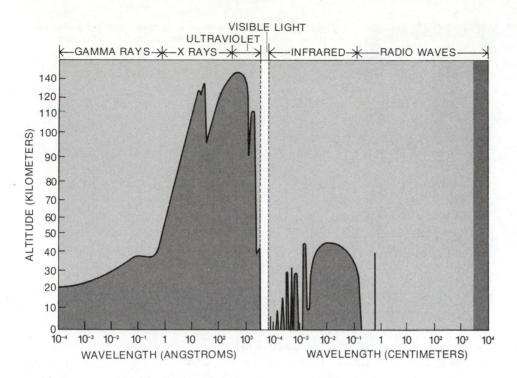

Figure 2–3 In a window of transparency in the terrestrial atmosphere, radiation can penetrate to the earth's surface. The curve specifies the altitude where the intensity of arriving radiation is reduced to half its original value. When this happens high in the atmosphere, little or no radiation of that wavelength reaches the ground. (From "Ultraviolet Astronomy" by Leo Goldberg. Copyright © 1969 by Scientific American. All rights reserved.)

Note also that light occupies only a very small portion of the entire electromagnetic spectrum. It is obvious that the new ability that astronomers have to study parts of the electromagnetic spectrum other than light waves enables us to increase our knowledge of celestial objects manyfold.

Only certain parts of the electromagnetic spectrum can penetrate the earth's atmosphere. We say that the earth's atmosphere has "windows" for the parts of the spectrum that can pass through it. The atmosphere is transparent at these windows and opaque at other parts of the spectrum. One window passes what we call "light," and what astronomers technically call *visible light,* or "the visible," or "the optical part of the spectrum." Another window falls in the radio part of the spectrum, and modern astronomy uses "radio telescopes" to detect that radiation (Fig. 2–3).

But we of the earth are no longer bound to our planet's surface; balloons, rockets, and satellites carry telescopes above the atmosphere to observe in parts of the electromagnetic spectrum that do not pass through the earth's atmosphere. By now, we have made at least some observations in each of the named parts of the spectrum. It seems strange, in view of the long-time identification of astronomy with visible observations, to realize that optical studies no longer dominate astronomy.

2.2 Spectral Lines

As early as 1666, Isaac Newton showed that sunlight is composed of all the colors of the rainbow. William Wollaston in 1804 and Joseph Fraunhofer in 1811 also studied sunlight as it was dispersed (spread out) into its rainbow of component colors (Fig. 2–4). Wollaston and Fraunhofer were able to see that at certain colors there were gaps that looked like dark lines across the spectrum at those colors (Color

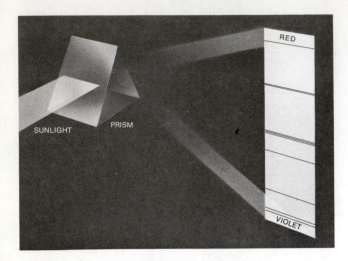

Figure 2–4 When a narrow beam of sunlight is dispersed by a prism, we see not only a continuous spectrum but also dark Fraunhofer lines.

Note that "absor**b**" ends with a "b" but "absor**p**tion" is spelled with a "p."

Stations on the radio, places in the spectrum where there is energy, represent spectral lines. In particular, they are emission lines.

We are not here giving examples of how continuous radiation is generated in a gas, though it usually results at relatively high pressure; different mechanisms apply at different temperatures and spectral ranges.

Though the exact German pronunciation is difficult for native speakers of English, the name Kirchhoff is normally pronounced in English with a "hard" *ch* (that is, like *k*): Kirk-hoff. Note the double "h."

Plate 5). These gaps are thus called *spectral lines;* the continuous radiation in which the gaps appear is called the *continuum* (pl: *continua*). The dark lines in the spectrum of the sun and in the spectra of stars are gaps that represent the diminution of electromagnetic radiation at those particular wavelengths. These dark lines are called *absorption lines;* they are also known (for the sun, in particular) as *Fraunhofer lines.* In contrast, an ordinary light bulb (an "incandescent" bulb) gives off only a continuum; it has no spectral lines.

It is also possible to have wavelengths at which there is somewhat more radiation than at neighboring wavelengths; these are called *emission lines.* We shall see that the nature of a spectrum, and whether we see emission or absorption lines, can provide considerable information about the nature of the body that was the source of light. We say that the lines are "in emission" or "in absorption."

It was discovered in laboratories on earth that patterns of spectral lines can be explained as the absorption or emission of energy at particular wavelengths by atoms of chemical elements in gaseous form. If a vapor (gaseous form) of any specific element is heated, it gives off a characteristic set of emission lines. That element, and only that element, has that specific set of spectral lines. If, on the other hand, a continuous spectrum radiated by a source of energy at a high temperature is permitted to pass through cooler vapor of any specific element, a set of absorption lines (Fig. 2–5) appears in the continuous spectrum at the same characteristic wavelengths as those of the emission lines of that element. Thus the vapor of an element through which light has passed has subtracted energy from the continuous spectrum at the set of wavelengths that is characteristic of that element (Color Plate 5). This was discovered by the German chemist Gustav Kirchhoff in 1859. If a continuous spectrum is directed first through the vapor of one absorbing element and then through the vapor of a second absorbing element, or through a mixture of the two gases, then the absorption spectrum that results will show the characteristic spectral absorption lines of both elements.

The same characteristic patterns of spectral lines that we detect on earth are observed in the spectra of stars, so we conclude basically that the same chemical elements are in the outermost layers of the stars. They absorb radiation from a continuous spectrum generated below them in the star, and thus cause the formation of absorption lines. Evidently, since each element has its own characteristic pattern of lines when it is in a certain range of conditions of temperature and density (a measure of how closely the star's matter is packed), the absorption spectrum of a star can be used (1) to identify the chemical constituents of the star's atmosphere, in other

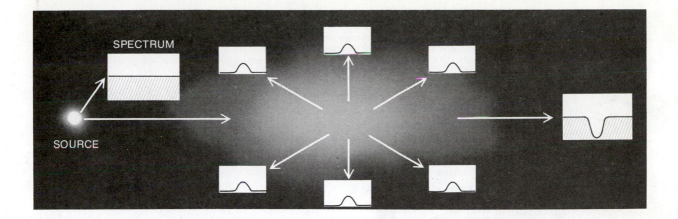

words, the types of atoms that make up the gaseous outer layers of the star, (2) to find the temperature of the surface of the star, and (3) to find the density of the radiating matter.

Not only individual atoms but also molecules, linked groups of atoms, exist in cooler bodies. (Such cooler bodies include the coolest stars, some of the gas between the stars, and the atmospheres of planets.) Molecules also have characteristic sets of absorption lines, and what we say for identifying elements goes for molecules as well.

Where does the radiation that is absorbed go? A very basic physical law called the *law of conservation of energy* says that the radiation cannot simply disappear. The energy of the radiation may be taken up in a collision of the absorbing atom with another atom. Alternatively, the radiation may be emitted again, sometimes at the same wavelength, but in random directions. Thus, fewer bits of energy proceed straight ahead at that particular wavelength than were originally heading in that direction. This leads to the appearance of an absorption line when we look from that direction.

Moreover, if one element or molecule is present in relatively great abundance, then its characteristic spectral lines will be especially strong. By observing the spectrum of a star or planet one can tell not only which kinds of atoms or molecules are present but also their relative abundances.

The method of spectral analysis is a powerful tool that can be used to explore the universe from our vantage point on earth. It tells us about the planets, the stars, and the other things in the universe. It also has many uses outside of astronomy. For example, by analyzing the spectrum, one can determine the presence of impurities in an alloy deep inside a blast furnace in a steel plant on earth. One can measure from afar the constituents of lava erupting from a volcano; indeed, colleagues and I have done so using the same spectrometer we used to study eclipses. Sensitive methods developed by astronomers trying to advance our knowledge of the universe often are put to practical uses in fields unrelated to astronomy.

Figure 2–5. When we view a source that emits a continuum through a vapor that emits emission lines, we may see absorption lines at wavelengths of the emission lines. Each of the inset boxes shows a graph of intensity (vertical axis) against wavelength (horizontal axis). Each graph is the one you would measure if you were at the tip of the arrow looking back along the arrow. Note that the view from the right shows an absorption line, while from any other angle the vapor in the center appears to be giving off an emission line. Only when you look through one source silhouetted against a hotter source do you see any absorption lines.

We shall say more about the formation of spectral lines in Section 3.3.

2.3 What a Telescope Is

How to define a telescope is a question with no simple answer, for a "telescope" to observe gamma rays may be a package of electronic sensors launched above the atmosphere, and a radio telescope may be a large number of small aerials strung over acres of landscape. We will begin, nevertheless, by discussing telescopes of the traditional types, which observe the radiation in the visible part of the spectrum. These optical telescopes are important because of the many things we have

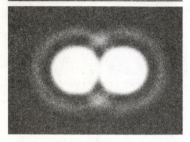

Figure 2–6 In addition to their primary use for gathering light from faint objects, telescopes are also often used to increase the resolution, our ability to distinguish details in an image. We see *(top)* a pair of point sources (sources so small or far away that they appear as points) that are not quite resolved; *(middle)* sources that have passed Dawes's limit, the condition in which they can barely be resolved; and *(bottom)* sources that are completely resolved.

learned over the years by studying visible radiation. And optical telescopes are the types of telescopes most frequently used by students and by amateur astronomers (some of whom are quite professional in their approach to the subject). Further, the principles of focusing and detecting electromagnetic radiation that were originally developed through optical observations are widely used throughout the spectrum.

Contrary to popular belief, the most important purpose for which most optical telescopes are used is to gather light. (For the next few sections I shall drop the qualifying word ''optical.'') True, telescopes can be used to magnify as well, but for the most part astronomers are interested in observing fainter and fainter objects and so must collect more light to make these objects detectable. There are certain cases where magnification is important—such as for observations of the sun or for observations of the planets—but stars are so far away that they appear as mere points of light no matter how much magnification is applied.

When we look at the sky with our naked eyes, several limitations come into play. First of all, our eyes see in only one part of the spectrum (the visible). Another limitation is that we can see only the light that passes through an opening of a certain diameter—the pupils of our eyes. In the dark, our pupils dilate so that as much light as possible can enter, but the apertures are still only a few millimeters across. Yet another limitation is that of time. Our brains distinguish a new image about 30 times a second, and so we are unable to store faint images over a long time in order to accumulate a brighter image. Astronomers overcome these additional limitations by using a telescope to gather light and a recording device, such as a photographic plate, to store the light. Other equipment may also be used, such as a spectrograph to analyze the spectral content of the light, or simple filters.

A further advantage of a telescope over the eye, or of a large telescope over a smaller telescope, is that of *resolution*, the ability to distinguish finer details in an image (Fig. 2–6). A telescope is capable of considerably better resolution than the eye, which is limited to a resolution of 1 arc min (the diameter of a dime at a distance of 60 m). (The larger the aperture accepting light, the better the resolution, and the eye's maximum opening is only a few mm across.) The aperture of each lens of binoculars is a few times larger, so binoculars can reveal details a few times finer. Thus binoculars or a telescope can distinguish the two components of a double star from each other in many cases where the unaided eye is unable to do so.

The earth's atmosphere limits us to seeing detail on celestial objects larger than roughly ½ arc sec across (the diameter of a dime at a distance of 7 km or of a human hair two football fields away). If a telescope with a collecting area 10 centimeters across can resolve double stars that are separated by 1 arc second, a telescope 20 centimeters across, twice the diameter, can resolve stars that are separated by half that angle, ½ arc sec. In principle, for light of a given wavelength, the size of the finest details resolved (the resolution) is inversely proportional to the diameter of the telescope's primary mirror or lens. By ''inversely proportional,'' we mean that as one quantity goes up, the other goes down by the same factor.

Even for large telescopes, the best resolution that can be achieved on the earth's surface on an average good night is limited by turbulence in the earth's atmosphere to about 1 arc sec. Thus increasing size no longer improves the resolution, even on the best of nights, though the advantages in terms of light-gathering power remain.

We will begin by considering the telescopes themselves, and then go on to consider the equally important devices that are used in conjunction with the telescopes.

For those of you who are photographers and who recognize the use of ''f-stops,'' the eye ranges from f/2.8 to f/22.

Actually, resolution depends not only on the aperture, but also on the wavelength of radiation being observed; the example we just considered was green light of approximately 5000 Å.

As the wavelength of radiation doubles, resolution is halved. For example, a telescope that can resolve two sources emitting 5000 Å (green) light that are 1 arc second apart could only resolve two 10,000 Å (infrared) sources if they were 2 arc secs apart. The limit of resolution for visible light, called Dawes's limit, is approximately 2×10^{-3} λ/d arc sec, when the wavelength, λ, is in Å and the diameter of a telescope, d, is in cm.

60 is a number that is conveniently divisible by many factors: 2, 3, 4, 5, 6, 10, 12, 15, 20, and 30.

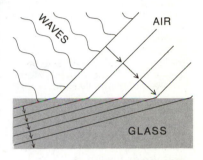

Figure 2–7 Light travels at a velocity in glass that is different from the velocity that it has in air. This leads to refraction.

2.3a Resolution

Let us consider waves of light hitting a block of glass at an angle, and draw an advancing plane of radiation (for example, the plane representing the first light that would arrive if the source were suddenly to become visible). This advancing plane is called a *wave front* (Fig. 2–7). (T.V. newscasters, similarly, talk of weather fronts.) The wave front is straight, so one side meets the glass slightly before the other side. It takes the other side of the wave front a very small additional amount of time to reach the glass, and in that time the first side has travelled a short distance through the glass. The speed of light is 50 per cent slower in the glass than it is in the air, so in that small additional time the part of the wave in the glass travels a shorter distance than the part of the wave still in the air. Because of this, by the time all the wave front has reached the glass, the wave front is moving in a different direction than it was before it started through the glass. This bending of the direction of travel of the light is called *refraction*. Put a straw in a glass of water (Fig. 2–8), a branch in a lake, or a foot in a bathtub to observe this phenomenon.

Figure 2–8 The flower stem appears both bent and displaced at the boundary between the water and the air.

2.3b Lenses and Telescopes

A suitable curved piece of glass can be made so that all the light that travels through it is bent, bringing all the rays of each given wavelength to a focus. Such a curved piece of glass is called a *lens*. The lens and cornea of your eye do a similar thing to make images of objects of the world on your retina—each point of the object is focused to a point on the image. An eyeglass lens helps your eye's lens and cornea accomplish this task.

The distance of the image from the lens on the side away from the object being observed depends in part on the distance of the object from the front side of the lens. Astronomical objects are all so far away that they are, for the purpose of forming images, as though they were infinitely far away. We say that they are "at infinity." The images of objects at infinity fall at a distance called the *focal length* behind the lens (Fig. 2–9).

A particular telescope forms an image of an area of the sky called its *field of view*. Objects that are at angles far from the center of the field of view may not be focused as well as objects in the center.

The technique of using a lens to focus faraway objects was, we think, developed in Holland in the first decade of the 17th century. It is not clear how Galileo heard of the process, but it is known that Galileo quickly bought or ground a lens and made

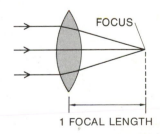

Figure 2–9 The focal length is the distance behind a lens to the point at which objects at infinity are focused. The focal length of the human eye is about 2.5 cm (1 in).

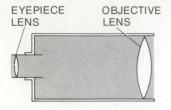

Figure 2–10 A simple refracting telescope consists of an objective lens and an eyepiece. The eyepiece shown here is a more modern type than the one used by Galileo. (Galileo's was concave—narrower at the center than at the edges—and this one is convex—wider at the center.) Modern eyepieces usually contain several pieces of carefully chosen shapes and materials. The telescope shown here, with a double convex eyepiece lens, gives an inverted image (that is, objects appear upside down).

a simple telescope that he demonstrated in 1610 to the Senate in Venice (Fig. 2–10). He put this small lens and another smaller lens at opposite ends of a tube. The second lens, called the *eyepiece*, is used to examine and to magnify the image made by the first lens, called the *objective*. Galileo's telescope was very small, but it magnified enough to impress the nobles of Venice, who had assembled to see the new invention.

Galileo turned his simple telescope on the heavens, and what he saw revolutionized not only astronomy but also much of seventeenth-century thought. He discovered, for example, that the Milky Way includes many stars, that there are mountains on the moon, that Jupiter has satellites of its own, and that Venus has phases. We shall be discussing the fundamental importance of these discoveries later on, in Section 18.5.

2.4 Refracting Telescopes

Refraction is the bending of light (or other electromagnetic radiation) when the light passes from one medium (e.g., transparent object, air, interstellar space) into another. A lens uses the property of refraction to focus light, and a telescope that has a lens as its major element is called a *refracting telescope*.

The current largest refracting telescope in the world has a lens 1 meter (40 inches) across. It is at the Yerkes Observatory in Williams Bay, Wisconsin (Figs. 2–11 and 2–12), and went into use during the 1890's.

Refracting telescopes suffer from several problems. For one thing, lenses suffer from *chromatic aberration*, the effect whereby different colors are focused at different points (Fig. 2–13). The speed of light in a substance—in glass, for example—depends on the wavelength of the light. As a result, light composed of different colors bends by different amounts, and not all wavelengths can be brought to a focus at the same point. Ingenious methods of making ''compound'' lenses of different glasses that have slightly differing properties have succeeded in reducing this problem. But even after these methods of reducing chromatic aberration have been used,

Figure 2–12 The opening of the 1-meter (40-inch) refractor of the Yerkes Observatory was the cause of much notice in the Chicago newspapers in 1893.

Figure 2–11 The Yerkes refractor, still the largest in the world. Note Albert Einstein at right of center in this 1921 picture.

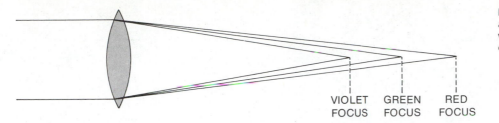

enough chromatic aberration remains to present a fundamental problem for the use of refracting telescopes.

With very large lenses, physical problems arise. It has been difficult to get a pure piece of glass sufficiently free of internal bubbles and sufficiently homogeneous throughout its volume to allow the construction of a lens larger than that at Yerkes. And a lens can be supported only from its rim where it is mounted in a telescope, since nothing may obstruct the aperture. Gravity causes the lens to sag in the middle, and this effect changes as the telescope is pointed in different directions. If the lens is made more rigid by making it thicker, this also makes it heavier and even more difficult to get a sufficiently pure lens blank. The 1-m telescope at Yerkes, for all of these reasons, remains the largest refracting telescope in the world.

2.5 Reflecting Telescopes

Reflecting telescopes are based on the principle of reflection, with which we are so familiar from ordinary household mirrors. A mirror reflects the light that hits it so that the light bounces off at the same angle at which it approached (Fig. 2–14). A flat mirror gives an image the same size as the object being reflected, but funhouse mirrors, which are not flat, cause images to be distorted.

One can construct a mirror in a shape so that it reflects incoming light to a focus. Let us consider a spherical mirror. If you were at the middle of a giant spherical mirror, in whatever direction you looked you would see your own image. What-

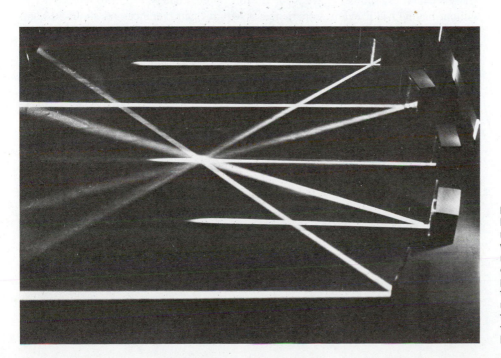

Figure 2–14 Each individual beam of light bounces off a flat mirror at the same angle at which it hits the mirror. By arranging several flat mirrors on a curve, as shown here, incoming parallel rays of light can be bent to a single point. A telescope mirror is made so that it is curved by the exact amount necessary to reflect all incoming parallel rays of light to a single point.

Figure 2–15 A spherical mirror focuses light that originates at its center of curvature back on itself, but suffers *spherical aberration* in that it does not perfectly focus light from infinity.

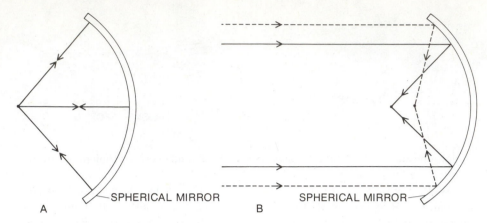

SPHERICAL MIRROR

A

SPHERICAL MIRROR

B

ever is at a spherical mirror's center is imaged at the same point, its center (Fig. 2–15). If we were to use just a portion of a sphere, it would still image whatever was at the center of its curvature back at that same point.

But the stars are far away, and so would not be at the center of curvature if we used a spherical mirror. Thus we can use a mirror that makes use of the fact that a parabola focuses *parallel light* to a point (Fig. 2–16). We say that the light from the star and planets is "parallel light" because the individual light rays are diverging by such an imperceptible amount by the time they reach us on earth that they are practically parallel (Fig. 2–17).

A parabola is a two-dimensional curve that has the property of focusing parallel rays to a point. In many cases, a telescope mirror is actually a *paraboloid*, which is the three-dimensional curve generated when a parabola is rotated around its axis of symmetry. Over a small area, a paraboloid differs from a sphere only very slightly, and one can ordinarily make a parabolic mirror by first making a spherical mirror and then deepening the center slightly.

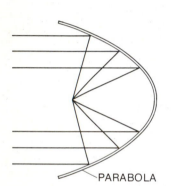

PARABOLA

Figure 2–16 A parabola focuses parallel light to a point.

The technique of making a small telescope mirror is not excessively difficult. Many amateur astronomers have made mirrors approximately 15 cm (6 inches) across without having had previous experience; it requires perseverance and about 50 hours of time. Making a large telescope mirror is another story, of course.

Reflecting telescopes have several advantages over refracting telescopes. The angle at which light is reflected does not depend on the color of the light, so chromatic aberration is not a problem for reflectors. One normally deposits a thin coat of a highly reflecting material on the front of a suitable ground and polished surface. Since light does not penetrate the surface, what is inside the telescope mirror does not matter too much. Further, one can have a network of supports all across the back of the telescope mirror, to prevent the mirror from sagging under the force of gravity. All these advantages allow reflecting telescopes to be made much larger than refracting telescopes. (Only reflecting telescopes can be used at wavelengths shorter than those of visible light, because most ultraviolet radiation does not pass through glass.)

Figure 2–17 Light rays from very distant objects are diverging so slightly by the time they reach us that we speak of *parallel light.*

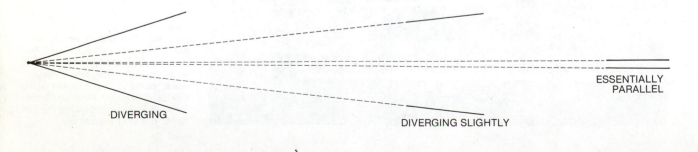

ESSENTIALLY
PARALLEL

DIVERGING

DIVERGING SLIGHTLY

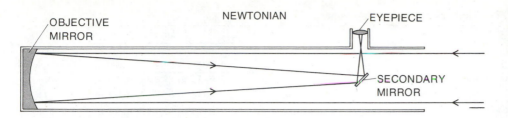

NEWTONIAN

OBJECTIVE MIRROR

EYEPIECE

SECONDARY MIRROR

Figure 2–18 A Newtonian reflector.

Figure 2–19 The replica of Newton's telescope in the visitor's gallery of the Isaac Newton telescope when it was at the Royal Greenwich Observatory at Herstmonceux, England, showing the side visible only to the astronomers.

A potential problem with small reflecting telescopes is the need to gather the reflected light without impeding the incoming light. If you put your head at the focal point of a small telescope mirror, the back of your head would block the incoming light and you would see nothing at all!

Isaac Newton got around this problem 300 years ago by putting a small diagonal mirror a short distance in front of the focal point to reflect the light so that the focus was outside the tube of the telescope. This type of apparatus, still in use today, is known as a *Newtonian* telescope (Fig. 2–18 and 2–19).

Two contemporaries of Newton invented alternative types of reflecting telescopes. G. Cassegrain, a French optician, and James Gregory, a Scottish astronomer, invented telescopes in which the light was reflected by the secondary mirror through a small hole in the center of the primary mirror. These designs are called *Cassegrainian* (or *Cassegrain*) and *Gregorian* telescopes (Fig. 2–20).

The largest telescopes have several interchangeable secondary mirrors built in so that different foci can be used at different times. The very largest telescopes are so huge that the observer can even sit in a "cage" suspended at the end of the telescope at the *prime focus*, the position of the direct reflection of the light from the primary mirror (Fig. 2–21). The largest reflecting telescopes today (Section 2.7) include the 6-m reflector in the Soviet Union (Color Plate 2), the 5-m reflector on Palomar Mountain in California, and 4-m telescopes at several locations including the Kitt Peak National Observatory in Arizona (Color Plate 1). We shall discuss in Section 2.7c some plans for even larger telescopes.

The fault of paraboloidal mirrors is that they focus light from a relatively narrow field of view, limiting astronomers' ability to study extended objects. The narrow field of view also makes it difficult at times for astronomers to find stars bright enough for "guiding," that is, stars to hold steady in view in order to photograph

The Hale telescope has been known for years as the "200-inch telescope," but the scientific journals are now converting even this sacred name to metric units. We will round off 5.08 meters to 5 meters in this book. (Note that the mirror is about 17 feet in diameter, the size of an ordinary room.) The hole in the mirror to allow use of the Cassegrain focus is 1 m (40 inches) across; it blocks only $(40/200)^2 = (1/5)^2 = 4$ per cent of the mirror's overall area. Even when the prime focus observing cage, which is 1.8 m (72 inches) across, is installed, only $(1.85)^2 = 13$ per cent of the incoming light is blocked.

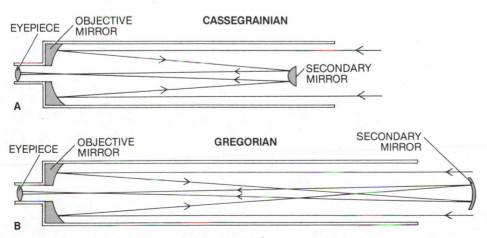

CASSEGRAINIAN

EYEPIECE

OBJECTIVE MIRROR

SECONDARY MIRROR

A

GREGORIAN

EYEPIECE

OBJECTIVE MIRROR

SECONDARY MIRROR

B

Figure 2–20 (*A*) A Cassegrain telescope has a convex secondary mirror. Cassegrains normally have short tubes relative to their mirror diameters. Most large telescopes have a Cassegrain focus. (*B*) A Gregorian telescope has a concave secondary that would be located beyond the focus. The Solar Maximum Mission, a spacecraft launched in 1980, has a Gregorian telescope.

Figure 2–21 The prime-focus cage of the 5-meter Palomar telescope. Note the observer sitting in it.

Table 2–1 Schmidt Telescopes

| Telescope or Institution | Location | Diameter | | Date Completed |
		Corrector (m)/(in)	Mirror (m)/(in)	
Karl Schwarzschild Obs.	East Germany	1.3 m/53 in	2.0 m/79 in	1960
Palomar Observatory	California	1.2 m/48 in	1.8 m/72 in	1948
U.K. Schmidt	Australia	1.2 m/48 in	1.8 m/72 in	1973
Tokyo Astronomical Obs.	Japan	1.1 m/40 in	1.5 m/60 in	1976
ESO Schmidt	Chile	1.0 m/39 in	1.6 m/63 in	1972

fainter or extended objects near them in the sky while compensating for the earth's rotation. Modern large reflectors use a version of the Cassegrain design that allows a relatively wide field of view compared to that of a paraboloid. The primary mirror has its center deepened from the shape of a paraboloid, and the secondary mirror has a complicated shape to compensate. This *Ritchey-Chrétien* design leads to a curved rather than a flat focus, but this problem can be overcome by using a small lens near the focus or by curving the film. Almost all the large telescopes constructed in the last 20 years are Ritchey-Chrétiens. So modern telescope mirrors are not usually paraboloids after all!

2.6 Schmidt Telescopes

Even Ritchey-Chrétiens are in focus over a field only about 1° across. The fact that such reflectors have only a very small field in good focus limits their usefulness in certain cases.

Bernhard Schmidt, working in Germany in about 1930, invented a type of telescope that combines some of the best features of both refractors and reflectors. In a *Schmidt camera* (Table 2–1), the main optical element is a large mirror, but it is spherical instead of being paraboloidal. Before reaching the mirror, the light passes through a thin lens, called a *correcting plate*, that distorts the incoming light in just the way that is necessary to have the spherical mirror focus it over a wide field (Fig. 2–22).

One of the largest Schmidt cameras in the world is on Palomar Mountain, in a dome near that of the 5-m telescope (Fig. 2–23). The Schmidt camera has a correcting plate 1.2 meters (48 inches) across, and a spherical mirror 1.8 meters (72 inches)

Figure 2–22 By having a nonspherical thin lens called a correcting plate, a Schmidt camera is able to focus a wide angle of sky onto a curved piece of film. Since the image falls at a location where you cannot put your eye, the image is always recorded on film. Accordingly, this device is often called a Schmidt camera rather than a Schmidt telescope.

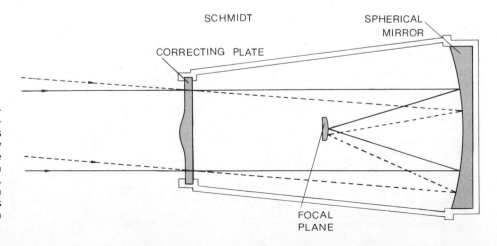

in diameter. Its field of view is about 7° across, which is much broader than the 5-m reflector's field of view (only about 2 minutes of arc across).

The field of view of the 5-m is so small that it would be hopeless to try to map the whole sky with it because it would take too long. But the 1.2-meter Schmidt has been used to map the entire sky that is visible from Palomar. The survey was carried out through both red and blue filters; 900 pairs of plates were taken over seven years, from 1949 to 1956. Sample plates from this National Geographic Society/Palomar Observatory Sky Survey are shown in Fig. 2–24. Blue, red, and infrared surveys will be carried out with the Palomar 1.2-m Schmidt as soon as a color-independent correcting plate is made. The photographs in the visible part of the spectrum will enable astronomers to compare the skies of the 80's with those of 30 years earlier. The new survey will reveal objects four times fainter than those in the original survey, primarily because of Kodak's progress in providing more sensitive and finer-grain emulsions.

Newer Schmidt cameras at the European Southern Observatory and at the observatory at Siding Spring, Australia, are now being used in a joint project to photograph the southern sky, including the one-quarter of the sky that cannot be seen from Palomar. The films they are using are also considerably improved from the ones used in the earlier Palomar/National Geographic survey.

Figure 2–23 Edwin P. Hubble, whose observational work is at the basis of modern cosmology, at the guide telescope of the 1.2-meter (48-inch) Schmidt camera on Palomar Mountain.

Figure 2–24 A 7° × 7° region of the sky in Orion taken with the 1.2-meter Schmidt camera as part of the National Geographic Society–Palomar Observatory Sky Survey. The left photograph shows the image with a blue-sensitive plate, and the right photograph shows the image with a red filter and a red-sensitive plate. The print is negative, so brighter areas appear blacker. The three bright stars at the lower right are Orion's belt; they are hot stars, which are brighter in the blue than in the red. The bright region to their left radiates mostly the red radiation characteristic of hydrogen, and so is brighter in the red than in the blue. The dust lanes, on the other hand, are more prominent in the blue. The Horsehead Nebula (also shown in Color Plate 59) is a region of dust obscuring part of the hydrogen emission that runs south of this region. Near it is a small nebula that is brighter in the blue than in the red, which leads us to conclude that we are seeing reflected starlight rather than hydrogen emission.

To the upper left is the brightest section of Barnard's ring, a ring of hydrogen-emitting gas that surrounds most of Orion. The three blotches at the upper left are artifacts called "ghost images" caused by internal reflections in the telescope of the three brightest stars.

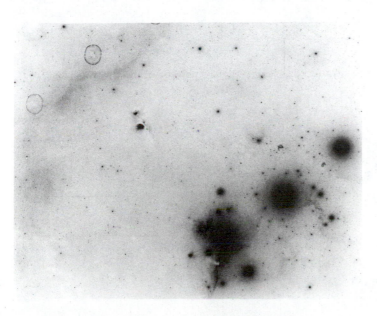
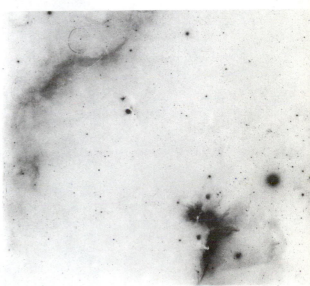

2.7 Optical Observatories

Once upon a time, over a hundred years ago, telescopes were put up wherever astronomers happened to be located. Thus all across the country and all around the world, on college campuses and near cities, we find old observatories. Later, sites that take advantage of still, dark mountain skies were used. Now all major observatories are built at remote sites chosen for the quality of their observing conditions rather than for their proximity to a home campus.

On Mount Wilson, near Los Angeles, George Ellery Hale built a 2.5-m telescope, with a mirror made of plate glass. It began operation in 1917 and was the largest telescope in the world for 30 years.

One of the limitations of any telescope, reflecting or refracting, is the length of time that the mirror or lens takes to reach its equilibrium shape when exposed to the temperature of the cold night air. In the 1930's, the Corning Glass Works invented Pyrex, a type of glass that is less sensitive to temperature variations than ordinary glass. They cast, with great difficulty, a mirror blank a full 5.08 meters (200 inches) across.

The construction of the telescope was held up by many factors, including the

---

Box 2.2 Converting Units

In order to know what to multiply and divide when converting from one system of units to another, some people find it convenient to multiply and divide the words representing the units as though they were numbers. They set up a chain of multiplication and division of known conversion factors so that the unit names from which they are trying to convert cancel out. Then they do the same series of multiplications and divisions with the numbers themselves.

For example, 1 inch = 2.54 cm. If we divide both sides of the equation by 2.54 cm, we have the following equation:

$$\frac{1 \text{ in}}{2.54 \text{ cm}} = 1.$$

Equivalently, we can divide both sides of "one inch = 2.54 cm" by 1 inch in order to get

$$\frac{2.54 \text{ cm}}{1 \text{ in}} = 1.$$

Now, we can always multiply any number by 1 without changing its value. Thus, to put, say, the size of the 200-inch telescope in units of centimeters, we multiply 200 inches by a term equal to 1, writing,

$$200 \text{ inches} \times \frac{2.54 \text{ cm}}{1 \text{ in}}.$$

The "inches" in the numerator and denominator cancel, and we have 508 cm as the converted figure. Similarly, to convert the size of a 4-meter telescope to inches,

$$4 \text{ m} \times \frac{100 \text{ cm}}{1 \text{ m}} \times \frac{1 \text{ in}}{2.54 \text{ cm}} = \frac{400 \text{ in}}{2.54} = 157.5 \text{ in.}$$

This method works just as well with units of time (months, weeks, days, hours, seconds), units of volume, and so on.

Figure 2–25 Periodically, the Palomar telescope's 5-m mirror is cleaned and a new aluminum coating is applied. Here, before its 1982 aluminizing, we can look through the mirror's front and see the ribbed structure of its back, designed to reduce weight and stress on the mirror itself.

Second World War. Only in 1949 did active observing begin with the telescope. It stands on Palomar Mountain in southern California, and for many years was the largest in the world. It is named the Hale telescope after George Ellery Hale, who was responsible for its construction (Fig. 2–25).

Materials for telescope mirrors exist that are even less sensitive to heat variations than Pyrex. They include fused quartz and some more newly developed ceramic materials. The mirrors of new telescopes for ground-based observatories as well as for space vehicles are made of these materials.

2.7a New Sites

Other telescopes have been built since the Palomar 5-m, but with a single exception they have all been substantially smaller. The largest telescope in the world is the Soviet Union's 6-m (236-inch) reflector opened in 1976 in the Caucasus (Fig. 2–26). After a considerable shakedown period, and replacement of the original mirror,

Figure 2–26 *(A)* The Soviet 6-meter (236-inch) telescope at the Special Astrophysical Observatory on Mount Pastukhov in the Caucasus. *(B)* A replacement 6-m mirror being installed.

A

B

Figure 2–27 An aerial view of Kitt Peak. The 4-meter telescope is at the right; the solar telescope is the sloping structure at the lower left. The white dome at extreme left is the radio telescope of the National Radio Astronomy Observatory, a separate organization.

it is now giving reasonable results. Its site, however, does not have the high quality of other major new telescope sites.

The National Science Foundation supports national optical and radio observatories. After a 1983 reorganization, the National Optical Astronomy Observatories (NOAO) includes the Kitt Peak National Observatory in Arizona, the Cerro Tololo Inter-American Observatory in Chile, and the National Solar Observatory (whose telescopes are on Kitt Peak in Arizona and on Sacramento Peak in New Mexico). The National Radio Astronomy Observatory, which we shall discuss further elsewhere, includes facilities in Virginia, West Virginia, Arizona, and New Mexico.

The Kitt Peak National Observatory's telescopes are located on Kitt Peak, a sacred mountain of the Papago Indians, about 80 km (50 miles) southwest of Tucson, Arizona (Fig. 2–27). Its headquarters are in Tucson.

Kitt Peak (as the Kitt Peak National Observatory is usually called) has been set up for use by all the optical astronomers in the United States. Any astronomer can apply for observing time at one of the telescopes at Kitt Peak. A 4-m (158-inch) telescope (shown at the opening of this chapter and in Color Plate 1) is the largest of 16 telescopes now there. Other institutions, including the University of Arizona and a Dartmouth College–University of Michigan–M.I.T. consortium, also have optical telescopes on Kitt Peak.

Astronomers of several campuses of the University of California use a 3-m (120-inch) reflector at the Lick Observatory on Mt. Hamilton in California. They are now planning a giant telescope, 10 m across. The University of Texas has its telescopes near Fort Davis in West Texas. They are planning a 7-m reflector.

Part of the sky is never visible from sites in the northern hemisphere, and many of the most interesting astronomical objects in the sky are only visible or best visible from the southern hemisphere. Therefore, in the last decades special emphasis has been placed on constructing telescopes at southern sites. Some of these sites are on coastal mountains west of the Andes range in Chile. The Cerro Tololo Inter-American Observatory there (Fig. 2–28) has a 4-m (158-inch) twin to the Kitt Peak 4-m telescope.

The Mount Wilson Observatory is jointly run with the Las Campanas Observatory, which includes a 2.5-m (101-inch) telescope on Cerro las Campanas, another Chilean peak. A consortium of European observatories and universities operates the European Southern Observatory, where a dozen telescopes including a 3.6-m (142-inch) telescope have been erected on still another Chilean mountain top. It is quite common in the astronomical community to go off to Chile for a while to ''observe.''

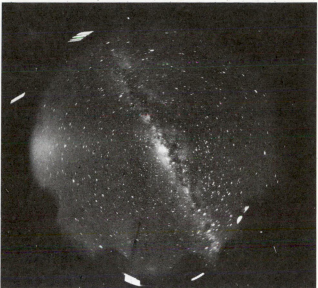

Australia and Great Britain remain at the forefront with the Anglo-Australian Telescope (AAT) of 3.9 meters (153 inches), at Siding Spring in eastern Australia.

Perhaps the most important new site is in Hawaii. The University of Hawaii's 2.2-m (88-inch) telescope, in use on Mauna Kea on the island of Hawaii, was joined in 1979 by the 3.8-m (150-inch) United Kingdom Infrared Telescope, the 3.6-m (140-inch) Canada–France–Hawaii Telescope, and by a 3.0-m (120-inch) Infrared Telescope Facility, built by NASA and operated by the University of Hawaii (Fig. 2–29). It is unprecedented for three such major telescopes to be erected at one place at one time. This site is at such a high altitude, 4200 m (13,800 ft), that the air above is particularly dry. Since water vapor, which blocks infrared radiation, is minimal, these telescopes are especially suitable for infrared observations. Also, the sky

Figure 2–28 Two views of the Cerro Tololo Inter-American Observatory taken from the same location with a fish-eye lens. *(left)* The camera was held fixed, and star trails and the overexposed image of Venus show above the domes. *(right)* The camera tracked the stars; the domes are blurred but the Milky Way shows.

Figure 2–29 The summit of Mauna Kea on the island of Hawaii now boasts a variety of large telescopes. The 3.6-m Canada-France-Hawaii Telescope is at far left. Then come the 0.6-m planetary patrol telescope and the 2.2-m telescope of the University of Hawaii, followed by the 3.8-m United Kingdom Infrared Telescope, and another 0.6-m telescope. The 3-m NASA/University of Hawaii Infrared Telescope Facility is in the right foreground.

is very dark, and the atmosphere is often exceptionally steady. Thus Mauna Kea is a prime candidate to be the site of giant telescopes of the next generation, including the 15-m National New Technology Telescope.

The United Kingdom, Spain, Sweden, Denmark, Holland, and Ireland are jointly constructing a new observatory, el Observatorio del Roque de los Muchachos, on the Canary Islands, Spain. The Isaac Newton telescope has been moved from the cloudy skies of the south of England to this site, with a new 2.5-m (100-inch) mirror installed. A 4.2-m (165-inch) telescope is being constructed alongside, to be opened in 1986. The world's third-largest single-mirror telescope, it is named after William Herschel, who discovered the planet Uranus in the 18th century.

2.7b New Directions in Telescope Making

Even though light-gathering power increases rapidly with the size of the telescope, cost goes up even more rapidly—with the cube of the diameter. Thus the funds have simply not been available in the United States to build a larger telescope than the 5-m (200-inch).

Alternative methods of getting more aperture are now under investigation. The Harvard-Smithsonian Center for Astrophysics and the University of Arizona have jointly built the Multiple Mirror Telescope (MMT) (Fig. 2–30), which has six 1.8-m (72-inch) paraboloids linked together to focus all the light at the same point. It is much cheaper to build even a half dozen 1.8-m telescopes than it is to build one larger telescope of 4.5 meters (176 inches), the equivalent total aperture. The MMT is on Mount Hopkins, south of Tucson, Arizona. Making a telescope with such an

Figure 2–30 Six 1.8-m mirrors are held in a single framework to make the Multiple Mirror Telescope at the Whipple Observatory in Arizona.

Table 2–2 The Largest Optical Telescopes

Telescope or Institution*	Location	Diameter (m)	Diameter (in)	Date Completed	Mirror
Soviet Special Astrophysical Obs.: BTA	Caucasus, U.S.S.R.	6.0	236	1976	Pyrex
Palomar Observatory: Hale Telescope	Palomar Mtn., California	5.0	200	1950	Pyrex
Multiple Mirror Telescope (MMT)	Mt. Hopkins, Arizona	4.5	176	1979	Fused silica
Roque de los Muchachos Obs.: Wm. Herschel Tel.	Canary Islands, Spain	4.2	165	1986	Cer-Vit
Cerro Tololo Inter-American Obs.	Cerro Tololo, Chile	4.0	156	1975	Cer-Vit
Anglo-Australian Telescope (AAT)	Siding Spring, Australia	3.9	153	1975	Cer-Vit
Kitt Peak National Obs.: Mayall Tel.	Kitt Peak, Arizona	3.8	150	1974	Quartz
United Kingdom Infrared Telescope (UKIRT)	Mauna Kea, Hawaii	3.8	150	1979	Cer-Vit
European Southern Observatory (ESO)	La Silla, Chile	3.6	142	1976	Fused silica
Canada-France-Hawaii Telescope (CFHT)	Mauna Kea, Hawaii	3.6	140	1979	Cer-Vit
German-Spanish Astronomical Center	Calar Alto, Spain	3.5	138	1983	Zerodur
Infrared Telescope Facility	Mauna Kea, Hawaii	3.0	120	1979	Cer-Vit
Lick Observatory: Shane Telescope	Mt. Hamilton, California	3.0	120	1959	Pyrex
McDonald Observatory	Mt. Locke, Texas	2.7	107	1968	Fused silica
Crimean Astrophysical Obs.: Shajn Tel.	Crimea, U.S.S.R.	2.6	102	1961	Pyrex
Byurakan Observatory	Armenia, U.S.S.R.	2.6	102	1976	Pyrex
Las Campanas Obs.: Irénée du Pont Tel.	Cerro Las Campanas, Chile	2.5	101	1977	Fused silica
Roque de los Muchachos Obs.: Isaac Newton Tel.	Canary Islands, Spain	2.5	101	1984	Zerodur
Mt. Wilson Observatory: Hooker Tel.	Mt. Wilson, California	2.5	100	1917	Plate glass
Edwin P. Hubble Space Telescope	Earth orbit	2.4	94	1986	Fused U.L.E.
Michigan/Dartmouth/MIT Astronomy Consortium	Kitt Peak, Arizona	2.4	93	1985	Cer-Vit
Wyoming Infrared Obs.	Jelm Mtn., Wyoming	2.3	92	1977	Cer-Vit
Mt. Stromlo and Siding Spring Obs.	Siding Spring, Australia	2.3	92	1984	Cer-Vit
Steward Observatory	Kitt Peak, Arizona	2.3	90	1969	Fused quartz
University of Hawaii	Mauna Kea, Hawaii	2.2	88	1970	Fused silica
German-Spanish Astronomical Center	Calar Alto, Spain	2.2	88	1979	Zerodur
European Southern Observatory	La Silla, Chile	2.2	88	1984	Zerodur

*7-, 10-, and 15-m telescopes being planned are discussed in Section 2.7c.

array, which requires constant adjustments of its elements, was an experiment that has turned out successfully. The original method of aligning the mirrors, using laser beams, didn't work out since moths kept flying through the beams, but a computer now adjusts the mirrors. Every 10 minutes, observing stops, the telescope is pointed at a star brighter than 16th magnitude, the six secondary mirrors are automatically tilted to spread apart the six images first in one direction and then in another, and the computer figures the positions necessary for correct alignment by analyzing the stellar images. The whole cycle takes only 20 seconds, and works well.

Similar multi-mirror techniques may be used in the future for "new technology telescopes" (Section 2.7c). Telescopes up to 15 meters across are being planned.

The continued building of optical observatories, with the current unprecedented spate of large telescopes, shows the vitality of ground-based optical astronomical research, even though new techniques involving space telescopes, radio astronomy, and other non-visual techniques are now providing important data at an increasing rate.

Though many of the world's large optical observatories have just been described, the current definition of observatory is no longer the place where the telescopes are located but rather the place where the astronomers are. Most of an astronomer's work is done by studying the data that have been recorded on film or on computer tape during a field trip to an observatory. Other astronomers—theoreticians—carry out their research without ever using a telescope. And new methods of communication and of telescope control are occasionally enabling telescopes to be remotely controlled. So it is still possible to have major observatories located in urban areas. All it takes are astronomers (and, perhaps, computers).

2.7c New Technology Telescopes

Telescopes in space will solve some scientific problems, but will also suggest many new kinds of observations that can best or most efficiently be made from the ground. The need for more large ground-based telescopes continues.

Several telescopes, any of which would be the largest in the world, are now planned for optical and infrared use by U. S. (Fig. 2–31), European Southern Observatory, Soviet, and Japanese astronomers. These telescopes will employ new technologies to provide large light-gathering power at relatively low cost compared to older standard designs. They will be at sites especially chosen for top-quality ''seeing'' (image steadiness and quality) and, to allow infrared use, for high altitude. Fund-raising and detailed planning are now under way.

2.8 Telescopes in Space

The largest astronomical telescope that has so far been sent into space was the 0.9-m (36-inch) telescope aboard the Copernicus satellite, the third Orbiting Astronomical Observatory, launched by NASA in 1973. The International Ultraviolet Explorer (IUE), launched by a joint effort of NASA and the British and European space organizations in 1978, carries a 0.45-m (18-inch) telescope with two spectrographs for studies of the ultraviolet. It carries a television-like system that records data much more efficiently than Copernicus and hence gives spectra more quickly. In Section 9.8, we describe what it is like to work with this important spacecraft.

Figure 2–31 *(A)* University of California: a 10-m telescope made of 36 contiguous hexagonal segments, planned for Mauna Kea in Hawaii but to be remotely controlled from California. *(B)* University of Texas: a 7.6-m telescope with a single thin mirror, for Mt. Fowlkes, a peak in West Texas. The mirror's shape would be maintained by varying the pressure behind it. *(C)* Kitt Peak and the Universities of Arizona, California, and Texas: a 15-m ''National New Technology Telescope.'' A single mirror assembled from smaller segments *(left)* was considered, but a multiple-mirror arrangement, with four 7.5-m mirrors, has been selected.

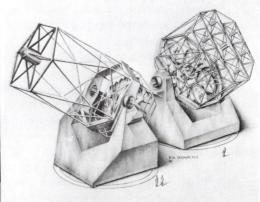

A   **B**   **C**

Figure 2–32 The Hubble Space Telescope is shown in this artist's conception. It will operate from a position above the earth's atmosphere, so it will be able to probe about 7 times deeper into space than ground-based telescopes. Space Telescope will give images with 0.1 arc second resolution, a few times better than can be obtained from the ground.

Space Telescope, with a mirror 2.4-m (94-inches) in diameter, is scheduled for launch in 1986. It will, in its original form, concentrate on visible and ultraviolet studies (Fig. 2–32). Free of the twinkling effects caused by turbulence in the earth's atmosphere and located above the atmosphere in perennially black skies, ST will be able to observe details of objects in space about seven times finer than can be observed with ground-based observatories. It will be ten billion times more sensitive than the human eye. Its sensitivity, the darker sky it will have, and the fact that images will be concentrated better on the detectors will allow it to see, on the whole, seven times farther into space. Since the volume of a box depends on the cube of the diameter, this will give us access to a volume of space about 7^3, or about 350 times, larger than we can now study to the same degree.

This orbiting observatory has been named the Edwin P. Hubble Space Telescope, after the scientist who, in the 1920's, discovered that there were other galaxies comparable in scale to our own (Part V) and, subsequently, discovered that the universe was expanding (Chapter 14). The Hubble Space Telescope will dramatically improve our understanding of many astronomical problems. For example, we should be able to improve our knowledge of distances to thousands of galaxies and thus improve our understanding of the overall cosmic scale of distances. We should be able to study clusters of galaxies and quasars so far back in time that conditions in the universe must have been very different from the way they are now. We will be able to image Jupiter and Saturn with as much detail as was measured by the spacecraft that flew close to them, and may be able to discover planets around other stars.

Space Telescope's primary mirror (Fig. 2–33), of the Ritchey-Chrétien design, will direct light to one of several analyzing instruments. The major instruments are a Wide Field/Planetary Camera, a Faint Object Spectrograph, a High Resolution Spectrograph, a faint object camera, and a High Speed Photometer/Polarimeter. The European Space Agency is cooperating with NASA by providing the solar array, one of the instruments, and some of the staff.

Space Telescope will be launched from a space shuttle and left in space. Astronauts will not fly with Space Telescope, because they would cause too much shaking. But astronauts can visit Space Telescope on a space shuttle to make repairs, carry out maintenance, and even change scientific instruments. Such visits are expected every 2½ years. ST can even be brought down to earth, refurbished, and launched again. It is designed to work for at least 15 years in orbit.

Space Telescope will be controlled from, and data will be analyzed at, the Space Telescope Science Institute. The STSI, upon the launch of Space Telescope, will

Work on the faintest objects will be very slow, for at the limit of its observing ability, Space Telescope will receive only 1 photon (i.e., one particle of light energy) per minute. So there will still be plenty for ground-based telescopes to do.

Figure 2–33 The 2.4-m mirror of Space Telescope, after it was ground and polished and covered with a reflective coating. The workers seen reflected in the mirror are actually far off to the right; their images appear magnified. The mirror is so smooth that were it to be blown up to the size of the U. S. from Maine to California, the biggest bump would be only a foot high.

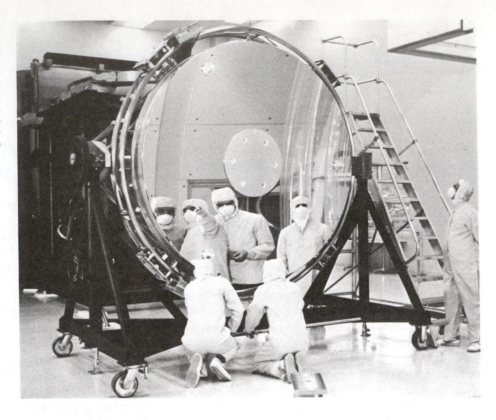

immediately become one of the major world institutions of astronomy. It is run by the same organization that runs Kitt Peak, and is located on the Homewood Campus of The Johns Hopkins University in Baltimore.

The space shuttles are also used to launch other telescopes. The European Space-lab (Fig. 2–34) first flew in 1983 with astronauts aboard operating an ultraviolet telescope. It flies periodically with different instruments aboard, though only two of the 70-odd experiments on its first flight were astronomical. The Hopkins Ultraviolet Telescope, Johns Hopkins' 0.9-m telescope, is also scheduled for future trips aboard a space shuttle, starting in 1986. It will extend Space Telescope's studies farther into the ultraviolet. In the late 1980's or early 1990's, space shuttles should carry aloft

Figure 2–34 Spacelab aboard a space shuttle. In the foreground, we see the tunnel through which astronauts move to the pressurized Spacelab module, where they work in shirtsleeves. Telescopes and antennas are in the vacuum of space behind them. A picture taken from Spacelab appears as Figure 10–6B.

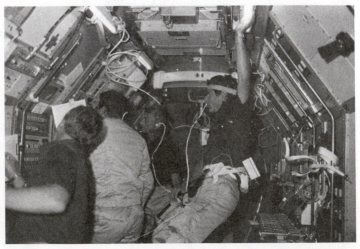

the Solar Optical Telescope, which will improve the resolution of solar studies by a factor of 10, and the Shuttle Infrared Telescope Facility, the heart of which will be cooled to 2 K ($-271°C$) to limit the amount of infrared radiation from the telescope and detectors themselves. Otherwise, this self-generated infrared radiation would overwhelm the infrared signal from celestial objects.

2.9 Light-Gathering Power and Seeing

The principal function of large telescopes is to gather light. The amount of light collected by a telescope mirror is proportional to the area of the mirror, πr^2. (It is often convenient to work with the diameter, which is twice the radius, since the diameter is the number usually mentioned. The formula for the area of the circle may, of course, also be written as $\pi(d/2)^2 = \pi d^2/4$.)

The ratio of the areas of two telescopes is thus the ratio of the square of their diameters (d_1^2/d_2^2), which is usually more easily calculated as the square of the ratio of their diameters $(d_1/d_2)^2$. For example, the 5-meter telescope has an area 4 times greater than the area of a 2.5-meter telescope, $(5/2.5)^2$ (Fig. 2–35). Thus with exposures of equal duration, the 5-m would collect four times as much light as a 2.5-m if all other things were equal. For the same exposure time, it would therefore record fainter stars.

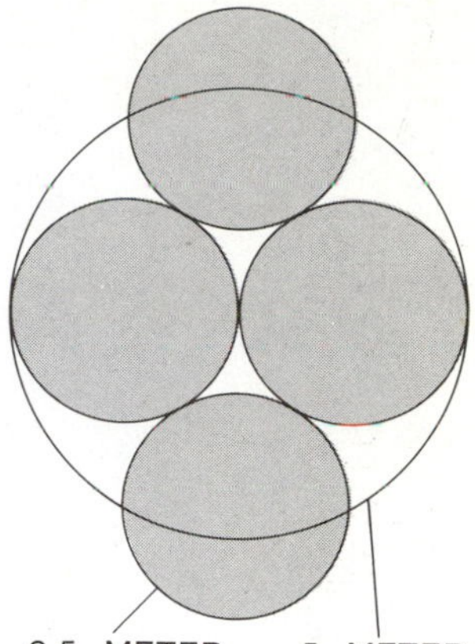

2.9a Seeing, Transparency, and Light Pollution

A telescope's light-gathering power (and its ability to concentrate light in as small an area as possible) is really all that is important for the study of individual stars, since the stars are so far away that no details can be discerned no matter how much magnification is used. The main limitation on the smallest image size that can be detected is caused not in the telescope itself, but rather by turbulence in the earth's atmosphere. The limitations are connected with the twinkling effect of stars. The steadiness of the earth's atmosphere is called, technically, the *seeing*. We say, when the atmosphere is steady, "The seeing is good tonight." "How was the seeing?" is a polite question to ask astronomers about their most recent observing runs. Another factor of importance is the *transparency*, or how clear the sky is. It is quite possible for the transparency to be good but the seeing to be very bad, or for the transparency to be bad (a hazy sky, for example), but the seeing excellent.

A problem of modern civilization is *light pollution* (Fig. 2–36).

2.5-METER MIRROR 5-METER MIRROR

Figure 2–35 A telescope twice the diameter of another has four times the collecting area.

Figure 2–36 The United States photographed from a satellite at night, showing how much light is escaping into the sky. The skies are no longer as dark as they used to be. Much of the problem of *light pollution* is caused by street lighting. Most street lights, for example, are designed so that some of their light goes up in the air instead of down toward the street. This stray light benefits nobody, and the simple addition of shields on street lights might help the light pollution problem considerably. Further, the type of street light is important; low-pressure sodium is best for astronomers, since the sodium emission lines from them can be filtered out.

2.9b Magnification

For some purposes, including the study of the planets and the resolution of double stars, the magnification provided by the telescope does play a role. Magnification can be important for the study of the sun, as well, and for *extended objects* (as distinguished from *point objects*, objects whose images are points) like nebulae or galaxies.

Magnification is the number of times larger an object's angular size is enlarged; the angular size is commonly expressed in degrees, minutes, or seconds of arc, and gives the angle across the sky that an object appears to cover in any one direction. Magnification works out to be equal to the focal length of the telescope objective (the primary mirror or lens) divided by the focal length of the eyepiece. By simply substituting eyepieces of different focal lengths you can get different magnifications. However, there comes a point where even though using an eyepiece of short focal length will enlarge the image, it will not give you any more visible detail. When using a telescope you should not exceed the maximum usable magnification, which depends on such factors as the diameter of the objective and the quality of the seeing.

A telescope of 1 meter focal length will give a magnification of 25 when used with a 40-millimeter eyepiece (that is, 1 m/40 mm = 25), an ordinary combination for amateur observing. If we substitute a 10-millimeter eyepiece, the same telescope has a magnification of 100. We would say that we were observing with "100 power."

2.10 Spectroscopy

We have discussed the collection of quantities of light by the use of large mirrors or lenses, and the focusing of this light to a point or suitable small area. Often one wants to break up the light into a spectrum instead of merely photographing the area of sky at which the telescope is pointed.

A prism breaks up light into its spectrum for the same reasons that light focused by a lens has chromatic aberration. A wave front is refracted as it hits the side of the prism, but it is refracted by different amounts at different wavelengths.

Today, astronomical spectra are no longer usually made by means of prisms. Light may be broken up into a spectrum, without need for a prism, by lines that are ruled on a surface very close together. And I do mean close together: one can have 10,000 lines ruled in a given centimeter (25,000 lines in a given inch)! Such a ruled surface is called a *diffraction grating*. Making diffraction gratings is a difficult art.

When a beam of parallel light falls directly on a prism or grating, the spectral lines that result are often indistinct because the spectrum from one place in the beam

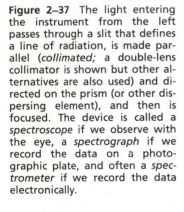

Figure 2–37 The light entering the instrument from the left passes through a slit that defines a line of radiation, is made parallel (*collimated;* a double-lens collimator is shown but other alternatives are also used) and directed on the prism (or other dispersing element), and then is focused. The device is called a *spectroscope* if we observe with the eye, a *spectrograph* if we record the data on a photographic plate, and often a *spectrometer* if we record the data electronically.

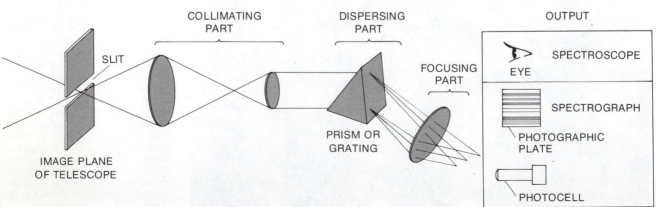

overlaps the spectrum from an adjacent spot in the beam. We can limit the blurring of spectral lines that results because of this by allowing only the light that passes through a long, thin opening—called a *slit*—to fall on the prism. Then the spectral lines are sharp and relatively narrow.

Scientists then use optics resembling a small telescope to examine the spectrum that results (Fig. 2–37). A device that makes a spectrum, usually including everything between the slit at one end and the viewing eyepiece at the other, is called a "spectroscope." When the resultant spectrum is not viewed with the eye but is rather recorded either with a photographic plate or electronically, the device is called a *spectrograph*.

2.11 Recording the Data

Astronomers usually want a more permanent record than is afforded by merely observing the image and more accuracy than can be guaranteed in a sketch. Traditionally, one puts a photographic film or "plate" (short for photographic plate, a layer of light-sensitive material on a sheet of glass) instead of an eyepiece at the observer's end of the telescope, so as to get a permanent record of the image or spectrum.

Basically, a photographic plate (or film) consists of a glass or plastic backing covered with an *emulsion*. The emulsion contains grains of a silver compound. When these grains are struck by enough light, they undergo a chemical change. The result, after development, is a "negative" image, with the darkest areas corresponding to the brightest parts of the incident image. Inspection of a photographic plate under high magnification shows the grainy structure of the image (Fig. 2–38).

Using film also significantly increases our ability to detect signals from faint sources. The eye and brain can process information for only a fraction of a second at a time and do not function cumulatively. A photographic plate, however, can be left exposed to radiation for a long period (Fig. 2–39), sometimes for hours. Just as a long exposure with an ordinary camera can record objects that are only dimly lit, a long exposure with a telescope can record fainter objects than would a shorter exposure.

Other methods of recording data have been developed. Some say that the day of film is past, though new emulsions now available are more sensitive and have finer grain than older emulsions. Still the future seems to lie in electronic devices.

The photocell is a basic electronic device used by astronomers. Certain materials give off an electric current when they absorb light. That current can be measured, and is often fed directly into a computer to be measured. (Materials that give off currents when struck by light may one day be widely used to convert solar energy into electricity for general use, but at present are much too expensive.)

The electrons that become available inside the photocell when light hits it can be directed to hit other special elements of the tube. Each of the elements can emit several electrons for each one hitting it. In this way, the original number of electrons

Figure 2–38 This enlargement of the spectrum of a distant galaxy shows clumps of grain on the photographic plate.

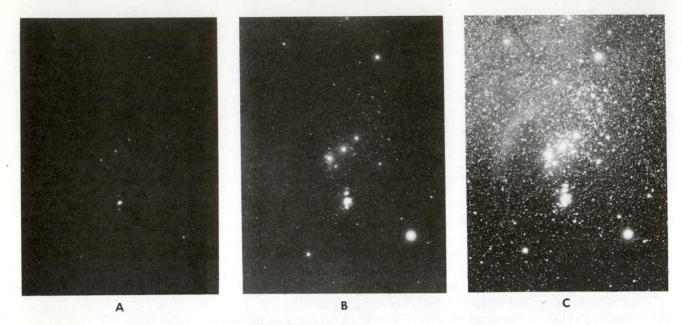

A   B   C

Figure 2–39 These three views of Orion (see also Fig. 10–2) show the effect of increased exposure time. The three stars representing the belt, shown also in Fig. 2–24, appear at the center of the photograph. The Orion Nebula (Color Plate 56) is below them. Though the red star Betelgeuse, at the upper left, appears bright to the eye, it does not appear very bright here because these photographs were taken on photographic plates that were sensitive only to the blue part of the spectrum. The longest exposure shows that nebulosity—glowing gas or the reflection of starlight off dust—covers almost the entire constellation.

can be multiplied, often by a substantial factor (perhaps 10^6). Such a device is called a *photomultiplier*.

One advantage of a photocell or photomultiplier is that its response is "linear"; that is, unlike film, it changes by twice as much electric current when twice as much light hits it. This makes it much easier to figure out how bright an object actually is or by what factor of brightness an object or spectrum is changing. One of the disadvantages of photocells and photomultipliers is that they measure only the total intensity of radiation falling on them from whatever direction; unlike film, they do not form an image.

For measuring the intensity of light of a star, photomultipliers are fine. Such measurements of intensity are called *photometry*. Often the intensity of objects is measured through each of several colored filters in turn, to find out how the continuous spectrum differs at different wavelengths.

Photomultipliers are much more sensitive to faint signals than is photographic film, and can also be made to be sensitive in a wider region of the spectrum. Photometry has long been carried out with photocells and photomultipliers and can be done very precisely. The procedure is often automated so that the strength of the current flowing from the photomultiplier is monitored by a computer. When enough light has been gathered to allow a measurement of a certain accuracy (say, to 1 per cent of the signal level), the computer ends the "exposure" and may even move the telescope to the next object.

2.11a *Electronic Imaging Devices*

For some years devices called *image tubes* have been used, sometimes to increase the sensitivity of film and sometimes to increase the wavelength coverage. Light from the telescope or spectrograph may fall on the face plate of a tube and generate internal electrons or light, but ultimately these secondary emissions still fall on a photographic plate or film. Since the light generated need not be at the same wavelength as the incident light, one can, say, have an infrared signal at a wavelength beyond the sensitivity of film fall on the image tube and generate green light to fall on ordinary film or on a television-type camera.

A new series of devices is now able to provide an electronic imaging capability.

The technology for making these is related to the new technology that has recently brought electronic calculators to their present versatility and inexpensive price. These devices can be used in the place of photographic plates to make either direct images of astronomical objects or of spectra. Rapid advances are being made in this field.

Astronomers have also been using ''silicon diode vidicons'' (Fig. 2–40), for which a series of small crystals, each only a few microns across, can actually be grown on a thin silicon wafer. Each point on the wafer acts as though it were a small photocell.

Still more recent electronic detectors are integrated-circuit devices built on ''chips.'' They sense light on a surface and simultaneously scan the image off that surface with internal circuitry. These chips come in several types, including Charge-Coupled Devices (*CCD's*; Fig. 2–41). The name ''CCD'' refers to the scanning method, which involves shifting the electrical signal from each point to its neighbor one space at a time, analogous to what happens with water in a bucket brigade (for fighting a fire) when buckets of water are passed from one individual to the next. A CCD's picture is created by scanning the picture elements line by line, as occurs in our televisions. CCD's are much more sensitive than film. Also, adjacent objects that differ greatly in brightness can be accurately compared on a CCD image, whereas film would not permit such a comparison.

CCD's are being developed for use on many large telescopes, and are the hottest technique in astronomy. They are becoming as commonplace at observatories as film. The imaging system on Space Telescope is a 1600×1600 ''pixel'' (picture element) array. Because current technology cannot build an array that large, it is actually a mosaic containing four CCD arrays, each 800 by 800 pixels.

*2.11b Fiber Optics

Many long-distance telephone lines are being converted from copper wires to optical fibers, over which signals are sent as flashes of tiny lasers. The laser beams travel basically straight along the fibers; if they should deviate by a small angle to the sides of the optical fibers, they are reflected off the sides so as to keep them travelling forward, even when the wire is bent (Fig. 2–42).

---

*Throughout this book, optional sections are starred. They can be skipped without loss of continuity.

Figure 2–40 The Cassegrain focus of the 1.5-m telescope at Cerro Tololo, shown here, contains the silicon vidicon spectrometer. The signals are sent to a nearby computer. After astronomer Karen Kwitter finishes her adjustments, she will run the equipment from a computer terminal on the observing floor.

CCD's are coming into widespread use in many fields, such as in tv cameras.

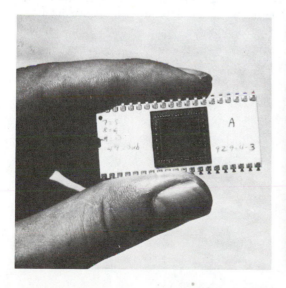

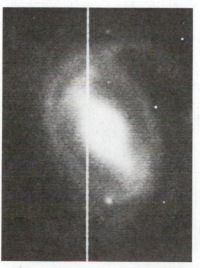

Figure 2–41 *(A)* CCD's, Charge-Coupled Devices, are now being used in many new telescopes. *(B)* A CCD image of a barred spiral galaxy (Arp 67). Each individual picture element—pixel—can be seen. The white streak is the result of one bad pixel. (Photograph taken with the Anglo-Australian Telescope)

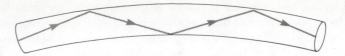

In astronomy, optical fibers are beginning to be used to make telescopes more efficient. For example, several optical fibers can be arranged so that each fiber accepts the light from a different star. The other ends of the fibers can be arranged so that they form a line, making the light coming out cover the whole slit of a spectrograph (Fig. 2–43). Each part of the length of the spectrograph slit thus represents light from a different star. The procedure allows a great gain in the efficiency of the spectrograph, for it now gathers several spectra at the same time instead of just one.

2.12 Observing at Short Wavelengths

In this section, we will first discuss the basic techniques of ultraviolet, x-ray, and gamma-ray astronomy. We will then go on, in the next section, to see how infrared and radio astronomy join with other observing methods to help us investigate space.

From antiquity until 1930, observations in a tiny fraction of the electromagnetic spectrum, the visible part, were the only way that observational astronomers could study the universe. Most of the images that we have in our minds of objects in our galaxy (e.g., the beautiful shapes of nebulae) are based on optical studies, since most of us depend on our eyes to discover what is around us.

If they had to make a choice, however, between observing only optical radiation or only everything else, many astronomers would choose "everything else." The rest of the electromagnetic spectrum carries more information in it than do a few thousand angstroms that we call visible light. Let us think of each factor of 2 in wavelength as an octave, similar to octaves in music. (In music, when two notes are separated by one octave, the wavelength of the sound represented by one note is twice the wavelength of the sound represented by the other note.) The visible spectrum takes up less than one octave, since the wavelength of red light is less than twice the wavelength of violet light. Just as a piano extends over several octaves, the overall electromagnetic spectrum extends over 30 octaves. Thus astronomers are now listening to the entire keyboards of several adjacent pianos, instead of merely the tinkling of a few notes in the center of one.

Thus far, we have talked of the spectrum in terms of the wavelengths of the radiation, which implies that radiation is a wave. But when light is absorbed, the energy is always transferred in discrete amounts that depend on the wavelength of

Figure 2–43 The "fiber optopus" [sic] *(left)* works as shown *(right)* by attaching optical fibers to a plate at the telescope's focal plane (top). The thin fibers, each of which starts at a star, are aligned at their other ends in a straight line, so that the light from several stars can be fed into the same straight spectrograph slit. The thick fibers carry light from stars used for guiding; a video system is used to keep the guide stars centered in these fibers. (Courtesy Anglo-Australian Telescope Board)

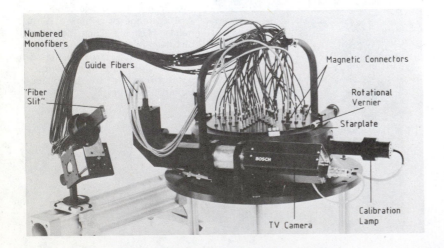

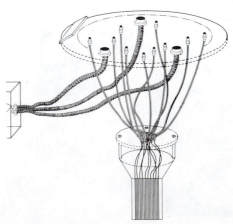

the light. Thus in absorption, radiation acts as though it were composed of particles rather than continuous waves. So it is often useful to think of light as particles of energy called *photons* instead of as waves. Photons that correspond to light of different wavelengths have different energies. We will come back to this idea in Section 4.3. But you will find it useful to have a picture in your mind of photons of high energy, and thus high penetrating power, corresponding to relatively short wavelengths (like x-rays). Similarly, photons corresponding to long wavelengths (radio waves or infrared) don't contain much energy, often not even enough to affect a photographic plate.

2.12a Ultraviolet and X-Ray Astronomy

Ultraviolet and x-ray photons have shorter wavelengths and greater energies than photons of visible light. Thus they too have enough energy to interact with the silver grains on film. Photographic methods can be used, therefore, throughout the x-ray, ultraviolet, and visible parts of the spectrum. Electronic devices like the ones that work in the visible work in the ultraviolet too.

At this shorter end of the spectrum, gamma rays, x-rays, and ultraviolet light do not come through the earth's atmosphere. Ozone, a molecule of three atoms of oxygen (O_3), is located in a broad layer between about 20 and 40 kilometers in altitude. The ozone prevents all the radiation at wavelengths less than approximately 3000 Å from penetrating. To get above the ozone layer, astronomers have launched telescopes and other detectors in rockets and in orbiting satellites. For example, the International Ultraviolet Observatory (Section 9.8) observes wavelengths down to 1000 Å from its orbit 150 km above the earth's surface.

The International Ultraviolet Explorer (IUE) spacecraft, for example, uses television-type devices (vidicons) to record the spectra that are imaged on them. The data can be read off the vidicons and radioed to earth.

2.12b X-Ray Telescopes

In the shortest wavelength regions, we cannot merely use mirrors to image the incident radiation in ordinary fashion, since the x-rays will pass right through the mirrors! Fortunately, x-rays can still be bounced off a surface if they strike the surface at a very low angle. This is called *grazing incidence*. This principle is similar to that of skipping stones across the water. If you throw a stone straight down at a lake surface, the stone will sink immediately. But if you throw a stone out at the surface some distance in front of you, the stone could bounce up and skip along a few times.

By carefully choosing a variety of curved surfaces that suitably allow x-radiation to ''skip'' along, astronomers can now make telescopes that actually make x-ray images. But the telescopes appear very different from optical telescopes (Figs. 2–44 and 2–45).

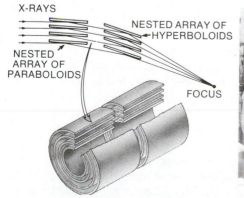

Figure 2–44 Four similar paraboloid/hyperboloid mirror arrangements, all sharing the same focus, were nested within each other to increase the area of telescope surface that intercepted x-rays in NASA's HEAO-2, the Einstein Observatory. The resolution of the system was 2 arc seconds, within a factor of about 4 of the best ground-based seeing.

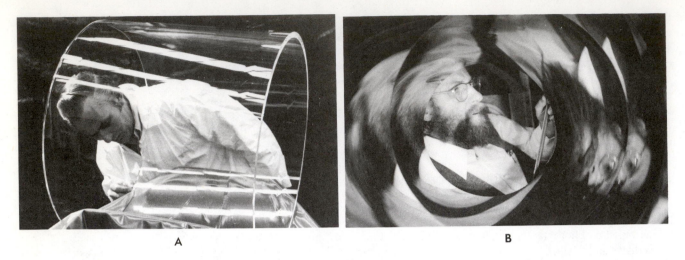

A   B

Figure 2–45 *(A)* The cylindrical mirror of the Einstein Observatory did not resemble mirrors for optical telescopes because it was used at grazing incidence. This was the outermost of four concentric mirrors. *(B)* Penn State scientist Gordon Garmire looks through a mirror polished for grazing incidence with parabolic and hyperbolic sections.

In order to form such high-energy photons, processes must be going on out in space that involve energies very much higher than most ordinary processes that go on at the surface layers of stars. The study of the processes that bring photons or particles of matter to high energies is called *high-energy astrophysics*. In the late 1970's NASA launched a series of three High-Energy Astronomy Observatories (HEAO's) to study high-energy astrophysics. HEAO-1 surveyed the x-ray sky with high sensitivity and excellent time resolution. HEAO-2, known as the Einstein Observatory, was able to image and study in detail the interesting objects mapped by HEAO-1. HEAO-3 studied cosmic rays (particles or matter moving with high energies) and gamma rays.

No American x-ray satellite is now aloft, but the European Space Agency's Exosat was launched in 1983. Though Exosat's images of objects are somewhat less detailed than those from the Einstein Observatory, Exosat has better spectral resolution than Einstein and can observe shorter wavelengths. Its orbit takes it in a path from which the moon blocks out objects in about 20 per cent of the sky. From the times at which the objects are first hidden and then uncovered by the moon, "lunar occultations," Exosat can make precision maps showing the objects' positions.

A small Japanese x-ray satellite, Tenma (which means Pegasus, the winged horse of Greek mythology), was also launched in 1983. Soon thereafter, a joint Soviet-French satellite, Astron, was launched carrying a 75-cm ultraviolet telescope and an x-ray spectrometer.

Larger x-ray satellites are planned, including the German-British Rosat, now scheduled for a 1987 launch aboard one of the U.S.'s space shuttles. It will, like the Einstein Observatory, use a nested array of grazing-incidence mirrors, and can observe in a longer-wavelength x-ray region of the spectrum than had been previously studied. It is expected to provide a survey of the entire sky at a sensitivity 1000 times higher than the last full-sky map.

In the early 1990's, we hope for the American AXAF—the "Advanced X-Ray Astrophysics Facility." AXAF will have four times the collecting area and eight times better spatial resolution than the Einstein Observatory, resulting in perhaps 100 times better sensitivity. AXAF received the highest priority from the "Field report," the 1982 result of a lengthy consideration of astronomy priorities for the 1980's by a national committee headed by George Field of the Harvard-Smithsonian Center for Astrophysics. Scientists hope that Congress and NASA will indeed produce AXAF as soon as possible.

X-ray satellites for studying the sun have different characteristics than the above satellites, and will be discussed in Section 7.5.

2.13 Observing at Long Wavelengths

2.13a Infrared Astronomy

Infrared photons have longer wavelengths and thus lower energies than visible photons. They do not have enough energy to interact with ordinary photographic plates. Some special films can be used at the very shortest infrared wavelengths, but for the most part astronomers have to employ methods of detection involving electronic devices in this region of the spectrum.

Another major limitation in the infrared is that there are very few windows of transparency in the earth's atmosphere. Most of the atmospheric absorption in the infrared is caused by water vapor, which is located at relatively low levels in our atmosphere compared to the ozone that causes the absorption in the ultraviolet. Thus we do not have to go as high to observe infrared as we do to observe ultraviolet. It is sufficient to send up instruments attached to huge balloons (Fig. 2–46).

NASA's Kuiper Airborne Observatory, an instrumented airplane, carries a 0.9-m telescope aloft (Fig. 2–47). It flies above most of the water vapor, so scientists use it especially to make infrared observations (Fig. 2–48). We will see in Chapter 13 that such observations are especially useful to study gas and dust clouds in space; they cannot be observed as well from the ground.

Many useful observations can be made in the infrared from high mountain tops. This has led to the construction of new infrared telescopes at such sites as Mauna Kea in Hawaii, and to the positioning of the Multiple Mirror Telescope on one of the higher Arizona peaks. Section 2.14 describes an observing run with the Infrared Telescope Facility on Mauna Kea. The Infrared Astronomical Satellite (IRAS) mapped the infrared sky in 1983; we have seen IRAS in the color essay following page 44, and we shall discuss it further in Section 13.8 along with how it is used to study stars in formation.

Figure 2–47 NASA'S Kuiper Airborne Observatory. The dark rectangle on top of the plane in front of the wing is the opening through which the 0.9-m telescope sees out.

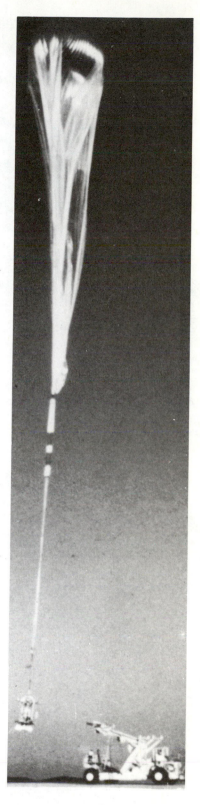

Figure 2—46 A balloon launch carries aloft a 102-cm infrared telescope.

Figure 2–48 Charles Townes *(left)*, the Nobel Prize–winning inventor of masers and lasers, sits at the control panel inside the Kuiper Airborne Observatory. He is studying the infrared spectrum of clouds of gas and dust. The 0.9-m telescope itself is through the hatch at extreme left.

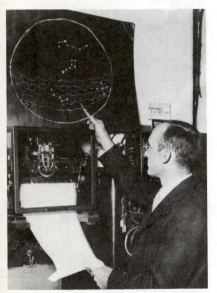

2.13b Radio Astronomy

At even longer wavelengths, in the radio region of the spectrum, is another window of transparency. The techniques of radio astronomy, now well established as one of the major branches of astronomy, will be discussed briefly here and further in various places in the book, especially in Parts IV and V.

One cannot always pinpoint the discovery of a whole field of research as decisively as that of radio astronomy. In 1931, Karl Jansky of the Bell Telephone Laboratories in New Jersey was experimenting with a radio antenna to track down all the sources of noise that might limit the performance of short-wave radiotelephone systems (Figs. 2–49 and 2–50). After a time, he noticed that a certain static appeared at approximately the same time every day. Then he made the key observation: the static was actually appearing four minutes earlier each day. This was the link between the static that he was observing and the rest of the universe. Jansky's static

Figure 2–50 Karl Jansky with the rotating antenna with which he discovered radio astronomy.

Figure 2–49 Karl Jansky lecturing about his discovery of a source of radio signals moving across the sky according to sidereal time.

IRAS —
The Infrared Astronomy Satellite

Beyond the rainbow of colors we can see, past the red, is the infrared. Most infrared is absorbed in the earth's atmosphere and does not reach us on earth. Further, our bodies, our telescopes, and even the sky radiate so strongly in the infrared part of the spectrum that infrared from celestial bodies is masked.

In 1983, the Infrared Astronomy Satellite (shown at right), IRAS, was launched as a joint American-Dutch-British venture. Located above the earth's atmosphere, and with its 60-cm telescope cooled to only 2 degrees above absolute zero, it could detect the faint infrared emitted by objects in space. Studying this part of the spectrum enables us to consider relatively cool bodies, those only a few tens or hundreds of degrees in temperature instead of the thousands of degrees typical of stars. The results are displayed here as false-color images, each color corresponding to a different temperature. (For the record, 12-micron radiation is blue, 25-micron radiation is yellow, 60-micron radiation is green, and 100-micron radiation is red.)

Below we see the Milky Way near the center of our own galaxy. The narrow band we see in the infrared view corresponds to dark absorbing dust we see when we look at the Milky Way in visible light. For this image, the warmest material (including warm dust in our solar system) is blue, while colder material is red. The middle-temperature yellow and green knots and blobs scattered along the band are giant clouds of interstellar gas and dust heated by nearby stars. Some are warmed by newly formed stars while others are heated by massive, hot stars thousands of times brighter than our sun.

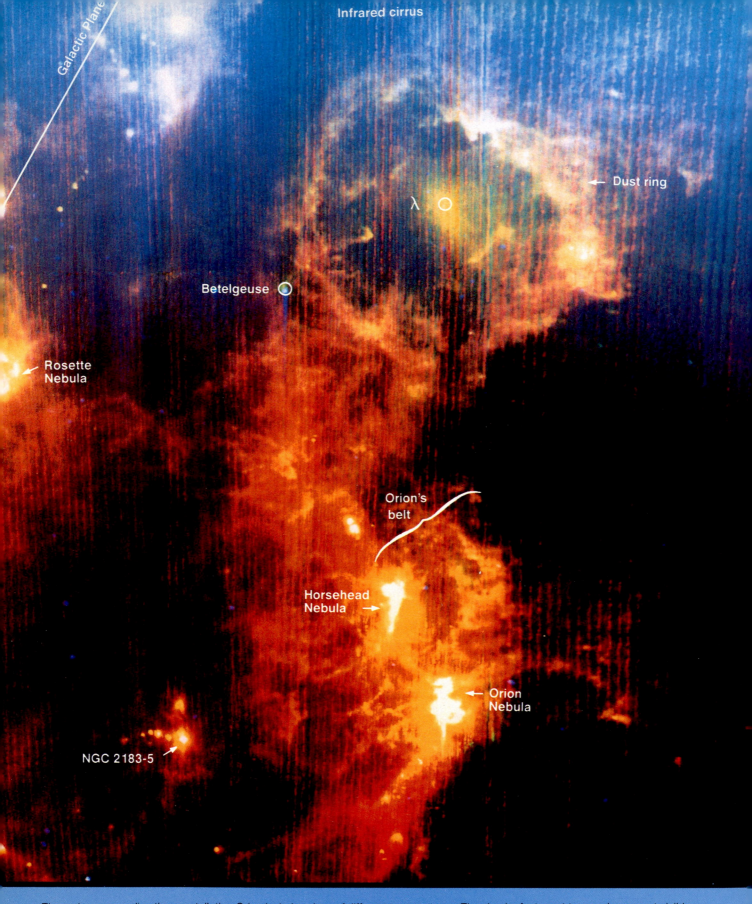

Galactic Plane

Infrared cirrus

Dust ring

λ ○

Betelgeuse ○

Rosette
Nebula

Orion's
belt

Horsehead
Nebula →

Orion
Nebula

NGC 2183-5 →

The region surrounding the constellation Orion includes signs of different temperatures. The circular feature at top can be seen at visible wavelengths, but appears different in intensity and size here in the infrared. It corresponds to hot hydrogen gas and dust heated by the star Bellatrix. The brightest regions correspond to regions where stars are forming.

Looking past the stripes that are an artifact of the data recording in the picture above, we see wispy structures extending roughly horizontally across the picture, and located about 30° above the Milky Way as seen from earth. This "infrared cirrus" was a surprise.

The four images of the center of our galaxy, shown below, use colors to show the intensity of radiation at a given wavelength. The image at left (12-micron radiation) records mainly stars. Most dust in our galaxy is too cool to be seen here, though we can see the dust that is near the galactic center or near a hot star as at top left. The next image (25-micron) also shows the dust only in hot, dense regions. In the next image (60-micron), we begin to see the interstellar dust well, and observe streamers of dust around the center of our galaxy. The final image (100-micron) clearly shows great clouds of interstellar dust.

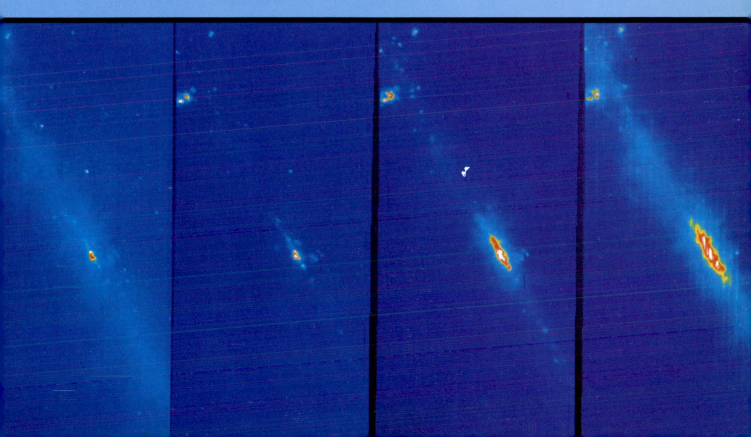

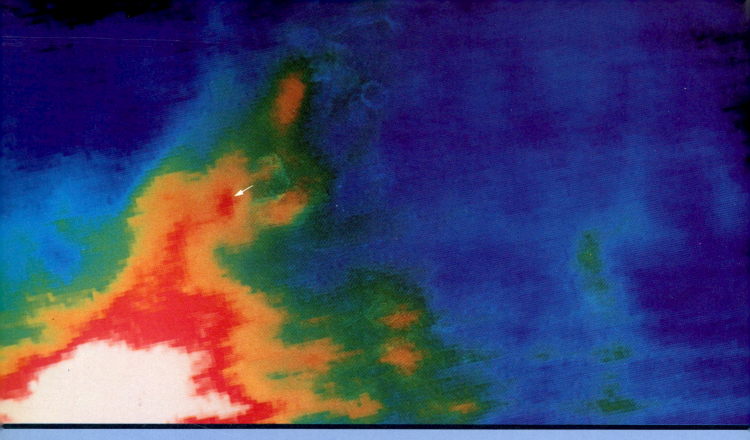

Above we see a star being born (red patch at end of arrow); it is embedded in a cloud of gas and dust known as Barnard 5, which is located only 1000 light years from us. The "protostar" is probably less than 100,000 years old, and has not reached the stage where it is shining by its own nuclear energy as do the sun and the other ordinary stars. The colors signify the intensity range in the image, taken at a wavelength of 100 microns. The white region at the bottom shows intense emission from relatively hot dust (at about our room temperature) in a region heated by a hot young star.

Below left we see Comet IRAS-Araki-Alcock, one of several comets discovered from IRAS. We see the "coma" of the comet, which is located in our solar system. The colors show intensities of 20-micron radiation from warm dust that has boiled off the solid central part of the comet and is then pushed by the pressure of light from the sun. Comets seem to be dustier than had been thought.

Below right we see an IRAS image of the Andromeda Galaxy, a spiral galaxy much like the one in which we live. Brighter areas show regions where star formation is especially active, including the galaxy's core and a ring many thousands of light years in diameter that corresponds to the inner part of the optical disk. When the liquid helium cooling IRAS's telescope ran out after 10 months (actually lasting longer than had been planned), the telescope lost its sensitivity and astronomers lost their best infrared observatory in space. The data will take years to analyze. We discuss the results at several places in this book.

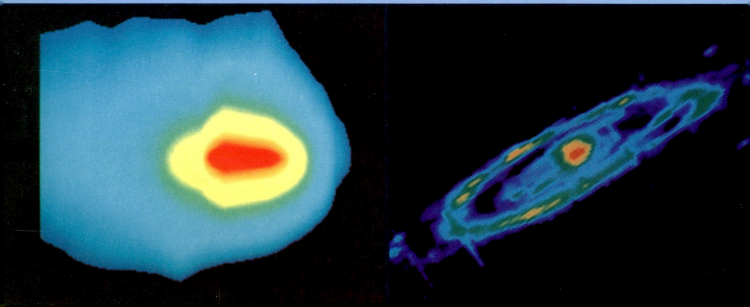

kept sidereal time, just as the stars do (see Section 3.3), and thus was coming from outside our solar system! Jansky was actually receiving radiation from the center of our galaxy.

The principles of radio astronomy are exactly the same as those of optical astronomy: radio waves and light waves are exactly alike—only the wavelengths differ. Both are simply forms of electromagnetic radiation. But different technologies are necessary to detect the signals. Radio waves cause electrical changes in antennas, and these faint electrical signals can be detected with instruments that we call radio receivers.

If we want to collect and focus radio waves just as we collect and focus light waves, we must find a means to concentrate the radio waves at a point at which we can place an aerial. Lenses to focus radio waves are impractically heavy, so refracting radio telescopes are not used. However, radio waves will bounce off metal surfaces. Thus we can make reflecting radio telescopes that work on the same principle as the 5-meter optical telescope on Palomar Mountain. The mirror, called a "dish" in common radio astronomy parlance, is usually made of metal rather than the glass used for optical telescopes. The dish needn't look shiny to our eyes, as long as it looks shiny to incoming radio waves (Fig. 2–51).

We want dishes as big as possible for two reasons: First, a larger surface area means that the telescope will be that much more sensitive. Second, a larger dish has better resolution. Since radio wavelengths are much longer than optical wavelengths, the resolution of any single radio telescope is thus far inferior to that of any single optical telescope (Section 2.3). The basic point is that measurements of the size of a telescope are most meaningful when they are in units of the wavelength of the radiation being observed (Fig. 2–52). Thus an optical telescope one meter across used to observe optical light, which is about one two-millionth of a meter in wavelength, is two million wavelengths across. A radio telescope, even one 100 meters across (the size of the currently largest fully steerable dish, at the Max Planck Institute for Radio Astronomy in Bonn, Germany), is only 1000 wavelengths across when used to observe radio waves 10 centimeters in wavelength (100 m/10 cm = 1000).

The overlap region representing the longest infrared and shortest radio waves, extending from about 0.3 mm (300 microns) to 1 mm, is known as the *submillimeter*. U.S. 10.4-m and British/Dutch 15-m telescopes, designed to operate in the submillimeter region, are being put in the dry, transparent air of Mauna Kea. They should

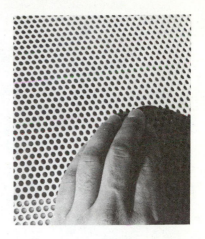

Figure 2–51 Even though the holes in this radio telescope may look large to the eye, they are much smaller than the wavelength of radio radiation being observed. Thus the telescope appears smooth and shiny to the incoming radio radiation. To appear shiny, the surface of a radio telescope has to be smooth to within a small fraction of the wavelength of radiation, one-twentieth or less.

Figure 2–52 It is more meaningful to measure the diameter of telescope mirrors in terms of the wavelength of radiation that is being observed than it is to measure it in terms of units like centimeters that have no particularly relevant significance. The radio dish at top is only 1¾ wavelengths across, while the mirror at bottom is 8 wavelengths across, making it effectively much bigger. The wavelengths are greatly exaggerated in this diagram relative to the size of any actual reflectors. For the Bonn radio telescope, 100 m diameter ÷ 0.1 m per wave = 1000 wavelengths.

Resolution was discussed in Section 2.3, where it was stated that resolution is proportional to wavelength and inversely proportional to telescope diameter.

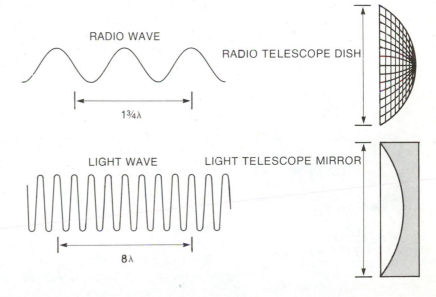

RADIO WAVE

RADIO TELESCOPE DISH

1¾λ

LIGHT WAVE

LIGHT TELESCOPE MIRROR

8λ

Figure 2–53 The beams of radio telescopes are ordinarily relatively large compared with radio sources. The Crab Nebula looks like a point to the radio telescope shown, even though it looks larger than a point to your eye.

Figure 2–54 The Infrared Telescope Facility on Mauna Kea, a 3-m telescope optimized for infrared observations.

be ready in 1986. At 1-mm wavelength, they are each over 10,000 wavelengths across. Their surfaces, though accurate enough for submillimeter waves, are not sufficiently accurate for visible light.

The larger the telescope, measured with respect to the wavelength of radiation observed, the narrower the width of the beam of radiation that the telescope receives. The reflecting radio telescope dish receives radiation only from within a narrow cone. This cone is called the *beam* (Fig. 2–53). Single dish radio telescopes have broad beams and cannot resolve any spatial details smaller than the size of their beams. The 100-m telescope at 10 cm has a beam that is four minutes of arc across; at longer wavelengths or with smaller telescopes one cannot even resolve an area as small as the angular size of the moon, 30 minutes of arc. Obviously, this is looking at the sky with less than the optimum resolution, given that large optical telescopes ordinarily resolve 1 arc sec. However, astronomers now use the technique known as interferometry to improve the resolution. This has led, for example, to the giant array of 27 radio telescopes in New Mexico called, prosaically, the VLA (Very Large Array). Both interferometry and the VLA will be discussed in Section 14.6.

*2.14 A Night at Mauna Kea

An observing run with one of the world's largest telescopes highlights the work of many astronomers. The construction of three of the world's ten largest telescopes at the top of Mauna Kea in Hawaii has made that mountain the site of the world's largest observatory.

Mauna Kea was chosen because of its outstanding observing conditions. Many of these conditions stem from the fact that the summit is so high—4200 m (13,800 ft) above sea level. It is an especially good site for observations. The top of Mauna Kea is above so much of the atmosphere's water vapor, all but 10 per cent, that observations can be made in many parts of the infrared inaccessible from other terrestrial observatories.

Mauna Kea has other general advantages as an observatory site. It is so isolated that the sky above is particularly dark, allowing especially faint objects to be seen. And the flow of air across the mountain top is often particularly smooth, leading to exceptional "seeing" for both visible and infrared observations. Moreover, the top of the mountain extends above most clouds that may cover the island below it. So Mauna Kea is perhaps the world's best site in that these advantages are linked with a large number of clear nights—about 75 per cent. Though there are taller mountains in the world, none has such favorable conditions for astronomy.

One of the new telescopes is the 3-m Infrared Telescope Facility (IRTF), sponsored by NASA and operated by the University of Hawaii (Fig. 2–54). Since it is a national facility, astronomers from all over the United States can propose observations. Every six months, a committee of infrared astronomers from all over the U.S. chooses the best proposals and makes out the observing schedule for the next half year. Priority for most of the telescope time is given to proposals closest to NASA's interests, especially planetary work, but a variety of other research is also included.

During the months before your observing run, you make detailed lists of objects to observe, prepare charts of the objects' positions in the sky based on existing star maps and photographs, and plan the details of your observing procedure. The air is very thin at the telescope. You may not think as clearly or rapidly as you do at lower altitudes. Therefore it is wise and necessary to plan each detail in advance.

The time for your observing run comes, and you fly off to Honolulu. Often you spend a day or two there at the Institute for Astronomy, consulting with the resident scientists, checking last-minute details, catching up with the time change, and perhaps giving a colloquium about your work. Then comes the brief plane trip to the island of Hawaii, the largest of the islands in the State of Hawaii. It is usually called simply "the Big Island."

Your plane may be met by one of the Telescope Operators, who will be assisting you with your observations. The Telescope Operators know the telescope and its systems very well, and are responsible for the telescope, its operation, and its safety. You drive together up the mountain, as the scenery changes from tropical to relatively barren, crossing dark lava flows. First stop is Hale Pohaku, the mid-level facility at an altitude of 2750 m (9000 ft). Hale Pohaku is Hawaiian for "Stone House," in honor of an original building still on the site. The astronomers and technicians sleep and eat at Hale Pohaku. Since the top of Mauna Kea is too high for people to sleep and work comfortably there for too long, everyone spends much time at the mid-level.

The rules say that you must spend a full 24 hours acclimatizing to the high altitude before you can begin your own telescope run. So you overlap with the last night of the run of the people using the telescope before you. This familiarizes you with the telescope and its operation. A combination of time and altitude adjustment sends you to bed early this first night.

The next night is yours. In the afternoon, you and the Telescope Operator may make a special trip up the mountain to install the systems you will be using to record data at the Cassegrain focus of the telescope. Since objects at normal outdoor or room temperatures give off enough radiation in the infrared to bother your observations, much of the instrumentation you install is cooled. The detectors used to observe wavelengths of 10 or 20 micrometers are bolometers, devices sensitive to small temperature changes that occur as a result of incoming radiation. The bolometers are cooled to a temperature of only 2 K ($-271°C$) by liquid helium. Surrounding the liquid helium is liquid nitrogen, at a temperature of 77 K ($-196°C$). These liquids have to be carefully transferred from large holding tanks into metal flasks that contain the bolometer and its accompanying optics (Figs. 2–55 and 2–56). Then the system has to be left for many hours to cool down.

After dinner, you return to the summit to start your observing. You dress warmly, since the nighttime temperature approaches freezing at this altitude, even in

Oh, Langley invented the
  bolometer,
A very good kind of
  thermometer.
It can measure degrees
On a polar bear's knees
At a distance of half a kilometer.

Figure 2–55 The telescope operator fills one of the bolometers with liquid nitrogen, which gives off a visible plume as it hits the air and partially evaporates.

Figure 2–56 Installing a bolometer at the Cassegrain focus. The back plate of the telescope is at the top; the 3-m mirror is on its other side.

Figure 2–57 A telescope operator at her console. The television screen shows that the telescope is pointing at Saturn.

the summertime. Since the sky is fairly dark at infrared wavelengths, you can start observing even before sunset. The telescope is operated by computer, and points at the coordinates you type in for the object you want to observe. A video screen (Fig. 2–57) displays an image of the object.

The first thing you look at is a known point object, a star with a known strength of infrared flux. You make the measurements by starting a computer program for that purpose. Later on, you compare your observations of this known object with what is known about the object, to calculate corrections that you should apply to all the data.

Everything is computer-controlled. You measure the strength of the object at the wavelength you are observing by starting a computer program on a console provided for the observer. In the infrared you usually measure the strength of your source by comparing it with the nearby sky. You adjust the telescope to automatically rock its secondary mirror back and forth perhaps 20 times each second, so that it sometimes looks at your object and sometimes at blank sky.

Let's say that your main object is a dark cloud of dust and gas in the sky, the subject of my observing run there. You may be looking to see if there are young stars inside, hidden to visible observations but detectable in the infrared because they make the dust glow. You have to search a small area of sky to see what you can find. You have to decide on the size of the area to be searched, the length of time to spend at each point, the frequency with which you rock the telescope mirror, the wavelength to use, and many other factors.

Throughout the night, you return periodically to your known point object. As you look through different angles at this object and at your main object, you look through different amounts of the earth's atmosphere. When reducing your data, you must take into account the dimming effect of looking through different amounts of air.

At some time during the night, you may go into the telescope dome to ponder the telescope at work. The telescope frame is very massive, so that it can point accurately and steadily in the right direction. It all but blocks your view of the sky.

Hardly anyone ever actually looks through the telescope; indeed, there is no good way to do so. You simply check your image on television, and record it with the bolometer or other measuring instrument. Still, you may sense a deep feeling for how the telescope is looking out into space (Fig. 2–58).

You usually quit when morning comes, since the Telescope Operator can only work a 12-hour shift, and you are exhausted anyway. You also have to stop because you have trouble finding your guide stars in visible light at dawn, though the infrared background is actually dark. For some projects, a daytime observing crew may take over.

You drive down the mountain in the early morning sun. The clouds you see may be a kilometer below you. The tiny pimples on the landscape that you first see turn out as you get closer to be giant cinder cones that tower above you. They are from past volcanic eruptions. The mountain around you is barren, littered with huge volcanic rocks, resembling surface views from the Viking landers on Mars more than any terrestrial view you have ever seen. Then some vegetation appears, and you reach Hale Pohaku, ready for breakfast and a day's sleep. The next night you start over again.

In the next few days, you may find time to visit some of the other telescopes on the mountain and talk to the scientists about their projects. The group at dinner is interesting, cosmopolitan, and international. In addition to the observers who come for special projects, the United Kingdom Infrared Telescope and the Canada-France-Hawaii Telescope each has scientists resident on the island who are often at the observatories.

When you leave Mauna Kea, it is a shock to leave the pristine air above the clouds. But you take home with you data about the objects you have observed. From the IRTF, the data are in the form of graphs, computer printouts, and sometimes computer tape. You may also have photographs that are the result of your work with other telescopes. It often takes you months to study the data you gathered in a few brief days at the top of the world.

Figure 2–58 A time exposure by moonlight showing the IRTF in the foreground and the Canada-France-Hawaii Telescope in the background.

Summary and Outline

The spectrum (Section 2.1)
> From short wavelengths to long wavelengths the various types of radiation are known as gamma rays, x-rays, visible, infrared, radio.
> Parts of visible spectrum are ROY G. BIV.
> All radiation travels at the ''speed of light''; in a vacuum this is 3×10^8 m/s.
> Windows of transparency of the earth's atmosphere exist for visible light and radio waves (plus some parts of the infrared).
> Continuous radiation is called the *continuum*.

Spectral lines (Section 2.2)
> Absorption lines are gaps in the continuum shown as narrow wavelength regions where the intensity is diminished from that of the neighboring continuum.
> Emission lines are narrow wavelength regions where the intensity is relatively greater than neighboring wave-

lengths (which are either continuum or zero in intensity).
> A given element has a characteristic set of spectral lines under certain conditions of temperature and pressure.

Optical telescopes (Sections 2.3 to 2.9)
> Resolution is inversely proportional to diameter (and proportional to wavelength).
> Refractors and reflectors: Ritchey-Chrétiens and Schmidt cameras have relatively wide fields.
> Optical observatories and their distribution across the world
> The Hubble Space Telescope is to be launched in 1986.
> Light-gathering power is proportional to diameter squared.

Spectroscopy and photometry (Sections 2.10 and 2.11)
> Film and electronic devices (especially CCD's) for gathering photons are described.
> Differences in observing in nonvisible parts of the spectrum

Fiber optics are making observing more efficient.
Ultraviolet and x-ray astronomy (Section 2.12)
  Video and grazing incidence techniques used
  Grazing incidence for HEAO-2

Infrared and radio astronomy (Section 2.13)
  Infrared: high altitude sites and electronic devices are necessary
  Radio: single-dish radio telescopes have low resolution

Key Words

radiation, speed of light, wavelength, ultraviolet, infrared, electromagnetic radiation, electromagnetic spectrum, electromagnetic waves, visible light, spectral lines, continuum (pl: continua), absorption lines, Fraunhofer lines, emission lines, law of conservation of energy, resolution, wave front, refraction, lens, focal length, field of view, eyepiece, objective, refraction, refracting telescope, chromatic aberration, reflecting telescope, parallel light, paraboloid, Newtonian, Cassegrain, Gregorian, prime focus, Ritchey-Chrétien, Schmidt camera, correcting plate, seeing, transparency, light pollution, extended objects, point objects, magnification, diffraction grating, slit, spectrograph, emulsion, photomultiplier, photometry, image tube, CCD, photons, grazing incidence, high-energy astrophysics, beam

Questions

1. What advantage does a reflecting telescope have over a refracting telescope?

2. What limits the resolving power of a 5-meter telescope?

3. Why does it matter whether a telescope is in the northern or southern hemisphere?

4. List the important criteria in choosing a site for an optical observatory meant to study stars and galaxies.

†5. Two reflecting telescopes have primary mirrors 2 m and 4 m in diameter. How many times more light is gathered by the larger telescope in any given interval of time?

†6. A 4-m telescope has a Cassegrain hole 1 m in diameter. What percentage of the total area is taken up by the hole?

7. Why might some stars appear double in blue light though they could not be resolved in red light?

†8. Use Dawes's limit to calculate how far apart two double stars have to be in order to be seen as separate with a 20-cm telescope.

9. For each of the following, identify whether it is a characteristic of a reflecting telescope, a refracting telescope, both, or neither. Give any limitations on the applicability of your answer.

   (a) Free of chromatic aberration.
   (b) Has more severe spherical aberration.
   (c) Requires aluminizing.
   (d) Can be used for photography.

   (e) Has an objective supported only by its rim.
   (f) Can be made in larger sizes.
   (g) Has a prime focus at which a person can work without blocking the incoming light.

10. Describe at least two methods that allow us to make large optical telescopes more cheaply than simply scaling up designs of previous large telescopes.

11. Why can't we use ordinary photographic plates to record infrared images?

†12. The Palomar–National Geographic Sky Survey made with the larger Palomar Schmidt telescope covers the sky in about 900 pairs of plates. Using the field of view given in the text, and comparing with a 30 arc min field of view for the Kitt Peak 4-m reflector, calculate how many red/blue pairs of photographs would be needed to cover the sky with the latter.

13. Why must we observe ultraviolet radiation using rockets or satellites, while balloons are sufficient for infrared observations?

14. Why can radio astronomers observe during the day, while optical astronomers are (for the most part) limited to nighttime observing?

15. What are the advantages of the Multiple Mirror Telescope? How does it compare with the 5-m telescope for studying faint objects?

16. What are the similarities and differences between making radio observations and using a reflector for optical observations? Compare the radiation path, the detection of signals, and limiting factors.

---

†This indicates a question requiring a numerical solution.

17. Why is it sometimes better to use a small telescope in orbit around the earth than it is to use a large telescope on a mountain top?

18. Why is it better for some purposes to use a medium-size telescope on a mountain instead of a telescope in space?

19. Compare the International Ultraviolet Explorer and the Space Telescope.

20. What are two reasons why the Space Telescope will be able to observe fainter objects than we can now study from the ground?

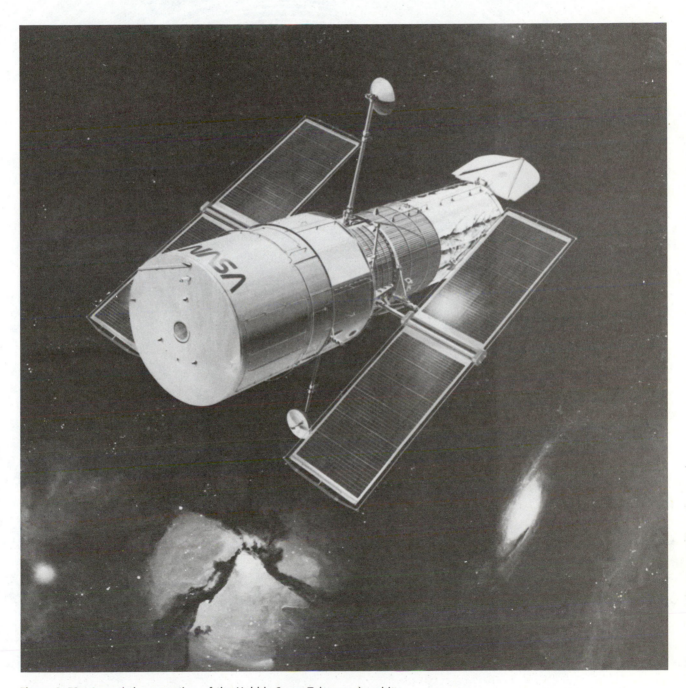

Figure 2–59 An artist's conception of the Hubble Space Telescope in orbit.

This 10-hour exposure shows southern star trails over the Anglo-Australian Telescope in Siding Spring, Australia.

Observing the Sky   3

Aims: To understand how astronomers locate objects in the sky, how objects appear to move in the sky, and time zones and calendars

When we look up at the sky on a dark night, from a location outside a city, we see a fantastic sight. If the moon is up, its splendor can steal the show; not only does its pearly white appearance draw our attention but also the light it gives off makes the sky so bright that we cannot see the other objects well. But when the moon is down, we see bright jewels in the inky sky. Generally, the brightest few shine steadily, which tells us that they are planets. The rest twinkle—sometimes gently, sometimes fiercely—which reveals them to be stars.

The Milky Way arches across the sky, and if we are in a good location on a dark night, it is quite obvious to the naked eye. (City dwellers may never see the Milky Way at all.) If you know where to look, from the northern hemisphere you can see a hazy spot that is actually a galaxy, rather than individual stars. This is the Andromeda Galaxy located in the constellation Andromeda. From the southern hemisphere, two other galaxies, the Large and Small Magellanic Clouds, can be easily seen.

Other objects are sometimes seen in the sky. If we are lucky, a bright comet may be there, but this happens rarely. More often, meteors—shooting stars—will dart across the sky above. And we may see the steady light of a spacecraft cross the sky in a few minutes. The lights of airplanes can be seen quite regularly.

3.1 The Constellations

Long, long ago, when Egyptian and other ancient astronomers were beginning to study and understand the sky, they divided the sky into regions containing fairly distinct groups of stars. The groups, called *constellations*, were given names, and stories were associated with them, perhaps to make them easier to remember.

Actually, the constellations are merely areas in the sky that happen to have stars in particular directions as we see them from the earth. There is no physical significance to the apparent groupings, nor are the stars in a given constellation necessarily associated with each other in any direct manner (Fig. 3–1).

The stories we now associate with the constellations come from Greek mythology (Fig. 3–2), though the names may have been associated with particular constellations more to honor Greek heroes than because the constellations actually looked like these people. Other civilizations (American Indians, for example) attached their own names, pictures, and stories to the stars (Fig. 3–3).

In his 1603 star atlas, Johann Bayer assigned Greek letters (Appendix 6) in alphabetical order to the stars in each constellation, roughly in order of brightness.

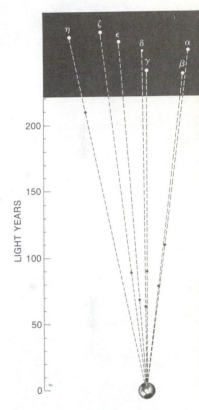

Figure 3–1 The stars we see as a constellation are actually differing distances from us. In this case we see the true distances of the stars in the Big Dipper, part of the constellation Ursa Major, in the lower part of the figure. Their appearance projected on the sky is shown in the upper part.

The International Astronomical Union put the scheme of constellations on a definite system in 1930. The sky was officially divided into 88 constellations (see appendix 9) with definite boundaries, and every star is now associated with one and only one constellation.

Figure 3–2 Pegasus, from the atlas of Hevelius. The horse appears reversed from what we see in the sky (in addition to being upside down), because the atlas is drawn as though we are looking at a globe of the sky from the outside instead of from the inside.

Ursa Major was originally the princess Callisto, an attendant of the goddess Juno, who became jealous of her. To protect Callisto, Jupiter turned her into a bear. However, when Callisto's son was about to kill the bear one day, Jupiter turned him into another bear and placed both of them in the sky.

Thus α (alpha) is usually the brightest star in a constellation, β (beta) is the second brightest, and so on. The Greek letters are used with the genetival form (''of . . .'') of the constellation name (Appendix 9), as in α Orionis, meaning ''alpha of Orion,'' which is Betelgeuse.

Sometimes familiar groupings in the sky do not make up a complete constellation. Such groupings are called *asterisms;* the Big and Little Dippers are examples, because they are really parts of the constellations Ursa Major (the Big Bear) and Ursa Minor (the Little Bear), respectively.

3.2 Twinkling

What about the twinkling? It is not a property of the stars themselves, but merely an effect of our earth's atmosphere. The starlight is always being bent by moving volumes of air in our atmosphere. The effect makes the images of the stars appear to be larger than points, to dance around slightly (such image motion is called *seeing,* as discussed in Section 2.9a) and to change rapidly in intensity (a property called scintillation). The change in intensity, scintillation, is what we non-technically call ''twinkling.''

The planets, unlike stars, do not usually seem to twinkle. They appear more steady because they are close enough to earth so that they appear as tiny disks large enough to be seen through telescopes. Though each point of the disk of light repre-

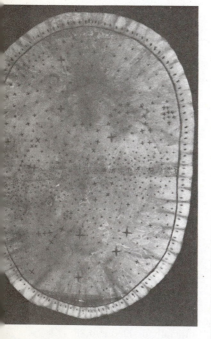

Figure 3–3 A sky chart from the Pawnee Indians who lived along the Platte River in what is now Nebraska.

Box 3.1

A set of four sky maps is bound into this book, one for each of the seasons. Choose the map that corresponds to your season of the year; the evening hours to which it corresponds are given.

Your horizon is one of the five curved lines closest to you on the map; the different horizons correspond to different latitudes, as is marked on them. Your zenith, the point directly overhead, is also marked on the map, in the upper center, for different latitudes. In designing these maps, specially drawn by Wil Tirion, we have allowed generous overlap beyond the zenith to make it easier to find constellations in those regions. Symbols describing the brightness of stars are shown in the legend.

senting a planet may change slightly in intensity, the disk is made of so many points of light that the total intensity doesn't change. To the naked eye the planets thus appear to shine more steadily than the stars. But when the air is especially turbulent, or when a planet is so low in the sky that we see it through a long path of air, even a planet may twinkle.

*3.3 Coordinate Systems

The stars and other astronomical objects that we study are at a wide range of distances from the earth. It is much harder to determine these distances than it is to determine the direction to these objects. Though we may have to know the distance to an object to understand how much energy it is giving off, we need to know only an object's direction to observe it. For this reason, it is useful to think of the astronomical objects as being at a common distance, all hung on the inside of a large (imaginary) sphere that we call the *celestial sphere*. In this section, we will see how to describe the positions of objects on the celestial sphere.

Both astronomers and geographers have established systems of coordinates—*coordinate systems*—to designate the positions of places in the sky or on the earth. The geographers' system is familiar to most of us: longitude and latitude. Lines (actually half-circles) of longitude called *meridians* run from the north pole to the south pole. The zero circle of longitude has been adopted, by international convention, to run through the former site of the Royal Greenwich Observatory in England (Fig. 3–4). We measure longitude by the number of degrees east or west an object is from the meridian that passes through Greenwich.

Latitudes are defined by parallel circles that run around the earth, all parallel to the equator. 0° of latitude corresponds to the equator; 90° of latitude corresponds to the poles.

The astronomers' system for the sky corresponds exactly to the geographers' system for the earth, except that the astronomers use the names *right ascension* for the celestial analogue of longitude and *declination* for the celestial analogue of latitude. Right ascension and declination are measured with respect to a *celestial equator*, which is the extension of the earth's equator into space, and *celestial poles*, which are on the extensions of the earth's axis of spin into space (Fig. 3–5).

Figure 3–4 The international zero circle of longitude in Greenwich, England.

Figure 3–5 The celestial equator is the projection of the earth's equator onto the sky, and the ecliptic is the sun's apparent path through the stars in the course of a year. The vernal equinox is one of the intersections of the ecliptic and the celestial equator and is the zero-point of right ascension. From a given location at the latitude of the United States, the stars nearest the north celestial pole never set and the stars nearest the south celestial pole never rise above the horizon.

Right ascension is measured along the celestial equator. Each hour of right ascension equals 15°. Declination is measured perpendicularly (−10°, −20°, etc.).

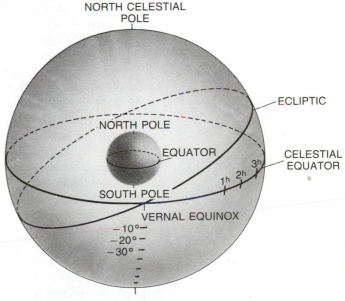

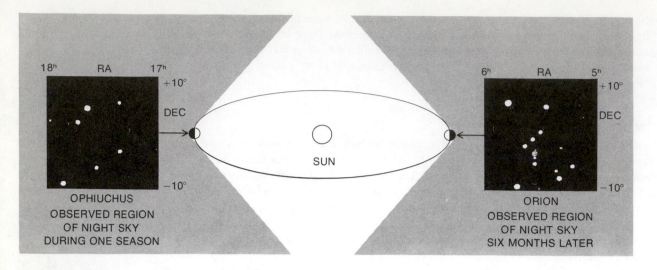

18ʰ RA 17ʰ
+10°
DEC
−10°
OPHIUCHUS
OBSERVED REGION
OF NIGHT SKY
DURING ONE SEASON

SUN

6ʰ RA 5ʰ
+10°
DEC
−10°
ORION
OBSERVED REGION
OF NIGHT SKY
SIX MONTHS LATER

Figure 3–6 For each time of the year, we can see the constellations that are in the direction away from the sun. Because the earth rotates much more rapidly than it revolves around the sun (24 hours compared with 365 days), all observers on the earth see the same constellations in a given season, even though it is daytime for some while it is nighttime for others.

Right ascension and declination form a coordinate system fixed to the stars. To observers on earth, the stars appear to revolve every 23 hours 56 minutes. The coordinate system thus appears to revolve at the same rate. Actually, of course, the earth is rotating and the stars and celestial coordinate system remain fixed.

Although the stars are fixed in their positions in the sky, the sun's position varies through the whole range of right ascension each year (Fig. 3–6). The path of the sun in the sky with respect to the stars is the *ecliptic*. The celestial equator is inclined by $23\frac{1}{2}°$ with respect to the ecliptic, since the earth's axis is tipped by that amount. The ecliptic and the celestial equator cross at two points. The sun crosses one of those points, the *vernal equinox*, on the first day of northern-hemisphere spring. The sun crosses the other intersection, the *autumnal equinox*, on the first day of autumn. The vernal equinox is the zero-point of right ascension.

Technically, we measure right ascension and declination with the use of *hour circles*, great circles running through the celestial poles. (A "great circle" on a sphere is a circle that is also on a plane that goes through the center of the sphere; it is the largest possible circle that can be drawn on the sphere's surface. On a sphere, the shortest distance between two points is on a great circle.) The hour circles cross the celestial equator perpendicularly. Right ascension is the angle to a body's hour circle, measured eastward along the celestial equator from the vernal equinox. Declination is the angle of an object north (+) or south (−) of the celestial equator along an hour circle.

Astronomers have set up a timekeeping system called *sidereal time* (sidereal means "by the stars"). Each location on earth has a unique *meridian*, the great circle linking the north and south poles and passing through the *zenith*, the point directly overhead. The sidereal time at any location is the length of time since the vernal equinox has crossed the meridian. The sidereal time is equal to the right ascension of any star on that place's meridian. Thus each location on the earth has a different sidereal time at each instant. Astronomers find this system convenient because they can consult clocks that are set to run on sidereal time. Thus by merely knowing the sidereal time, they can tell whether a star is favorably placed for observing.

Sidereal time and solar time differ, though both are caused by the rotation of the earth on its axis. A *sidereal day* (a day by the stars) is the length of time that it takes the vernal equinox to return to the celestial meridian. A *solar day* is the length of time that the sun takes to return to your meridian. Since the earth revolves around the sun once a year, by the time a day has passed, the earth has moved 1/365 of the way around the sun. Thus after the earth has turned far enough for the stars to return to the same apparent positions in the sky, the earth must still turn an additional 1/365 of 24 hours (24 hours/365 = 4 minutes) for the sun to return to your meridian (Fig. 3–7). A solar day is thus approximately 4 minutes (actually 3 minutes 56 seconds of time) longer than a sidereal day.

Every observatory has both sidereal clocks, for the astronomers to tell when to observe their stars, and solar clocks, for the astronomers to gauge when sunrise will come and to know when to go to dinner. A solar clock and a sidereal clock show the same time (on a 24-hour system) only one instant each year, the autumnal equinox. (An equinox is both a point in the sky and the time when the sun passes that point.) The next day the sidereal clock is 4 minutes ahead, the second day afterward it is 8 minutes ahead, and so on. Six months later the two clocks differ by 12 hours and the stars that were formerly at their highest at midnight are then at their highest at noon, when the sun is out. As a result, they may not be visible at all at that season.

Though the coordinates are basically fixed to the stars, a small effect called *precession* causes a slow drift of the coordinate system with respect to the stars with a 26,000-year period (Fig. 3–8). Precession takes place because the earth's axis doesn't always point exactly at the same spot in the sky; the axis rather traces out a small circle. (The effect is like the wobbling of a spinning top.) The axis takes approximately 26,000 years to return to the same orientation. About half-way through the cycle from now—in, say, A.D. 14,985—the north star will be Vega. But don't worry—Polaris will be our north star again in about 26,000 years.

Because the pole moves, the celestial equator—which is always 90° from the pole—moves. Hence the equinoxes, which are the intersections of the celestial equator and the ecliptic, precess—that is, they apparently move slowly along the ecliptic. Because of this *precession of the equinoxes,* the right ascension and declination of objects in the sky change slowly. (The formulas for computing these changes are given in Appendix 2.)

As a result of precession, one has to make small corrections in any catalogue of celestial positions to update them to the present time. Precession is a small effect—the change in celestial coordinates is less than one minute of arc (one sixtieth of a degree) per year, and is much less for some parts of the sky. Thus it need not be taken into account for casual observing, though good star maps and catalogues are now being recalculated and redrawn for "epoch 2000.0."

3.4 Motions in the Sky

At the latitudes of the United States, which range from +25° for the tip of the Florida Keys up to +49° for the Canadian border, and down to +19° in Hawaii and up to +67° in Alaska, the stars rise and set at angles to the horizon. In order to

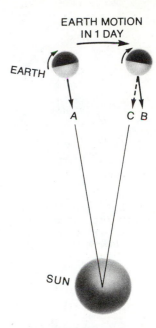

Figure 3–7 While the earth rotates once on its own axis with respect to the stars (one sidereal day), it also moves slightly in its orbit around the sun. Thus after one sidereal day, arrow A becomes arrow B. But one solar day has passed only when arrow B rotates a little farther and becomes arrow C. This takes an additional 4 minutes, making a solar day 4 minutes longer than a sidereal day.

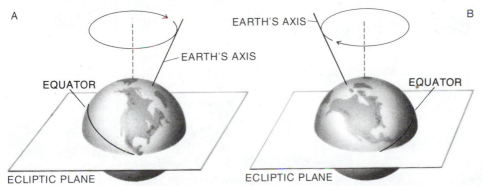

Figure 3–8 The earth's axis precesses with a period of 26,000 years. The two positions shown are separated by 13,000 years.

As the earth's pole precesses, the equator moves with it (since the earth is a rigid body). The celestial equator and the ecliptic will always maintain the 23½° angle between them, but the points of intersection, the equinoxes, will change. Thus, over the 26,000-year precession cycle, the vernal equinox will move through all the signs of the zodiac. It is now in the constellation Pisces and approaching Aquarius (and thus the celebration in the musical "Hair" of the "Age of Aquarius").

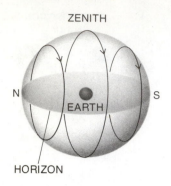

Figure 3–9 From the equator, the stars rise straight up, pass right across the sky, and set straight down.

Figure 3–10 From the pole, the stars move around the sky in circles parallel to the horizon, never rising or setting.

understand the situation, it is best first to visualize simpler cases. These concepts involve spherical geometry, which many individuals find difficult to visualize without practical experience. But a little experience with a telescope or in a planetarium can make right ascension and declination seem very easy.

If we were standing on the equator, the stars would rise perpendicularly to the horizon (Fig. 3–9). The north celestial pole would lie exactly on the horizon in the north, and the south celestial pole would lie exactly on the horizon in the south. Each star would rise somewhere on the eastern half of the horizon; each would remain ''up'' for twelve hours, and then would set. By waiting long enough, we would be able to see all stars, no matter what their declination. The sun, no matter what its declination, would also rise, be up for twelve hours, and then set, so day and night would each last twelve hours.

If, on the other hand, we were standing on the north pole, the north celestial pole would be directly overhead, and the celestial equator would be on the horizon (Fig. 3–10). All the stars would move around the sky in circles parallel to the horizon. Since the celestial equator would be on the horizon, we could see only the stars with northern declinations. The stars with southern declinations would never be visible.

Let us consider a latitude between the equator and the north pole, say $+40°$. There the stars seem to rise out of the horizon at oblique angles. The north celestial pole (with the north star nearby it) is always visible in the northern sky, and is at an *altitude* of 40° above the horizon. The star Polaris, a 2nd-magnitude star, happens to be located within one degree of the north celestial pole, and so is called the *pole star*. If you can see the north celestial pole, then the south celestial pole must be hidden; the celestial pole you can see is the only fixed point in the sky (Figs. 3–11 and 3–12).

If you point a camera at the sky and leave its lens open for a long time—many minutes or hours—the stars appear as trails. Those near the south celestial pole (see the figure opening this chapter) or the north pole (Fig. 3–13) move in relatively small and obvious circles around the poles. The circles followed by stars farther from the celestial poles are so large that sections of them seem straight (Fig. 3–14).

When astronomers want to know if a star is favorably placed for observing, they must know both its right ascension and declination (Section 3.3). By seeing if the right ascension is reasonably close to the sidereal time, they can tell if it is the best time of year at which to observe. But they must also know the star's declination to know how long it will be above the horizon each day.

Figure 3–11 Cartoon by Charles Schulz. © 1970 United Feature Syndicate, Inc.

© 1970 United Feature Syndicate, Inc.

Figure 3–12 Cartoon by Charles Schulz. © 1970 United Feature Syndicate, Inc.

© 1970 United Feature Syndicate, Inc.

Figure 3–13 Star trails over the 4-m telescope of the Kitt Peak National Observatory. (Richard E. Hill)

Figure 3–14 Near the celestial equator, the star circles are so large that they appear almost straight in this view past the University of California's Lick Observatory.

Telescopes are often mounted at an angle such that one axis—the *polar axis*— points directly at the north celestial pole. Since all stars move across the sky in circles centered at the pole, the telescope must merely turn about that axis to keep up with stellar motions. The other axis of the telescope is used to point the telescope in declination.

Since motion around only one axis is necessary to track the stars, one need have only a single small motor set to rotate once every 24 sidereal hours. This motor turns the polar axis in the direction opposite to the rotation of the earth. The principle is the same for a small telescope in your backyard as for the 5-meter telescope at Palomar. Arrangements of this type are called *equatorial mounts* (Fig. 3–15), since one axis rotates perpendicularly to the celestial equator.

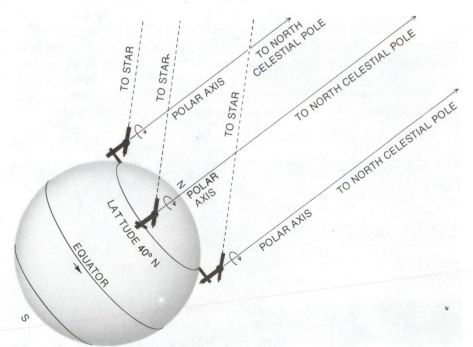

Figure 3–15 An equatorially mounted telescope need rotate on only one axis to keep pointing at a star. This axis is called the *polar axis;* it is fixed for telescopes in the northern hemisphere so that it points at the north celestial pole. Rotation of the telescope around this axis keeps up with the rotation of the earth on its axis. The axis perpendicular to the polar axis sweeps out a circle of declination in the sky and is called the *declination axis.*

Figure 3–16 Alt-azimuth mounts, with one "up-down" motion *(altitude)* and one "around" motion *(azimuth),* are used for mounting binoculars at public viewpoints.

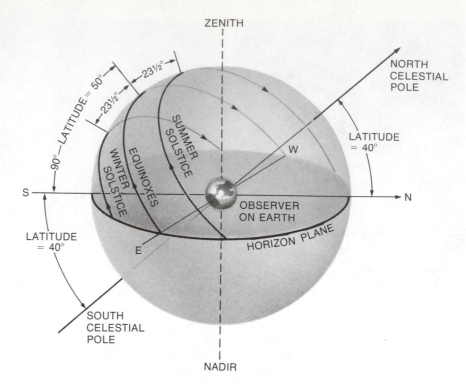

Figure 3–17 The path of the sun at different times of the year. Around the summer solstice, June 21, the sun is at its highest declination, rises highest in the sky, stays up longer (because, as shown, more of its path is above the horizon), and rises and sets farthest to the north. The opposite is true near the winter solstice, December 21. The diagram is drawn for latitude 40 degrees.

The alternative to this system is to mount a telescope such that one axis goes up-down (altitude) and the other goes around (azimuth). The azimuth motion is parallel to the horizon and the altitude motion is perpendicular. Binoculars on stands at scenic overlooks are mounted this way (Fig. 3–16). To track a star, continual adjustments have to be made in both axes, which used to be very inconvenient to do smoothly enough for photography. But now, computers make the necessary calculations that allow large new telescopes to be mounted using this *alt-azimuth* system.

A large alt-azimuth mount is often less expensive to construct then an equatorial mount of the same size, and can be housed in a smaller, less expensive dome. The

Figure 3–18 In this series taken in June from northern Norway, above the Arctic Circle, one photograph was taken each hour for an entire day. The sun never set, a phenomenon known as the *midnight sun.* Since the site was not at the north pole, the sun and stars move somewhat higher and lower in the sky in the course of a day.

6-m Soviet telescope, the Multiple Mirror Telescope in Arizona, and Britain's new 4.2-m Herschel telescope in the Canary Islands have alt-azimuth mounts for these reasons. Some amateur observers are using large (for amateurs), thin telescope mirrors in wooden or composition alt-azimuth mounts that have plastic bearings. They merely nudge the telescope along to follow the stars. These *Dobsonian* telescopes are a relatively inexpensive way to obtain large apertures for visual observing, though they cannot be used for photography because they do not track the stars.

*3.4a Positions of the Sun, Moon, and Planets

As the sun moves along the ecliptic each year, it crosses the vernal equinox on approximately March 21st and the autumnal equinox on approximately September 23rd, respectively. On these days, the sun's declination is 0°; the sun's declination varies over the year from $+23\frac{1}{2}°$ to $-23\frac{1}{2}°$ (Fig. 3–17).

These points are called *equinoxes* ("equal nights"; *nox* is Latin for night) because the daytime and the nighttime are supposedly equal on these days. Actually, because refraction (bending) of light by the earth's atmosphere makes the sun appear to rise a little early and set a little late, and the fact that the top of the sun rises ahead of the middle of the sun, the daytime exceeds the nighttime at U.S. latitudes by about 10 minutes on the days of the equinoxes. The days of equal daytime and nighttime precede the vernal equinox and follow the autumnal equinox by a few days.

If we were at the north pole, whenever the sun had a northern declination, it would be above our horizon and we would have daytime. The sun would move in a circle all around us, moving essentially parallel to the horizon. From day to day it would appear slightly higher in the sky for 3 months, and then move lower. The date when it is highest in the sky is the summer *solstice*. It occurs on approximately June 21st each year. On that date the sun is $23\frac{1}{2}°$ above the horizon because its declination is $+23\frac{1}{2}°$. The time of the year when the sun never sets is known as the time of the *midnight sun* (Fig. 3–18 and Color Plate 8), which lasts six months at the poles and shorter times at locations other than the poles. The sun crosses the celestial equator and begins to have northern declination at the vernal equinox. The sun crosses the celestial equator in the other direction six months later at the autumnal equinox.

Since the sun goes $23\frac{1}{2}°$ above the celestial equator, the midnight sun is visible at some time anywhere within $23\frac{1}{2}°$ of the north pole, a boundary at $66\frac{1}{2}°$ latitude known as the Arctic Circle. The midnight sun within $23\frac{1}{2}°$ of the south pole, within the Antarctic Circle, is six months out of phase with that near the north pole.

When the sun is at the summer solstice, it is at its greatest northern declination

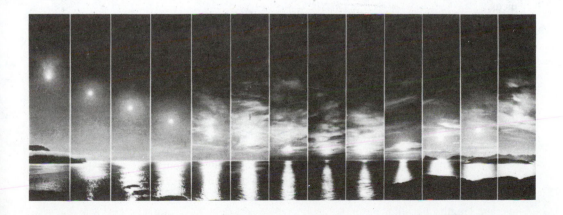

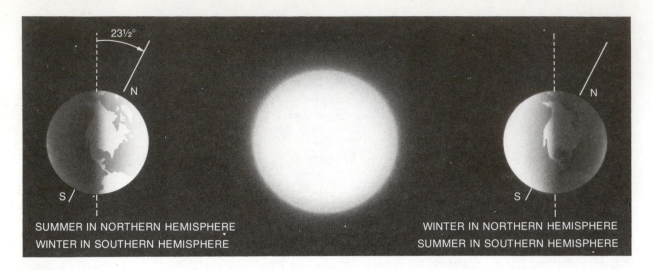

SUMMER IN NORTHERN HEMISPHERE
WINTER IN SOUTHERN HEMISPHERE

WINTER IN NORTHERN HEMISPHERE
SUMMER IN SOUTHERN HEMISPHERE

Figure 3–19 The seasons occur because the earth's axis is tipped with respect to the plane of the orbit in which it revolves around the sun. The dotted line is drawn perpendicularly to the plane of the earth's orbit. When the northern hemisphere is tilted toward the sun it has its summertime; at the same time, the southern hemisphere is having its winter. At both locations of the earth shown, the earth rotates through many 24-hour day-night cycles before its motion around the sun moves it appreciably. The diagram is not to scale.

and is above the horizon of all northern hemisphere observers for the longest time each day. Thus daytimes in the summer are longer than daytimes in the winter, when the sun is at its lowest declinations. In the winter, the sun not only is above the horizon for a shorter period each day but also never rises very high in the sky. As a result, the weather is colder. The instant of the sun's lowest declination is the winter solstice. Winter in the northern hemisphere corresponds to summer in the southern hemisphere.

The seasons (Fig. 3–19), thus, are caused by the variation of declination of the sun, which in turn is caused by the fact that the earth's axis of spin is tipped by $23\frac{1}{2}°$ with respect to the perpendicular to the plane of the earth's orbit around the sun.

The moon goes around the earth once each month, and so the moon's right ascension changes through the entire range of ascension once each month. Since the moon's orbit is inclined to the celestial equator, the moon's declination also varies.

The planets' motions in the sky are less easy to categorize, but they also change their right ascension and declination from day to day. Their positions are listed or graphed in *A Field Guide to the Stars and Planets,* in monthly magazines and bulletins, and in *The Astronomical Almanac* published each year under government auspices.

*3.5 Time and the International Date Line

Every city and town on earth used to have its own time system, based on the sun, until widespread railroad travel made this inconvenient. In 1884, an international conference agreed on a series of longitudinal time zones. Now all localities in the same zone have a standard time (Fig. 3–20). Since there are twenty-four hours in a day, the 360° of longitude around the earth are divided into 24 standard time zones, each 15 degrees wide. Each time zone is centered on a meridian of longitude exactly divisible by 15. Because the time is the same throughout each zone, the sun is not directly overhead at noon at each point in a given time zone, but in principle is less than about a half-hour off. Standard time is based on a *mean solar day,* the average length of a solar day.

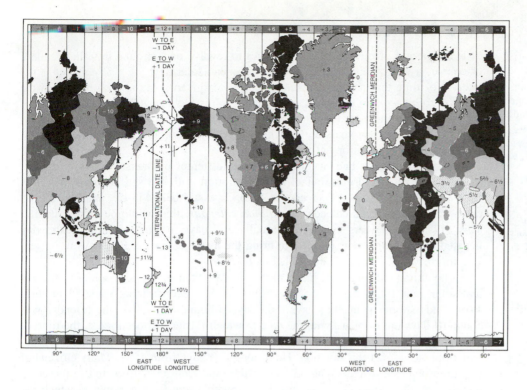

Figure 3–20 Although in principle the earth is neatly divided into 24 time zones, in practice, political and geographic boundaries have made the system much less regular. The existence of daylight-saving time in some places and not in others further confuses the time zone system. Other countries, shown with stripes, have time zones that differ by one-half hour from a neighboring zone. India and Nepal actually differ by 10 minutes! At the international date line, the date changes. In the U.S., most states have daylight-saving time for six months a year—from the last Sunday in April until the last Sunday in October. But Arizona, Hawaii, and parts of Indiana have standard time year round. Alaska's time zones were changed in 1983, placing almost all of the state in a single zone, one hour earlier than Pacific time.

As the sun seems to move in the sky from east to west, the time in any one place gets later. We can visualize noon, and each hour, moving around the world from east to west, minute by minute. We get a particular time back 24 hours later, but if the hours circled the world continuously the date would not have changed. So we specify a north-south line and have the date change there. We call it the *international date line*. England won for Greenwich, then the site of the Royal Observatory, the distinction of having the basic line of longitude, 0°. Realizing that the international date line would disrupt the calendars of those who crossed it, that line was put as far away from the populated areas of Europe as possible—near or along the 180° longitude line. The international date line passes from north to south through the Pacific Ocean and actually bends to avoid cutting through continents or groups of islands, thus providing them with the same date as their nearest neighbor.

In the summer, in order to make the daylight last into later hours, many countries have adopted daylight-saving time. Clocks are set ahead 1 hour on a certain date in the spring. Thus if darkness falls at 6 p.m. E.S.T., that time is called 7 p.m. E.D.T., and most people have an extra hour of daylight after work. In most places, that hour is taken away in the fall, though some places have adopted daylight-saving time all year. The phrase to remember to help you set your clocks is "fall back, spring ahead." Of course, daylight-saving time is just a bookkeeping change in how we name the hours, and doesn't result from any astronomical changes.

*3.6 Calendars

The period of time that the earth takes to revolve once around the sun is called, of course, a *year*. This period is about 365¼ mean solar days. A *sidereal year* is the interval of time that it takes the sun to return to a given position with respect to the stars. A *solar year* (in particular, a *tropical year*) is the interval between passages of the sun through the vernal equinox, the point where the ecliptic crosses the celestial equator. Since tropical years are growing shorter by about half a second per century, we refer to a standard tropical year: 1900.

Roman calendars had, at different times, different numbers of days in a year, so the dates rapidly drifted out of synchronization with the seasons (which follow solar years). Julius Caesar decreed that 46 B.C. would be a 445-day year in order to catch up, and defined a calendar, the *Julian calendar,* that would be more accurate. This calendar had years that were normally 365 days in length, with an extra day inserted every fourth year in order to bring the average year to 365¼ days in length. The fourth years were, and are, called *leap years.*

The Julian calendar was much more accurate than its predecessors, but the actual solar year 1985 will be 365 days 5 hours 45.6 seconds long, some 11 minutes 14 seconds shorter than 365¼ days (365 days 6 hours). By 1582, the calendar was about 10 days out of phase with the date at which Easter had occurred at the time of a religious council 1250 years earlier, and Pope Gregory XIII issued a bull—a proclamation—to correct the situation. He dropped 10 days from 1582. Many citizens of that time objected to the supposed loss of the time from their lives and to the commercial complications. Does one pay a full month's rent for the month in which the days were omitted, for example? ''Give us back our fortnight,'' they cried.

In the Gregorian calendar, years that are evenly divisible by four are leap years, except that three out of every four century years, the ones not divisible evenly by 400, have only 365 days. Thus 1600 was a leap year; 1700, 1800, and 1900 were not; and 2000 will again be a leap year.

Although many countries adopted the Gregorian calendar as soon as it was promulgated, Great Britain (and its American colonies) did not adopt it until 1752, when 11 days were skipped. As a result, we celebrate George Washington's birthday on February 22nd, even though he was born on February 11. Actually, since the year had begun in March instead of January, 1752 was cut short. Washington was born in February 1731, often then written February 1731/32, but we now refer to his date of birth as February 22nd, 1732 (Fig. 3–21). The Gregorian calendar is the one in current use. It will be over 3000 years before this calendar is as much as one day out of step.

The name of the fifth month, formerly Quintillis, was changed to honor Julius Caesar; in English we call it July. The year then began in March; the last four months of our year still bear names from this system of numbering. Augustus Caesar, who carried out subsequent calendar reforms, renamed August after himself. He also transferred a day from February in order to make August last as long as July.

Figure 3–21 In George Washington's family bible his date of birth is given as 1731/2. Some contemporaries would have said 1731; we now say 1732.

Summary and Outline

The constellations (Section 3.1)
    Stars in them are not necessarily physically grouped.

Twinkling (Section 3.2)
    Stars appear to twinkle; planets usually do not.
    Twinkling is caused in the earth's atmosphere.

Coordinate systems (Section 3.3)
    Celestial longitude and latitude are right ascension and declination.
    The celestial equator and celestial poles are the points in the sky that are on the extensions of the earth's equator

and poles into space. The ecliptic and the celestial equator cross at the equinoxes.

Sidereal time is time by the stars.

Sidereal day: a given right ascension returns to your meridian

Solar day: the sun returns to your meridian

Precession is the slow drift in the coordinate system; 26,000-year period

Motions in the sky (Section 3.4)
    Stars rise and set; at the earth's equator we would see them do so perpendicularly to the horizon.

At the earth's poles, we would see the stars move around parallel to the horizon.

At or near the poles, the sun and moon rise and set only when they change sufficiently in declination; the sun goes through this sequence once a year—thus ''midnight sun.''

The seasons are caused by the sun's variations in declination.

Telescopes can be mounted so that motion on one axis allows the earth's rotation to be counteracted; such mounts are called ''equatorial.'' Alt-azimuth mounts have separate motions in altitude and azimuth; it takes both motions to follow the stars.

Time (Section 3.5)

Standard time is based on a mean solar day.

The date changes at the international date line.

Daylight-saving time: ''fall back, spring ahead''

Calendars (Section 3.6)

Leap years are needed to keep up with the 365¼-day year.

The Julian calendar, introduced by Julius Caesar, was a reasonably good calendar, but, over the centuries, the days drifted.

We now use the Gregorian calendar, set up in 1582.

Key Words

constellations, asterisms, seeing, celestial sphere*, coordinate systems*, meridians*, right ascension*, declination*, celestial equator*, celestial poles*, ecliptic*, vernal equinox*, autumnal equinox*, hour circles*, sidereal time*, meridian*, zenith*, sidereal day*, solar day*, precession*, altitude, pole star, polar axis, equatorial mounts, alt-azimuth, Dobsonian, solstice*, midnight sun*, mean solar day*, international date line*, year*, sidereal year*, solar year (tropical year)*, Julian calendar*, leap years*

---

*These terms are found in optional sections.

Questions

1. Explain why we cannot tell by merely looking in the sky that stars in a given constellation are at different distances, while in a room we can easily tell that objects are at different distances from us. What is the difference between the two situations?

2. Why is the Big Dipper only an asterism while the Big Bear is a constellation?

3. Can a star in the sky not be part of a constellation? Explain.

4. Can you reason that all the stars in the constellation Pegasus are close together, so that they must have formed at about the same time? Explain.

5. What is the difference between ''seeing'' and ''twinkling''?

6. If you look toward the horizon, are the stars you see likely to be twinkling more or less than the stars overhead? Explain.

7. Is the planet Uranus, which is in the outskirts of the solar system, likely to twinkle more or less than the nearby planet Venus?

8. Explain how it is that some stars never rise in our sky, while others never set.

9. By comparing their right ascensions and declinations (Appendix 5, the brightest stars), describe whether Sirius and Canopus, the two brightest stars, are close together or far apart in the sky. Explain.

†10. When Arcturus (Appendix 5) is due south of you, what is the sidereal time where you are?

†11. Between the vernal equinox, March 21st, and the autumnal equinox, about 6 months later, ignoring precession,
   (a) by how much does the right ascension of the sun change?

   (b) by how much does the declination of the sun change?
   (c) by how much does the right ascension of Sirius change?
   (d) by how much does the declination of Sirius change?

12. (a) How does the declination of the sun vary over the year?
   (b) Does its right ascension increase or decrease from day to day? Justify your answer.

13. We normally express longitude on the earth in degrees. Why would it make sense to express longitude in units of time?

†14. By how many hours do sidereal clocks and solar clocks differ at the summer solstice?

†15. Divide the number of minutes in a day by 365 to find out by how many minutes each day the sidereal day drifts with respect to the solar day, as the earth goes 1/365 of the way around the sun. Show your work.

16. If a planet always keeps the same side toward the sun, how many sidereal days are there in a year on that planet?

†17. When it is 6 p.m. on October 1st in New York City, what time of day and what date is it in Tokyo?

†18. When it is noon on April 1st in Los Angeles, describe how to use Figure 3–20 to find the date and time in China. Follow through both going westward across the international date line and going eastward across the Atlantic, and describe why you get the same answer.

19. What is the advantage of an equatorial mount?

20. What has changed that has led large telescopes to be placed on alt-azimuth mounts?

---

†This indicates a question requiring a numerical solution.

Part II The Stars

When we look up at the sky at night, most of the objects we see are stars. In the daytime, we see the sun, which is itself a star. The moon, the planets, even a comet may give a beautiful show, but, however spectacular, they are only minor actors on the stage of observational astronomy.

From the center of a city, we may not be able to see very many stars, because city light scattered by the earth's atmosphere makes the light level of the sky brighter than most stars. All together, about 6000 stars are bright enough to be seen with the unaided eye under good observing conditions.

Even a cursory glance at the sky shows patterns in the distribution of stars—the **constellations.** The names that the ancients gave to these areas, associating them with figures and objects from their myths and religions, are still used to this day. It takes a great imagination to see Orion (the Hunter) or Draco (the Dragon) in the sky above, but it is nice to be like Saint-Exupéry's Little Prince (Fig. 4–1), and not lose our sense of wonder.

The stars we see defining a constellation may be at very different distances from us. One star may appear bright because it is relatively close to us even though it is intrinsically faint; another star may appear bright even though it is far away because it is intrinsically very luminous. When we look at the constellations, we are observing only the directions of the stars and not whether the stars are physically very close together. Nor do the constellations tell us anything about the nature of the stars. Astronomers tend not to be interested in the constellations because the constellations do not give useful information about how the universe works. After all, the constellations merely tell the directions in which stars lie. Most astronomers want to know *why* and *how.* Why is there a star? How does it shine? Such studies of the workings of the universe are called *astrophysics.* Almost all modern day astronomers are astrophysicists as well, for they not only make observations but also think about their meaning.

Some properties of the stars that can be seen with the naked eye tell us about the natures of the stars themselves. For example, some stars seem blue-white in the sky, while others appear slightly reddish. From information of this nature, astronomers are able to determine the temperatures of the stars. Chapter 4 is devoted to the basic properties of stars and some of the ways we find out such information. Chapter 5 describes some of the ingenious ways we use basic information about stars. We can sort stars into categories and tell how far away they are and how they move in space.

Most stars occur in pairs or in larger groups, and we shall discuss types of groupings and what we learn from their study in Chapter 6. Binary stars are important, for example, in finding the masses of stars.

Some stars vary in brightness, a property that astronomers use to tell us the scale of the universe itself. The study of star clusters leads to our understanding of the ages of stars and how they evolve.

In Chapter 7 we study an average star in detail. It is the only star we can see close up—the sun. The phenomena we observe on the sun take place on other stars as well, as has recently been verified by telescopes in space.

In recent years, the heavens have been studied in other ways besides observing the ordinary light that is given off by many astronomical objects. Other forms of radiation—radio waves, x-rays, gamma rays, ultraviolet, and infrared—and interstellar particles called cosmic rays are increasingly studied from the earth's surface or from space. Many a contemporary astronomer—even many who consider themselves observers (who mainly carry out observations) rather than theoreticians (who do not make observations, but rather construct theories)—has never looked through an optical telescope. Still, astronomy began with optical studies, and our story begins there. This part of the book will tell us how we study stars, and what they teach us. Part III will take up the life stories of stars.

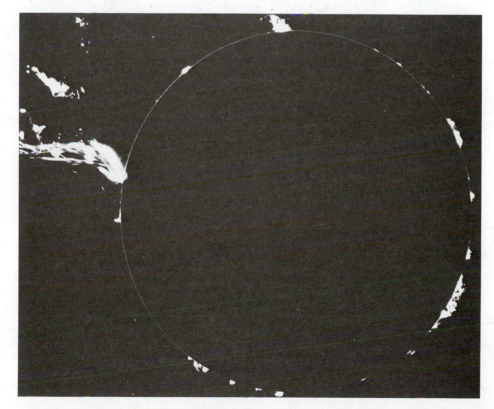

Many prominences appear around the sun, and a giant eruptive prominence ejects matter westward (toward the left). This hydrogen-light photograph, taken from a high mountain in Hawaii, looks past a dark "occulting disk" that hides the sun's surface.

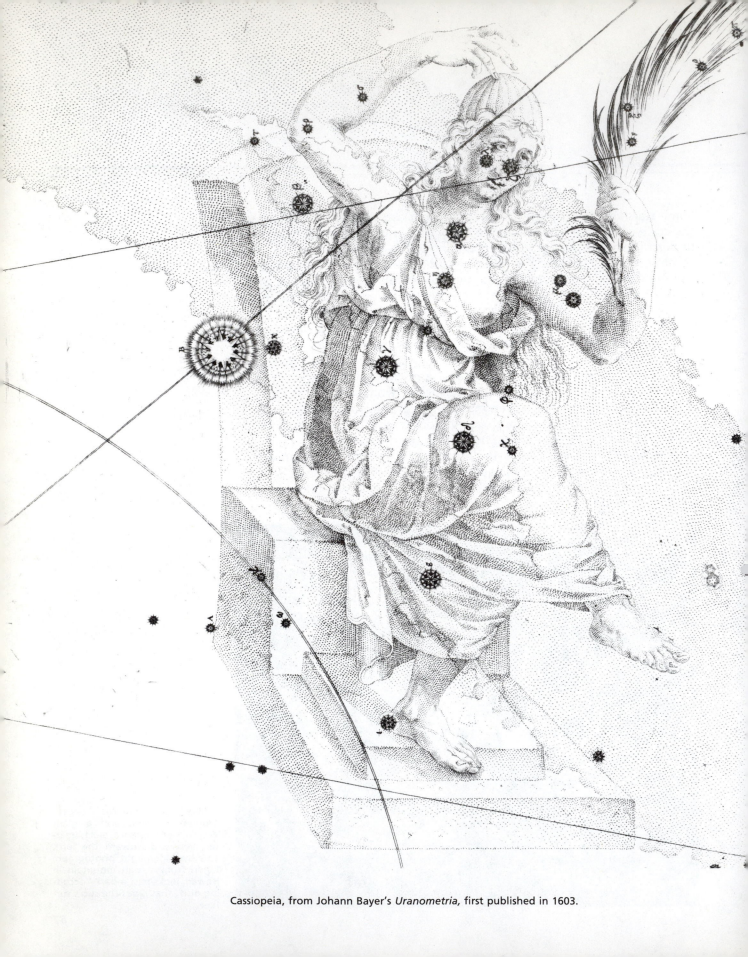

Cassiopeia, from Johann Bayer's *Uranometria,* first published in 1603.

Ordinary Stars   4

Aims: To study the colors and spectra of ordinary stars, and to see how this tells us about the temperatures of the surfaces of the stars

What are these stars that we see as twinkling dots of light in the nighttime sky? The stars are luminous balls of gas scattered throughout space. All the stars we see are among the trillion stellar members of a collection of stars and other matter called the Milky Way Galaxy. The sun, which is the star that gives most of the energy to our planet, is so close to us that we can see detail on its surface, but it is just an ordinary star like the rest.

Stars are balls of gas held together by the force of gravity, the same force that keeps us on the ground no matter where we are on the earth. Long ago, gravity compressed large amounts of gas and dust into dense spheres that became stars. The dust vaporized, and balls of gas remained.

Stars generate their own energy and light. It has been only 50 years since the realization that the stars shine by nuclear fusion. Now we are looking to the nearest star, our sun, to be used either as a direct source of energy in the form of its light or as a laboratory to study the physics of hot gases in magnetic fields. The latter study may tell us how to tame the fusion process and recreate it on earth.

We see only the outer layers of the stars; the interiors are hidden from our view. The stars are gaseous through and through; they have no solid parts. The outer layers do not generate energy by themselves, but glow from energy transported outward from the stellar interiors. So when we study light from the stars we are not observing the processes of energy generation directly. We will study these processes in Chapter 8, where we start to study the life cycles of stars. In this chapter, we shall concentrate on how we study the stars.

4.1 The Colors of Stars

In Section 2.1 we saw that the different colors we perceive result from radiation at certain ranges of wavelength. For example, an object appears blue because most of its radiation (or at least most of its visible radiation) is at wavelengths between roughly 4300 and 5000 Å. We will now see that how a star's radiation is distributed in wavelength depends on the temperature of the surface of that star. By measuring the color of a star, we can determine the wavelength distribution and thus deduce the star's temperature.

To measure a star's color, we can graph the intensity of light from the star at each wavelength. We can measure the intensity by passing the light through a spectrograph and measuring the intensity at each wavelength. More simply, we can measure the light that emerges from filters, each of which allows only a certain group of wavelengths to pass. As few as three filters are enough to define the shape of the curve.

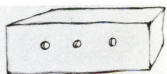

Figure 4–1 (A) The Little Prince is shown at top on Asteroid B-612, his home. (B) When asked by the Little Prince to draw a sheep, the author drew the above. "This is only his box. The sheep you asked for is inside." The Little Prince replied that the picture was exactly the way he wanted it, and asked for quiet because the sheep had gone to sleep. We too must not allow our knowledge to cause us to lose our sense of wonder about the stars.
From *The Little Prince* by Antoine de Saint-Exupéry.

Box 4.1 Temperature

We must choose the units that we will use to discuss temperature. In America, until recently, we have used for most everyday purposes the Fahrenheit temperature scale. On this scale, water freezes at 32°F and boils at 212°F, a span of 180°F between freezing and boiling. Absolute zero (the minimum temperature theoretically possible for any material thing) is −459.7°F (Fig. 4–2). The Fahrenheit scale is not used in the physical sciences. Most of the rest of the world uses degrees centigrade; a modern minor adjustment of the scale is called degrees Celsius. The centigrade and Celsius scales differ by less than 0.1° centigrade or Celsius.) The United States is changing over gradually to this system; most scientists already have for their professional work. On the centigrade scale, water freezes at 0°C and boils at 100°C, a span of 100°C; a Celsius degree (which is defined in terms of absolute zero) is essentially the same size. Thus 180 Fahrenheit degrees are equivalent to 100 Celsius degrees; simplifying gives 1.8°F per 1°C.

On the Celsius scale absolute zero is about −273°C. For us on earth the temperatures with which we deal conveniently range around zero or a few tens of degrees. But the freezing and boiling points of water don't have much relevance to the stars. Astronomers choose to use the Kelvin temperature scale, which has its zero point at absolute zero instead of at the freezing point of water. The size of a Kelvin degree (now officially called one kelvin and abbreviated K instead of °K) is the same as the size of a Celsius degree. Thus absolute zero is 0 K, water freezes at +273 K, and water boils at +373 K.

It is simple to change from Celsius degrees to kelvins; simply add 273 to the Celsius temperature. To change from Fahrenheit to Celsius is more complicated. One must first subtract 32°F, to align the freezing points, and then divide by 1.8 to transform the smaller Fahrenheit degrees into the larger Celsius degrees. The formula is °C = (°F − 32) ÷ 1.8.

Stellar temperatures, whether expressed in Fahrenheit or Celsius degrees or in kelvins, are very high—in the thousands or tens of thousands (Fig. 4–3).

An easy way to convert from degrees Celsius to degrees Fahrenheit is to take the Celsius number, double it, subtract 10 per cent of the result, and add 32. Example: take 30°C, double it to get 60°, subtract 10 percent of 60° = 6° from 60° to get 54°, and add 32° to get the answer of 86°F.

When we examine a set of these graphs, we can see that, spectral lines aside, the radiation follows a fairly smooth curve (Fig. 4–4). The radiation "peaks" in intensity at a certain wavelength (that is, has a maximum in its intensity at that wavelength), and decreases in amount more slowly on the long wavelength side of the peak than it does on the short wavelength side.

Figure 4–2 Temperature scales.

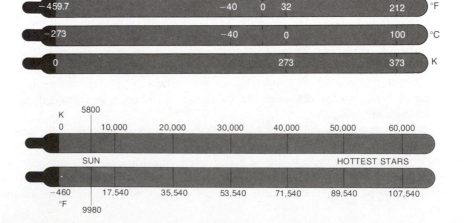

Figure 4–3 Stellar temperature scales.

Table 4–1 Wien's Displacement Law

Temperature	Wavelength of Peak	Spectral Region of Wavelength of Peak
3 K	.97 mm	infrared-radio
3,000 K	9660 Å	infrared
6,000 K	4830 Å	green
12,000 K	2415 Å	ultraviolet
24,000 K	1207 Å	ultraviolet

The graphs can be best understood by first considering the radiation that represents different temperatures. When you first put an iron poker in the fire, it glows faintly red. Then as it gets hotter it becomes redder. If the poker could be even hotter without melting, it would turn yellow, white, and then blue-white. White hot is hotter than red hot.

A set of physical laws governs the sequence of events when material is heated. As the material gets hotter, the peak of radiation shifts toward shorter wavelengths (the blue). This shift of wavelength of the peak is known as *Wien's displacement law* (Box 4.2 and Table 4–1).

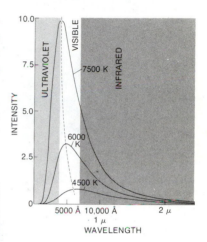

---

*Box 4.2 Wien's Displacement Law

$\lambda_{max}T$ = constant,

where λ_{max} is the **wavelength** at which the energy given off is at a **maximum,** and T is the temperature. The numerical value of the constant is given in Appendix 2.

Also, as the material gets hotter, the total energy of the radiation grows quickly. Notice in Fig. 4–4 how the total energy, which is the area under each curve, is much greater for higher temperatures. The energy follows the *Stefan-Boltzmann law* (Box 4.3). This law says that the total energy emitted from each square centimeter of a source in each second grows as the fourth power of the temperature ($T \times T \times T \times T$). If we compare two square centimeters of gas at different temperatures, the Stefan-Boltzmann law tells us that the ratio of the energies emitted is the fourth power of the ratio of the temperatures. For example, consider the same gas at 10,000 K and at 5000 K. Because 10,000 K is twice 5000 K, the 10,000 K gas gives off $2^4 = 2 \times 2 \times 2 \times 2 = 16$ times more energy than does the same amount of gas at 5000 K. (We must measure our scale of degrees from absolute zero, the minimum temperature possible.) Note how even a small increase in temperature can make a star much brighter.

The Stefan-Boltzmann law relates the amount of energy a gas gives off to its temperature. Also, a larger volume of gas at the same temperature gives off more energy.

Figure 4–4 The intensity of radiation for different stellar temperatures, according to Planck's law (Section 4.2). The wavelength scale is linear, i.e., equal spaces signify equal wavelength intervals. The dotted line shows how the peak of the curve shifts to shorter wavelengths as temperature increases; this is known as Wien's displacement law. The rate of growth of the total energy emitted at a given temperature—the area under the curve—is given by the Stefan-Boltzmann law.

Note that we are discussing, for the moment, only the continuous part of the spectrum; it is possible to observe both a continuous spectrum and superimposed spectral lines in a single spectrum.

---

*Box 4.3 Stefan-Boltzmann Law

Energy = σT^4.

Energy *(E)* is the "strength" of the radiation. The numerical value of σ, the Stefan-Boltzmann constant, is given in Appendix 2.

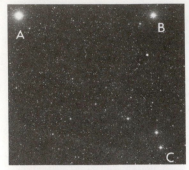

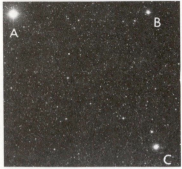

Figure 4–5 The top photograph is a view of several stars in blue light, and the bottom photograph is a view of the same stars in red light. Star A appears about the same brightness in both. Star B, a relatively hot star, appears brighter in the blue photograph. Star C, a relatively cool star, appears brighter in the red photograph.

Figure 4–6 Max Planck *(left)* presented an equation in 1900 that explained the distribution of radiation with wavelength. Five years later, Albert Einstein, in suggesting that light acts as though it is made up of particles, found an important connection between his new theory and the earlier work of Planck. The photograph was taken years later.

The overall color a star appears in the sky depends on the color of the peak of a star's energy distribution. Thus, from Wien's displacement law, we see that the colors of stars in the sky tell us the stars' temperatures. The reddish star Betelgeuse, for example, is a comparatively cool star, while the blue-white star Sirius is very hot. By simply measuring the intensity of a star with a set of filters at different wavelengths, we can find out the temperature of the star (Fig. 4–5). Astronomers measure colors directly at the telescope using a procedure discussed in Section 5.3a.

4.2 Planck's Law and Black Bodies

The laws of distribution of energy from a heated gas are much more general than their particular application to the stars. In principle, if any gas is heated, its continuous emission follows a certain peaked distribution called *Planck's law,* suggested in 1900 by the German physicist Max Planck (Fig. 4–6). Wien's displacement law and the Stefan-Boltzmann law, which had been discovered earlier from study of experimental data, can be derived from Planck's law. Actual gases may not, and usually don't, follow Planck's law exactly. For example, Planck's law governs only the continuous spectrum; any line emission or absorption does not follow Planck's law.

Because of these difficulties, scientists consider an idealized case: a fictional object called a *black body,* which is defined as something that absorbs all radiation that falls on it. Note that the "black" in a black body is an idealized black that absorbs 100 per cent of the light that hits it. Ordinary black paint, on the other hand, does not absorb 100 per cent of the light that hits it. Further, something that looks black in visible light might not seem very black at all in the infrared, and so would not be a real black body.

An important physical law that governs the radiation from a black body says that anything that is a good absorber is a good emitter too. The radiation from a black body follows Planck's law. As a black body is heated, the peak of the radiation shifts toward the short wavelength end of the spectrum (Wien's displacement law) and the total amount of radiation grows rapidly (Stefan-Boltzmann law). The radiation from a black body at a higher temperature is greater at every wavelength than the radiation from a cooler black body, but as the temperature increases, the intensity of radiation at shorter wavelengths increases. Note that the curves of energies emitted by black bodies at different temperatures in Fig. 4–4 do not cross.

To a certain extent, the atmospheres of stars appear as black bodies, in that over a broad region of the spectrum that includes the visible, the continuous radiation from stars follows Planck's law fairly well.

4.3 The Formation of Spectral Lines

Spectral lines arise when there is a change in the amount of energy present in any given atom. An atom (Fig. 4–7) can be thought of as consisting of particles in its core, called the *nucleus,* surrounded by orbiting particles, called *electrons.* Examples of nuclear particles are the *proton,* which carries one unit of electric charge, and the *neutron,* which has no electric charge. (Protons and neutrons are, in turn, made up of particles called *quarks,* as we discuss further in Chapter 17.) An electron, which has one unit of negative electric charge, is only 1/1800 as massive as a proton or a neutron. Normally, an atom has the same number of electrons as it has protons. The negative and positive charges balance, leaving the atom electrically neutral.

The amount of energy that an atom can have is governed by the laws of *quantum mechanics,* a field of physics that was developed in the 1920's and whose applications to spectra were further worked out in the 1930's. According to the laws of quantum mechanics, light (and other electromagnetic radiation) has the properties of waves in some circumstances and the properties of particles under other circumstances. One cannot understand this dual set of properties of light intuitively, and some of the greatest physicists of the time had difficulty adjusting to the idea. Still, quantum mechanics has been worked out to explain experimental observations of spectra, and is now thoroughly accepted.

According to quantum mechanics, an atom can exist only in a specific set of energy states, as opposed to a whole continuum (continuous range) of energy states. An atom can have only discrete values of energy; that is, the energy cannot vary continuously. For example, the energy corresponding to one of the states might be 10.2 electron volts (eV). The next energy state allowable by the quantum mechanical rules might be 12.0 eV. We say that these are *allowed states*. The atom **simply cannot have** an energy between 10.2 and 12.0 eV. The energy states are discrete (separate and distinct); we say that they are "quantized," and hence the name "quantum mechanics." The discrete energy states are called *energy levels*.

When an atom drops from a higher energy state to a lower energy state without colliding with another atom, the difference in energy is sent off as a bundle of radiation. This bundle of energy, a *quantum*, is also called a *photon*, which may be thought of as a particle of electromagnetic radiation (Fig. 4–8). Photons always travel at the speed of light. Each photon has a specific energy, which does not vary.

A link between the particle version of light and the wave version is seen in the equation that relates the energy of the photon with the wavelength it has: $E = hc/\lambda$. In this equation, E is the energy, h is a constant named after Planck, c is the speed of light (in a vacuum, as always), and λ (the Greek letter lambda) is the wavelength. Since λ is in the denominator on the right hand side of the equation, a small λ corresponds to a photon of great energy. (We are dividing by a small number, and thus get a large result.) When, on the other hand, λ is large, the photon has less energy. For example, an x-ray photon of wavelength 1 Å has much more energy than does a photon of visible light of wavelength 5000 Å.

4.3a Emission and Absorption Lines

In a cold gas, most of the atoms are on the lowest possible energy level. When the gas is heated, many of its atoms are raised from this lowest energy level to higher energy states. From the higher energy levels, they then spontaneously drop back to their lowest energy levels, emitting photons as they do so. These new photons represent energies at certain wavelengths. Since photons did not necessarily previously exist at those wavelengths, the new photons appear on the spectrum as wavelengths brighter than neighboring wavelengths. These are the **emission** lines (Fig. 4–9).

Much of what we know about stars comes from studying their spectral lines. First we will discuss spectral lines themselves; we will go on to their appearance in stars in Section 4.5.

An electron volt (eV) is an extremely small unit of energy. It takes billions of billions of eV to light a 100-watt light bulb for one second.

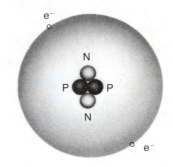

Figure 4–7 A helium atom contains two protons and two neutrons in its nucleus and two orbiting electrons.

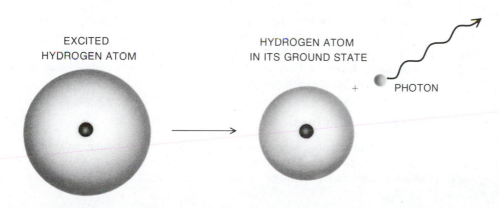

EXCITED HYDROGEN ATOM   HYDROGEN ATOM IN ITS GROUND STATE   PHOTON

Figure 4–8 Since the hydrogen atom has only a single electron, it is a particularly simple case to study. The lowest possible energy state of an atom is called its *ground state*. All other energy states are called *excited states*. When an atom in an excited state gives off a photon, it drops back to a lower energy state, perhaps even to the ground state. We see the photons as an emission line.

Figure 4–9 When photons are emitted, we see an emission line; a continuum may or may not be present (Cases A and B). An absorption line must absorb radiation from something. Hence an absorption line necessarily appears in a continuum (Case C). The top and bottom horizontal rows are graphs of the spectrum before and after passing by an atom; a schematic diagram of the atom's energy levels (only two levels are shown to simplify the situation) is shown in the center row.

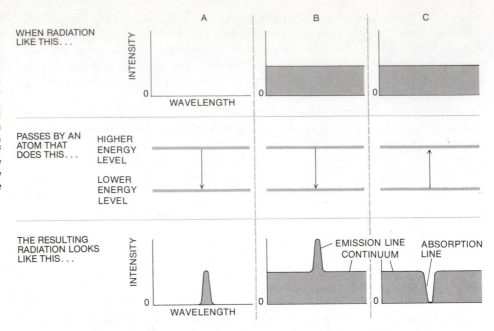

As we saw in Section 2.2, when we allow continuous radiation from a body at some relatively high temperature to pass through a cooler gas, the atoms in the gas can take energy out of the continuous radiation at certain discrete wavelengths. We now see that the atoms in the gas are changed into higher energy states that correspond to these wavelengths. Because energy is taken up in the gas at these wavelengths, less radiation remains to travel straight ahead at these particular wavelengths. These wavelengths are those of the **absorption** lines (Fig. 4–9). In the visible part of the spectrum, these wavelengths appear dark when we look back through the gas. (The energy absorbed is soon emitted in other directions, which explains why we can see emission lines when we look from the side, as was shown in Fig. 2–5.)

Note that emission lines can appear without a continuum (Case A in Fig. 4–9) or with a continuum (Case B in Fig. 4–9), since they are merely the addition of energy to the radiation field at certain wavelengths. But absorption lines must be absorbed **from** something (Case C in Fig. 4–9), namely, the continuum, which is just the name for a part of the spectrum where there is some radiation over a continuous range of wavelengths.

A normal stellar spectrum is a continuum with absorption lines. This simple fact leads us to the important conclusion that the temperature in the outer layers of the star is decreasing with distance from the center of the star. The presence of absorption lines tells us that we have been looking through cooler gas at hotter gas. As a result of this temperature trend with height in the stellar atmosphere, stellar spectra rarely show emission lines, which result from the presence of gas hotter than any background continuum. Emission lines can appear, however, in some cases, such as those of stars surrounded by hot shells of gas.

4.3b Excited States and Ions

We can think of an atom's energy states as resulting from a change in energy of an atom's electrons. (The electrons are orbiting the nucleus. The production of spectral lines does not involve changes in the nucleus itself.) If a gas were at a

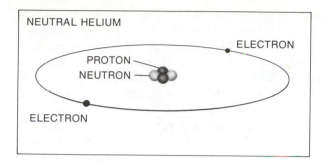
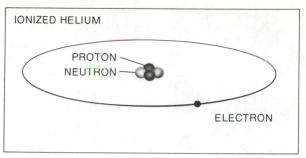

Figure 4–10 An atom missing one or more electrons is *ionized*. Neutral helium is He I and singly ionized helium is He II.

temperature of absolute zero, all the electrons in it would be in the lowest possible energy levels. The lowest energy level that an electron can be on is called the *ground state* of that atom.

As we add energy to the atom we can "excite" one or more electrons to higher energy levels. If we were to add even more energy, some of the electrons would be given not only sufficient energy to be excited but also enough to escape entirely from the atom. We then say that the atom is *ionized*. The remnant of the atom with less than its quota of electrons is called an *ion* (Fig. 4–10). Since the remnant has more positive charges in its nucleus than it has negative charges on its electrons, its net charge is positive, and it is sometimes called a *positive ion*.

Before the atom was ionized, it had the same number of protons and electrons, and so was electrically neutral. Such an atom is called a neutral atom, and is denoted with the Roman numeral I. For example, neutral helium is denoted He I. When the atom has lost one electron (is singly ionized), we use the Roman numeral II. Thus when helium has lost one electron, it is called He II (read "helium two"). If an atom is doubly ionized, that is, has lost two electrons, it is in state III, and so on.

Atoms can be excited or ionized either by collisions or by radiation. In a star the former is the case when the density is sufficiently high that collisions of the atoms with electrons are frequent. The latter case occurs when the gas is "heated," for "heating" simply means that the individual atoms are given more energy. The atoms thus collide more frequently and exchange more energy with each other. Thus in a "hotter" gas, more atoms are excited or ionized.

4.4 The Hydrogen Spectrum

Hydrogen gas emits, and absorbs, a set of spectral lines that falls across the visible spectrum in a distinctive pattern (Fig. 4–11). The strongest line is in the red, the second strongest is in the blue, and the other lines continue through the blue and ultraviolet, with the spacing between the lines getting smaller and smaller.

Figure 4–11 The Balmer series, representing transitions down to or up from the second energy state of hydrogen. The strongest line in this series, Hα (H-alpha), is in the red.

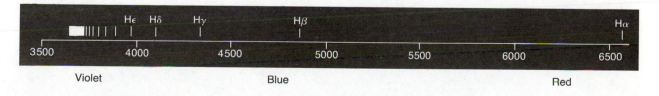

It is easy to visualize how the hydrogen spectrum is formed by using the picture that Niels Bohr (Fig. 4–12), the Danish physicist, laid out in 1913. In the *Bohr atom* (Fig. 4–13), electrons can have orbits of different sizes that correspond to different energy levels. Only certain orbits are allowable, which is the same as saying that the energy levels are quantized. This simple picture of the atom has been superseded by the development of quantum mechanics, but the notion that the energy levels are quantized remains.

We use the letter n to label the energy levels. It is called the *principal quantum number*. We call the energy level for $n = 1$ the *ground level* or *ground state*, as it is the lowest possible energy state. The hydrogen atom's series of transitions from or to the ground level is called the *Lyman series,* after the American physicist Theodore Lyman. The lines fall in the ultraviolet, at wavelengths far too short to pass through the earth's atmosphere. Lyman's observations of what we now call the Lyman lines were made in a laboratory in a vacuum tank. At present, telescopes aboard earth satellites, such as the International Ultraviolet Explorer (IUE), enable us to observe Lyman lines from the stars.

The series of transitions with $n = 2$ as the lowest level, the *Balmer series,* is in the visible part of the spectrum (Box 4.4). The bright red emission line in the visible part of the hydrogen spectrum, known as the Hα line, arises from the transition from level 3 to level 2. The transition from $n = 4$ to $n = 2$ causes the Hβ line, and so on. Because the series falls in the visible where it is so well observed, we usually call the lines H alpha, etc., instead of Balmer alpha, etc.

Figure 4–12 Niels Bohr and his wife, Margrethe.

Figure 4–13 The representation of hydrogen energy levels known as the Bohr atom.

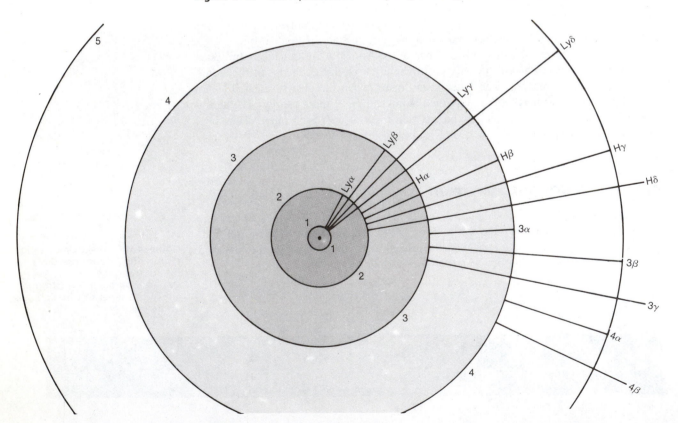

Box 4.4 The Balmer Series and the Bohr Atom

The fact that the spectral lines from hydrogen—one in the red, one in the blue, and then others increasing in number farther into the blue and violet—showed a recognizable pattern was long known, but it was not at first obvious how to explain that pattern mathematically. In 1885, a Swiss schoolteacher named Johann Balmer announced that he had found by trial and error a mathematical pattern that fit the Balmer series. If the first line in the series, hydrogen alpha (Hα) in the red, is assigned the number 3; the second line in the series, hydrogen beta (Hβ) in the blue, is assigned the number 4; the third line, hydrogen gamma (Hγ), is assigned 5; etc., then the wavelengths of the lines fit the following formulas:

$$1/\text{wavelength (H}\alpha) = \text{constant} \times (1/2^2 - 1/3^2),$$
$$1/\text{wavelength (H}\beta) = \text{constant} \times (1/2^2 - 1/4^2),$$
$$1/\text{wavelength (H}\gamma) = \text{constant} \times (1/2^2 - 1/5^2),$$

and so on for the farther hydrogen spectral lines. These formulas are so simple—they contain only the squares of the smallest integers—that there must be some simple regularity in the hydrogen atom.

Eventually, Theodore Lyman at Harvard, experimenting in the laboratory far in the ultraviolet beyond where the eye can see, discovered another series of hydrogen lines. For this set of lines, now called the *Lyman series,* each line corresponded (with the same constant as for the Balmer series) to

$$1/\text{wavelength (Ly}\alpha) = \text{constant} \times (1 - 1/2^2),$$
$$1/\text{wavelength (Ly}\beta) = \text{constant} \times (1 - 1/3^2).$$

Other scientists soon discovered further series, with $1/3^2$, $1/4^2$, etc., in place of $1/1^2$ and $1/2^2$ of the Lyman and Balmer series.

Thus a general formula involving small integers explains all the hydrogen lines. If n is a number assigned to each series (1 for the Lyman series, 2 for the Balmer series, etc.) and m signifies all the integers greater than n, the hydrogen lines are all

$$1/\text{wavelength} = \text{constant} \times (1/n^2 - 1/m^2).$$

The fact that each line could be explained by the difference between two such simple numbers indicated that something fundamental was involved. The reason why this simple subtraction explains all the hydrogen lines was found by Niels Bohr in 1913.

In this *Bohr atom,* we think of each of the allowable energy states as having a fixed energy E_1, E_2, E_3, etc. Each E is equal to a constant times the speed of light (c) times $1/(\text{subscript})^2$. Thus the Lyman series has $1/\text{wavelength} = E_1 - E_m$, the Balmer series has $1/\text{wavelength} = E_2 - E_m$, etc. (The m represents the integers higher than the subscript in the first term.)

Lyman went so far as to climb Mount McKinley in 1915, hoping that from that great altitude there might happen to be a window of transparency that would pass Lyman alpha. Unfortunately, there is no such window, and his photographs of the solar spectrum were blank in the region where ultraviolet radiation would have appeared.

There are many other series of lines, each corresponding to transitions to a given lower level (Fig. 4–14).

Since the higher energy levels have greater energy, a spectral line caused by transition from a higher level to a lower level yields an emission line. When, on the other hand, continuous radiation falls on cool hydrogen gas, some of the atoms in the gas can be raised to higher energy levels. Absorption lines result in this case.

All transitions between the same two energy states always cause a spectral line at the same wavelength. When the transition is from the higher energy state to the

Figure 4–14 The energy levels of hydrogen and the series of transitions among the lowest of these levels.

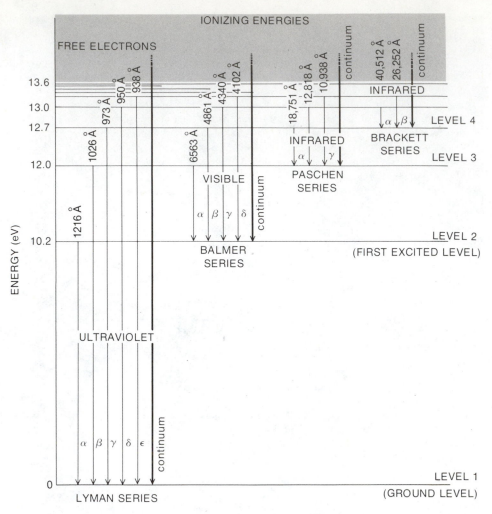

The hydrogen atom, with its lone electron, is a very simple case. More complicated atoms have more complicated sets of energy levels and thus their spectral lines have less simple patterns.

lower one, the spectral line is in emission. When the transition is from the lower energy state to the higher one, the line is in absorption.

Each of the chemical elements has its own set of energy levels in a given stage of ionization, and thus has its own set of spectral lines. (The Lyman and Balmer series are only for hydrogen; the lines of most other elements do not form such a regular pattern.) Whether the lines appear in absorption or in emission depends on external factors.

4.5 Spectral Types

In the last decades of the 19th century, spectra (almost always absorption lines) of thousands of stars were photographed. Differences existed among spectra from different stars, and classifications of the different types of spectra were developed. The most famous worker at the vital task of classifying spectra was Annie Jump Cannon (Fig. 4–15).

At first, the stellar spectra were classified only by the strength of certain absorption lines from hydrogen, and were lettered alphabetically: A for stars with the strongest hydrogen lines, B for stars with slightly weaker lines, and so on. These categories are called *spectral types* or *spectral classes* (Fig. 4–16). It was later real-

Figure 4–15 Part of the computing staff of the Harvard College Observatory in about 1917. Annie Jump Cannon, fifth from the right, classified over 500,000 spectra in the decades following 1896. Her catalogue, called the Henry Draper catalogue after the benefactor who made the investigation possible, is still in use today. Many stars are still known by their HD (Henry Draper catalogue) numbers, such as HD 176387. Also in this photograph is Henrietta S. Leavitt, fifth from the left, whose work on variable stars we shall meet in Sections 6.4 and 14.4.

ized that the types of spectra varied primarily because of differing temperatures of the stellar atmospheres (which, you recall, can independently be measured by observing the color of the star). The hydrogen lines were strongest in stars in the middle range of stellar temperatures and were weaker at both higher and lower temperatures. In the hottest stars, too much hydrogen is ionized for the lines to be strong. In stars cooler than spectral type A, too few hydrogen atoms have enough energy for electrons to be on levels above the ground state. When we now list the spectral types of stars in order of decreasing temperature, they are no longer in alphabetical order. From hottest to coolest, the spectral types are O B A F G K M.

O stars (that is, stars of type O) are the hottest, and the temperatures of M stars are more than 10 times lower. Generations of American students and teachers have remembered the spectral types by the mnemonic: Oh, Be A Fine Girl, Kiss Me. The spectral types have been subdivided into 10 subcategories each. For example, the

Spectral lines arise from transition of electrons between energy states. When hydrogen is ionized, since it has only one electron to begin with, it then has no electrons at all, and thus has no spectral lines.

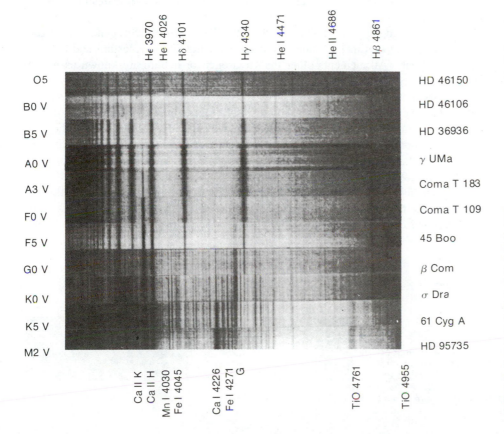

Figure 4–16 Spectral types. Note that all the stellar spectra shown have absorption lines. The Roman numeral "V" indicates that the stars are normal ("dwarf") stars, which will be defined in Section 5.4. The numbers following the letters in the left-hand column represent a subdivision of the spectral types into tenths; the step from B8 to B9 is equal to the adjacent step from B9 to A0. The abbreviations of the elements forming the lines and the individual wavelengths (in angstroms) of the most prominent lines appear at the top and the bottom of the graph.

Historically, the hotter stars are called early types and the cooler stars are called late types. Thus, a B star is "earlier" than an F star, although stars do not change spectral type during the long stable period that takes up most of their lives.

hottest B stars are B0, followed by B1, B2, B3, and so on. Spectral type B9 is followed by spectral type A0. It is fairly easy to tell the spectral type of a star by mere inspection of its spectrum.

The actual assignment of spectral types follows a particular set of stellar spectra that serve as standards. Stellar spectra are compared with these standards to find the closest match.

4.5a Stellar Spectra

When we observe the light from a star, we are seeing only the radiation from a thin layer of that star's atmosphere. The interior is hidden within. As we discuss stellar spectra, let us keep this fact in mind.

As we consider stars of different spectral types (Fig. 4–16) ranging from the hottest to the coolest, we can observe the effect of temperature on the spectra. For example, the hottest stars, those of type O, have such high temperatures that there is enough energy to remove the outermost electron or electrons from most of the atoms. Since most of the hydrogen is ionized, relatively few neutral hydrogen atoms are left and the hydrogen spectral lines are weak. Helium is not so easily ionized—it takes a large amount of energy to do so—yet even helium appears not in its neutral state but in its singly ionized state in an O star. Temperatures range from 30,000 K to 60,000 K in O stars. O stars are relatively rare since they have short lifetimes; none of the nearest stars to us are O stars. The system really begins with spectral types O3 and O4, of which only a handful are known at any distance. No O0, O1, or O2 stars apparently exist. (In the late stages of the evolution of a star, some even hotter objects result, but we are discussing only ordinary stars here.)

Rigel and Spica are familiar B stars.

B stars are somewhat cooler, 10,000 K to 30,000 K. The hydrogen lines are stronger than they are in the O stars, and lines of neutral helium instead of ionized helium are present.

Sirius, Deneb, and Vega are among the bright A stars in the sky.

A stars are cooler still, 7500 K to 10,000 K. The lines of hydrogen are strongest in this spectral type. Lines of singly ionized elements like magnesium and calcium begin to appear. (All elements other than hydrogen or helium are known as *metals* for this purpose.) O, B, and A stars are all bluish in color.

Canopus, the second brightest star in the sky (not visible from the latitudes of most of the United States), is a prominent F star, as is Polaris.

F stars have temperatures of 6000 K to 7500 K. Hydrogen lines are weaker in F stars than they are in A stars, but the lines of singly ionized calcium are stronger. Singly ionized calcium has a pair of lines that are particularly conspicuous; they are easy to pick out and recognize in the spectrum. These lines are called H and K (Fig. 4–17), from their alphabetical order in an extended version of Fraunhofer's original list (Section 2.2 and Color Plate 5).

Besides the sun, Alpha Centauri (actually the brightest of the three stars that together make up the bright point in the sky that we call Alpha Centauri) is also a G star.

G stars, the spectral type of our sun, are 5000 K to 6000 K. They are yellowish in color, since the peak in their spectra falls in the yellow/green part of the spectrum. The hydrogen lines are visible, but the H and K lines are the strongest lines in the

Figure 4–17 The H and K lines of singly ionized calcium are the strongest lines in the visible part of the solar spectrum, and are prominent in spectra of stars of spectral types F, G, K, and M. These classes are known as *late spectral types,* while classes O, A, and B are known as *early spectral types,* although no connotations of age are meant.

> **Box 4.5**
>
> Do not confuse:
> K stars: a spectral class
> K: a kelvin, a unit of temperature
> K: a strong line in many stellar spectra
> K: the element potassium

spectrum. The H and K lines are stronger in G stars than they are in any other spectral type.

K stars are relatively cool, only 3500 K to 5000 K. The spectrum is covered with many lines from neutral metals, a strong contrast to the spectra of the hottest stars, which show few spectral lines.

M stars are cooler yet, with temperatures less than 3500 K. Their atmospheres are so cool that molecules can exist without being torn apart, and the spectrum shows many molecular lines. (Hotter stars do not usually show molecular lines.) Lines from the molecule titanium oxide are particularly numerous.

For the very coolest stars, there are alternative spectral types that reflect differences in the relative abundances of various elements. For example, stars that are relatively rich in carbon are called *carbon stars* or *C-type*. (These were originally called R and N stars.) *S-type* stars have, among other things, relatively strong zirconium oxide lines, as opposed to titanium oxide lines. Both C- and S-type stars have similar temperatures to M stars.

We have listed the ordinary spectral types of stars, each of which shows an absorption spectrum, that is, a continuum crossed with dark lines. There are many unusual stellar spectra, though. *Wolf-Rayet stars* are O stars that not only show emission lines but also have the strange feature that the emission is particularly broad; that is, a given line may cover several angstroms of spectrum instead of a fraction of an angstrom. The emission in the spectra of Wolf-Rayet stars comes from shells of material that the star had ejected into the space surrounding it and results from strong outflowing "stellar winds." Peter Conti, Catharine Garmany, and colleagues have concluded that Wolf-Rayet stars are descendants of stars whose initial masses were over 40 times that of the sun.

Other types of stars, like Oe, Be, Ae, and Me stars (where the "e" stands for emission) also have emission lines, though these are relatively rare.

In one of the first applications of astrophysics to the study of stars, in 1925, Cecilia Payne-Gaposchkin (then Cecilia Payne) graphed the strength of certain absorption lines versus spectral type. She used the graph together with theoretical calculations to determine the stellar temperature scale for each spectral type.

Arcturus and Aldebaran, both visible as reddish points in the sky, are K stars.

Betelgeuse is an example of such a reddish star of spectral class M.

*4.5b Spectra Observed from Space

In the ultraviolet and x-ray regions of the spectrum, we often see radiation from higher levels of the star than the level that gives off the visible spectrum. These higher levels are hotter, as was originally learned in studies of the sun. For example, we discovered from Einstein Observatory observations that stars of all spectral types give off x-rays. Though it had been expected that the coolest stars, the M stars, would be too cool to be so strong in x-rays, they are actually as much as 10 per cent

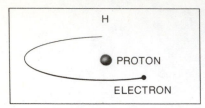 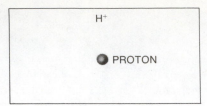 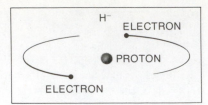

Figure 4–18 Neutral hydrogen, the hydrogen ion, and the negative hydrogen ion.

as strong in the x-ray spectrum as they are in the visible. But this doesn't say that we were wrong about the temperature of these stars' surfaces. The x-rays come from a hot *stellar corona* surrounding the stars at a temperature of over a million kelvins.

Ultraviolet spectra of even cool stars taken with the International Ultraviolet Explorer show spectral lines from gas at temperatures of up to about 100,000 kelvins. These spectral lines are formed in a *stellar chromosphere* between the star's ordinary surface and its corona, or in a *transition zone* between the chromosphere and corona. Only since the launch of these spacecraft could we study chromospheres and coronas in stars other than the sun.

*4.5c The Continuous Spectrum

Where does the continuous spectrum come from? In order for there to be an absorption line, there must be some continuous radiation to be absorbed. The particular mechanism that causes the continuous emission is different for different spectral types. In the sun and other stars of similar spectral types, the continuous emission is caused by a strange type of ion: the *negative hydrogen ion* (Fig. 4–18). Only a tiny fraction of the hydrogen atoms are in this state, but the sun is so overwhelmingly (90 per cent) hydrogen that enough negative hydrogen atoms are present to be important.

When a neutral hydrogen atom takes up an extra electron, the second electron may temporarily become part of the atom, resident on an energy level. But the energy **difference** between the extra electron's energy and its energy as part of the atom is not limited to discrete values, because the electron could have had any amount of energy before it joined the atom (Fig. 4–19). Alternatively, the extra

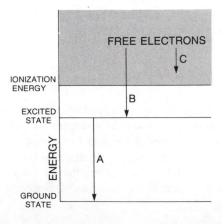

Figure 4–19 A bound-bound transition *(A)* would go from one discrete level to another (that is, from a bound level to another bound level), analogous to jumping on a staircase from one step to another. A free-bound transition *(B)* would be analogous to jumping from some arbitrary height above the staircase (as from the sky with a parachute) to some particular step (that is, from being free to a bound level). In a free-free transition *(C)* the electron never joins the atom.

electron starts free of the atom, is affected by the atom, but winds up still free of the atom. In this case too, the amount of energy involved is not limited to discrete values. Thus a continuous spectrum is formed. The case is thus different from the one in which an electron in an atom jumps between two energy levels, because the electron has a fixed amount of energy in each level, and a fixed number subtracted from another fixed number gives a fixed number, and cannot lead to a continuous range of values.

In all stars, the processes of continuous emission and spectral-line absorption take place together throughout the outer layers of the stars rather than in separate layers. The detailed study of the processes of emission and absorption is called the study of *stellar atmospheres*. Nowadays supercomputers are able to handle complicated sets of equations that describe the transfer of radiation from the outer layers of a star into space. Sometimes the largest computers, such as the Cray computer at the National Center for Atmospheric Research (NCAR) in Boulder, Colorado, may run for hours to calculate a model of the atmosphere of a single star.

Summary and Outline

Stars are balls of gas held together by gravitation and generating energy by nuclear fusion.

Colors of stars and of other gas radiating a continuum (Sections 4.1 and 4.2)

As gas gets hotter, it goes from red hot to bluish hot; the peak of the radiation moves to shorter wavelengths (*Wien's displacement law*):
$\lambda_{max}T = $ constant.

As gas gets hotter, the total energy radiated grows rapidly, with the fourth power of the temperature (the *Stefan-Boltzmann law*):
$E = \sigma T^4$.

Planck's law, discovered after the above laws, gives the amount of radiation at each wavelength, given the temperature. Wien's displacement law and the Stefan-Boltzmann law are really included within Planck's law.

> Wien's displacement law:
> $\lambda_{max}T = $ constant
> Stefan-Boltzmann law: $E = \sigma T^4$
> Planck's law relates λ, E_λ, and T.

Matter whose radiation follows Planck's law precisely is called a *black body*.

The structure of an atom (Section 4.3)

Protons and neutrons in the nucleus with electrons orbiting the nucleus.

Spectral lines (Sections 4.3 and 4.4)

Quantum theory showed that atoms can exist only in discrete states of energy. When an atom drops from a higher energy state to a lower energy state it emits the difference in energy as a photon of a particular wavelength λ and energy hc/λ. Such photons make an emission line. If, on the other hand, an atom takes up enough energy to raise itself from a lower to a higher state, then such atoms lead to the formation of an absorption line in radiation that has been hitting them.

Hydrogen has a series of spectral lines that falls in a distinct pattern across the visible; this is the Balmer series, and includes Hα and Hβ.

The Bohr atom, with each energy level having a different size, is a good way to visualize the hydrogen atom. The Balmer series corresponds to transitions between level 2 and higher levels. The Lyman series, which falls in the ultraviolet (and thus does not pass through the earth's atmosphere), involves transitions to the more basic level, the first level.

Spectral types (Section 4.5)

Hottest to coolest: O B A F G K M

Hydrogen spectra are strongest in type A.

Calcium H and K lines are strongest in type G.

Molecular spectra begin to appear in type M.

All types have absorption spectra. Only a few kinds of stars with emission lines exist.

From space we can observe stellar chromospheres and coronas.

The sun's continuous spectrum comes from the negative hydrogen ion (H^-).

Key Words

Wien's displacement law, Stefan-Boltzmann law, Planck's law, black body, nucleus, electrons, proton, neutron, quarks, quantum mechanics, allowed states, energy levels, quantum, photon, ground state, ionized, ion, positive ion, Bohr atom, principal quantum number, ground level (ground state), Balmer series, Lyman series, spectral types (spectral classes), metals, carbon stars, C-type, S-type, Wolf-Rayet stars, stellar atmosphere*, stellar corona*, stellar chromosphere*, transition zone*, negative hydrogen ion*

---

*This term is found in an optional section.

Questions

1. What are two differences between a star and a planet?

†2. Room temperature is now about 65°F. (a) What is this in °C? (b) in kelvins?

†3. The sun's surface is about 5800 K. (a) What is this in °C? (b) In °F?

†4. The sun's spectrum peaks at 5600 Å. Would the spectrum of a star whose temperature is twice that of the sun peak at a longer or a shorter wavelength? How much more energy than a square centimeter of the sun would a square centimeter of the star give off?

†5. The sun's temperature is about 5800 K and its spectrum peaks at 5600 Å. An O star's temperature may be 40,000 K. At what wavelength does its spectrum peak?

6. For the O star of question 5, in what part of the spectrum does its spectrum peak? Can the peak be observed with the Palomar telescope? Explain.

†7. One black body peaks at 2000 Å. Another, of the same size, peaks at 10,000 Å. Which gives out more radiation at 2000 Å? Which gives out more radiation at 10,000 Å? What is the ratio of the total radiation given off by the two bodies?

†8. A coolish B star has twice the temperature of the sun in kelvins. (a) How many times more than the sun does each square centimeter of the B star's surface radiate? (b) The B star's diameter is 5 times the sun's. How many times more than the sun does the whole star radiate, given that the surface area of a sphere increases as the radius squared ($A = 4\pi r^2$)?

9. Which contains more information, Wien's displacement law or Planck's law? Explain.

†10. What is the ratio of energy output for a bit of surface of an average O star and a same-sized bit of surface of the sun?

11. Star A appears to have the same brightness through a red and a blue filter. Star B appears brighter in the red than in the blue. Star C appears brighter in the blue than in the red. Rank these stars in order of increasing temperature.

12. (a) From looking at Figure 4–14, draw the Lyman series and the Balmer series on the same wavelength axis. (b) Why is the Balmer series the most observed spectral series of atomic hydrogen?

13. What is the difference between the continuum and an absorption line? The continuum and an emission line? Draw a continuum with absorption lines. Can you draw absorption lines without a continuum? Can you draw emission lines without a continuum? Explain.

†14. Consider a hypothetical atom in which the energy levels are equally spaced from each other; that is, the energy of level n is n. (a) Draw the energy level diagram for the first five levels. (b) Indicate on the diagram the transition from level 4 to level 2 and from level 4 to level 3. (c) What is the ratio of energies of these two transitions? (d) If the 4→2 transition has a wavelength of 4000 Å, what is the wavelength of the 3→2 transition?

15. Does the spectrum of the solar surface show emission or absorption lines? Compare and/or distinguish between the sun's spectrum and the spectrum of a B star.

16. (a) Why are singly ionized helium lines detectable only in O stars? (b) Why aren't neutral helium lines prominent in the solar spectrum?

17. Compare Planck curves for stars of spectral types O, G, and M.

18. Why don't we see strong molecular lines in the sun?

19. Make up your own mnemonic for the spectral types.

†20. Using Balmer's formula and the fact that the wavelength of Hα is 6563 Å, calculate the wavelength of Hβ. Show your work.

21. What factors determine spectral type? What instrument would you use to determine the spectral type of a star?

22. List spectral types of stars in approximate order of strength, from strongest to weakest, of (a) hydrogen lines, (b) ionized calcium lines, (c) titanium oxide lines.

---

†This indicates a question requiring a numerical solution.

23. If we are examining a stellar spectrum, explain how and why comparing the relative intensity of the absorption line at about 3970 Å with the intensity of the calcium K line allows us to judge the star's spectral type.

24. If we take two stones, one twice the diameter of the other, and put them in an oven until they are heated to the same temperature and begin glowing, what will be the relationship between the total energy in the light given off by the stones? (The surface area of a sphere is proportional to the square of the radius.)

25. Why does the negative hydrogen atom have a continuous rather than a line spectrum?

Stars in the Milky Way in the constellation Sagittarius, looking toward the center of our galaxy. Two globular clusters are visible.

Stellar Brightness, Distances, and Motions

5

Aims: To describe the absolute and apparent magnitude scales, the method of trigonometric parallax and how it tells us the distances to the stars, the Hertzsprung-Russell diagram and what it tells us about types of stars and their distances, and the Doppler effect and how it tells us about stellar motions

We have seen in Chapter 4 how we can analyze the spectrum of a star to tell us about the outer layer of that star; for example, we can tell how hot it is. In this chapter, we see how information about the spectra of many stars can be put together to tell us about the overall properties of stars.

We begin by describing the scale in which astronomers give brightness. We also see how we can fairly directly measure the distances to the nearest stars, and that these distances, together with classification by spectral type, can give us distances to farther stars. We see how graphing the temperatures and brightnesses of stars gives us an important tool: the Hertzsprung-Russell diagram. Then we study the Doppler effect, which enables us to tell how stars are moving.

Later in this book we shall use this same method to discover that indeed the whole universe is expanding, with galaxies moving away from us in all directions.

5.1 Light from the Stars

The most obvious thing we notice about the stars is that they have different brightnesses. Over 2000 years ago, the Greek astronomer Hipparchus divided the stars that he could see into classes of brightness. His work was extended by Ptolemy in Alexandria in about A.D. 140. In their classification, the brightest stars were said to be of the first magnitude, a sensible and reasonable idea. Somewhat fainter stars were said to be of the second magnitude, and so on down to the sixth magnitude, which represented the faintest stars that could be seen with the naked eye.

5.1a Apparent Magnitude

In the nineteenth century, when astronomers became able to make quantitative measurements of the brightnesses of stars, the magnitude scale was placed on an accurate basis. It was discovered that the brightest stars that could be seen with the naked eye were about 100 times brighter than the faintest stars. This corresponded to a difference of 5 magnitudes on Hipparchus' scale, a number that was used to set up the present magnitude system. Since this type of magnitude tells us how bright a star **appears,** it is called *apparent magnitude.*

The new system is based on the definition that stars that, on the magnitude scale, show a **difference** of five magnitudes are different in intensity by a **factor** of exactly 100 times. Thus a star of first magnitude is exactly 100 times brighter than a star of sixth magnitude.

The brightnesses of the stars that had been classified by the Greeks were placed on this new magnitude scale (Fig. 5–1). Many of the stars that had been of the first magnitude in the old system were indeed "first magnitude stars" in the new system. But a few stars were much brighter, and thus corresponded to "zeroth" magnitude, that is, a number less than one. A couple, like Sirius, the brightest star in the sky, were brighter still. The numerical magnitude scale was easily extended to negative numbers. The new scale, being numerical, admits fractional magnitudes. Sirius is actually magnitude −1.4, for example. Physiologically, this type of measurement makes sense because the human eye happens to perceive equal **factors** of luminosity (such a factor is the energy coming from one object divided by the energy coming from another) as roughly equal **intervals** (additive steps) of brightness, which is just how the magnitude scale is set up.

The magnitude scale was extended not only to stars brighter than first magnitude, but also to stars fainter than 6th magnitude. Since 5 magnitudes corresponds to a hundred times in brightness, 11th magnitude is exactly 100 times fainter than 6th

Figure 5–1 The apparent magnitude scale is shown on the vertical axis. At the left, sample intervals of 1, 5, 10, and 15 magnitudes are marked and translated into multiplicative factors.

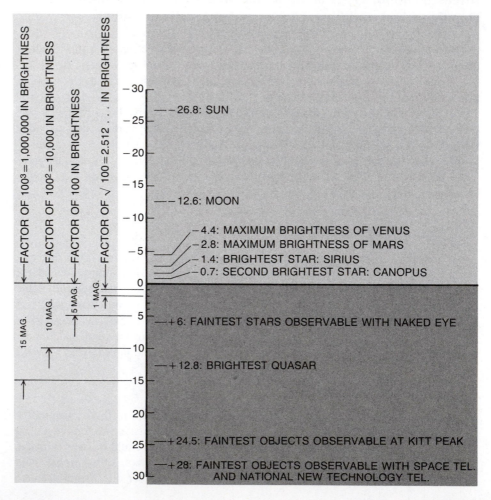

magnitude (that is, 1/100 times as bright); 16th magnitude, in turn, is exactly 100 times fainter than 11th magnitude. The faintest stars that can be photographed from the ground are fainter than 24th magnitude.

5.1b Working with the Magnitude Scale

We have seen how a difference of 5 magnitudes corresponds to a factor of 100 in brightness. What does that indicate about a difference of 1 magnitude? Since an increase of 1 in the magnitude scale corresponds to a decrease in brightness by a certain factor, we need a number that, when multiplied by itself 5 times, will equal 100. This number is just $\sqrt[5]{100}$. Its value is 2.512..., with the dots representing an infinite string of other digits. For many practical purposes, it is sufficient to know that it is approximately 2.5.

Thus a second magnitude star is about 2.5 times fainter than a first magnitude star. A third magnitude star is about 2.5 times fainter than a second magnitude star. Thus, a third magnitude star is approximately $(2.5)^2$ (which is about 6) times fainter than a first magnitude star. (We could easily give more decimal places, writing 6.3 or 6.31, but the additional figures would not be meaningful. We started with only one digit being significant when we wrote down ''2nd magnitude,'' so our data were given to only 1 *significant figure*. Our result is thus only accurate to 1 significant figure.)

In similar fashion, using a factor of 2.5 or 2.512 for the single magnitudes and a factor of 100 for each group of 5 magnitudes, we can very simply find the ratio of intensities of stars of any brightness.

> **EXAMPLE:** By what factor do the brightness of stars of magnitudes −1 and 6 differ (that is, what is the brightness ratio between the brightest and faintest stars, respectively, that are seen with the naked eye)?
>
> **ANSWER:** Since the stars differ by 7 magnitudes, the factor is (100)(2.5)(2.5), which is about 600. The star of −1st magnitude is 600 times brighter than the star of 6th magnitude.

One unfortunate thing about the magnitude scale is that it operates in the opposite sense from a direct measure of brightness. Thus the brighter the star, the lower the magnitude (a second magnitude star is brighter than a third magnitude star), which is sometimes a confusing convention. This kind of problem comes up occasionally in an old science like astronomy, for at each stage astronomers have made their new definitions so as to have continuity with the past.

5.2 Stellar Distances

We can tell a lot by looking at a star or by examining its radiation through a spectrograph. But such observations do not tell us directly how far away the star is. Since all the stars are but points of light in the sky even when observed through the biggest telescopes, we have no reference scale to give us their distances (Fig. 5–2).

The best way to find the distance to a star is to use the principle that is also used in rangefinders in cameras. Sight toward the star from different locations, and see how the direction toward the star changes with respect to a more distant background of stars. You can see a similar effect by holding out your thumb at arm's length. Examine it first with one eye closed and then with the other eye closed. Your thumb seems to change in position as projected against a distant background; it appears to move across the background by a certain angle, about 3°. This is because your eyes are a few centimeters apart from each other, so each eye has a different point of view.

½ mag	= 1.585 times
1 mag	= 2.512 times
2 mag	= 6.310 times
3 mag	= 15.85 times
4 mag	= 39.81 times
5 mag	= 100 times
6 mag	= 251.2 times
7 mag	= 631.0 times
8 mag	= 1585 times
9 mag	= 3981 times
10 mag	= 10^4 times
15 mag	= 10^6 times
20 mag	= 10^8 times

We must be careful not to fool ourselves that we can gain in accuracy in an arithmetical process; the number of significant figures we put in is the number of significant figures in the result.

Figure 5–2 ''Excuse me for shouting—I though you were further away.'' Without clues to indicate distance, we cannot properly estimate size. (Reproduced by special permission of PLAYBOY Magazine; copyright © 1971 by Playboy.)

In closeup photography, "parallax error" can cause you to mistakenly leave someone's head out of the picture, if you don't take account of the parallax caused by the fact that the lens and your viewfinder are in different places.

Devices to measure the positions of stars on photographic plates are called *measuring engines;* the recent development of automatic measuring engines has enabled trigonometric parallaxes to be measured for stars five times more distant than could previously be determined. New ground-based techniques using electronic detectors will soon improve the situation further.

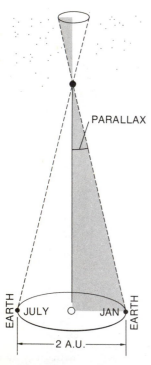

Figure 5–3 The nearer stars seem to be slightly displaced with respect to the farther stars when viewed from different locations in the earth's orbit.

Now hold your thumb up closer to your face, just a few centimeters away. Note that the angle your thumb seems to jump across the background as you look through first one eye and then through the other is greater than it was before. Exactly the same effect can be used for finding the distances to nearby stars. The nearer the star is to us, the farther it will appear to move across the background of distant stars, which are so far away that they do not appear to move with respect to each other.

To maximize this effect, we want to observe from two places that are separated from each other as much as possible. For us on earth, that turns out to be the position of the earth at intervals of six months. In that six-month period, the earth moves to the opposite side of the sun. In principle, we observe the position of a star in the sky (photographically), with respect to the background stars, and several months later, when the earth has moved part way around the sun, we repeat the observation. (Six months later, the star would be up in the daytime and could not be seen.)

The straight line joining the points in space from which we observe is called the *baseline*. The distance from the earth to the sun is called an *Astronomical Unit* (A.U.). In the case above, we have observed with a baseline 2 A.U. in length. The angle across the sky that a star seems to move (with respect to the background of other stars) between two observations made from the ends of a baseline of 1 A.U. (half the maximum possible baseline) is called the *parallax* of the star (Fig. 5–3). From the parallax, we can use simple trigonometry to calculate the distance to the star. The process is thus called the method of *trigonometric parallax*.

The basic limitation to the method of trigonometric parallax is that the farther away the star is, the smaller is its parallax. It turns out that this method can only be used for about ten thousand of the stars closest to us; most other stars have parallaxes too small for us to measure. (These other stars, with negligible parallaxes, are the distant stars that provide the unmoving background against which we compare.) Even Proxima Centauri, the star nearest to us beyond the sun, has a parallax of only $\frac{3}{4}$ arc sec, the angle subtended (taken up across our vision) by a dime at a distance of 2 km! Clearly we must find other methods to measure the distances to most of the stars. But for the cases in which it can be used, the method of trigonometric parallaxes gives the most accurate answers.

Parallaxes are measured by comparing pairs of photographs taken several months apart. Measurements are repeated over several years to determine the effect of actual motions of stars and to reduce observational errors. Trigonometric parallaxes can now be measured for stars out to about 300 light years from the sun, which is less than 1 per cent of the diameter of our galaxy. Only stars brighter than 20th magnitude can be measured.

A European Space Agency satellite devoted entirely to astronomy is planned. It is named Hipparcos (**Hi**gh **Pr**ecision **Par**allax **Co**llecting **S**atellite), close to the name of the astronomer Hipparchus. Hipparcos will give higher accuracy than we now have for a huge number of stars, hundreds of thousands. It will thus lead to great improvement in *astrometry,* the study of the positions of stars and their apparent motion across the sky. New astrometric measurements are also necessary for providing accurate positions for faint optical objects so they can be identified with emitters in other parts of the spectrum, such as those identified in radio astronomy or with x-ray satellites.

When Space Telescope is launched in 1986, it will be able to use its very high resolution to pinpoint the positions of stars even more accurately than Hipparcos. We will be able to measure parallaxes still farther into space, and the parallaxes remeasured for closer objects will be much more accurate. Among the many projects that will become possible is a thorough search for dark companions (that is, companions that are not shining or are shining very faintly)—perhaps even planets—of nearby stars. But Space Telescope will have time to observe only a small number of paral-

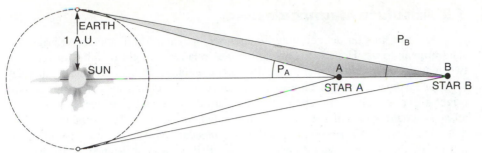

Figure 5–4 Star B is farther from the sun than star A and thus has a smaller parallax. The parallax angles are marked P$_A$ and P$_B$. To *subtend* is to take up an angle; as seen from point A, 1 A.U. subtends an angle P$_A$. As seen from point B, the same 1 A.U. subtends an angle P$_B$.

laxes. Still, though both Hipparcos and Space Telescope will improve parallaxes for close sources, Space Telescope will uniquely be able to tie in these parallaxes with the positions of objects, like galaxies and quasars, that are so far away that they form an unmoving background.

5.2a Parsecs

Since parallax measures have such an important place in the history of astronomy, a unit of distance was defined in terms that relate to these measurements. If we were outside the solar system, and looked back at the earth and the sun, they would appear to us to be separated in the sky. We could measure the angle by which they are separated. From twice as far away as Pluto, for example, when the earth and the sun were separated by the maximum amount (1 A.U.), they would be approximately 1° apart. As we go farther and farther away from the solar system, 1 A.U. subtends a smaller and smaller angle. When we go about 60 times farther away, 1 A.U. subtends only 1 arc min (60 arc min = 1°). From 60 times still farther, 1 A.U. subtends 1 arc sec (60 arc sec = 1 arc min). Note that the **angle** subtended by the astronomical unit is a measure of the **distance** we have gone (Fig. 5–4), similar to the way we measure an object's parallax.

The distance at which 1 A.U. subtends only 1 arc sec, we call *1 parsec* (Fig. 5–5). A star that is 1 parsec from the sun has a parallax of one arc sec. One parsec is a long distance. Most astronomers tend to use parsecs instead of light years, the distance that light travels in a year (= 9.5×10^{12} km), when talking about stellar distances. It takes light about 3.26 years to travel 1 parsec; thus 1 parsec = 3.26 light years. Note that parsecs and light years are both **distances,** just like kilometers or miles, even though their names contain references to their definitions in terms of angles or time.

*5.2b Computing with Parsecs

The advantage of using "parsecs" instead of "light years," "kilometers," or any other distance unit is that the distance in parsecs is equal to the inverse of the parallax angle in seconds of arc:

$d(\text{parsecs}) = 1/p$ (seconds of arc).

For example, a star with a parallax angle of 0.5 arc sec is 2 parsecs away (1/0.5 = 2) and would thus be one of the nearest stars to the sun. A star with a parallax of 0.01 is 1/0.01 = 100 pc away. Since the parallax angle is the quantity that is measured directly at the telescope, it was convenient to choose a distance unit very closely related to the parallax angle.

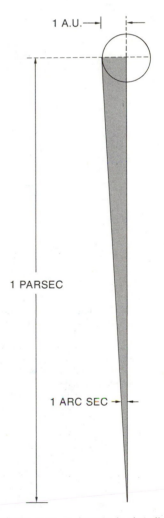

Figure 5–5 A *parsec* is the distance at which 1 A.U. subtends an angle of 1 arc sec. The drawing is not to scale. 1 arc sec is actually a tiny angle, the width of a dime at a distance of 3 km.

5.3 Absolute Magnitudes

We have thus far defined only the *apparent magnitudes* of stars, how bright the stars **appear** to us. However, stars can appear relatively bright or faint for either of two reasons: they could be intrinsically bright or faint, or they could be relatively close or far away. We can remove the distance effect, giving us a measure of how bright a star actually is, by choosing a standard distance and considering how bright all stars would appear if they were at that standard distance. The standard distance that we choose is 10 parsecs. We define the *absolute magnitude* of a star to be the magnitude that the star would appear to have if it were at a distance of 10 parsecs. Absolute magnitude gives the intrinsic brightness of a star, its *luminosity,* the total amount of energy the star gives off each second.

We normally write absolute magnitude with a capital *M*, and apparent magnitude with a small *m*. Sometimes a subscript signifies that we have observed through a special filter: M_V is the absolute magnitude through a special ''visual'' filter.

If a star happens to be exactly 10 parsecs away from us, its absolute magnitude is exactly the same as its apparent magnitude. If we were to take a star that is farther than 10 parsecs away from us and somehow move it to be 10 parsecs away, then it would appear brighter to us than it does at its actual position (since it is closer). Since the star would be brighter, its absolute magnitude would be a lower number (for example, 2 rather than 6) than its apparent magnitude.

On the other hand, if we were to take a star that is closer to us than 10 parsecs, and move it to 10 parsecs away, it would then be farther away and therefore fainter. Its absolute magnitude would be higher (more positive, for example, 10 instead of 6) than its apparent magnitude.

To assess just how much brighter or fainter that star would appear at 10 parsecs than at its real position, we must realize that the intensity of light from a star follows the *inverse-square law* (Fig. 5–6). That is, the intensity of a star varies inversely with the square of the distance of the star from us. (This law holds for all point sources of radiation, that is, sources that appear as points without length or breadth.) If we could move a star 2 times as far away, it would grow 4 times fainter. If we could move a star 9 times as far away, it would grow 81 times fainter. Of course, we are not physically moving stars (we would burn our shoulders while pushing),

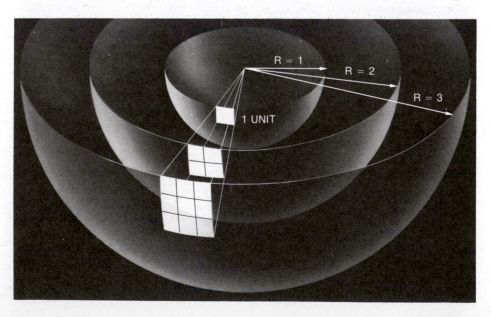

Figure 5–6 The inverse-square law. Radiation passing through a sphere twice as far away as another sphere has spread out so that it covers $2^2 = 4$ times the area; *n* times farther away it covers n^2 times the area.

but merely considering how they would appear at different distances. In all this, we are assuming that there is no matter in space between the stars to absorb light. Unfortunately, this is not always a good assumption (as we shall see in Chapter 13), but we make it anyway.

Astronomers normally speak in terms of magnitudes, so let us see how to interpret the inverse-square law in the magnitude scale. Remember that a difference of one magnitude is a factor of approximately 2.5 in brightness. Thus, if a star is 2.5 times fainter (that is, gives off 2.5 times less light) than another star, it is 1 magnitude fainter. If it is 3 times fainter, it is slightly more than 1 magnitude fainter. If it is 6 times fainter, it is slightly less than 2 magnitudes fainter.

> **EXAMPLE:** A star is 20 parsecs away from us, and its apparent magnitude is +4. What is its absolute magnitude?
>
> **ANSWER:** If the star were moved to the standard distance of 10 parsecs away, it would be twice as close as it was at 20 parsecs and, therefore, by the inverse-square law, would appear four times brighter. Since 2.5 times is one magnitude, and $(2.5)^2 = 6.25$ is two magnitudes, it would be approximately $1\frac{1}{2}$ magnitudes brighter. Since its actual apparent magnitude is 4, its absolute magnitude would be $4 - 1\frac{1}{2} = 2\frac{1}{2}$, equivalent to the apparent magnitude it would have if it were 10 parsecs from us. (We see that we must subtract the magnitude difference, since the star would be closer and therefore brighter at 10 parsecs.)

We have followed the following steps, knowing one type of magnitude and the distance, and wanting to find the other type of magnitude: (1) Find the ratio of the distances (the distance of the star and the 10 parsec standard distance; make sure the star's distance is also in parsecs); (2) square to find the factor by which the brightness is changed; (3) convert this factor to a difference in magnitudes (using each 1 magnitude = a factor of 2.5; each 5 magnitudes = a factor of 100); (4) add or subtract the magnitude difference to change between absolute and apparent magnitude.

> **EXAMPLE:** A star has apparent magnitude 10 and absolute magnitude 5. How far away is it?
>
> **ANSWER:** The star would grow brighter by 5 magnitudes were it to be moved to the standard distance of 10 parsecs. It would thus brighten by 100 times. By the inverse-square law, it would do so if it were 10 times closer than its real position. Its distance must therefore be 10 times farther away than the standard distance of 10 parsecs, and 10×10 parsecs is 100 parsecs.

Note that to find the distance in light years, we must multiply by 3.26 parsecs/ light year. The distance here is 326 light years.

We have followed the following steps, knowing both types of magnitudes and wanting to find the distance: (1) Find the difference in magnitudes, by subtracting; (2) convert this difference in magnitudes into a factor by which the brightness is changed (using a factor of 2.5 = 1 magnitude; a factor of 100 = 5 magnitudes); (3) find the change in distance by taking the square root of the factor of brightness; (4) multiply or divide the standard distance by this factor of distance, multiplying if the star's apparent magnitude is fainter than its absolute magnitude (and, if necessary, converting from parsecs to light years).

> **EXAMPLE:** A star has apparent magnitude 10 and absolute magnitude 3. How far away is it? (This is an example in which the numbers don't work out quite as smoothly.)
>
> **ANSWER:** The star would grow brighter by 7 magnitudes if we were to move it to the standard distance, making a factor of brightness of $(2.5)^2 \times 100 = 600$ (approximately). By the inverse-square law, it is thus growing

> ### *Box 5.1 Linking Apparent Magnitude, Absolute Magnitude, and Distance*
>
> Astronomers have a formula that allows you to calculate one of the terms apparent magnitude, absolute magnitude, and distance, given the other two. But the formula merely does numerically what we have just carried out logically. The formula is $m - M = 5 \log_{10} r/10$, where m is apparent magnitude, M is the absolute magnitude, and r is the distance in parsecs. Note that if $r = 10$, then $r/10 = 1$, $\log 1 = 0$, and $m = M$. If $r = 20$, as in the first example above, $m - M = 5 \log 2 = 5 \times 0.3 = 1.5$ magnitudes. Be careful not to get carried away using the formula without understanding the point of the manipulations.

closer by the square root of 600, which is about 25. (After all, $20^2 = 400$, and $30^2 = 900$, so the answer must be in between 20 and 30.) The star must thus actually be 25 times 10 parsecs, or 250 parsecs away.

5.3a Photometry

The actual value of either absolute or apparent magnitude can depend on the wavelength region in which we are observing. Let us consider a blue star and a red star, each of which gives off the same amount of energy. The blue star can seem much brighter than the red star if we observe them both through a blue filter, while the red star can seem much brighter than the blue if we observe them both through a red filter. Astronomers often measure a *color index* by subtracting the magnitude in one spectral range from the magnitude in another. Since the color of a star gives its temperature, color index is a measure of temperature.

A standard set of filters (Fig. 5–7) has been defined with one filter in the ultraviolet (called U), one filter in the blue (called B), and one filter in the yellow (called V for visual). Sets of equivalent filters exist at observatories all over the world. Hundreds of thousands of stars have had their colors measured with this *UBV* set of filters; we call the process *three-color photometry*. Other filter sets exist, as do filters for other spectral ranges; a star's magnitude measured in the infrared, for example, may be very different from its magnitude measured in the blue.

The quantity of fundamental importance is not the magnitude as observed through any given filter but rather the magnitude that corresponds to the total amount of energy given off by the star over all spectral ranges. This is called the *bolometric magnitude*, M_{bol}. Our ability to observe from space now allows us to measure magnitudes over a much wider range than previously.

"Bolometer," the instrument used to measure the total amount of energy arriving in all spectral regions, derives its name from "boli," Greek for "beam of light."

Figure 5–7 The U, B, and V curves represent the standard set of filters used by many astronomers. The response of the eye under normal conditions and the response of the dark-adapted eye are also shown.

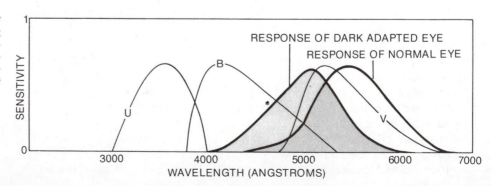

5.4 The Hertzsprung-Russell Diagram

In about 1910, Ejnar Hertzsprung (Fig. 5–8) in Denmark and Henry Norris Russell (Fig. 5–9) at Princeton University in the United States independently plotted a new kind of graph (Fig. 5–10). On the horizontal axis, the *x*-axis, each graphed a quantity that measured the temperature of stars. On the vertical axis, the *y*-axis, each graphed a quantity that measured the brightness of stars. They found that all the points that they plotted fell in limited regions of the graph rather than being widely distributed over the graph.

There are several possible ways to plot a measure of the temperature without plotting the temperature itself. One can plot the spectral type, for example, or even the color of a star (usually in the form of color index, as mentioned in Section 5.3a).

There are also several possible ways to plot the brightness. The simplest is merely to plot the apparent magnitudes of stars, with the brightest stars (i.e., those with the most negative magnitudes) at the top. But remember that a star can appear bright either by really being intrinsically bright or, alternatively, by being very close to us. One way to get around this problem is to plot only stars that are at the same distance away from us.

How do we find a group of stars at the same distance? Luckily, there are clusters of stars in the sky (described in more detail in Chapter 6) that are really groups of

Figure 5–8 Ejnar Hertzsprung in the 1930's.

Figure 5–9 Henry Norris Russell and his family circa 1917.

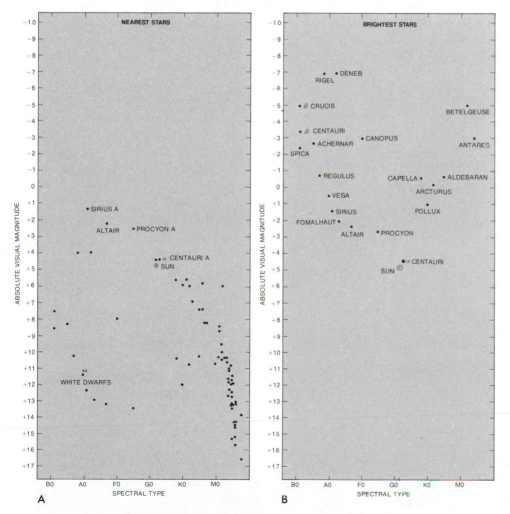

Figure 5–10 Hertzsprung-Russell diagrams *(A)* for the nearest stars in the sky and *(B)* for the brightest stars in the sky. The brightness scale is given in *absolute magnitude*. Because the effect of distance has been removed, the intrinsic properties of the stars can be compared directly on such a diagram.

Note that none of the nearest stars is intrinsically very bright. Also, the brightest stars in the sky are, for the most part, intrinsically very luminous, even though they are not usually the very closest to us.

Figure 5–11 A physical grouping of stars in space. The globular cluster ω (omega) Centauri, photographed with the Anglo-Australian telescope.

Figure 5–12 The Hertzsprung-Russell diagram, with the stars of Figure 5–10 included. The spectral type axis *(bottom)* is equivalent to the temperature axis *(top)*. The absolute magnitude axis *(left)* is equivalent to the luminosity axis *(right)*.

stars in space at approximately the same distance away from us. We do not have to know what the distance is to plot a diagram of the type plotted by Hertzsprung; all we have to know is that the stars are really clustering in space (Fig. 5–11).

If we somehow knew the absolute magnitudes of the stars, that would also be a good thing to plot. When we know the distances (for relatively close stars, we can get this from trigonometric parallax measurements, which is what Russell originally did), we can calculate the absolute magnitudes, which are actually what is plotted in Fig. 5–10.

Such a plot of temperature versus brightness is known as a *Hertzsprung-Russell diagram,* or simply as an *H-R diagram*. Note that since H-R diagrams were sometimes originally plotted by spectral type, from O to M, the hottest stars are on the left side of the graph. Thus temperature increases from right to left. Also, since the brightest stars are on the top, magnitude decreases toward the top. In some sense, thus, both axes are plotted backwards from the way a reasonable person might choose to do it if there were no historical reasons for doing it otherwise.

When plotted on a Hertzsprung-Russell diagram, the stars lie mainly on a diagonal band from upper left to lower right (Fig. 5–12). Thus the hottest stars are normally brighter than the cooler stars. Most stars fall very close to this band, which is called the *main sequence*. Stars on the main sequence are called *dwarf stars,* or *dwarfs*. There is nothing strange about dwarfs; they are the normal kind of stars. The sun is a type G dwarf. Some dwarfs are quite large and bright; the word ''dwarf'' is used only in the sense that these stars are not a larger, brighter kind of star that we will define below as giant.

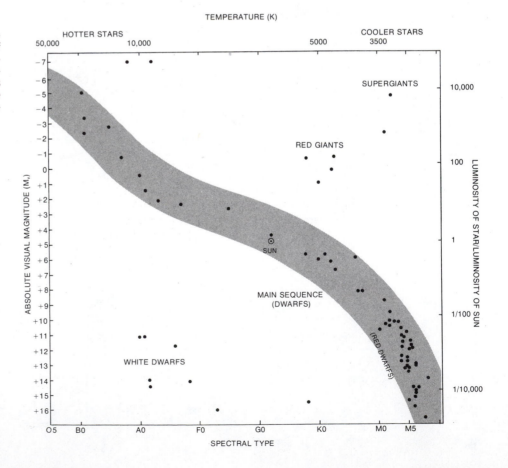

Some stars lie above and to the right of the main sequence. That is, for a given spectral type the star is intrinsically brighter than a main-sequence star. These stars are called *giants,* because their luminosities are large compared to dwarfs of their spectral type. Some stars, like Betelgeuse, are even brighter than normal giants, and are called *supergiants.* Since two stars of the same spectral type have the same temperature and, according to the Stefan-Boltzmann law (Section 4.1), the same amount of emission from each area of their surfaces, the brighter star must be bigger than the fainter star of the same spectral type.

Stars in a class of faint hot objects, called *white dwarfs,* are located below and to the left of the main sequence. They are smaller and fainter than main-sequence stars (ordinary dwarfs) of the same spectral type.

The use of the Hertzsprung-Russell diagram to link the spectrum of a star with its brightness is a very important tool for stellar astronomers. The H-R diagram provides, for example, another way of measuring the distances to stars, as we shall see below.

5.5 *Spectroscopic Parallax*

When we take a group of stars whose distances we can measure directly, by some method like that of trigonometric parallax, we can plot a Hertzsprung-Russell diagram. From this standard diagram we can read off the absolute magnitude that corresponds to any star on the main sequence.

We can apply the standard Hertzsprung-Russell diagram to find the distance to any star whose spectrum we can observe, no matter how faint, if we know that the star is on the main sequence. Examining the spectrum in detail reveals to trained eyes whether or not a star is on the main sequence (Fig. 5–13).

For a main-sequence star, we first find the absolute magnitude that corresponds to its spectral type. We must also know the apparent magnitude of the star, but we can get that easily and directly by simply observing it. Once we know both the apparent magnitude and the absolute magnitude, we have merely to figure out how far the star has to be from the standard distance of 10 parsecs to account for the difference $m - M$. (We get this from the inverse-square law.)

EXAMPLE: We see a G2 star on the main sequence. Its apparent magnitude is $+8$. How far away is it?

ANSWER: From the H-R diagram, we see that $M = +5$. The star appears fainter than it would be if it were at 10 parsecs. It is approximately $(2.5)^3 = 6 \times 2.5 = 15$ times fainter. (More accuracy in carrying out this calculation would not be helpful because the original data were not more accurate.) By the inverse-square law, this means that it is approximately 4 times farther away (exactly 4 times would be a factor of $4^2 = 16$ in brightness). Thus the star is 4 times 10 parsecs, or 40 parsecs, away.

We are measuring a distance, and not actually a parallax, but by analogy with the method of trigonometric parallax for finding distance, the method using the H-R diagram is called finding the *spectroscopic parallax.*

5.6 *The Doppler Effect*

The Doppler effect is one of the most important tools that astronomers can use to understand the universe. Without having to measure the distance to an object, they can use the Doppler effect to determine its *radial velocity,* its speed toward or away from us (on the radius of an imaginary sphere centered at us). The Doppler effect in sound is familiar to most of us, and its analogue in electromagnetic radiation, including light, is very similar.

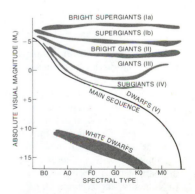

Figure 5–13 One can tell whether a star is a dwarf, giant, or supergiant by looking closely at its spectrum, because slight differences exist for a given spectral type. This procedure is called luminosity classification. Once we know a star's luminosity class, we can place the star on the H-R diagram and find its spectroscopic parallax.

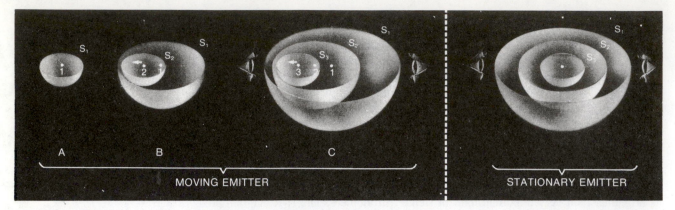

Figure 5–14   An object emits waves of radiation that can be represented by spheres representing the peaks of the wave, each centered on the object and expanding. In the left side of the drawing, the emitter is moving in the direction of the arrow. In part *A*, we see that the peak emitted when the emitter was at point 1 becomes a sphere (labelled S) around point 1, though the emitter has since moved toward the left. In part *B*, some time later, sphere S has continued to grow around point 1 (even though the emitter is no longer there), and we also see sphere S_2, which shows the position of the peaks emitted when the emitter had moved to point 2. Sphere S_2 is thus centered on point 2, even though the emitter has continued to move on. In *C*, still later, yet a third peak of the wave has been emitted, S_3, this time centered on point 3, while spheres S_1 and S_2 have continued to expand.

For the case of the moving emitter, observers who are being approached by the emitting source (those on the left side of the emitter, as shown on the left side of Case *C*) see the three peaks drawn coming past them relatively bunched together (that is, at shorter intervals of time, which is the same as saying at a higher frequency), as though the wavelength were shorter, since we measure the wavelength from one peak to the next. This corresponds to a wavelength farther to the blue than the original, a *blueshift*. Observers from whom the emitter is receding (those on the right side of the emitter, as shown on the right side of Case *C*) see the three peaks coming past them with decreased frequency (at increased intervals of time), as though the wavelength were longer. This corresponds to a color farther to the red, a *redshift*.

Contrast the case at the extreme right, in which the emitter does not move and so all the peaks are centered around the same point. No redshifts or blueshifts arise.

Note that once the light wave is emitted, it travels at a constant speed ("the speed of light," 3×10^{10} cm/s, equivalent to seven times around the earth in a second), so that a shorter (lower) wavelength corresponds to a higher frequency, and a longer (higher) wavelength corresponds to a lower frequency.

5.6a Blueshifts and Redshifts

You may be familiar with how the sound of a train whistle, a jet engine, or a motorcycle motor changes in pitch as the train, plane, or motorcycle first approaches you and then passes you and begins to recede. As the object that is emitting the sound waves approaches, it has moved closer to you by the time it emits a second wave than it had been when it emitted the first wave. Thus waves arrive more frequently than they would if the source were not moving. The wavelengths seem compressed, and the pitch is higher. After the emitting source has passed you, the wavelengths are stretched, and the pitch of the sound is lower (Fig. 5–14).

With light waves, the effect is similar. As a body emitting light, or other electromagnetic radiation, approaches you, the wavelengths become slightly shorter than they would be if the body were at rest. Visible radiation is thus shifted slightly in the direction of the blue (it doesn't actually have to become blue, only be shifted in that direction). We say that the radiation is *blueshifted*. Conversely, when the emitting object is receding, the radiation is said to be *redshifted* (Fig. 5–15). We generalize these terms to types of radiation other than light, and say that radiation is blueshifted whenever it changes to shorter wavelengths, and redshifted whenever it changes to longer wavelengths.

The point of all this is that we can measure a Doppler shift for any object we

The Doppler shift allows us to measure only velocities toward or away from us and not side to side. Since these velocities are along a radius extending outward from earth, we call them *radial velocities*. If the object is moving at some angle to the radius, then the Doppler shift measures only the part of the velocity (technically, the *component* of the velocity) in the radial direction.

Even the Doppler shift of a car's 30 km/hr velocity, 1/3600 times less than the speed of the star in the example, is easily measurable by police radar, as many have discovered to their chagrin.

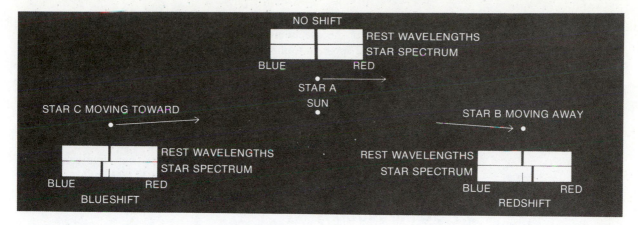

can see that has a spectral line or some other feature in its spectrum for which we can measure a wavelength. We can then tell how fast the object is moving along a radius toward or away from us. If we were moving at the same speed and in the same direction as a source of radiation or sound, it would have no net velocity with respect to us. Then no Doppler shift would be observed.

Figure 5–15 The Doppler effect. In each pair of spectra, the position of the spectral line in the laboratory is shown on top and the position observed in the spectrum of the star is shown below it. Lines from approaching stars appear blueshifted, lines from receding stars appear redshifted, and lines from stars that are moving transverse to us are not shifted because the star has no velocity toward or away from us. A short vertical line marks the unshifted position on the spectra showing shifts.

*5.6b Working with Doppler Shifts

The wavelength when the emitter is at rest is called the *rest wavelength*. Let us consider a moving emitter. The fraction of the rest wavelength that the wavelength of light is shifted is the same as the fraction of the speed of light at which the body is travelling (or, for a sound wave, the fraction of the speed of sound).

We can write

$$\frac{\text{change in wavelength}}{\text{original wavelength}} = \frac{\text{velocity of emitter}}{\text{speed of light}}$$

or

$$\frac{\Delta\lambda}{\lambda_o} = \frac{v}{c}$$

where $\Delta\lambda$ is the change in wavelength (the Greek delta, Δ, usually stands for the change), λ_o is the original (rest) wavelength, v is the velocity of the emitting body, and c is the velocity of light ($= 3 \times 10^5$ km/s). We define positive velocities as velocities of recession (redshifts) and negative velocities as velocities of approach (blueshifts). The new wavelength, λ, is equal to the old wavelength plus the change in wavelength, $\lambda_o + \Delta\lambda$.

EXAMPLE: A star is approaching at 30 km/s (i.e., $v = -30$ km/s). At what wavelength do we see a spectral line that was at 6000 Å (which is in the orange part of the spectrum) when the radiation left the star?

ANSWER:

$$\frac{\Delta\lambda}{\lambda_o} = \frac{v}{c} = \frac{-30 \text{ km/s} \times 10^5 \text{ cm/km}}{3 \times 10^{10} \text{ cm/s}} = \frac{-3 \times 10^6 \text{ cm/s}}{3 \times 10^{10} \text{ cm/s}} = -10^{-4}.$$

$\Delta\lambda = -10^{-4}\lambda_o = -10^{-4} \times 6000$ Å $= -0.6$ Å. (Since the star is approaching, this change of wavelength is a blueshift, and the new wavelength is slightly shorter than the original wavelength.) The new wavelength, λ, is thus $\lambda_o + \Delta\lambda = 6000$ Å -0.6 Å $= 5999.4$ Å. It is still in the orange part of the spectrum. The change would be easily measurable, but would not be apparent to the eye.

Note that since there are proportions on both sides of the equation, if we take care to use v and c in the same units (e.g., km/s, cm/s, or whatever), the $\Delta\lambda$ will be in the same units that λ is in, no matter whether that is angstroms, centimeters, or whatever.

The 30 km/s given in the example is typical of the random velocities that stars have with respect to each other. These velocities are small on a universal scale, much too small to change the overall color that the eye perceives, but large compared to terrestrial velocities (30 km/s $=$ 30 km/s $\times$ 3600 s/hr $=$ 108,000 km/hr).

5.7 Stellar Motions

On the whole, the network of stars in the sky is fixed. But radial velocities can be measured for all stars, and some of the stars are seen to move slightly across the sky with respect to the more distant stars. The actual velocity of a star in 3-dimensional space, with respect to the sun, is called its *space velocity,* but when we detect a star's change in position in the sky, we know only through what angle it moved. We cannot tell how far it moved in linear units (like km or light years) unless we also happen to know the distance to the star.

We usually deal separately with the part of the velocity of a star that is toward or away from us (the radial velocity), and the angular velocity of the star across the sky (how fast the object is moving across the sky in units of angle). The angular velocity is called the *proper motion*. Radial velocity and proper motion are perpendicular to each other.

The stars that show proper motions are generally the closest stars to us. If more distant stars were moving at the same velocities as measured in km/s, their angular movement would be imperceptible.

Proper motion is ordinarily very small, and is accordingly very difficult to measure. Usually we must compare the positions of stars taken at intervals of decades to detect a sufficient amount of proper motion to allow a measurement of any accuracy to be made. Only a few hundred stars have proper motions as large as 1 arc sec per year; proper motions of other stars are sometimes measured in seconds of arc per century! Space Telescope and Hipparcos will soon improve the accuracy of proper motion measurements by a considerable factor.

Figure 5–16 Two photographs taken 11 months apart. In making this combined print, one of the negatives was shifted slightly. The proper motion of Barnard's star is clearly visible.

The star with the largest proper motion was discovered in 1916 by E. E. Barnard, and is known as Barnard's star (Fig. 5–16). It moves across the sky by $10\frac{1}{4}$ arc

Figure 5–17 If we know a star's proper motion and its distance, we can compute its linear velocity through space in the direction across our field of view. The Doppler effect gives us its linear velocity toward or away from us. These two velocities can be combined to tell us the star's actual velocity through space, its *space velocity*.

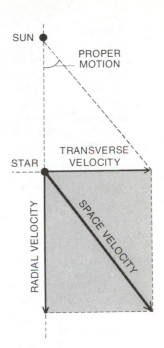

sec per year. Some call it ''Barnard's runaway star'' because of this proper motion, which, though the highest for any known star, is still minuscule. Barnard's star moves so rapidly across the sky because it is relatively close to us. Only 1.8 parsecs away, it is the nearest star to us (Appendix 7) after the sun, Proxima Centauri, and the rest of the Alpha Centauri system (of which Proxima is a part). Since the moon appears half a degree across, Barnard's star moves across the sky by the equivalent of the diameter of the moon in only 180 years.

If we know the distance to a star, that distance when combined with the proper motion gives us the star's actual velocity across space from side to side, its *transverse velocity*. From the transverse velocity and the radial velocity (obtained from the Doppler shift) together, we can calculate the star's *space velocity* (Fig. 5–17).

Summary and Outline

Magnitudes (Section 5.1)
Lower (or negative) numbers are the brightest objects.
5 magnitudes difference = 100 times in brightness; adding or subtracting magnitudes corresponds to dividing or multiplying brightness.
Brightest star is magnitude −1.4; faintest stars visible to the naked eye are about sixth magnitude.

Trigonometric parallax (Section 5.2)
We triangulate, using the earth's orbit as a baseline.
We find the angle through which a nearby star appears to shift; the more the shift, the nearer the star.
This method works only for the nearest stars; Space Telescope and Hipparcos will improve our capabilities.
A *parsec* is the distance of a star from the earth when the radius of the earth's orbit subtends 1 arc sec when viewed from the star.

Absolute magnitude (Section 5.3)
If we know how bright a star **appears,** and how far away it is, we can calculate how intrinsically bright it is. This can be placed on the same scale as that of apparent magnitudes, and is known as the *absolute magnitude*. The absolute magnitude is defined as the magnitude a star would appear to have if it were at a distance of 10 parsecs.
The apparent brightness of a point object follows the inverse square law: brightness decreases with the square of the distance. We can use the inverse square law to relate apparent magnitude, absolute magnitude, and distance.

The Hertzsprung-Russell diagram (Section 5.4)
A plot of brightness versus temperature
Stars fall in only limited regions of the graph.
The *main sequence* contains most of the stars; these stars are called *dwarfs*. *Giants* are brighter (and bigger) than dwarfs; *supergiants* are brighter and bigger still. *White dwarfs* are fainter than dwarfs, and so fall below the main sequence.

Method of *spectroscopic parallax* (Section 5.5)
Observing the spectrum of a star tells us where it falls on an H-R diagram; this tells us its absolute magnitude. Since we can easily observe its apparent magnitude, we can derive its distance.

The Doppler effect (Section 5.6)
Radiation from objects that are receding is shifted to longer wavelengths: redshifted. Radiation from objects that are approaching is shifted to shorter wavelengths: blueshifted.

Stellar motions (Section 5.7)
The angular velocity from side to side is measured by observing the *proper motion*—the motion of the star across the sky. This can be observed only for the nearest stars. If we know both distance and proper motion, we can derive the star's *transverse velocity*. Angular velocity is usually measured in arc sec/year and radial velocity in km/s.
The radial velocity and the transverse velocity together give us a star's *space velocity*, its actual motion through space with respect to the sun.

Key Words

apparent magnitude, significant figure, baseline, Astronomical Unit, parallax, trigonometric parallax, astrometry, parsec, apparent magnitude, absolute magnitude, luminosity, inverse-square law, color index*, UBV*, three-color photometry*, bolometric magnitude*, Hertz-sprung-Russell diagram, H-R diagram, main sequence, dwarf stars, dwarfs, giants, super-giants, white dwarfs, spectroscopic parallax, radial velocity, blueshifted, redshifted, compo-nent, rest wavelength*, space velocity, proper motion, transverse velocity, measuring engines, subtend

---

*These terms are found in optional sections.

Questions

1. Which star is visible to the naked eye: one of the 4th magnitude or one of 8th magnitude?

2. Which is brighter: Mars when it is magnitude +0.5 or Spica, a star whose magnitude is 0.9 (Appendix 5)?

††3. Venus can reach magnitude −4.4, while the bright-est star, Sirius, is magnitude −1.4. How many times brighter is Venus at its maximum than is Sirius?

†4. Venus can be brighter than magnitude −4. Antares is a first-magnitude star (m = +1). How many times brighter is Venus at magnitude −4 than Antares?

†5. Pluto is about 14th magnitude at most. How many times fainter is it than Venus? (m_{Venus} = −4, approximately)

†6. If a variable star brightens by a factor of 15, by how many magnitudes does it change?

†7. If a variable star starts at 5th magnitude and bright-ens by a factor of 60, at what magnitude does it appear?

†8. The variable star Mira ranges between magnitudes 9 at minimum and 3 at maximum. How many times brighter is it at maximum than at minimum?

†9. Star A has magnitude +11. Star B appears 10,000 times brighter. What is the magnitude of star B? Star C ap-pears 10,000 times fainter than star A. What is its magnitude?

†10. Star A has magnitude +10. The magnitude of star B is +5 and of star C is +3. How much brighter does star B appear than star A? How much brighter does C appear than B?

†11. How much brighter is a 0th-magnitude star than a +3rd-magnitude star?

12. You are driving a car, and the speedometer shows that you are going 80 km/hr. The person next to you asks why you are going only 75 km/hr. Explain why you each saw dif-ferent values on the speedometer.

13. Would the parallax of a nearby star be larger or smaller than the parallax of a more distant star? Explain.

†14. A star has an observed parallax of 0.2 arc sec. An-other star has a parallax of 0.02 arc sec. (a) Which star is farther away? (b) How much farther away is it?

†15. (a) What is the distance in parsecs to a star whose parallax is 0.05 arc sec? (b) What is the distance in light years?

†16. Estimate the farthest distance for which you can de-tect parallax by alternately blinking your eyes and looking at objects at different distances. (You may find it useful to look through a window to outdoor objects, so that you can see them silhouetted against a very distant background.) To what angle does this correspond?

†17. Vega is about 8 parsecs away from us. What is its parallax?

18. What are the two fundamental quantities that are being plotted on the axes of a Hertzsprung-Russell diagram?

19. What is the significance of the existence of the main sequence?

20. What is the observational difference between a dwarf and a white dwarf?

†21. Two stars have the same apparent magnitude and are the same spectral type. One is twice as far away as the other. What is the relative size of the two stars?

†22. Two stars have the same absolute magnitude. One is ten times farther away than the other. What is the difference in apparent magnitudes?

†23. A star has apparent magnitude of +5, and is 100 parsecs away from the sun. If it is a main-sequence star, what is its spectral type? (Hint: refer to Fig. 5–12.)

†24. A star is 30 parsecs from the sun and has apparent magnitude +2. What is its absolute magnitude?

††25. A star has apparent magnitude +9 and absolute magnitude +4. How far away is it?

†26. The nearest star, alpha Centauri, is apparent magni-tude 0 and absolute magnitude +4.4. How far away is it in parsecs? In light years?

†27. Betelgeuse has apparent magnitude 0.4 and absolute magnitude −5.6. How far away is it in parsecs? In light years?

†28. The first quasar to be discovered, 3C273 (Chapter

---

†This indicates a question requiring a numerical solution.

††Answers: **3.** $(2.512)^3$ = 15 times   **25.** 100 pc

15) was identified with what appeared to be a star of magnitude 13. It turned out to be 1 billion parsecs away. When the discoverer did the calculation, what absolute magnitude did it turn out to be? How many times brighter is it than the sun, whose absolute magnitude is about +5? The result astonished scientists.

29. If a star is moving away from the earth at very high speed, will the star have a continuous spectrum that appears hotter or cooler than it would if the star were at rest? Explain.

30. (a) What does the proper motion of Barnard's star indicate about its distance from us? (b) What would Barnard's star's proper motion be if it were twice as far away from us as it is?

†31. A star has a proper motion of 10 arc sec per century. Can you tell how fast it is moving in space in km/s? If so, describe how.

†32. Two stars have the same space velocity in the same direction, but star A is 10 parsecs from the earth and star B is 30 parsecs from the earth. (a) Which has the larger radial velocity? (b) Which has the larger proper motion?

33. How might Space Telescope help in determining proper motions?

34. What is the only direct way to determine the distance to a star?

M13, a globular cluster in Hercules.

Doubles, Variables, and Clusters

6

Aims: To discuss double stars, variable stars, and stellar clusters, and to draw conclusions about the distances to stars and about the ages of stars

We often think of stars as individual objects that shine steadily, but many stars vary in brightness and most stars actually have companions close by. Also, many stars appear as members of groupings called clusters. By studying the effects that the stars have on each other, or by studying the nature of all the stars in a star cluster, we can learn much that we could not discover by studying the stars one at a time. This information even leads us to a general understanding of the life history of the stars.

6.1 Binary Stars

Most of the objects in the sky that we see as single "stars" really contain two or more component stars. Sometimes a star appears double merely because two stars that are located at different distances from the sun appear in the same line of sight. Such systems are called *optical doubles,* and will not concern us here. We are more interested in stars that are physically associated with each other. We will use the terms *double star* or *binary star* interchangeably to mean two or more stars held together by the gravity they have between each other. (Even systems with three or more stars are usually called "double stars.")

The easiest way to tell that more than one star is present is by looking through a telescope of sufficiently large aperture. Stars that appear double when observed directly are called *visual binaries*. The resolution of small telescopes may not be sufficient to allow the components of a double star to be "separated" from each other; larger telescopes can thus distinguish more double stars. When the stars have different colors, they are particularly beautiful to observe even with small telescopes. Five to ten per cent of the stars in the sky are visual binaries.

We are not always fortunate enough to have the two components of a double star appear far enough apart from each other to be observed visually. Sometimes a star appears as a single object through a telescope, but one can see that the spectrum of that "object" actually consists of the overlapping spectra of at least two objects. If we can detect the presence of two spectra of different types—of say, one hot star and one cool star—then we say that the object has a *composite spectrum*.

We can tell that a second star is present even when an image appears single if the spectral lines we observe change in wavelength with time. We know of no other

Albireo (β Cygni) contains a B star and a K star, which make a particularly beautiful pair because of their different colors.

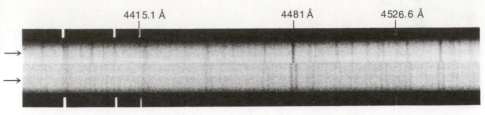

Figure 6–1 Two spectra of Mizar (ζ Ursae Majoris) taken 2 days apart show that it is a spectroscopic binary. The lines of both stars are superimposed in the upper stellar absorption spectrum *(top arrow),* but are separated in the lower spectrum *(bottom arrow)* by 2 Å, which corresponds to a relative velocity of 140 km/sec. Emission lines from a laboratory source are shown at the extreme top and extreme bottom to provide a comparison with a source at rest that has lines at known wavelengths.

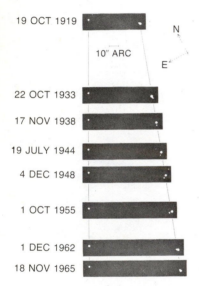

Figure 6–3 The two components of the visual binary Krüger 60 are seen to orbit each other with a period of about 44 years in this series of photographs taken at the Leander McCormick Observatory in Virginia and at the Sproul Observatory in Pennsylvania. We also see the proper motion of the Krüger 60 system, as it moves farther away from the single star at the left.

way such wavelength changes can occur except for finding variable Doppler shifts, which indicate that the speed of an object is moving along a line linking it with us. We deduce that two or more stars are present, and are revolving around each other. Such an object is called a *spectroscopic binary* (Fig. 6–1). As the stars in the system orbit each other, unless we are looking straight down on the orbit from above, each spends half of its orbit approaching us and the other half receding from us, relative to its average space motion. The variations of velocity in spectroscopic binaries are periodic; the spectrum of each component varies separately in wavelength. Note that even if the spectrum of one of the stars is too faint to be seen, we can still tell if the star is a spectroscopic binary from the variations in radial velocity of the visible component (Fig. 6–2).

Careful spectroscopic studies have shown that two-thirds of all solar-type stars have stellar companions. Though the presence of companions of stars of other spectral types has not been studied in such detail, it seems that about 85 per cent of all stars are members of double-star systems. Few stars are single, an idea that is verified by noticing that many of the nearest stars to our sun (Appendix 7), stars we can study in detail, are double. In recent years, we have realized that in some circumstances mass can flow from one member of a binary system to the other, changing the evolution of each. We shall discuss such effects on stellar evolution in Section 9.7.

We can detect visual binaries most easily when the stars are relatively far apart. Then the period of the orbit is relatively long, over 100,000 years in most cases. So we know little about the motions in such systems because of our short human lifetime, even though most double stars discovered are visual binaries. It is harder to detect spectroscopic binaries so we know fewer of them. But for this group, most of whose periods are between one day and one year, we are able to determine such important details of the orbit as size and period (Fig. 6–3).

Sometimes the components of a double star pass in front of each other, as seen from our viewpoint on the earth. The "double star" then changes in brightness periodically, as one star cuts off the light from the other. Such a pair of stars is called an *eclipsing binary* (Fig. 6–4). The easiest to observe is Algol, β Persei (beta of Perseus), in which the eclipses take place every 69 hours, dropping the total bright-

Figure 6–2 Two spectra of Castor B (α Geminorum B) taken at different times show a Doppler shift. Thus the star is a spectroscopic binary, even though lines from only one of the components can be seen. The comparison spectrum of a laboratory source appears at the top and bottom. Note that emission lines from this laboratory source are in the same horizontal positions at extreme top and bottom, while the absorption lines of the stellar spectra (shown with arrows) are shifted laterally (that is, in wavelength) with respect to each other.

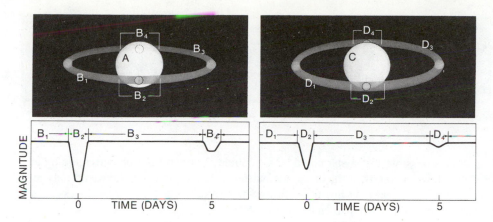

Figure 6–4 The shape of the light curve of an eclipsing binary depends on the sizes of the components and the angle from which we view them. At lower left, we see the light curve that would result for star B orbiting star A, as pictured at upper left. When star B is in the positions shown with subscripts at top, the regions of the light curve marked with the same subscripts result. The eclipse at B_4 is total. At right, we see the appearance of the orbit and the light curve for star D orbiting star C, with the orbit inclined at a greater angle than at left. The eclipse at D_4 is partial. From earth, we observe only the light curves, and use them to determine what the binary system is really like, including the inclination of the orbit and the sizes of the objects.

ness of the system from magnitude 2.3 to 3.5. (A third star is present in the Algol system. It orbits the other two every 1.86 years, and does not participate in the eclipse.)

Note that the way a binary star appears to us depends on the orientation of the two stars not only with respect to each other, but also with respect to the earth. If we are looking down at the plane of their mutual orbit (Fig. 6–5), then we might see a composite spectrum; under the most favorable conditions the star might be seen as a visual binary. But in this orientation we would never be able to see the stars eclipse. We would also not be able to see the Doppler shifts typical of spectroscopic binaries, since only radial velocities contribute to the Doppler shift.

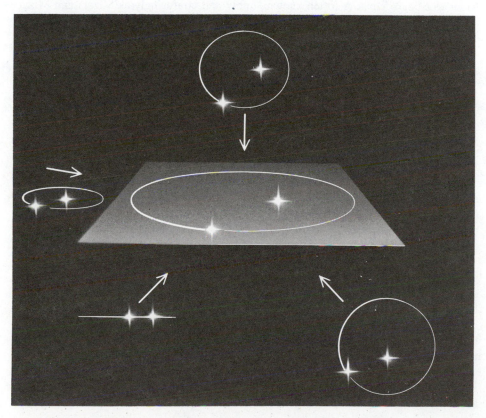

Figure 6–5 The appearance and the Doppler shift of the spectrum of a binary star depend on the angle from which we view the binary. From far above or below the plane of the orbit, we might see a visual binary, as shown on the top and at the lower right of the diagram. From close to but not exactly in the plane of the orbit, we might see only a spectroscopic binary, as shown at left. (The stars appear closer together, so might not be visible as a visual binary.) From exactly in the plane of the orbit, we would see an eclipsing binary, as shown at lower left.

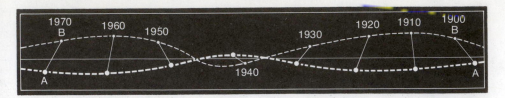

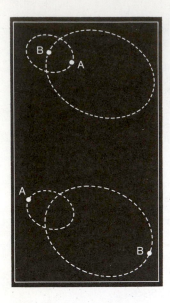

Figure 6–6 Sirius A and B, an astrometric binary. From studying the motion of Sirius A (often called, simply, Sirius) astronomers deduced the presence of Sirius B before it was seen directly. Sirius B's orbit is larger because Sirius A is a more massive star. They contain 2.14 and 1.05 times as much mass as the sun, respectively.

Example: Castor C is not only a spectroscopic binary but also an eclipsing binary, and is sometimes known as YY Gem. From the fact that eclipses occur, we know that the orbits of the stars lie in a plane parallel to our line of sight, as in the bottom left case of Figure 6–5. Thus we can use measurement of the Doppler shift to derive the speed in its orbit of each component, and from this and its measured period can find the size of its orbit. From this we can derive (we use a form of Kepler's third law, which we discuss in Sections 18.7 and 18.8) the sum of the masses of the components. Since we can tell from the spectra that the masses of each are the same, we simply divide this sum by two to get the mass of each star by itself. Each turns out to contain 0.6 times as much mass as the sun.

It is possible that a star could be a ''double,'' but still not be detectable by any of the above methods. Sometimes, the existence of a double star shows up only as a deviation from a straight line in the proper motion of the ''star'' across the sky. Such stars are called *astrometric binaries* (Fig. 6–6).

A double star may fall in more than one category. For example, two stars that eclipse each other could well be a spectroscopic binary too.

Many of the celestial sources of x-rays that have been observed in recent years from orbiting telescopes turn out to be binary systems. Matter from one member of the pair falls upon the other member, heats up, and radiates x-rays. We will discuss several such systems in the chapter on neutron stars.

Though one usually speaks of ''double stars,'' stars can exist in threes, or fours, or with even more components. For example, the second star from the end of the handle of the Big Dipper, Mizar, has an apparent fainter companion called Alcor. Alcor and Mizar, which are separated in the sky by about one-third the diameter of the moon, are an optical double. Many people can see them as double with the naked eye. The American Indians knew these two stars as a horse and rider.

A telescope shows that two stars are present in Mizar, as well. Mizar's components (called Mizar A and Mizar B) are separated by 1/50 the angular distance from Alcor to Mizar. Mizar A (often known simply and confusingly as Mizar) and Mizar B are, in turn, each spectroscopic binaries. Mizar A is a double-line spectroscopic binary (Fig. 6–1), and Mizar B shows only a single set of lines.

Another interesting star is Castor, α Geminorum. Through a telescope it appears as a triple star: Castor A and Castor B are relatively close to each other and Castor C is somewhat farther off. Castor A, B, and C are each spectroscopic binaries; we saw the spectrum of Castor B in Fig. 6–2. Thus Castor is a sextuple star.

6.2 Stellar Masses

The study of binary stars is of fundamental importance in astronomy because it allows us to determine stellar masses. If we can determine the orbits of the stars around each other, we can calculate theoretically the masses of the stars necessary to produce the gravitational effects that lead to those orbits. For a star that is a visual binary with a sufficiently short period (only 20 or even 100 years, for example), we are able to determine the masses of both of the components.

We are more limited if a star is only a spectroscopic binary, even one with spectral lines of both stars present. If the star is not also a visual binary, we can find only the lower limits for the masses—that is, we can say that the masses must be larger than certain values. This limitation occurs because we cannot usually tell the angle by which the plane of the stars' orbit is inclined. Only if the spectroscopic binary is also an eclipsing binary do we know the angle of inclination, allowing us to find the individual masses.

So we do not always know as much as we would like about the masses of stars in binary systems. Of course, we are better off than we are for stars outside binary systems, for we cannot directly measure their masses at all! From studies of the

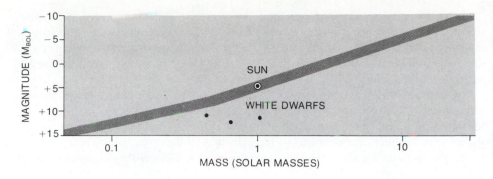

Figure 6–7 The mass-luminosity relation, measured from binary stars. Most of the stars fit within the shaded band. The relation is defined by straight lines of different slopes for stars brighter than and fainter than magnitude +7.5. We do not know if there is a real difference between the brighter and fainter stars that causes this or whether it is an effect introduced when the data are reduced, at which stage we try to take account of the energy that the stars radiate outside the visible part of the spectrum.

Note that white dwarfs, which are relatively faint for their masses, lie below the mass-luminosity relation. Red giants lie above it. Thus we must realize that the mass-luminosity relation holds only for main-sequence stars.

several dozen binaries for which we can accurately tell the masses, astronomers have graphed the luminosities (intrinsic brightnesses) of the stars on one axis and the masses on the other axis. Most of the stars turn out to lie on a narrow band, the *mass-luminosity relation* (Fig. 6–7).

The mass-luminosity relation is valid only for stars on the main sequence. The more massive a main-sequence star is, the brighter it is. The mass, in fact, is the prime characteristic that determines where on the main sequence a star will settle down to live its lifetime.

The most massive stars we know are about 50 times more massive than the sun, and the least massive stars are about 7 per cent the mass of the sun. A dwarf star (i.e., main-sequence star) of 10 solar masses is a B star. A dwarf star of 2 solar masses is an A star. The sun, which of course has 1 solar mass, is a G star. A dwarf of half a solar mass is a K star, and a dwarf of four-tenths of a solar mass or less is an M star.

There are many more stars of low mass than there are of high mass. In fact, the number of stars of a given mass in any sufficiently large volume of space increases as we consider stars of lower and lower mass. Thus most of the mass in a cluster of stars, or in a galaxy, comes from stars at the lower end of the H-R diagram. Even though the many faint stars contain most of the total mass in a group of stars, most of the light that we receive from a group of stars, on the contrary, comes from the few very brightest stars. A single O star can outshine a thousand main-sequence K stars.

How massive can a star be? Both observational evidence and theoretical reasoning have led to an upper limit of 60 to 100 solar masses before unstable oscillations rip the star apart. But there is some evidence that a few *supermassive stars* may exist, containing thousands of times the mass of the sun. The prime example is the bright central object of the Tarantula nebula in the Large Magellanic Cloud (Color Plate 65), known as R136a (from a catalogue of the Radcliffe Observatory). Though most astronomers have considered it to be an unresolved dense cluster of dozens or hundreds of O stars, other astronomers now think on the basis of IUE and optical observations that it is a single supermassive star containing 3,000 solar masses. Each school of thought has marshalled interesting evidence for its position and against the other; observational and theoretical research continues.

*6.3 Stellar Sizes

Stars appear as points to the naked eye and as small, fuzzy disks through large telescopes. Distortion by the earth's atmosphere blurs the stars' images into disks, and hides the actual size and structure of the surface of the stars. The sun is the only star whose angular diameter we can measure easily and directly.

Astronomers use an indirect method to find the sizes of most stars. If we know the absolute magnitude of a star (from some type of parallax measurement) and the temperature of the surface of the star (from measuring its spectrum), then we can tell the amount of surface area the star must have. The extent of surface area, of course, depends on the radius.

One type of direct measurement works only for eclipsing binary stars. As the more distant star is hidden behind the nearer star, one can follow the rate at which the intensity of radiation from the further star declines. From this information together with Doppler measurements of the velocities of the components, one can calculate the sizes of the stars. One can sometimes even tell how the brightness varies across a star's disk.

It is much more difficult to measure the size of a single star directly. One way of measuring the diameter is by *lunar occultation.* As the moon moves with respect to the star background, it occults (hides) stars. By studying the light from a star in the fraction of a second it takes for the moon to completely block it, we can deduce the size of the stellar disk. Unfortunately, the moon only passes over 10 per cent of the sky in the course of a year.

Other methods for measuring the diameters of single stars use a principle called *interferometry,* a technique that measures incoming radiation at two different locations and then combines the two signals. This gives the effect, for the purpose of determining the resolution, of a single very large telescope. The method overcomes many of the problems of blurring by the earth's atmosphere.

The first such measurements were carried out at Mt. Wilson 50 years ago; 7 of the stars that subtended the largest angles in the sky had their diameters measured. These were the largest of the closest stars, that is, nearby red giants and supergiants. A modern variation of this technique has been worked out at the University of Maryland.

The above interferometric methods work best for large, cool stars. For the last 25 years, R. Hanbury Brown and his associates in Australia have used another type of stellar interferometer (Fig. 6–8) that works best for hot, bright stars. They have determined the diameters of three dozen stars whose diameters cannot be measured with other techniques.

Only a few dozen stellar diameters have been measured directly, and they confirm the indirect measurements (Fig. 6–9). Direct and indirect measurements show that the diameters of main-sequence stars decrease as we go from hotter to cooler; that is, O dwarfs are relatively large and F dwarfs are relatively small. Indirect mea-

Figure 6–8 Hanbury Brown's intensity interferometer in Australia, used to measure the diameters of hot stars. The two sets of mirrors can be placed up to 188 meters apart from each other.

Box 6.1 Speckle Interferometry

A relatively new technique called *speckle interferometry* uses high-speed imaging to overcome the blurring of earth's atmosphere to a certain extent. At any one instant, the image of a star through a large telescope looks speckled because different parts of the image are affected by different small turbulent areas in the earth's atmosphere. A long exposure blurs these speckles and gives us the fuzzy disk that normally appears on a photograph. Speckle interferometry involves short electronic or photographic exposures—perhaps 1/50 s—and mathematical techniques with computer assistance to deduce the properties of the starlight that entered the telescope. In this manner, from many short exposures of an image, scientists can measure a star's diameter or find out if two normally unresolved objects are present.

surements alone show that this trend continues for K and M stars, which are smaller still. As for stars that are not on the main sequence, red giants are indeed giant in size as well as in brightness.

6.4 Variable Stars

Some stars vary in brightness with respect to time. The most basic property of a star's variation is the *period*. The period is the time it takes for a star to go through its entire cycle of variation.

One way for a star to vary in brightness, as we have seen, is for the star to be an eclipsing binary. Sometimes we get the light simultaneously from both members of the binary, and sometimes the light from one of the members is at least partially blocked by the other member. But it is possible for individual stars to vary in brightness all by themselves. Thousands of such stars are known in the sky. The periods of the variations can be seconds for some types of stars or years for others. Plots of the brightness of a star (usually in terms of its magnitude) versus time are called *light curves*.

Besides ordinary variables of various types, some stars occasionally briefly flare up. These stellar flares often make a bigger difference in parts of the spectrum other than the visible. Fig. 6–10 shows an x-ray flare in the nearest star to us, Proxima Centauri.

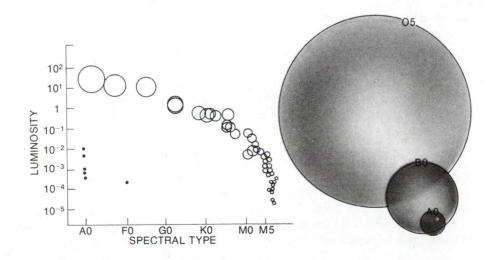

Figure 6–9 The H-R diagram for stars within 17 light years of the sun. The open circles indicate the relative diameters of main-sequence stars; white dwarfs are shown as dots. In addition, relative sizes of O, B, and A stars are shown at right.

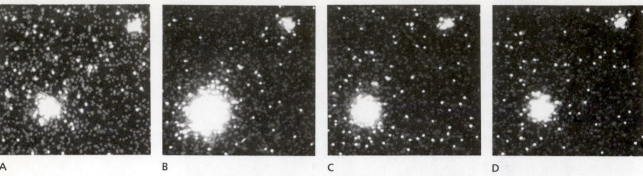

A   B   C   D

Figure 6–10 An x-ray flare on Proxima Centauri, the nearest star, was observed with the Einstein Observatory in this series of 35-minute exposures taken on 20 August 1980. *(A)* is before the flare, *(B)* is near the flare's maximum, and *(C)* and *(D)* show the flare's decay. The steady x-ray flux probably comes from the star's corona. The flare's x-ray flux was about that of a bright solar flare. A second x-ray source appears at upper right.

*Box 6.2 Naming Variable Stars

A naming system has been adopted for variable stars (often simply called "variables") that helps us to recognize them on any list of stars. The first variable to be discovered in a constellation is named R, followed by the genitival form ("of . . .") of the Latin name of the constellation (see Appendix 9), e.g., R Coronae Borealis (R of Corona Borealis). One continues with S, T, U, V, W, X, Y, Z, then RR, RS, etc., up to RZ, then SS (not SR) up to SZ, and so on up to ZZ. Then the system starts over with AA up to AZ, BB (not BA) and so on up to QZ. The letter J was omitted when the method was set up to avoid confusion with I in the German script of that time. This system covers the first 334 variables; after that, one numbers the stars beginning with V for variable (V335 . . .). It should be noted that variable stars with more commonly known names, such as Polaris and δ Cephei, retain their common names instead of being included in the lettering system.

We will limit our discussion to three types of variables, one of which is especially numerous, and the other two of which have provided important information about the scale of distance in the universe.

6.4a Mira Variables

A type of variable star is often named after its best-known or brightest example. A star named Mira in the constellation Cetus (the star is also known as o Ceti, omicron of the Whale) fluctuates in brightness with a long period (Fig. 6–11), about a year. Red stars that share this characteristic are called *Mira variables*. The period of a given Mira-type star can be from three months up to about two years. The period of an individual star is not strictly regular; it can vary from the average period.

Mira itself is sometimes of apparent magnitude 9, and is thus invisible to the naked eye. However, it brightens fairly regularly by about six magnitudes, a factor of 250, with a period of about 11 months. At maximum brightness it is quite noticeable in the sky. As it brightens, its spectral type changes from M5 to M9. Thus real changes are taking place at the surface of the star that result in a change of temperature.

Mira stars are giants of spectral type M, and are about 700 times the diameter of the sun. Such a star would extend beyond Mars if one were in our solar system. Mira stars emit most of their radiation in the infrared. They are also the source of strong radio spectral lines from water vapor.

These stars, which are the most numerous type of variable star in the sky, are also known as *long-period variables*.

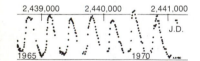

Figure 6–11 The light curve for Mira, the prototype of the class of long-period variables. Dates are given at top in Julian days (J.D.), elapsed time since January 1, 4713 B.C.

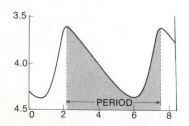

Figure 6–12 The light curve for δ Cephei, the prototype of the class of Cepheid variables.

6.4b *Cepheid Variables*

The most important variable stars in astronomy are the *Cepheid variables* (cef'-e-id). The prototype is δ Cephei (delta of Cepheus). Cepheid variables have very regular periods that, for individual Cepheids (as they are called), can be from 1 to 100 days. δ Cephei itself varies between apparent magnitudes 3.6 and 4.3 with a period of 5.4 days (Fig. 6–12). Astronomers can tell Cepheid variables apart from other variables, even when the periods are the same, by the shape and regularity of their light curves.

Cepheids are relatively rare stars in our galaxy; only about 700 are known. Some have been detected in other galaxies.

Cepheids are important because a relation has been found that links the periods of their light changes, which are simple to measure, with their absolute magnitudes. For example, if we measure that the period of a Cepheid is 10 days, we need only look at Fig. 6–13 to see that the star is of absolute magnitude −3. We can then compare its absolute magnitude to its apparent magnitude, which gives us (by the inverse-square law of brightness) its distance from us.

The study of Cepheids is the key to our current understanding of the distance scale of the universe and has allowed us to determine that the objects in the sky that we now call galaxies (discussed in Chapter 14) are giant systems comparable to that of our own galaxy (discussed in Chapter 12).

The story started out 75 years ago with Henrietta Leavitt (Fig. 6–14) who, at the Harvard College Observatory, was studying the light curves of variable stars in the southern sky. In particular, she was studying the variables in the Large and Small Magellanic Clouds (Fig. 6–15), two hazy areas in the sky that were discovered by the crew of Magellan's expedition around the world when they sailed far south. Whatever the Magellanic Clouds were—we now know them to be galaxies, but Leavitt did not know this—they looked like concentrated clouds of material. It thus seemed clear that, for each cloud, all its stars were at approximately the same distance away from the earth. Thus even though she could plot only the apparent magnitudes, the relation of the absolute magnitudes to each other was exactly the same as the relation of the apparent magnitudes.

By 1912, Leavitt had established the light curves and determined the periods for two dozen stars in the small Magellanic Cloud. She plotted the magnitude of the stars (actually a median brightness, a value between the maximum and minimum brightness) against the period. She realized that there was a fairly strict relation between the two quantities, and that the Cepheids with longer periods were brighter than the Cepheids with shorter periods. By simply measuring the periods, she could determine the magnitude of one star relative to another; each period uniquely corresponds to a magnitude.

Henrietta Leavitt could measure **apparent** magnitude but she did not know the distance to the Magellanic Clouds so could not determine the **absolute** magnitude (intrinsic brightness) of the Cepheids. To find the distance to the Magellanic Clouds, we first had to be able to find the distance to any Cepheid—even one not in the Magellanic Clouds—to tell its absolute magnitude. A Cepheid of the same period but located in the Magellanic Clouds would presumably have the same absolute magnitude.

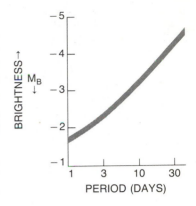

Figure 6–13 The period-luminosity relation for Cepheid variables.

Figure 6–14 Henrietta S. Leavitt, who worked out the period-luminosity relation, at her desk at the Harvard College Observatory in 1916.

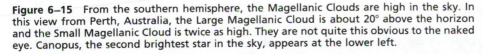

Figure 6–15 From the southern hemisphere, the Magellanic Clouds are high in the sky. In this view from Perth, Australia, the Large Magellanic Cloud is about 20° above the horizon and the Small Magellanic Cloud is twice as high. They are not quite this obvious to the naked eye. Canopus, the second brightest star in the sky, appears at the lower left.

To find the absolute magnitude of a Cepheid it would seem easiest to start with the one nearest to us. Unfortunately, not a single Cepheid is close enough to the sun to allow its distance to be determined by the method of trigonometric parallax. More complex, statistical methods had to be used to study the relationships between stellar motions and distances. This gave the distance to a nearby Cepheid. Once we have the distance to and thus the absolute magnitude of a nearby Cepheid, we know the absolute magnitude of all Cepheids of that same period in the Magellanic Clouds, since all Cepheids of the same period have the same absolute magnitude.

From that point on it is easy to tell the absolute magnitude of Cepheids of **any** period in the Magellanic Clouds. After all, if a Cepheid has an **apparent** magnitude that is, say, 2 magnitudes brighter than the apparent magnitude of our Cepheid of known intrinsic brightness, then its **absolute** magnitude is also 2 magnitudes brighter. After this process, we have the period-luminosity relation in a more useful form—period vs. absolute magnitude. We call this process ''the calibration'' of the period-luminosity relation.

Thus Cepheids can be employed as indicators of distance: First we identify the star as a Cepheid (by studying its spectrum and the shape of its light curve). Then we measure its period. Third, the period-luminosity relation gives us the absolute magnitude of the Cepheid. And last, we calculate (with the inverse-square law) how far a star of that absolute magnitude would have to be moved from the standard distance of 10 parsecs to appear as a star of the apparent magnitude that we observe.

When the calibration of the period-luminosity relation was worked out quantitatively by the American astronomer Harlow Shapley in 1917, the distance to the Magellanic Clouds could be calculated. They were very far away, a distance that we now know means that they are not even in our galaxy! Instead, they are galaxies by themselves, two small irregular galaxies that are companions of our own, larger galaxy. (Shapley's values have been superseded by later measurements, but the conclusion remains valid.)

Cepheids can also be seen not only in our own galaxy and in the Magellanic Clouds but also in more distant galaxies. They are bright stars (giants and supergiants, in fact) and so can be seen at quite a distance. The use of Cepheids is the prime method of establishing the distance to all the nearer galaxies.

As the Cepheids were further studied, astronomers could make a model of what a Cepheid was like, based on their light curves and on their spectra. The absorption lines in the spectrum of a Cepheid show Doppler shifts that prove that the star is actually pulsating—expanding and contracting—as it changes in brightness. The size of a Cepheid variable may change by 5 to 10 per cent as it goes through its pulsations. The surface temperature also varies. The star is brightest roughly at its hottest phase. A detailed theory of stellar pulsations that explains these effects has been worked out.

In the 1950's, when overlapping methods of finding distances were applied to some relatively nearby stars and clusters of stars, it was realized that there was a second type of Cepheid whose light curves looked the same but which were about 1.5 magnitudes fainter at each given period than the original type. The distances to many galaxies had to be recalculated because this distinction between ordinary Cepheids and Type II Cepheids had previously not been made. Our estimates of the distances to many distant galaxies doubled.

6.4c RR Lyrae Variables

Many stars are known to have short regular periods, less than one day in duration. Certain of these stars, no matter what their period, have light curves of a certain distinctive shape (Fig. 6–16). All these stars have the same average absolute magnitude.

Such stars are called *RR Lyrae stars* after the prototype of the class. Since many of these stars appear in globular clusters (which will be described in Section 6.6), RR Lyrae stars are also called *cluster variables*.

Once we detect an RR Lyrae star by the shape of its light curve, we immediately know its absolute magnitude, since all the absolute magnitudes are the same (about

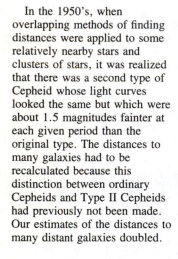

Figure 6–16 The light curve of RR Lyrae.

0,6) Just as before, we measure the star's apparent magnitude and can thus easily calculate its distance, which is also the distance to the cluster.

6.5 Clusters and Stellar Populations

Even aside from the hazy band of the Milky Way, the distribution of stars in our sky is not uniform. There are certain areas where the number of stars is very much higher than the number in adjacent areas. Such sections of the sky are called *star clusters* (Color Plates 47 and 48). All the stars in a given cluster are essentially the same distance away from us, and were also formed at essentially the same time.

One type of star cluster appears only as an increase in the number of stars in that limited area of sky. Such clusters are called *open clusters*, or *galactic clusters* (Fig. 6–17). The most familiar example of a galactic cluster is the Pleiades, a group of stars visible in the winter sky. The unaided eye sees at least six stars very close together. With binoculars or the smallest telescopes, dozens more can be seen. A larger telescope reveals hundreds of stars. Another galactic cluster is called the Hyades, which forms the ''V'' that outlines the face of Taurus, the bull, a constellation best visible in the winter sky. More than a thousand such clusters are known, most of them too faint to be seen except with telescopes.

All the stars in a galactic cluster are packed into a volume not more than 10 parsecs across.

Stars in galactic clusters seem to be representative of stars in the spiral arms of our galaxy and of other galaxies. When the spectra of stars in galactic clusters are analyzed to find the relative abundances of the chemical elements in their atmospheres, we find that over 90 per cent of the atoms are hydrogen, most of the rest are helium, and less than 1 per cent are elements heavier than helium. This is similar to the composition of the sun. Such stars are said to belong to stellar *Population I*.

The second major type of star cluster appears in a small telescope as a small, hazy area in the sky. Observing with larger telescopes distinguishes individual stars, and reveals that these clusters are really composed of many thousands of stars packed together in a very limited space. The clusters are spherical, and are known as *globular clusters* (Fig. 6–18).

Globular clusters can contain 10,000 to one million stars, in contrast to the 20 to several hundred stars in a galactic cluster. A globular cluster can fill a volume up to 30 parsecs across. It has fewer stars at its periphery and more closely packed stars toward the center.

Globular clusters are typically found only in regions above and below the galactic plane. We say that they are in the galactic *halo*. (We do not see many in the plane of the galaxy because they are hidden by interstellar dust there.) The abundances of the elements heavier than helium in globular clusters are much lower, by

Figure 6–17 The Pleiades, a galactic cluster. The six brightest stars are visible to the naked eye.

The Pleiades are often known as the Seven Sisters, after the seven daughters of Atlas who were pursued by Orion and who were given refuge in the sky. That one is missing—the Lost Pleiad—has long been noticed. Of course, a seventh star is present (and hundreds of other as well), although too faint to be plainly seen with the naked eye. The Pleiades seem to be riding on the back of Taurus.

We need only know here that ordinary hydrogen contains one nuclear particle and so is the least massive element, while ordinary helium contains 4 nuclear particles and is the second least massive element. (See Section 8.3.)

Figure 6–18 Four views of the globular cluster NGC 288. From left to right, we see a plate taken with a blue-sensitive emulsion, a plate taken with a red-sensitive emulsion, a plate taken with an infrared-sensitive emulsion, and a plate taken with a large glass prism in front of the telescope's objective to produce a spectrum for each stellar image.

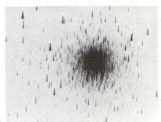

> ### Box 6.3 Star Clusters
>
> | Galactic Clusters | Globular Clusters |
> |---|---|
> | No regular shape (also called ''open clusters'') | Shaped like a ball, stars more closely packed toward center |
> | Many young stars | All old stars |
> | H-R diagrams have long main sequences | H-R diagrams have short main sequences |
> | Where stars leave the main sequence tells the cluster's age | All clusters have the same H-R diagram and thus the same age |
> | Stars have similar composition to sun, Population I | Stars have lower abundances of elements than sun, Population II |
> | Hundreds of stars per cluster | 10,000–1,000,000 stars per cluster |
> | Found in galactic plane | Found in galactic halo |

a factor of 10 or more, than their abundances in the sun. Such stars are said to belong to *Population II*.

The fact that galactic clusters and globular clusters are distributed throughout our galaxy in different fashions has been used to determine the makeup and structure of our galaxy. We shall return to this point in more detail in Chapter 12.

6.5a H-R Diagrams and the Ages of Galactic Clusters

The Hertzsprung-Russell diagram for several galactic clusters is shown in Fig. 6–19. For the purposes of this discussion, it is most important to note that the horizontal axis is a measure of temperature and the vertical axis is a measure of brightness. Fig. 6–19 is actually a set of many individual diagrams like the two on the right, laid on top of each other.

Since all the stars in a cluster were formed at the same time out of the same gas, we can presume that they have similar chemical compositions and differ only in mass. The difference between one cluster and another is principally that the two clusters were formed at different times.

The H-R diagrams for different galactic clusters appear to be similar over the lower part of the main sequence but diverge at the upper part. Comparison of the diagrams for the different individual clusters has given us a picture of how stars evolve.

We see on the figure that the H-R diagrams for some clusters lie almost entirely along the main sequence, while others have only their lowest portions on the main sequence. Since stars spend most of their lifetimes on the main sequence, we can deduce that the cluster whose stars lie mostly on the main sequence must be the youngest. They would not have had time for many stars to have evolved enough to move off the main sequence. Thus NGC 2362 and the pair of clusters known as h and χ Persei (Fig. 6–20) are the youngest clusters in the composite diagram. When a star finishes its main-sequence lifetime, its surface becomes larger and cooler. Thus the star moves upward and to the right on the H-R diagram.

At the telescope, we measure temperatures as a ''color index'' (Section 5.3a).

The stars that finish their main-sequence lifetimes most rapidly, and move off the main sequence, are the most massive ones. The most massive stars are the O and B stars on the extreme upper left of the H-R diagram. They are more luminous and use up their nuclear fuel at a faster rate than the cooler, more numerous, ordinary stars like the sun.

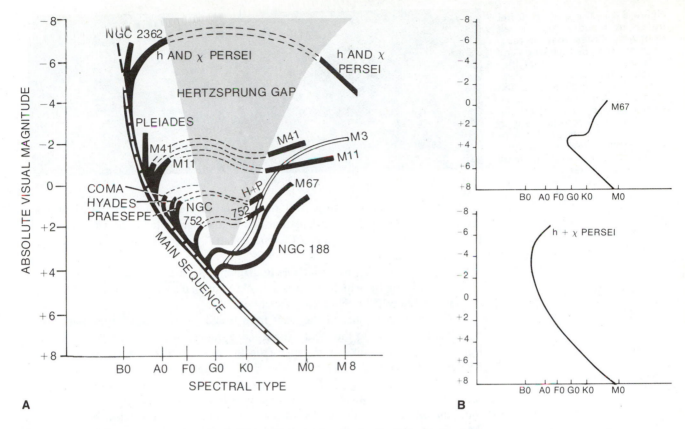

Figure 6–19 *(A)* Hertzsprung-Russell diagrams of several galactic clusters, showing the overlay of several individual diagrams. We see that the fainter stars of galactic clusters are on the main sequence, while the brighter stars are above and to the right of the main sequence. Almost all stars in the younger clusters, like h and χ Persei, follow the main sequence, while the hotter and more massive members of older clusters, like M67, have had time to evolve toward the red giant region. By observing the point where the cluster turns off the main sequence, we deduce the length of time since its stars were formed, i.e., the age of the cluster. The presence of the *Hertzsprung gap (shaded)*, a region in which few stars are found, indicates that stars evolve rapidly through this part of the diagram.

Part of the H-R diagram for the globular cluster M3 *(hollow bar)* is graphed for comparison.

(B) Two of the H-R diagrams of individual galactic clusters are shown separately here to illustrate how several of these are put together to make the composite diagram shown in *A*.

In NGC 2362 and h and χ Persei, only the most massive stars have lived long enough to die and move off the main sequence. Theoretical calculations tell us that stars of the spectral type of the point where these two clusters leave the main sequence (about B0) have masses such that their main-sequence lifetimes are 10^7 years. Thus these two clusters must be about 10^7 years old. The stars that live for less time than 10^7 years have died and moved off to the right. The stars that live for longer than 10^7 years are still on the main sequence.

The Pleiades, on the other hand, must be older than 10^7 years since the stars with lifetimes of 10^7 years have already died. Calculations tell us that stars of the spectral type of the Pleiades' turnoff from the main sequence (about B5) live 10^8 years, so the Pleiades must be about 10^8 years old. Similarly, the Hyades must be

Figure 6–20 The double cluster in Perseus, h and χ Persei, a pair of galactic clusters that are readily visible in a small telescope and close enough together that they appear in the same field of view. Study of the H-R diagram reveals that they are relatively young. Perseus is a northern constellation that is most prominent in the winter sky. In Greek mythology, Perseus slew the Gorgon Medusa and saved Andromeda from a sea monster.

I←DIAMETER OF MOON→I

older than the Pleiades, since still more stars have had time to live their main-sequence lifetimes and die. Calculations tell us that stars at the turnoff for the Hyades (about A0) have main-sequence lifetimes of 10^9, so the Hyades must have been formed 10^9 years ago.

We can thus read a cluster's H-R diagram like a clock, telling how long the cluster has lived by which of its stars (observationally) have turned off the H-R diagram and how old (theoretically) such a star is.

Note also that some of the stars from h and χ Persei have even had time to move quite far over to the right. These move quite fast since they are so massive. But notice also that these stars appear on the other side of a zone of the diagram where we do not detect any stars—the Hertzsprung gap. If stars kept evolving upward and to the right at a steady rate, we would expect a uniform number of stars across this zone. From the fact that some stars have reached the right side though no stars appear in the zone, we can deduce that stars have properties that would put them in the zone for only a very brief time. This is equivalent to saying that these stars must evolve quite quickly at those points.

In about 10^9 years, the stars at the extreme right of the h and χ Persei graph will have evolved downward on the diagram, since h and χ Persei will then be the same age that the Hyades are now. Main-sequence stars in h and χ Persei will have died and peeled off the main sequence, until the graph resembles the one for the Hyades today.

So think of the H-R diagram of a cluster as a string that extends along the main sequence's entire length. Points representing stars lie along the string. Now imagine slowly pulling the string's top end off to the upper right and then eventually downward. The longer you pull, the more of the string is pulled off the main sequence. By simply looking at any time to see where the string leaves the main sequence, anyone can tell how long you have been pulling.

6.5b H-R Diagrams for Globular Clusters

The H-R diagram for a globular cluster is shown in Fig. 6–21. H-R diagrams for all globular clusters look essentially identical. They all have stubby main sequences like this one.

From the fact that all globular clusters have identical H-R diagrams, we can conclude that they are all about the same age. From the fact that they all have H-R diagrams with short main sequences, we can conclude that the globular clusters must be very old. Detailed studies assign an age of about 18 billion years. If we assume that the globular clusters formed when or soon after our galaxy formed, then our galaxy must be about 18 billion years old.

The H-R diagrams for globular clusters have prominent *horizontal branches*. The horizontal branch goes leftward from the stars on the right side of the diagram

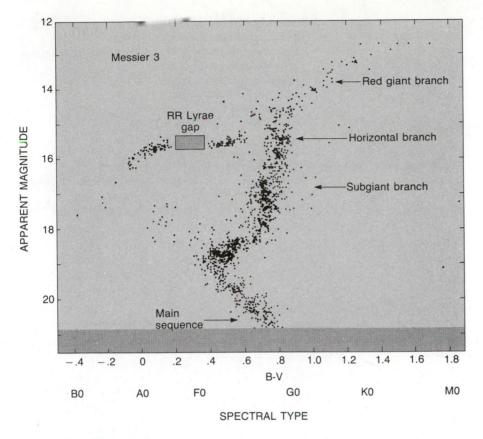

Figure 6–21 The Hertzsprung-Russell diagram for the globular cluster M3, part of which was included in Fig. 6–19A. There is a gap in the horizontal branch where no stars of constant brightness are found. The RR Lyrae variables fall here, some 200 for this cluster (though most clusters contain few of them). We can observe only the members of the cluster that are brighter than a certain limit. Thus no stars appear in the shaded region at the bottom.

The positions of stars on the horizontal axis were actually determined from measurements of the colors of the stars made at the telescope by comparing their brightness in the blue and in the yellow (which gives a *"color index"* B-V). The transformation to spectral type is approximate.

that have long since turned off the main sequence. It represents very old stars that have evolved past their giant or supergiant phases and are returning leftward. No galactic cluster has stars that old. The horizontal branch has a gap where no stars of constant brightness are found. Only RR Lyrae stars, which vary in brightness, are found there.

Summary and Outline

Binary stars (Section 6.1)
   Optical doubles
   Visual binaries
   Binaries with composite spectra
   Spectroscopic binaries

Eclipsing binaries
Astrometric binaries
Determination of stellar masses
   (Section 6.2)
   Mass-luminosity relation

Determination of stellar sizes (Section 6.3)
   Indirect: calculate from absolute magnitude or measure in eclipsing binaries
   Direct: interferometry
Variable stars (Section 6.4)
   Mira variables (Section 6.4a)
   Cepheid variables (Section 6.4b)
     Period-luminosity relation
     Uses for determining distances
   RR Lyrae variables (Section 6.4c)
     All of approximately the same absolute magnitude
     Uses for determining distances

Clusters and stellar populations (Section 6.5)
   Galactic (open) clusters (Section 6.5a)
     Population I: relatively high abundance of elements heavier than helium
     Representative of spiral arms
     20 to several hundred members
     Turn-off on H-R diagram gives age
   Globular clusters (Section 6.5b)
     Population II: relatively low abundance of metals
     Representative of galactic halo
     10^4 to 10^6 members
     Old enough to have H-R diagrams with horizontal branches

Key Words

optical doubles, double star, binary star, visual binaries, composite spectrum, spectroscopic binary, eclipsing binary, astrometric binaries, mass-luminosity relation, supermassive stars, lunar occultation*, interferometry*, speckle interferometry*, period, light curves, Mira variables, long-period variables, Cepheid variables, RR Lyrae stars, cluster variables, star clusters, open clusters, galactic clusters, Population I, globular clusters, halo, Population II, horizontal branch, Hertzsprung gap

*These terms are found in optional sections.

Questions

1. Sketch the orbit of a double star that is simultaneously a visual, an eclipsing, and a spectroscopic binary.

2. Define briefly and contrast an astrometric binary and an eclipsing binary.

3. (a) Assume that an eclipsing binary contains two identical stars. Sketch the intensity of light received as a function of time. (b) Sketch to the same scale another curve to show the result if both stars were much larger while the orbit stayed the same.

†4. How much brighter than the sun is a main-sequence star whose mass is 10 times that of sun?

5. What does the mass-luminosity relationship tell us about the mass of the supergiant star Betelgeuse (spectral type M2)?

†6. A main-sequence star is 3 times the mass of the sun. What is its luminosity relative to that of the sun?

†7. (a) Use the mass-luminosity relation to determine about how many times brighter than the sun are the most massive main-sequence stars of which we know. (b) How many times fainter are the least massive main-sequence stars?

8. When we look at the Andromeda galaxy, from what mass range of star does most of the light come? What mass range of star provides most of the mass of the galaxy?

9. When we consider the gravitational effects of a distant galaxy on its neighbors, are we measuring the effects of mostly low- or high-mass stars? Explain.

†10. A Cepheid variable has a period of 30 days. What is its absolute magnitude?

†11. What is the ratio of apparent brightness of two Cepheid variables in the Large Magellanic Cloud, one with a 10-day period and the other with a 30-day period?

†12. An RR Lyrae star has an apparent magnitude of 6. How far is it from the sun?

13. A Cepheid variable with a period of 10 days has an apparent magnitude of 8. How bright would an RR Lyrae star be if it were in a globular cluster near the Cepheid?

14. An astronomer observes a galaxy and notices that a star in it brightens and dims every 11 days. Sketch the light curve, labelling the axes, and explain how to find the distance to the galaxy.

†15. What is the absolute magnitude of an RR Lyrae star with an 18-hour period?

---

†This indicates a question requiring a numerical answer.

16. Briefly distinguish Population I from Population II stars.

17. Cluster X has a higher fraction of main-sequence stars than cluster Y. Which cluster is probably older?

18. What is the advantage of studying the H-R diagram of a cluster, compared to that of the stars in the general field?

19. Which galactic cluster from Fig. 6–19 is about 8 billion years old?

†20. From the position of the RR Lyrae gap and your knowledge of RR Lyrae stars, how far away is M3, the globular cluster whose H-R diagram is shown in Fig. 6–21?

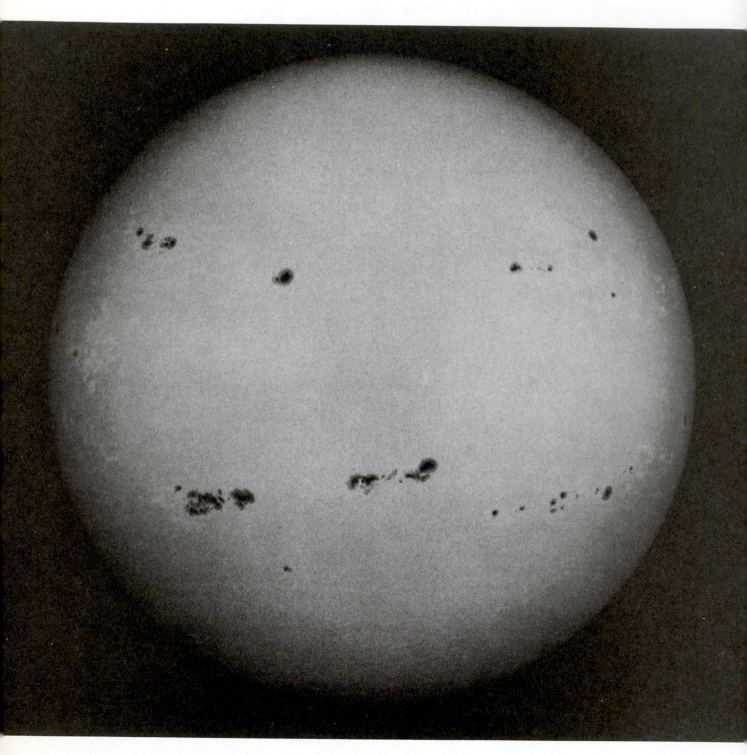

The sun, observed in white light on November 9, 1980, near the most recent solar maximum.

The Sun   7

Aims: To study the sun, which is the nearest star and an example of all the other stars whose surfaces we cannot observe in such detail

We have discussed a range of individual stars of different spectral classes and have discussed groupings of stars in close physical proximity to each other. Studying these distant stars has allowed us to learn a lot about the properties of stars and how they evolve. But not all stars are far away; one is close at hand. By studying the sun, we not only learn about the properties of a particular star but also can study the details of processes that undoubtedly take place in more distant stars as well. We will first discuss the *quiet sun,* the solar phenomena that appear every day. Afterwards, we will discuss the *active sun,* solar phenomena that appear non-uniformly on the sun and vary over time.

7.1 Basic Structure of the Sun

We think of the sun as the bright ball of gas that appears to travel across our sky every day. We are seeing only one layer of the sun, part of its atmosphere; the properties of the solar interior below that layer and of the rest of the solar atmosphere above that layer are very different. The outermost parts of the solar atmosphere even extend through interplanetary space beyond the orbit of the earth.

The layer that we see is called the *photosphere,* which simply means the sphere from which the light comes (from the Greek *photos,* light). As is typical of many stars, about 94 per cent of the atoms and nuclei in the outer parts are hydrogen, about 5.9 per cent are helium, and a mixture of all the other elements make up the remaining one-tenth of one per cent. The overall composition of the interior is not very different.

The sun is an average star, since stars much hotter and much cooler, and stars intrinsically much brighter and much fainter, exist. Radiation from the photosphere peaks (is strongest) in the middle of the visible spectrum; after all, our eyes evolved over time to be sensitive to that region of the spectrum because the greatest amount of the solar radiation occurred there. If we lived on a planet orbiting an object that emitted mostly x-rays, we, like Superman, might have x-ray vision.

The solar photosphere is about 1.4 million km (1 million miles) across. The disk of the sun (the apparent surface of the photosphere) takes up about one-half a degree across the sky; we say that it *subtends* one-half degree. This angle is large enough for us to see detailed structure on the solar surface, and we shall soon describe some of it.

Beneath the photosphere is the solar *interior.* All the solar energy is generated there at the solar *core,* which is about 10 per cent of the solar diameter at this stage of the sun's life. The temperature there is about 15,000,000 K. In Chapter 8 we will discuss how energy is generated in the sun and other stars.

The photosphere is the lowest level of the *solar atmosphere* (Fig. 7–1). Though the sun is gaseous through and through, with no solid parts, we still use the term

Figure 7–1 The parts of the solar atmosphere and interior. The solar surface is depicted as it appears through light equally filtered across the spectrum (called *white light*), and through filters that pass only light of certain elements in certain temperature stages. These specific wavelengths, counter-clockwise from the top, are the Hα line of hydrogen (which appears in the red), the K line of ionized calcium (which appears in the part of the ultraviolet that passes through the earth's atmosphere and can be seen with the naked eye), and the line of ionized helium at 304Å in the extreme ultraviolet (which can be observed only from rockets and satellites).

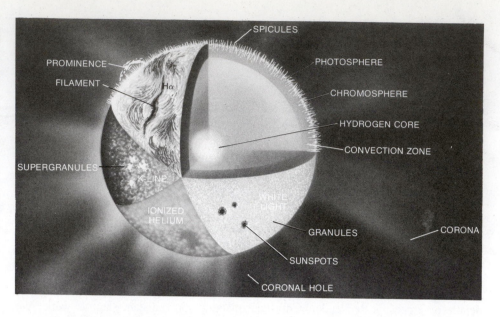

atmosphere for the upper part of the solar material. The parts of the atmosphere above the photosphere are very tenuous, and contribute only a small fraction to the total mass of the sun. These upper layers are very much fainter than the photosphere, and cannot be seen with the naked eye except during a solar eclipse, when the moon blocks the photospheric radiation from reaching our eyes directly. Now we can also study these upper layers with special instruments on the ground and in orbit around the earth.

Just above the photosphere is a jagged, spiky layer about 10,000 km thick, only about 1.5 per cent of the solar radius. This layer glows colorfully pinkish when seen at an eclipse, and is thus called the *chromosphere* (from the Greek *chromos,* color). Above the chromosphere, a ghostly-white halo called the *corona* (from the Latin, crown) extends tens of millions of kilometers into space. The corona is continually expanding into interplanetary space and in this form is called the *solar wind.* We shall discuss all these phenomena in succeeding sections.

7.2 The Photosphere

The sun is a normal star of spectral type G2, which means that its surface temperature is about 5800 K. But the sun is the only star close enough to allow us to study its surface in detail. In recent years, we have become able to verify directly from the ground and from space that many of its features indeed also appear on other stars.

7.2a High-Resolution Observations of the Photosphere

One major limitation in observing the sun is the turbulence in the earth's atmosphere, which also causes the twinkling of stars. The problem is even more serious for studies of the sun, since the sun is up in the daytime when the atmosphere is heated by the solar radiation and so is more turbulent than it is at night.

Only at the very best observing sites, specially chosen for their steady solar observing characteristics, can one see detail on the sun subtending an angle as small as 1 second of arc. This corresponds to about 700 km on the solar surface, the distance from Boston to Washington, D.C. Occasionally, objects $\frac{1}{2}$ arc second across can be seen. Since atmospheric turbulence causes bad "seeing" that limits our ability to observe small-scale detail (see Section 2.9), it does little good to build solar telescopes larger than about 50 cm (20 inches) in order to increase resolution, even though larger telescopes are inherently capable of resolving finer detail.

Sometimes we observe the sun in *white light*—all the visible radiation taken together. When we study the solar surface in white light with 1 arc second resolution, we see a salt-and-pepper texture called *granulation* (Fig. 7–2). The effect is similar to that seen in boiling liquids on earth, which are undergoing *convection*. Convection, the transport of energy by moving matter, is one of the basic ways in which energy can be transported, and carries energy to your boiling eggs. Conduction and radiation are the other major methods.

Granulation on the sun is an effect of convection. Each granule is only about 1000 km across, and represents a volume of gas that is rising from and falling to a shell of convection, called the *convection zone,* located below the photosphere. The granules are convectively carrying energy from the hot solar interior to the base of the photosphere.

But the granules are about the same size as the limit of our resolution, so are difficult to study. NASA plans to launch SOT, the Solar Optical Telescope, about 1990; it will allow astronomers to resolve features as small as 0.1 second of arc. This tenfold improvement in resolution should allow us to see clearly the smallest structures that the sun is theoretically predicted to have.

It was discovered in the early 1960's that areas of the upper photosphere are oscillating up and down with a 5-minute period—basically, ringing like a gong. This surprising phenomenon is now thought to be caused by waves of energy coming from the convection zone. We have since discovered that the whole sun's surface is oscillating with periods ranging from minutes up to hours. Some of the longer periods have been studied with a telescope at the earth's south pole (Fig. 7–3), where the sun is up for months without setting. We hope that interpreting the oscillations will tell us what the inside of the sun is like.

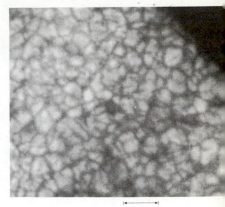

10 ARC SEC

Figure 7–2 The picture was taken during the partial phase of the 1973 solar eclipse so that the sharp edge of the moon, visible at upper right, shows the blurring caused by the earth's atmosphere. The smallest features visible—granules—are about 1 arc sec across.

Figure 7–3 (*A*) A few of the thousands of possible modes of the sun's oscillations. Solid lines represent zones of expansion and dotted lines represent zones of contraction. (*B*) Astronomers erecting a solar telescope at the South Pole to get a long run of uninterrupted sunlight, in order to study solar oscillations with periods of hours. Their longest run was 115 consecutive hours, during the South Pole's summer when the sun never sets. (AURA, Inc., National Solar Observatory)

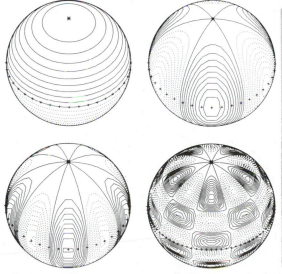

A

B

Figure 7–4 The visible part of the solar spectrum.

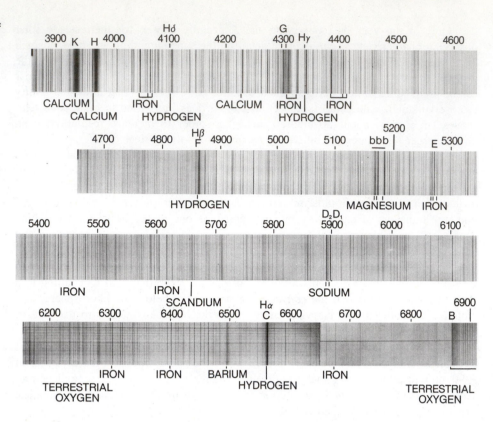

7.2b The Photospheric Spectrum

The spectrum of the solar photosphere, like that of other G stars, is a continuous spectrum crossed by absorption lines (Fig. 7–4). Hundreds of thousands of these absorption lines, which are also called Fraunhofer lines, have been photographed and catalogued. They come from most of the chemical elements, although some of the elements have many lines in their spectra and some have very few. Iron has many

Table 7–1 Solar Abundances of the Most Common Elements

	Symbol	Atomic Number
For each 1,000,000 atoms of hydrogen, there are	H	1
63,000 atoms of helium	He	2
690 atoms of oxygen	O	8
420 atoms of carbon	C	6
87 atoms of nitrogen	N	7
45 atoms of silicon	Si	14
40 atoms of magnesium	Mg	12
37 atoms of neon	Ne	10
32 atoms of iron	Fe	26
16 atoms of sulfur	S	16
3 atoms of aluminum	Al	13
2 atoms of calcium	Ca	20
2 atoms of sodium	Na	11
2 atoms of nickel	Ni	28
1 atom of argon	Ar	18

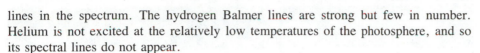

lines in the spectrum. The hydrogen Balmer lines are strong but few in number. Helium is not excited at the relatively low temperatures of the photosphere, and so its spectral lines do not appear.

Fraunhofer, in 1814, labelled the strongest of the absorption lines in the solar spectrum with letters from A through H. His C line, in the red, is now known to be the first line in the Balmer series of hydrogen, and is called Hα (H alpha). (We have seen in Section 4.4 that H alpha comes from a transition between the 2nd and 3rd energy levels.) We still use Fraunhofer's notation for some of the strong lines: The D lines, a pair of lines close together in the yellow part of the spectrum, are caused by neutral sodium (Na I). The H line and the K line, both in the part of the violet spectrum that is barely visible to the eye, are caused by singly ionized calcium (Ca II).

None of the elements other than hydrogen and helium makes up as much as one tenth of one per cent of the number of hydrogen atoms (Table 7–1). Nevertheless, the D, H, and K lines are among the strongest in the visible spectrum. Why? Sodium and calcium are among the more abundant of the elements other than hydrogen and helium. Also, most of the absorbing power of sodium and calcium at the temperatures of the photosphere and chromosphere is concentrated in just a few lines. Furthermore, these lines occur in the visible part of the spectrum.

The places where the continuous spectrum (called the *continuum*) and most of the absorption lines are formed are mixed together throughout the photospheric layers. The energy we receive on earth in the form of solar radiation has been transported upward to these layers from the solar interior.

7.3 The Chromosphere

7.3a The Appearance of the Chromosphere

Under high resolution, we see that the chromosphere is not a spherical shell around the sun but rather is composed of small spikes called *spicules* (Fig. 7–5). The spicules rise and fall, and have been compared in appearance to blades of grass or burning prairies.

Spicules are more-or-less cylinders of about 1 arc second in diameter and perhaps ten times that in height, which corresponds to about 700 km across and 7000 km tall. They seem to have lifetimes of about 5 to 15 minutes, and there may be approximately half a million of them on the surface of the sun at any given moment.

Studies of velocities on the sun showed the existence of large organized convection cells of matter on the surface of the sun called *supergranulation*. Supergranulation cells look somewhat like polygons of approximately 30,000 km diameter. Supergranulation is an entirely different phenomenon from granulation. Each supergranulation cell may contain hundreds of individual granules. Supergranules can be seen in Fig. 7–6.

Matter seems to well up in the middle of a supergranule and then slowly move horizontally across the solar surface to the supergranule boundaries. The matter then sinks back down at the boundaries. This slow circulation of matter seems to be a basic process of the lower part of the solar atmosphere. The network of supergranu-

Figure 7–5 Spicules at the solar limb, observed through a filter passing a narrow band of wavelengths centered 1 Å redward of Hα.

Fraunhofer's original spectrum is shown in Color Plate 5.

We are discussing here only the part of the solar spectrum in the visible. The formation of the Fraunhofer lines was discussed in Section 4.3. Note that to absorb is spelled with a ''b,'' but absorption is spelled with a ''p.''

Figure 7–6 Supergranulation is best visible in an image like this one, in which the velocity field of the sun is shown. Dark areas are receding and bright areas are approaching us. The supergranules are each less than 1 mm across in this reproduction. Because of the flow of matter from the center to the edge of each supergranule, the Doppler shift makes one side appear dark and the other bright on this velocity image.

Figure 7–7 The flash spectrum 0.6 sec after totality at the African solar eclipse of 1973. We see the chromospheric spectrum.

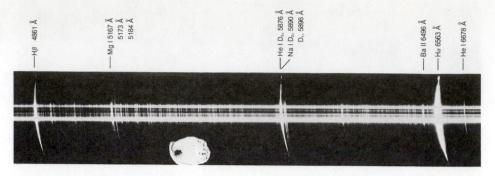

lation boundaries, called the *chromospheric network,* is visible in the radiation of hydrogen alpha or the H and K lines of calcium.

Chromospheric matter appears to be at a temperature of approximately 15,000 K, somewhat higher than the temperature of the photosphere. New ultraviolet spectra of distant stars recorded by the International Ultraviolet Explorer have shown unmistakable signs of chromospheres in stars of spectral types like the sun. Thus by studying the solar chromosphere we are also learning what the chromospheres of other stars are like.

7.3b The Chromospheric Spectrum at the Limb

During the few seconds at an eclipse of the sun that the chromosphere is visible, its spectrum can be taken. This type of observation has been performed ever since the first spectroscopes were taken to eclipses in 1868.

Since the chromosphere then appears as hot gas silhouetted against dark sky, the chromospheric spectrum then consists of emission lines.

The chromospheric emission lines appear to flash into view at the beginning and at the end of totality, so the visible spectrum of the chromosphere is known as the *flash spectrum* (Fig. 7–7 and Color Plate 13).

Astronomers have been able to study the chromospheric spectrum in the ultraviolet using telescopes on spacecraft; the spectrum in the ultraviolet contains emission lines, even when viewed at the center of the disk.

7.4 The Corona

7.4a The Structure of the Corona

During total solar eclipses, when first the photosphere and then the chromosphere are completely hidden from view, a faint white halo around the sun becomes visible. This *corona* (Fig. 7–8) is the outermost part of the solar atmosphere, and extends throughout the solar system. Close to the solar limb, the corona's temperature is about 2,000,000 K.

Even though the temperature of the corona is so high, the actual amount of energy in the solar corona is not large. The temperature quoted is actually a measure of how fast individual particles (electrons, in particular) are moving. There aren't very many coronal particles, even though each particle has a high velocity. The corona has less than one billionth the density of the earth's atmosphere, and would be considered to be a very good vacuum in a laboratory on earth. For this reason, the corona serves as a unique and valuable celestial laboratory in which we may study gaseous plasmas in a near vacuum.

Photographs of the corona show that it is very irregular in form. Beautiful long *streamers* extend away from the sun in the equatorial regions. At the poles, delicate thin *plumes* are suspended above the surface. The shape of the corona varies continuously and is thus different at each successive eclipse. The structure of the corona is maintained by the magnetic field of the sun.

*7.4b Other Ground-Based Observations of the Corona

Figure 7–9 From a few mountain sites, the innermost corona can be photographed without need for an eclipse. The coronagraph at the Haleakala Observatory of the University of Hawaii on Maui took these observations on March 8, 1970, the day after a total eclipse, with this coronagraph. Five exposures were superimposed to improve the image.

The corona is normally too faint to be seen except at an eclipse of the sun because it is fainter than the everyday blue sky. But at certain locations on mountain peaks on the surface of the earth, the sky is especially clear and dust-free, and the innermost part of the corona can be seen (Fig. 7–9). Special telescopes called *coronagraphs* study the corona from such sites. Coronagraphs are built with special attention to low scattering of light, since their object is not to gather a lot of light but to prevent the strong solar radiation from being scattered about within the telescope.

Since so many of the problems in observing the faint solar corona are caused by the earth's atmosphere, it is obvious that we want to observe the corona from above the atmosphere. If we stood on the moon, which has no air, we would see the corona rising each lunar morning a bit ahead of the sun.

But it is much less expensive to observe the corona from a satellite in orbit around the earth than it is to do so from the moon. Several manned and unmanned spacecraft have used coronagraphs to photograph the corona hour by hour in visible light (Color Plates 16 to 19). These satellites studied the corona to much greater distances from the solar surface than can be studied with coronagraphs on earth. Among the major conclusions of the research is that the corona is much more dynamic than we had thought. For example, many blobs of matter were seen to be ejected from the corona into interplanetary space (Fig. 7–10), often in connection with solar flare activity (which will be discussed in Section 7.7). In 1984, Space Shuttle astronauts repaired the Solar Maximum Mission, a spacecraft bearing a coronagraph that was launched in 1980 but failed a few months later (Fig. 7–23).

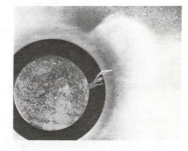

Figure 7–10 This example of solar ejection was photographed from Skylab. Here we see two pictures superimposed. The eruption at the extreme right, photographed with the coronagraph, is the response of the corona to a lower-level eruption, photographed in the ultraviolet and superimposed on the central dark region occulted by the coronagraph.

*7.4c The Coronal Spectrum

The visible region of the coronal spectrum, when observed at eclipses, shows a continuum, and also both absorption lines and emission lines. The emission lines do not correspond to any normal spectral lines known in laboratories on earth or on other stars, and for many years their identification was one of the major problems in solar astronomy. The lines were even given the name of an element: coronium. In the late 1930's, it was discovered that they arose in atoms that were multiply ionized. This was the major indication that the corona was very hot (and that coronium doesn't exist). By contrast, in the photosphere we find atoms that are neutral, singly ionized, or doubly ionized (Ca I, Ca II, and Ca III, for example). In the corona, on the other hand, we find ions that are ionized approximately a dozen times (Fe XIV, for example, iron that has lost 13 of its normal quota of 26 electrons). The corona must be very hot indeed, millions of degrees, to have enough energy to strip that many electrons off atoms.

By multiply (mul-ti-plē) we mean more than once, that is, twice, three times, and so on.

7.5 The Heating of the Corona

The gas in the corona is so hot that it emits mainly x-rays, photons of high energy. The photosphere, on the other hand, is too cool to emit x-rays. As a result, when photographs of the sun are taken in the x-ray region of the spectrum, they show the corona and its structure (Fig. 7–11).

X-ray astronomy, which is now able to study distant stars, galaxies, and quasars, began with studies of the sun. In the 1940's, scientists launched rockets with x-ray devices for observing the sun. Techniques of making high-quality x-ray images were developed in the 1960's and 1970's and climaxed with an orbiting x-ray telescope on Skylab in 1973 and 1974. Three successive crews of astronauts visited Skylab, for periods of 1 to 3 months each.

The x-ray pictures show structure that varies with the 11-year cycle of solar activity, which we shall discuss in Section 7.7. The brightest regions visible in x-rays are part of active regions that can also be seen in white light. (In white light, these active regions include sunspots.) X-ray images show a higher layer of the solar atmosphere than the white-light images show.

Detailed examination of the x-ray images shows that most, if not all, the radiation appears in the form of loops of gas joining points separated from each other on the solar surface. The latest thinking is that the corona is composed entirely of these loops. One important line of thought that follows from this new idea is that we must understand the physics of coronal loops in order to understand how the corona is heated. It is not sufficient to think in terms of a uniform corona, since the corona is obviously so non-uniform.

The x-ray image also shows a very dark area at the sun's north pole and extending downward across the center of the solar disk. (As the sun rotates from side to side, obviously, the part extending across the center of the disk will not usually be facing us.) These dark locations are *coronal holes* (Fig. 7–12), regions of the corona that are particularly cool and quiet. The density of gas in those areas is lower than the density in adjacent areas.

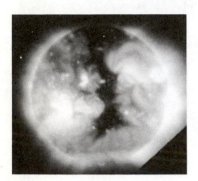

Figure 7–11 An x-ray photograph of the sun taken from Skylab on June 1, 1973. The filter passed wavelengths 2–32 Å and 44–54 Å. The dark region across the center is a coronal hole.

Figure 7–12 This sequence, taken from Skylab on August 14, 1973, shows a coronal hole, a phenomenon that usually appears over one or both poles of the sun and often at other latitudes as well. Ionized helium (*A*) shows chromospheric temperatures, Ne VII (*B*) shows the transition zone, and Mg IX (*C*) shows the corona.

A   He II 304 Å   B   Ne VII 465 Å   C   Mg IX 368 Å

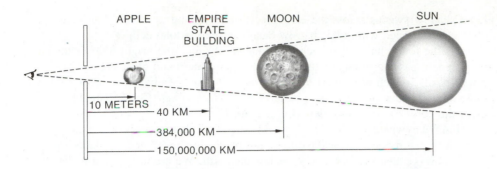

Figure 7–13 An apple, the Empire State Building, the moon, and the sun are very different from each other in size, but they subtend the same angle in the picture because they are shown at different distances from us.

There is usually a coronal hole at one or both of the solar poles. Less often, we find additional coronal holes at lower solar latitudes. The regions of the coronal holes seem very different from other parts of the sun.

Solar physicists used to think that the corona was heated to millions of degrees by shock waves that carried energy upward from underneath the photosphere. Though this theory had seemed very well established and had been worked on continuously since 1946, it was discarded in 1980 on the basis of new evidence. Some of the important new evidence was a set of observations from a spacecraft in the late 1970's. The data showed directly that the shock waves were carrying, as they passed through the chromosphere, 1000 times too little energy to heat the corona.

It was also discovered from the Einstein Observatory x-ray spacecraft that even type O and B stars give off enough x-rays that they must have hot coronas, even though their internal structure is such that they cannot generate shock waves. The x-ray observations show that different mechanisms are necessary to heat the coronas of other stars. Some of these different mechanisms probably heat the solar corona, too. Though the old ideas about coronal heating have been discarded, there is no agreement about what the real mechanism is for heating the corona. Probably the energy is generated in connection with the solar magnetic or electric fields. This would explain the Skylab x-ray pictures, which show that the corona is hotter above active regions.

*7.6 Solar Eclipses

Since the solar chromosphere and corona are visible to the eye only at the time of a total solar eclipse, eclipses have played a major role in solar physics.

Solar eclipses arise because of a happy circumstance: though the moon is 400 times smaller in diameter than the solar photosphere, it is also 400 times closer to the earth. Because of this, the sun and the moon subtend almost exactly the same angle in the sky—about $\frac{1}{2}°$ (Fig. 7–13).

The moon's position in the sky, at certain points in its orbit around the earth, comes close to the position of the sun. This happens approximately once a month at the time of the new moon. Since the lunar orbit is inclined with respect to the earth's orbit, the moon usually passes above or below the line joining the earth and the sun. But occasionally the moon passes close enough to the earth-sun line that the moon's shadow falls upon the surface of the earth.

At a total solar eclipse, the lunar shadow barely reaches the earth's surface (Fig. 7–14). As the moon moves through space on its orbit, and as the earth rotates, this

Figure 7–14 A solar eclipse with the earth, moon, and the distance between them shown to actual scale.

lunar shadow sweeps across the earth's surface in a band up to 300 km wide. Only observers stationed within this narrow band can see the total eclipse.

As the total phase of the eclipse—totality—begins, the bright light of the solar photosphere passing through valleys on the edge of the moon glistens like a series of bright beads, which are called *Baily's beads*. The last bead seems so bright that it looks like the diamond on a ring—the *diamond ring effect* (Color Plate 11). About then, the corona comes into view (Color Plate 12).

From anywhere outside the band of totality, one sees only a partial eclipse. Sometimes the moon, sun, and earth are not precisely aligned and the darkest part of the shadow—called the *umbra*—never hits the earth. We are in the intermediate part of the shadow, which is called the *penumbra*. Only a partial eclipse is visible on earth under these circumstances. As long as the slightest bit of photosphere is visible, even as little as 1 per cent, one cannot see the important eclipse phenomena—the chromosphere and corona.

The last total eclipse to cross the continental U.S. and Canada occurred in 1979. The next total eclipse in the U.S. will cross one of Alaska's Aleutian Islands in 1990, and a long total eclipse (almost 7 minutes of totality) will be visible from Hawaii in 1991. The next total eclipse in the continental United States won't be until 2017. Canada won't see another total eclipse until 2024.

On the average, a solar eclipse occurs somewhere in the world every year and a half. The band of totality, though, usually does not cross populated areas of the earth, and astronomers often have to travel great distances to carry out their observations. Table 7–2 and Fig. 7–15 show sites and dates of future eclipses. Astrono-

Figure 7–15 The paths of the moon's shadow during solar eclipses occurring between 1979 and 2017. Anyone standing along such a path on the dates indicated will see a total solar eclipse. The annular eclipse of 1984 is shown as a dotted line.

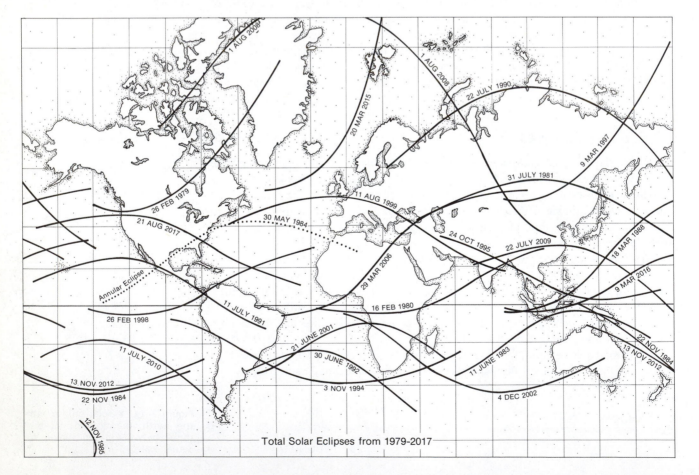

Total Solar Eclipses from 1979-2017

Table 7–2 Total Solar Eclipses

Date	Maximum Duration	Location
1984 November 22	1ᵐ 59ˢ	Papua New Guinea, South Pacific Ocean
1985 November 12	1ᵐ 59ˢ	Ocean near Antarctica
1988 March 18	3ᵐ 46ˢ	Indonesia, Philippines
1990 July 22	2ᵐ 33ˢ	Finland, U.S.S.R., U.S.A. (Alaska)
1991 July 11	6ᵐ 54ˢ	U.S.A. (Hawaii), Central and South America

mers took advantage of the relatively long 1980 and 1983 eclipses by sending large-scale expeditions to India and Indonesia, respectively (Fig. 7–16). The next eclipse for which favorable circumstances are expected will take place in 1988, with observing expected from Sumatra, Indonesia.

Sometimes the moon subtends a slightly smaller angle in the sky than the Sun, because the moon is on the part of its orbit that is relatively far from the earth. When a well-aligned eclipse occurs in such a circumstance, the moon doesn't quite cover the sun. An annulus—a ring—of photospheric light remains visible, so we call this an *annular eclipse* (Fig. 7–17). Such an annular eclipse crossed the southeastern United States on May 30, 1984. The rest of the United States saw a partial eclipse.

In these days of orbiting satellites, is it worth travelling to observe eclipses? There is much to be said for the benefits of eclipse observing. Eclipse observations are a relatively inexpensive way, compared to space research, of observing the chromosphere and corona to find out such quantities as the temperature, density, and magnetic field structure. Even realizing that some eclipse experiments may not work out because of the pressure of time or because of bad weather, eclipse observations are still very cost-effective. And for some kinds of observations, those at the highest resolutions, space techniques have not yet matched ground-based eclipse capabilities. Space coronagraphs, even though they can study the outer part of the corona day to day, are not able to observe the inner and middle corona that we observe at eclipses. Coronal studies have been made much more cheaply during an eclipse than the cost of a space experiment would have been, even if it had been mounted.

Figure 7–16 Mid-totality at the Indian eclipse of 1980. The corona is visible at upper left. The Sun was in the clear, although there were clouds around it in the sky. The sky at the horizon is bright because we are seeing light from outside the cone of the Moon's shadow. The Williams College expedition is in the foreground. The corona shows at lower left on a television screen; we used the image for maintaining the accurate tracking of the telescopes. Students, staff, and equipment appear in silhouette.

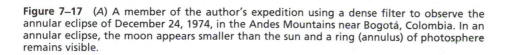

Figure 7–17 (*A*) A member of the author's expedition using a dense filter to observe the annular eclipse of December 24, 1974, in the Andes Mountains near Bogotá, Colombia. In an annular eclipse, the moon appears smaller than the sun and a ring (annulus) of photosphere remains visible.

Figure 7–17 (*B*) The U.S. annular eclipse of May 30, 1984. The 1984 eclipse was so close to total—99.8% coverage at maximum—that the annulus was always broken by the mountains on the moon. This photograph, taken without a filter, shows some Baily's beads. The bright crescent at upper right was the photosphere. The thin curve at lower left was the reddish chromosphere. The path of annularity was clear from New Orleans through Atlanta to North Carolina, and cloudy farther east. The rest of the country had a partial eclipse, though it was cloudy in the northeast. I took the picture in Picayune, Mississippi.

*7.6a A Solar Eclipse Expedition

Just as heart surgeons would travel to the ends of the earth each 18 months if the only place and time they could operate was there and then, eclipse astronomers often travel great distances to study total eclipses. Then the corona and other solar phenomena outside the solar limb are best visible. Some eclipses are more favorable than others, for the sun may be higher in the sky, totality longer, and the weather forecasts better. The U.S. National Science Foundation sponsored national expeditions to the 1980 eclipse in India and the 1983 eclipse in Indonesia.

The 11 June 1983 eclipse path crossed the island of Java. The eclipse was long—5 minutes—and weather forecasts were excellent. The people from the National Center for Atmospheric Research who handled logistics for the U.S. team selected a site on the north coast, overlooking the Java Sea. Not only are solar images often steadier when seen over water but also a few degrees off the 95°F temperature or a few per cent off the almost equal humidity would help the astronomers work more comfortably.

Scientists from half a dozen U.S. colleges, universities, and observatories were members of the official team. My own group, from Williams College, was studying a proposed mechanism for heating the solar corona as a result of waves involving magnetic fields. If the mechanism was valid, then we might expect to see coronal loops oscillating with a period of about 1 second. At the Indian eclipse, we had found indications of these oscillations, and we brought improved equipment this time. The National Geographic Society also supported my efforts. Groups from the Sacramento Peak Observatory, the Kitt Peak National Observatory, and from research institutes in India (who were set up nearby) were studying temperatures and velocities in the corona. University of Hawaii scientists were taking spectra of prominences. A group from Iowa State University was searching for dust in interplanetary space. High Altitude Observatory researchers studied the structure and polarization of the corona. Groups from East Carolina University and North Carolina State University observed the eclipse's effect on the earth's atmosphere. And a U.S. Naval Observatory scientist led a team to time the Baily's beads precisely, to measure an accurate size for the sun in order to see if it were changing over the years.

The logistics people had been working for over a year to arrange shipping and customs clearance; our equipment had been shipped out months before. My group, and some others, each had two tons of equipment. Our job on site for the two weeks before the eclipse was to set up, test, and align the equipment. Adjusting telescopes and mirrors so that they accurately track the sun is not usually done on such a short time scale. Carpenters and masons built sturdy bases for our equipment, and bamboo huts to shield it and us from the sun (Fig. 7–18).

Figure 7–18 I demonstrate our partially installed equipment to the Governor of the Indonesian state of East Java. Light from the telescope fell onto a ring that we could rotate to choose isolated parts of the sun; we used fiber optics to carry light from that focal plane to our photomultipliers. At this point, the electronics and computer were not yet installed. Note the concrete bases and the bamboo and woven-mat protective building. (Courtesy of Daniel Fischer)

Of course, we had time to see the small town around us. We had passed briefly through the big cities of Jakarta and Surabaya en route here. People thronged our site to see what was going on.

Bit by bit, scientists checked equipment out, and tried to fix what was broken. My group actually had to have someone fly down from Boston at the last minute with a replacement digital tape recorder. By eclipse day, all was working.

But the day before, the weather had stopped cooperating. Two weeks of sun had turned into pelting rain. When eclipse morning dawned, the rain had stopped but the clouds were heavy. We couldn't tell if we would see the eclipse or not, but the optimists among us never gave up hope.

Indeed, the clouds parted, and it was pretty clear by the time the partial phases started. The times can now be predicted accurately, to within seconds. As the solar crescent became smaller, the light turned eerie and shadows sharpened. Then the Baily's beads appeared on the edge of the sun, the last one gleaming. ''Diamond ring, lens caps off,'' someone called. I monitored and adjusted where on the sun our telescope was pointing, one colleague ran the measuring instruments, and another colleague monitored the microcomputer he had built to digitize the data. We stood in the middle of totality, the moon's shadow around us and a reddish glow on the horizon in every direction.

High in the sky above us was the glorious corona, looking ghostly white (Fig. 7–19). The corona seemed especially bright this time; we could read dials and notes with ease in its light. Every now and then during the five minutes of totality, we could look up and around to enjoy the view. But all too soon it was over; the second diamond ring came and the sky grew rapidly brighter. We capped our equipment, and carried out calibration tests.

All experiments seemed to run well, and the data are being studied at our various home institutions. In a year or so, reports will appear in scientific journals and will be delivered at meetings of the American Astronomical Society. In these ways, the results of the eclipse will be made available to the astronomical community at large.

Figure 7–19 The eclipsed sun high in the sky and the equipment of the author's expedition, part of the U.S. expedition to Indonesia for the 1983 total eclipse, in the foreground.

7.7 Sunspots and Other Solar Activity

A host of other time-varying phenomena are superimposed on the basic structure of the sun. Many of them, notably the sunspots, vary with an 11-year cycle, which is called the *solar activity cycle*.

In this section we discuss the *active* sun.

7.7a Sunspots

Sunspots (Fig. 7–20) are the most obvious sign of solar activity. They are areas of the sun that appear relatively dark when seen in white light. Sunspots appear dark because they are giving off less radiation than the photosphere that surrounds them. This implies that they are cooler areas of the solar surface, since cooler gas radiates less than hotter gas. Actually, if we could somehow remove a sunspot from the solar surface and put it off in space, it would appear bright against the dark sky; a large one would give off as much light as the full moon.

A sunspot includes a very dark central region, called the *umbra* from the Latin for ''shadow'' (plural: *umbrae*). The umbra is surrounded by a *penumbra* (plural: *penumbrae*), which is not as dark (just as during an eclipse the umbra of the shadow is the darkest part and the penumbra is less dark).

To understand sunspots we must understand magnetic fields. When iron filings are put near a simple bar magnet on earth, the filings show a pattern that is illustrated

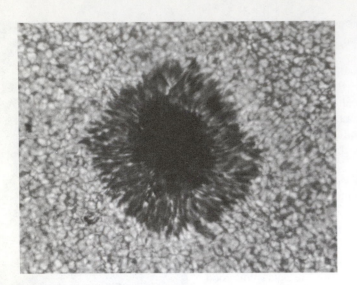

Figure 7–20 A sunspot, showing the dark *umbra* surrounded by the lighter *penumbra*. Granulation is visible in the surrounding photosphere.

in Fig. 7–21. The magnet is said to have a north pole and a south pole, and the magnetic field linking them is characterized by what we call *magnetic lines of force,* or *magnetic field lines* (after all, the iron filings are spread out in what look like lines). The earth (as well as some other planets) has a magnetic field that has many characteristics in common with that of a bar magnet. The structure seen in the solar corona, including polar plumes and equatorial streamers, results from matter being constrained by the solar magnetic field.

We can measure magnetic fields on the sun by using a spectroscopic method. In the presence of a magnetic field, certain spectral lines are split into a number of components, and the amount of the splitting depends on the strength of the magnetic field.

Measurements of the solar magnetic field were first made by George Ellery Hale. He showed, in 1908, that the sunspots are regions of very high magnetic field strength on the sun, thousands of times more powerful than the earth's magnetic field or than the average solar magnetic field. Sunspots usually occur in pairs, and often these pairs are part of larger groups. In each pair, one sunspot will have a polarity typical of a north magnetic pole and the other will have a polarity typical of a south magnetic pole.

Magnetic fields are able to restrain matter—this is the property we are trying to exploit on earth to contain superheated matter sufficiently long to allow nuclear fu-

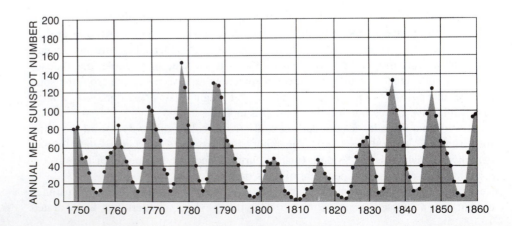

Figure 7–21 (*A*) Lines of force from a bar magnet are outlined by iron filings. One end of the magnet is called a north pole and the other is called a south pole. Similar poles ("like poles")—a pair of norths or a pair of souths—repel each other and unlike poles (1 north and 1 south) attract each other. Lines of force go between opposite poles. (*B*) The magnetic field on November 9, 1980, the day that the white light photograph that opens this chapter was taken. One polarity appears as relatively dark and the opposite polarity as relatively light. Note how the regions of strong fields correspond to the positions of sunspots. Notice also how the two members of a pair of sunspots have polarity opposite to each other, and how which polarity comes first as the sun rotates is different above and below the equator.

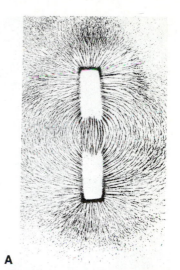

A

B

sion for energy production to take place. The strongest magnetic fields in the sun occur in sunspots. The magnetic fields in sunspots restrain the motions of the matter there, and in particular keep convection from carrying energy to photospheric heights from lower, hotter levels. This results in sunspots being cooler and darker, though exactly why they remain so for weeks is not known. The parts of the corona above active regions are hotter and denser than the normal corona. Presumably the energy is guided upward by magnetic fields. These locations are prominent in radio or x-ray maps of the sun.

Sunspots were discovered in 1610, independently by Galileo in Italy, Fabricius and Christopher Scheiner in Germany, and Thomas Harriot in England. In about 1850, it was realized that the number of sunspots varies with an 11-year-cycle, as is shown in Fig. 7–22. This is called the *sunspot cycle,* although we now realize that many related signs of solar activity vary with the same period.

Besides the specific magnetic fields in sunspots, the sun seems to have a weak overall magnetic field with a north magnetic pole and a south magnetic pole, which may entirely result from the sum of the weak magnetic fields from vanished sunspots. Every 11-year-cycle, the north magnetic pole and south magnetic pole on the sun reverse polarity; what had been a north magnetic pole is then a south magnetic pole and vice versa. This occurs a year or two after the number of sunspots has reached its maximum. For a time during the changeover, the sun may even have two north magnetic poles or two south magnetic poles! But the sun is not a simple bar magnet, so this strange-sounding occurrence is not prohibited. Because of the changeover, it is 22 years before the sun returns to its original configuration, so the real period of the solar activity cycle is 22 years.

The last maximum of the sunspot cycle—the time when there is the greatest number of sunspots—took place in late 1979. Shortly thereafter, NASA launched a satellite called the Solar Maximum Mission to study solar activity at that time (Fig. 7–23 and Color Plates 16 and 18 to 21).

A type of variable star—colloquially called RS Can Ven stars after its prototype—apparently has giant starspots on its surface. Many amateur and professional astronomers are now following the resulting changes in the brightness of such stars, and are taking spectra with IUE and with ground-based telescopes.

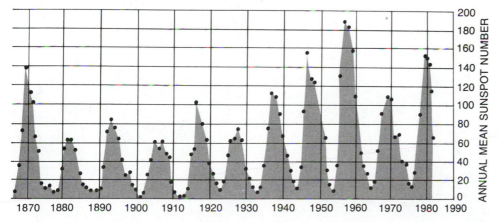

Figure 7–22 The 11-year sunspot cycle is but one manifestation of the solar activity cycle.

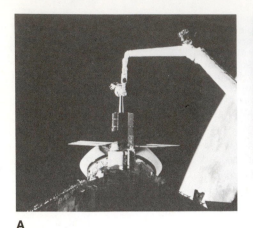

A

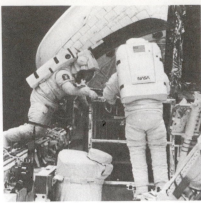

B

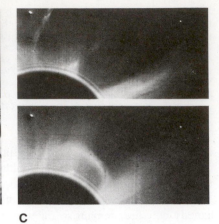

C

PHOTOSPHERE

N

EQUATOR

A

S

N

S N
N S

B

S

Figure 7–24 A leading model to explain sunspots suggests that the solar differential rotation winds tubes of magnetic flux around and around the sun. When the tubes kink and penetrate the solar surface, we see the sunspots that occur in the areas of strong magnetic field.

Figure 7–23 (*A*) Space-shuttle astronauts retrieved the Solar Maximum Mission spacecraft with the remote arm and attached it to the cargo bay. (*B*) The astronauts were able to repair or replace almost all the defective systems. (*C*) The eruption of a large prominence led to this transient event in the corona, observed from the coronagraph aboard Solar Maximum Mission. 56 minutes elapsed between the two frames. (Courtesy of Lewis House, Ernest Hilder, William Wagner, and Constance Sawyer—HAO/NCAR/NSF and NASA)

Box 7.1 The Origin of Sunspots

Although the details of the formation of sunspots are not yet understood, a general picture was suggested in 1961. In this model, just under the solar photosphere the magnetic field lines are bunched in tubes of magnetic field that wind around the sun (Fig. 7–24).

The sun rotates approximately once each earth month. Different latitudes on the sun rotate at different speeds. Though a solid ball like the earth rotates at a constant rate at all latitudes (both Tampa and New York rotate in 24 hours), a gaseous ball like the sun can rotate differentially (Fig. 7–25). Because of this *differential rotation*, if a line of sunspots started at the same longitude on the sun at a given time, the ones closest to the equator would make a full revolution faster than the ones farthest from the equator. Gas at the equator rotates in 25 days, gas at 40 degrees latitude rotates in about 28 days, and gas nearer the poles rotates even more slowly. The differential rotation can be measured spectrographically from Doppler shifts.

A line of force that may have started out in the direction from north to south on the solar surface is wrapped around the sun by the action of the differential rotation. These lines collect in the equivalent of tubes located not far beneath the solar photosphere. Under some circumstances, buoyant forces carry part of a tube upward until the tube sticks up through the solar photosphere. Where the tube emerges we see a sunspot of one magnetic polarity, and where the tube returns through the surface we see a sunspot of the other polarity. The north and south polarities are connected by magnetic lines of force that extend above the sun.

Because of the differential rotation, the spiral winding of the magnetic lines of force is tighter at higher latitudes on the sun than at lower latitudes. Thus the instability that allows part of a tube to be carried to the surface arises first at higher solar latitudes. As the solar cycle wears on, the differential rotation continues and the tubes rise to the surface at lower and lower latitudes. This explains why spots form at higher latitudes earlier in the sunspot cycle than they do later in the cycle.

The effects that lead to the formation of sunspots arise from the interaction of the solar differential rotation with turbulence and motions in the convective zone. Dynamos in factories on earth also depend on the interaction of rotation and magnetic fields; thus these theories for the solar activity are called *dynamo theories*, and one says that the sunspots are generated by the *solar dynamo*.

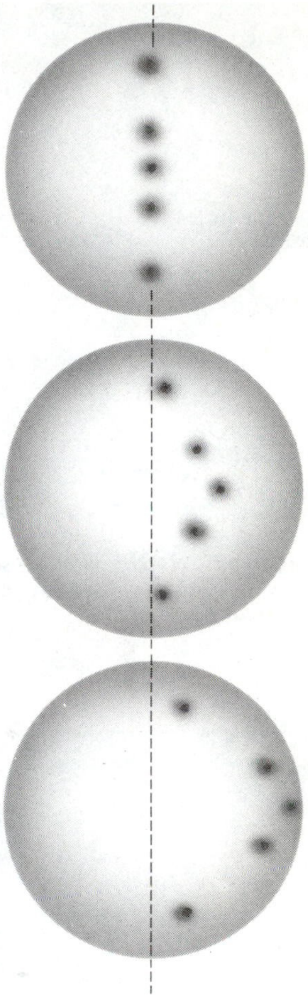

Figure 7–25 The notion that the sun rotates differentially is illustrated by this series, which shows the progress of a schematic line of sunspots month by month. The equator rotates faster than the poles by about 4 days per month.

7.7b Flares

Violent activity sometimes occurs in the regions around sunspots. Tremendous eruptions called *solar flares* (Figs. 7–26 and 7–27) can eject particles and emit radiation from all parts of the spectrum into space. These solar storms begin in a few seconds and can last up to four hours. A typical flare lifetime is 20 minutes. Temperatures in the flare can reach 5 million kelvins, even hotter than the quiet corona. Flare particles that are ejected reach the earth in a few hours or days and can cause disruptions in radio transmission, cause the aurorae (Color Plate 36)—the *aurora borealis* is the "northern lights" and the *aurora australis* is the "southern lights"—and even cause surges on power lines. Because of these solar-terrestrial relationships, high priority is placed on understanding solar activity and being able to predict it. The U.S. government even has a solar weather bureau to forecast solar storms, just as it has a terrestrial weather bureau.

Solar flares also emit x-rays, which have been studied from satellites. Observing flares in the ultraviolet and x-ray region was one of the Solar Maximum Mission's major goals. The radio emission of the sun also increases at the time of a solar flare.

Figure 7–26 (*A*) One of the largest solar flares in decades occurred on August 7, 1972, and led to power blackouts, short-wave radio blackouts, and aurorae. The whole sun is shown at the peak of the flare, 15:30 U.T. (Universal Time, approximately corresponding to Greenwich Time). (*B*) The development of the flare with time and its subsidence is shown in this sequence from the Big Bear Solar Observatory. It was hours before the flare completely died down. Other strong flares also occurred in this active region during that week. Labels show Universal Time in hours and minutes. All photographs are through a filter passing only Hα.

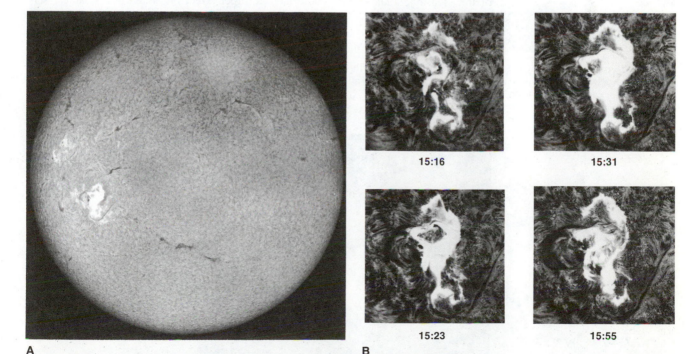

15:16 15:31

15:23 15:55

A B

No specific model is accepted as explaining the eruption of solar flares. But it seems that a tremendous amount of energy is stored in the solar magnetic fields in sunspot regions. Something unknown triggers the release of the energy.

7.7c Plages, Filaments, and Prominences

Studies of the solar atmosphere in Hα radiation also reveal other types of solar activity. Bright areas called *plages* (from the French word for beach, pronounced "plahges") surround the entire sunspot region (Fig. 7–28). Dark *filaments* are seen threading their way across the sun in the vicinity of sunspots. The longest filaments can extend for 100,000 km. They mark the locations of zero magnetic field that separate regions of positive and negative magnetic polarities. When filaments happen to be on the limb of the sun, they can be seen to project into space, often in beautiful shapes. Then they are called *prominences* (Fig. 7–29). Prominences can be seen with the eye at solar eclipses, and glow pinkish at that time because of their emission in Hα and a few other spectral lines. They can be observed from the ground even without an eclipse, if an Hα filter is used. Prominences appear to be composed of matter in a condition of temperature and density similar to matter in the quiet chromosphere, somewhat hotter than the photosphere.

Sometimes prominences can hover above the sun, supported by magnetic fields, for weeks or months. They are then called *quiescent prominences,* and can extend tens of thousands of kilometers above the limb. Other prominences can seem to undergo rapid changes (Color Plates 14 and 15).

In sum, the regions around sunspots show many kinds of solar activity, and so are called *active regions*.

7.8 Solar-Terrestrial Relations

Careful studies of the solar activity cycle are now increasing our understanding of how the sun affects the earth. Although for many years scientists were skeptical of the idea that solar activity could have a direct effect on the earth's weather, scientists presently seem to be accepting more and more the possibility of such a relationship.

An extreme test of the interaction may be provided by the interesting probability that there were no sunspots at all on the sun from 1645 to 1715! The sunspot cycle may not have been operating for that 70-year period (Fig. 7–30). This period, called the *Maunder minimum*, was known to the British astronomer Walter Maunder and

Figure 7–27 A flare near the solar limb, photographed at Big Bear on August 20, 1971.

Figure 7–28 The sun, photographed in hydrogen radiation at the last solar maximum, shows bright areas called plages and many dark filaments around the many sunspot regions.

Figure 7–29 A solar prominence.

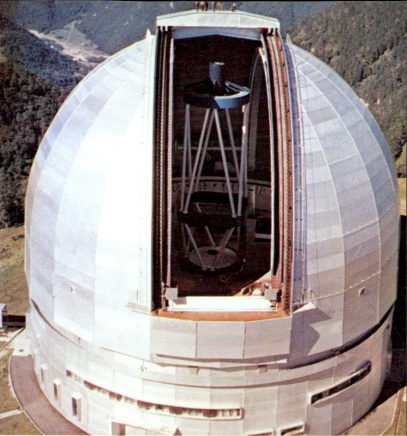

Color Plate 1 (top, left): The 4-m Mayall optical telescope on Kitt Peak, near Tucson, Arizona. (Kitt Peak National Observatory photo)

Color Plate 2 (top, right): The Soviet 6-m telescope, the largest in the world. Note the figures standing on top of the dome. (Courtesy of J.M. Kopylov, Special Astrophysical Observatory)

Color Plate 3 (bottom, left): The Greenbelt, Maryland, control room of the International Explorer Spacecraft. The image displayed on the screen shows a stellar spectrum in the ultraviolet. (Photo by the author)

Color Plate 4 (bottom, right): The 67-m radio telescope at the Australian National Radio Observatory at Parkes, N.S.W. (Photo by the author)

Color Plate 5 (bottom): Fraunhofer's original spectrum from 1814. Only the D and H lines retain their notation from that time. The C line is Hα

Color Plate 6 (top): HEAO-1 (High-Energy Astronomy Observatory 1), launched in 1977 to study x-rays and gamma rays. (NASA and TRW Systems Group photo)

Color Plate 7 (bottom): The VLA (Very Large Array), the aperture synthesis radio telescope near Socorro, New Mexico. It is composed of 27 dishes, each 26 m in diameter, arranged in the shape of a "Y" over a flat area 27 km in diameter. The third arm of the "Y" extends to the right. (National Radio Astronomy Observatory photo)

Color Plate 8 (top): Multiple exposures at hourly intervals taken in Alaska. The upper section shows the phenomenon of the midnight sun (which never dips below the horizon) on the day of the summer solstice, and the lower section shows the sun on the shortest day of the year. (Photo by Mario Grassi)

Color Plate 9 (bottom): Lightning flashes illuminate the Kitt Peak National Observatory in this 45-second exposure. (Photo ©1972 Gary Ladd)

Color Plate 10 (top): Two of the partial phases of the June 30, 1973, total solar eclipse. Note the clouds that covered part of the crescent shortly before totality.

Color Plate 11 (bottom): The diamond ring effect at the beginning of totality at the June 30, 1973, eclipse. These photographs were taken from Loiengalani, Kenya, by the author's expedition.

Color Plate 12 (top): The solar corona at the 1980 total eclipse, photographed by the author's expedition to India. Note the many streamers typical of solar maximum.

Color Plate 13: (top) The flash spectrum of the solar corona at the 1977 total solar eclipse. The crescents are chromospheric spectral lines. The horizontal streak is an overexposed spectrum of a Baily's bead. (bottom) The spectrum during totality shows the green and red emission lines from the corona and the emission lines of prominences dotted around the solar limb. (Dennis di Cicco photos; Williams College expedition)

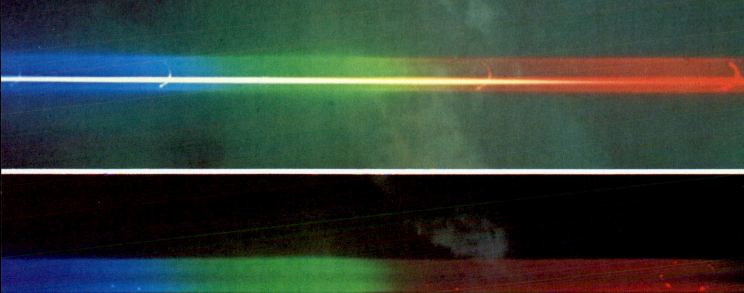

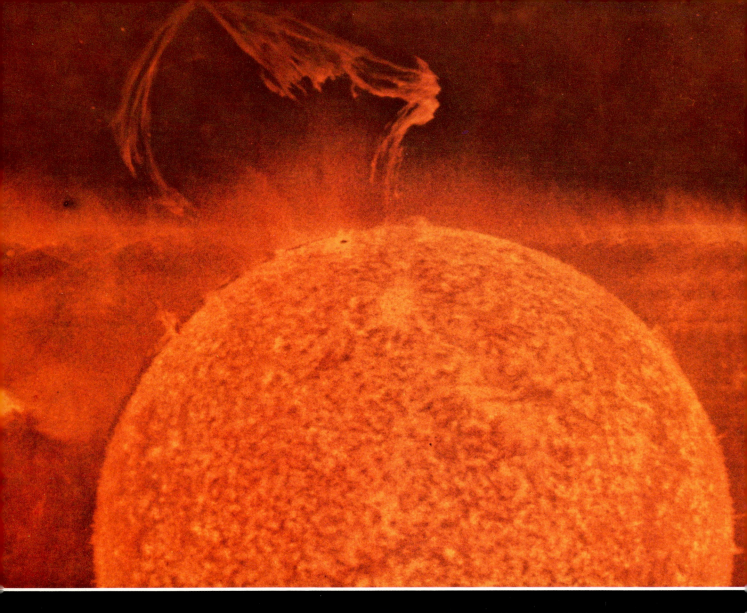

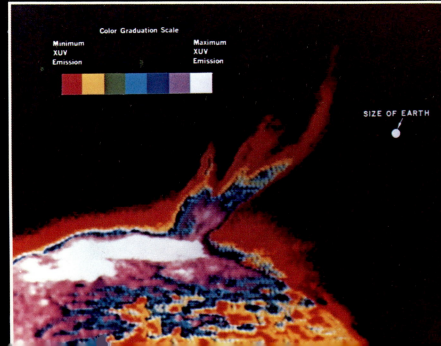

Color Plate 14 (top): An erupting prominence on the sun, photographed on August 9, 1973, in the ultraviolet radiation of ionized helium at 304 Å with the Naval Research Laboratory's slitless spectroheliograph aboard Skylab. Because of the technique used, radiation at nearby wavelengths appears as a background. A black dot visible on the solar surface slightly to the left of the base of the prominence may be the site of the flare that caused the eruption. Supergranulation is visible on the solar disk in the light of He II, and a macrospicule shows on the extreme right. (NRL/NASA photo)

Color Plate 15 (bottom): Contours of equal intensity computed for another eruptive prominence observed in the radiation of ionized helium with the NRL instrument aboard Skylab. Each color represents a different strength of emission. This view shows the prominence 90 minutes after its eruption, when the gas had moved 500,000 km from the solar surface. (NRL/NASA photo)

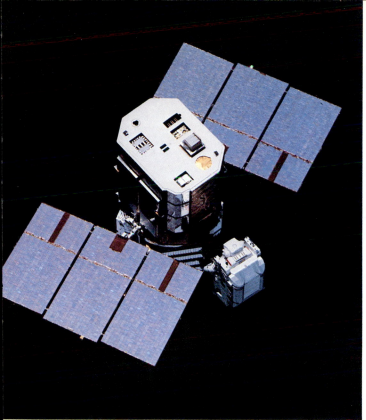

Color Plate 16: A Space Shuttle astronaut made an excursion to the damaged Solar Maximum Mission, but could not attach himself to it (left). Later, the spacecraft was retrieved by the remote arm and attached to the Space Shuttle, where it was repaired (right).

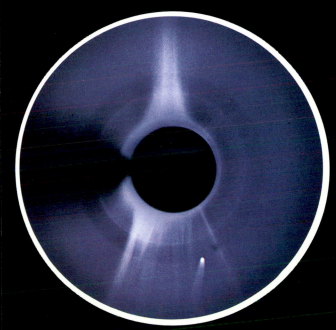

Color Plate 17 On the few days when Comet Kohoutek was closest to its perihelion, it was invisible to observers on Earth. The Skylab astronauts, fortunately, were able to sketch and photograph it. Here we see the image of Comet Kohoutek in the High Altitude Observatory's coronagraph on December 27, 1973. (HAO/NASA photo)

Color Plate 18: An eruption of the sun observed with the coronagraph aboard Solar Maximum Mission, launched in 1980. The image is 6 solar radii across. (HAO/NCAR/NSF and NASA)

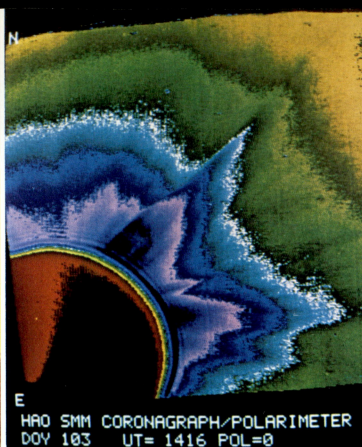

HAO SMM CORONAGRAPH/POLARIMETER
DOY 103   UT= 1837

HAO SMM CORONAGRAPH/POLARIMETER
DOY 103   UT= 1416 POL=0

Color Plate 19: Two false-color views of the corona in visible light with the coronagraph aboard Solar Maximum Mission. The different colors show the intensities, which correspond roughly to the density of the gas present. Blue is the densest and yellow is the least dense. The occulting disk, which is the quarter-circle in the corner of each image, is 1.75 times the diameter of the sun. It is immediately surrounded by a series of colored rings that are an optical effect in the system. The image on the right shows a sharp-edged narrow feature known as a coronal spike. (Courtesy of Lewis House, Ernest Hildner, William Wagner, and Constance Sawyer—HAO/NCAR/NSF and NASA)

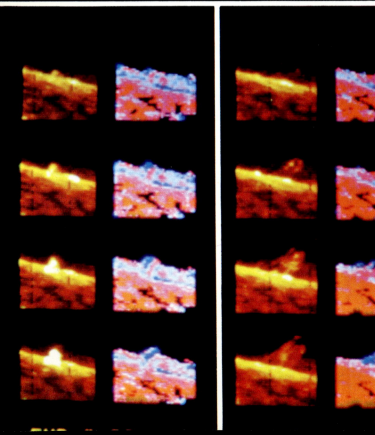

Color Plate 20 (left): A solar flare observed on April 30, 1980 from the Solar Maximum Mission. The sequence on the left, at 2.5-minute intervals shows the radiation of three-times ionized carbon at 1548 Å The sequence on the right shows line-of-sight velocities found from the Doppler shift. Red represents material moving away and blue represents approaching material.

We see a jet of hot gas rising from the base of a magnetic loop and filling the top of the loop. Some of the gas may even go over the top of the loop. The velocity picture supports this last idea since it shows material coming toward us at the top of the loop. A bright flare was detected in x-rays 3 minutes after the last frame shown.

Color Plage 21 (right): A similar sequence of events observed from the Solar Maximum Mission on its next orbit 90 minutes later. Each picture here is separated by 7.5 minutes. Although at first glance the material appears to be rising, detailed measurement gives a different picture. Actually, features are not moving but new ones are appearing successively higher up. This may occur because hotter gas is cooling to the 200,000 degrees necessary to be visible this temperature. (Color Plates 20 and 21 are courtesy of Bruce E. Woodgate, Einar Tandberg-Hanssen and colleagues at NASA's Marshall Space Flight Center)

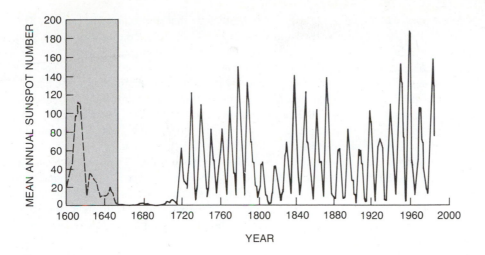

Figure 7–30 The Maunder minimum (1645–1715), when sunspot activity was negligible for decades, may indicate that the sun does not have as regular a cycle of activity as we had thought. Activity has been extrapolated into the shaded regions with measurements made by indirect means (such as the frequency of auroras).

others in the early years of this century but was largely forgotten until its importance was recently noted and stressed by John A. Eddy of the High Altitude Observatory. Although no counts of sunspots exist for most of that period, there is evidence that people were looking for sunspots; it seems reasonable that there were no counts of sunspots because they were not there and not just because nobody was observing. A variety of indirect evidence has also been brought to bear on the question. For example, the solar corona appeared very weak when observed at eclipses during that period.

It may be significant that the anomalous sunspotless period coincided with a "Little Ice Age" in Europe and with a drought in the southwestern United States. An important conclusion from the existence of this sunspotless period is that the solar activity cycle may be much less regular than we had thought.

The evidence for the Maunder minimum is indirect, and has been challenged. For example, the Little Ice Age could have as much or more to do with volcanic dust in the air or with changing patterns of land use than with sunspots. Old auroral records may show activity, though this is controversial. It should come as no particular surprise that several mechanisms affect the earth's climate on this time scale, rather than only one.

*7.9 The Solar Wind

At about the time of the launch of the first earth satellites, in 1957, it was realized that the corona must be expanding into space. This phenomenon is called the *solar wind*. The expansion causes comet tails (Fig. 7–31) always to point away

Figure 7–31 The solar wind causes the wavy streaming of the tails of comets, as in this view of Comet Mrkos in 1957.

from the sun, which had previously led to suspicions that the solar wind existed. The solar wind apparently emanates from coronal holes. The gas takes about 10 days to reach the earth.

The density of particles in the solar wind, always low, decreases with distance from the sun. There are only about 5 particles in each cubic centimeter at the earth's orbit. The particles are a mixture of ions and electrons. The solar wind extends into space far beyond the orbit of the earth, possibly even beyond the orbit of Pluto. The Pioneer and Voyager spacecraft are now exploring the region beyond Saturn, and have not yet come to the end of the solar wind.

The earth's outer atmosphere is bathed in the solar wind. Thus research on the nature of the solar wind and on the structure of the corona is necessary to understand our environment in the solar system. A European spacecraft is to be launched in about 1985 to fly over one of the solar poles, to give information about the solar wind above the plane in which the planets all lie. A companion American spacecraft was cancelled for budgetary reasons.

*7.10 The Solar Constant

The solar constant is the amount of energy per second that would hit each square centimeter of the earth at its average distance from the sun if the earth had no atmosphere. 99% of solar energy is in the range from 2760 Å to 49,000 Å (nearly 5 microns), and 99.9% is between 217 Å and 10.94 microns.

Every second a certain amount of solar energy passes through each square centimeter of space at the average distance of the earth from the sun. This quantity is called the *solar constant*. Accurate knowledge of the solar constant is necessary to understand the terrestrial atmosphere, for to interpret our atmosphere completely we must know all the ways in which it can gain and lose energy. Further, knowledge of the solar constant enables us to calculate the amount of energy that the sun itself is giving off, and thus gives us an accurate measurement on which to base our quantitative understanding of the radiation of all the stars.

Ground-based measurements give a value of about 2 calories/cm^2/min, which is equivalent to about 135 milliwatts/cm^2 (where calories are a unit of energy and watts a unit of power). The value has been determined from the ground to an accuracy of 1.5 per cent. High accuracy—0.1 per cent—was obtained with the Solar Maximum Mission. Short-term variations of about 0.2 per cent appeared. These dips in the sun's overall intensity turn out to be correlated with large sunspots crossing the sun (Fig. 7–32). Thus sunspots really prevent energy from escaping from the solar surface, at least temporarily.

One interesting aim is to determine if there are long-term changes in the solar constant, perhaps arising from changes in the luminosity of the sun. Certainly our climate would be profoundly affected by small long-term changes. Of course, if the solar constant changed with time, it wouldn't really be a constant after all.

Makers of solar-energy systems for heating and generating electricity start their calculations with the solar constant. But much of the sun's energy is lost passing through the earth's atmosphere and in conversion to a usable form, not to mention the effect of cloudy days. For all these reasons, less solar energy is available to us than the solar constant itself.

7.11 The Sun and the Theory of Relativity

The intuitive notion we have of gravity corresponds to the theory of gravity advanced by Isaac Newton in 1687. We now know, however, that Newton's theory and our intuitive ideas are not sufficient to explain the universe in detail. Theories advanced by Albert Einstein in the first decades of this century now provide us with a more accurate understanding.

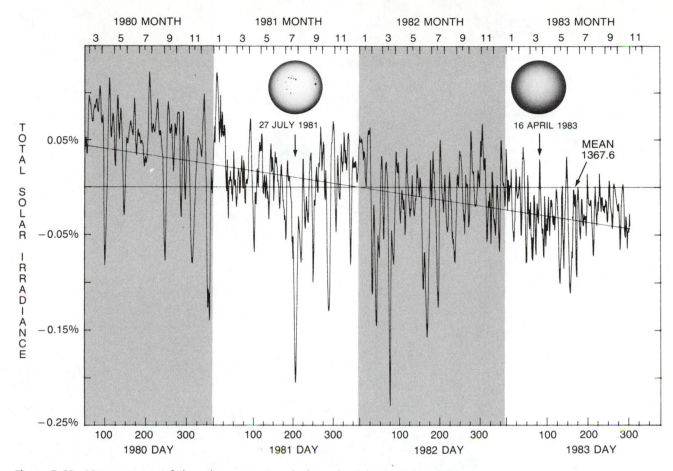

Figure 7–32 Measurements of the solar constant made from the Solar Maximum Mission, with comparison photos showing the area of the sun covered with sunspots. Dips in the solar constant correspond nicely to large sunspots crossing the face of the sun. (AURA, Inc., National Solar Observatory)

The sun, as the nearest star to the earth, has been very important for testing some of the predictions of Albert Einstein's theory of gravitation, which is known as the general theory of relativity. The theory, which Einstein advanced in final form in 1916, made three predictions that depended on the presence of a large mass like the sun for experimental verification. These predictions included (1) the gravitational deflection of light, (2) the advance of the perihelion of Mercury, and (3) the gravitational redshift; comparing the predictions with observations provides basic tests of the theory. We shall discuss the gravitational redshift in Section 9.5 and discuss the other two tests here. The theory has recently been further verified by discovery of changes in a binary pulsar (Section 10.9) and of multiple images of a distant quasar (Section 15.9).

Einstein's theory predicts that the light from a star would act as though it were bent toward the sun by a very small amount (Fig. 7–33). We on earth, looking back, would see the star from which the light was emitted as though it were shifted slightly away from the sun (Fig. 7–34). Only a star whose radiation grazed the edge of the sun would seem to undergo the full deflection; the effect diminishes as one considers stars farther away from the solar limb. To see the effect, one has to look near the sun at a time when the stars are visible, and this could be done only at a total solar eclipse (Fig. 7–35).

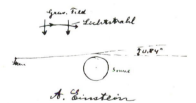

Figure 7–33 The prediction in Einstein's own handwriting of the deflection of starlight by the sun, taken from a letter from Einstein to Hale at Mt. Wilson. "Lichtstrahl" is "light ray." Einstein asked if the effect could be measured without an eclipse, to which Hale replied negatively. The drawing came from a date when Einstein had developed only an early version of the theory, which gave a predicted value half that of his later predicted value.

Figure 7–34 Under the general theory of relativity, the presence of a massive body essentially warps the space nearby. This can account for both the bending of light near the sun and the advance of the perihelion point of Mercury by 43 arc sec per century more than would otherwise be expected. The diagram shows how a two-dimensional surface warped into three dimensions can change the direction of a ''straight'' line that is constrained to its surface; the warping of space is analogous, although with a greater number of dimensions to consider. The effect is similar to a golfer putting on a warped green. Though the ball is hit in a straight line, we see it appear to curve.

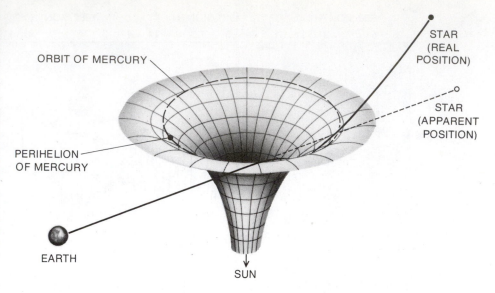

Box 7.2 The Special and General Theories of Relativity

In 1905, Einstein advanced a theory of relative motion that is called the *special theory of relativity*. A basic postulate is that nothing can go faster than the speed of light. Further, strangely, the speed of light is the same even if we are moving toward or away from the object that is emitting the light that we are observing. Einstein's theory shows that the values that we measure for length, mass, and the rate at which time advances depend on how fast we are moving relative to the object we are observing. One consequence of the special theory is that mass and energy are equivalent and that they can be transformed into each other, following a relation $E = mc^2$. The consequences of the special theory of relativity have been tested experimentally in many ways, and the theory has long been well established.

''Special relativity'' was limited in that it did not take the effect of gravity into account. Einstein proceeded to work on a more general theory that would explain gravity. Isaac Newton in 1687 had shown the equations that explain motion caused by the force of gravity, but nobody had ever said how gravity actually worked. Einstein's *general theory of relativity* links three dimensions of space and one dimension of time to describe a four-dimensional space-time. It explains gravity as the effect we detect when we observe objects moving in curved space. Imagine the two-dimensional analogy to curved space of a billiard ball rolling on a warped table. The ball would tend to curve one way or the other. The effect would be the same if the table were flat but there were objects with the equivalent of gravity spaced around the table. In four-dimensional space-time, masses warp the space near them and objects or light change their path as a result, an effect that we call gravity. Thus Einstein's theory of general relativity explains gravity as an observational artifact of the curvature of space. (See also Box 16.1.)

In Einstein's own words, ''If you will not take the answer too seriously, and consider it only as a kind of joke, then I can explain it as follows. It was formerly believed that if all material things disappeared out of the universe, time and space would be left. According to the relativity theory, however, time and space disappear together with the things.''

Figure 7–35 The photographic plate from the 1922 eclipse expedition to test Einstein's theory; the confirming results first found in 1919 were verified. The circles mark the positions of the stars used in the data reduction; the stars themselves are too faint to see in the reproduction.

Figure 7–36 Albert Einstein visiting California in the 1930's.

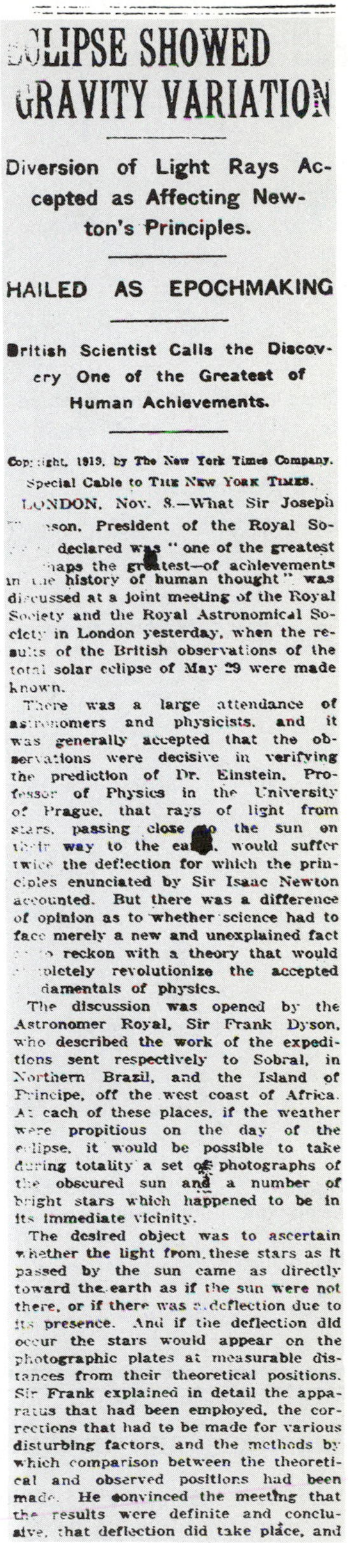

The British astronomer Arthur Eddington and other scientists observed the total solar eclipse of 1919 from sites in Africa and South America. The effect for which they were looking was tiny, and it was not enough merely to observe the stars at the moment of eclipse. One had to know what their positions were when the sun was not present in their midst, so the astronomers had already made photographs of the same field of stars six months earlier when the same stars were in the nighttime sky. The duration of totality was especially long, and the sun was in a rich field of stars, the Hyades, making it a particularly desirable eclipse for this experiment. Even though his observations had been limited by clouds, Eddington detected that light was deflected by an amount that agreed with Einstein's revised predictions. The scientists hailed this confirmation of Einstein's theory, and from the moment of its official announcement, Einstein was recognized by scientists and the general public alike as the world's greatest scientist (Figs. 7–36 through 7–38).

The experiment has been repeated, but is a very difficult one. Attempts at the eclipses in Mexico in 1970 and in India in 1980 failed because of a power failure and because of overexposure of the photographic plates, respectively. A University of Texas 1973 expedition to Africa, hampered by a dust storm, confirmed Einstein's theory, but only to 10 per cent accuracy. A restudy of the original 1919 photographic plates carried out for the 1979 Einstein centennial confirmed the predictions with a precision of only about 6 per cent.

Though the data agree with Einstein's prediction, they are not accurate enough to distinguish between Einstein's theory and newer, more complicated, rival theories of

ECLIPSE SHOWED GRAVITY VARIATION

Diversion of Light Rays Accepted as Affecting Newton's Principles.

HAILED AS EPOCHMAKING

British Scientist Calls the Discovery One of the Greatest of Human Achievements.

Copyright, 1919, by The New York Times Company.

Special Cable to THE NEW YORK TIMES.

LONDON, Nov. 8.—What Sir Joseph Thomson, President of the Royal Society, declared was " one of the greatest—perhaps the greatest—of achievements in the history of human thought " was discussed at a joint meeting of the Royal Society and the Royal Astronomical Society in London yesterday, when the results of the British observations of the total solar eclipse of May 29 were made known.

There was a large attendance of astronomers and physicists, and it was generally accepted that the observations were decisive in verifying the prediction of Dr. Einstein, Professor of Physics in the University of Prague, that rays of light from stars, passing close to the sun on their way to the earth, would suffer twice the deflection for which the principles enunciated by Sir Isaac Newton accounted. But there was a difference of opinion as to whether science had to face merely a new and unexplained fact or to reckon with a theory that would completely revolutionize the accepted fundamentals of physics.

The discussion was opened by the Astronomer Royal, Sir Frank Dyson, who described the work of the expeditions sent respectively to Sobral, in Northern Brazil, and the Island of Principe, off the west coast of Africa. At each of these places, if the weather were propitious on the day of the eclipse, it would be possible to take during totality a set of photographs of the obscured sun and a number of bright stars which happened to be in its immediate vicinity.

The desired object was to ascertain whether the light from these stars as it passed by the sun came as directly toward the earth as if the sun were not there, or if there was a deflection due to its presence. And if the deflection did occur the stars would appear on the photographic plates at measurable distances from their theoretical positions. Sir Frank explained in detail the apparatus that had been employed, the corrections that had to be made for various disturbing factors, and the methods by which comparison between the theoretical and observed positions had been made. He convinced the meeting that the results were definite and conclusive, that deflection did take place, and

Figure 7–37 The first report of the events that made Einstein world-famous. From *The New York Times* of November 9, 1919.

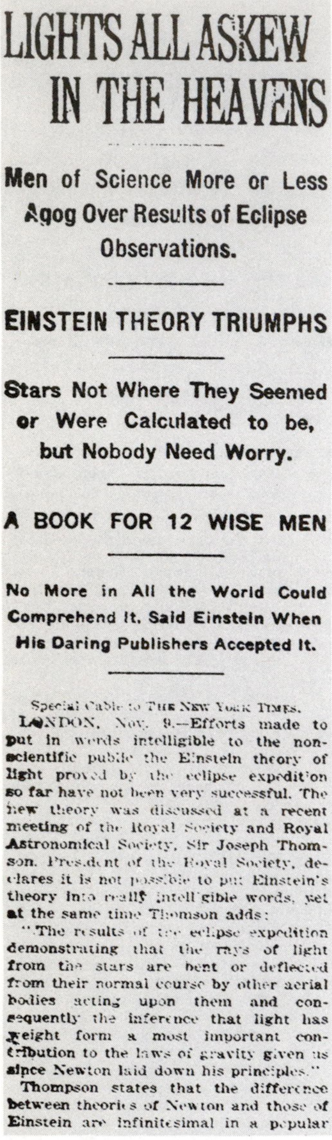

LIGHTS ALL ASKEW IN THE HEAVENS

Men of Science More or Less Agog Over Results of Eclipse Observations.

EINSTEIN THEORY TRIUMPHS

Stars Not Where They Seemed or Were Calculated to be, but Nobody Need Worry.

A BOOK FOR 12 WISE MEN

No More in All the World Could Comprehend It. Said Einstein When His Daring Publishers Accepted It.

Special Cable to THE NEW YORK TIMES.
LONDON, Nov. 9.—Efforts made to put in words intelligible to the non-scientific public the Einstein theory of light proved by the eclipse expedition so far have not been very successful. The new theory was discussed at a recent meeting of the Royal Society and Royal Astronomical Society. Sir Joseph Thomson, President of the Royal Society, declares it is not possible to put Einstein's theory into really intelligible words, yet at the same time Thomson adds:

"The results of the eclipse expedition demonstrating that the rays of light from the stars are bent or deflected from their normal course by other aerial bodies acting upon them and consequently the inference that light has weight form a most important contribution to the laws of gravity given us since Newton laid down his principles."

Thompson states that the difference between theories of Newton and those of Einstein are infinitesimal in a popular

Figure 7–38 These reports captured the public's fancy. From *The New York Times* of November 10, 1919.

gravitation. Fortunately, the effect of gravitational deflection is constant through the electromagnetic spectrum and the test can now be performed more accurately by observing how the sun bends radiation from radio sources, especially quasars. The results agree with Einstein's theory to within 1 per cent, enough to make the competing theories very unlikely. The double quasars (Section 15.9) provide strong verification of Einstein's theory.

A related test involves not deflection but a delay in time of signals passing near the sun. This can now be performed with signals from interplanetary spacecraft, most recently with a Viking orbiter near Mars, and these data also agree with Einstein's theory.

Another of the triumphs of Einstein's theory, even in its earliest versions, was that it explained the "advance of the perihelion of Mercury." The orbit of Mercury, like the orbits of all the planets, is elliptical; the point at which the orbit comes closest to the sun is called the *perihelion*. The elliptical orbit is pulled around the sun over the years, mostly by the gravitational attraction on Mercury by the other planets, so that the perihelion point is at a different orientation in space. Each century (!) the perihelion point appears to move around the sun by approximately 5600 seconds of arc (which is less than 2°). Subtracting the effects of precession (the changing orientation of the earth's axis in space, described in Section 3.3) and the gravitational effects of the other planets leaves 43 seconds of arc per century whose origin had not been understood on the basis of Newton's theory of gravitation.

Einstein's general theory predicts that the presence of the sun's mass warps the space near the sun. Since Mercury is sometimes closer to and sometimes farther away from the sun, it is sometimes travelling in space that is warped more than the space it is in at other times. This should have the effect of changing the point of perihelion by 43 seconds of arc per century. The agreement of this prediction with the measured value was an important observational confirmation of Einstein's theory. More recent refinements of solar system observations have shown that the perihelions of Venus and of the earth also advance by the even smaller amounts predicted by the theory of relativity. And the verification has since been completely established by studying a pulsar in a binary system (Section 10.9).

Historically, in explaining the perihelion advance, general relativity was explaining an observation that had been known. On the other hand, in the bending of light it actually predicted a previously unobserved phenomenon.

As a general rule, scientists try to find theories that not only explain the data that are at hand, but also make predictions that can be tested. This is an important part of the *scientific method*. Because the bending of electromagnetic radiation by a certain amount was a prediction of the general theory of relativity that had not been anticipated, the verification of the prediction was a more convincing proof of the theory's validity than the theory's ability to explain the perihelion advance.

Summary and Outline

Parts of the sun: interior (15,000,000 K), photosphere including granulation (5,800 K), chromosphere and spicules (15,000 K), supergranulation, corona (2,000,000 K), coronal holes (perhaps 1,500,000 K)
Spectra
    Photospheric spectrum: continuum with Fraunhofer lines; abundances of the elements (Section 7.2b)

Chromospheric spectrum: emission at the limb during eclipses in the visible part of the spectrum (flash spectrum) (Section 7.3b)
Coronal spectrum: identification of coronium with highly ionized ions shows the high temperature (Section 7.4c)
Heating of the corona, change to theories involving solar magnetism stemmed from space observations: x-ray

pictures show link with magnetic field (Section 7.5)
(Section 7.5)

Eclipse phenomena: diamond ring, totality (Section 7.6)

Solar activity cycle: sunspots, flares, plages, filaments, prominences all linked to magnetic field structure (Section 7.7)

Solar-terrestrial relations (Section 7.8)

Possible effect of solar activity on terrestrial weather

May not have been any solar activity during the Maunder minimum

Solar wind flows from coronal holes (Section 7.9)

Solar constant and its possible variability (Section 7.10)

Using the sun to test Einstein's general theory of relativity (Section 7.11)

Predicted in advance of detection of gravitational bending at eclipses; now also tested in radio region of spectrum

Advance of the perihelion of Mercury; phenomenon was known in advance of Einstein's theory

Now also verified for the binary pulsar and the multiple quasar

Key Words

quiet sun, active sun, photosphere, subtends, interior, core, solar atmosphere, chromosphere, corona, solar wind, white light, granulation, convection, convection zone, continuum, spicules, supergranulation, chromospheric network, flash spectrum, streamers, plumes, coronagraphs, active regions, coronal holes, umbra, penumbra, Baily's beads, diamond ring effect, annular eclipse, solar activity cycle, sunspots, magnetic lines of force, magnetic field line, differential rotation, dynamo theories, solar dynamo, sunspot cycle, solar flares, aurora borealis, aurora australis, plages, filaments, prominences, quiescent prominences, Maunder minimum, solar constant, special theory of relativity, general theory of relativity, perihelion, scientific method

Questions

1. Sketch the sun, labelling the interior, the photosphere, the chromosphere, the corona, sunspots, and prominences. Give the approximate temperature of each.

2. What elements make most of the lines in the Fraunhofer spectrum? What elements make the strongest lines? Why?

3. Explain why the photospheric spectrum is an absorption spectrum and the chromospheric spectrum seen at an eclipse is an emission spectrum.

4. Sketch the strongest lines of the solar spectrum, including H alpha, H beta, H gamma, H and K, and the D lines.

5. Explain why helium lines can be seen in the chromospheric spectrum but not in the photospheric spectrum.

6. Graph the temperature of the interior and atmosphere of the sun as a function of distance from the center.

7. Define and contrast a prominence and a filament.

8. List three phenomena that vary with the solar activity cycle.

9. (a) Why isn't there a solar eclipse once a month? (b) Whenever there is a total solar eclipse, a lunar eclipse occurs either two weeks before or two weeks after. Explain.

10. Why can't we observe the corona every day from any location on earth?

11. Describe why we cannot see the solar corona during the partial phases of an eclipse.

12. Why does the chromosphere appear pinkish at an eclipse?

13. How do we know that the corona is hot?

14. Describe relative advantages of ground-based eclipse studies and of satellite studies of the corona.

15. Describe the sunspot cycle and a mechanism that explains it.

16. What are the differences between the solar wind and the normal "wind" on earth?

17. (a) What is the solar constant? (b) Why is it difficult to measure precisely?

18. Of what part of the spectrum is it most important to make accurate measurements in order to determine the solar constant?

†19. Referring to Appendix 3 for data about Mars, what would the solar constant be if we lived there?

20. In what tests of general relativity does the sun play an important role? Describe the current status of these investigations.

_____
†This indicates a question requiring a numerical answer.

Topic for Discussion

Discuss the relative importance of solar observations (a) from the ground, (b) at eclipses, (c) from unmanned satellites, and (d) from manned satellites.

Part III Stellar Evolution

We have seen how the Hertzsprung-Russell diagram for a cluster of stars allows us to deduce the age of the cluster and the ages of the stars themselves (Section 6.4). Our human lifetimes are very short compared to the billions of years that a typical star takes to form, live its life, and die. Thus our hope of understanding the life history of an individual star depends on studying large numbers of stars, for presumably we will see them at different stages of their lives.

Though we can't follow an individual star from cradle to grave, studying many different stars shows us enough stages of development to allow us to write out a stellar biography. Chapters 8 through 11 are devoted to such life stories. Similarly, we could study the stages of human life not by watching someone's aging, but rather by studying people of all ages who are present in, say, a city on a given day.

In Chapter 8, we shall study the birth of stars, and observe some places in the sky where we expect stars to be born very soon, possibly even in our own lifetimes. We shall then consider the properties of stars during the long, stable phase in which they spend most of their histories.

In the following chapters, we go on to consider the ways in which stars end their lives. Stars like the sun sometimes eject shells of gas that glow beautifully; the part of such a star that is left behind then contracts until it is as small as the earth (see Chapter 9). Sometimes when a star is newly visible in a location where no star was previously known to exist—a "nova"—we are seeing the interaction of a dead solar-type star with a companion.

Sometimes newly visible stars rival whole galaxies in brightness. Then we may be seeing the spectacular death throes of massive stars (see Chapter 10). The study of what happens to stars after that explosive event is currently a topic of tremendous interest to many astronomers. Some of the stars become pulsars (also discussed in Chapter 10). Other stars, we think, may even wind up as black holes, objects that are invisible and therefore difficult—though not impossible—to detect (see Chapter 11).

When we write a stellar biography from the clues that we get from observation, we are acting like detectives in a novel. As new methods of observation become available, we are able to make better deductions, so the extension of our senses throughout the electromagnetic spectrum has led directly to a better understanding of stellar evolution. Our work in the x-ray part of the spectrum, for example, has told us more about tremendously hot gases like those that result from an exploding star. X-ray astronomy is also intimately connected with our current exciting search for a black hole.

In these chapters, we shall see that an important new tool has also been added: the computer. Calculations that would have taken years or centuries to carry out—and indeed which never would have been carried

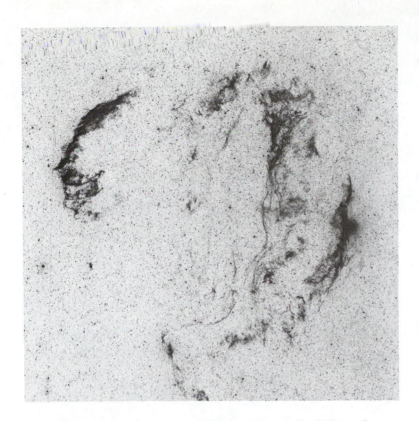

The Cygnus loop, the remnant of a supernova. The remainder of the star that explodes as a supernova contracts to become either a neutron star or a black hole, depending on how much mass is left. The print is a negative.

out because of the time involved or the probability of error—can now be routinely made in minutes. As computers grow faster and cheaper, as they do every year, our capabilities grow.

The importance of this Part of the book is based on a theorem that states that all stars of the same mass and of the same chemical composition evolve in the same way, if we neglect the effects of rotation and magnetic field. Since the chemical differences among stars are usually not overwhelming, it is largely the mass that is the chief determinant of stellar evolution. Because of this fact, we need only study a few groups of stars to gain pictures of the evolution of most stars.

We can set up divisions, using terminology (suggested by Martin Schwarzschild of Princeton) from the sport of boxing, depending on the mass of stars. We use the name *featherweight star* for collapsing gas that is less than about 7 per cent as massive as the sun; it never reaches stardom, and becomes a black dwarf. Stars more massive than that but containing less than about four solar masses eventually lose some of their mass and become white dwarfs; we call these *lightweight stars*. *Heavyweight stars*, which contain greater than eight solar masses, explode as one of the types of supernovae. Some wind up as neutron stars, which we may detect as pulsars or in x-ray binaries. Other heavyweight stars, after they become supernovae, wind up withdrawing from the universe in the form of black holes. The fate of *middleweight stars*, between 4 and 8 solar masses, is less clear. Note that in all cases, we are really discussing mass (which is an intrinsic property) rather than weight (which is the force of gravity on a mass).

Let us start in the next chapter with the formation and the main lifetime of stars. The following three chapters will then discuss the death of stars.

149

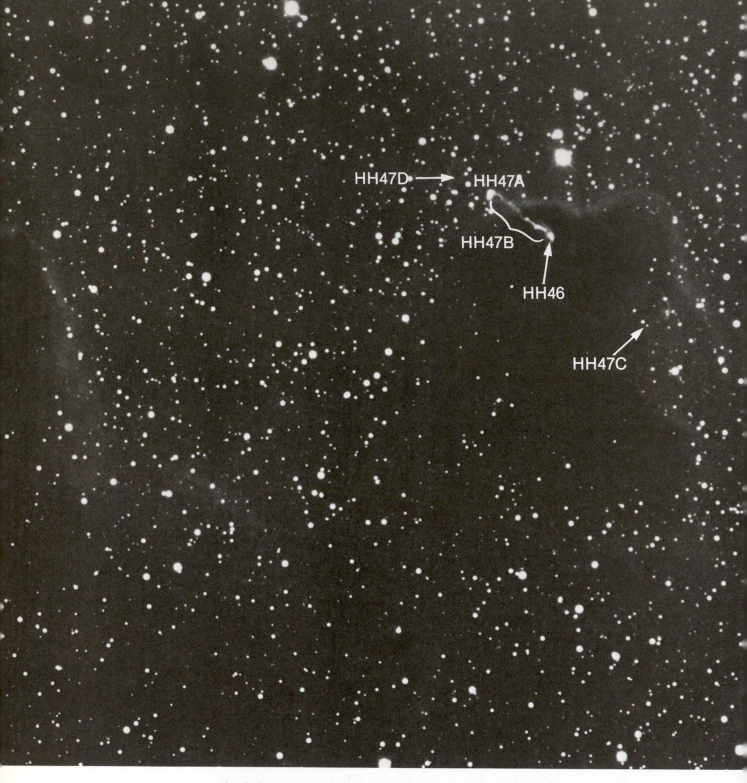

The dark regions are Bok globules, each about 3 arc min across, which are so opaque that they block our view of the stars behind. The one at right is ESO 210–6A; crossing its boundary at its top left are Herbig-Haro objects, the results of interactions of gas ejected from a young star with its surrounding material. HH47A is the knot just outside the boundary and HH46 is the knot just inside. The star that ejected them, visible only in the infrared, is near HH46. It has apparently ejected HH46 and HH47A&B in a line slanting toward us. Another H-H object, HH47C, is just outside the right (west) boundary of the Bok globule on the continuation of the line and has a velocity showing that it is receding from us. We are seeing a "bipolar ejection" from a young star of the type known as T Tauri. Bok took this "Valentine's night" (14 February 1978) photo at Cerro Tololo.

Young and Middle-Aged Stars

8

Aims: To study the formation of stars, their main-sequence lifetimes, and how they generate their energy

Even though individual stars shine for a relatively long time, they are not eternal. Stars are born out of gas and dust that may exist within a galaxy; they then begin to shine brightly on their own. Though we can observe only the outer layers of stars, we can deduce that the temperatures at their centers must be millions of kelvins. We can even deduce what it is deep down inside that makes the stars shine.

In this chapter we will discuss the birth of stars, then the processes that go on in a stellar interior during a star's life on the main sequence, and then begin the story of the evolution of stars when they finish this stage of their lives. The next three chapters will continue the story of what is called *stellar evolution*.

8.1 Stars in Formation

The process of star formation starts with a region of gas and dust. The dust—tiny solid particles—may have been given off from the outer atmospheres of giant stars. A region may become of slightly higher density than its surroundings, perhaps from a random fluctuation in density or because a "density wave" in our galaxy compressed it, as we shall discuss in Section 12.5c. Or a star may explode nearby—a "supernova"—sending out a shock wave that compresses gas and dust. In any case, once a minor density enhancement occurs, gravity keeps the gas and dust contracting. As they contract, energy is released, and it turns out that half of that energy heats the matter, causing it to give off an appreciable amount of radiation.

Such a not-quite-yet-formed star is called a "protostar" (from the prefix of Greek origin meaning "primitive"). The "track" of this protostar, the path its properties take when plotted on a Hertzsprung-Russell diagram, is shown in Fig. 8–1. At first, the protostar brightens, and the track bends upward and toward the left on the diagram. While this is occurring, the central part of the protostar continues to contract and the temperature rises. The higher temperature results in a higher pressure, which pushes outward more and more strongly. Eventually a point is reached where this outward force balances the inward force of gravity for the central region.

By this time, the dust has vaporized and the gas has become opaque, so that energy emitted from the central region does not escape directly. The outer layers continue to contract. Since the surface area is decreasing, the luminosity decreases and the track of the protostar begins to move downward on the H-R diagram. As the protostar continues to heat up, it also continues to move toward the left on the H-R diagram (that is, it gets hotter).

The gas and dust from which stars are forming is best observed in the infrared and radio regions of the spectrum. We shall have more to say about these regions when we discuss the Milky Way Galaxy in Chapter 13.

We can plot the position of a star on an H-R diagram for any particular instant of its life. Each of the pairs of luminosity and temperature corresponds to a point on the H-R diagram. Connecting the points representing the entire lifetime of the star gives an *evolutionary track*. A particular star, of course, is only at one point of its track at any given time. And note that an evolutionary track shows how the properties of a star evolve, not how or whether the star is physically moving in space.

151

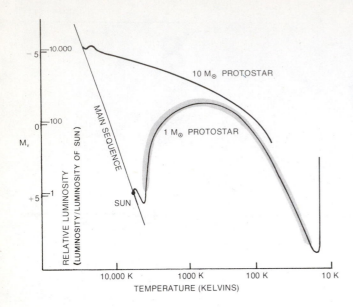

Figure 8–1 Evolutionary tracks for two protostars, one of 1 solar mass and the other of 10 solar masses. ($M_\odot$ is the symbol for the mass of the sun. By saying "a star of 10 solar masses," we mean simply a star that has 10 times the mass of the sun.) The evolution of the shaded portion of the track for the 1 solar-mass star is described in the text. These stages of its life last for about 50 million years. More massive stars whip through their protostar stages more rapidly; the corresponding stages for a 10 solar-mass star may last only 200,000 years. These very massive stars wind up being very luminous.

On the vertical axis, absolute visual magnitude is plotted to the left of the line. An alternative scale showing the star's luminosity relative to that of the sun is plotted to the right of the line.

Note that a given star does not move **along** the main sequence; it stays for a long time at essentially the same place on the H-R- diagram. The main sequence shows up only when we plot the properties of a lot of stars.

The time scale of this gravitational contraction depends on the mass of gas in the protostar. Very massive stars contract to approximately the size of our solar system in only ten thousand years or so. These massive objects become O and B stars, and are located at the top of the main sequence. They are sometimes found in groups, which are called *O and B associations*.

Less massive stars contract much more leisurely. A star of the same mass as the sun may take tens of millions of years to contract, and a less massive star may take hundreds of millions of years to pass through this stage.

Theoretical analysis shows that the dust surrounding the stellar embryo we call a protostar should absorb much of the radiation that the protostar emits. The radiation from the protostars should heat the dust to temperatures that produce primarily infrared radiation. Infrared astronomers have found many objects that are especially bright in the infrared but that have no known optical counterparts. The IRAS satellite, in orbit in 1983, has discovered so many of these that we now think that about one star forms each year in our Milky Way Galaxy. These objects seem to be located in regions where the presence of a lot of dust and gas and other young stars indicates that star formation might be going on.

In the visible part of the spectrum, several classes of stars that vary erratically are found. One of these classes, called *T Tauri stars*, includes stars of spectral types G, K, and M. Their visible radiation can vary by as much as several magnitudes. Astronomers work in the infrared to study the dust grains that surround T Tauri; observations have included some made with the Kuiper Airborne Observatory farther into the infrared than can be detected from the ground. Presumably, T Tauri stars have not quite settled down to a steady and reliable existence. They fall slightly above the main sequence of the H-R diagram, which is consistent with their being very young.

The optical spectra of T Tauri stars show strong emission lines, probably showing that they have chromospheres. Ultraviolet observations from IUE confirm this, and show spectral lines formed at temperatures as hot as 100,000 K. This may, in turn, indicate that a corona is present too, though no coronal lines are in the spectral region that IUE can study. In $\frac{1}{3}$ of the T Tauri stars it examined, the Einstein Observatory detected x-rays that may be from such coronas.

T Tauri stars are found in close proximity to each other; these groupings are known as *T associations*. The T Tauri stars are embedded in dark dust clouds. Also

Figure 8–2 *(A)* T Tauri itself is embedded in a Herbig-Haro object. The H-H object has changed in brightness considerably in the last century, first fading from view and then brightening considerably. This probably results from changes in the angle at which T Tauri illuminated the cloud. The H-H object has now been constant in brightness for decades. The image of a second H-H object is merged with the image of T Tauri on this plate. Herbig-Haro objects are named after George Herbig of the Lick Observatory and Guillermo Haro of the Mexican National Observatory.

The frame shown is about 1 arc min across. It is magnified by 100 times less than the high-resolution radio image below, all of which would fit into the white photographic image of T Tauri itself. The H-H object is on the inner rim of the nebula visible.

(B) A radio map of T Tauri and surrounding matter. Note that the outer contours are not circular. The map was made at a wavelength of 6 cm with the VLA, the Very Large Array of radio telescopes in New Mexico that allows such high-resolution mapping. A subsequent surprising discovery was that T Tauri is a double star with components too close together to resolve. The companion is hidden by dust, and can be observed and distinguished only in the infrared, though it is also a radio source. (The infrared observations were made at Mauna Kea by Mel Dyck, Ted Simon, and Ben Zuckerman.)

(C) With the high resolution of the VLA, operating at the radio wavelength of 6 cm, the T Tauri radio source is resolved into these components separated by 0.54 arc sec. The weaker source is T Tau itself and the lower source is the infrared companion. (Courtesy of P. R. Schwartz, Theodore Simon, B. Zuckerman, and R. R. Howell)

A

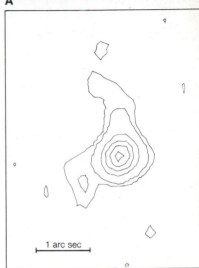

1 arc sec

B

in these clouds are bright nebulous regions of gas and dust called *Herbig-Haro objects*. T Tauri itself, the prototype of its class of variable stars, is near a typical Herbig-Haro object (Fig. 8–2).

The Herbig-Haro objects are the results of gas ejected from very young T Tauri stars. The supersonic jet-like gas flows interact with surrounding material to produce flowing shock waves, which appear as Herbig-Haro objects.

Radio maps of T Tauri (Fig. 8–2*B*) made with the VLA (Section 14.6) show that T Tauri's stellar wind is not spherical. A strong magnetic field is probably present, which could explain why observations of some T Tauri stars seem to show that material has been sent out in beams. Similarly, the H-H objects are often aligned as though they were ejected to opposite sides of the T Tauri star (Fig. 8–3). From time

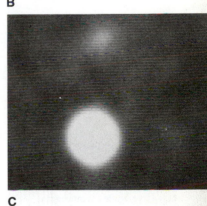

C

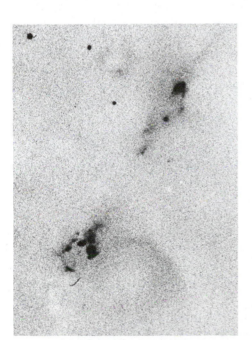

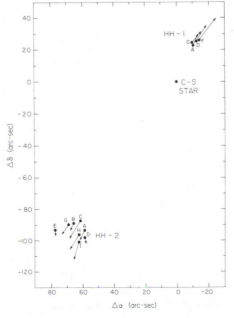

Figure 8–3 An infrared object known as the C-S star (after its discoverers, M. Cohen and R. D. Schwartz) has apparently ejected the knots of gas (Herbig-Haro objects) known as HH1 and HH2 in opposite directions. The arrows show their projected motion after 100 years, based on observed proper motions.

**H¹ NUCLEUS
PROTON**

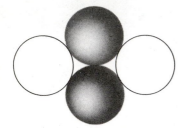

**He⁴ NUCLEUS
ALPHA PARTICLE**

Figure 8–4 The nucleus of hydrogen's most common form is a single proton while the nucleus of helium's most common form consists of two protons and two neutrons.

In the interiors of stars, we are dealing with nuclei instead of the atoms we have discussed in stellar atmospheres, because the high temperature strips the electrons off the nuclei. The electrons are mixed in with the nuclei through the center of the star; there can be no large imbalance of positive and negative charges, or else strong repulsive forces would arise inside the star.

to time later in this book, we will run across other places where astronomers now think that magnetic fields may be beaming matter.

It came as a surprise in 1982 when T Tauri turned out to have a companion observable only in the infrared. The companion, whose temperature is only 800 K, is the radio source. One reasonable interpretation is that the companion is embedded inside and taking up matter from a disk of gas around T Tauri. The companion might even be too small to be a star, which would shine on its own by the fusion process we discuss in the next section. If so, it would be more like a giant planet in formation. It is, in any case, 30 times farther from T Tauri than Jupiter is from our sun.

8.2 Stellar Energy Generation

All the heat energy in stars that are still contracting toward the main sequence results from the gravitational contraction itself. If this were the only source of energy, though, stars would not shine for very long on an astronomical time scale—only about 30 million years. Yet we know that even rocks on earth are older than that, since rocks over 4 billion years old have been found. We must find some other source of energy to hold the stars up against their own gravitational pull.

The gas in the protostar will continue to heat up until the central portions become hot enough for *nuclear fusion* to take place. Using this process, which we will soon discuss in detail, the star can generate enough energy inside itself to support it during its entire lifetime on the main sequence. The energy makes the particles in the star move around rapidly. For short, we say that the particles have a ''high temperature.'' The particles exert a *thermal pressure* pushing outward, providing a force that balances gravity's inward pull.

The basic fusion processes in most stars fuse four hydrogen nuclei into one helium nucleus, just as hydrogen atoms are combined into helium in a hydrogen bomb here on earth. In the process, tremendous amounts of energy are released.

A hydrogen nucleus is but a single proton. A helium nucleus is more complex. It consists of two protons and two neutrons (Fig. 8–4). The mass of the helium nucleus that is the final product of the fusion process is slightly less than the sum of the masses of the four hydrogen nuclei that went into it. A small amount of the mass ''disappears'' in the process: 0.7 per cent of the mass of the four hydrogen nuclei.

The mass does not really simply disappear, but is rather converted into energy according to Albert Einstein's famous formula $E = mc^2$. Now c, the speed of light, is a large number, and c^2 is even larger. Thus even though m is only a small fraction of the original mass, the amount of energy released is prodigious. The loss of only .007 of the central part of the sun, for example, is enough to allow the sun to radiate as much as it does at its present rate for a period of at least ten billion (10^{10}) years. This fact, not realized until 1920 and worked out in more detail in the 1930's, solved the long-standing problem of where the sun and the other stars got their energy.

All the main-sequence stars are approximately 90 per cent hydrogen (that is, 90 per cent of the atoms are hydrogen), so there is lots of raw material to stoke the nuclear ''fires.'' We speak colloquially of ''nuclear burning,'' although, of course, the processes are quite different from the chemical processes that are involved in the ''burning'' of logs or of autumn leaves. In order to be able to discuss these processes, we must first discuss the general structure of nuclei and atoms.

8.3 Atoms

An atom consists of a small *nucleus* surrounded by *electrons*. Most of the mass of the atom is in the nucleus, which takes up a very small volume in the center of the atom. The effective size of the atom, the chemical interactions of atoms to form molecules, and the nature of spectra are determined by the electrons.

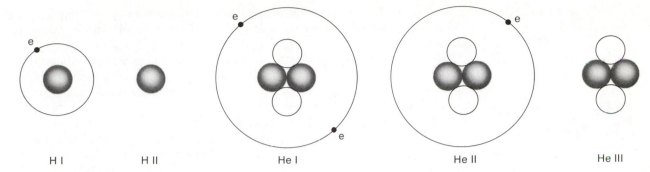

H I   H II   He I   He II   He III

Figure 8–5 Hydrogen and helium ions. The sizes of the nuclei are greatly exaggerated with respect to the sizes of the orbits of the electrons.

The nuclear particles with which we need be most familiar are the *proton* and *neutron*. Both these particles have nearly the same mass, 1836 times greater than the mass of an electron, though still tiny (Appendix 2). The neutron has no electric charge and the proton has one unit of positive electric charge. The electrons, which surround the nucleus, have one unit each of negative electric charge. When an atom loses an electron, it has a net positive charge of 1 unit for each electron lost. The atom is now a form of *ion* (Fig. 8–5 and Section 4.3b). Astronomers use Roman numerals to show the stage of ionization: I for the neutral state, II for the state in which one electron is lost, etc. Thus Fe I is neutral iron, Fe II is the state of Fe^+ ions, Fe III is the state of Fe^{++} ions, etc. (The Roman numeral is one higher than the number of electrons removed.)

The number of protons in the nucleus determines the quota of electrons that the neutral state of the atom must have, since this number determines the charge of the nucleus. Each *element* (sometimes called "chemical element") is defined by the specific number of protons in its nucleus. The element with one proton is hydrogen, that with two protons is helium, that with three protons is lithium, and so on.

Though a given element always has the same number of protons in a nucleus, it can have several different numbers of neutrons. (The number is always somewhere between 1 and 2 times the number of protons. Hydrogen, which need have no neutrons, and helium are the only exceptions to this rule.) The possible forms of an element having different numbers of neutrons are called *isotopes*.

For example, the nucleus of ordinary hydrogen contains one proton and no neutrons. An isotope of hydrogen (Fig. 8–6) called deuterium (and sometimes "heavy hydrogen") has one proton and one neutron. Another isotope of hydrogen called tritium has one proton and two neutrons.

Most isotopes do not have specific names, and we keep track of the numbers of protons and neutrons with a system of superscripts and subscripts. The subscript **before** the symbol denoting the element is the number of protons (called the *atomic number*), and a superscript **following** the symbol is the total number of protons and neutrons together (called the *mass number*). For example, $_1H^2$ is deuterium, since deuterium has one proton, which gives the subscript, and an atomic mass of 2, which gives the superscript. Deuterium has atomic number equal 1 and mass number equal

Many other nuclear particles have been detected; some only show up for tiny fractions of a second under extreme conditions in giant particle accelerators ("atom smashers"). The nuclear particles are apparently made up of even more fundamental particles called *quarks*.

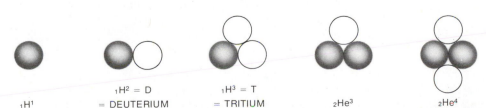

$_1H^1$ $_1H^2 = D$ $_1H^3 = T$ $_2He^3$ $_2He^4$
                     = DEUTERIUM   = TRITIUM

Figure 8–6 Isotopes of hydrogen and helium. $_1H^2$ (deuterium) and $_1H^3$ (tritium) are much rarer than the normal isotope, $_1H^1$. $_2He^3$ is much rarer than $_2He^4$.

Figure 8–7 Enrico Fermi, Werner Heisenberg, and Wolfgang Pauli enjoy an outing on Lake Como in 1927. Pauli predicted on theoretical grounds that the neutrino would exist. Fermi further developed the theory and gave the particle its name. Heisenberg, one of the originators of quantum mechanics, is most widely known for his *uncertainty principle*. The principle shows that there are fundamental limits in how accurately we can pinpoint the position or momentum of a nuclear particle (or other object) because the very act of observing it disturbs it.

Figure 8–8 The proton-proton chain; e^+ stands for a positron, ν (nu) is a neutrino, and γ (gamma) is radiation at a very short wavelength.

In the first stage, two nuclei of ordinary hydrogen fuse to become a deuterium (heavy hydrogen) nucleus, a positron (the equivalent of an electron, but with a positive charge), and a neutrino. The neutrino immediately escapes from the star, but the positron soon collides with an electron. They annihilate each other, forming gamma rays. (A positron is an anti-electron, an example of antimatter; whenever a particle and its antiparticle meet, they annihilate each other.)

Next, the deuterium nucleus fuses with yet another nucleus of ordinary hydrogen to become an isotope of helium with two protons and one neutron. More gamma rays are released.

Finally, two of these helium isotopes fuse to make one nucleus of ordinary helium plus two nuclei of ordinary hydrogen. The protons are numbered to help you keep track of them.

2. Similarly, $_{92}U^{238}$ is an isotope of uranium with 92 protons (atomic number = 92) and mass number of 238, which is divided into 92 protons and $238 - 92 = 146$ neutrons.

Each element has only certain isotopes. For example, most naturally occurring helium is in the form $_2He^4$, with a lesser amount as $_2He^3$. Sometimes an isotope is not stable, in that after a time it will spontaneously change into another isotope or element; we say that such an isotope is *radioactive*.

During certain types of radioactive decay, a particle called a *neutrino* is given off (Fig. 8–7). A neutrino is a neutral particle (its name comes from the Italian for "little neutral one"). It has long been thought that neutrinos have no "rest mass," the mass they would have if they were at rest. There is now some experimental evidence that a neutrino may in fact have some small amount of rest mass, about 1/100,000,000 that of a proton. If true, then most of the mass of the universe would be in the form of neutrinos (see Section 17.2). But the results have not been confirmed, and many scientists are skeptical about them; we will just have to wait to see.

Neutrinos have a very useful property for the purpose of astronomy: they do not interact very much with matter. Thus when a neutrino is formed deep inside a star, it can usually escape to the outside without interacting with any of the matter in the star. This means that the neutrino heads right out of the star, without bumping into any of the matter in the star along the way. Electromagnetic radiation, on the other hand, does not escape from inside a star so easily. A photon of radiation can travel only about 1 cm in a stellar interior before it is absorbed, and it is millions of years before a photon zigs and zags its way to the surface.

The elusiveness of the neutrino makes it a valuable messenger—indeed, the only possible direct messenger—carrying news of the conditions inside the sun at the present time, but it also makes it very difficult for us to detect on earth. Later in this chapter we shall discuss the experiment that is being performed to study solar neutrinos.

8.4 Stellar Energy Cycles

Several chains of reactions have been proposed to account for the fusion of four hydrogen atoms into a single helium atom. Hans Bethe, now at Cornell University, suggested some of these procedures during the 1930's. The different possible chains that have been proposed are important at different temperatures, so chains that are

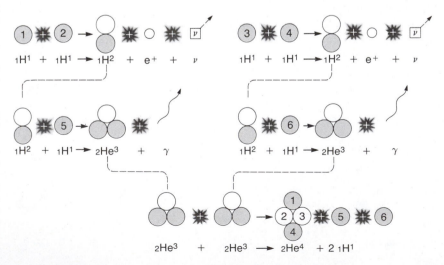

dominant in the centers of very hot stars may be different chains from the ones that are dominant in the centers of cooler stars.

When the center of a star is at a temperature less than 15×10^6 K, the *proton-proton* chain (Fig. 8–8) dominates. In the chain, we put in six hydrogens one at a time, and wind up with one helium plus two hydrogens, a net transformation of four hydrogens into one helium. But the six protons contained more mass than do the final single helium plus two protons. The small fraction of mass that disappears in the process is converted into an amount of energy that we can calculate with the formula $E = mc^2$.

For stellar interiors hotter than that of the sun, the *carbon-nitrogen (CN) cycle* (Fig. 8–9) dominates. The CN cycle begins with the fusion of a hydrogen nucleus with a carbon nucleus. After many steps, and the insertion of four hydrogen nuclei, we are left with one helium nucleus plus a carbon nucleus. Thus as much carbon remains at the end as there was at the beginning, and the carbon can start the cycle again. Again, four hydrogens have been converted into one helium, 0.007 of the

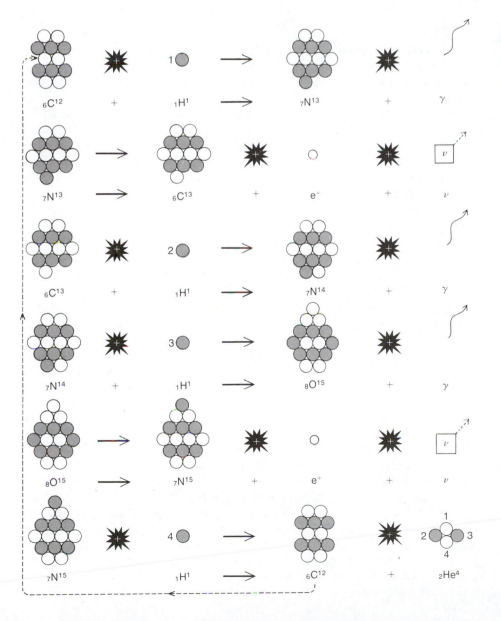

Figure 8–9 The carbon-nitrogen cycle, also called the *carbon cycle.* The 4 hydrogen atoms are numbered. Note that the carbon is left over at the end, ready to enter into another cycle.

$_6C^{12}$ + $_1H^1$ ⟶ $_7N^{13}$ + γ

$_7N^{13}$ ⟶ $_6C^{13}$ + e^+ + ν

$_6C^{13}$ + $_1H^1$ ⟶ $_7N^{14}$ + γ

$_7N^{14}$ + $_1H^1$ ⟶ $_8O^{15}$ + γ

$_8O^{15}$ ⟶ $_7N^{15}$ + e^+ + ν

$_7N^{15}$ + $_1H^1$ ⟶ $_6C^{12}$ + $_2He^4$

Figure 8–10 The triple-alpha process, which takes place only at temperatures above about 10^8 K. Beryllium, $_4\text{Be}^8$, is but an intermediate step.

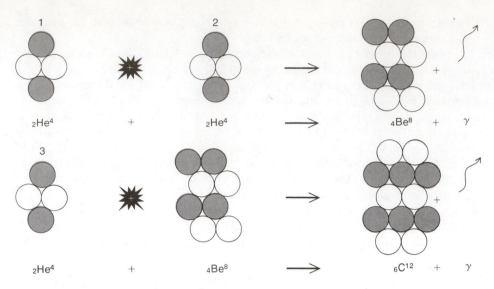

The Greek letters derive from a former confusion of radiation and particles. We know now that α particles, formerly called α rays, are helium nuclei; β particles, formerly called β rays, are electrons, and γ rays, as we have seen, are radiation of short wavelength.

mass has been transformed, and an equivalent amount of energy has been released according to $E = mc^2$. Later, it was found that alternative cycles involving different oxygen isotopes sometimes occur. These processes are often called the *carbon-nitrogen-oxygen (CNO) tri-cycle.*

Stars with even higher interior temperatures, above 10^8 K, fuse helium nuclei to make carbon nuclei. The nucleus of a helium atom is called an "alpha particle" for historical reasons. Since three helium nuclei ($_2\text{He}^4$) go into making a single carbon nucleus ($_6\text{C}^{12}$), the procedure is known as the *triple-alpha process* (Fig. 8–10). A series of other processes can build still heavier elements inside stars. This process is called *nucleosynthesis*.

The theory of nucleosynthesis can account for the abundances we observe of the elements heavier than helium. Currently, we think that the synthesis of isotopes of hydrogen and helium took place in the first few minutes after the origin of the universe (Section 17.1), and that the heavier elements were formed, along with additional helium, in stars or in supernova explosions (Section 10.2). William A. Fowler of Caltech shared the 1983 Nobel Prize in Physics for his work on nucleosynthesis, including the measurements of the rates of the nuclear reactions that make the stars shine.

8.5 The Stellar Prime of Life

Now that we have discussed basic nuclear processes, let us return to an astronomical situation. We last discussed a protostar in a collapsing phase, with its internal temperature rapidly rising.

One of the most common definitions of temperature describes the temperature as a measure of the velocities of individual atoms or other particles. Since this type of temperature depends on kinetics (motions) of particles, it is called the *kinetic temperature*. A higher kinetic temperature (we shall call it simply "temperature" from now on) corresponds to higher particle velocities.

For a collapsing protostar, the energy from the gravitational collapse goes into giving the individual particles greater velocities; that is, the temperature rises. For nuclear fusion to begin, atomic nuclei must get close enough to each other so that the force that holds nuclei together, the *strong nuclear force* (Section 17.3), can play its part. But all nuclei have positive charges, because they are composed of protons (which bear positive charges) and neutrons (which are neutral). The positive charges on any two nuclei cause an electrical repulsion between them, and this force tends to prevent fusion from taking place.

However, at the high temperatures typical of a stellar interior, some nuclei have enough energy to overcome this electrical repulsion and to come sufficiently close to each other for the strong nuclear force to take over. The electrical repulsion that must be overcome is the reason why hydrogen nuclei, which have net positive charges of 1, will fuse at lower temperatures than will helium nuclei, which have net positive charges of 2.

Once nuclear fusion begins, the thermal pressure of the star's gas (the pressure resulting from temperature) provides a force that pushes outward strongly enough to balance gravity's inward pull. In the center of a star, the fusion process is self-regulating. If the nuclear energy production rate increases, then an excess pressure is generated that would tend to make the star grow larger. However, this expansion in turn would cool down the gas and slow down the rate of nuclear fusion. Thus the star finds a temperature and size at which it can remain stable for a very long time. This balance between thermal pressure pushing out and gravity pushing in character-izes the main-sequence phase of a stellar lifetime. These fusion processes are well regulated in stars. When we learn how to control fusion in power-generating stations on earth, which currently seems decades off, our energy crisis will be over.

The more mass a star has, the hotter its core becomes before it generates enough pressure to counteract gravity. The hotter core leads to a higher surface luminosity, explaining the mass-luminosity relation (Section 6.2). Thus more massive stars use their nuclear fuel at a much higher rate than less massive stars, and even though the more massive stars have more fuel to burn, they go through it relatively quickly. The next three chapters continue the story of stellar evolution by discussing the fate of stars when they have used up their hydrogen.

8.6 The Solar Neutrino Experiment

Astronomers can apply the equations that govern matter and energy in a star, and model the star's interior and evolution in a computer. Though the resulting model can look quite nice, nonetheless it would be good to confirm it observationally. Hap-pily, the models are consistent with the conclusions on stellar evolution that one can make from studying different types of Hertzsprung-Russell diagrams. Still, it would be nice to observe a stellar interior directly.

Since the stellar interior lies under opaque layers of gases, we cannot observe directly any electromagnetic radiation it might emit. Only neutrinos escape directly from a stellar interior. Neutrinos interact so weakly with matter that they are hardly affected by the presence of the rest of the solar mass. Once formed, they zip right out into space at the speed of light. Raymond Davis, Jr., of the Brookhaven National Laboratory has spent many years studying the neutrinos formed in the solar interior.

Neutrinos formed in the sun's normal proton-proton chain do not have enough energy for Davis's apparatus to detect them. So Davis's experiment can detect only the neutrinos that occur in branches of the chain that are followed less than 1 per cent of the time. Thus theoretical calculations of how often these branches are fol-lowed are an important part of the study.

How do we detect the neutrinos? Neutrinos, after all, pass through the earth and sun, barely affected by their mass. At this instant, neutrinos are passing through your body. Davis makes use of the fact that very occasionally a neutrino will interact with the nucleus of an atom of chlorine, and transform it into an isotope of the noble gas argon.

This transformation takes place very rarely, so Davis needs a large number of chlorine atoms. He found it best to do this by filling a large tank with liquid cleaning

Figure 8–11 The neutrino tele-scope, deep underground in the Homestake Gold Mine in Lead, South Dakota, consists mainly of a tank containing 400,000 liters of perchloroethylene. The tank is now surrounded by water.

Figure 8–12 Davis tried out the water that helps the kilometer of earth overhead shield his neu-trino telescope from particles other than neutrinos.

Davis needs 50 tons of gallium, roughly equivalent to the entire world's production for a whole year. Nonetheless, world gallium production is increasing since this element is used in many solid-state devices. The gallium, of course, would not be damaged in the course of the experiment and could be used for other purposes later on. It might even increase in value while being used.

fluid, C_2Cl_4, where the subscripts represent the number of atoms of carbon and chlo-rine in the molecule, which is called perchloroethylene. One fourth of the chlorine is the Cl^{37} isotope, which is able to interact with a neutrino. Davis now has a large tank containing 400,000 liters (100,000 gallons) of this cleaning fluid (Figs. 8–11 and 8–12).

Even with this huge tankful of chlorine atoms, calculations show that Davis should expect only about one neutrino-chlorine interaction every day. This experi-ment is surely one of the most difficult ever attempted. When an interaction occurs, a radioactive argon atom is formed. Any such argon atoms can be removed from the tank by standard chemical techniques involving bubbling helium gas through the fluid. Then the radioactivity of the few resulting argon atoms can be measured, also by standard means.

It is fair to say that Davis's results have astounded the scientific community, which had confidently expected to hear reports of a number of neutrino interactions in agreement with theoretical predictions. But Davis is finding fewer neutrinos than expected. His measurement is at the minimum level that standard theories of nuclear fusion in stellar interiors are able to predict.

Several questions immediately come to mind. The first deals with Davis's ap-paratus, and whether there are some experimental effects that could explain the re-sults in some normal manner. Davis has carried out a series of careful checks of his apparatus, and most astronomers and chemists are convinced that the experimental setup is not the cause of the problem.

Next, perhaps we do not understand the basic physics of nuclear reactions or of neutrinos as well as we thought. Davis's apparatus can detect only the form of neu-trinos that the sun emits. Perhaps a neutrino changes in type during the 8 minutes it takes to travel at the speed of light from the sun to the earth. Some evidence was advanced for this possibility in 1980, but has not been confirmed.

Another interesting possibility is that our understanding of stellar interiors is not satisfactory. If, for example, stars are cooler inside than we have predicted, fewer neutrinos would be produced.

Another possibility is that the sun is simply not generating neutrinos at the mo-ment! The neutrino experiment tells us what the sun is doing now, or really what it did eight minutes earlier. On the other hand, the sun shines because energy generated inside it 10 million years earlier has just reached the surface. Perhaps ice ages in the past resulted from other periods long ago when the sun was turned "off"; perhaps the current episode will lead to ice ages hundreds of thousands of years from now.

Though the discrepancy between theory and observations seems serious—the theoretical value is more than three times the observational value—both values have inherent uncertainties. The uncertainties involved, especially in the theoretical value, must be considerably reduced before we can definitely even say that there is a major problem.

Davis plans to extend his search by using gallium, a material more sensitive to neutrinos than chlorine (Fig. 8–13). Gallium is sensitive to neutrinos of much lower energy than those with which chlorine interacts. It would detect essentially all solar neutrinos, instead of only the high-energy one the current experiment can detect. Another experiment, using lithium, is also being developed.

8.7 Dying Stars

We shall devote the next three chapters to the various end stages of stellar evolution. The mass of the star determines its fate. First we shall discuss the less massive stars, such as the sun. In Chapter 9, we shall see how such stars swell in

Figure 8–13 A pilot gallium experiment at Brookhaven, with a tank containing 1.3 tons of gallium. Gallium turns into germanium from an interaction with a neutrino, and Davis is now working on chemical means to separate the few germanium atoms from the gallium, which is liquid at room temperature.

size to become giants, possibly become planetary nebulae, and then end their lives as white dwarfs.

More massive stars come to more explosive ends. In Chapter 10, we shall see how some stars are blown to smithereens, and how strange objects called "pulsars" are the remnants. Other remnants appear to be part of x-ray–emitting binary systems.

In Chapter 11, we shall discuss the strangest kind of stellar death of all. The most massive stars may become "black holes," and effectively disappear from view.

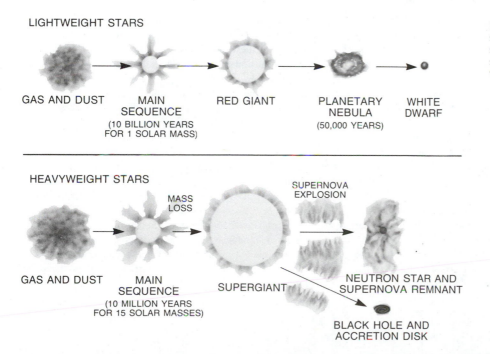

LIGHTWEIGHT STARS

GAS AND DUST   MAIN SEQUENCE (10 BILLION YEARS FOR 1 SOLAR MASS)   RED GIANT   PLANETARY NEBULA (50,000 YEARS)   WHITE DWARF

HEAVYWEIGHT STARS

GAS AND DUST   MAIN SEQUENCE (10 MILLION YEARS FOR 15 SOLAR MASSES)   MASS LOSS   SUPERGIANT   SUPERNOVA EXPLOSION   NEUTRON STAR AND SUPERNOVA REMNANT   BLACK HOLE AND ACCRETION DISK

Figure 8–14 A summary of the stages of stellar evolution for stars of different masses; these stages will be described in the following three chapters.

Summary and Outline

Protostars (Section 8.1)
    Their evolutionary tracks on the H-R diagram bring them
        in from the right; they form when a cloud of gas and
        dust is compressed; they can be observed in the in-
        frared
T Tauri stars (Section 8.1)
    Irregular variations in magnitude; may not have reached the
        main sequence; near H-H objects
Stellar energy generation (Sections 8.2, 8.3, and 8.4)
    Nuclear fusion provides the energy for stars to shine.
    Proton-proton chain dominates in cooler stars, including
        the sun.

Carbon cycle is for hotter stars
Triple-alpha process takes place at high temperatures
Stellar nucleosynthesis accounts for most of the elements
The main sequence: balance of pressure and gravity (Sec-
    tion 8.5)
A test of the theory: the neutrino experiment (Section 8.6)
    The theoretical prediction is 3 times the observational re-
        sult; possible resolutions; uncertainty in the theoretical
        result must be lowered; further, more definitive exper-
        iments planned
Mass is the determining factor in evolution (Section 8.7)

Key Words

featherweight stars, lightweight stars, middleweight stars, heavyweight stars, stellar evolution, evolutionary track, O and B associations, T Tauri stars, T associations, Herbig-Haro objects, nuclear fusion, thermal pressure, $E = mc^2$, nucleus, electrons, proton, neutron, ion, quarks, element, uncertainty principle, isotopes, atomic number, mass number, radioactive, neutrino, proton-proton chain, carbon cycle, carbon-nitrogen cycle, CNO cycle, triple-alpha process, nucleosynthesis, kinetic temperature, strong nuclear force

Questions

1. Since individual stars can live for billions of years, how can observations taken at the current time tell us about stellar evolution?

2. What is the source of energy in a protostar? At what point does a protostar become a star?

3. What is the *evolutionary track* of a star?

4. Describe the evolutionary track of the sun during the 10 billion years it is on the main sequence.

5. Arrange the following in order of development: OB associations; T Tauri stars; dark clouds; sun; pulsars.

†6. Give the number of protons, the number of neutrons, and the number of electrons in: ordinary hydrogen ($_1H^1$), lithium ($_3Li^6$), iron ($_{26}Fe^{56}$).

7. (a) If you remove one neutron from helium, the remainder is what element? (b) Now remove one proton. What is left? (c) Why is He IV not observed?

8. (a) Explain why nuclear fusion takes place only in the centers of stars rather than on their surfaces as well. (b) What is the major fusion process that takes place in the sun?

9. If you didn't know about nuclear energy, what is one possible energy source you might suggest for stars? What is wrong with these alternative explanations?

10. What forces are in balance for a star to be on the main sequence?

†11. Use the speed of light and distance between the earth and sun given in Appendix 2 to verify that it takes neutrinos 8 minutes to reach the earth from the sun.

12. What does it mean for the temperature of a gas to be higher?

†13. (a) If all the hydrogen in the sun were converted to helium, what fraction of the solar mass would be lost? (b) How many times the mass of the earth would that be? (Consult Appendix 2.)

14. (a) How does the temperature in a stellar core determine which nuclear reactions will take place? (b) Why do more massive stars have shorter main-sequence lifetimes?

15. In what form is energy carried away in the proton-proton chain?

---

†This indicates a question requiring a numerical solution.

16. In the proton-proton chain, the products of some reactions serve as input for the next and therefore don't show up in the final result. Identify these intermediate products. What are the **net** input and output of the proton-proton chain?

17. What do you think would happen if nuclear reactions in the sun stopped? How long would it be before we noticed?

18. Why do neutrinos give us different information about the sun than does light?

19. Why are the results of the solar neutrino experiment so important?

20. Why will the gallium experiment for solar neutrinos be superior to the chlorine experiment?

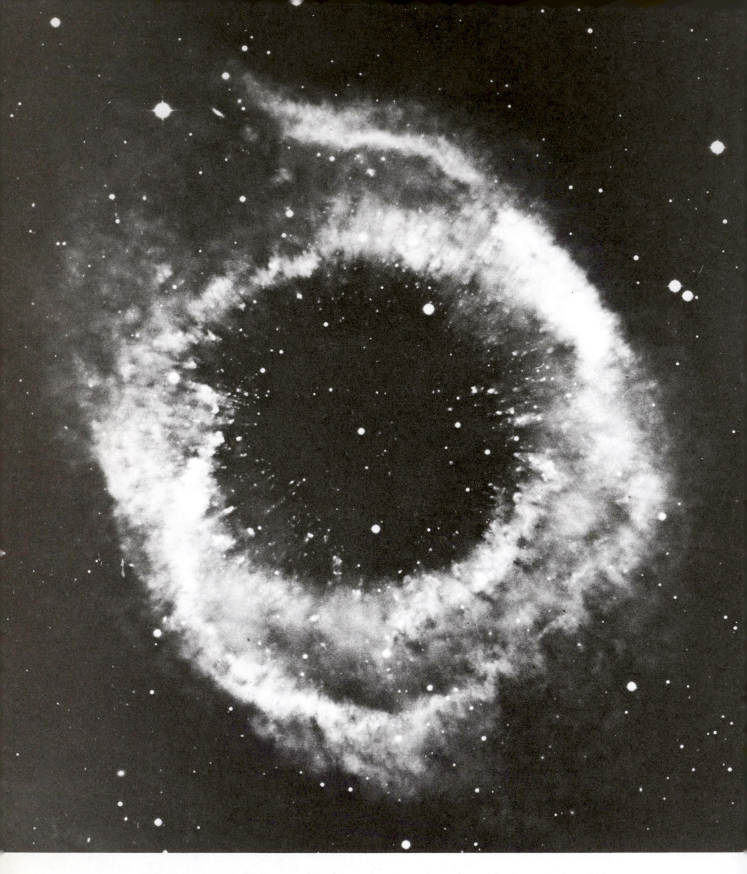

The Helix Nebula, NGC 7293, the nearest planetary nebula to us at a distance of 400 light years. Long exposures show that its diameter in the sky is about the same as the moon's. (Photo with the Anglo-Australian 4-m telescope)

The Death of Stars Like the Sun

9

Aims: To understand what happens to stars of up to about 4 solar masses when they have finished their time on the main sequence of the H-R diagram

Let us remember that a star is a continual battleground between gravity pulling inward and pressure pushing outward. In main-sequence stars, thermal gas pressure resulting from energy provided by nuclear fusion balances gravity. Here we will see what happens when fusion stops and so no longer provides that pressure.

The sun is just an average star, in that it falls in the middle of the main sequence. But the majority of stars have less mass than the sun. So when we discuss the end of the main-sequence lifetimes of low-mass stars, including not only the sun but also all stars containing up to a few times its mass, we are discussing the future of most of the stars in the universe. In this chapter we will discuss the late (post–main-sequence) stages of evolution of all stars that, when they are on the main sequence, contain up to about 4 solar masses. We shall call them *lightweight stars*. Let us consider a star like the sun, in particular, remembering that all lightweight stars go through similar stages but at different rates.

9.1 Red Giants

During the main-sequence phase of lightweight stars, as with all stars, hydrogen in the core gradually fuses into helium. By about 10 billion (10^{10}) years after a one-solar-mass star first reaches the main sequence, no hydrogen is left in its core, which is thus composed almost entirely of helium. Hydrogen is still undergoing fusion in a shell around the core.

Since no fusion is taking place in the core, no ongoing nuclear process is replacing the heat that flows out of this hot central region. The core no longer has enough pressure to hold up both itself and the overlying layers against the inward force of gravity. As a result, the core then begins to contract under the force of gravity (we say it contracts **gravitationally**). This gravitational contraction not only replaces the heat lost by the core, but also in fact heats the core up further. Thus, paradoxically, soon after the hydrogen burning in the core stops, the core becomes hotter than it was before because of the gravitational contraction.

As the core becomes hotter, the hydrogen-burning shell around the core becomes hotter too, and the nuclear reactions proceed at a higher rate. A very important phenomenon then occurs: part of the increasing energy production goes into expand-

Half the energy from gravitational contraction always goes into kinetic motion in the interior. An example of such increased kinetic motion is the rise in temperature that we call heat. This is an important theorem, the ''virial theorem,'' which we met before in Section 8.1 discussing the contraction of protostars.

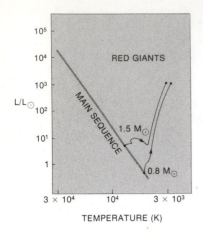

TEMPERATURE (K)

Figure 9–1 An H-R diagram, showing the evolutionary tracks of 0.8 and 1.5 solar mass stars, as the stars evolve from the main sequence to become red giants. For each star, the first dot represents the point where hydrogen burning starts and the second dot the point where helium burning starts. The vertical axis is the luminosity of the star, in units of the sun's luminosity; the horizontal axis is the temperature of the star's surface.

ing the outer parts of the star. There isn't enough energy to go around at the same rate in this expanded surface area, and the surface temperature of the star decreases.

In sum, as soon as a main-sequence star forms a substantial core of helium, the outer layers grow slowly larger and redder. The star winds up giving off more energy (because the nuclear reaction rates are increasing). The star's total luminosity, which is the emission per unit area times the total area, is increased. The luminosity increases even though each bit of the surface is less luminous than before (as a result of its decreasing temperature) because the total surface area that is emitting increases rapidly.

The process proceeds at an ever accelerating pace, simply because the hotter the core gets, the faster heat flows from it and the more rapidly it contracts and heats up further. The hydrogen shell gives off more and more energy. The layers outside this hydrogen-burning shell continue to expand.

The sun is presently still at the earlier, main-sequence stage, where changes are stately and slow. But in a few billion years, when the hydrogen in its core (about 10 per cent of the hydrogen in the whole sun) is exhausted, the time will come for the sun to brighten and redden faster and faster until it eventually swells and engulfs Mercury and Venus. At this point, the sun will no longer be a dwarf but will rather be a *red giant* (Fig. 9–1). Its surface will be so close to earth (Fig. 9–2) it will sear and char whatever is then here.

While the outer layers are expanding, the inner layers continue to heat up, and eventually reach 10^8 K. At this point, the triple-alpha process (Section 8.4) begins for all but the least massive lightweight stars, as groups of three helium nuclei (alpha particles) fuse into single carbon nuclei. For stars about the mass of the sun, the onset of the triple-alpha process happens rapidly. The development of this *helium flash* produces a very large amount of energy in the core, but only for a few years. This input of energy at the center reverses the evolution that took place just prior to the helium flash: now the core expands and the outer portion of the star (which is no longer receiving as much energy from the core as it was before the helium flash) contracts. As the star adjusts to this situation it becomes smaller and less luminous, moving back down and to the left on the H-R diagram to the horizontal branch. Helium now burns steadily and over a million years the helium in the core is completely converted to carbon. The star is now stabilized, burning hydrogen and helium in separate shells; its core is made of carbon.

Computer models show that different amounts of different molecules form in red giants depending on the relative amounts of carbon and oxygen. But red giants of the same temperature have three spectral types (M, S, and C) because different molecules dominate their spectra. This fact verifies that red giants have different relative amounts of carbon, presumably because they are making new carbon.

9.2 Planetary Nebulae

When the carbon core forms in a red giant, it contracts and heats up as did the helium core before it. This drives up the rate at which hydrogen is being converted to helium in the shell surrounding the core. The star's parameters then fit a point further to the upper right on the H-R diagram; the star expands and becomes a red giant again. This time, however, the swelling and cooling continue until the outer layers grow sufficiently cool that the nuclei and electrons combine to form neutral atoms. Each atom must release some energy when the previously free electrons recombine with the nuclei, since free electrons have more energy than bound ones.

Since the temperature is on the borderline for this recombination to occur, the energy supply is unsteady. The star begins pulsating and ejects a small fraction of its outer layers with each pulse. The pulsations become unstable. In a very short time (perhaps 1000 years) the outer layers have been entirely ejected. On this timescale, the ejected gas has moved several hundred astronomical units away from the star, and has spread into a shell thin enough to be transparent. A typical temperature for such a shell is 10,000 K. We see it shining because it is ionized by ultraviolet radiation from the central star.

We know of a thousand such objects in our galaxy that can be explained by the above theoretical picture, and there may be 10,000 more. They were named *planetary nebulae* because two hundred years ago, when they were discovered, they appeared similar to the planet Uranus when viewed in a small telescope. Both the planet and "planetary nebulae" appeared as small, greenish disks. The nebulae's greenish color is caused by the presence of certain strong emission lines of multiply (pronounced "mul-ti-plee") ionized oxygen (that is, oxygen that has lost more than one electron) and other elements. We can determine the chemical composition of the nebula by studying these emission lines.

An alternative picture also explains how planetary nebulae may arise. The expanding outer layers of the red giant form a stellar wind, an analogy to the solar wind. As the core is bared, the wind speed increases; the inner, faster moving material then piles up against the outer, slower moving material. The dense shell of gas that results from this snowplow effect, ionized by ultraviolet radiation from the central star, is the planetary nebula that we observe.

Planetary nebulae are exceedingly beautiful objects (Figs. 9–3 and 9–4, and Color Plates 49, 50, and 51). These nebulae are actually semitransparent shells of gas. When we look at their edges, we are looking obliquely through the shells of gas, and there is enough gas along our line of sight to be visible. But when we look through the centers of the shells, there is less gas along our line of sight, and the nebulae appear transparent. In the middle of most planetary nebulae, we see the *central star* from which the nebula was ejected. Central stars are very hot, some as hot as 100,000 K, so appear high up on the left side of the H-R diagram.

Astronomers are particularly interested in planetary nebulae because they want to study the various means by which stars can eject mass into interstellar space. The material given off by a planetary nebula contains heavy elements that had been "cooked" inside the original star. Thus the study of planetaries tells astronomers both how a star can reduce its own mass and also something about the origin of

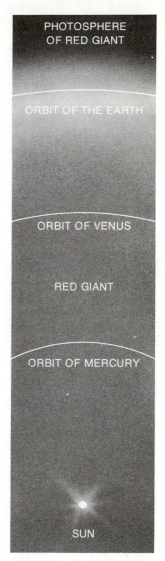

Figure 9–2 A red giant swells so much it can be the diameter of the earth's orbit.

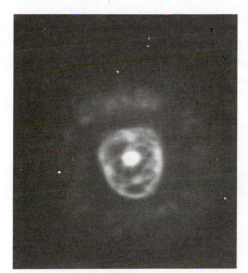

Figure 9–3

Figure 9–4

Figure 9–3 NGC 2392, a planetary nebula in the constellation Gemini, the Twins. This object is sometimes known as the Eskimo Nebula for obvious reasons.

Figure 9–4 NGC 3587, the Owl Nebula, a planetary nebula in the constellation Ursa Major.

interstellar matter and how it is enriched with heavy elements, since some planetary-nebula shells include material that underwent nuclear processing inside the progenitor star. Each planetary nebula represents the ejection of 10 to 20 per cent of a solar mass, which is only a small fraction of the mass of the star. However, statistically about one planetary nebula forms in our galaxy per year; hence, within just one century 10 to 20 solar masses of processed material enrich the interstellar medium.

We can measure the ages of some planetary nebulae by tracing back the shells at their current rate of expansion and calculating when they would have been ejected from the star. Ages derived in this way have large uncertainties, since many effects could cause the expansion to speed up or slow down. Estimates for ages also come from observations of central stars, interpreted by comparing their measured positions on the H-R diagram with theoretical calculations of how the positions there vary with age. Most of the planetary nebulae are less than 50,000 years old. The data fit with our picture: after a longer time, the nebulae will have expanded so much that the gas will be invisible and the central star may have cooled off enough that it is now unable to ionize gas and so cause the nebula to glow. Then only the central contracted hot star is left to see. We identify it as a white dwarf star, as we will discuss next.

9.3 White Dwarfs

Consider what happens later on for a lightweight star. We have seen it transform all the hydrogen in its core to helium and then all the helium in its core into carbon. The star does not then heat up sufficiently to allow the carbon to fuse into still heavier elements. There comes a time when nuclear reactions are no longer generating energy to maintain the internal pressure that balances the force of gravity.

All stars that at this stage in their lives contain less than 1.4 solar masses have the same fate. Many or all stars in this low-mass group come from the red giant phase, and may pass through the phase of being central stars of planetary nebulae. Those that do not go through a planetary nebula stage lose parts of their mass in some other way, perhaps during the giant stage. In any case, they lose large fractions of their original masses. Stars that contained four solar masses when they were on the main sequence probably now contain only 1.4 solar masses.

When their nuclear fires die out for good, the stars with less than 1.4 solar masses remaining shrink in size, and reach a stable condition that we shall describe below. As they shrink they grow very faint (in the opposite manner to that in which red giants grew brighter as they grew bigger). Whatever their actual color, all these stars are called *white dwarfs*. The white dwarfs occupy a region of the Hertzsprung-Russell diagram that is below and to the left of the main sequence (Fig. 9–5).

White dwarfs represent a stable phase in which stars of less than 1.4 solar masses live out their old age. The value of 1.4 solar masses is known as the *Chandrasekhar limit* after the astronomer S. Chandrasekhar (known far and wide as "Chandra"). Chandra derived the concept and the value when he was 19 years old, and was on a boat from India to England to go to college; he was later on the faculty of the University of Chicago. He shared in the 1983 Nobel Prize in Physics for the discovery.

Chandrasekhar reasoned that something must be holding up the material in the white dwarfs against the force of gravity; nuclear reactions generating thermal pressure no longer take place in their interiors. The property that holds up the white dwarfs is a condition called *electron degeneracy*; we thus speak of "degenerate white dwarfs."

Electron degeneracy is a condition that arises in accordance with certain laws of quantum mechanics (and is not something that is intuitively obvious). As the star contracts and the electrons get closer together, there is a continued increase in their

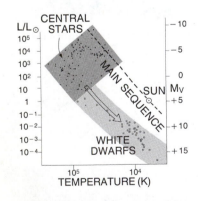

Figure 9–5 When the planetary nebula expands around a red giant, the core of the star becomes visible at the center of the nebula. Since this *central star* has a high temperature, it appears to the left of the main sequence. These stars eventually shrink and cool to become white dwarfs, as shown on this H-R diagram.

resistance to being pushed even closer. This shows up as a pressure. At very great densities, the pressure generated in this way exceeds the normal thermal pressure. When this pressure from the degenerate electrons is sufficiently great, it balances the force of gravity and the star stops contracting.

Thus, the effect of the degenerate electron pressure is to stop the white dwarf from contracting; the gas is then in a very compressed state. In a white dwarf, a mass approximately that of the sun is compressed into a volume only the size of the earth (Fig. 9–6). (This great collapse is possible because atoms are mostly empty space: the nucleus takes up only a very small part of an atom.) A single teaspoonful of a white dwarf weighs 10 tons; it would collapse a table if you somehow tried to put some there. A white dwarf contains matter so dense that it is in a truly incredible state.

What will happen to the white dwarfs with the passage of time? The pressure from degenerate electrons doesn't depend on temperature, so the stars are stabilized even though no more energy is ever generated within them. Because of their electron degeneracy, they can never contract further. Still, they have some heat stored in the nuclear particles present, and that heat will be radiated away over the next billions of years. Then the star will be a burned-out hulk called a *black dwarf*, though it is probable that no white dwarfs have yet lived long enough to reach that final stage. It will be billions of years before the sun becomes a white dwarf and then many billions more before it reaches the black dwarf stage.

9.4 Observing White Dwarfs

White dwarfs are very faint and thus are difficult to detect. We find them by looking for bluish (hot) stars with high proper motions or, in binary systems, by their gravitational effect on the companion stars.

In the former case, by studying proper motions we find the stars that are close to the sun. Some are fainter than main-sequence stars would be at their distances; they must be white dwarfs. (White dwarfs at greater distances would be too faint for us to see.) To understand the latter case, we must realize that for any system of mass orbiting each other, we can define an imaginary point, called the *center of mass*, which moves in a straight line across the sky. The individual bodies move around the center of mass, however, so the path in the sky of any of the individual bodies appears wavy. We have already discussed such astrometric binaries in Section 6.1.

At least three of the 40 stars within 5 parsecs of the sun (Appendix 7: The Nearest Stars)–Sirius (Fig. 9–7), 40 Eridani, and Procyon—have white dwarf companions. Another nearby object, known as van Maanen's star, is a white dwarf, although not in a multiple system. So even though we are not able to detect white dwarfs at great distances from the sun, there seems to be a great number of them. Hundreds of white dwarfs are known.

Even though the theory that explains white dwarfs has seemed consistent with actual observations, their size had never before been measured directly until 1975, when a set of observations was made in the ultraviolet from the Copernicus satellite. The ultraviolet part of the spectrum of Sirius B could be distinguished from that of

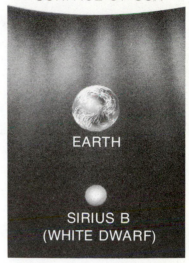

SURFACE OF SUN

EARTH

SIRIUS B (WHITE DWARF)

Figure 9–6 The sizes of the white dwarfs are not very different from that of the earth. A white dwarf contains about 300,000 times more mass than does the earth, however.

Do not confuse the term "white dwarf" with the term "dwarf." The former refers to the dead hulks of stars in the lower left of the H-R diagram, while the latter refers to normal stars on the main sequence.

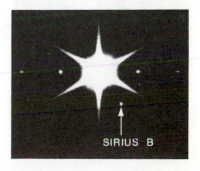

SIRIUS B

Figure 9–7 Sirius A appears as a bright overexposed image with six spikes caused in the telescope system; fainter additional images of Sirius A appear at regular spacing to both sides. Sirius B appears to the lower right of the brightest image of Sirius A; its additional images are too faint to see. Sirius A and B differ by 10 magnitudes, a factor of 10,000, and are only a few arc seconds apart, making them very difficult to photograph on the same picture. This led to the use of the special technique in which much of the overexposed light from Sirius A is caused to fall in the six spikes so that Sirius B can be seen between the spikes.

its much brighter neighbor, and was found to peak at 1100 Å. This corresponds to a temperature of under 30,000 K. Since the absolute magnitude of Sirius B is known, its surface area and thus its radius can be found. The radius of Sirius B is indeed only 4200 km, smaller than the earth's. Many white dwarfs are now being observed with the International Ultraviolet Explorer spacecraft. IUE observations confirm that the temperature of Sirius B is about 26,000 K. At the end of this chapter, we shall discuss what it is like to observe with IUE.

White dwarfs have been observed with temperatures as low as 4000 K and as high as 85,000 K. Since white dwarfs are the cores of stars, revealed as the outer layers were lost, they are made of helium, carbon, and heavier elements. Some have atmospheres of hydrogen, but others have atmospheres entirely of helium.

*9.5 White Dwarfs and the Theory of Relativity

One special reason to study white dwarfs is to make use of them as a laboratory to test extreme physical conditions. It is impossible to create such strong gravity in a laboratory on earth. We have already discussed two tests of general relativity in Section 7.11: the deflection of starlight by the sun, and the advance of the perihelion of the planet Mercury.

A third test, the *gravitational redshift* of light, is best carried out with white dwarfs. Einstein's theory predicts that light leaving a mass would be redshifted by an amount that depends on the amount of mass.

The effect is minuscule on the sun, though it has been detected. Furthermore, turbulence in the solar photosphere, such as the rise and fall of granules, distorts and confuses the solar results. Gravity on a white dwarf's surface is much stronger than gravity on the surface of the sun, since the surface of a white dwarf is so much closer to the star's center.

Unfortunately, one must know the mass and size of the white dwarf to perform the test accurately, and the masses and sizes of white dwarfs are not known very well. Still, strong redshifts are found. The results agree with the predictions of the theory of relativity to within the possible error that results from our uncertain knowledge of the masses and sizes of white dwarfs. The test has best been carried out with 40 Eridani B.

9.6 Novae

Although the stars were generally thought to be unchanging on a human time scale, occasionally a "new star," a *nova* (plural: *novae*, from the Latin for "new"), became visible. Such occurrences have been noted for thousands of years; ancient Oriental chronicles report many such events. Only in recent years have we found out that white dwarfs are at the center of the nova phenomenon.

A nova is a newly visible star rather than actually a new star. A nova is actually a brightening of a star by 5 to 15 magnitudes or more, making it visible to the eye or to the telescope.

A nova (Fig. 9–8) may brighten within a few days or weeks. It ordinarily fades drastically within months, and then continues to fade gradually over the years. Besides these "classical" novae, "dwarf novae" brighten at intervals of months by smaller factors.

Novae occur in binary systems in which one member has evolved into a white dwarf while another member is on the path toward becoming a red giant. The two are separated by about the distance between the earth and the moon. We have mentioned that the outer layers of a red giant are not held very strongly by the star's gravity. If a white dwarf is nearby, some of the matter originally from the red giant goes into orbit around the white dwarf rather than falling straight onto its surface.

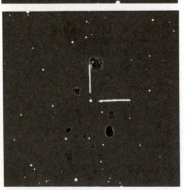

Figure 9–8 Nova Herculis 1934 (DQ Her), showing its rapid fading from 3rd magnitude on March 10, 1935, to below 12th magnitude on May 6, 1935.

August 28, 1975   11:30 U.T.   August 30, 1975   6:45 U.T.

Figure 9–9 These photographs are part of a unique series of observations covering the eruption of Nova Cygni 1975 during the period of its brightening. A Los Angeles amateur astronomer, Ben Mayer, was repeatedly photographing this area of the sky at the crucial times to search for meteors. When he heard of the nova, he retrieved his meteor-less film from the waste basket. Never before had a nova's brightening been so well observed. On August 28 *(left)*, no star was visible at the arrow. By August 30 *(right)*, the nova had reached 2nd magnitude, the brightness of Deneb, which is seen at the right.

Mayer has now organized an international amateur effort, Problicom (*Pro*jection *Bli*nk *Com*parator), in which each participant photographs a region of sky every few weeks and checks over the film using a simple combination of two ordinary slide projectors. Volunteers are welcome.

The orbiting gas forms a disk around the white dwarf. The disk gives off ultraviolet and x-rays that we detect from satellites, and causes the light from the system to flicker rapidly.

In a "classical" nova, some of this orbiting material falls onto the white dwarf's surface. After a time, enough builds up and enough heat is deposited to trigger nuclear reactions on the star's surface for a brief time. This runaway hydrogen burning causes the brightening we see as a nova; it lasts until all the hydrogen is consumed. Years after the outburst of light, the shell of gas blown off sometimes becomes detectable through optical telescopes (Fig. 9–10).

A dwarf nova, while still involving a white dwarf, takes place by a different mechanism: some of the orbiting material suddenly falls onto the white dwarf's surface. The energy from the gravitational collapse is released as heat and radiation. Outbursts of this type are smaller than "classical" nova eruptions, recur many times, and do not eject a shell of gas. Amateur astronomers, through the American Association of Variable Star Observers, provide many of the observations of dwarf novae, such as SS Cygni. SS Cyg is particularly fun to monitor (through a telescope) because it brightens by a quite noticeable 4 magnitudes fairly often—about every 50 days.

An old observational category—"recurrent novae"—includes examples of both classical novae (on which we are seeing runaway nuclear burning) and dwarf novae (on which we are seeing the result of disk collapse).

The European Exosat is observing x-ray pulsations with a period of a few minutes in some novae, indicating an effect of the white dwarf's magnetic field on the disk of gas.

9.7 The Evolution of Binary Stars

A single star evolves in an orderly fashion and lives on the main sequence for a length of time that depends only on its mass and chemical composition. But steady evolution takes place only for stars whose mass does not change substantially during their main-sequence lifetimes. Observations from satellites have led us to the realization that many of the x-ray sources are binary systems, and so have spurred new interest in studying the evolution of binaries. The importance of the investigation is underscored by the long-known fact that most stars are members of binary systems. We have just seen that the long-known novae are really examples of advanced states of stellar evolution in binary systems.

Figure 9–10 In 1951, a shell of gas could be seen to surround Nova Herculis 1934. Nova shells expand at 100 times the rate of planetary nebulae.

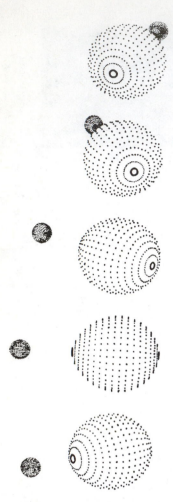

Figure 9–11 A computer simulation of an eclipse in a short period binary. The five phases shown here are separated by $\frac{1}{10}$ of a period. The light we would measure from such a system would appear to vary in part because of the eclipses and in part because of the distortion of the components from being round.

In a binary system, the mutual gravitational attraction the two stars have for each other can sometimes pull off mass, which then flows from one star to the other. This, in turn, can have a drastic effect on the evolution of both stars.

In order for one star to pull mass off another star, the force of gravity from the first star must be attracting some part of the surface of the second star more strongly than that star's own gravity is attracting it. This happens most readily when the second star is very large (for example, a red giant, which has swelled considerably over its main-sequence size). Then its outer layers are much further from its center. Since a star's gravity acts as though all the star's mass were concentrated at the center of the star, gravity does not bind the outer layers of a red giant very strongly.

The effect of the first star's gravity on the second star varies from point to point because the different points are at different distances from the center of the first star. The second star may be distorted (Figure 9–11) because of this *differential effect*, which is called a *tidal force* because the tides of the earth's oceans also arise from such a differential effect, in that case chiefly from the moon's gravity (Section 19.3).

Let us consider a binary system made up of a widely separated main-sequence star (a dwarf) and a red giant. The parts of the red giant that are closer to the dwarf feel a stronger pull of gravity from the dwarf than the parts of the red giant that are farther from the dwarf. The tidal force pulls the outer layers of the giant component until it is more egg-shaped than round. In fact, we can draw a figure-8–shaped curve (Fig. 9–12) that marks the edge of the volumes in which each star's own gravity dominates, which we call the *Roche lobes* of the two stars. We categorize double stars according to the relation of each of the components to its Roche lobe (Fig. 9–13). The size of the Roche lobes is calculated theoretically from knowledge of the stars' masses. In *contact binaries*, the stars fill their Roche lobes; in *detached binaries*, they do not.

A dwarf is much smaller than its Roche lobe. But a giant may be so large that it fills its Roche lobe. When this happens, material can flow freely through the "neck" of the figure-8 and fall onto the other star. In this way, the dwarf gains mass while the giant star loses mass. This takes place rapidly at first. Sometimes the fact that the mass is newly transferred shows up in unusual abundances of the chemical elements. Angular momentum, the quantity that describes the tendency to keep spinning, is transferred too. The angular momentum causes the transferred mass to spiral down onto the dwarf instead of falling straight in.

If the recipient star gains a lot of mass, its evolution can speed up very considerably. It can carry on nuclear fusion in its interior at a greatly increased rate, and will change spectral type. The relative brightness of the two stars in the binary system can change completely, or even invert. During this "common envelope phase," the orbit shrinks considerably and a system with an orbital period of several hundred days can turn into one with a period of 1 day or less.

Figure 9–12 The Roche lobes for a binary system in which the primary starts with 80 percent of the mass and the secondary starts with 20 percent of the mass. As the mass is transferred, the separation between the two stars diminishes until the two are equal in mass and then the separation increases. By the end of the transfer, the ratio of the masses is reversed. Numbers below each frame of the sequence show the fraction of the total mass in the left component.

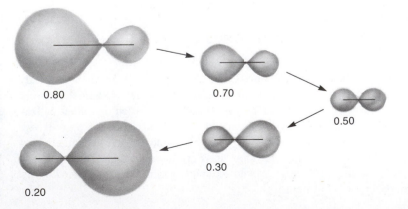

0.80   0.70   0.50

0.20   0.30

Once the masses become about equal, the rate of mass transfer slows down considerably. Stars in this phase are known as Algol binaries. Algol (beta Persei) itself is interesting to observe as it varies between magnitudes 2.2 and 3.5 every 2.9 days. This stage ends when the original donor can no longer sustain nuclear burning and shrinks until it is a white dwarf (or smaller).

Eventually, the original recipient star evolves toward the red giant stage, although the orbit is not big enough for it to become a red giant. By this time, the period has shrunk to a few hours. When mass transfer to the white dwarf begins, we have a nova (Section 9.6). In the next chapter, we discuss what happens when the recipient star is more condensed yet.

Since it is very difficult or even impossible to gain an accurate idea of the amount of mass flowing between stars in a binary system, stellar evolution is considerably more complicated to understand for a star in a binary system than it is for an isolated star.

*9.8 Observing with the IUE Spacecraft

The most widely used telescope now available to astronomers is not on a mountain top, but is rather hovering over earth. Known as IUE, the International Ultraviolet Explorer, it has supplied data to hundreds of astronomers in the United States and around the world since its launch in 1978.

IUE (Fig. 9–14) carries a 45-cm (18-inch) telescope. It uses a vidicon (that is, a television-type device) for a detector, and records the data in a particularly efficient way so that it can survey the spectrum of a star through the entire ultraviolet in an hour or so. It has studied many white dwarfs, since they emit most of their radiation in the ultraviolet. IUE is so sensitive that it can also be used to observe much fainter objects, including galaxies and even quasars.

IUE is in an orbit that carries it around the earth once every 24 hours. Since the earth rotates underneath at the same rate, the spacecraft hovers within view of NASA's Goddard Space Flight Center in Greenbelt, Maryland, all the time—it is in a *synchronous orbit*. It actually traces out an ellipse on the ground that brings it within view of a European data center in Spain for 8 hours a day. European astronomers there control it during this time.

American astronomers prepare proposals that describe what they want to observe with IUE and why, and send them periodically to NASA. For the successful proposers, the observing time follows some months later. Three days or a week of 16-hour shifts, for example, might be available for a given project. Let us say that our proposal to observe the outer atmospheres of stars is accepted. We take our leave from home for a week or so and set out for the Goddard Space Flight Center, which is in a suburb of Washington, D.C.

We arrive at an office building, not at all like a telescope dome. In a small room on an ordinary office corridor, we find the American control room for this fabulous telescope in space.

Before we arrive, we have carefully figured out exactly what objects we want to observe, and indeed have a list of specific objects approved by the IUE team. We meet the Telescope Operator, the person who actually sends commands from the control room. We are really there in an advisory capacity.

Acquiring the star in the field of view might take fifteen minutes. Then an image of a field of view 16 minutes of arc across—about half the diameter of the moon—

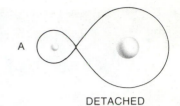

DETACHED

SEMI-DETACHED

CONTACT

COMMON ENVELOPE (OVERCONTACT)

Figure 9–13 Depending on whether one or both stars fill its Roche lobe, we can categorize double stars as *(A)* detached, *(B)* semi-detached, or *(C)* contact binaries. When both stars overfill their Roche lobes, we have an overcontact binary *(D)*.

Figure 9–14 The International Ultraviolet Explorer spacecraft, IUE, before its launch. The telescope, of the Ritchey-Chrétien design, extends out the top. Solar cells to provide power are visible at both sides. The spectrographs are in the middle section.

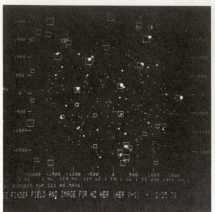

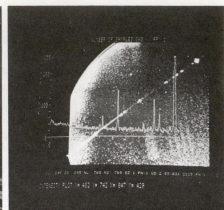

Figure 9–15 Within a couple of minutes, IUE displays its field of view so that the astronomers can specify which object to observe and to use as a guide star.

Figure 9–16 The control room contains video screens to display the data *(left)* and spacecraft housekeeping equipment *(right),* plus a large projection tv screen *(top)* for a second view of the data. The high dispersion spectrum is dispersed by a kind of spectrograph in which many short strips of spectrum, each only 20 Å wide, are displayed one above the other. This is particularly convenient for a vidicon detector, which has a round face, because it fills up much of the viewing surface, making observing pretty efficient.

Figure 9–17 Within seconds after the observer's request, the computer can plot on its screen a graph of any part of the spectrum that the astronomer designates.

IUE observations confirmed that Nova Cygni 1975 was caused by nuclear fusion in material that fell onto a white dwarf.

appears on a television screen before us (Fig. 9–15). Our star is probably within the field of view, because the telescope points pretty accurately where we tell it to look.

If there are several stars close together in the field, however, we can have the computer automatically display on the same screen a star map in which each star is represented by a box, the size of the box representing the brightness of a star. We then simply have the telescope slide around until the stars in the field of view fit within the boxes. This is certainly a far cry from having to locate stars from their position in a constellation!

Now we tell the Telescope Operator to take the first exposure. We have to choose which of two possible wavelength bands to use: a relatively short one from 1150 Å (below the wavelength of hydrogen's Lyman α line) up to 2000 Å, or else a relatively long ultraviolet range from 1900 Å up to 3200 Å. Only the upper tip of the upper range comes through the earth's atmosphere, so just what is to be found in these spectra is often a tremendous surprise.

After our hour-long exposure, the data are radioed back from the spacecraft to an antenna at Goddard, and soon appear on our video screen. The spectrum is displayed both on a television screen on the console and also on the large screen overhead (Fig. 9–16 and Color Plate 3). We have the Telescope Operator begin the next exposure, perhaps the other wavelength band for the same star.

As soon as these instructions are radioed to the spacecraft, we can make use of one of the nicest features of IUE—an interactive computer. We can tell the computer to display an enlarged region of the spectrum on the screen.

We can even have the computer display, in a few seconds, a graph of a part of the spectrum (Fig. 9–17). No more do we have to develop photographic plates and then have the densities on the emulsion traced laboriously. Within seconds, we have a spectrum in view, and can even lay a plastic overlay on it to make a few measurements right away. So much of the spectrum of objects is completely unknown in the ultraviolet that important results have regularly been found in this way.

Another advantage of the interactive computer system is that by the time our next exposure is complete, we have already gained a good idea about the kind of results we might hope for from the first spectrum. We would also know whether it was well exposed or has to be retaken.

A rest at our motel—our shifts may well run through the night—brings us up to the time to observe again. We see where the telescope is pointing, figure out what to observe first, and start again. A week of this is exhausting but rewarding. We even have preliminary data to take home with us, though computer tapes with final treatments of the data will follow us by mail a few weeks later.

IUE—the first international observatory in space—has been a tremendous success. It has already sent back data on stars, planets, nebulae, galaxies, and quasars, and we look for continued productivity in the years to come.

Summary and Outline

Red giants (Section 9.1)

They follow the end of hydrogen burning in the core. The core contracts gravitationally; the core and the hydrogen-burning shell become hotter; the star is red, so each bit of surface has relatively low luminosity but the surface area is very greatly increased, so the total luminosity is greatly increased

Helium flash: rapid onset of triple-alpha process

Planetary nebulae (Section 9.2)

Ages: less than 50,000 years old; the gas is blown off when it absorbs photons; mass loss: 0.1 or 0.2 solar mass

White dwarfs (Sections 9.3, 9.4, and 9.5)

End result of lightweight stars; less than 1.4 solar masses

remaining; supported by electron degeneracy; detected by their proper motion or by their presence in a binary system

They provide the best test of the gravitational redshift

Novae (Section 9.6)

Interaction of a red giant and a white dwarf

Evolution of binary stars (Section 9.7)

The exchange of mass between components of binary stars can change their evolution drastically; a star gives off mass to its companion when it fills its Roche lobe

International Ultraviolet Explorer (IUE) (Section 9.8)

A NASA/European spacecraft with an international set of observers can study faint objects in the ultraviolet

Key Words

lightweight stars, gravitationally, red giant, helium flash, planetary nebulae, central star, white dwarfs, Chandrasekhar limit, electron degeneracy, black dwarf, center of mass, gravitational redshift, nova (novae), differential effect, tidal force, Roche lobes, contact binaries, detached binaries, synchronous orbit

Questions

1. What event signals the end of the main-sequence life of a star?

2. When hydrogen burning in the core stops, the core contracts and heats up again. Why doesn't hydrogen burning start again?

3. When the core first starts contracting, what halts the collapse?

4. How can a star reach its red giant stage twice?

5. Why, physically, is there a Chandrasekhar limit?

6. If you are outside a spherical mass, the force of gravity varies inversely as the square of the distance from the center. What is the ratio of the force of gravity at the surface of the sun to what it will be when the sun has a radius of one astronomical unit? (Use data in Appendix 2.)

7. Why is helium ''flash'' an appropriate name?

8. If you compare a photograph of a nearby planetary nebula taken 80 years ago with one taken now, how would you expect them to differ?

9. Why is the surface of a star hotter after the star sheds a planetary nebula?

10. What keeps a white dwarf from collapsing further?

11. What are the differences between the sun and a one-solar-mass white dwarf?

12. When the sun becomes a white dwarf, approximately how much mass will it have? Where will the rest of the mass have gone?

13. Which has a higher surface temperature, the sun or a white dwarf?

14. Compare the surface temperatures of the hottest white dwarf with an O star. What spectral type of normal dwarf has the same surface temperature as the coolest white dwarf?

15. Sketch an H-R diagram indicating the main sequence. Show the evolution of the sun starting at the time it leaves the main sequence.

16. Compare the desirability of the sun and of 40 Eridani B for testing the gravitational redshift prediction of Einstein's general theory of relativity.

17. When the proton-proton chain starts at the center of a star, it continues for billions of years. When it starts at the surface (as in a nova) it only lasts a few weeks. How can you explain the difference?

18. Compare IUE with the Palomar Observatory, listing two advantages and two disadvantages.

19. Why is it harder to predict the evolution of a binary star than of a single star?

20. Describe the Roche lobes and their relation to binary-star evolution.

Filaments at the north end of the supernova remnant known as the Cygnus Loop.

Supernovae, Neutron Stars, and Pulsars

10

Aims: To see how stars become supernovae, some from exploding white dwarfs and others as very massive stars whose cores sometimes collapse to become neutron stars; to study how we observe neutron stars as pulsars and as x-ray binaries

We have seen how the run-of-the-mill lightweight stars end, not with a bang but a whimper. Still, when the white dwarfs that result are in binary systems, huge explosions called supernovae may yet result.

More massive stars, which contain more than about 8 solar masses when they are on the main sequence, put on a dazzling display more directly. These heavyweight stars cook the heavy elements deep inside. They then blow themselves almost to bits as supernovae, forming still more heavy elements in the process. The matter that is left behind settles down to even stranger states of existence than that of a white dwarf.

In this chapter and the next one we shall discuss supernovae. Then we shall see that some of the massive stars, after their supernova stage, become neutron stars, which we detect as pulsars and x-ray binaries. In the following chapter, we shall discuss the death of the most massive stars, which become black holes.

10.1 Red Supergiants

Stars that are much more massive than the sun whip through their main-sequence lifetimes at a rapid pace. These prodigal stars use up their store of hydrogen very quickly. A star of 15 solar masses may take only 10 million years from the time it first reaches the main sequence until the time when it has exhausted all the hydrogen in its core. This is a lifetime a thousand times shorter than that of the sun. When the star exhausts the hydrogen in its core, the outer layers expand and the star becomes a red giant.

For these massive stars, the core can then gradually heat up to 100 million degrees, and the triple-alpha process begins to transform helium into carbon. After its ignition, the helium then burns steadily, unlike the helium flash of less massive stars.

By the time helium burning is concluded, the outer layers have expanded even further, and the star has become much brighter than even a red giant. We call it a *red supergiant* (Fig. 10–1); Betelgeuse, the star that marks the shoulder of Orion, is the best-known example (Fig. 10–2). Supergiants are inherently very luminous stars, with absolute magnitudes of up to −10, one million times brighter than the sun. A

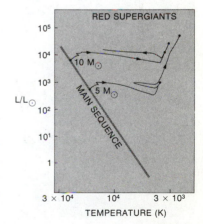

Figure 10–1 An H-R diagram (as in Fig. 9–1) showing the evolutionary tracks of 5- and 10-solar-mass stars as they evolve from the main sequence to become red supergiants. For each star, the first dot represents the point where hydrogen burning starts, the second dot represents the point where helium burning starts, and the third dot the point where carbon burning starts.

Figure 10–2 The red supergiant Betelgeuse is the star labelled α in the shoulder of Orion, the Hunter, shown here in Johann Bayer's *Uranometria*, first published in 1603. Betelgeuse, visible in the winter sky, appears reddish to the eye.

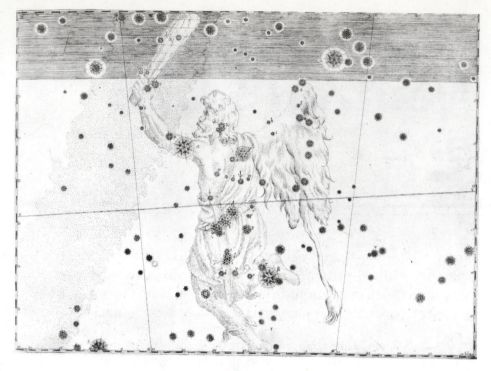

Figure 10–3 Views of the central part of the galaxy NGC 5253, taken in 1959 and in 1972. The supernova that appeared in 1972 was nearly as bright as the rest of the galaxy.

⊙ is the symbol of the sun. 8M⊙, read "eight solar masses," means 8 times as much mass as the sun has.

supergiant's mass is spread out over such a tremendous volume, though, that its average density is less than one millionth that of the sun.

The carbon core of a supergiant contracts, heats up, and begins fusing into still heavier elements. Eventually, even iron builds up. The iron core is surrounded by layers of elements of different mass, with the lightest toward the periphery.

10.2 Supernovae

A dying star can explode in a glorious burst (Figs. 10–3 and 10–4) called a *supernova* (plural: *supernovae*). Two quite different situations apparently lead to supernovae. The best current models indicate that *Type I supernovae* represent the incineration of a white dwarf (which results from a lightweight star). On the other hand, *Type II supernovae* take place in heavyweight stars, and show that stellar evolution can run away with itself and go out of control. (To help remember which is which, recall that the less massive star leads to the lower Roman number—I instead of II.)

The two types can be distinguished by their spectra and by the rate at which their brightnesses change. Rudolph Minkowski pointed out the distinction in the 1940's based on the observation that some supernovae—Type II—have prominent hydrogen lines when they are at maximum light, while hydrogen lines are absent in the others—Type I. Type I are more often seen because they are 1.5 magnitudes (4 times) brighter and take place in all types of galaxies. Type II supernovae are seen only in spiral galaxies. Indeed, they are usually found near the spiral arms, which endorses the idea that they come from massive stars. (Since the massive stars are short lived, they are found in the spiral arms near where they were born.) In both cases, Doppler shifted absorption lines show that gas is expanding rapidly: about 10,000 km/s for Type I's and half that for Type II's.

In one new model for Type II supernovae, a substantial core of heavy elements has formed in a massive star (perhaps 8–12 $M_\odot$), and begins to shrink and heat up.

Figure 10–4 Views of the supernova that erupted in 1979 in the galaxy M100. Less spiral structure appears in the later view, chiefly because of a different exposure time, yet the supernova appears prominently at lower left. At 12th magnitude (apparent), it was the brightest supernova since 1970. IUE spectra were the first ever of a supernova in the ultraviolet. (Photo by Paul Griboval with an electronic camera and the 0.76-m telescope of the McDonald Observatory, University of Texas)

The heavy elements represent the ashes of the previous stages of nuclear burning. Oxygen and magnesium ignite when they capture electrons, and "burn" to form iron. The conditions in the center of the star change so quickly that it becomes very difficult to model them satisfactorily in sets of equations or on computers, but new powerful computers have enabled astronomers to keep up with the rapidly changing conditions; whether the following model ultimately proves correct is uncertain. In this model, the temperature becomes high enough for iron to form and then to undergo nuclear reactions. The stage is now set for disaster because iron nuclei have a fundamentally different property from other nuclei when undergoing nuclear reactions. Unlike other nuclei, iron absorbs rather than produces energy in order to undergo either fusion or fission. So processes that release energy no longer can build up heavier nuclei as the core shrinks this time. Iron cores are formed in stars more massive than 12 $M_\odot$ by other processes, and these more massive stars then join the iron-collapse scenario.

The temperature climbs beyond billions of kelvins, and the iron is broken up into alpha particles (helium nuclei), protons, and neutrons by the high-energy photons of radiation that are generated. As a result, the central pressure diminishes, and the core begins to collapse. The process goes out of control. Within milliseconds—a fantastically short time for a star that has lived for millions and millions of years—the inner core collapses and heats up catastrophically.

As the outer core falls in upon the collapsing inner core, electrons and protons combine to make neutrons and neutrinos (Sections 8.3 and 8.6). The core collapses and (calculations show) bounces outward, since its gas has a natural barrier at nuclear densities against being compressed too far. The neutrinos may also play a role in blowing off the outer layers or at least in stopping the core from collapsing completely.

The collision of the rebounding inner core with the outer core sends off shock waves that cause heavy elements to form and that throw off the outer layers. With this tremendous explosion, the star is destroyed. Only the core may be left behind.

The heavy elements formed either near the center of the star or in the supernova explosion are spread out into space, where they enrich the interstellar gas. Though most Type II supernovae come from stars containing between 8 and 12 solar masses, they don't supply as much of the heavy elements as more massive stars do. When a star forms out of gas enriched by supernovae of both types, these heavy elements are present. (It is strange, though, that studies of existing supernova remnants don't find as much enhancement of heavy elements as one might expect.) Our understanding of element formation in Type II supernovae is circumstantial, since the outburst itself is

Spectroscopic studies of the star η (eta) Carinae (Color Plate 57) show that a gas condensation spewed out 150 years ago contains a relatively large abundance of nitrogen. This indicates that the star has processed material in the carbon cycle, brought it to the surface as the star's material mixed up, and ejected it. This is expected from a massive star that will become a supernova soon—though "soon" could be next year or in 10,000 years.

Eta Carinae is 9000 light years away. A supernova there would be too far away to affect us, but would appear brighter than Venus.

hidden from our view by the outer layers of the massive star; we can only study the transformed material later on, when it has spread out considerably.

In contrast to Type II supernovae, a Type I supernova apparently takes place in a binary system containing a white dwarf when mass from the second star (perhaps also a degenerate object) falls onto the white dwarf. Type I supernovae thus take place in the late stages of lightweight stars. The extra mass puts the white dwarf up to the 1.4-solar-mass Chandrasekhar limit, which leads to the star's thermonuclear explosion. The light fades drastically for a month, and then begins a slower decline. The energy comes (according to theoretical models) from radioactive decay of material produced in the supernova explosion. In particular, nickel-56, which was formed, decays into cobalt-56 and then iron-56, releasing energy. The spectrum of the long, slow decline can be explained as a composite of overlapping lines of ionized iron. NASA's Gamma Ray Observatory, hoped for in the 1990's, should be sensitive enough to test this model by observing Type I supernovae in galaxies some distance away from us, detecting the gamma rays emitted as the iron, cobalt, and nickel are formed.

The star is apparently entirely disrupted, spreading heavy elements through the galaxy but leaving no core behind. The fact that Type I supernovae don't show hydrogen in their spectra fits with the model, since the white dwarf may have shed its entire outer atmosphere before its incineration.

Since Type I and Type II supernovae are the source of heavy elements in the universe, they provide the heavy elements that are necessary for life to arise. The heavy atoms in each of us have been through such supernova explosions.

The name "supernova" persists from a time when they were thought to be merely unusually bright novae. The supernova in the Andromeda Galaxy in 1885 (known by the variable-star name of S Andromedae), since it was thought to have the same brightness as a nova, gave a value for the distance to that galaxy that misled astronomers for years. They thought the Great Nebula in Andromeda was too close to be an independent galaxy. But in the 1930's, Walter Baade and Fritz Zwicky realized the distinction between novae and supernovae. A supernova explosion represents the death of a star and the scattering of most of its material, while a nova uses up only a small fraction of a stellar mass and can recur. One new similarity does exist: both novae and Type I supernovae are phenomena of binary systems in which one member is a collapsed star.

10.2a Detecting Supernovae

Whether Type I or Type II, in the weeks following the explosion, the amount of radiation emitted by the supernova can equal that emitted by the rest of its entire galaxy. It may brighten by over 20 magnitudes, a factor of 10^8 in luminosity. Doppler-shift measurements show high velocities, about 10,000 km/s, that prove that an explosion has taken place.

In our galaxy, the only supernovae thoroughly reported were in A.D. 1572 and 1604. Tycho's supernova, observed by the great astronomer (Section 18.6) in 1572, is the bright object shown at the left edge of Cassiopeia's chair in the drawing that opens Chapter 4.

Kepler's supernova, observed by that great astronomer (Section 18.7) in 1604, was slightly farther away from earth than Tycho's, though in another direction. Though these two supernovae were seen only 32 years apart, one actually exploded about 10,000 years after the other. The light curves measured then were accurate enough for us to tell that both were probably Type I.

There is some debate as to whether a star that the first Astronomer Royal in Britain, John Flamsteed, detected in 1680 was a different supernova, Cassiopeia A. We see nothing now in the sky in the location where he drew an object on a star map: optical, radio, and x-ray observations now show that a supernova took place there. If what he drew was the supernova, it would have had to be fainter than normal. The optical spectra fit with the enhanced abundances that we would expect from a 20-solar-mass star. This corresponds to a Type II supernova, perhaps fainter than usual because the massive star had already lost its outer atmosphere. Optical studies of the faint remnant show an 8,000 km/s expansion, and verify the date of the explosion.

In A.D. 1054, chronicles in China, Japan, and Korea recorded the appearance of a "guest star" in the sky that was sufficiently bright that it could be seen in the daytime. No one is certain why no Western European records of the supernova were made, for the Bayeux tapestry illustrates a comet and thus shows that even in the Middle Ages people were aware of celestial events; perhaps a Western philosophical mind set played a role. A reference to a sighting of the supernova in Constantinople was discovered a few years ago. Certain cave and rock paintings made by Indians in the American southwest may show the supernova (Fig. 10–5), though this interpretation is controversial. The light-curve was not well enough observed even to determine whether it was a Type I or a Type II supernova. (It may even be an example of some rare additional type.) We shall discuss it further in the next section.

The sightings of A.D. 185, 393, 1006, and 1181 are now also thought to have probably been supernovae. The 1006 supernova, in the southern constellation Lupus, was apparently the brightest ever, perhaps apparent magnitude −10. It remained visible, even in northern locations, for at least two years.

Not a single supernova has been immediately noticed in our galaxy since the invention of the telescope. Nor do we usually find supernovae in other galaxies before they have passed their peak. We would obviously like to detect a supernova right away, so that we could bring to bear all our modern telescopes in various parts of the spectrum. An attempt a decade ago to set up an automated search program was a bit too far in advance of the technology available. But two new search programs got under way in 1984, both taking advantage of the sensitivity of CCD's (Section 2.11a) as detectors.

The program of the Lawrence Berkeley Laboratory is set up at the 0.9-m telescope of MIRA (the Monterey Institute for Research in Astronomy) in California. It is programmed to survey a set of 7,500 galaxies repeatedly over a period of 3 nights of observing. Developing the computer software to allow the CCD image of each galaxy to be compared with an image stored the previous week has been a formidable task.

The researchers hope to find 75 supernovae a year. They think they can detect supernovae in nearby galaxies at only 1 per cent of maximum light, which would be a week or so before maximum. Most supernovae are now discovered at or after maximum, when there is little or no time to make a wide range of observations. And they would have pre-discovery observations at intervals of a few days, instead of the few-year interval we saw in Figure 10–3.

The program of the Steward Observatory of the University of Arizona is set up at a new 1.8-m telescope on Kitt Peak. The telescope points at a fixed direction, so a narrow strip of sky drifts across it. The data are read out from the CCD's every minute, which allows observation down to 22nd magnitude. Ten to thirty thousand galaxies of all types will be included in the survey each night. The same strip of sky, changing only by the drift of nighttime hours over the year, will be surveyed over and over again. Novae and supernovae in our galaxy as well as brightness variations of many other types of objects will also be detected.

Figure 10–5 An American Indian cave painting discovered in northern Arizona that may depict the supernova explosion 900 years ago that led to the Crab Nebula. It had been wondered why only Oriental astronomers reported the supernova, so searches have been made in other parts of the world for additional observations.

10.2b Supernova Remnants

Optical astronomers have photographed two dozen of the stellar shreds that are left behind, which are known as *supernova remnants* (Fig. 10–6, the photograph that opens this chapter, and Color Plates 54 and 55), in our galaxy alone, and others in nearby galaxies. Strong radio radiation and x-radiation can also be detected. Over 100 radio supernova remnants are known. The comparison of radio and x-ray observations of supernova remnants (Fig. 10–7 and Color Plate 53) tells us about the supernova itself and about its interaction with interstellar matter.

We find the most famous remnant in the location where Chinese astronomers reported their "guest star" in A.D. 1054. When we look at the reported position in

Figure 10–6 *(A)* S147, the remnant of a supernova explosion in our galaxy. The long delicate filaments shown cover an area over 3° × 3°, about 40 times the area of the moon. *(B)* The Cygnus Loop supernova remnant is seen here in far-ultraviolet radiation (1500–1700 Å), which does not penetrate the earth's atmosphere; the image was obtained from Spacelab I aboard a space shuttle. The radiation that made the image was spectral lines of twice- and three-times–ionized oxygen and carbon. It can be interpreted to give information about the shock wave that the supernova gave off. (Stuart Bowyer, U. of Cal. Berkeley)

A

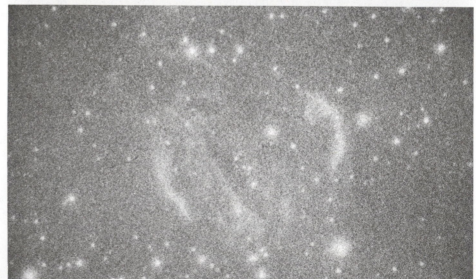

B

the sky, in the constellation Taurus, we see an object—the Crab Nebula—that clearly looks as though it is a star torn to shreds (Color Plate 54). The Crab appears to be about ⅕ of the distance from the sun to the center of our galaxy. We find its age by noting the rate at which the filaments in the Crab Nebula are expanding. Tracing the filaments back in time shows that they were at a single point approximately 900 years ago. The Crab's age, location, and current appearance leave us little doubt that it is the remnant of the supernova whose radiation reached earth twelve years before William the Conqueror invaded England. The discovery that the filaments are rich in helium and are expanding at thousands of km/s as from an explosion further confirms that the Crab is a supernova remnant. And most of the gas in the filaments that emits spectral lines is helium, a product of evolution. We will have more to say about the Crab when we discuss pulsars.

From a study of the rate at which supernovae appear in distant galaxies, we estimate that supernovae should appear in our galaxy about every 30 years. Interstellar matter obscures much of the galaxy from our optical view. Still, we wonder why we haven't seen one in so long. It's about time. Radiation from hundreds of distant supernovae is on its way to us now. The light from a nearby supernova may be on its way too. Maybe it will get here tonight.

10.3 Cosmic Rays

Most of the information we have discussed thus far in this book has been gleaned from the study of electromagnetic radiation, which can be thought of as sometimes having the effects of waves and sometimes having the effects of particles called photons. But certain high-energy particles of matter travel through space in addition to the photons. These particles are nuclei of atoms moving at tremendous velocities, and are called *cosmic rays*.

The cosmic rays accelerated to high velocities deep in space, called *primary cosmic rays*, often interact with atoms in the earth's atmosphere, at which time *secondary cosmic rays* are given off. It is possible to detect many such secondary rays without leaving the earth's surface.

The origin of the primary cosmic rays has long been a matter of controversy. It seems likely that most cosmic rays are particles accelerated to tremendous velocities in supernova explosions in our galaxy or by shock waves—the equivalent of sonic booms—in gas near the supernova. Even though they are particles, not rays, we retain the historic name of "cosmic rays."

To best capture cosmic rays from outer space, one must travel above most of the earth's atmosphere in a rocket, satellite, or balloon. (Only a few of the most energetic cosmic rays reach the ground.) Balloons bear stacks of suitable plastics to altitudes of 50 kilometers and now the space shuttles carry similar materials higher; the cosmic rays leave marks as they travel through the plastic, and a three-dimensional picture of the track can be built up because of the three-dimensional nature of the stack. The third High-Energy Astronomy Observatory (HEAO-3) carried two cosmic ray experiments into space in 1979. Cosmic rays were even detected by emulsion chambers mounted for two months in the baggage compartment of a supersonic Concorde as it flew around. And in 1984, a space shuttle carried cosmic-ray detectors aloft to remain in space for 10 months on NASA's Long Duration Exposure Facility (LDEF).

Special ground-based telescopes are now able to see streaks of light in the upper atmosphere formed by cosmic rays. The highest energy cosmic rays seem to come from the direction of the nearest large cluster of galaxies. Scientists suspect that a peculiar galaxy there, M87, may be giving off these particles, perhaps from a giant black hole.

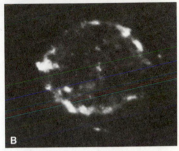

60 ARC SEC

Figure 10–7 The Cassiopeia A supernova remnant. *(A)* An x-ray image made with the Einstein Observatory. *(B)* A radio image made with the Very Large Array. The source is a relatively recent supernova remnant, but the supernova itself was not observed. We can date it to the late 17th century by studying its expansion. Both images show a bright shell of emission, presumably the supernova matter encountering the interstellar medium. Analysis of the x-ray spectrum shows that the abundances of the elements silicon, sulfur, and argon are higher than the normal solar abundance, demonstrating that heavy elements are indeed formed in supernovae. Other supernova remnants have even greater enhancements of these elements. (The x-ray image is courtesy of Steven Murray et al. of the Harvard-Smithsonian Center for Astrophysics. The radio image was made by A. R. Thompson et al., and is courtesy of R. Hjellming.)

Figure 10–8 *(A)* A University of Hawaii professor holds one of the giant photomultipliers being deployed in tests of Project DU-MAND. *(B)* A Soviet test photomultiplier array is deployed at sea.

A

B

An ingenious proposal to study a variety of cosmic-ray particles and neutrinos is in the planning and preliminary testing stage. Project DUMAND (Deep Underwater Muon and Neutrino Detection) would set up chains of giant photomultipliers (Fig. 10–8) in the clear ocean water near Hawaii. The photomultipliers would detect flashes of light from secondary cosmic-ray particles. The flashes arise as the cosmic-ray particles slow down when they pass through water.

Figure 10–9 A neutron star may be only 20 kilometers in diameter (10 km in radius), the size of a city, even though it may contain a solar mass or more. A neutron star might have a solid, crystalline crust about a hundred meters thick. Above these outer layers, its atmosphere probably takes up only another few **centimeters**. Since the crust is crystalline, there may be irregular structures like mountains, which would only poke up a few centimeters through the atmosphere.

If more than about three solar masses remain, then the force of gravity will overwhelm even neutron degeneracy. We shall discuss that case in Chapter 11.

10.4 Neutron Stars

We have discussed the fate of the outer layers of a massive star that explodes as a Type II supernova. Now let us discuss the fate of the core.

As iron fills the core of a massive star, the temperatures are so high that the iron nuclei begin to break apart into smaller units like helium nuclei. The pressure is too low to counteract gravity. The core collapses.

As the density increases, the electrons are squeezed into the nuclei and react with the protons there to produce neutrons and neutrinos. The neutrinos escape, thus stealing still more energy from the core, and may also help eject the outer layers if enough neutrinos interact as they pass through. In any case, a gas composed mainly of neutrons is left behind in the dense core as the outer layers explode as a supernova.

Following the explosion, the core may contain as little as a few tenths of a solar mass or possibly as much as two or three solar masses. This remainder is at an even higher density than that at which electron degeneracy holds up a white dwarf. At this density an analogous condition called *neutron degeneracy,* in which the neutrons cannot be packed any more tightly, appears. The pressure caused by neutron degeneracy balances the gravitational force that tends to collapse the core, and as a result the core reaches equilibrium as a *neutron star.*

Whereas a white dwarf packs the mass of the sun into a volume the size of the earth, the density of a neutron star is even more extreme. A neutron star may be only 20 kilometers or so across (Fig. 10–9), in which space it may contain the mass of about two suns. In its high density, it is like a single, giant nucleus. A teaspoonful of a neutron star could weigh a billion tons.

Before it collapses, the core (like the sun) has only a weak magnetic field. But as the core collapses, the magnetic field is concentrated. It becomes much stronger than any we can produce on earth.

When neutron stars were discussed in theoretical analyses in the 1930's, there seemed to be no hope of actually observing one. Nobody had a good idea of how to look. Let us jump to consider events of 1967, which will later prove to be related to the search for neutron stars.

10.5 The Discovery of Pulsars

By 1967, radio astronomy had become a flourishing science. But one problem had thus far prevented a major discovery from being made: Like radio receivers in our living rooms, radio telescopes are subject to static—rapid variations in the strength of the signal. It is difficult to measure the average intensity of a signal if the signal strength is jumping up and down many times a second, so radio astronomers usually adjusted their instruments so that they did not record any variation in signal shorter than a second or so. But in 1967 the only rapid variations that astronomers expected besides terrestrial static were those that correspond to the twinkling of stars.

An astronomer at Cambridge University in England, Antony Hewish, wanted to study this "twinkling" of radio sources, which is called *scintillation*. The light from stars twinkles because of effects of our earth's atmosphere. The radio waves from radio sources scintillate not because of terrestrial effects but because radio signals are affected by clouds of electrons in the solar wind. Hewish therefore built a radio telescope (Fig. 10–10) uniquely able to detect faint, rapidly varying signals.

One day in 1967, Jocelyn Bell (now Jocelyn Burnell), then a graduate student working on the project, noticed that a set of especially strong variations in the signal appeared in the middle of the night, when scintillations caused by the solar wind are usually weak. In her own words, there was "a bit of scruff" on the tracing of the signal. After a month of observation, it became clear to her that the position of the source of the signals remained fixed with respect to the stars rather than constant in terrestrial time, a sure sign that the object was not terrestrial or solar.

Detailed examination of the signal showed that, surprisingly, it was a rapid set of pulses, with one pulse every 1.3373011 seconds. The pulses were very regularly spaced (Fig. 10–11). Soon, Burnell located three other sources, pulsing with regular periods of 0.253065, 1.187911, and 1.2737635 seconds, respectively.

When the discovery was announced to an astonished astronomical community, it was immediately apparent that the discovery of these pulsating radio sources, called *pulsars*, was one of the most important astronomical discoveries of the decade. But what were they?

Figure 10–10 The radio telescope—actually a field of aerials—with which pulsars were discovered at Cambridge, England. The total collecting area was large, so that it could detect faint sources, and the electronics was set so that rapid variations could be observed.

One immediate thought was that the signal represented an interstellar beacon sent out by extraterrestrial life on another star. For a time the source was called an LGM, for Little Green Men, and the possibility led to the Cambridge group's withholding the announcement for a time. The sources were briefly called LGM 1, LGM 2, LGM 3, and LGM 4, but soon it was obvious that the signals had not been sent out by extraterrestrial life. It seemed unlikely that there would be four such beacons at widely spaced locations in our galaxy. Besides, any being would probably be on a planet orbiting a star, and no effect of a Doppler shift from any orbital motion was detected.

Figure 10–11 This pulsar, PSR 0329+54, has a period of 0.7145 second. (Observations made in 1979 at the 91-m radio telescope of the National Radio Astronomy Observatory by Joseph H. Taylor, Marc Damashek and Peter Backus of the University of Massachusetts at Amherst)

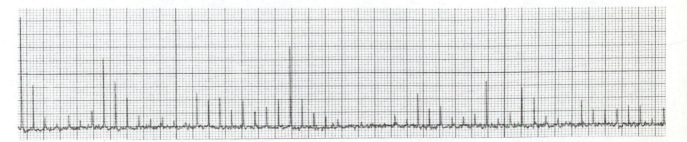

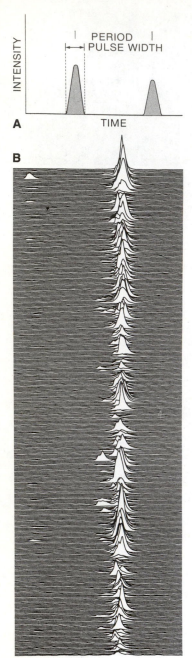

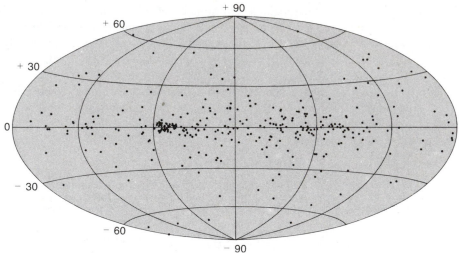

Other observatories turned their radio telescopes to search the heavens for new pulsars. Dozens of new pulsars were found. A typical pulse shape is shown in Figure 10–12*A*. The time from one pulse to the next pulse is the *period*. The pulse itself lasts only a small fraction of the period. The periods of known pulsars range from thousandths of a second to four seconds. A series of pulses appears in Fig. 10–12*B*.

When the positions of all the known pulsars are plotted on a chart of the heavens, it can easily be seen that they are concentrated along the plane of our galaxy (Fig. 10–13). Thus they are clearly objects in our galaxy, for if they were located outside our galaxy, we would expect them to be distributed uniformly or even for fewer to be visible where the Milky Way obscures the universe beyond.

10.6 What Are Pulsars?

Theoreticians went to work to try to explain the source of the pulsars' signals. The problem had essentially two parts. The first part was to explain what supplies the energy of the signal. The second part was to explain what causes the signal to be regularly timed, that is, what the "clock" mechanism is.

From studies of the second part of the problem, the choices were quickly narrowed down. The signals could not be coming from a pulsation of a normal dwarf star because such a star would be too big. If the sun, for example, were to turn off at one instant, we on earth would not see it go dark at once. The point nearest to us would disappear first, and then the darkening would spread farther back around the side of the sun (Fig. 10–14). The sun's diameter is 1.4 million km, so it would take 5 s to go dark. Pulsars pulsed too rapidly to be the size of a normal dwarf star.

That left white dwarfs and neutron stars as candidates. Remember, at that time, neutron stars were merely objects that had been predicted theoretically but had never been detected observationally. Could the pulsars be the more ordinary objects, namely, special kinds of white dwarfs?

Fiugre 10–12 *(A)* A schematic series of pulses. *(B)* A series of 400 consecutive pulses from pulsar PSR 0950+08. Each line from bottom to top represents the next 0.253 seconds. We can see structure within the individual pulses, and an "interpulse" about halfway between some consecutive pulses.

Figure 10–13 The distribution of most of the 369 known pulsars on a projection that maps the entire sky, with the plane of the Milky Way along the zero degree line on the map. The concentration of pulsars near 60° galactic longitude on this map merely represents the fact that this section of the sky has been especially carefully searched for pulsars because it is the area of the Milky Way best visible from the Arecibo Observatory. Thirty-four of the pulsars were newly reported in 1983. From the concentration of pulsars along the plane of our galaxy, we can conclude that pulsars are members of our galaxy; had they been extragalactic, we would have expected to see as many near the poles of this map. We can extrapolate that there are at least 100,000 pulsars in our galaxy.

Astronomers could conceive of two basic mechanisms as possibilities for the clock. The pulses might be coming from a star that was actually oscillating in size and brightness, or they might be from a star that was rotating. A third alternative was that the emitting mechanism involved two stars orbiting around each other in a binary system. However, it was soon shown that systems of orbiting stars would give off energy, making the period change. This was not observed, so orbiting stars were ruled out. This left the rotation or oscillation of white dwarfs or neutron stars.

Theoreticians have found that the denser a star, the more rapidly it oscillates. A white dwarf oscillates in less than a minute. However, it could not oscillate as rapidly as once every second, as would be required if pulsars were oscillating white dwarfs. Neutron stars, however, would indeed oscillate more rapidly, but they would oscillate once every 1/1000 second or so, too rapidly to account for pulsars. Thus oscillating neutron stars were ruled out as well.

What about rotation? Consider a lighthouse whose beacon casts its powerful beam many miles out to sea. As the beacon goes around, the beam sweeps past any ship very quickly, and returns again to illuminate that ship after it has made a complete rotation. Perhaps pulsars do the same thing: they emit a beam of radio waves that sweeps out a path in space. We see a pulse each time the beam passes the earth (Fig. 10–15).

Figure 10–14 Even if a large star were to turn off all at one time, the size of the star is such that it would appear to us to darken over a measurable period of time because radiation travels at the finite speed of light. Pulsars pulse so rapidly that we know that they cannot be large objects flashing.

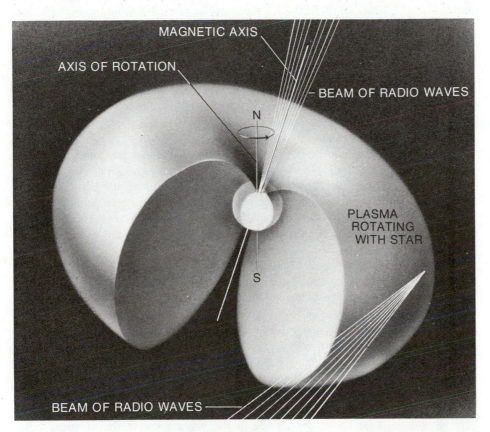

Figure 10–15 In the lighthouse model for pulsars, which is now commonly accepted, a beam of radiation flashes by us once each pulsar period, just as a lighthouse beam appears to flash by a ship at sea. It is believed that the generation of a pulsar beam is related to the neutron star's magnetic axis being aligned in a different direction than the neutron star's axis of rotation. The mechanism by which the beam is generated is not currently understood. Two possible variations of the lighthouse theory are shown here. The beam may be emitted along the magnetic axis, as shown in the top half of the figure. Alternatively, the beam could be generated on the surface of a doughnut of magnetic field (shown cut away) that surrounds and rotates with the neutron star, as shown in the bottom half of the figure.

Table 10–1 Possible Explanations of Pulsars

Hypothesis	Probability
Regular dwarf or giant	Ruled out
System of orbiting white dwarfs	Ruled out
System of orbiting neutron stars	Ruled out
Oscillating white dwarf	Ruled out
Oscillating neutron star	Ruled out
Rotating white dwarf	Ruled out
Rotating neutron star	Most likely

But what type of star could rotate at the speed required to account for the pulsations? We expect a collapsed star to be rotating relatively rapidly, just as ice skaters rotate faster when they pull in their arms to bring all parts of their bodies closer to their axes of rotation (Section 18.9). If the sun, which rotates once a month, would shrink by a factor of 100 to make it the size of the earth, it would speed up by 100 times, to 8 hours per rotation. If it shrank another 500 times, to make it the size of a neutron star, it would rotate every minute or so. And many stars rotate more rapidly than the sun. So rotation periods of 1 s are in the range we would expect for collapsed stars.

Could a white dwarf be rotating fast enough? Recall that a white dwarf is approximately the size of the earth. If an object that size were to rotate once every $\frac{1}{4}$ second, the centrifugal force—the outward force resulting from the rotation—would overcome even the immense inward gravitational force of a white dwarf. The outer layers would begin to be torn off. Thus pulsars were probably not white dwarfs; if a pulsar with a period even shorter than $\frac{1}{4}$ second were to be found, then the white dwarf model would be completely ruled out.

Neutron stars, on the other hand, are much smaller, so the centrifugal force would be weaker and the gravitational force would be stronger. As a result, neutron stars can indeed rotate four times a second. There is nothing to rule out their identification with pulsars. Since no other reasonable possibility has been found, astronomers accept the idea that pulsars are in fact neutron stars that are rotating. This is called the *lighthouse model*. Thus by discovering pulsars, we have also discovered neutron stars.

Note that we argued from elimination. For such an argument to be complete and convincing, we must be certain that we started with a complete list (Table 10–1).

About 300 pulsars have thus far been discovered. But the lighthouse model implies that there are many more, for we can see only those pulsars whose beams happen to strike the earth. If a pulsar were spinning at most orientations, then we would not know that it was there.

We have dealt above with only the second part of the explanation of the emission from pulsars: the clock mechanism. We understand much less well the mechanism by which the radiation is actually emitted in a beam. Presumably it has something to do with the extremely powerful magnetic field of the neutron star. Energy may be generated far above the surface of the neutron star in the magnetic field, perhaps near the location where the magnetic field lines, which are swept along with the star's rotation, attain velocities that approach that of light. Or the radio waves may be generated by charged particles that have escaped from the neutron star near its magnetic poles. These magnetic poles may not coincide with the neutron star's poles of rotation. After all, the earth's north magnetic pole is not at the north pole, but rather is near Hudson Bay, Canada (Fig. 10–16).

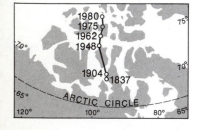

Figure 10–16 The magnetic north pole is the point on the earth's surface, about 1400 km from the north geographic pole, to which a magnetic compass would lead you. Interestingly, the magnetic pole is in continuous motion, and its mean position moves about 0.1° per year. It is believed that the same fluid motions of the earth's core that create the geomagnetic field are responsible for the movement of the magnetic pole.

Figure 10–17 The Crab Nebula, the remnant of a supernova explosion that became visible on earth in A.D. 1054. The pulsar, the first pulsar detected to be blinking on and off in the visible part of the spectrum, is marked with an arrow. In the long exposure necessary to take this photograph, the star turns on and off so many times that it appears to be a normal star. Only after its radio pulsation had been detected was it observed to be blinking in optical light. It had long been suspected, however, of being the leftover core of the supernova because of its unusual spectrum, which has only a continuum and no spectral lines.

10.7 The Crab, Pulsars, and Supernovae

Several months after the first pulsars had been discovered, strong bursts of radio energy were discovered coming from the direction of the Crab Nebula (Fig. 10–17). The bursts were sporadic rather than periodic, but astronomers hoped that they were seeing only the strongest pulses and that a periodicity could be found.

Within two weeks of feverish activity at several radio observatories, it was discovered that the pulsar in the Crab Nebula (named NP 0532) had a period of 0.033 second, the shortest period by far of all the known pulsars. The fastest pulsar previously known pulsed four times a second, while NP 0532 pulsed at the very rapid rate of thirty times a second. No white dwarf could possibly rotate that fast.

Furthermore, the Crab Nebula is a supernova remnant, and theory predicts that neutron stars should exist at the centers of at least some supernova remnants. Thus the discovery of a pulsar there, exactly where a neutron star would be expected, clinched the identification of pulsars with neutron stars.

Soon after NP 0532 was discovered, three astronomers at the University of Arizona decided to examine its position to look for an optical pulsar, even though earlier optical searches had failed to detect a pulsar at other sites. They turned their telescope toward a faint star in the midst of the Crab Nebula, and it very soon became apparent (Fig. 10–18*A*) that the star was pulsing! They had found an optical pulsar.

In all this time, only one other object has been found that pulses in the visible spectrum. Several objects are known that pulse x-rays or gamma rays, but the mechanism by which these pulses are generated is different from the run-of-the-mill radio-pulsing pulsar. When you say "pulsar," you mean that the star pulses radio waves.

X-ray observations from the Einstein Observatory revealed the structure of hot gas in the Crab Nebula, and clearly showed the pulsar blinking on and off (Fig. 10–18*B*). But Einstein observations of other supernova remnants often don't show bright central objects.

In one case, an x-ray source at the center of a young supernova remnant is apparently black-body radiation from a neutron star whose surface is 2,000,000 K. No x-ray or radio pulsations are seen.

Figure 10–18 *(A)* A sequence of pictures taken by adding light from different phases of the pulsar's period shows the pulsar apparently blinking on and off. Other stars in the field remain constant in intensity. Observations like this one clinched the identification of this star with the pulsar. (Observation by H. Y. Chiu, R. Lynds, and S. P. Maran at the Kitt Peak National Observatory) *(B)* High-resolution x-ray images of the Crab Nebula and its pulsar, displayed in 10 time intervals covering the 33-millisecond pulse period. A total of 10 hours of observation is included. The main pulse occurs in the second interval, and an in-between pulse called the "interpulse" falls in the sixth. (Einstein Observatory data from F. R. Harnden, Jr., High Energy Astrophysics Division of the Harvard-Smithsonian Center for Astrophysics)

10.8 *Slowing Pulsars and the Fast Pulsar*

When pulsars were first discovered, their most prominent feature was the extreme regularity of the pulses. It was hoped that they could be used as precise timekeepers, perhaps even more exact than atomic clocks on earth. After they had been observed for some time, however, it was noticed that the pulsars were slowing down very slightly. Presumably, the interaction of their magnetic field with interstellar matter slows the pulsars down.

The filaments in the Crab Nebula still glow brightly, even though over 900 years have passed since the explosion. Furthermore, the Crab gives off tremendous amounts of energy across the spectrum from x-rays to radio waves. It had long been wondered where the Crab got this energy. Theoretical calculations show that the amount of rotational energy lost by the Crab's neutron star as its rotation slows down is just the right amount to provide the energy radiated by the entire nebula. Thus the discovery of the Crab pulsar solved a long-standing problem in astrophysics: where the energy originates that keeps the Crab Nebula shining.

The Crab pulsar, pulsing 30 times a second, seemed to be spinning incredibly fast. And since it is a young pulsar (only 900 years old) and is slowing down more rapidly than other pulsars, astronomers assumed that the shorter the period, the younger the pulsar and the greater the slow-down rate. But in 1982, a pulsar spinning 20 times faster—642 times per second—was discovered. Even a neutron star rotating at this speed would be on the verge of being torn apart. The "fast pulsar," also known as the "millisecond pulsar" since its period is 1.5 milliseconds (0.0015 s), was found by a University of California team who chose to study in detail a strong radio source whose spectrum resembled that of a pulsar.

Since this pulsar's pulse rate (and thus speed of rotation) is changing only very slightly, its magnetic field must be relatively small for a pulsar. So it must be an old pulsar, in spite of its short period. A second millisecond pulsar since discovered shares its properties. This second source is in a binary system, one of only four such pulsars known. Though the first millisecond pulsar is not now in a binary system, theoretical models can explain its period if it was once in a binary system. It is the interaction between the members of the system that has speeded up the rotation rate of the pulsar; we think it was formed over a million years ago and its rotation had decayed to a much slower rate.

The first millisecond pulsar's period is changing so slightly and at such a steady rate that we may be able to use it to tell time more accurately over long periods than with atomic clocks. It seems to give us time to within a millionth of a second per year.

*10.9 The Binary Pulsar

All pulsars are odd, but some pulsars are odder than others. In 1974, Joseph H. Taylor, Jr., and Russell Hulse, then both at the University of Massachusetts at Amherst and now at Princeton, found a pulsar whose period of pulsation (approximately 0.059 second, one of the shortest) did not seem very regular. They finally realized that its variation in pulse period could be explained if the pulsar were orbiting another star about every 8 hours. If so, when the pulsar was approaching us, the pulses would be jammed together a bit, and when it was receding the pulses would be spread apart in time in a type of Doppler effect. The object is a *binary pulsar* and this particular case is "**the** binary pulsar."

Interpreting the binary pulsar system on the basis of many years of pulse arrival-time data has told us much. The pulses arrive 3 seconds earlier at some times than at others, indicating that the pulsar's orbit is 3 light seconds (1,000,000 km) across, approximately the diameter of the sun. Thus each of the objects has to be smaller than a normal dwarf. The companion object is probably a neutron star too. No pulses have been detected from it, but if it is giving off a radio beacon, it could simply be oriented at a different angle.

The binary pulsar has allowed us to derive the components' masses, which fall in the range we expected for neutron stars. But the main interest in the binary pulsar has come about because it provides powerful tests of the general theory of relativity. In Section 7.11, we discussed the excess advance of the perihelion of Mercury, which is only 43 seconds of arc per **century.** The binary pulsar, which is moving in a more elliptical orbit and in a much stronger gravitational field than is Mercury, should have its periastron advance by 4° per **year** (Fig. 10–19). (Periastron for a star corresponds to "perihelion," the closest approach of a planet's orbit to the sun.) Thus the periastron of the binary pulsar moves as far in a day as Mercury moves in a century. Observations of the binary pulsar show that its periastron indeed advances 4° per year, and so provide a strong confirmation of Einstein's general theory of relativity. In the next section, we will see how the binary pulsar has verified a previously untested prediction of Einstein's theory: gravity waves.

Two more binary pulsars have since been discovered, but the sizes and shapes of their orbits are not as favorable for testing relativity.

10.10 Gravitational Waves

Almost all astronomical data come from studies of the electromagnetic spectrum, with a small additional contribution of cosmic rays. But Einstein's general theory of relativity predicts the existence of still another type of signal: *gravitational*

Taylor's data are so sensitive that they show time on earth running up to 1.6 microseconds ahead of or behind the average rate each month as the earth and moon revolve around each other. The earth then moves slightly farther inward or outward on the warp of space caused by the sun's gravity.

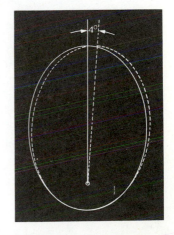

Figure 10–19 The periastron of the binary pulsar PSR 1913+16 advances by 4° per year, providing a strong endorsement of Einstein's general theory of relativity. The angle is shown for the apastron, the farthest point of the pulsar from the other object.

A   B

Figure 10–20 *(A)* Joseph Weber and one of his original gravitational-wave detectors from the 1960's, a large aluminum cylinder, weighing 4 tons, delicately suspended so that gravitational waves would set it vibrating 1660 times per second. Though Weber originally detected many "events" simultaneously affecting bars in both Maryland and Illinois, which could apparently be caused only by a global event like a gravity wave because they were separated by so great a distance, most of the newer, more sensitive experiments have not found events caused by gravity waves. Events recently affecting widely separated bars in Italy seem to occur too often to be from gravity waves. In any case, Weber's efforts have stimulated interest in detecting gravity waves, and even more sensitive detectors are currently in use or under construction at Louisiana State and the Universities of Maryland, Rochester, Rome, and Tokyo. *(B)* MIT's laser interferometer to measure gravity waves; similar apparatus is under development at Caltech. A laser beam enters the vertical tube A. It is then split into 2 beams that bounce back and forth between A and B and between A and C. The beams are then combined, which allows us to determine if the 2 different paths have changed in length. A passing gravity wave would change the effective length of one of the paths. The technique, which is similar to that used in the 1880's by A. A. Michelson to study the speed of light in various directions, can measure changes of a fraction of the wavelength of the laser light. It is believed that future larger versions would be sensitive enough to detect gravity waves from supernovae in nearby galaxies. Scientists at several institutions here and abroad are developing other laser detectors.

waves. Any accelerating mass will cause these waves, which are ripples in the curvature of space-time.

Whereas electromagnetic waves affect only charged particles, gravitational waves should affect all matter. But gravitational waves would be so weak that we could hope to detect them only in situations where large masses were being accelerated rapidly. Two neutron stars revolving in close orbits around each other, as in the binary pulsar, is one possible example. Another would be a supernova or the collapse of a massive star to become a black hole.

Joseph Weber of the University of Maryland was the first to build sensitive apparatus to search for gravitational waves. The detector was a large metal cylinder (Fig. 10–20), delicately suspended from wires and protected from local motion. When a gravitational wave hit the cylinder, it should have begun vibrating; the vibrations could be detected with sensitive apparatus. His apparatus started working in 1969; though it sometimes started vibrating, the vibrations seem to be from causes other than gravitational waves. Other scientists have since built more sensitive gravitational-wave apparatus, and no gravitational waves have been detected.

In the meantime, the binary pulsar turned out to be a quicker way to show that gravitational waves exist. The gravitational waves emitted by the system carry away energy. This makes the orbits shrink, which results in a slight decrease of the period of revolution. By 1978 Taylor and colleagues had detected such a slight speedup. The speedup has continued at precisely the rate predicted by the general theory of relativity. The pulsar now reaches periastron more than two seconds earlier than it would if its period had remained constant since it was discovered. The agreement of observations with predictions provides strong though indirect evidence for the existence of gravitational waves.

*10.11 X-Ray Binaries

Telescopes in orbit that are sensitive to x-rays have detected a number of strong x-ray sources, some of which are pulsating. Several of these objects pulse only in the x-ray region of the spectrum. One of the most interesting is Hercules X-1 (the first x-ray source to be discovered in the constellation Hercules). It is pulsating with a 1.24-second period.

Hercules X-1 and the other pulsating x-ray sources are apparently examples of the type of evolution of binary systems that we discussed in Section 9.7. Theoreticians think that the x-ray sources are radiating because mass from the companion is being funneled toward the poles of the neutron star by the neutron star's strong magnetic field. Unlike the slowing down of the pulse rate of pulsars (which give off pulses in the radio region of the spectrum), the pulse rate of the binary x-ray sources usually speeds up. The period of Hercules X-1, for example, is growing shorter.

Thus we know of two types of systems in which neutron stars are located. First, isolated neutron stars may give off pulses of radio waves, and we detect these stars as pulsars. Second, neutron stars in binary systems tend to give off pulses of x-rays.

*10.12 SS433: The Star that Is Coming and Going

SS433, one of the most exciting single objects in astronomy, contains spectral lines that change in wavelength by a tremendous amount. Its shifts of hundreds of angstroms correspond to Doppler shifts from velocities up to 50,000 km/sec. Such large shifts, about 15 per cent of the speed of light, are entirely unprecedented for an object in our galaxy. Furthermore, spectral lines appear simultaneously that are shifted to the red and to the blue by that same tremendous amount. Something seems to be coming and going at the same time.

SS433 was identified as a potentially interesting object because its optical spectrum showed emission lines. Then it turned out to be in the midst of an interstellar cloud thought to be a radio supernova remnant; it also seemed to correspond to an x-ray source. Following these discoveries, Bruce Margon and colleagues observed that a set of SS433's optical spectral lines apparently change in wavelength from night to night. It took a lengthy series of observations before they realized that the wavelengths varied in a regular fashion (Fig. 10–22), though large changes take place from night to night. The wavelengths have a period of 164 days.

''SS'' indicated that it is from a catalogue of stars with hydrogen emission lines compiled in the 1960's by C. Bruce Stephenson and Nicholas Sanduleak of Case Western Reserve University.

Figure 10–21 A computer simulation of the rotation of the disk of gas accumulated by the gravity of Hercules X-1—its *accretion disk*.

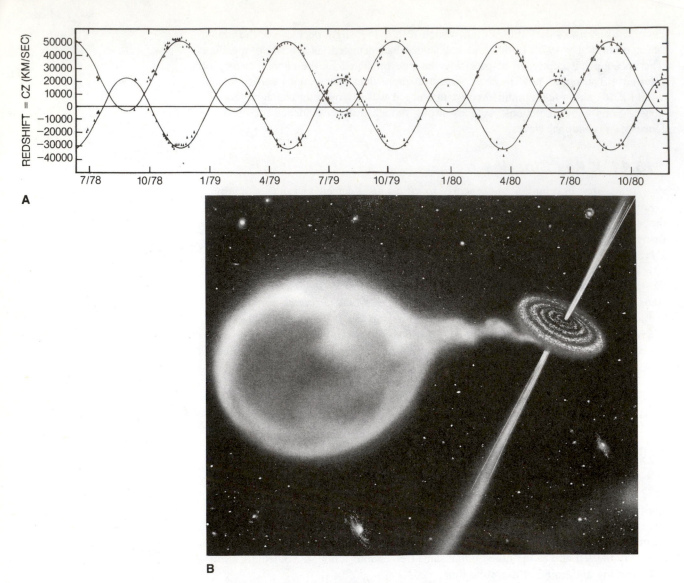

Figure 10–22 *(A)* The redshifts and blueshifts of spectral lines in SS433 vary in a regular fashion with a 164-day period. *(B)* A model of SS433 in which the radiation emanates from two narrow beams of matter that are given off by the accretion disk. The model follows ideas of Mordechai Milgrom of the Weizmann Institute in Israel, Bruce Margon, now at the University of Washington, and Jonathan Katz and George Abell of UCLA. The velocity of gas emitted is 80,000 km/s in each direction, though we see a somewhat lower velocity because we do not see the jets head on. In some models, variations in pressure and density of the disk can cause the emission.

It was later discovered that SS433 is in orbit with a 13-day period around another object. It is thus an x-ray binary, though its x-ray emission is relatively weak for a binary x-ray source. The object is apparently a massive star, probably of spectral type O or B.

The best models indicate that SS433 is a collapsed object—a neutron star or a black hole—surrounded by a disk of material gathered from its companion. Such a disk of accreted material is known as an *accretion disk;* astronomers have recently been finding signs of accretion disks in many types of objects. In the leading model (Fig. 10–22), material is being ejected from SS433's accretion disk along a line.

Because of the precession of the accretion disk, the line traces out a cone. (Precession is the wobbling motion, as of a child's top, that we have already met for the earth's rotation in Section 3.3.) If we assume that SS433 is oriented with its spin axis more-or-less but not completely perpendicular to the direction in which we are looking, we see radiation given off by the material ejected upward as redshifted and radiation given off by the material ejected downward as blueshifted. The redshift and blueshift vary as the line traces out the cone. The extended series of observations give the angles, and show that the material is ejected at 26 per cent the speed of light. X-ray (Fig. 10–23) and high-resolution radio (Fig. 10–24) observations seem to show the jets, confirming the model.

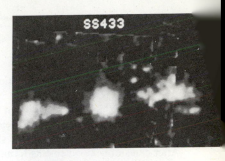

Figure 10–23 An x-ray image of SS433 from the Einstein Observatory. Two jets (or, at least, features that resemble two jets) can be seen extending to the sides of the bright central image. This central peak is actually a point source, though it is spread on this image because of the exposure necessary to emphasize the jets. SS433 is not an especially strong source; some of the other binary x-ray sources are 100 times stronger. (Einstein image courtesy of M. Watson, R. Willingale, J. Grindley, and F. Seward.)

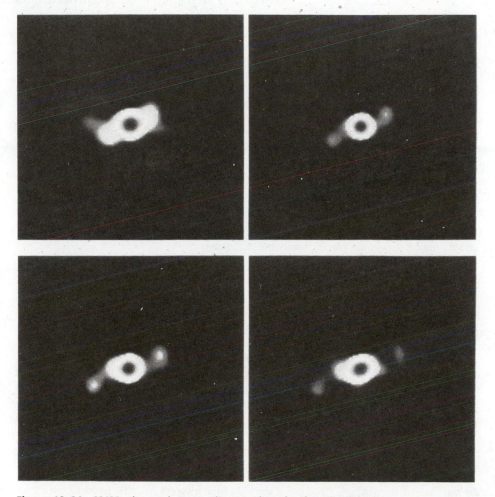

Figure 10–24 SS433 observed at a radio wavelength of 5 cm with the Very Large Array (VLA). SS433 is in the center of a supernova remnant known as W50. The images are taken at intervals of one month. Slight changes are visible in the structure and orientation of what appear to be the jets in accordance with the model we show above. (Courtesy of R. Hjellwing and K. Johnston, VLA of NRAO)

Summary and Outline

Entire chapter concerns the evolution of stars of more than 8 solar masses.

When they exhaust their hydrogen in the core, they become red giants. Later, heavier elements are built up in the core, and the stars become supergiants (Section 10.1).

After cores of heavy elements form, the stars explode as Type II supernovae. Type I supernovae come from less massive stars, and may be the incineration of a white dwarf. Only a few optical supernovae have been seen in our galaxy. Supernova remnants can also be studied with radio and x-ray astronomy (Section 10.2).

Cosmic rays, high-energy particles in space, probably come from supernovae (Section 10.3).

After a star's fusion stops, neutron degeneracy can support a remaining mass of up to 2 or 3 solar masses. A neutron star may be only 10 km in radius, and so is fantastically dense (Section 10.4).

Pulsars were discovered when a radio telescope was built that did not mask rapid time variations in the signal. Pulsars have very regular signals, and are slowing down very slightly. Pulse periods range from 1.5 milliseconds to 4 sec (Section 10.5).

Scientists have identified pulsars with rotating neutron stars by a process of elimination (Section 10.6), which requires complete, logical reasoning.

A pulsar is observed in the center of the Crab Nebula, a supernova remnant, where we would expect to observe a neutron star.

Pulsars emit mainly in the radio spectrum, although the Crab and Vela pulsars also pulse light, x-rays, and γ-rays. Other neutron stars pulse x-rays (Section 10.7).

Pulsars are gradually slowing down as they grow older (Section 10.8).

Several binary pulsars have been discovered. The periastron of the Hulse-Taylor binary advances $4°$/year, confirming general relativity (Section 10.9).

General relativity predicts the existence of gravitational waves. Attempts to detect them have probably been unsuccessful. The orbital period of the binary pulsar is slowing down by just the amount that is expected to correspond to the emission of gravitational waves from that system, confirming Einstein's theory (Section 10.10).

Many x-ray sources are caused by the infall of material on a neutron-star member of the binary system from its companion (Section 10.11).

An object in our galaxy, SS433, has spectral lines that shift periodically to the red and to the blue by about 15 per cent of the speed of light. The Doppler-shifted radiation may come from beams ejected from an accretion disk about the neutron star in an x-ray binary (Section 10.12).

Key Words

supernova, red supergiant, supernova remnant, cosmic rays (primary, secondary), neutron degeneracy, neutron star, scintillation, pulsar, period, lighthouse model, binary pulsar*, gravitational wave, accretion disk

Questions

1. Why are red supergiants so bright?

2. What are the basic differences between a nova and a supernova?

3. If we see a massive main-sequence star (a heavyweight star), what can we assume about its age, relative to most stars? Why?

4. What do we know about the core of a star when it leaves the main sequence?

5. What is special about iron in the core of a star?

†6. In a supernova explosion of a 20-solar-mass star, about how much material is blown away?

†7. A supernova can brighten by 20 magnitudes. By what factor of brightness is this? Show your calculations.

8. How do we distinguish observationally between Type I and Type II supernovae?

9. How do the physical models of Type I and Type II supernovae differ?

†10. A typical galaxy has a luminosity 10^{11} times that of the sun. If a supernova equals this luminosity, and if it began as a B star, by what factor did it brighten? By how many magnitudes?

11. How does eta Carinae endorse our model of supergiants?

12. Would you expect the appearance of the Crab Nebula to change in the next 500 years? How?

†13. If the light from a supernova 2000 parsecs from us was received in A.D. 1054, when did the star actually explode?

---

*This word is found in an optional section.

†This indicates a question requiring a numerical solution.

14. What are cosmic rays?

15. Where are secondary cosmic rays formed?

16. In your own words, fill in Table 10–1 with the reasons why each of the explanations for pulsars involving single stars was ruled out except for the case of rotating neutron stars.

17. In view of current theories about supernovae and pulsars, list the pieces of evidence that indicate that the Crab pulsar is a young one.

†18. A pulsar has a period of 1 second and a pulse lasts 5 per cent of the cycle. What is the pulse's width in seconds? What might that say about the size of the object emitting the pulse on (a) the oscillating and (b) the rotating model?

19. Why did it seem strange that the fast pulsar is slowing down so slowly? Explain.

20. Explain two important consequences of the discovery of the binary pulsar.

21. Why is the discovery of gravitational waves a test of the general theory of relativity?

22. Sketch a model for an x-ray binary and explain how it leads to x-ray emission.

23. A shift of the Hα line in SS433 of 40,000 km/sec corresponds to a shift of how many Å in wavelength? (Ignore any effects of relativity.)

24. Describe why the precession of an accretion disk in SS433 can cause varying Doppler shifts.

25. Explain why the true velocities of the jets in SS433 may be even greater than the velocities we measure directly from Doppler shifts.

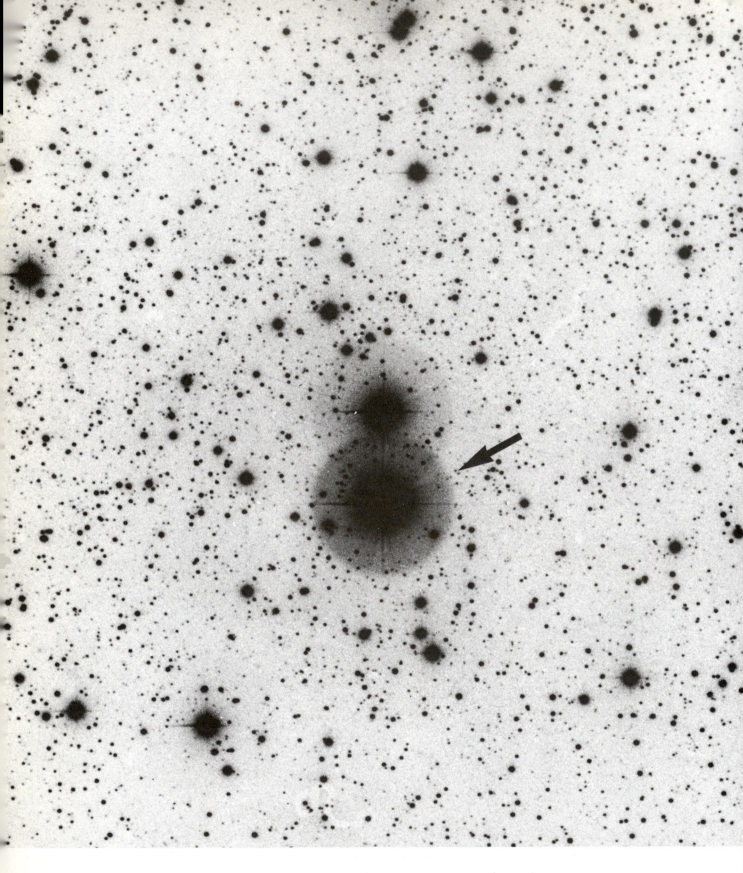

The overexposed dark object in the center of this negative print is the blue supergiant star HDE 226868, which is thought to be the companion of the first black hole to be discovered, Cygnus X-1.

Black Holes 11

Aims: To understand what black holes are, how they form, and how they might be detected

The strange forces of electron and neutron degeneracy support dying lightweight, middleweight, and some heavyweight stars against gravity. The strangest case of all occurs at the death of the most massive stars, which contained much more than 8 and up to about 60 solar masses when they were on the main sequence. They generally undergo supernova explosions like less massive heavyweight stars, but some of these most massive stars may retain cores of over 2 or 3 solar masses. Nothing in the universe is strong enough to hold up the remaining mass against the force of gravity. The remaining mass collapses, and continues to collapse forever. The result of such a collapse is one type of *black hole,* in which the matter disappears from contact with the rest of the universe. Later, we shall discuss the formation of black holes in processes other than those that result from the collapse of a star.

11.1 The Formation of a Stellar Black Hole

When nuclear fusion ends in the core of a star, gravity causes the star to contract. We have seen in Chapter 9 that a star that has a final mass of less than 1.4 solar masses will end its life as a white dwarf. In Chapter 10 we saw that a heavier star will become a supernova, and that if the remaining mass of the core is less than 2 or 3 solar masses, it will wind up as a neutron star. We are now able to observe both white dwarfs and neutron stars and so can study their properties directly.

It now seems reasonable that in some cases more than 2 or 3 solar masses remain after the supernova explosion. The star collapses through the neutron star stage, and we know of no force that can stop the collapse. In some cases, the matter may even have become so dense as the star collapsed that no explosion took place and no supernova resulted.

The value for the maximum mass that a neutron star can have is a result of theoretical calculation. We must always be aware of the limits of accuracy of any calculation, particularly of a calculation that deals with matter in a state that is very different from states that we have been able to study experimentally. Still, the best modern calculations show that the limit of a neutron star mass is 2 or 3 solar masses.

We may then ask what happens to a 5- or 10- or 50-solar-mass star as it collapses, if it retains more than 3 solar masses. It must keep collapsing, getting denser and denser. We have seen that Einstein's general theory of relativity predicts that a strong gravitational field will redshift radiation, and that this prediction has been verified both for the sun and for white dwarfs (as discussed in Section 9.5). Also, radiation will be bent by a gravitational field, or at least appear to us on earth as though it were bent (Section 7.11). Further, the general theory of relativity indicates that pressure as well as mass produces gravity. This effect compounds the gravitational contraction of a very massive star.

As the mass contracts and the star's surface gravity increases, radiation is continuously redshifted more and more, and radiation leaving the star other than perpen-

Astronomers long assumed that the most massive stars would somehow lose enough mass to wind up as white dwarfs. When the discovery of pulsars ended this prejudice, it seemed more reasonable that black holes could exist.

Compressed matter, perhaps like that in a neutron star, was formed in a terrestrial laboratory for the first time in 1984.

Figure 11–1 As the star contracts, a light beam emitted other than radially outward will be bent.

dicularly to the surface is bent more and more. Eventually, when the mass has been compressed to a certain size, radiation from the star can no longer escape into space. The star has withdrawn from our observable universe, in that we can no longer receive radiation from it. We say that the star has become a *black hole*.

Why do we call it a black hole? We think of a black surface as a surface that reflects none of the light that hits it. Similarly, any radiation that hits the surface of a black hole continues into the black hole and is not reflected. In this sense, the object is perfectly black. (A ''black'' piece of paper on earth, in contrast, may radiate in the infrared, and is not truly black.)

11.2 The Photon Sphere

Let us consider what happens to radiation emitted by the surface of a star as it contracts. Although what we will discuss affects radiation of all wavelengths, let us simply visualize standing on the surface of the collapsing star while holding a flashlight.

On the surface of a supergiant star, we would note only very small effects of gravity on the light from our flashlight. If we shine the beam at any angle, it seems to go straight out into space.

As the star collapses, two effects begin to occur. Although we on the surface of the star cannot notice them ourselves, a friend on a planet revolving around the star could detect the effects and radio back information to us about them. For one thing, our friend could see that our flashlight beam is redshifted. Second, our flashlight beam would be bent by the gravitational field of the star (Fig. 11–1). If we shined the beam straight up, it would continue to go straight up. But if we shined it away from the vertical, the beam would be bent even farther away from the vertical. When the star reaches a certain size, a horizontal beam of light would not escape (Fig. 11–2).

From this time on, only if the flashlight is pointed within a certain angle of the vertical does the light continue outward. This angle forms a cone, with its apex at the flashlight, and is called the *exit cone* (Fig. 11–3). As the star grows smaller yet, we find that the flashlight has to be pointed more directly upward in order for its light to escape. The exit cone grows smaller as the star shrinks.

When we shine our flashlight upward in the exit cone, the light escapes. When we shine our flashlight in a direction outside the exit cone, the light is bent sufficiently that it falls back to the surface of the star. When we shine our flashlight exactly along the side of the exit cone, the light goes into orbit around the star, neither escaping nor falling onto the surface (Fig. 11–3).

The sphere around the star in which the light can orbit is called the *photon sphere*. Its size can be calculated theoretically. It is 13.5 km in radius for a star of 3 solar masses, and its size varies linearly with the mass; that is, a star of 6 solar masses is 27 km in radius, and so on.

As the star continues to contract, the theory shows that the exit cone gets narrower and narrower. Light emitted within the exit cone still escapes. The photon sphere remains at the same height even though the matter inside it has contracted further, since the total amount of matter within has not changed.

11.3 The Event Horizon

If we were to depend on our intuition, which is based on classical physics as advanced by Newton, we might think that the exit cone would simply continue to get narrower. But when we apply the general theory of relativity, we find that when

Figure 11–2 Light can be bent so that it falls back onto the star.

the star contracts beyond a certain size the cone vanishes. Light no longer can escape into space, even when it is travelling straight up.

The solutions to Einstein's equations that predict this were worked out by Karl Schwarzschild in 1916, shortly after Einstein advanced his general theory. The radius of the star at the time at which light can no longer escape is called the *Schwarzschild radius* or the *gravitational radius,* and the spherical surface at that radius is called the *event horizon* (Fig. 11–4). The event horizon is not a physical surface. The radius of the photon sphere is exactly 3/2 times this Schwarzschild radius.

We can visualize the event horizon in another way, by considering a classical picture based on the Newtonian theory of gravitation. The picture is essentially that conceived in 1796 by Laplace, the French astronomer and mathematician. You must have a certain velocity, called the *escape velocity,* to escape from the gravitational pull of another body. For example, we have to launch rockets at 11 km/sec (40,000 km/hr) in order for them to escape from the earth's gravity. For a more massive body of the same size, the escape velocity would be higher. Now imagine that this body contracts. We are drawn closer to the center of the mass. As this happens, the escape velocity rises. When all the mass of the body is within its Schwarzschild radius, the escape velocity becomes equal to the speed of light. Thus even light cannot escape. If we begin to apply the special theory of relativity, we might then reason that since nothing can go faster than the speed of light, nothing can escape. Now let us return to the picture according to the general theory of relativity.

The size of the Schwarzschild radius depends linearly on the amount of mass that is collapsing. A star of 3 solar masses, for example, would have a Schwarzschild radius of 9 km. A star of 6 solar masses would have a Schwarzschild radius of twice 9 km, or 18 km. One can calculate the Schwarzschild radii for less massive stars as well, although the less massive stars would be held up in the white dwarf or neutron-star stages and not collapse to their Schwarzschild radii (although conceivably some objects could be sufficiently compressed in a supernova). Since one couldn't, theoretically, put a tape measure into and out of a black hole to measure its radius, we would actually measure its circumference and divide by 2π.

Note that anyone or anything on the surface of a star as it passed its event horizon would not be able to survive. An observer would be torn apart by the tremendous difference in gravity between head and foot. (This is called a tidal force, since this kind of difference in gravity also causes the tides on earth.) If the tidal force could be ignored, though, the observer on the surface of the star would not notice anything particularly wrong as the star passed its event horizon. If we chose to stay with the observer at that time, we could still be pointing our flashlights up into space trying to signal our friends on the earth. But once we passed the event horizon, no answer would ever come because our signal would never get out.

Once the star passes inside its event horizon, it continues to contract. Nothing can ever stop its contraction. In fact, the mathematical theory predicts that it will contract to zero radius, a situation that seems impossible to conceive of. The point at which it will have zero radius (it has infinite density there) is called a *singularity.* Strange as it seems, theory predicts that a black hole contains a singularity.

Even though the mass that causes the black hole has contracted further, the event horizon doesn't change. It remains at the same radius forever, as long as the amount of mass inside doesn't change.

*11.4 Rotating Black Holes

Once matter is inside a black hole, it loses its identity in the sense that from outside a black hole, all we can tell is the mass of the black hole, the rate at which it is spinning, and what total electric charge it has. These three quantities are suffi-

Figure 11–3 When the star has contracted enough (the inner sphere), only light emitted within the exit cone escapes. Light emitted on the exit cone goes into the photon sphere. The further the star contracts within the photon sphere, the narrower the exit cone becomes.

The sun's Schwarzschild radius is 3 km. The Schwarzschild radius for the earth is only 9 mm; that is, the earth would have to be compressed to a sphere only 9 mm in radius in order to form an event horizon and be a black hole. Its surface would be so much closer to its center that its surface gravity would be 5×10^{13}—50 trillion times higher than it is now.

Figure 11–4 When the star becomes smaller than its Schwarzschild radius, we can no longer observe it. We say that it has passed its *event horizon,* by analogy to the statement that we cannot see a thing on earth once it has passed our horizon.

cient to completely describe the black hole. Thus, in a sense, black holes are simple objects to describe physically, because we only have to know three numbers to characterize each one. (By contrast, for the earth we have to know shapes, sizes, densities, motions, and other parameters for the interior, the surface, and the atmosphere.) The theorem that describes the simplicity of black holes is often colloquially stated by astronomers active in the field as ''a black hole has no hair.'' All other properties, such as size, can be derived from the three basic properties.

Most of the theoretical calculations about black holes, and the Schwarzschild solutions to general relativity, in particular, are based on the assumption that black holes do not rotate. But this assumption is only a convenience; we think, in fact, that the rotation of a black hole is one of its important properties. It was not until 1963 that Roy P. Kerr solved Einstein's equations for a situation that was later interpreted in terms of the notion that a black hole is rotating. In this more general case, an additional special boundary—the *stationary limit*—appears, with somewhat different properties from the original event horizon. At the stationary limit, space-time is flowing at the speed of light, so particles that would otherwise be moving at the speed of light would remain stationary. Within the stationary limit, no particles can remain at rest even though they are outside the event horizon.

"Now here, you see, it takes all the running you can do, to keep in the same place. If you want to get somewhere else, you must run twice as fast as that.''
The Red Queen,
Through the Looking Glass
By Lewis Carroll, 1871

Let us consider a series of non-moving (''static'') points at distances close to a black hole and visualize what happens to a sphere representing expanding wavefronts of light emitted from these points. In Fig. 11–6*A,* we see the wavefronts a standard time (say, 1 second) after they are emitted from each dot. For a point inside the event horizon, all the emitted waves go inward instead of outward, and no light escapes. In Fig. 11–6*B,* we see top views of a series of points near a rotating black hole. The rotation carries the wavefront to the side, and for a point at the stationary limit, some of the wavefront is outside the limit. These light waves can escape. Only for points within the event horizon do all wavefronts move inward.

The equator of a stationary limit of a **rotating** black hole has the same diameter as the event horizon of a **non-rotating** black hole of the same mass. But a rotating black hole's stationary limit is squashed. The event horizon touches the stationary limit at the poles. Since the event horizon remains a sphere, it is smaller than the event horizon of a non-rotating black hole (Fig. 11–6*C*).

The region between the stationary limit and the event horizon is the *ergosphere*. An object can be shot into the ergosphere of a rotating black hole at such an angle that, after it is made to split in two, one part continues into the event horizon while the other part is ejected from the ergosphere with more energy than the incoming

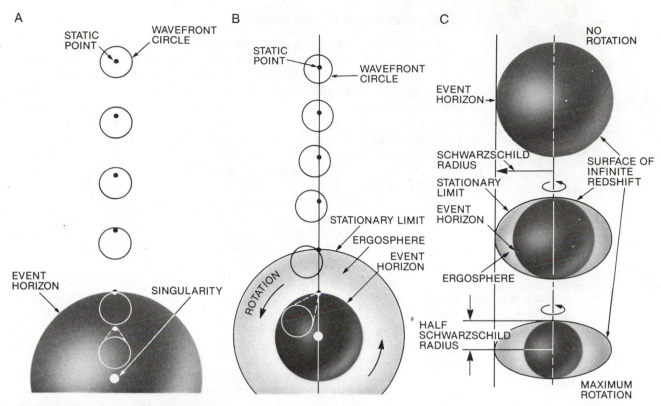

Figure 11–6 *(A)* A static point that emits a pulse of light in all directions, and the circle showing the wavefront a standard time later for locations near a non-rotating black hole. The circle is dragged more toward the black hole the closer the static point is to the black hole. For a static point at the event horizon, the wave circles are entirely inside and no light escapes. *(B)* Near a rotating black hole, the wave circles in this top view are displaced not only inward but also in the direction of rotation. For a static point on the stationary limit, part of the circle is outside the stationary limit, so some light escapes. Some light can escape even from inside the stationary limit. *(C)* Side views of black holes of the same mass and different rotation speeds. The region between the stationary limit and the event horizon of a rotating black hole is called the *ergosphere* (from the Greek word *ergon,* work) because in principle work can be extracted from it.

particle had. This energy must have come from somewhere; it could only come from the rotational energy of the black hole. Thus by judicious use of the ergosphere, we could, in principle, tap the energy of the black hole. We are always looking for new energy sources and this is the most efficient known, although it is obviously very far from being practical.

A black hole can rotate up to the speed at which a point on the event horizon's equator is travelling at the speed of light. The event horizon's radius is then half the Schwarzschild radius. If a black hole rotated faster than this, its event horizon would vanish. Unlike the case of a non-rotating black hole, for which the singularity is always unreachably hidden within the event horizon, in this case distant observers could receive signals from the singularity. Such a point would be called a *naked singularity* and, if one exists, we might have no warning—no photon sphere or orbiting matter, for example—before we ran into it. Most theoreticians assume the existence of a law of "cosmic censorship," which requires all singularities to be "clothed" in event horizons, that is, not naked. Since so much energy might erupt from a naked singularity, we can conclude that there are none in our universe from the fact that we do not find signs of them.

11.5 Detecting a Black Hole

What if we were watching a faraway star collapse? Photons that were just barely inside the exit cone when emitted would leak very slowly away from the black hole. These few photons would be spread out in time, and the image would become very faint. As the star went through its event horizon, the last photons to reach us would be extremely redshifted, so the star would get very red and then blink out. It would do this in a fraction of a second, so the odds are unfavorable that we would actually see a collapsing star as it went through the event horizon.

But all hope is not lost for detecting a black hole, even though we can't hope to see the photons from the moments of collapse. The black hole disappears, but it leaves its gravity behind. It is a bit like the Cheshire Cat from *Alice's Adventures in Wonderland,* which fades away, leaving only its grin behind (Fig. 11–7).

The black hole attracts matter, and the matter accelerates toward it. Some of the matter will be pulled directly into the black hole, never to be seen again. But other matter will go into orbit around the black hole, and will orbit at a high velocity. This accreted matter forms an *accretion disk* (Fig. 11–8).

It seems likely that the gas in orbit will be heated. Theoretical calculations of the friction that will take place between adjacent filaments of gas in orbit around the black hole show that the heating will be so great that the gas will radiate strongly in the x-ray region of the spectrum. The inner 200 km should reach hundreds of millions of kelvins. Thus, though we cannot observe the black hole itself, we can hope to observe x-rays from the gas surrounding it.

In fact, a large number of x-ray sources are known in the sky. They have been surveyed from NASA's Uhuru and HEAO's (High-Energy Astronomy Observatories), and now from the Japanese Hakucho x-ray satellite and the European Space Agency's Exosat. Some of these sources have been identified with galaxies or quasars. Others pulse regularly, like Hercules X-1, and are undoubtedly neutron stars (Section 10.11). Some of the rest, which pulse sporadically, may be related to black holes.

looked goodnatured, she thought: still it had *very* long claws and a great many teeth, so she felt it ought to be treated with respect.

"Cheshire Puss," she began, rather timidly, as she did not at all know whether it would like the

Figure 11–7 Lewis Carroll's Cheshire Cat, from *Alice's Adventures in Wonderland,* shown here in John Tenniel's drawing, is analogous to a black hole in that it left its grin behind when it disappeared, while a black hole leaves its gravity behind when its mass disappears. Alice thought that the Cheshire Cat's persisting grin was "the most curious thing I ever saw in my life!" We might say the same about the black hole and its persisting gravity.

Figure 11–8 (A) The optical appearance of a rotating black hole surrounded by a thin accretion disk. The drawing shows, on the basis of computer calculations, how curves of equal intensity are affected by the mass of the black hole. The observer is located 10° above the plane of the accretion disk. The asymmetry in appearance is caused by the rotation of the black hole. (B) A supercomputer simulation shows how a black hole's gravity affects the orbiting gas in the accretion disk that surrounds it. Effects close to the event horizon show for the first time; it turns out that gas can sometimes even bounce back out. (John F. Hawley and Larry Smarr, University of Illinois at Urbana-Champaign)

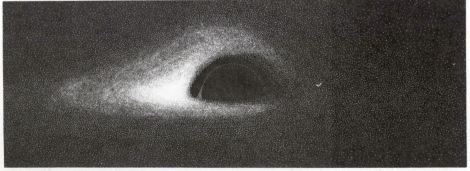

A

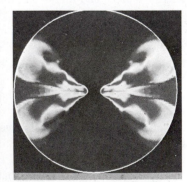

B

It is not enough to find an x-ray source that gives off sporadic pulses, for one can think of other mechanisms besides matter revolving around a black hole that can lead to such pulses. One would like to show that a collapsed star of greater than 3 solar masses is present. We can determine masses only for certain binary stars. When we search the position of the x-ray sources, we look for a spectroscopic binary (that is, a star whose spectrum shows a Doppler shift that indicates the presence of an invisible companion). Then, if we can show that the companion is too faint to be a normal, main-sequence star, it must be a collapsed star. If, further, the mass of the unobservable companion is greater than 3 solar masses, it must be a black hole.

The most persuasive case is named Cygnus X-1, an x-ray source that varies in intensity on a time scale of milliseconds. In 1971, radio radiation was found to come from the same direction. A 9th magnitude star called HDE 226868 was found at its location (Fig. 11–9).

HDE 226868 has the spectrum of a blue supergiant, and thus has a mass of about 15 times that of the sun. Its spectrum is observed to vary in radial velocity with a period of 5.6 days, indicating that the supergiant and the invisible companion are orbiting each other with that period. From the orbit, it is deduced that the invisible companion must certainly have a mass greater than 4 solar masses; the best estimate is 8 solar masses. Because this is so much greater than the limit of 2 or 3 solar masses above which neutron stars cannot exist, it seems that even allowing for possible errors in measurement or in the theoretical calculations, too much mass is present to allow the matter to settle down as a neutron star. Thus many, and probably most, astronomers believe that a black hole has been found in Cygnus X-1.

Another promising candidate is LMC X-3 (the third x-ray source to be found in the Large Magellanic Cloud). Since we know the distance to the Large Magellanic Cloud and thus to anything in it much more accurately than we know the distance to Cygnus X-1, our calculations for the invisible object's mass may be more accurate. Again, the invisible object seems to have about 8 solar masses of matter, and so must be a black hole.

Matter in the accretion disk around the black hole (Fig. 11–10) would orbit very quickly. If a hot spot developed somewhere on the disk, it might beam a cone of radiation into space, similar to the lighthouse model of pulsars. If the hot spot lasted for several rotations, we could detect a pulse of x-rays every time the cone swept past the earth. The period of the x-ray pulses would be extremely short, only a few

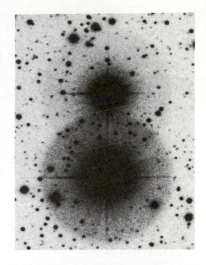

Figure 11–9 A section of the image reproduced in the photograph opening this chapter of the blue supergiant star HDE 226868. The black hole Cygnus X-1 that is thought to be orbiting the supergiant star is not visible. Note the fact that the image of the supergiant appears so large because it is overexposed on the film. The image does not represent the actual angular diameter subtended by the star, which is too small to resolve from the earth.

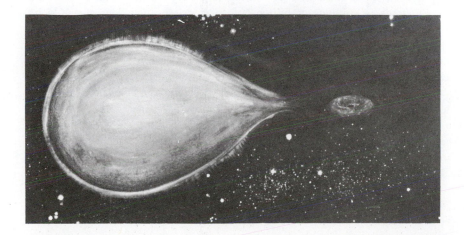

Figure 11–10 An artist's conception of the disk of swirling gas that would develop around a black hole like Cygnus X-1 *(right)* as its gravity pulled matter off the companion supergiant *(left)*. The x-radiation would arise in the disk. (Painting by Lois Cohen, courtesy of the Griffith Observatory)

Figure 11–11 An x-ray view of Cygnus X-1, the first picture taken with HEAO-2, the Einstein Observatory. The resolution of the telescope system aboard was about 4 arc sec, whereas the orbiting gas around the black hole—the "accretion disk"—is 100,000 times smaller, according to calculations, so that a photograph like this one mainly shows that the telescope optics are focusing properly. The object was allowed to drift into the field of view for this first image, hence the appearance of a "tail." (Courtesy of Riccardo Giacconi, then at the Harvard-Smithsonian Center for Astrophysics)

milliseconds. Searches for the short-period pulses have been made with x-ray satellites. Cygnus X-1 and LMC X-3 do indeed flicker in x-rays, though current theories do not get the periods exactly right.

*11.6 Non-stellar Black Holes

We have discussed how black holes can form by the collapse of massive stars. But theoretically a black hole should result if a mass of any amount is sufficiently compressed. No object containing less than 2 or 3 solar masses will contract sufficiently under the force of its own gravity in the course of stellar evolution. But the density of matter was so high at the time of the origin of the universe (see Chapter 17) that smaller masses may have been sufficiently compressed to form what are called mini black holes.

Stephen Hawking, an English astrophysicist, has suggested the existence of mini black holes, the size of pinheads (and thus with masses equivalent to those of asteroids). There is no observational evidence for a mini hole, but they are theoretically plausible. Hawking has deduced that small black holes can seem to emit energy in the form of elementary particles (neutrinos and so forth). The mini holes would thus evaporate and disappear. This may seem to be a contradiction to the concept that mass can't escape from a black hole. But when we consider effects of quantum mechanics, the simple picture of a black hole that we have discussed up to this point is not sufficient. Hawking suggests that a black hole so affects space near it that a pair of particles—a nuclear particle and its antiparticle—can form simultaneously. The antiparticle disappears into the black hole, and the remaining particle reaches us. Photons, which are their own antiparticles, appear too.

Emission from a black hole is significant only for the smallest mini black holes, for the amount of radiation increases sharply as we consider less and less massive black holes. Only mini black holes up to the mass of an asteroid—far short of stellar masses—would have had time to disappear since the origin of the universe. Hawking's ideas set a lower limit on the size of black holes now in existence, since we think the mini black holes were formed only in the first second after the origin of the universe by the tremendous pressures that existed then.

On the other extreme of mass, we can consider what a black hole would be like if it contained a very large number, i.e., thousands or millions, of solar masses. Thus far, we have considered only black holes the mass of a star or smaller. Such black holes form after a stage of high density. But the more mass involved, the lower the density needed for a black hole to form. For a very massive black hole, one containing hundreds of millions or billions of solar masses, the density would be fairly low when the event horizon formed, approaching the density of water. For even higher masses, the density would be lower yet.

Thus if were traveling through the universe in a spaceship, we couldn't count on detecting a black hole by noticing a volume of high density. We could pass through the event horizon of a high-mass black hole without even noticing. We would never be able to get out, but it would be hours before we would notice that we were being drawn into the center at an accelerating rate.

Where could such a supermassive black hole be located? The center of our galaxy may contain a black hole of a million solar masses. Though we would not observe radiation from the black hole itself, the gamma-rays, x-rays, and infrared radiation we detect would be coming from the gas surrounding the black hole. Active galaxies and quasars (Section 15.4) are other probable locations for massive black holes.

Summary and Outline

Gravitational collapse occurs if more than 2 or 3 solar masses remain (Section 11.1).

Critical radii of a black hole

Photon sphere: exit cones form (Section 11.2).

Event horizon: exit cones close (Section 11.3).

Schwarzschild radius (gravitational radius) defines the limit of the black hole, the event horizon.

Singularity

Rotating black holes (Section 11.4)

Can get energy out of the ergosphere

Detecting a black hole ((Section 11.5)

Flickering x-radiation expected from accretion disks

Detection in a spectroscopic binary, as an invisible high-mass companion

Cygnus X-1 and LMC X-3: black holes?

Non-stellar black holes (Section 11.6)

Mini black holes could have formed in the big bang.

Very massive black holes do not have high densities.

Key Words

black hole, exit cone, photon sphere, Schwarzschild radius, gravitational radius, event horizon, escape velocity, singularity, stationary limit*, naked singularity*, ergosphere*, accretion disk

Questions

1. Why doesn't electron or neutron degeneracy prevent a star from becoming a black hole?

2. Why is a black hole blacker than a black piece of paper?

3. (a) Is light acting more like a particle or more like a wave when it is bent by gravity?

   (b) Explain the bending of light as a property of a warping of space (Section 7.11).

4. (a) How does the escape velocity of the moon compare with the escape velocity of the earth? Would a larger rocket engine be necessary to escape from the gravity of the earth or from the gravity of the moon?

   (b) How will the velocity of escape from the surface of the sun change when the sun becomes a red giant? A white dwarf?

†5. (a) What is the Schwarzschild radius for a 10-solar-mass star?

   (b) What is your Schwarzschild radius?

†6. What are radii of the photon sphere and the event horizon of a non-rotating black hole of 18 solar masses?

†7. What is the size of the stationary limit and of the event horizon of an 18-solar-mass star that is rotating at the maximum possible rate?

8. What is the relation in size of the photon sphere and the event horizon? If you were an astronaut in space, could you escape from within the photon sphere of a rotating black hole? From within its ergosphere? From within its event horizon?

9. (a) How could the mass of a black hole that results from a collapsed star increase?

   (b) How could mini black holes, if they exist, lose mass?

10. Would we always notice when we reached a black hole by its high density? Explain.

11. Could we detect a black hole that was not part of a binary system?

12. Under what circumstances does the presence of an x-ray source associated with a spectroscopic binary suggest to astronomers the presence of a black hole?

---

*These words are found in an optional section.

†This indicates a question requiring a numerical solution.

Part IV The Milky Way Galaxy

The Origin of the Milky Way by Tintoretto, circa 1578. (The National Gallery, London)

On the clearest nights, when we are far from city lights, we can see a hazy band of light stretched across the sky. This band is the *Milky Way*—the aggregation of dust, gas, and stars that makes up the galaxy in which the sun is located.

Don't be confused by the terminology: the Milky Way itself is the band of light that we can see from the earth, and the Milky Way Galaxy is the whole galaxy in which we live. Like other galaxies, our Milky Way Galaxy is composed of perhaps a trillion stars plus many different types of gas, dust, planets, etc. The Milky Way is that part of the Milky Way Galaxy that we can see with the naked eye in our nighttime sky.

The Milky Way appears very irregular in form when we see it stretched across the sky—there are spurs of luminous material that stick out in one direction or another, and there are dark lanes or patches in which nothing can be seen. This patchiness is simply due to the splotchy distribution of dust, stars, and gas.

Here on earth, we are inside our galaxy together with all of the matter we see as the Milky Way. Because of our position, we see a lot of matter when we look in the plane of our galaxy. On the other hand, when we look "upwards" or "downwards" out of this plane, our view is not obscured by matter, and we can see past the confines of our galaxy. We are able to see distant galaxies only by looking in parts of our sky that are away from the Milky Way, that is, by looking out of the plane of the galaxy.

The gas in our galaxy is more or less transparent to visible light, but the small solid particles that we call "dust" are opaque. So the distance we can see through our galaxy depends mainly on the amount of dust that is present. This is not surprising: we can see great distances through our gaseous air on earth, but if a small amount of particulate material is introduced in the form of smoke or dust thrown up from a road, we find that we can no longer see very far. Similarly, the dust between the stars in our galaxy dims the starlight by absorbing it or by scattering it in different directions.

The abundance of dust in the plane of the Milky Way Galaxy actually prevents us from seeing very far toward its center—with visible light, we can see only $\frac{1}{10}$ of the way toward the galactic center itself. The

dark lanes across the Milky Way are just areas of dust, obscuring any emitting gas or stars. The net effect is that we can see just about the same distance in any direction we look in the plane of the Milky Way. These direct optical observations fooled scientists at the turn of the century into thinking that the earth was near the center of the universe.

We shall see in the next chapter how the American astronomer Harlow Shapley in the 1920's realized that our sun was not in the center of the Milky Way. This fundamental idea took humanity one step further away from thinking that we were at the center of the universe. Later in this book, we will see how Copernicus in 1543 had already made the first step in removing the earth from the center of the universe.

Some of the dust and gas in our galaxy takes a pretty shape, and may glow, reflect light, or be visible in silhouette. These shapes are the *nebulae.* Another class of objects visible in our sky was once known as "spiral nebulae," since they looked like glowing gas with "arms" spiralling away from their centers. (We still speak of the Great Nebula in Andromeda, Color Plate 68.) In the first decades of this century, astronomers debated whether these objects were part of our own galaxy or were independent "island universes." We shall see in Chapter 14, Part V, of this book how we learned that the island-universe idea was correct. Our own Milky Way turned out to be part of a galaxy equal in stature to these other objects, which are now known to be spiral galaxies in their own right.

In recent years astronomers have been able to use wavelengths other than optical ones to study the Milky Way Galaxy. In the 1950's and 60's especially, radio astronomy gave us a new picture of our galaxy. In the 1980's, we have benefited from infrared observations, most recently from the IRAS spacecraft. Infrared and radio radiation can pass through the galaxy's dust, and allow us to see our galactic center and beyond.

In this part of the book, we shall first discuss the types of objects that we find in the Milky Way Galaxy. Chapter 12 is devoted to the general structure of the galaxy, and its major parts. In Chapter 13, we describe the matter between the stars, and what studying this matter has told us about how stars form.

The Milky Way in Sagittarius. The Great Rift across the center may be an overlap of many giant molecular clouds.

The Structure of the Milky Way Galaxy

12

Aims: To describe the basic components of the Milky Way Galaxy, and to understand why spiral structure forms

We have now described the stars, which are important constituents of any galaxy, and how they live and die. And we have seen that many or most stars exist in pairs and clusters. In this chapter, we describe the gas and dust that accompany the stars and that we see in the optical part of the spectrum as nebulae. We also discuss the overall structure of the Milky Way Galaxy and how, from our location inside it, we detect this structure.

12.1 Nebulae

Not all the gas and dust in our galaxy has coalesced into stars. A nebula is a cloud of gas and dust that we see in visible light. When we see the gas actually glowing in the visible part of the spectrum, we call it an *emission nebula*. Sometimes we see a cloud of dust that obscures our vision in some direction in the sky. When we see the dust appear as a dark silhouette against other glowing material, we call the object a *dark nebula* (or, sometimes, an *absorption nebula*). The photo of the Milky Way on the facing page shows both emission nebulae (some of the brightest parts) and dark nebulae (some of the dark parts where relatively few stars can be seen). In the heart of the Milky Way, shown in the picture, we also see the great *star clouds* of the galactic center, the bright regions where the stars are too close together to tell them apart.

The clouds of dust in Figure 12–1 or surrounding some of the stars in the Pleiades (Color Plate 48) are examples of *reflection nebulae*—they merely reflect the starlight toward us without emitting visible radiation of their own. Reflection nebulae usually look bluish because blue light scatters around better than red. Whereas an emission nebula has its own spectrum, as does a neon sign on earth, a reflection nebula shows the spectrum of the nearby star or stars whose light is being reflected.

The Great Nebula in Orion (Color Plate 56) is an emission nebula. In the winter sky, we can readily observe it through even a small telescope, but only with long photographic exposures or large telescopes can we study its structure in detail. In the center of the nebula are four closely grouped bright stars called the Trapezium, which provide the energy to make the nebula glow. In the region of the Orion Nebula we think stars are being born this very minute.

The North America Nebula (Color Plate 60), gas and dust that has a shape in the sky similar to the shape of North America on the earth's surface, is an example of an object that is simultaneously an emission and an absorption nebula. The reddish

Nebula is Latin for fog or mist. The plural is usually *nebulae* rather than ''nebulas.''

Figure 12–1 A reflection nebula, NGC 7129, in Cepheus.

Figure 12–2 The Horsehead Nebula, IC 434, in Orion.

emission comes from glowing gas spread across the sky; its strongest emission is in the red hydrogen-alpha line. This gas is not completely opaque to visible light; of the stars that are visible in the nebula, some are between the sun and the nebula, but others are behind the nebula.

The boundary of the North America Nebula as we see it is not just the point where the glowing gas stops; dust absorbs light in the area that corresponds to the Gulf of Mexico. We can tell this because we see fewer stars there.

The Horsehead Nebula (Fig. 12–2 and Color Plate 59) is another example of an object that is both an emission and an absorption nebula simultaneously. It also looks red because the hydrogen-alpha line is so strong. A bit of absorbing dust intrudes onto emitting gas, outlining the shape of a horse's head. We can see in the pictures that the horsehead is a continuation of a dark area in which very few stars are visible.

We have already discussed some of the most beautiful nebulae in the sky, composed of gas thrown off in the late stages of stellar evolution. They include planetary nebulae and supernova remnants.

12.2 The New Subdivision of Our Galaxy

It was not until the 1920's that the American astronomer Harlow Shapley realized that we were not in the center of the galaxy. He was studying the distribution of globular clusters and noticed that they were all in the same general area of the sky as seen from the earth. They mostly appear above or below the galactic plane and thus are not obscured by the dust. When he plotted their distances and directions (using the methods described in Chapter 6), he saw that they formed a spherical halo around a point thousands of light years away from us (Fig. 12–3). Shapley's touch of genius was to realize that this point must be the center of the galaxy.

The picture that we have of our own galaxy has changed in the last few years. In the 1970's, we knew of three components of our galaxy; now we speak of four. We have known of the first three for half a century: (1) the nuclear bulge, (2) the disk, and (3) the halo that contains the globular clusters. Now we also know of an additional component: (4) the galactic corona, though we don't know what it is made of. Let us now discuss these four parts of the galaxy.

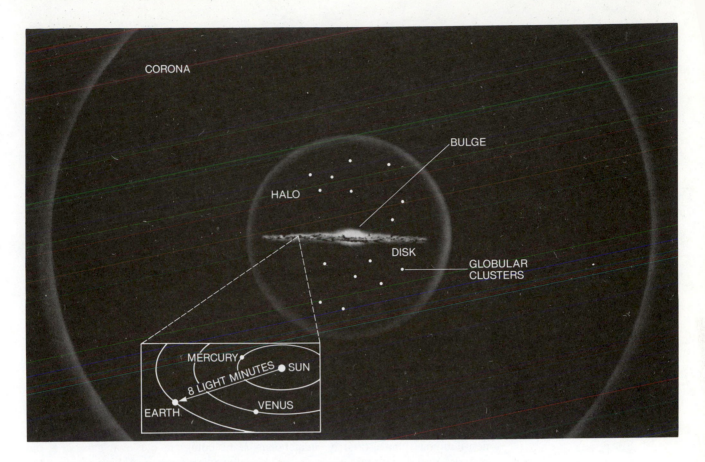

Figure 12–3 The drawing shows the *nuclear bulge* surrounded by the *disk,* which contains the spiral arms. The globular clusters are part of the *halo,* which extends above and below the disk. From the fact that most of the clusters appear in less than half of our sky, Shapley deduced that the galactic center is in the direction indicated. Extending even farther than the halo is the *galactic corona.*

(1) *The nuclear bulge:* Our galaxy has the general shape of a pancake with a bulge at its center. This *nuclear bulge* is about 5000 parsecs (16,000 light years) in radius, with the galactic *nucleus* at its midst. The nucleus itself is only about 5 parsecs across. The nuclear bulge has the shape of a flattened sphere and does not show spiral structure. It contains densely packed old stars as well as interstellar dust and gas. We shall say more about the nucleus in the next section.

(2) *The disk:* The part of the pancake outside the bulge is called the galactic *disk.* It extends out 15,000 parsecs (15 kiloparsecs or 50,000 light years) or so from the center of the galaxy. The sun is located about half-way out, about 8,500 parsecs (8.5 kpc) from the nucleus. The disk is very thin, 2 per cent of its width, like a phonograph record. It contains all the young stars and interstellar gas and dust. It is slightly warped at its ends, perhaps by interaction with our satellite galaxies, the Magellanic Clouds. So, in the words of Leo Blitz of the University of Maryland, Michael Fich of Berkeley, and Anthony Stark of Bell Labs, our galaxy looks a bit like a "fedora hat" with a turned-down brim.

It is very difficult for us to tell how the material is arranged in our galaxy's disk, just as it would be difficult to tell how the streets of a city were laid out if we could only stand still on one street corner without moving. Still, other galaxies have similar properties to our own, and their disks are filled with great *spiral arms,* regions of dust, gas, and stars in the shape of a pinwheel (see, for example, Color Plates 66, 68, 69, and 71). So we assume the disk of our galaxy has spiral arms too, though in Section 12.5 we will see that the evidence is ambiguous.

The disk contains relatively young stars of Population I (Section 6.4).

(3) *The halo:* Older stars (including the globular clusters) and interstellar matter form a galactic *halo* around the disk. This halo is at least as large across as the disk, perhaps 20 kpc (65,000 light years) in radius. It extends far above and below the plane of our galaxy in the shape of a flattened spheroid. Spectra from the International Ultraviolet Explorer spacecraft show that gas in the halo is hot, 100,000 K. IUE discovered the spectra of this hot gas when pointing at more distant objects. The gas in the halo contains only about 2 per cent of the mass of the gas in the disk.

The halo contains relatively old stars of Population II (Section 6.4).

(4) *The galactic corona:* We shall see in the next chapter (Section 13.6) how studies of the rotation of material in the outer parts of galaxies tell us how much mass is present. These studies have told us of the existence of a lot of mass we had overlooked before because we couldn't see it. This mass extends out 60 or 100 kpc (200,000 or 300,000 light years). Believe it or not, this galactic corona contains 5 or 10 times as much mass as the nucleus, disk, and halo together. And it makes our galaxy 3 or 5 times larger across than we had thought. (The subject is so new that the vocabulary is not settled; "galactic corona" and "outer halo" are among the names for this material.)

The existence of this extra material had been suspected from older gravitational studies, which had led to the "missing-mass problem" (which we will discuss further in Section 17.2). But only recently have we accepted the actual presence of so much non-luminous matter (that is, matter that isn't shining) in our own galaxy. If the material in the galactic corona was of an ordinary type, we would have seen it directly, if not in visible light, then in radio waves, x-rays, infrared, etc. But we don't see it at all! We only detect its gravitational properties.

What is the galactic corona made of? We just don't know. A tremendous number of very faint stars is a possibility, though it seems unlikely, extrapolating from the numbers of the faintest stars that we can study. A large number of small black holes has been suggested. And another possibility is a huge number of neutrinos, the subatomic particles of the type we discussed in Section 8.6 on the solar neutrino problem, if neutrinos have mass. Work is now going on to find out if neutrinos have any mass at all.

If the preceding paragraphs—stating that perhaps 95 per cent of our galaxy's mass is in some unknown form—seem unsatisfactory to you, you may feel better by knowing that astronomers find the situation unsatisfactory too. But all we can do is go out and do our research, and try to find out more. We just don't know the answers . . . yet.

12.3 The Center of Our Galaxy and Infrared Studies

We cannot see the center of our galaxy in the visible part of the spectrum because our view is blocked by interstellar dust. We did not even know that the center of our galaxy (Fig. 12–4) lies in the direction of the constellation Sagittarius as seen from the earth until Shapley deduced the fact from his study of the distribution of globular clusters.

In recent years, observations in other parts of the electromagnetic spectrum have become increasingly important for the study of the Milky Way Galaxy. The center

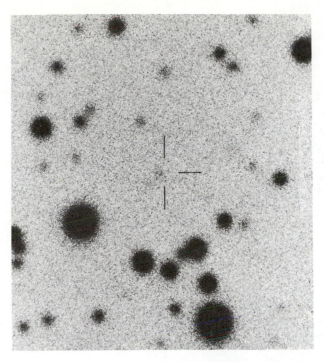

Figure 12–4 The galactic center region shown in a negative print. The position of the center of our galaxy is marked.

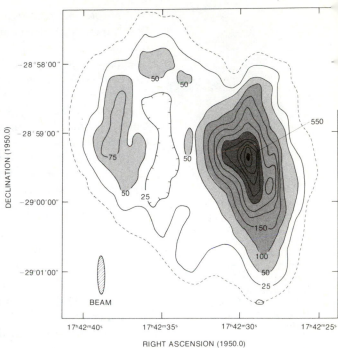

Figure 12–5 A radio map of the galactic center. The intense source Sgr A at right, marked 550 in intensity units, coincides with the center.

of our galaxy is a prominent source in each of these spectral regions. In the radio region of the spectrum, for example, Sagittarius A (Fig. 12–5) has long been known as one of the strongest sources in the sky. Both radio waves and infrared penetrate the dust that fills our galactic disk.

12.3a Infrared Observations

For many years the sky has been intensively studied at optical wavelengths up to about 8500 Å (0.85 micron) and to a lesser extent up to 1.1 microns, which is in the near infrared. The sky has also been studied for decades at radio wavelengths down to one or two centimeters (10,000 or 20,000 microns). But until the last few years, the sky has been studied very little at the wavelengths in between. We saw in Section 2.13a that the lack of film sensitivity in the infrared, whose photons contain relatively low energy, is a major reason for this limitation. Neither have especially sensitive electronic detection devices been available in the infrared. Atmospheric limitations, caused by the presence of only a few windows of transparency, have been another major factor. Since one can observe better in the infrared from locations where there is little water vapor overhead, the new emphasis on infrared telescopes at sites like Mauna Kea in Hawaii should improve the situation considerably.

Another difficulty is the fact that the heat radiation is in the infrared. The earth's atmosphere radiates conspicuously in the infrared, so the radiation coming into a telescope includes infrared radiation from the atmosphere, from the source being observed, and from the telescope itself. To limit the telescopic contribution, equipment is usually bathed in liquid nitrogen, or even in liquid helium, which is colder and more expensive.

Astronomers working in the infrared usually use the unit of microns for wavelength. One micron is 1/1,000,000 m, and is 10,000 Å.

Figure 12–6 The IRAS spacecraft, shown here with its cooled telescope, gave 1000 times better results at 12- and 25-micron wavelengths than ground-based observations. Aloft for 10 months in 1983, it was sensitive enough to detect the heat from a 20-watt light bulb at the distance of the planet Pluto.

See the Color Essay on IRAS following page 44.

A 1969 sky survey in the 2.2-micron window revealed 20,000 infrared sources. IRAS, the NASA/Netherlands/UK **I**nfrared **A**stronomy **S**atellite (Fig. 12–6), with an 0.6-m telescope, mapped the sky at much longer wavelengths in 1983. With its detectors cooled by liquid helium to only 2 K (2°C above absolute zero) and its telescope cooled to only slightly more than that by keeping the whole assembly in a Thermos-like bottle, it was 1000 times more sensitive at 12 and 25 microns than past observations, and made the first full survey at 60 and 100 microns. IRAS sent back data for 10 months, until its liquid helium was exhausted. It discovered hundreds of thousands of new sources—so much data that it will take years to interpret.

Many of the infrared sources are cool stars but many others do not coincide in space with known optical sources. The map of the radio sky doesn't look like the map of the optical sky, and the map of the infrared sky doesn't resemble either of the others. Some of the infrared sources without optical counterparts may be galaxies too distant to see optically. Most of the identified infrared-emitting objects, however, unlike the radio objects, are in our galaxy. Many of the more interesting infrared objects turn out to be intimately connected with stars in formation, which we shall consider in Section 13.8.

IRAS mapped the entire galaxy; its view of the disk penetrated to the galactic center (Fig. 12–7). Another of IRAS's discoveries was that the sky is covered with infrared-emitting material, probably outside our solar system but in our galaxy. Since its shape resembles terrestrial cirrus clouds, the material is being called "infrared cirrus."

In the early 1990's, NASA plans to carry aloft an 0.85-m infrared telescope with a Space Shuttle. This space Infrared Telescope Facility (SIRTF) will be about 1000 times more sensitive than IRAS, in large part because infrared detectors have improved. The cooled telescope will focus infrared radiation on several instruments to allow both photometry and spectroscopy. Also for the early 1990's, the European Space Agency is planning Infrared Space Observatory (ISO) with an 0.6-m telescope. The European ISO will fly freely for perhaps 18 months. NASA's spacecraft was originally supposed to stay on its shuttle and so be aloft for only a week or two each flight, but may wind up as a free-flier. NASA and ESA have promised to coordinate their infrared missions.

Figure 12–7 An IRAS view of 45° across the sky, showing the Milky Way. The warm dust that IRAS images is close to the plane of the galaxy. The bright spots are regions where stars are forming; we see dust heated by the stars. The bright region at the core is about 25 K. Infrared cirrus seems to stream from the plane.

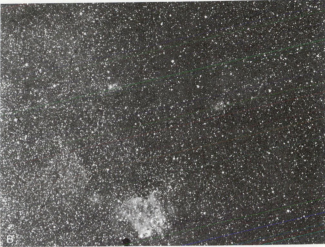

12.3b The Galactic Nucleus

One of the brightest infrared sources in our sky, known before IRAS, is located at the position of the radio source Sagittarius A (Fig. 12–8), which marks the center of our galaxy. Since the amount of scattering by dust varies with wavelength, we can see further through interstellar space in the infrared than we can in the visible. In particular, in the infrared we can see to the center of our galaxy.

This infrared source subtends 1 arc min, and so is about 3 parsecs (about 10 light years) across. This makes it a very small source for the prodigious amount of energy it emits: as much energy as if there were 80 million suns radiating. It is also a strong and variable x-ray source.

We don't know just what it is in the center of our galaxy that causes this radiation, but this tiny region gives off as much as 0.1 per cent of the total radiation from our galaxy. Thus the center of our galaxy is qualitatively different from the outer parts. (Other galaxies have even more prominent infrared, radio, and x-ray sources in their nuclei.)

Several models have been advanced to account for the radiation from the galactic nucleus. In one model, bright sources of ultraviolet or visible radiation, such as a dense group of stars, are present there. These sources would be surrounded by shells of dust that heat up to about 50 K, thus providing the infrared luminosity. The equivalent of about 10 solar masses of dust, not an inconceivable amount, would have to be present to make this model work.

Another model is based on the observation that both matter and antimatter are annihilating each other. This liberates a tremendous amount of energy in the form of gamma rays. HEAO-3's gamma-ray spectrometer confirmed the detection of the gamma-ray spectral line that results from electrons and positrons annihilating each other. The region emitting the line is too small for the gamma rays to arise from cosmic rays interacting with the interstellar medium or from a distribution of many supernovae, novae, or pulsars.

In the very center of the nucleus, radio and infrared astronomers have discovered an extremely narrow source. It is only about 10 astronomical units across, smaller than the orbit of Jupiter. It is giving off a great deal of energy (though not as much as the nuclei of some distant galaxies). The leading model for the galactic center at the moment, with black holes in fashion, is that a high-mass black hole is present there. This model explains particularly well why the radio and infrared source

Figure 12–8 (*A*) A high-resolution map of the center of the Milky Way, made from the ground at a wavelength of 2.2 microns, which is in the infrared. At this wavelength, we see stars rather than dust. Dust becomes apparent at longer wavelengths. (*B*) A photograph in visible light taken in the direction of the center of our galaxy, showing exactly the same region observed in the infrared in *A*. In visible light, interstellar dust completely hides the galactic center, and we see only relatively nearby stars and gas.

The resolution is about 1.2 arc min, and the height of the entire image is about 1.1°.

One kind of matter-antimatter annihilation occurs when an electron meets its antiparticle, a *positron*. The two annihilate each other completely, releasing energy in the amount $E = 2mc^2$, where m is the mass of each of the two particles.

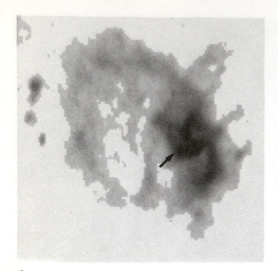

A

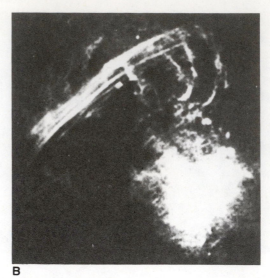

B

Figure 12–9 (*A*) The nucleus of our galaxy at the center of Sagittarius A, observed at the radio wavelength of 6 cm with the VLA. The resolution is 8 arc sec and the field of view is 2 arc min square. The brightest spot, at the center of the spiral, is the image of the nucleus, though the point radio source is known to be 100 times smaller than it appears here. The spirals may be the matter of stars being drawn into the giant black hole in the nucleus.

(*B*) In a field of view 10 times larger across, also centered near the nucleus, the VLA showed Mark Morris of UCLA and Farhad Yusef-Zadeh and Don Chance of Columbia that a vast arc of parallel filaments stretched over 130 light years perpendicularly to the plane of the galaxy. Resembling a solar quiescent prominence, the structure seems to indicate that a strong magnetic field is present.

at the galactic center is so small. The black hole would contain millions of times the mass of the sun. Interstellar gas and dust spiralling in toward the black hole would heat up and give off the large amount of energy that we detect. Infrared studies of the motions of gas near the center endorse the idea that a very massive source is present.

The high-resolution radio maps of our galactic center, now made with the Very Large Array, show a small bright spot that could well be the central giant black hole (Fig. 12–9).

12.4 High-Energy Sources in Our Galaxy.

The study of our galaxy provides us with a wide range of types of sources. Many of these have been known for many years from optical studies (Fig. 12–10). We have just seen how the infrared sky looks quite different. The radio sky provides still a different picture. Technological advances have enabled us to study sources in our galaxy in the x-ray and gamma-ray region of the spectrum as well.

Non-solar x-ray astronomy began in 1962, when Riccardo Giacconi and colleagues discovered x-rays from a source in the constellation Scorpius (and named it Scorpius X-1). Herbert Friedman and colleagues soon carried out additional x-ray work. The 1960's research was carried out with rockets rather than with satellites. A few dozen sources, including Scorpius X-1, the Crab Nebula, and the Virgo cluster of galaxies, were found.

The first reasonable map of the x-ray sky was made with the Uhuru satellite, which observed hundreds of x-ray sources in about 1970. Most of our current knowledge of x-ray sources came from the U.S. **H**igh-**E**nergy **A**stronomy **O**bservatories. HEAO-1 mapped 1500 x-ray sources (Fig. 12–11). Scientists using HEAO-1 and HEAO-2 (Einstein) studied many of these sources over extended periods of time.

NASA launched the Uhuru satellite (SAS-1) in 1970, as part of the Small Astronomy Satellite series. We have already mentioned (in Section 11.5 on black holes) some of the results provided by the Uhuru satellite.

Other satellites to study x-rays have been launched by the U.S., by the Soviet Union, by Britain from a launch pad in Kenya, and by the Netherlands from the U.S. Current x-ray satellites aloft include the Japanese Tenma, the European Exosat, and the Soviet-French Astron. Though American x-ray expertise was shown to great advantage while the HEAO's were up, there is now a fear that American work in the field may die out almost completely because of the current lack of NASA funding in the field.

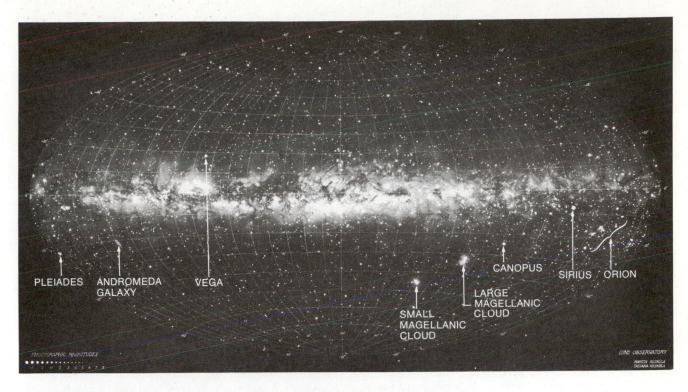

Figure 12–10 A drawing of the Milky Way, made under the supervision of Knut Lundmark at the Lund Observatory in Sweden. 7000 stars plus the Milky Way are shown in this panorama, which is in coordinates such that the Milky Way falls along the equator.

Figure 12–11 An x-ray map of the Milky Way showing the objects observed by the first High Energy Astronomy Observatory, the HEAO A-1 All-Sky X-Ray Source Map, drawn in the same "galactic coordinates" as the preceding figure. A wide variety of objects shows up. (Kent S. Wood/Naval Research Laboratory)

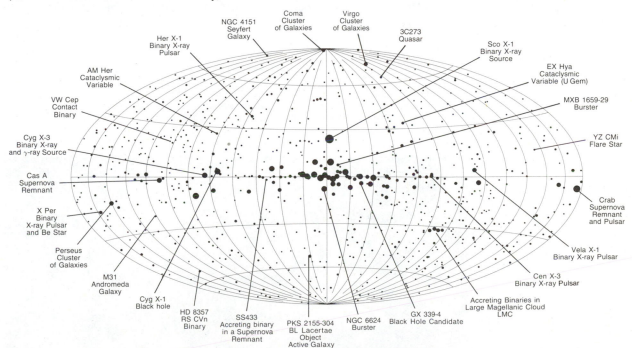

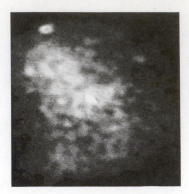

Figure 12–12 This x-ray image shows a bright point at the galactic center. This field of view is identical to that shown in Figure 12–8, except that the scale of this image is 3/4 that of the infrared and optical images. (Einstein Observatory image by Michael Watson, Paul Hertz, and colleagues, Harvard-Smithsonian Center for Astrophysics)

The situations are opposite for novae and for bursters. In a nova, we have mass transfer onto a white dwarf. The energy released in a collapse of an accretion disk (a dwarf nova) is much less than that released by runaway hydrogen burning on the white dwarf's surface (a classical nova). In a burster, we have mass transfer onto a neutron star. The much higher gravity on the surface of a neutron star compared to that on the surface of a white dwarf makes the disk-collapse situation more energetic, and accounts for the rapid burster. Runaway burning, of helium in this case, can account for the ''normal'' bursters.

Among the many discoveries were the x-ray bursts from globular clusters and from other locations in the sky. These *bursters,* which give off bursts of x-rays on a time scale of seconds or hours, turn out to be x-ray binaries, some of which are located in globular clusters.

Another x-ray discovery was the bright x-ray source at the galactic center (Fig. 12–12), which fits in with the model that a giant black hole is present there.

Although many of the x-ray sources are in our galaxy, others are extragalactic. The Andromeda Galaxy previously had been only barely detected as a single x-ray source, but the Einstein Observatory found many sources distributed through its central region (Fig. 12–13). Since it is easier to understand galaxies seen from outside than it is from the perspective of our interior view of the Milky Way Galaxy, these observations can help us understand our own galaxy.

Although an occasional strange object, like the Crab Pulsar, can be detected at every wavelength of the spectrum, astronomers know of very few gamma-ray objects. The European COS-B spacecraft has been mapping the sky in gamma rays since 1975. It has discovered 25 sources emitting gamma rays of relatively high energy. Only four of these sources correspond to well-known objects: the Crab and Vela pulsars, a nearby quasar (3C 273), and a complex of interstellar clouds near

Figure 12–13 An x-ray image of the central region of M31, the Andromeda Galaxy, taken with the Einstein Observatory. Twenty x-ray sources can be seen in this image. (Einstein Observatory image by Leon Van Speybroeck and colleagues, Harvard-Smithsonian Center for Astrophysics)

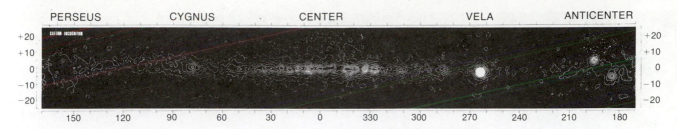

| PERSEUS | CYGNUS | CENTER | VELA | ANTICENTER |

Figure 12–14 This gamma-ray map made by the European COS-B spacecraft shows the plane of the Milky Way and several individual gamma-ray sources. The map includes the region between 25° above and 25° below the plane of the Milky Way.

ρ (rho) Ophiuchi. With a single exception, the other sources lie close to the plane of our galaxy. Extrapolating from the observations, several hundred such sources may exist near the plane of our galaxy, but we do not know what they are. U.S. satellites in the Vela series, whose prime purpose is to detect nuclear explosions, and HEAO-3 have also made gamma-ray observations.

A general background of gamma rays is concentrated along the plane of our galaxy (Fig. 12–14). This diffuse background is not concentrated in particular objects, not even in the direction of the galactic center. Presumably, these gamma rays are caused by the interaction of cosmic rays with interstellar matter. Even though the cosmic rays are already formed with high energies at such sites as supernovae, many of them are brought up to even higher energies as they move through our galaxy.

Another gamma-ray observation is the occurrence of mysterious bursts, which come from seemingly random locations in the sky; about 150 are discovered each year. One of the problems in the past has been limited observational ability to pinpoint the location of the bursts. An international set of sensors put in space by the U.S., Europe, and the U.S.S.R. has discovered a new type of gamma-ray burst ten times more powerful than any previous bursts. The burst seems to come from a supernova remnant associated with the Large Magellanic Cloud, an object that is hundreds of times farther away from us than we had assumed the sources of gamma-ray bursts would be. This identification is not universally accepted. Models for the gamma-ray bursts include thermonuclear explosions on neutron stars, with the explosions made more powerful because of the high magnetic fields present.

The most studied gamma-ray object is Geminga, a gamma-ray–emitting region in the constellation Gemini that has been observed for over a decade. Gamma-ray observations do not pinpoint positions very accurately; searches of the region have not turned up special optical or radio objects to identify with the gamma-ray source. A possible x-ray identification has been made. If the identification is correct, then Geminga may be an old pulsar whose radio emission has stopped. The Italian scientists studying the source chose its name, which is pronounced with a hard second "g," for **Gemin**i **ga**mma ray source; they found the name suitable since the source was unidentified for so long and since "geminga" means "it does not exist" in Milanese dialect.

Studies of electromagnetic radiation like x-rays and gamma rays and of rapidly moving cosmic-ray particles are part of the new field of *high-energy astrophysics*. With the termination of the HEAO program, x-ray observing has fallen on hard times. Hope remains that an Advanced X-Ray Astrophysics Facility (AXAF) will be launched in the early 1990's. The project was given the highest priority by the recent review committee of American astronomers. An elaborate Gamma Ray Observatory (GRO) is also planned.

A powerful burst was observed on March 5, 1979. The initial burst, only 1/5000 second in duration, was followed for a few minutes by pulses with an 8-second period. The periodicity may indicate the presence of a neutron star. During the burst, the object gave off energy at a rate greater than that of the entire Milky Way Galaxy. A burst two years earlier has also been discovered by subsequent analysis of HEAO-1 data. Reverberations with a 4.2-s period followed. The source's direction is unknown.

Figure 12–15 The positions of young galactic clusters and H II regions in our own galaxy are projected on a photograph of the spiral galaxy NGC 1232, scaled to the size of our galaxy. The sun is along the line labelled 0°.

12.5 The Spiral Structure of the Galaxy

12.5a Bright Tracers of the Spiral Structure

When we look out past the boundaries of the Milky Way Galaxy, usually by observing in directions above or below the plane of the Milky Way, we can see a number of galaxies with arms that appear to spiral outward from near their centers. In this section we shall discuss some of the evidence that our own Milky Way Galaxy also has spiral structure.

It is always difficult to tell the shape of a system from a position inside it. Think, for example, of being somewhere inside a maze of tall hedges. We might be able to see through some of the foliage and be reasonably certain that layers and layers of hedges surrounded us, but we would find it difficult to trace out the pattern. If we could fly overhead in a helicopter, though, the pattern would become very easy to see.

Similarly, we have difficulty tracing out the spiral pattern in our own galaxy, even though the pattern would presumably be apparent from outside the galaxy. Still, by noting the distances and directions to objects of various types, we can tell about the Milky Way's spiral structure.

Galactic clusters are good objects to use for this purpose, for they are always located in the spiral arms. We have found the distance to 200 galactic clusters by studying their color-magnitude diagrams.

We think that spiral arms are regions where young stars are found. Some of the young stars are the O and B stars; their lives are so short we know they can't be old. But since our methods of determining the distances to O and B stars from their spectra and colors are uncertain to 10 per cent, they give a fuzzy picture of the distant parts of our galaxy.

Other signs of young stars are the presence of regions of ionized hydrogen known as H II regions (pronounced ''H two'' regions). We know from studies of other galaxies that H II regions are preferentially located in spiral arms. In studying the locations of the H II regions, we are really again studying the locations of the O stars and the hotter B stars, since it is ultraviolet radiation from these hot stars that provides the energy for the H II regions to glow. In Section 12.5c we shall discuss a theory that explains why O and B stars should be found in spiral arms.

When the positions of the galactic clusters, the O and B stars, and the H II regions of known distances are studied (by plotting their distances and directions as seen from earth), they appear to trace out bits of three spiral arms (Fig. 12–15). The spacings between the arms and the widths of the arms appear consistent with spacings and arm widths that we can observe in other galaxies.

These observations are carried out in the visible part of the spectrum. Even when we are observing the O and B stars, which are very bright and therefore can be seen at great distances, the optical map shows only regions of our galaxy close to the sun. Interstellar dust prevents us from studying parts of our galaxy farther away from the sun. Another very valuable method of studying the spiral structure in our own galaxy involves spectral lines of hydrogen and of carbon monoxide in the radio part of the spectrum (Sections 13.4 and 13.5). Radio waves penetrate the interstellar dust, and we are no longer limited to studying the local spiral arms.

*12.5b Differential Rotation

The latest calculations indicate that the sun is approximately 8.5 kiloparsecs from the center of our galaxy. From spectroscopic observations of the Doppler shifts of globular clusters, which do not participate in the galactic rotation, or of distant galaxies, we can tell that the sun is revolving around the center of our galaxy at a speed of approximately 250 kilometers per second. At this velocity, it would take the sun about 250 million years to travel once around the center; this period is called the *galactic year*. But not all stars revolve around the galactic center in the same period of time. The central part of the galaxy rotates like a solid body. Beyond the central part, the stars that are farther out have longer galactic years than stars closer in.

If this system of *differential rotation,* with differing rotation speeds at different distances from the center, has persisted since the origin of the galaxy, we may wonder why there are still only a few spiral arms in our galaxy and in the other galaxies we observe. The sun could have made fifty revolutions during the lifetime of the galaxy, but points closer to the center would have made many more revolutions. Thus the question arises: why haven't the arms wound up very tightly?

*12.5c Density-Wave Theory

The leading current solution to this conundrum is a theory first suggested by the Swedish astronomer B. Lindblad and elaborated mathematically by the American astronomers C. C. Lin at M.I.T. and Frank Shu, now at Berkeley. They say, in effect, that the spiral arms we now see are not the same spiral arms that were previously visible. In their model, the spiral-arm pattern is caused by a spiral *density wave,* a wave of increased density that moves through the stars and gas in the galaxy. This density wave is a wave of compression, not of matter being transported. It rotates more slowly than the actual material, and causes the density of material to build up as it passes. A shock wave—a moving surface at which a compression takes place, as in a sonic boom from an airplane—is also built up. The shock wave heats the gas.

So, in the density-wave model, the spiral arms we see at any given time do not represent the actual motion of individual stars in orbit around the galactic center. We can think of the analogy of a crew of workers painting a white line down the center of a busy highway. A bottleneck occurs at the location of the painters. If we were observers in an airplane, we would see an increase in the number of cars at that place. As the line painters continued slowly down the road, we would seem to see the place of increased density move down the road at that slow speed. We would see the bottleneck move along even if our vision were not clear enough to see the individual cars, which could still speed down the highway, slow down briefly as they cross the region of the bottleneck, and then resume their high speed.

Similarly, we might be viewing only some galactic bottleneck at the spiral arms. The gas is compressed there, both from the increased density itself and, to a greater extent, from the shock wave formed by other gas piling into this slowed-up gas from behind. The compression causes heating, and also leads to the formation of protostars that collapse to become stars. As the density wave passes a given point, stars begin to form there because of the increased density. The distribution of new stars can take a spiral form (Fig. 12–16). The new stars heat the interstellar gas so that it becomes visible. In fact, we do see young, hot stars and glowing gas outlining the spiral arms, which are checks of this prediction of the density-wave theory.

Some astronomers do not accept the density-wave theory; one of their major objections is the problem of explaining why the density wave forms. One alternative, very different theory says that stars are produced by a chain reaction and are then spread out into spiral arms by the differential rotation of the galaxy. The chain reaction begins when high-mass stars become supernovae. The expanding shells from the supernovae trigger the formation of stars in nearby regions. Some of these new stars become massive stars, which become supernovae, and so on.

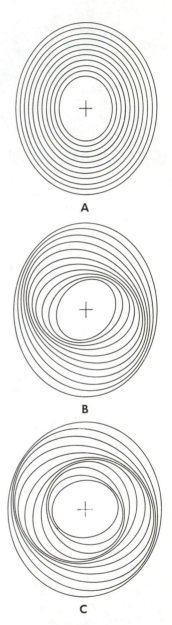

A

B

C

Figure 12–16 Each part of the figure includes the same set of ellipses; the only difference is the relative alignment of their axes. Consider that the axes are rotating slowly and at different rates. The compression of their orbits takes a spiral form, even though no actual spiral exists. The spiral structure of a galaxy may arise from an analogous effect.

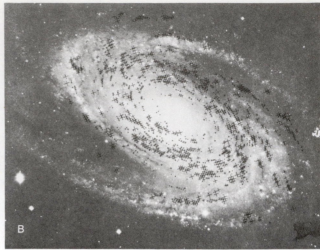

Figure 12–17 The model galaxies—plotted as +'s—superimposed upon photographs of the galaxies. The models were computed using observed values of rotational velocities for each distance from the center. (*A*) M101 (that is, the 101st object in Messier's Catalogue, Appendix 8. (*B*) M81, including the effect of projecting the model at a 58° angle. (Courtesy of Humberto Gerola and Philip E. Seiden, IBM Watson Research Center)

Computer modelling of the *supernova-chain-reaction model* gives values that seem to agree with the observed features of spiral galaxies (Fig. 12–17). Each galaxy has a differential velocity, with regions at different distances from the center rotating at different speeds. At some distance from the center, the speed is a maximum. The degree of winding depends on the value for this maximum velocity.

Summary and Outline

Nebulae (Section 12.1)
   Emission; dark (absorption); reflection; planetary; supernova remnants
   The reddish glow results from H-alpha emission.
Our galaxy has a nuclear bulge surrounding its nucleus, a disk, a halo, and a corona (Section 12.2).
Infrared observations (Section 12.3)
   Limited by atmospheric transparency and technology
   IRAS mapped the entire sky at relatively long wavelengths.
   Radiation from the galactic nucleus indicates a lot of energy is generated in a small volume there; a very massive black hole may be present.
High-energy astronomy (Section 12.4)
   The Uhuru and HEAO satellites observed x-rays.
   Gamma-ray background radiation is the result of cosmic-ray interaction with matter.
   Next stage, in early 1990's, will be NASA's AXAF (Advanced X-Ray Astrophysics Facility) and GRO (Gamma Ray Observatory).
Spiral structure of the galaxy (Section 12.5)
   Bright tracers of the spiral structure (Section 12.5a)
      Difficult to determine shape of our galaxy because of obscuring gas and dust and because of our vantage point
      Galactic clusters, O and B stars, and H II regions are used to determine the locations of spiral arms.
   Differential rotation (Section 12.5b)
      Sun is approximately 8.5 kpc from center of galaxy.
      Central part of galaxy rotates as a solid body.
      Outer part, including the spiral arms, is in differential rotation.
   Theories for spiral structure (Section 12.5c)
      Density wave: we see the effect of a wave of compression; the distribution of mass itself is not spiral in structure.
      Chain of supernovae

Key Words

nebula, emission nebula, dark nebula, absorption nebula, reflection nebula, star cloud, nuclear bulge, nucleus, disk, halo, galactic corona, spiral arms, positron, bursters, high-energy astrophysics, galactic year, differential rotation, density wave*, supernova-chain-reaction model.

---

*These words are found in an optional section.

Questions

1. Why do we think our galaxy is a spiral?

2. How would the Milky Way appear if the sun were closer to the edge of the galaxy?

3. Sketch (a) a side view and (b) a top view of the Milky Way Galaxy, showing the shapes and relative sizes of the nuclear bulge, the disk, and the halo. Mark the position of the sun. Indicate the location of the galactic corona.

4. Compare (a) absorption (dark) nebulae, (b) reflection nebulae, and (c) emission nebulae.

5. How can something be both an emission and an absorption nebula? Explain and give an example.

6. Sketch the North America Nebula, indicating on the sketch where you see H-alpha radiation and where you see dust.

7. If you see a red nebula surrounding a blue star, is it an emission or a reflection nebula? Explain.

8. How do we know that the galactic corona isn't made of ordinary stars like the sun?

9. Why may some infrared observations be made from mountain observatories while all x-ray observations must be made from space?

10. What radio source corresponds to the center of our galaxy?

11. Illustrate, by sketching the appropriate Planck radiation curves, why cooling an infrared telescope from room temperature to 4 K greatly reduces the background noise.

12. Describe infrared and radio results about the center of our galaxy.

13. What are three tracers that we use for the spiral structure of our galaxy? What are two reasons why we expect them to trace spiral structure?

14. Since Einstein's result was that $E = mc^2$, explain how, if m_e is the mass of an electron, $2m_ec^2$ of energy results when an electron and a positron annihilate each other.

15. Why will x-rays expose a photographic plate while infrared radiation will not?

16. What types of objects give off strong infrared radiation?

17. What types of objects give off x-rays?

18. Discuss how observations from space have added to our knowledge of our galaxy.

19. Comment on our understanding of the gamma-ray bursts.

20. Why does the density-wave theory lead to the formation of stars?

21. If a spacecraft could travel at 10% of the speed of light, how long would it have to travel to get far enough out to be able to take a photograph showing the spiral structure of our galaxy?

The 100-meter radio telescope at Effelsburg, near Bonn, Germany, the largest fully steerable radio telescope in the world. It is often used for studies of the interstellar medium.

The Interstellar Medium 13

Aims: **To discuss the interstellar medium, to describe the techniques (especially radio astronomy) used to study it, and to understand the relation of interstellar clouds to star formation**

The gas and the dust between the stars is known as the *interstellar medium*. The nebulae represent regions of the interstellar medium in which the density of gas and dust is higher than average. The interstellar medium contains the elements in what we call their *cosmic abundances,* that is, the overall abundances they have in the universe (the cosmos).

In this chapter we shall see that the studies of the interstellar medium are vital for understanding the structure of the galaxy and how stars are formed. We shall also discuss the recent realization that units of the interstellar medium called ''giant molecular clouds'' are basic building blocks of our galaxy.

13.1 H I and H II Regions

For many purposes, we may consider interstellar space as being filled with hydrogen at an average density of about 1 atom per cubic centimeter, although individual regions may have densities departing greatly from this average. Regions of higher density in which the atoms of hydrogen are predominantly neutral are called *H I regions* (pronounced ''H one regions''; the roman numeral ''I'' refers to the first, or basic, state). Where the density of an H I region is high enough, pairs of hydrogen atoms combine to form molecules (H_2). The densest part of the gas associated with the Orion Nebula might have a million or more hydrogen molecules per cubic centimeter. So hydrogen molecules (H_2) are often found in H I regions.

A region of ionized hydrogen, with one electron missing, is known as an *H II region* (from ''H two''; the second state). Since hydrogen, which makes up the overwhelming proportion of interstellar gas, contains only one proton and one electron, a gas of ionized hydrogen contains individual protons and electrons. Wherever a hot star provides enough energy to ionize hydrogen, an H II region (Fig. 13–1) results. Emission nebulae are such H II regions. They glow because the gas is heated. Emission lines appear.

Studying the optical and radio spectra of H II regions and planetary nebulae tells us the abundances of several of the chemical elements (especially helium, nitrogen, and oxygen). How these abundances vary from place to place in our galaxy and in other galaxies helps us choose between models of element formation and of galaxy formation. The variation of the relative abundances from the center of a galaxy to its outer regions—the gradient of abundances—differs for different galaxies. The abundances of the elements known as ''metals'' are greatest in the locations where the most star formation took place; this indicates that these elements were indeed produced in stars.

The cosmic abundance of elements is roughly 90 per cent hydrogen atoms, 9 per cent helium atoms, and less than one per cent of heavier atoms, though it is important to note that essentially all the heavier atoms are represented. (These percentages are for number of atoms, not for the relative masses.)

Even the nebulae with the highest densities are still many orders of magnitude less dense than the best vacuums that can be made in laboratories on earth, so that the nebulae provide scientists with a way of studying the basic properties of gases under conditions that are unobtainable on earth.

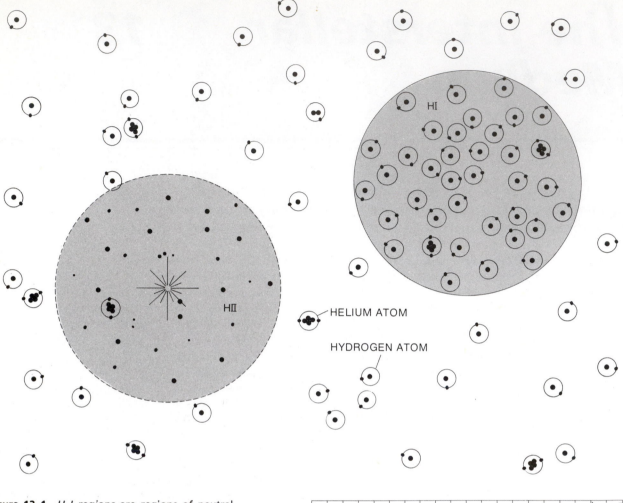

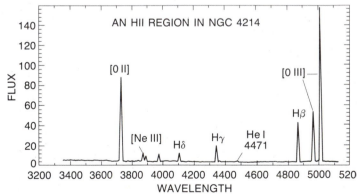

Figure 13–1 *H I regions* are regions of neutral hydrogen of higher density than average, and *H II regions* are regions of ionized hydrogen. The protons and electrons that result from the ionization of hydrogen by ultraviolet radiation from a hot star, and the neutral hydrogen atoms, are shown schematically. The larger dots represent protons or neutrons and the smaller dots represent electrons. The star that provides the energy for the H II region is shown. Though all hydrogen is ionized in an H II region, only some of the helium is; heavier elements have often lost 2 or 3 electrons. (Bottom) The emission lines in an H II region, including some of the members of hydrogen's Balmer series, helium, and "forbidden lines" of ionized oxygen and neon (enclosed in square brackets to show that they are transitions that happen relatively rarely).

13.2 Interstellar Reddening and Extinction

Distant stars in the plane of our galaxy are obscured from our vision. In addition, many stars that are still close enough to be visible are partially obscured. The amount of obscuration varies with the wavelength at which we observe. The blue light is *scattered* by dust in space more efficiently than the redder light is; for a given distance through the dust, more of the blue light has been bounced around in every

Box 13.1 Why Is the Sky Blue?

A similar scattering by the electrons in air molecules in the terrestrial atmosphere makes the sky blue. Further, when the sun is near the horizon, we have to look diagonally through the earth's layer of air. Our line of sight through the air is then longer than a line of sight straight up. Thus when the sun is low in the sky, most of the blue light is scattered out before it reaches us. Relatively more red light reaches us. This accounts for the reddish color of sunsets. The light that reaches us from the part of the sky away from the sun is scattered sunlight, and is therefore predominantly blue. Electrons scatter visible wavelengths of light much more efficiently than dust does.

direction. Thus less of the blue light comes through to us than the red light, and the stars look redder—we say that the stars are *reddened*. This reddening is thus a consequence of the scattering properties of the dust. It has nothing to do with "redshifts," since the spectral lines are not shifted in wavelength by reddening.

The amount of scattering together with the amount of actual absorption of visible radiation by dust is known as the *extinction*. In the blue part of the spectrum, the total extinction in the 8.5 kpc between the sun and the center of the galaxy is about 25 magnitudes. Most of this takes place far from the sun, in regions with a high dust content. Even a tremendously bright object located near the center of our galaxy would be dimmed too much—25 magnitudes—to be seen from the earth in the visible part of the spectrum.

What is the dust made of? We know the particles are tiny—smaller than the wavelength of light—or else they would not scatter light in this way, with the shorter wavelengths scattered more efficiently. From spectral studies, we know that at least some of the particles are carbon in the form of graphite. Other particles may be silicates or ices.

Interstellar dust is heated a bit by radiation, thus causing the dust to radiate. The dust never gets very hot, so its radiation peaks in the infrared. The radiation from dust scattered among the stars is too faint to detect, but the radiation coming from clouds of dust surrounding stars has been observed from the ground and from the IRAS spacecraft. IRAS found infrared radiation from so many stars forming in our galaxy that we now think that about one star forms in our galaxy each year, a higher value than previously realized.

Similarly, since the interstellar gas is "invisible" in the visible part of the spectrum (except at the wavelengths of certain weak spectral lines), special techniques are needed to observe the gas in addition to observing the dust. Radio astronomy is the most widely used technique, so we will now discuss its use for mapping our galaxy.

13.3 Radio Observations of our Galaxy

The rest of the electromagnetic spectrum carries more information in it than do the few thousand angstroms that we call visible light. We will first discuss some basic techniques of radio astronomy (Section 2.13b), and then go on to see how radio astronomy joins with other observing methods to investigate interstellar space.

13.3a Continuum Radio Astronomy

All radio-astronomy studies in the early days were of the continuum. In radio astronomy, as in optical astronomy, studying the continuum means that we consider the average intensity of radiation at a given frequency without regard for variations in intensity over small frequency ranges. In short, we ignore any spectral lines. When radio ''continuum spectra'' were first measured, the continuum levels were measured at widely separated frequencies. These continuum values, and the spectra derived from them by joining the values by straight lines or other simple smooth curves, provided information about the mechanisms that cause the continuum emission.

As radio astronomical observations got under way, it was immediately apparent that the brightest objects in the radio sky, that is, the objects that give off the most intense radio waves, are not identical with the brightest objects in the optical sky. The radio objects were named with letters and with the names of their constellations. Thus Taurus A is the brightest radio object in the constellation Taurus; we now know it to be the Crab Nebula. Sagittarius A is the center of our galaxy; Sagittarius B is another radio source nearby, whose emission is caused by clouds of gas near the galactic center.

*13.3b Synchrotron Radiation

Continuum radio radiation can be generated by several processes. One of the most important is *synchrotron emission* (Fig. 13–2), the process that produces the radiation from Taurus A, a supernova remnant. Lines of magnetic field extend throughout the visible Crab Nebula and beyond. Electrons, which are electrically charged, tend to spiral around magnetic lines of force. Electrons of high energy spiral very rapidly, at speeds close to the speed of light. We say that they move at ''relativistic speeds,'' since the theory of relativity must be used for calculations when the electrons are going that fast. Under these conditions, the electrons radiate very efficiently. (This is the same process that generates the light in electron synchrotrons in some kinds of atom-smashing physics laboratories on earth, hence the name synchrotron radiation.)

The suggestion that the synchrotron mechanism causes radiation from various astronomical sources was first made by several Soviet theoreticians in about 1950. Synchrotron radiation is highly polarized, and the discovery a few years later that

Many of the radio objects have been discovered to lie outside our galaxy, and we shall discuss them in subsequent chapters. Quasars (Chapter 15), for example, are an important group of extragalactic (outside our galaxy) objects that radiate strongly in the radio region of the spectrum.

Figure 13–2 Electrons spiralling around magnetic lines of force at velocities near the speed of light (we say ''at relativistic velocities'') emit radiation in a narrow cone. This radiation, which is continuous and highly polarized, is called synchrotron radiation. Synchrotron radiation has been observed in both optical and radio regions of the spectrum.

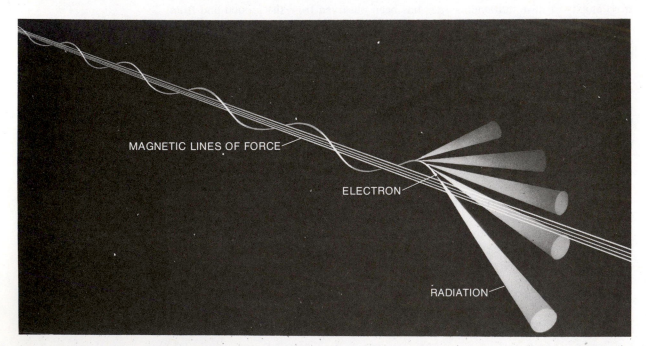

MAGNETIC LINES OF FORCE

ELECTRON

RADIATION

the optical radiation from the Crab Nebula is highly polarized (Fig. 13–3) was an important confirmation of this suggestion. The radio radiation from the Crab is also highly polarized.

The intensity of synchrotron radiation is related not to the temperature of the astronomical body that is emitting the radiation, but rather to the strength of its magnetic field and to the number and energy distribution of the electrons caught in that field. Since the temperature of the object cannot be derived from knowledge of the intensity of the radiation, we call this radiation *non-thermal radiation*. Synchrotron radiation is but one example of such non-thermal processes. The synchrotron process can work so efficiently that a relatively cool astronomical body can give off a tremendous amount of such radiation, perhaps so much that it would have to be heated to a few million degrees before it would radiate as much *thermal radiation* at a given frequency.

Figure 13–3 Photographs of the Crab Nebula taken through filters that pass visible light polarized at the angles shown with the arrows. A non-polarized source would appear the same when viewed at any angle of polarization. The pictures at different polarization angles look very different from each other. This shows that the light from the Crab is highly polarized, which implies that it is caused by the synchrotron mechanism.

By *thermal radiation* we mean continuous radiation whose spectrum is directly related to the temperature of the gas.

13.4 The Radio Spectral Line from Interstellar Hydrogen

In about 1950, though radio astronomers were very busy with continuum work, there was still a hope that a radio spectral line might be discovered. This discovery would allow Doppler-shift measurements to be made.

What is a radio spectral line? Remember that an optical spectral line corresponds to a wavelength (or frequency) in the optical spectrum that is more (for an emission line) or less (for an absorption line) intense than neighboring wavelengths or frequencies. Similarly, a radio spectral line corresponds to a frequency (or wavelength) at which the radio noise is slightly more, or slightly less, intense.

A radio station is an emission line on our home radios.

13.4a The Hydrogen Spin-Flip

The most likely candidate for a radio spectral line that might be discovered was a line from the lowest energy levels of interstellar hydrogen atoms. This line was predicted to be at a wavelength of 21 cm. Since hydrogen is by far the most abundant element in the universe, it seems reasonable that it should produce a strong spectral line. Furthermore, since most of the interstellar hydrogen has not been heated by stars or otherwise, it is most likely that this hydrogen is in its state of lowest possible energy.

This line at 21 cm comes not from a transition to the ground state from one of the higher states, nor even from level 2 to level 1 as does Lyman alpha, but rather from a transition between the two sublevels into which the ground state of hydrogen is divided (Fig. 13–5).

Figure 13–4 Radio astronomers often speak in terms of frequency instead of wavelength. Since all electromagnetic radiation travels at the same speed (the speed of light) in a vacuum, fewer waves of longer wavelength *(top)* pass an observer in a given time interval than do waves of a shorter wavelength *(bottom)*. If the wavelength is half as long, twice as many waves pass, i.e., the frequency is twice as high. The wavelength λ times the frequency *ν* is constant, with the constant being the speed of light *c:* λ*ν* = *c*.

Frequency is given in hertz (Hz), formerly called cycles per second. The 21-cm line is at 1420 MHz.

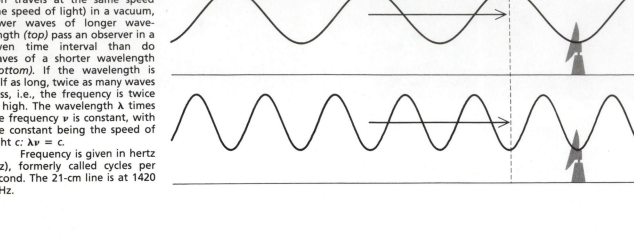

FIXED POINT

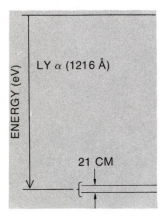

Figure 13–5 21-cm radiation results from an energy difference between two sublevels in the lowest principal energy state of hydrogen. The energy difference is much smaller than the energy difference that leads to Lyman α. So (because $E = h\nu = hc/\lambda$) the wavelength is much longer.

If we were to watch any particular group of hydrogen atoms, we would find that it would take 11 million years before half of the electrons had undergone spin-flips; we say that the *half-life* is 11 million years for this transition. But even though the probability of a transition taking place in a given interval of time is very low, there are so many hydrogen atoms in space that enough 21-cm radiation is given off to be detected.

For this astronomical discussion, it is sufficient to think of a hydrogen atom as an electron orbiting a proton. Both the electron and the proton have the property of spin; each one has angular momentum (Section 18.9) as if it were spinning on its axis.

The spin of the electron can be either in the same direction as the spin of the proton or in the opposite direction. The rules of quantum mechanics prohibit intermediate orientations. If the spins are in opposite directions, the energy state of the atom is very slightly lower than the energy state occurring if the spins are in the same direction. The energy difference between the two states is equal to a photon of 21-cm radiation.

If an atom is sitting alone in space in the upper of these two energy states, with its electron and proton spins aligned in the same direction, it has a certain small probability that the spinning electron would spontaneously flip over to the lower energy state and emit a photon. We thus call this a *spin-flip* transition (Fig. 13–6). The photon of hydrogen's spin-flip corresponds to radiation at a wavelength of 21 cm — the *21-cm line*.

We have just described how an emission line can arise at 21 cm. But what happens when continuous radiation passes through neutral hydrogen gas? In this case, some of the electrons in atoms in the lower state will absorb a 21-cm photon and flip over, putting the atom into the higher state. Then the radiation that emerges from the gas will have a deficiency of such photons and will show the 21-cm line in absorption (Fig. 13–6).

In 1944, H. C. van de Hulst, a Dutch astronomer (then a graduate student), predicted that the 21-cm emission would be strong enough to be observable as soon as the proper equipment was developed. It took seven more years for the instrumental capability to be built up. In 1951, scientists both in Holland at Leiden and in the United States at Harvard were building equipment to detect the 21-cm radiation. An unfortunate fire set back the Dutch effort many months. In the meantime, the work at Harvard went on. Electronic equipment to observe in the direction of the galactic center was built in a physics lab with a small antenna stuck out a window (into which passing undergraduates occasionally lobbed snowballs). Finally, the Harvard team, consisting of a graduate student named Harold Ewen and his advisor, Edward M. Purcell, went ''on the air'' and succeeded in observing the 21-cm line in emission (Fig. 13–7). Soon the Dutch group and then a group in Australia confirmed the detection of the 21-cm line. Spectral-line radio astronomy had been born.

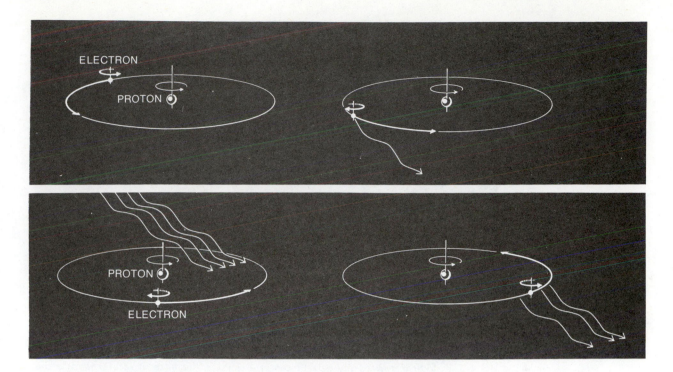

Figure 13–6 When the electron in a hydrogen atom flips over so that it is spinning in the opposite direction from the spin of the proton *(top)*, an emission line at a wavelength of 21 cm results. When an electron takes energy from a passing beam of radiation, causing it to flip from spinning in the opposite direction from the proton to spinning in the same direction *(bottom)*, then a 21-cm line in absorption results.

13.4b Mapping Our Galaxy

21-cm hydrogen radiation has proved to be a very important tool for studying our galaxy because this radiation passes unimpeded through the dust that prevents optical observations very far into the plane of the galaxy. Using 21-cm observations, astronomers can study the distribution of gas in the spiral arms. We can detect this radiation from gas located anywhere in our galaxy, even on the far side, whereas light waves penetrate the dust clouds in the galactic plane only about 10 per cent of the way to the galactic center.

But here again we come to the question that bedevils much of astronomy: how do we measure the distances? Given that we detect the 21-cm radiation from a gas cloud (since the cloud contains neutral hydrogen, it is an H I region), how do we know how far away the cloud is from us?

The answer can be found by using a model of rotation for the galaxy, that is, a description of how each part of the galaxy rotates. As we have already learned, the outer regions of galaxies rotate differentially; that is, the gas nearer the center rotates faster than the gas farther away from the center.

Figure 13–8 shows a simplified version of differential rotation. Because of the differential rotation, the distance between us and point A is decreasing. Therefore, from our vantage point at the sun, point A has a net velocity toward us. Thus its 21-cm line is Doppler-shifted toward shorter wavelengths. If we were talking about light, this shift would be in the blue direction; even though we are discussing radio waves, we say "blueshifted" anyway. If we look from our vantage point toward gas cloud C, we see a redshifted 21-cm line, because its higher speed of rotation is carrying C away from us. But if we look straight toward the center, clouds B_1 and

Figure 13–7 Harold Ewen with the horn radio telescope he and Edward Purcell used to discover the 21-cm line.

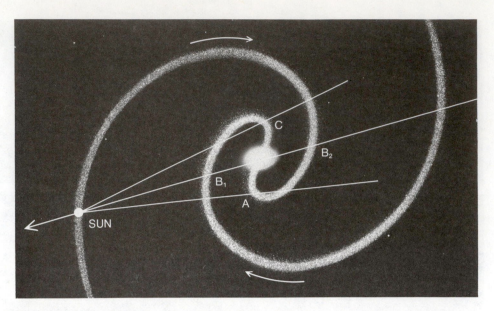

Figure 13–8 Because of the differential rotation, the cloud of gas at point A appears to be approaching the sun, and the cloud of gas at point C appears to be receding. Objects at points B_1 or B_2 have no net velocity with respect to the sun, and therefore show no Doppler shifts in their spectra.

The closer in toward the galactic center a cloud is, the faster it rotates, and so the larger its Doppler shift along any line of sight (other than that toward the center) when observed from the sun. (The difference between the earth's and sun's velocity is known and can be accounted for.)

B_2 are both passing across our line of sight in a path parallel to that of our own orbit. They have no net velocity toward or away from us. Thus this method of distance determination does not work when we look in the direction of the center, nor indeed in the opposite direction.

Once we measure the Doppler shift of an H I region, we can deduce its net velocity: the speed with which it is coming toward or moving away from the sun. In the inner part of the galaxy, we must first deduce the pattern of differential rotation. Since rotation speed is less the farther an object is from the center of our galaxy, we can tell the location of individual H I regions when two or more are in the same line of sight. (Clouds farther from the galactic center take longer to revolve than do clouds closer to the center, in a manner similar to Kepler's third law, discussed in Section 18.7.) The highest velocity cloud along any line of sight must be the cloud that is closest to the center.

Once we work out the law of differential rotation for the whole galaxy by looking along various lines of sight, then we can apply our knowledge to any particular cloud we observe. From its observed velocity, we can tell how far it is from the center of the galaxy: we figure out where along our line of sight in the direction we are looking we meet the circle of gas having the proper velocity to match the observations. By observing in different directions, we can build up a picture of the spiral arms.

Each line of sight along which we look inward from the sun crosses a circle of gas at a given distance from the center at two points. Thus an ambiguity arises, because a cloud at either point would have the same Doppler velocity with respect to the sun. We have to resort to methods other than 21-cm studies to resolve this ambiguity. We know, for example, the average size of H I regions, and so can often tell whether the emitting cloud is at the farther or nearer point by its angular size in the sky. Alternatively, we can sometimes tell by noting whether the 21-cm radiation seems relatively weak or strong. The problem, though it can be partially overcome, gives a basic uncertainty in our knowledge of our galaxy's rotation.

Unfortunately, gas clouds have not only a velocity of revolution around the center of the galaxy but also random velocities to and fro. When we look at the center of the galaxy, we see that there are some clouds of gas moving outward with an average velocity of 50 km/s. These motions, both systematic and random, place a fundamental uncertainty on the conclusions drawn from this method. But this is the best that we can do.

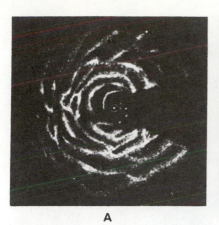

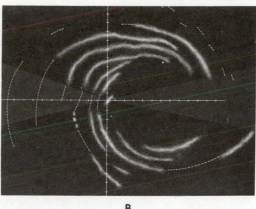

A   B

Figure 13–9 Two artists' impre
sions of the structure of our ga
axy based on 21-cm data. Be
cause hydrogen clouds located i
the directions either toward o
away from the galactic cente
have no radial velocity with respect to us, we cannot find their
distances. Differences between
the drawings result from uncertainties in the observations or in
their interpretation.

The earliest 21-cm maps showed many narrow arms (Fig. 13–9) but no clear pattern of a few broad spiral arms like those we see in other galaxies. The question emerged: is our galaxy really a spiral at all? Only in recent years, with the additional information from studies of molecules in space that we describe in the next section, have we made further progress. Although questions remain, it is clear that the 21-cm radiation is at the base of our mapping efforts.

13.5 Radio Spectral Lines from Molecules

For several years after the 1951 discovery of 21-cm radiation, spectral-line radio astronomy continued with just the one spectral line. Astronomers tried to find others. One prime candidate was OH, hydroxyl, a molecule which should be relatively abundant because it is a combination of the most abundant of the remaining elements, oxygen. OH has four lines close together at about 18 cm in wavelength, and the relative intensities expected for the four lines had been calculated.

It wasn't until 1963 that other radio spectral lines were discovered, and four new lines were indeed found at 18 cm. But the intensity ratios were all wrong to be OH, according to the predicted values, and for a time we spoke of the "mysterium" lines. Mysterium turned out to be OH after all, but with the process of "masering" affecting the excitation of the energy levels of OH and thus amplifying certain of the lines at the expense of others, which are weakened.

Masers and *lasers* are of great practical use on the earth. (Maser is an acronym for *m*icrowave *a*mplification by *s*timulated *e*mission of *r*adiation, and lasers are their analogue using *l*ight instead of *m*icrowaves.) For example, masers are used as sensitive amplifiers. Masers were "invented" on earth not long before they were found in space. For maser action to occur, a large number of electrons are pushed into a higher energy state in which they tend to stay. When "triggered" by a photon on the proper frequency, the electrons jump down together to a lower energy level. (Their emission is "stimulated" by the trigger, leading to the name "maser.") The original radiation is thus amplified, since there are now many photons at that wavelength instead of just one.

Since the interstellar abundance of OH seemed very much lower than that of isolated hydrogen or oxygen atoms (only one OH molecule for every billion H atoms), it seemed quite unlikely that the quantities of any molecules composed of three or more atoms would be great enough to be detected. The chance of three atoms getting together in the same place should be very small.

In 1968, however, Berkeley's Charles Townes and colleagues observed the radio frequencies that were predicted to be the frequencies of water (H_2O) and ammo-

A few interstellar optical and ultraviolet lines from atoms and simple molecules have been studied from the ground and from space satellites.

The idea of discovering a new element was not unprecedented—after all, unknown lines at the solar eclipse in 1868 had been assigned to an unknown element, "helium," because they occurred only (as far as was known at that time) on the sun. But the periodic table of elements has been filled in during the last 100 years, so we can't expect to find new elements in that way anymore.

Figure 13–10 An overall view of the National Radio Astronomy Observatory's field site at Green Bank, West Virginia. Three dishes that are used together as an interferometer are at right. The 91-m dish is in front of the 43-m dish at extreme left, which was used for the discovery of interstellar formaldehyde.

DUST GRAIN

H

H

H

5000 YEARS LATER

H_2

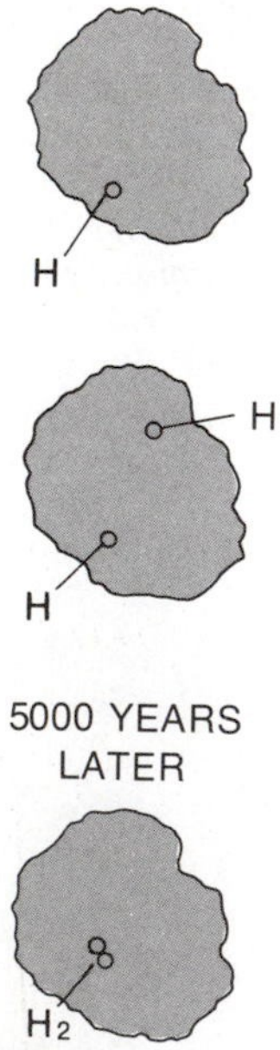

Figure 13–11 Hydrogen molecules are formed in space with the aid of dust grains at an intermediate stage.

nia (NH_3). The spectral lines of these molecules proved surprisingly strong, and were easily detected.

Soon afterwards, another group of radio astronomers used a telescope of the National Radio Astronomy Observatory (Fig. 13–10) to discover interstellar formaldehyde (H_2CO) at a wavelength of 6 cm. This discovery (by Ben Zuckerman, now of UCLA; Patrick Palmer, now of the University of Chicago; David Buhl, now of NASA; and Lewis Snyder, now of the University of Illinois) was the first molecule that contained two ''heavy'' atoms, that is, two atoms other than hydrogen.

By this time, it was apparent that the earlier notion that it would be difficult to form molecules in space was wrong. There has been much research on this topic, but the mechanism by which molecules are formed has not yet been satisfactorily determined. For some molecules, including molecular hydrogen, it seems that the presence of dust grains is necessary. In this scenario, one atom hits a dust grain and sticks to it (Fig. 13–11). It may be thousands of years before a second atom hits the same dust grain, and even longer before still more atoms hit. But these atoms may stick to the dust grain rather than bouncing off, which gives them time to join together. Complex reactions may take place on the surface of the dust grain. Then, somehow, the molecule must get off the dust grain, thus being released into space as part of a gas. Perhaps either incident ultraviolet radiation or the energy released in the formation of the molecule allows the molecule to escape from the grain surface.

Though hydrogen molecules form on dust grains, a strong body of opinion holds that most of the other molecules are formed in the interstellar gas without need for grains. Recent theoretical and laboratory studies indicate that reactions between neutral molecules and ionized molecules may be particularly important. Many of these chains start with molecular hydrogen, formed on grains, being ionized by cosmic rays. There are still many gaps in the theory. It is likely that no single process forms all the interstellar molecules.

After the discoveries of water, ammonia, and formaldehyde in interstellar space, further discoveries came one after another. Many different radio astronomers looked up the wavelengths of likely spectral lines from molecules containing abundant elements, and were able to observe the radiation. Sometimes the lines were in absorption and sometimes in emission. Eventually, astronomers and chemists started measuring frequencies especially for use in searches at the telescope.

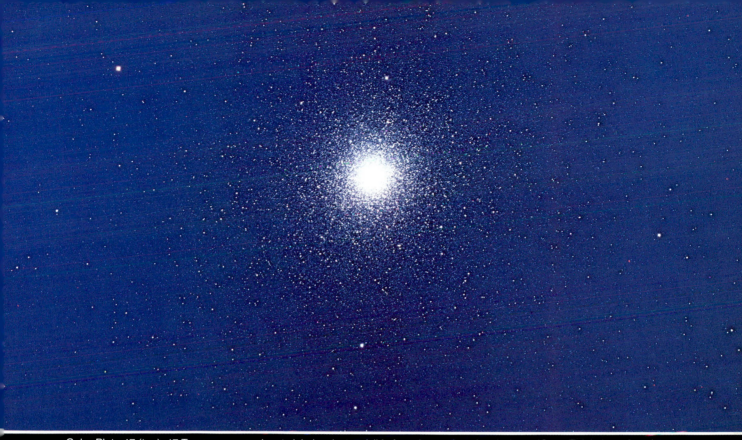

Color Plate 47 (top): 47 Tucanae, a prominent globular cluster visible from the southern hemisphere. (Cerro Tololo Inter-American Observatory photo with the 4-m telescope)

Color Plate 48 (bottom): The Pleiades. M45, is a galactic cluster in the constellation Taurus, the bull. Reflection nebulae are visible around the brightest stars. (Palomar Observatory, California Institute of Technology photo with the 1.2-m Schmidt camera)

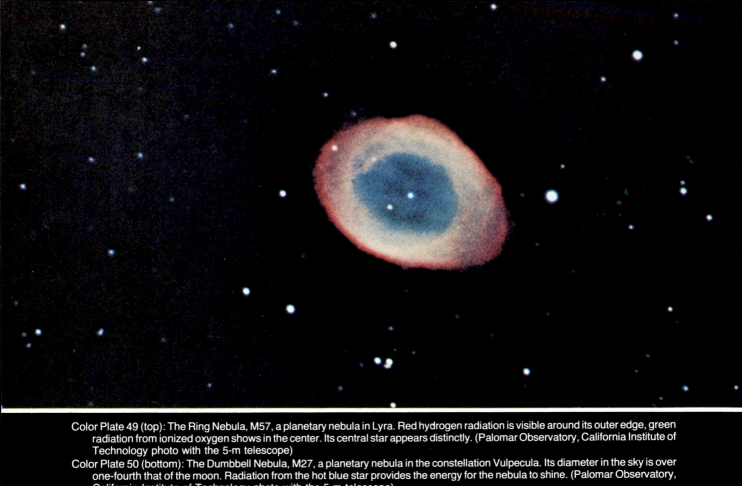

Color Plate 49 (top): The Ring Nebula, M57, a planetary nebula in Lyra. Red hydrogen radiation is visible around its outer edge, green radiation from ionized oxygen shows in the center. Its central star appears distinctly. (Palomar Observatory, California Institute of Technology photo with the 5-m telescope)

Color Plate 50 (bottom): The Dumbbell Nebula, M27, a planetary nebula in the constellation Vulpecula. Its diameter in the sky is over one-fourth that of the moon. Radiation from the hot blue star provides the energy for the nebula to shine. (Palomar Observatory, California Institute of Technology photo with the 5-m telescope)

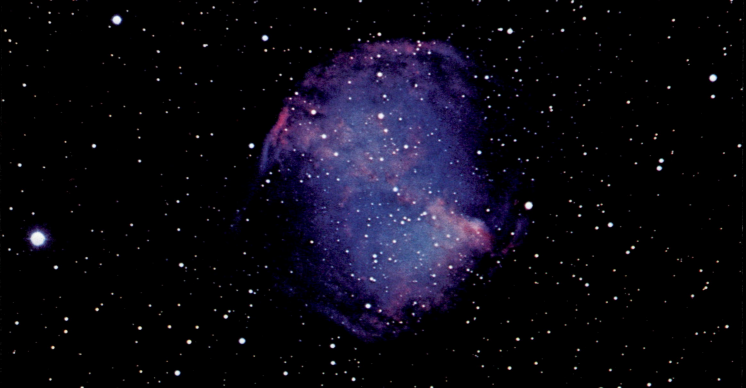

Color Plate 51 (top): A planetary nebula, NGC 6781, in the constellation Aquila. (Palomar Observatory, California Institute of Technology photo with the 1.2-m Schmidt camera)

Color Plate 52 (bottom, left): A radio image of the Cassiopeia A supernova remnant, observed at a wavelength of 20 cm with the VLA. (Courtesy of Philip E. Angerhofer, Richard A. Perley, Bruce Balick, and Douglas Milne with the VLA or NRAO)

Color Plate 53: (bottom, right): An x-ray image of the Cassiopeia A supernova remnant, observed with the Einstein Observatory. The bright ring is thought to be the region associated with the expanding shock front. No pulsar has been detected. (Courtesy of S. S. Murray and colleagues at the Harvard-Smithsonian Center for Astrophysics)

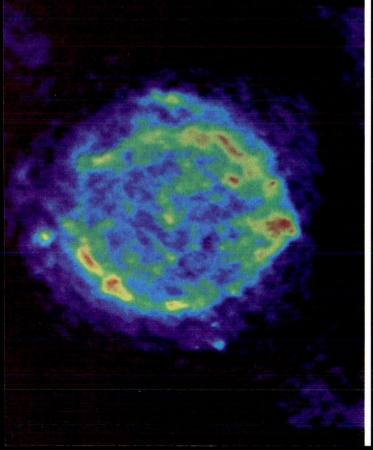

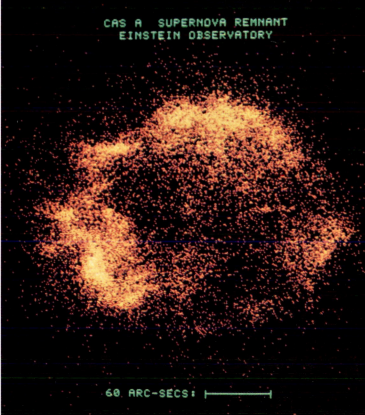

Color Plate 54: The Crab Nebula, MI, in Taurus is the remnant of the supernova of 1054 A.D. The red filaments radiate in the hydrogen lines; the white continuum is from synchrotron radiation. (Palomar Observatory, California Institute of Technology photo with the 5-m telescope)

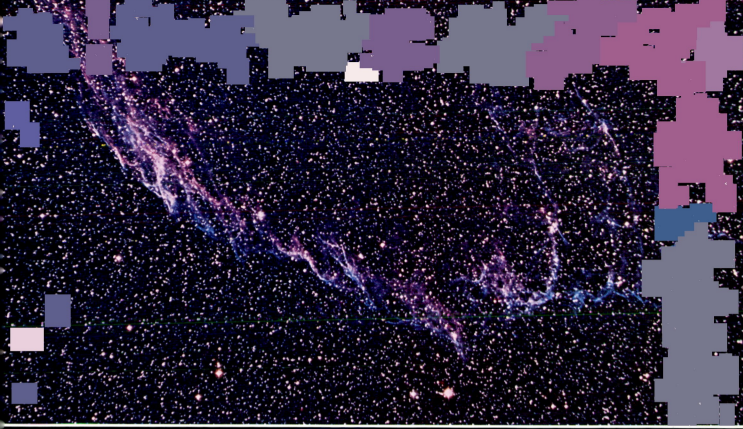

Color Plate 55 (top): The Veil Nebula, NGC 6992, part of the Cygnus loop, a supernova remnant. (Palomar Observatory, California Institute of Technology photo with the 1.2-m Schmidt camera)

Color Plate 56 (bottom): The Orion Nebula, M42, is glowing gas excited by the Trapezium, four hot stars. The nebula contains stars in formation; it is 25 ly across and 400 parsecs away. (Palomar Observatory, California Institute of Technology photo with the 5-m telescope)

Color Plate 57: The Eta Carinae Nebula, NGC 3372, in the southern constellation Carina. Visually, this is the brightest part of the Milky Way. The central dark cloud superimposed on the brightest gas is the Keyhole Nebula.(GK. "eta") Carinae is the brightest star left of the Keyhole. (Cerro Tololo Inter-American Observatory photo).

Color Plate 58: Clouds of gas and dust around ρ (rho) Ophiuchi (in the blue reflection nebula). The bright star Antares is surrounded by yellow and red nebulosity. M4, the nearest globular cluster to us, also shows. (© 1979 Royal Observatory, Edinburgh)

Color Plate 59: The Horsehead Nebula and ζ (zeta) Orionis. (© 1980 Royal Observatory, Edinburgh)

The list of molecules discovered expanded gradually from three-atom molecules like ammonia and water, and four-atom molecules like formaldehyde, to even more complex molecules. Over 50 molecules have now been discovered in interstellar space—one of the most complex is $HC_{11}N$, which has 12 "heavy" atoms and only 1 hydrogen. Hundreds of spectral lines remain unidentified, some of which are undoubtedly from still other molecules.

Studying the spectral lines provides information about physical conditions—temperature, densities, and motion, for example—in the gas clouds that emit the lines. For example, formaldehyde radiates only when the density is roughly ten times that of the gas that radiates carbon monoxide. The clouds that emit molecular lines are usually so dense that hydrogen atoms have combined into hydrogen molecules and very little 21-cm is emitted.

Studies of molecular spectral lines have been used together with 21-cm observations to improve the maps of the spiral structure of our galaxy. Observations of carbon monoxide (CO) in particular have provided better information about the parts of our galaxy farther out than the distance of the sun from the galaxy's center. The result is a four-armed spiral (Fig. 13–12). But there are still differences to be resolved between the different spirals that have been suggested. In one of the models, the data from the outer part of the galaxy are made to connect with data from interior regions, though the interior regions seem clumpier. This difference between inside and outside may or may not turn out to be real. There are signs, probably real, of short partial arms or spurs, in addition to the four major arms.

13.6 Measuring the Mass of our Galaxy

We can measure the mass of our galaxy by studying the velocity of rotation of gas clouds. All the mass inside the radius of the cloud's orbit acts as though it were concentrated at one point; the more mass that is present (and thus the stronger the gravity), the faster a gas cloud has to revolve around the center to keep itself from falling inward. (Objects "revolve" around something, but "rotate" as part of something that is turning around; a gas cloud in a galaxy is an intermediate case, and one should pay attention to when it is suitable to speak of it "revolving" and when "rotating.")

Figure 13–12 *(A)* This four-armed spiral has been fit to 21-cm hydrogen observations and to carbon-monoxide observations to map the spiral structure of our galaxy. The solar circle marks the distance of the sun from the center of our galaxy. The dashed lines show the extrapolation of the solid lines inward to 4 parsecs from the center. Three arms and a spur are based on observations and are given names corresponding to the constellations in which they appear; the fourth arm is drawn assuming a symmetric galaxy. (Courtesy of Leo Blitz, University of Maryland) *(B)* The structure of the inner part of our galaxy based primarily on southern-hemisphere CO observations from Australia and northern-hemisphere CO observations from the University of Massachusetts. The star marks the sun and the + marks the galactic center. The expanding arm at a radius of 3.6 kpc is marked with arrows. The dashed segment is based on 21-cm hydrogen observations, since CO is faint there. Most structure corresponds to four spiral arms; other emission regions appear as spurs and bifurcations of the arms. (Courtesy of B. J. Robinson, J. B. Whiteoak, R. N. Manchester, C.S.I.R.O., Australia, and W. H. McCutcheon, Univ. of British Columbia)

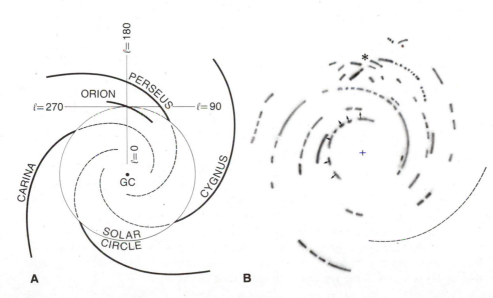

A   B

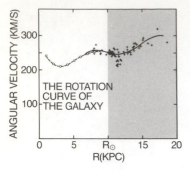

Figure 13–13 The rotation curve of our galaxy. The inner part of the curve *(dotted line)* is based on 21-cm hydrogen observations. It had been anticipated that beyond about 10 parsecs the velocity would decline following Kepler's third law, because there would be no additional mass added as we go further out. But by observing carbon monoxide we are newly able to make direct measurements of the outer regions *(shaded)*. The observations *(crosses)* show that the velocity of rotation doesn't decline. This must mean that there is more mass in the outer regions of the galaxy than we had anticipated. (Courtesy of Leo Blitz, University of Maryland)

The gravitational effect of the mass outside the cloud's orbit turns out to balance out (assuming spherical symmetry); the mass has no net gravity, and thus does not affect the cloud's velocity. Conversely, the cloud's velocity gives us no information about this mass, so the results of this section do not include the galactic corona (Section 12.2).

From the velocity with which a gas cloud is orbiting, we can thus calculate how much mass is inside the orbit of the cloud. (We shall see in Chapter 18 that the method is called Kepler's third law, and was first worked out for the planets.)

It is thus important to measure the velocity of rotation of gas clouds in our galaxy as far out from the center as possible. Velocities have been measured inside the sun's orbit from measurements of the 21-cm line, using the method discussed in the previous section. Leo Blitz of the University of Maryland has now extended these measurements out twice as far using radiation from carbon monoxide instead of hydrogen.

A graph of the velocity of rotation vs. distance from the center is called a *rotation curve* (Fig. 13–13). It had been expected up until about 1980 that our galaxy's rotation curve would stop increasing beyond the sun's distance from the galactic center and begin to decrease, since the galaxy essentially ended. Then no more mass would be included inside as we went to larger radii. But the curve does not stop increasing and begin to decrease, which indicates that the galaxy is larger and contains more mass than we had thought. It now seems that our galaxy is twice as massive as the Andromeda Galaxy, and contains perhaps 10^{12} solar masses; this is at least twice as massive as had previously been calculated. If we assume that an average star has 1 solar mass, then our galaxy contains about 10^{12} stars.

13.7 Molecular Hydrogen

While individual atoms of hydrogen, whether neutral or ionized, have been extensively observed in the interstellar gas, hydrogen molecules (H_2) have been observed only more recently. Even though astronomers had long thought that molecular hydrogen could be a major constituent of the interstellar medium, it was simply not possible to observe hydrogen in molecular form. At the low temperature of interstellar space, only the lowest energy levels of molecular hydrogen are excited, and the lines linking these levels fall in the far ultraviolet. Because our atmosphere prevents these lines from reaching us on earth, we had to wait to observe from space in order to observe lines from H_2. H_2 was first observed from a rocket in 1970.

In 1972, a 90-cm telescope was carried into orbit aboard NASA's third Orbiting Astronomical Observatory, named "Copernicus" in honor of that astronomer's 500th birthday (which occurred a year later). The telescope, operated by scientists from the Princeton University Observatory, was largely devoted to observing interstellar material. The observers could point the telescope at a star and look for absorption lines caused by gas in interstellar space as the light from the star passed through the gas en route to us. Since it is easiest to pick up interstellar absorption lines if the star itself has no lines of its own, the scientists observed in the direction of B stars, which have few lines and are very bright (Fig. 13–14).

Since the hydrogen molecule is easily torn apart by ultraviolet radiation, the observers did not find much molecular hydrogen in most directions. But whenever they looked in the direction of highly reddened stars, they found a very high fraction of hydrogen in molecular form: more than 50 per cent. Presumably, in these regions some of the dust that causes the reddening shields the hydrogen from being torn apart by ultraviolet radiation. There are also theoretical grounds for believing that the molecular hydrogen is formed on dust grains, so it seems reasonable that the high fraction of H_2 is found in the regions with more dust grains.

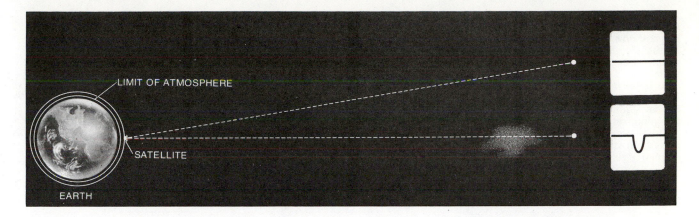

13.8 The Formation of Stars

Most radio spectral lines seem to come only from a very limited number of places in the sky—molecular clouds. (Carbon monoxide is the major exception, for it is widely distributed across the sky.) Infrared and radio observations together have provided us with an understanding of how stars are formed from these dense regions of gas and dust.

Giant molecular clouds are 50 to 100 parsecs across. There are a few thousand of them in our galaxy. The largest giant molecular clouds are about 100 parsecs across and contain about 100,000 to 1,000,000 times the mass of the sun. Their internal densities are about 100 times that of the interstellar medium around them. Since giant molecular clouds break up to form stars, they only last 10 million to 100 million years.

Carbon monoxide observations reveal the giant molecular clouds, but it is molecular hydrogen (H_2) rather than carbon monoxide that is significant in terms of mass. It is difficult to detect the molecular hydrogen directly, though there is over 100 times more molecular hydrogen than dust. So we must satisfy ourselves with observing the tracer, carbon monoxide, whose spectral lines are excited by collisions with hydrogen molecules and so shows us where the hydrogen molecules are.

The Great Rift in the Milky Way, the extensive dark area that can be seen in the midst of the photograph opening Chapter 12, is made up of many overlapping giant molecular clouds.

A giant molecular cloud is fragmented into many denser bits of 1 parsec in size. These bits must become much smaller and denser yet to form a star or, in many cases, several stars.

Sometimes we see smaller objects, called globules, or *Bok globules*. (They are named after Bart Bok, who has extensively studied these objects.) Some of the smaller globules are visible in silhouette against H II regions, as in Color Plate 61 (top right of picture, for example). Others are larger, and appear isolated against the stellar background (Fig. 13–15). Two hundred large globules are known to be within 500 parsecs of the sun (5 per cent of the way to the center of the galaxy), and there may be 25,000 in our galaxy. It has been suggested that the globules may be on their way to becoming stars. Molecular-line observations of the larger globules indicate that they may contain sufficient mass for gravitational collapse to be taking place. And though there had been some doubt whether stars were forming in smaller globules, IRAS discovered infrared radiation from at least some that indicate the presence of star formation inside. For example, a few stars, each about the mass of the sun

Figure 13–14 From above the atmosphere of the earth, the Copernicus satellite looked toward a star that provides a more-or-less continuous spectrum. (A rapidly rotating B star was usually chosen, because B stars have few lines and those lines are washed out by Doppler shifts resulting from the velocity of rotation.) When a cloud of gas is in the path, then absorption or emission from molecules or atoms in the cloud results.

Radio studies of interstellar molecules and infrared studies have recently led to a major improvement in our understanding of stars in formation. The discussion in this section thus completes the cycle we began in the chapters on stellar evolution, for the older stars have been born, died, and spread out their material through interstellar space.

Figure 13–15 The Bok globule Barnard 335. The small cloud of gas and dust appears as a "hole" in the distribution of stars. We assume that the stars are roughly uniformly distributed and that such holes result because of the extinction caused by dust in the line of sight. Emission from carbon monoxide has been observed from this region, signifying that something is really present. The angular size of this globule is about 4 arc minutes.

and each only a few hundred thousand years old, were discovered in the globule called Barnard 5 (as we saw in the Color Essay on IRAS).

Extrapolating from the region near the sun that IRAS studied best, to the galaxy as a whole, about one one-solar-mass star is forming in our galaxy each year. The result indicates that stars are forming more often than had been realized.

Observations of interstellar molecules have shown that gas flows rapidly outward from some newly formed stars in two opposing streams. These high-velocity *bipolar flows* typically carry more than a solar mass of material outward and transport a lot of energy. They are thus clearly very important for the evolution of young stars. The gas flow extends only about 1 light-year, which can be traversed by gas at the high velocities measured in only about 10,000 years. In at least one case, Herbig-Haro objects (Section 8.1) are located at the end of one of the jets; the motion of the H-H objects can be projected back to the same location at which the jets originate. In this and in most cases, an infrared source (with often no image in the visible part of the spectrum) is located at the center. We have already met a jet of gas in the model of SS433 and we will learn about other jets in radio galaxies and in quasars. Astrophysicists are finding that signs of violent activity like jets are more common than had been suspected.

13.8a A Case Study: The Orion Molecular Cloud

Many radio spectral lines have been detected only in a particular cloud of gas located in the constellation Orion, not very far from the main Orion Nebula. This *Orion Molecular Cloud* (Fig. 13–16), which contains about 500 solar masses of material, is itself buried deep in nebulosity. It is relatively accessible to our study because it is only about 500 parsecs from us. Even though less than 1 percent of the cloud's mass is dust, that is still a sufficient amount of dust to prevent ultraviolet light from nearby stars from entering and breaking the molecules apart. Thus molecules can accumulate.

We know that young stars are found in this region—the Trapezium, a group of four hot stars readily visible in a small telescope, is the source of ionization and of energy for the Orion Nebula. The Trapezium stars are relatively young, about 100,000 years old. The Orion Nebula (Color Plate 56), prominent as it is in the visible, is an H II region located along the near side of the molecular cloud (Fig. 13–17). The nebula contains much less mass than does the molecular cloud. The ultraviolet radiation from the H II region causes a sharp front of ionization at the edge of the molecular cloud. A shock wave forms and concentrates gas and dust in the molecular cloud to make protostars.

The properties of the molecular cloud can be deduced by comparing the radiation from its various molecules and by studying the radiation from each molecule

Figure 13–16 The contours show the molecular cloud associated with the Orion Nebula. The molecular cloud is actually on the far side of the glowing gas of the nebula, but the radio waves from the molecules penetrate the nebula. The contours of radio emission correspond roughly to regions of different density. The densest part (the smallest region) also includes a region of infrared radiation called the Kleinmann-Low Nebula and a strong point infrared emitter called the Becklin-Neugebauer object. The shape of the outermost contour indicates that the lower H II region may be expanding against the molecular cloud.

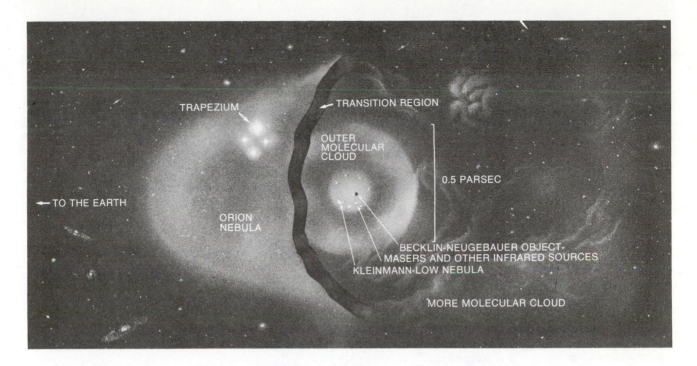

TRAPEZIUM

TRANSITION REGION

OUTER
MOLECULAR
CLOUD

0.5 PARSEC

TO THE EARTH

ORION
NEBULA

BECKLIN-NEUGEBAUER OBJECT
MASERS AND OTHER INFRARED SOURCES
KLEINMANN-LOW NEBULA

MORE MOLECULAR CLOUD

individually. The density, 1000 particles per cubic centimeter in the outer limits at which the cloud is visible to us, increases toward the center. The cloud may actually be as dense as 10^6 particles/cm^3 at its center. This is still billions of times less dense than our earth's atmosphere, and 10^{16} times less dense than the star that may eventually form, though it is substantially denser than the average interstellar density of about 1 particle per cm^3.

One of the brightest of all the infrared sources in the sky, an object in the Orion Nebula that was discovered by Eric Becklin, now at the University of Hawaii, and Gerry Neugebauer of Caltech, is right in the midst of the Orion Molecular Cloud. This *Becklin-Neugebauer object* (Fig. 13–18), also known for short as the *B-N object,* is about 200 astronomical units across. Its temperature is 600 K. The B-N object is the prime candidate for a star about to be born.

The B-N object is behind so much dust as seen from earth that it appears very faint. Comparison of certain of the spectral lines indicated how much the B-N object

Figure 13–17 The structure of the Orion Nebula and the Orion Molecular Cloud, proposed by Ben Zuckerman, now of UCLA.

Figure 13–18 Infrared images of the Kleinmann-Low Nebula in Orion, a region of active star formation. The brightest part of the image at right is the Becklin-Neugebauer object, probably one of the youngest stars known. A map of equal-intensity lines appears at left. (Courtesy of Robert D. Gehrz, J. A. Hackwell, and Gary Grasdalen, Wyoming Infrared Observatory)

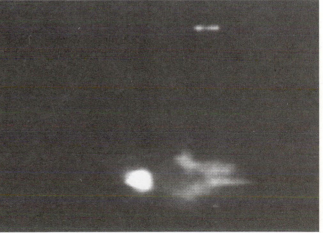

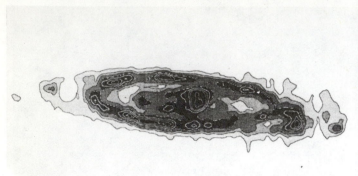

A

B

Figure 13–19 *(A)* The Androm-eda Galaxy (Color Plate 68) ob-served at the long infrared wave-length of 100 microns from IRAS. The darkest regions in the image, corresponding to the strongest radiation, are regions where there are young stars. A ring, cor-responding to the dust lane of the visual photograph, shows. *(B)* An optical view, observed with an amateur (20-cm) tele-scope, for comparison.

Another example of star formation in a molecular cloud is M17, shown in Color Plate 62. The relatively sharp edge at right marks the left edge of the molecular cloud, which contains both infrared objects and maser sources. M8 (Color Plate 64) is also an example.

If the Becklin-Neugebauer object is indeed a star at the distance of the Orion Nebula, we can calculate how much its light would have to be reddened to appear to us so bright in the infrared yet so faint in the visible. The extinction in the visible region of the spectrum would be at least 50 magnitudes (a factor of 10^{20}).

is reddened. Donald Hall, now of the University of Hawaii, and the other scientists involved concluded that it is a very young star of spectral type B, a massive, hot star. It is already, but just barely, operating on the basis of nuclear fusion.

Other infrared sources are present nearby the B-N object, and probably also contain young stars or stars in formation. Small intense sources of maser radiation from various molecules also exist nearby. Such interstellar masers can only exist for small sources, about the size of our solar system, and are also signs of stars in formation.

13.8b IRAS and Star Formation

Not only has IRAS discovered stars forming in globules in our own galaxy but also it has detected signs of similar star formation in other galaxies. For example, its scan of the Andromeda galaxy (Fig. 13–19) showed a ring of bright infrared emission as well as a bright infrared region near the galaxy's nucleus that indicate the recent formation of stars in those locations. We don't yet know why stars should form in such a ring. (See the Color Essay following page 44.)

IRAS observations of the Tarantula Nebula, which glows brightly in reddish H-alpha in the Large Magellanic Cloud (Color Plate 65), showed many stars in forma-tion. Each infrared wavelength observed tended to show a different stage of evolution (Fig. 13–20).

*13.9 At a Radio Observatory

What is it like to go observing at a radio telescope? First, you decide just what you want to observe, and why. You have probably worked in the field before, and your reasons might tie in with other investigations under way. Perhaps a newly avail-able set of radio frequencies received in the mail from a colleague convinced you to observe a particular molecule.

Then you decide at which telescope you want to observe; let us say it is one of the telescopes of the National Radio Astronomy Observatory. You first send in a proposal to the NRAO, describing what you want to observe and why. You also list recent articles you have published, providing yourself with a track record. After all, even the best observations don't help anyone if you don't study your data and report your results.

Your proposal will be read by a panel of scientists. If the proposal is approved, it is placed in a queue waiting for observing time. You might be scheduled to observe for a five-day period to begin six months to a year after you have submitted your proposal.

Figure 13–20 An IRAS scan of part of the Large Magellanic Cloud, including the Tarantula Nebula. The longest infrared wavelengths observed show the molecular clouds about to collapse, the medium wavelengths show the matter collapsing and becoming denser, and the shortest infrared wavelengths show the newly formed stars.

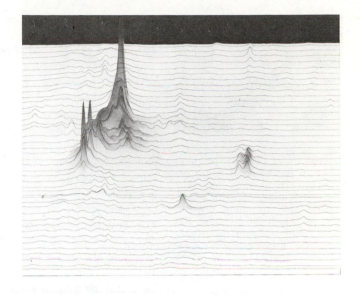

At the same time you might apply (usually to the National Science Foundation) for support to carry out the research. Your proposal possibly contains requests for some support for yourself, perhaps for a summer, and support for a student or students to work on the project with you. It might also contain requests for funds for computer time at your home institution, some travel support, and funds to support the eventual study of the data and the publication of the research.

The NRAO has telescopes at Green Bank in West Virginia, for study of wavelengths longer than about 1 cm, at Kitt Peak in Arizona, for study of millimeter wavelengths (Figs. 13–21 and 13–22), and an array of telescopes (the Very Large Array, Section 14.6) west of Socorro, New Mexico, for making high-resolution images. All the observing locations have dormitories and dining halls so that astronomers can stay near the telescopes, which are in isolated locations.

The NRAO does not charge you directly for telescope time—that cost is covered in the overall budget of the observatory itself.

Unlike the optical sky, which is bright (and blue) in the daytime, the radio-sky background at long wavelengths remains dark even when the sun is up. As long as they don't point their telescopes to within a few degrees of the sun, radio astronomers working at long enough wavelengths can observe anywhere in the sky at any hour of the day or night.

Figure 13–21 The 11-m telescope of the National Radio Astronomy Observatory, seen here through a fish-eye lens, was on Kitt Peak in Arizona. It was used for observations at millimeter wavelengths, and has been used to discover many interstellar molecules.

Figure 13–22 The 12-m telescope of the National Radio Astronomy Observatory on Kitt Peak in Arizona, used for observations at millimeter wavelengths. Much of its time has been spent studying interstellar molecules. This 12-m surface replaced an early 11-m dish in the same dome, and has been in use since 1983. It is more accurate in shape than the earlier dish.

At the observatory, you will meet the other members of your team if they have come from different places. If this is your first time at this particular telescope, an astronomer experienced at observing there will help you get started. The discussions you will have with other astronomers in your field are professionally valuable and personally interesting. Wherever they go in the world, astronomers have colleagues to talk to about research of mutual interest.

The astronomers sit at a computer console monitoring the data as they come in from the telescope or array of telescopes (Fig. 13–23). A trained observing assistant actually runs the mechanical aspects of the telescope. You give the observing assistant the coordinates of the point in the sky that you want to observe, and the telescope is pointed for you. These observing assistants are regular employees of the observatory and work in shifts. The astronomers arrange their own schedules so that they can observe around the clock—one doesn't want to waste any observing time. This is unlike optical astronomers who can work all night and sleep for part of the day.

The electronics that are used to treat the incoming signal from the feed are particularly advanced and are as important as the dish itself. A computer can display the incoming data on a video screen. You can even use the computer to manipulate the data a bit, perhaps adding together the results from different five-minute exposures that you have made. At the Very Large Array, huge computers combine the output from two dozen telescopes and show you a color-coded image, with each color corresponding to a different brightness level (Color Plate 52). The data are stored on magnetic computer tape. You take home with you both copies made on paper or film and the computer tapes themselves for further study.

Depending on what you are observing, your results may or may not be immediately apparent to you on the computer screen. Some spectral lines are so intense in certain sources that you can see an emission line on the screen in just a few minutes. Some lines are so faint that you may have to observe for many hours, or even days or months, to reach an acceptable level of sensitivity.

When you have finished observing one source, perhaps because it has set below the horizon, you ask the operator to point the telescope to another source, and off you go again.

At the end of your observing run, you take the data back to your home institution to complete the analysis. You are expected to publish the results as soon as possible in one of the scientific journals, probably after you have given a paper about the results at a professional meeting, such as that of the American Astronomical Society.

Often a research paper deals with some special aspect of an object or set of objects. In order to generalize and to draw basic conclusions, many sets of observations may be necessary. The conclusions now being reached about how stars are formed, for example, make use of observations made by many individuals and groups.

Figure 13–23 The taking of spectral-line data is controlled by a scientist or observing assistant at a keyboard linked to a computer. A screen on which the spectra can be continually seen is at top left. The view here is inside the control room of the 64-m telescope at Parkes, Australia.

Summary and Outline

Line occurs at 21 cm through a spin-flip transition, corresponding to a change between energy subdivisions of the ground level of hydrogen.

Both emission and absorption have been detected.

21-cm radiation used to map the galaxy

Distances measured using differential rotation and the Doppler effect

Measuring mass of our galaxy (Section 13.5)

Revised measurements have doubled the galaxy's previously accepted size.

Molecular hydrogen (Section 13.6)

Interstellar ultraviolet absorption lines from H_2, with abundances up to 50 per cent, are observed from above the earth's atmosphere.

Radio spectral lines from molecules (Section 13.7)

''Mysterium'' lines discovered at 18 cm—later identified as OH affected by masering process

Masers first developed artificially on earth, but later discovered to exist in space

Three dozen molecules have been discovered in space.

Analysis of molecular lines tells us physical conditions (e.g., temperature, density, and motions).

The formation of stars (Section 13.8)

Molecular clouds contain gas that is collapsing to become stars.

Molecules are associated with dark clouds, such as the Orion Molecular Cloud, where the molecules are shielded by dust from being torn apart by ultraviolet radiation.

The Becklin-Neugebauer object, a compact infrared source deep inside the Orion Molecular Cloud, has been revealed through infrared spectral observations to be a young star.

At a radio observatory (Section 13.9)

Stages of conceiving the project, applying for time, observing, reducing data, publishing results

Key Words

interstellar medium, cosmic abundances, H I region, H II region, scattered, reddened, extinction, synchrotron radiation*, non-thermal radiation*, thermal radiation*, spin-flip, half-life, 21-cm line, differential rotation, masers, lasers, rotation curve, giant molecular cloud, Bok globule, bipolar flows, Orion Molecular Cloud, B-N object

---

*These words are found in an optional section.

Questions

1. List two relative advantages and disadvantages of radio astronomy and optical astronomy.

2. What is the ratio of the cosmic abundance of hydrogen to the cosmic abundance of helium?

3. Though cool gas does not give off much continuum radiation, we can detect 21-cm radiation from cool interstellar gas. Explain.

4. Describe the relation of hot stars to H I and H II regions.

5. Briefly define and distinguish between the redshift of a gas cloud and the reddening of that cloud.

6. What determines whether the 21-cm lines will be observed in emission or absorption?

7. Is 21-cm radiation thermal or non-thermal? Explain.

8. If our galaxy rotated like a rigid body, with each point rotating with the same period no matter what the distance from the center, would we be able to use the 21-cm line to determine distances to H I regions? Explain.

9. Describe how a spin-flip transition can lead to a spectral line, using hydrogen as an example. Could deuterium also have a spin-flip line?

10. Why did it take so long to discover interstellar hydrogen molecules?

11. What was ''mysterium''? Why was it thought to be strange?

12. How does a maser work?

13. Why did the abundance of heavy molecules predicted from observations of hydroxyl give too low a value?

14. Why are dust grains important for the formation of interstellar molecules?

15. Which molecule is found in the most locations in interstellar space?

16. Explain what property of young stars and/or the region of space nearby was observed with the IRAS spacecraft.

17. Describe the relation of the Orion Nebula and the Orion Molecular Cloud.

18. Describe the Becklin-Neugebauer object. Why do scientists find it interesting?

19. How do we detect star formation in another galaxy? Describe an example.

20. Optical astronomers can observe only at night. In what time period can radio astronomers observe? Why?

Topics for Discussion

1. What does the discovery of fairly complex molecules in space imply to you about the existence of extraterrestrial life?

2. What does the formation of stars like the sun so commonly in our galaxy imply to you about the existence of planets in our galaxy?

Part V Galaxies and Beyond

The individual stars that we see with the naked eye are all part of the Milky Way Galaxy, discussed in the preceding two chapters. But we cannot be so categorical about the conglomerations of gas and stars that can be seen through telescopes. Once they were all called "nebulae," but we now restrict the meaning of this word to gas and dust in our own galaxy. Some of the objects that were originally classed as nebulae turned out to be huge collections of gas, dust, and stars located far from our Milky Way Galaxy and of a scale comparable to that of our galaxy. These objects are galaxies in their own right, and are both fundamental units of the universe and the stepping stones that we use to extend our knowledge to tremendous distances.

In the 1770's, a French astronomer named Charles Messier was interested in discovering comets. To do so, he had to be able to recognize whenever a new fuzzy object appeared in the sky. He thus compiled a list of about 100 diffuse objects that could always be seen. To this day, these objects are commonly known by their *Messier numbers.* Messier's list contains the majority of the most beautiful objects in the sky, including nebulae, star clusters, and galaxies.

Soon after, William Herschel, in England, compiled a list of 1000 nebulae and clusters, which he expanded in subsequent years to include 2500 objects. Herschel's son John continued the work, incorporating observations made in the southern hemisphere. In 1864, he published the *General Catalogue of Nebulae.* In 1888, J. L. E. Dreyer published a still more extensive catalogue, *A New General Catalogue of Nebulae and Clusters of Stars,* the *NGC,* and later published two supplementary *Index Catalogues, IC's.* The 100-odd non-stellar objects that have Messier numbers are known by them, and sometimes also by their numbers in Dreyer's catalogue. Thus the Great Nebula in Andromeda is very often called M31, and is less often referred to as NGC 224. The Crab Nebula = M1 = NGC 1952. Objects without M numbers are known by their NGC or IC numbers, if they have them.

When larger telescopes were turned to the Messier objects, especially by Lord Rosse in Ireland in about 1850, some of the objects showed traces of spiral structure, like pinwheels. They were called "spiral nebulae." But where were they located? Were they close by or relatively far away?

When such telescopes as the 0.9-m reflector at Lick in 1898, and later the 1.5-m and 2.5-m reflectors on Mount Wilson, began to photograph the "spiral nebulae," they revealed many more of them. The shapes and

The large reflector, with a mirror 1.8 meters across, built by the Earl of Rosse in England in 1845. Problems with maintaining an

accurate shape for the mirror, which was made of metal, led to the telescope's abandonment.

motions of these "nebulae" were carefully studied. Some scientists thought that they were merely in our own galaxy, while others thought that they were very far away, "island universes" in their own right, so far away that the individual stars appeared blurred together. (The name "island universes" had originated with the philosopher Immanuel Kant in 1755.)

The debate raged, and an actual debate on the scale of our galaxy and the nature of the "spiral nebulae" was held on April 16, 1920, as an after-dinner event of the National Academy of Sciences. Harlow Shapley (pronounced to rhyme with "map lee") argued that the Milky Way Galaxy was larger than had been thought, and thus implied that it could contain the spiral nebulae. Heber Curtis argued for the independence of the "spiral nebulae" from our galaxy. This famous *Shapley-Curtis debate* is an interesting example of the scientific process at work. (It has recently been pointed out that most reports of the Shapley-Curtis debate are based on the published transcript, while the actual words spoken on that evening were less thorough. Shapley, for one, had prepared a low-level introductory talk for this audience.)

Shapley's research on globular clusters had led him to correctly assess our own galaxy's large size. But he also argued that the "spiral nebulae" were close by because proper motion had been detected in some of them by another astronomer. These observations were subsequently shown to be incorrect. He also reasoned that an apparent nova in the Andromeda Galaxy, S Andromedae, had to be close or else it couldn't have been as bright; nobody knew about supernovae then. Curtis's conclusion that the "spiral nebulae" were external to our galaxy was based in large part on an incorrect notion of our galaxy's size. He treated S Andromedae as an anomaly and considered only "normal" novae.

So Curtis's conclusion that the "spiral nebulae" were comparable to our own galaxy was correct, but for the wrong reasons. Shapley, on the other hand, came to the wrong conclusion but followed a proper line of argument that was unfortunately based on incorrect and inadequate data.

The matter was settled in 1924, when observations made at the Mount Wilson Observatory by Edwin Hubble proved that there were indeed other galaxies in the universe besides our own. (The European astronomers K. Lundmark and E. Opik had earlier derived a good distance for the Andromeda Galaxy, and Hubble's evidence clinched the case.) In fact, we think of galaxies and clusters of galaxies as fundamental units in the universe. The galaxies are among the most distant objects we can study. Many quasars are even farther away, and turn out to be certain types of galaxies seen at special times.

Galaxies and quasars can be studied in most parts of the spectrum. Radio astronomy, in particular, has long proved a fruitful method of study. The study across the spectrum of galaxies and quasars provides tests of physical laws at the extremes of their applications and links us to cosmological consideration of the universe on the largest scale. Our notion of the past and future of our universe has changed recently, and we describe this new research.

An edge-on view of a galaxy of type Sb, NGC 4565, in the constellation Coma Berenices. Its nuclear bulge and dust lane show clearly. Several other galaxies also appear on this picture, though they are much farther away and thus much smaller. The tiny irregular objects about 2 cm upward from the center of NGC 4565, for example, are distant galaxies.

Galaxies 14

Aims: To discuss the different types of galaxies, to see that galaxies are fundamental units of the universe, to study the expansion of the universe, and to consider how interferometry is now allowing radio astronomy to make significant advances in the study of galaxies

The question of the distance to the "spiral nebulae"—the spiral-shaped regions observable in the sky with telescopes—was settled only in 1924 by Edwin Hubble (who was shown in Fig. 2–23). He used the Mount Wilson telescopes to observe Cepheid variables in three of the "spiral nebulae." He concluded (following the line of argument we described in Section 6.3b) that the "spiral nebulae" were outside our own galaxy; from their observed angular sizes and their distances from us, it followed that they are not overwhelmingly different from the Milky Way Galaxy in size.

Since Hubble's work, there has been no doubt that the spiral forms we observe in the sky are galaxies like our own. For the rest of the book we shall strictly use the term *spiral galaxies;* the currently incorrect, historical term "spiral nebula," often hangs on in certain contexts, chiefly when we discuss the "Great Nebula in Andromeda" (Color Plate 68), which is actually a spiral galaxy.

14.1 Types of Galaxies

Hubble used the Mount Wilson telescopes to study the different types of galaxies. Actually, spiral galaxies are in the minority; many galaxies have elliptical shapes and others are irregular or abnormal in appearance. In 1925, Hubble set up a system of classification of galaxies that we still use today; we normally describe a galaxy by its *Hubble type,* as we shall discuss below.

14.1a Elliptical Galaxies

Most galaxies are elliptical in shape (Fig. 14–1). The largest of these *elliptical galaxies* contain 10^{13} solar masses and are 10^5 parsecs across (approximately the diameter of our own galaxy); these *giant ellipticals* are rare. Much more common are *dwarf ellipticals,* which contain "only" a few million solar masses and are only 2000 parsecs across.

Elliptical galaxies range from nearly circular in shape, which Hubble called *type E0,* to very elongated, which Hubble called *type E7*. The spiral Andromeda Galaxy, M31 (Color Plate 68), is accompanied by two elliptical companions of types E2 (for the galaxy closer to M31) and E5, respectively. It is obvious on the photograph that the companions are much smaller than M31 itself.

We assign types based on the optical appearance of a galaxy rather than how elliptical it actually is. After all, we can't change our point of view for such a far-off object. But even a very elliptical galaxy will appear round when seen end on. (Just picture looking straight at the end of an egg or a cigar; they look round.) So

Figure 14–1 M87 (NGC 4486), a galaxy of Hubble type E0(pec) in the constellation Virgo. The jet that makes M87 a peculiar elliptical galaxy doesn't show in this view. Globular clusters can be seen in the outer regions.

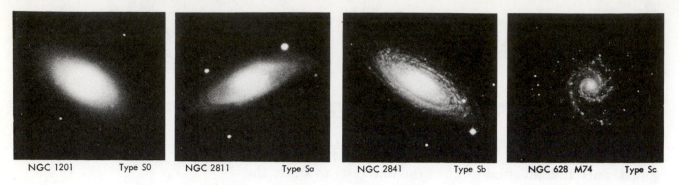

NGC 1201   Type S0   NGC 2811   Type Sa   NGC 2841   Type Sb   NGC 628 M74   Type Sc

Figure 14–2 Normal spiral galaxies.

the ellipticals are, in actuality, at least as elliptical as their Hubble type (which is based only on their appearance) shows; they may actually be more elliptical than they appear.

14.1b Spiral Galaxies

Although spiral galaxies, with arms unwinding gracefully from the central regions, are a minority of all the galaxies in the universe, they form a majority in certain particular groups of galaxies. Also, since they are brighter than the more abundant small ellipticals, we tend to see the spirals even though the ellipticals may make up the majority.

Sometimes the arms are tightly wound around the nucleus; Hubble called this type *Sa,* the S standing for "spiral" (Fig. 14–2). Spirals with their arms less and less tightly wound (that is, looser and looser) are called *type Sb* (Color Plates 71 and 72) and *type Sc* (Color Plate 66). The nuclear bulge as seen from edge-on is less and less prominent as we go from Sa to Sc. On the other hand, the dust lane—obscuring dust in the disk of the galaxy—becomes more prominent. Spectroscopic measurements from Doppler shifts indicate that galaxies rotate in the sense that the arms trail.

Spiral galaxies can be 25,000 to 80,000 light years across. They contain 10^9 to over 10^{12} solar masses. Since most stars are of less than 1 solar mass, this means that spirals contain over 10^9 to over 10^{12} stars—we now think our own galaxy has perhaps 10^{12} (1000 billion = 1 trillion).

In about one-third of the spirals, the arms unwind not from the nucleus but rather from a straight *bar* of stars, gas, and dust that extends to both sides of the nucleus. These are similarly classified in the Hubble scheme from *a* to *c* in order of increasing openness of the arms, but with a *B* for "barred" inserted: *SBa, SBb,* and *SBc.*

Gerard de Vaucouleurs of the University of Texas has expanded Hubble's classification scheme. In addition to Shapley's class Sd with extremely open arms, he has further subdivided barred spirals. Those whose arms come off a ring are given an "(r)" while those whose arms come off the ends of the central bar are given an "(s)." There is actually a continuous range of intermediate types from ordinary spiral galaxies (called SA by de Vaucouleurs for parallel notation to SB) to barred spirals (Fig. 14–3), so "normal" spirals and barred spirals may not really be distinct types.

14.1c Irregular Galaxies

A few per cent of galaxies show no regularity. The Magellanic Clouds, for example, are basically irregular galaxies (Fig. 14–4). Irregular galaxies are classified as *Irr.*

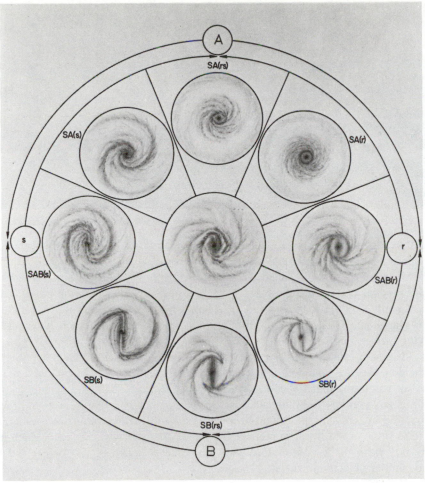

Figure 14–3 This set of photographs by Gerard de Vaucouleurs of regular spirals (SA) and barred-spiral galaxies (SB) shows that there is no clear demarcation between those classes. We see here arms typical of subclass "b," as in SAb (formerly Sb) and SBb.

Figure 14–4 The Large Magellanic Cloud *(left)* and the Small Magellanic Cloud *(right),* photographed with the new British 1.2-m Schmidt telescope at Siding Spring, Australia, as part of the current project to extend the National Geographic Society–Palomar Observatory Sky Survey to the southern hemisphere.

The LMC is about 50 kpc and the SMC about 65 kpc from us, and are about 20 kpc apart. A few globular clusters, considered part of our Milky Way Galaxy, and some dwarf elliptical galaxies are also that far away. The LMC contains about $\frac{1}{100}$ and the SMC about $\frac{1}{1,000}$ the mass of our galaxy.

The Clouds were first reported to Europe by 15th-century Portuguese navigators who had travelled to the Cape of Good Hope at the southern tip of Africa. They were named a few decades later in honor of Magellan, who was then circumnavigating the world by that route.

Sometimes traces of regularity—perhaps a bar—can be seen. De Vaucouleurs' classification scheme defines class Sm = Magellanic spiral class beyond Sd, with the Large Magellanic Cloud (see also Color Plate 65) as the prototype. Since the LMC has a bar, it is class SBm.

Radio-astronomy studies of the velocities of hydrogen gas in the Small Magellanic Cloud have revealed two regions with motions (calculated from Doppler shifts)

Figure 14–5 M82, a most unusual galaxy that is a powerful source of radio radiation. It was once thought to be exploding, but then gentler processes were thought to cause its form and nonthermal radiation. Now there is new evidence that it may be exploding after all.

so different that we may really be seeing two separate galaxies superimposed. So in the same direction as the Small Magellanic Cloud, we may also be seeing through the SMC to what is being called the Mini-Magellanic Cloud. The Small Magellanic Cloud may have split into these two parts in a near collision with the Large Magellanic Cloud 200,000,000 years ago. The Small Magellanic Cloud Remnant and the Mini-Magellanic Cloud are now 30,000 light years apart from each other and are moving farther apart.

Also from studies of 21-cm radiation, Cornell scientists (including Stephen Schneider, George Helou, Ed Salpeter, and Yervant Terzian) have found a large cloud of gas in the midst of a cluster of distant galaxies. The cloud contains at least a billion and perhaps 10 billion solar masses of material. We don't know yet whether it is a true intergalactic gas cloud or a galaxy in which for some reason stars didn't begin to shine or are fewer in number and fainter than we would expect. Much more work will have to be done before we know what to make of this object and how many similar objects may exist.

14.1d Peculiar Galaxies

In some cases, as in M82 (Fig. 14–5), it appears at first look as though an explosion has taken place in what might have been a regular galaxy. Another possibility, though, is that we are seeing light from the galaxy's nucleus scattered toward us by dust in the filaments. The ring galaxy shown in Figure 14–6A probably resulted from the passage of one galaxy through another. For other *peculiar galaxies*—usually

Figure 14–6 (A) A group of galaxies photographed at the prime focus of the Soviet 6-m telescope. One component of this group is a ring-like galaxy without a nucleus. It contains much gas and many hot stars. The nearby galaxy in the upper left corner, NGC 4513, has a ring of gas around it. The print is a negative to bring out the faint structure. (B) NGC 2685, a peculiar galaxy, type S0(pec), in Ursa Major. It seems to be wrapped with helical filaments around a second axis of symmetry; the reasons for this are not understood.

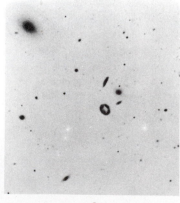

A

B

spiral or elliptical galaxies with some peculiarity—such as those shown in Figure 14–6*B*, we have little idea of what might have gone on.

Peculiar galaxies are classified as the corresponding Hubble type followed by (pec): for example, Sa (pec).

14.1e The Hubble Classification

Hubble drew out his scheme of classification in a *tuning-fork diagram* (Fig. 14–7). The transition from ellipticals to spirals is represented by *type S0*. Galaxies of this transition type resemble spirals in having the shape of a disk, but do not have spiral arms. The farther from M31 of its two elliptical companions (Color Plate 68) has outer parts whose intensity decreases as an S0 galaxy, and so may be a transition case from ellipticals to the transition type S0.

It has since been shown, from optical observations and from studies of the 21-cm hydrogen line, that the amount of gas between the stars in galaxies depends on the type of galaxy. Elliptical galaxies have essentially no gas or dust, while spiral galaxies have a lot. The relative amount of gas increases from types Sa (or SBa) to Sc (or SBc). The interstellar medium in an irregular galaxy is usually even denser. Though no star formation is found in elliptical or S0 galaxies, the amount of star formation increases towards type Sc. Only in the galaxies with a substantial gas and dust content—mostly the Sc, SBc, and Irr galaxies—are the O and B stars to be found. Since these stars have short lifetimes on a stellar scale, they must have been formed comparatively recently, within the last several million years.

At first it was thought that the arrangement of galaxies in the Hubble classification, and the differing amounts of gas in different types, might indicate that one type of galaxy evolves into another, but that is no longer generally thought to be the case. Most astronomers believe that the differences in shape and gas content result from differing conditions at the time of formation of the galaxies.

New theoretical work has been undertaken to explain the huge faint shells of gas recently observed on long exposures to surround some elliptical galaxies. Australian researchers have found that they can explain the shells as stars thrown out by collisions of the ellipticals with spiral galaxies. Thus in this model, some elliptical galaxies are of a later stage of galactic evolution than spirals.

Modern studies have shown that all E7 galaxies are in fact S0. E6 is the most elongated elliptical.

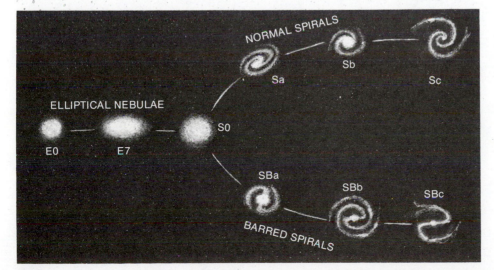

Figure 14–7 The Hubble tuning-fork classification of galaxies. But there are intermediate types between the arms, as we saw in Fig. 14–3.

14.1f IRAS Observations of Galaxies

1983's infrared spacecraft IRAS observed many galaxies, mostly spirals. Since the infrared radiation observed presumably comes from dust clouds heated by young stars, the infrared brightness should tell us how fast stars are being created. About half the energy emitted by our own Milky Way Galaxy is in the infrared, which indicates that a lot of stars are being formed. But we don't know why the Andromeda Galaxy, which in optical radiation resembles the Milky Way, emits only 3 percent of its energy in the infrared.

Some galaxies give out 10 or more times as much infrared as optical energy, which indicates that they are creating stars especially rapidly. Early indications are that many of these galaxies have distorted optical images; the gravitational interaction of a nearby galaxy that distorted the shape may have caused gas clouds to collapse to form stars.

Some bright sources discovered by IRAS don't correspond to anything visible. Some of them could be distant galaxies that emit many times more infrared than optical radiation, but are too far away for us to see them optically. It cannot yet be ruled out, though, that they are nearby dwarf galaxies. The Space Telescope and the next generation of infrared satellites may answer these questions.

14.2 The Origin of Galactic Structure

In Section 12.5c, we saw how spiral structure can be maintained in a single galaxy by density waves or by a chain of supernovae. In some cases, structure can arise from a gravitational interaction of two galaxies. Some scientists have used computers to follow the evolution of a system over time, using the computer to follow the gravitational interaction between many particles, with each particle interacting by gravity with the center of mass of the other galaxy. Long arms, often called "tails," are drawn out by tidal forces (Fig. 14–8). The spin of the original galaxies contributes to the graceful curvature. The shapes can be made to match known galaxies (Fig. 14–9; see also Color Plate 67).

The above example involved galaxies of equal mass. If one galaxy is much more massive than the other, the structure can resemble that of an ordinary spiral galaxy. Note that the Milky Way Galaxy has close companions—the Magellanic Clouds— and the Andromeda Galaxy has companions as well, so it is possible that there was a gravitational interaction contribution to the spiral structure of our own galaxy and other galaxies.

An alternate class of theories to explain the persistence of spiral structure involves magnetic fields. The magnetic field in our galaxy is only one-millionth the strength of the earth's magnetic field and is aligned along the spiral arms. (We measure this magnetic field by studying how starlight is polarized.) Even though galactic magnetic fields are weak, magnetic and other theories cannot yet be ruled out as possible contributors to spiral structure.

14.3 Clusters of Galaxies

Careful study of the positions of galaxies and their distances from us has revealed that most galaxies are part of groups or clusters. Groups have just a handful of members, while *clusters of galaxies* may have hundreds or thousands.

14.3a The Local Group

The two dozen or so galaxies nearest us form the *Local Group*. The Local Group contains a typical distribution of types of galaxies and extends over a volume 1

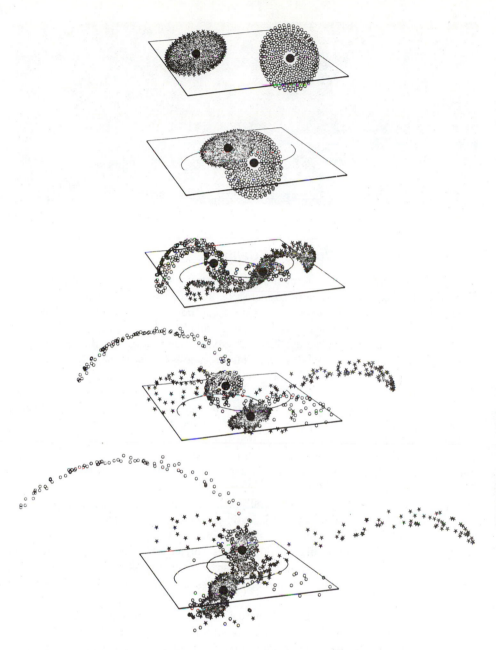

Figure 14–8 Drawings made on a computer by Alar and Juri Toomre, showing a time sequence of the very close encounter of two identical model galaxies. The large dot represents the center of mass of each galaxy; the mass and thus the gravitational force of each galaxy are concentrated at its center for purposes of the calculation. A disk of 350 particles is associated with each galaxy. The drawings are separated by an interval of 200 million years in the evolution of the galaxies.

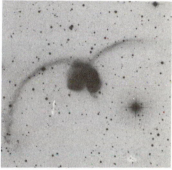

Figure 14–9 A negative print of the pair of galaxies NGC 4038 and NGC 4039, known as The Antennae. It closely matches the computer drawings of Fig. 14–8 when they are displayed from the proper angle.

megaparsec in diameter. It contains three spiral galaxies, each at least 15 to 50 kiloparsecs across—the Milky Way, Andromeda, and M33. There are four irregular galaxies, each 3 to 10 kiloparsecs across, including the Large and Small Magellanic Clouds. At least a dozen other dwarf irregulars are known. The rest of the galaxies are ellipticals, including four regular ellipticals, two of which are companions to the Andromeda Galaxy. The others are dwarf ellipticals.

Previously unknown candidates for membership in the Local Group are occasionally found. Some of these newly discovered galaxies have been difficult to discover even though they are so close because they lie in the plane of our galaxy and are thus hidden from our view by dust (Fig. 14–10).

The Andromeda Galaxy at 2.2 million light years and perhaps M33 at 2.4 million light years are the farthest objects you can see with your unaided eye; they appear as fuzzy blobs in the sky if you know where to look.

14.3b More Distant Clusters of Galaxies

There are apparently other small groups of galaxies in the vicinity of the Local Group, containing only a dozen or so members. The nearest cluster of many galaxies (a *rich cluster,* as opposed to a *poor cluster*) can be observed in the constellation Virgo and surrounding regions of the sky; it is called the Virgo Cluster. It covers a region in the sky over 6° in radius, 12 times greater than the angular diameter of the moon. The Virgo Cluster contains hundreds of galaxies of all types. It is about 2 million parsecs across, and most of it is located about 20 million parsecs away from us.

Other rich clusters are known at greater distances, including the Coma Cluster in the constellation Coma Berenices (Berenice's Hair). The Coma Cluster has spherical symmetry; its galaxies are concentrated toward its center, not unlike the distribution of stars in a globular cluster (which is a cluster of stars, and is therefore on a much smaller scale). Thus the Coma Cluster is a *regular cluster,* as opposed to an *irregular cluster.*

Rich clusters of galaxies are generally x-ray sources. Studies with the Einstein Observatory have revealed a hot intergalactic gas containing as much mass as is in the galaxies themselves. The temperature of the gas is 10–100 million K. The gas is clumped in some clusters while in others it is spread out more smoothly with a concentration near the center. This may be an evolutionary effect, with gas being ejected from individual galaxies in younger clusters and spreading out as the clusters age.

X-ray spectra from British and American satellites showed that the intergalactic gas contains iron with an abundance that approximates that found in the sun. This implies that it had been ejected from galaxies, in which nucleosynthesis in stars formed the iron. This also indicates that matter apparently flows out of most of the galaxies in a cluster.

The matter is pulled by gravity toward the center of the cluster, although it may be heated and not fall to the center. A giant elliptical galaxy, like M87 at the center of the Virgo Cluster, may be so large because it has gobbled up gas and other galactic debris.

There may be as many as 10,000 galaxies in a cluster, and the density of galaxies near the center of a rich cluster may be higher than that near the Milky Way by a factor of one thousand to one million. Thousands of clusters of galaxies are known (Figs. 14–11 and 14–12).

Stephen A. Gregory of Bowling Green State University and Laird Thompson of the University of Hawaii have concluded that every nearby very rich cluster is located in a cluster of clusters, a *supercluster.* A new catalogue of superclusters includes 16

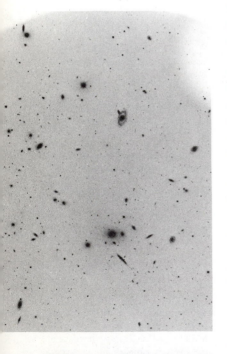

Figure 14–11 Many galaxies of all types are visible in this photo of a cluster of galaxies, A1367.1. (Kitt Peak 4-m photo by A. Oemler, courtesy of L. Thompson)

within $z = 0.1$ (2 billion light years) of us. The Local Group, the several similar groupings nearby, and the Virgo Cluster form the *Local Supercluster*. This cluster of clusters contains 100 member clusters roughly in a pancake shape, on the order of 100 million light years across and 10 million light years thick. Superclusters are apparently separated by giant voids. We discuss the relation of these observations to theory in Section 14.7.

Does the clustering continue in scope? Are there clusters of clusters of clusters, and clusters of clusters of clusters of clusters, and so on? The evidence at present is that this is not so.

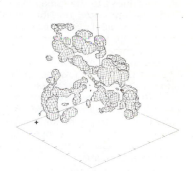

14.4 *The Expansion of the Universe*

In the decade before the problem of the location of the ''spiral nebulae'' was settled, Vesto M. Slipher of the Lowell Observatory took many spectra that indicated that the spirals had large redshifts. This work was to lead to a profound generalization. In 1929 Hubble announced that galaxies in all directions are moving away from us, and that the distance of a galaxy from us is directly proportional to its redshift (that is, when the redshift we observe is greater by a certain factor, the distance is greater by the same factor; Fig. 14–13A). The proportionality between redshift and distance is known as *Hubble's law*. Hubble, in collaboration at Mt. Wilson with Milton L. Humason, went on during the 1930's to establish the relation more fully (Fig. 14–13B). The redshift is presumably caused by the Doppler effect (which was described in Section 5.6). The law is usually stated in terms of the velocity that corresponds (by the Doppler effect) to the measured wavelength, rather than in terms of the redshift itself.

Hubble's law states that the velocity of recession of a galaxy is proportional to its distance. It is written

$$v = H_0 d,$$

where v is the velocity, d is the distance, and H_0 is the present-day value of the constant of proportionality (simply, the constant factor by which you multiply d to get v), which is known as *Hubble's constant* (Figs. 14–14 and 14–15).

Allan Sandage of the Mt. Wilson and Las Campanas Observatories is Hubble's intellectual heir in using the Mt. Wilson and Palomar telescopes to study the distances and redshifts of the farthest galaxies. Working over many years, Sandage and Gustav Tammann of the University of Basel in Switzerland have derived that H_0 is 50 km/s/Mpc, as we shall describe in Section 14.4b. This is about 10 times lower than the value that Hubble originally announced, but Sandage and Tammann have used new techniques for finding the distance to distant galaxies, incorporating also earlier corrections to the distance scale. In recent years, other scientists have made similar sets of observations, and most have derived larger values. Gerard de Vaucouleurs at Texas, Marc Davis at Berkeley, John Huchra of the Harvard-Smithsonian Center for Astrophysics, Marc Aaronson of the University of Arizona, Jeremy Mould of Caltech and others have found values closer to 100 km/s/Mpc. Let us use 50 for the rest of this book for convenience, as many astronomers do; there are strong arguments for each value.

Figure 14–12 A computer plot of the distribution of galaxies in space, plotted on the basis of many radial velocity measurements made with new electronic techniques. Long chains of galaxies, showing their grouping into clusters, are clearly visible in this three-dimensional display. A cross marks our galaxy. Tick marks show intervals of 50 million light-years. A+ marks the position of our Milky Way Galaxy. (Courtesy of Frenk and White, and the CfA survey)

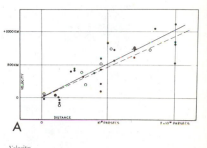

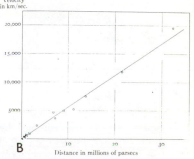

Figure 14–13 *(A)* Hubble's original diagram from 1929. Dots are individual galaxies; open circles are from groups of galaxies. The scatter to one side of the line or the other is substantial. *(B)* By 1931, Hubble and Humason had extended the measurements to greater distances, and Hubble's law was well established. All the points shown in the 1929 work appear bunched near the origin of this graph. $v = H_0 d$ represents a straight line of slope H_0. These graphs use older distance measurements than we now use, and so give different values for H_0 than we now derive.

Figure 14–14 Spectra are shown at right for the galaxies at left, all reproduced to the same scale. Distances are based on Hubble's constant = 50 km/sec/Mpc. Notice how the farther away a galaxy is, the smaller it looks. The arrow below each horizontal streak of spectrum shows how far the H and K lines of ionized calcium are redshifted. The spectrum of an emission-line source located inside the telescope building appears as vertical lines above and below each galactic spectrum to provide a comparison with a redshift known to be zero.

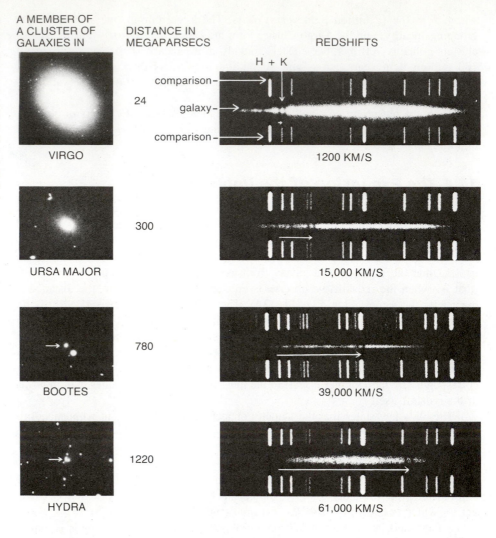

A MEMBER OF A CLUSTER OF GALAXIES IN

DISTANCE IN MEGAPARSECS

REDSHIFTS

VIRGO — 24 — 1200 KM/S

URSA MAJOR — 300 — 15,000 KM/S

BOOTES — 780 — 39,000 KM/S

HYDRA — 1220 — 61,000 KM/S

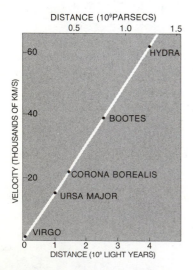

Figure 14–15 The Hubble diagram for the galaxies shown in Fig. 14–14.

Hubble's constant is given in units that may appear strange, but they merely state that for each megaparsec (3.3×10^6 light years) of distance from the sun, the velocity increases by 50 km/s (thus the units are 50 km/s per Mpc). From Hubble's law, we see that a galaxy at 10 Mpc would have a redshift corresponding to 500 km/s; at 20 Mpc the redshift of a galaxy would correspond to 1000 km/s; and so on. The redshift for a given galaxy is the same no matter in which part of the spectrum we observe.

Note that if Hubble's constant is 100 instead of 50, then a galaxy whose redshift is measured to be 500 km/s would be (500 km/s)/(100 km/s/Mpc) = 5 Mpc, half the distance than was derived for the smaller Hubble's constant. So the debate over the size of Hubble's constant has broad effect on the size of the universe. The debate is often heated, and sessions of scientific meetings at which the subject is discussed are well attended.

14.4a The Expanding Universe

The major import of Hubble's law is that objects in the universe in all directions are moving away from us; the universe is expanding. Since the time when Copernicus moved the earth out of the center of the universe (and the time when Shapley moved the earth and sun out of even the center of the Milky Way Galaxy), we have

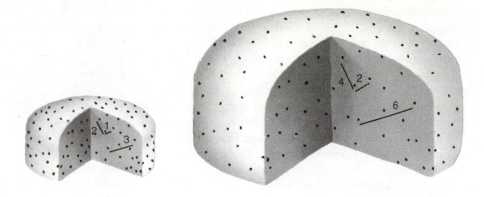

Figure 14–16 From every raisin in a raisin cake, every other raisin seems to be moving away from you at a speed that depends on its distance from you. This leads to a relation like the Hubble law between the velocity and the distance. Note also that each raisin would be at the center of the expansion measured from its own position, yet the cake is expanding uniformly. For a better analogy with the universe, consider an infinite cake; unlike the finite analogy pictured, there is then no center to its expansion.

not liked to think that we could be at the center of the universe. Fortunately, Hubble's law can be accounted for without our having to be at any such favored location, as we see below.

Imagine a raisin cake (Fig. 14–16) about to go into the oven. The raisins are spaced a certain distance away from each other. Then, as the cake rises, the raisins spread apart from each other. If we were able to sit on any one of those raisins, we would see our neighboring raisins move away from us at a certain speed. It is important to realize that raisins farther from us would be moving away faster, because there is more cake between them and us to expand. No matter in what direction we looked, the raisins would be receding from us, with the velocity of recession proportional to the distance.

The next important point to realize is that it doesn't matter which raisin we sit on; all the other raisins would always seem to be receding. Of course, the raisin cake is finite in size, while the universe may have no limit so that we would never see an edge. The fact that all the galaxies appear to be receding from us does not put us in a unique spot in the universe; there is no center to the universe. Each observer at each location would observe the same effect.

Note that in our analogy the size of the raisins themselves is not changing; only the separations are changing. In the universe, the galaxies themselves and the clusters of galaxies are not expanding; only the distances between the clusters (or, perhaps, the superclusters) are increasing.

Note also that individual stars in our galaxy can appear to have small redshifts or blueshifts, caused either by their peculiar velocities or by the differential galactic rotation. Also, even some of the nearer galaxies (such as M31 in Andromeda) have random velocities of sufficient size, or velocities less than our rotational velocity in our galaxy, so that they are approaching us. But except for these few nearby cases, all the galaxies are receding.

14.4b The Distance Scale of the Universe

The major problem for setting the Hubble law on the firmest footing is finding the distances to the galaxies for which redshifts are measured. Let us first discuss the distance indicators used by Sandage and Tammann in deriving their value of 50 km/s/Mpc for Hubble's constant.

We can't measure trigonometric parallaxes beyond the nearest region of our galaxy. Only for the nearest galaxies can we detect *primary distance indicators:* Cepheid or RR Lyrae variable stars, for which we derive the distance directly by comparing absolute magnitude (from the period-luminosity relation for Cepheids or the known value for RR Lyrae stars) with the observed apparent magnitude.

Beyond those nearest galaxies, we use such *secondary distance indicators* as supergiant stars or H II regions, assuming that the magnitudes of supergiants or sizes of H II regions are more or less the same as they are in our own galaxy. We then calculate for a supergiant star its spectroscopic parallax; for an H II region we calculate the distance at which its observed angular size would correspond with the linear dimensions we know for those objects in the Milky Way.

At still greater distances, we must resort to *tertiary distance indicators*. We compare the maximum brightnesses of supernovae, assuming that they all reach the same maximum brightness. Or we assume that the brightest member in a cluster of galaxies has the same absolute magnitude as all other brightest members of other clusters. Sometimes, when one or two members are exceptionally bright, we consider instead the third-brightest member of a cluster, to lessen the possibility of one odd object being the brightest. If this seems a weak way to measure distances, you are right, but it is all we have.

In his separate study of the distance scale, de Vaucouleurs used an overlapping but wider-ranging set of primary, secondary, and tertiary distance indicators. For example, he adopted the brightnesses of novae as an additional primary indicator. An additional secondary indicator was the distribution of internal velocities in H II regions. (Greater internal velocities seem—observationally—to correspond to larger H II regions; the distance is derived by comparing the measured and observed diameters.) An additional tertiary distance indicator is the diameter of the observed ring in barred-spiral galaxies whose arms unwind from such a ring. The result of these investigations indicates that Hubble's constant is 100 km/s/Mpc, making all distances based on Hubble's law smaller by a factor of two than they are for the Sandage-Tammann value of the Hubble constant.

De Vaucouleurs and many other scientists rely also on the Tully-Fisher relation, a new and powerful relation discovered by Brent Tully of the University of Hawaii and J. Richard Fisher of the National Radio Astronomy Observatory. The relation is based on a link between the rotational velocities of galaxies measured from their 21-cm hydrogen lines, and the absolute magnitudes of the galaxies (with the best relation obtained when magnitudes are measured in the infrared).

Some but not all of the difference in measured velocities stems from the fact that the earlier measurements were all made in the northern hemisphere, while we

Following the style of the late George Gamow, I can thus write my address as:

Jay M. Pasachoff
Williamstown
Massachusetts
United States of America
North America
Earth
Solar System
Milky Way Galaxy
Local Group
Local Supercluster
Universe

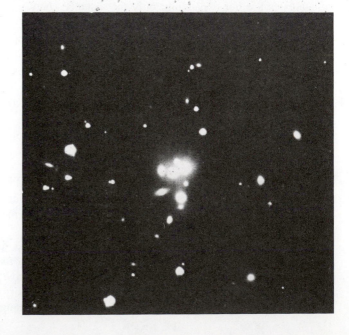

Figure 14–17 Is this a galactic cannibal with an undigested supper? This galaxy contains 9 small nuclei within a halo of stars. The nucleus may have originally been independent galaxies; their outer layers would have been stripped off. Computer simulations seem to show that the nuclei will fuse within a billion years, and the surrounding stars will take the shape of a giant elliptical galaxy. It should then be comparable to other giant ellipticals in the centers of clusters. Described by Donald Schneider and James Gunn as "the most nightmarish known multiple-nucleus system," it is the central galaxy in the poor cluster of galaxies known as V Zw 311 (in Zwicky's catalogue) or as Abell 407.

now have many more southern hemisphere observations available. Our Local Group is moving toward the Virgo Cluster with an appreciable velocity, and this velocity must be taken into account when computing Hubble's constant. The effect of this velocity is different for galaxies observed in the northern and southern hemispheres.

The methods we have to use grow less precise as we get farther from the sun. In particular, at the very farthest distances we are seeing galaxies that emitted their light very long ago. A galaxy may well have then had a very different brightness than the nearby galaxies to which we are comparing it. For example, there is some evidence that the brightest galaxies in clusters may be devouring other galaxies (Fig. 14–17) and so growing brighter; this process is called *galactic cannibalism*. One radio galaxy that is not rotating contains a rapidly rotating cloud of ionized hydrogen, something that seems easiest to explain if the cloud were the remnant of another galaxy that was ''eaten.''

Beyond a certain range, we can no longer independently measure distances, and our only method of assessing distance is applying the Hubble law with our current ''best value'' for Hubble's constant to the observed redshifts. What is the best value for Hubble's constant? The two camps—those deriving 50 and those deriving 100— don't agree. Each seems expert, and the methods will ultimately stand or fall on the basis of the validity of the methods used for each step of the chain. Each value has some inherent uncertainty. For the value of 50, for example, the ''standard deviation'' is 7, and most scientists believe that the true value of a quantity falls within three standard deviations of the average value. Thus the allowable range is really 30–70, a factor of over 2.

A larger Hubble constant corresponds to a more rapid expansion, and so to a younger universe since the galaxies would have needed less time to reach their current spacing. If the globular clusters are really as old as 18 billion years, as some recent work shows, then a Hubble constant as high as 100 is excluded. Studies of the ages of the radioactive elements seem to agree with the 18-billion-year age. But the one thing that I can guarantee is that the discussion isn't over.

> The link between Hubble's law and the age of the universe assumes that we know how the expansion rate changes with time, a question that is still being discussed (Section 16.2).

14.4c The Most Distant Galaxies

Hyron Spinrad and colleagues at Berkeley have been looking for the farthest galaxies, to study how they evolve and what they tell us about the early times of our universe. A handful of galaxies have thus far been detected with redshifts larger than 0.75. Among these are the giant elliptical galaxies 3C 324 and 3C 427.1 (Fig. 14–18). The objects are over 10 billion light years away. We are seeing the galaxies as they

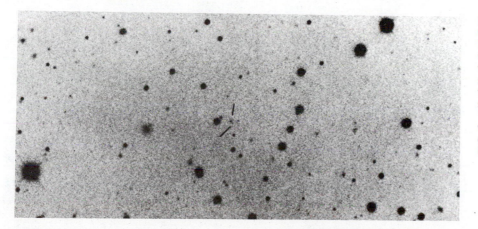

Figure 14–18 One of the farthest known galaxies, 3C 427.1, is barely observable even with a very long exposure, as shown in this negative print. It has a redshift of 1.175, which means that the spectral lines that are observed are shifted to the red by 117.5 per cent of their original values. The size of the redshift indicates that the object is over 10 billion light years away.

The redshift of 3C 324 is 1.21 and that of 3C 472.1 is 1.175, so great that the formula used for the Doppler shift must be the version that takes the special theory of relativity into account. This redshift says only that each wavelength of 3C 324 is shifted by an amount equal to 1.21 times its rest wavelength. At such great redshifts, the redshifts no longer correspond to v/c; the velocity is still less than the speed of light. We delay our discussion of the formula for such high redshifts until Section 15.2, since they are more typical of quasars than of galaxies.

were 10 billion years ago. We say that the *look-back time* for these galaxies is 10 billion years. New observational techniques, like the use of CCD's, have reduced drastically the time to take the spectra necessary for this research.

Even though we are seeing the galaxy 3C 427.1 as it was 10 billion years ago, its spectrum shows signs that the galaxy was at that time already 6 billion years old. Thus the galaxy formed about 16 billion years ago, and the universe must be at least that old. Studies of such distant galaxies also tell us how the properties of galaxies evolved with time.

Our knowledge of the most distant galaxies is being helped greatly by the development of CCD imaging devices (Section 2.11). A CCD system being developed for Space Telescope is playfully known as 4Shooter, because a mosaic of four detectors allows larger coverage than a single CCD can. It is being used at the Palomar 5-m telescope to image or take spectra of extremely distant objects. When Space Telescope itself is launched in 1986, much of its time will be devoted to studying distant galaxies in order to study the universe as far back in time as possible.

14.5 Active Galaxies

Most of the objects that we detect in the radio sky turn out not to be located in our galaxy. The study of these *extragalactic radio sources* is a major subject of this section.

The core of our galaxy, the radio source we call Sagittarius A, is one of the strongest radio sources that we can observe in our galaxy. But if the Milky Way Galaxy were at the distance of other galaxies, its radio emission would be very weak.

Some galaxies emit quite a lot of radio radiation, many orders of magnitude (that is, many powers of ten) more than "normal" galaxies. We shall use the term *radio galaxy* to mean these relatively powerful radio sources. They often appear optically as peculiar giant elliptical galaxies. Radio galaxies, and galaxies that similarly radiate much more strongly in x-rays than normal galaxies, are called *active galaxies*.

The first radio galaxy to be detected, Cygnus A (Fig. 14–19), radiates about a million times more energy in the radio region of the spectrum than does the Milky

Figure 14–19 *(A)* A radio map of Cygnus A, with shading indicating the intensity of the radio emission. A negative image of the faint optical object or objects observable is superimposed at the proper scale. *(B)* The especially high resolution on this new image of Cygnus A reveals a third, less prominent feature between the two main structures. This central region contains the central radio source. The resolution was improved over past images by using a computer-controlled monitor of the seeing and telescope guiding to form an image with ⅔-arc sec resolution on a CCD at Mauna Kea. The overall structure again makes the source look like galaxies in collision. (Radio image from the VLA; optical image by Laird Thompson, U. Hawaii)

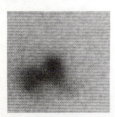

B

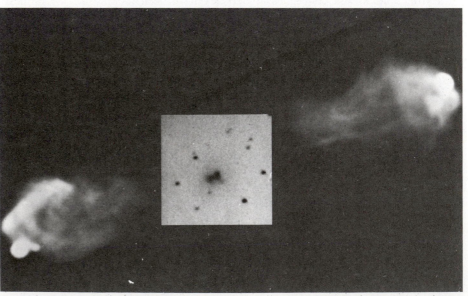

A

Box 14.1 The Peculiar Galaxy M87

The radio galaxy M87 is a giant elliptical galaxy, one of the brighter members of the Virgo Cluster. The galaxy turns out to have an odd optical appearance on short exposures, for a jet of gas can be seen. The galaxy corresponds to the powerful radio galaxy Virgo A. A radio jet 2 kiloparsecs long has been discovered.

High-resolution radio observations have shown that the nucleus of M87 is only 0.01 arc sec across. Even at its distance of 50 million light years, this makes it a very small source in which to generate so much energy, which is emitted across the spectrum from x-rays to radio waves. Indeed, this galaxy gives off much more energy than do other galaxies of its type; the central region is as bright as 10^8 suns.

Optical studies of the motion of matter circling M87's nucleus have allowed astronomers to estimate the mass of the nucleus. The stars are moving so fast that a huge mass must be present to hold them in. It turns out that 5 billion solar masses of matter must be there. Other optical studies disclosed the presence of an extremely bright point of light in the center of the galaxy. Astronomers continue to gather spectral and spatial information to determine if the source is a giant black hole or is matter under more conventional conditions. Higher-resolution observations from Space Telescope should tell us more about this exotic source.

Way Galaxy. Cygnus A, and dozens of other radio galaxies, emit radio radiation mostly from two zones, called *lobes,* located far to either side of the optical object. Such *double-lobed structure* is typical of many radio galaxies.

The optical object that corresponds to Cygnus A—a fuzzy, divided blob or perhaps two fuzzy blobs—has been the subject of much analysis, but its makeup is not yet understood. Perhaps we see a single object partly obscured by dust.

Often the optical images that correspond to radio sources show peculiarities. For example, on short exposures of M87 (Fig. 14–20), which corresponds to the powerful radio galaxy Virgo A, we see (optically) a jet of gas. Light from the jet is

At first, the optical object in Cygnus A was thought to be two galaxies in collision. This idea was discarded when it was realized that collisions between galaxies would not be frequent enough to account for the many similar radio sources that had by then been discovered. One distinguished astronomer lost a bet of a bottle of whiskey to a similarly distinguished astronomer on the subject. But now there is evidence that the material that fuels the giant black holes at the centers of radio galaxies may come from collisions of galaxies. (See also Section 15.6c.) The suggestion has thus been made that a bottle of whiskey might be returned (posthumously). And computer processing of a photograph to improve resolution has made it look even more as though Cygnus A is galaxies in collision after all.

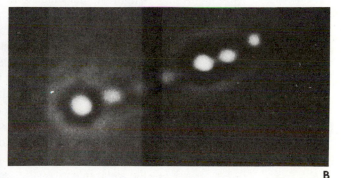

B

A

Figure 14–20 *(A)* The galaxy M87, which corresponds to the radio source Virgo A, as photographed with the Kitt Peak 4-m telescope. Two small background galaxies appear at the "5 o'clock" position, falsely mimicking a jet of gas. The inset shows a shorter exposure reproduced to the same scale, on which a real jet of gas is visible. *(B)* Computer enhancement of a photograph taken with the Palomar 5-m telescope brings out details of the jet. North is at top and East is at left in all photographs.

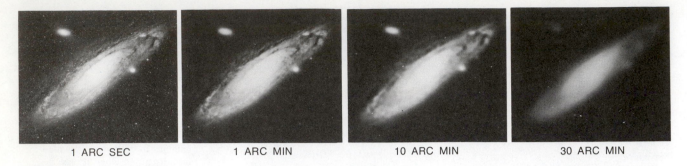

| 1 ARC SEC | 1 ARC MIN | 10 ARC MIN | 30 ARC MIN |

Figure 14–21 The Andromeda Galaxy as it would appear at different resolutions.

polarized, which confirms that the synchrotron process is at work here. The powerful radio source Centaurus A has an optical counterpart (Color Plate 70) that shows unusual wrapping by dust. In recent years, both x-ray and optical jets, aligned with the radio lobes, have been discovered. Perhaps a radio jet would be discovered as well if a high-resolution, high-sensitivity telescope were available in the southern hemisphere, where it could properly observe this southern source.

As our observational abilities in radio astronomy have increased, especially with the techniques described in the next section, lobes and jets aligned with them have become commonly known. The best current model is that a giant rotating black hole in the center of a radio galaxy is accreting matter. Twin jets carrying matter at a high velocity are given off almost continuously along the poles of rotation. These jets carry energy into the lobes. (We may only see one, depending on the alignment and Doppler shifts.) Calculations show that a not-very-hungry giant black hole would provide the right amount of energy to keep the lobes shining.

*14.6 Radio Interferometry

14.6a Radio Interferometers

The resolution of single radio telescopes is very low, because of the long wavelength of radio radiation. Single radio telescopes may be able to resolve structure only a few minutes of arc or even a degree or so across. The techniques of interferometry are now used in radio astronomy to detect fine detail. Arrays of radio telescopes can now map the sky with resolutions far higher than the 1 arc sec or so that we can get with optical telescopes (Fig. 14–21). Let us first describe how these radio interferometers work, and then discuss some of the high-resolution results.

The resolution of a single-dish radio telescope at a given frequency depends on the diameter of the telescope. (A single reflecting surface of a radio telescope is known as a "dish.") If we could somehow retain only the outer zone of the dish (Fig. 14–22), the resolution would remain the same. (The collecting area would be decreased, though, so we would have to collect the signal for a longer time to get the same intensity.)

Figure 14–22 A large single mirror *(A)* can be thought of as a set of smaller mirrors *(B)*. Since the resolution for radiation of a certain wavelength depends only on the telescope's aperture, retaining only the outermost segments *(C)* matches the resolution of a full-aperture mirror. We can use a property of radiation called *interference* to analyze the incoming radiation. The device is then called an *interferometer*.

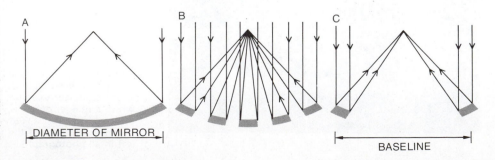

DIAMETER OF MIRROR

BASELINE

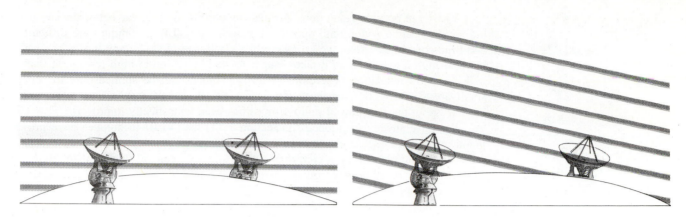

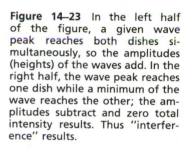

Let us picture radiation from a distant source as coming in wavefronts, with the peaks of the waves in step (Fig. 14–23); we say that the waves are "coherent." If we can maintain our knowledge of the relative arrival times of the wavefront at each of two dishes, we can retain the same resolution as though we had one large dish of the spacing of the two small dishes shown. For a single dish, the maximum spacing of the two most distant points from which we can detect radiation is the "diameter." For a two-dish interferometer, we call it the *baseline*.

We study the signals by adding together the signals from the two dishes; we say the signals "interfere," hence the device is an "interferometer." Since the delay in arrival time of a wavefront at the two dishes depends on the angular position of an object in the sky with respect to the baseline, by studying the time delay one can figure out angular information about the object.

If the source were made of two points close together, the wavefronts from the two sources would be at slight angles to each other, and the time interval between the source reaching the two dishes would be slightly different. Thus an interferometer can tell if an object is double, even if it is unresolved by each of the dishes used alone.

The first radio interferometers were separated by hundreds or thousands of meters. The signals were sent over wires to a central collecting location, where the signals were combined.

Figure 14–23 In the left half of the figure, a given wave peak reaches both dishes simultaneously, so the amplitudes (heights) of the waves add. In the right half, the wave peak reaches one dish while a minimum of the wave reaches the other; the amplitudes subtract and zero total intensity results. Thus "interference" results.

14.6b Very-Long-Baseline Interferometry

The breakthrough in timekeeping came with the invention of atomic clocks, which drift only three hundred-billionths of a second in a year. A time signal from an atomic clock can be recorded by a tape recorder on one channel of the tape, while the celestial radio signal is recorded on an adjacent tape channel. Since a wide band of frequencies must be recorded, video tape recorders are used, and their development was another necessary event. The radio signal recorded can be compared at any later time with the signal from the other dish, synchronized accurately through comparison of the clock signals.

The ability to record the time so accurately freed radio astronomers of the need to have the dishes in direct contact with each other during the period of observation. Now all that is necessary is that the two telescopes observe the same object at the same period of time; the comparison of the signals can take place in a computer weeks later. With this ability, astronomers can make up an interferometer of two or more dishes very far apart, even thousands of kilometers (Fig. 14–24). This technique is called *very-long-baseline interferometry (VLBI)*.

Figure 14–24 VLBI techniques with a baseline that is the diameter of the earth allow radio sources to be studied with extremely high resolution.

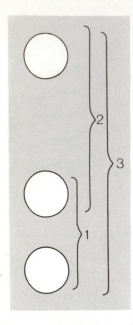

Figure 14–25 The lines joining the several dishes making up an interferometer represent pairs of dishes that give results equivalent to having several baselines simultaneously. The same results could be gathered with only two dishes, one of which was moved around, but it would take longer. With the three dishes shown, one can simultaneously make measurements with three different baselines.

VLBI can provide resolutions as small as 0.0001 arc sec, far better than resolutions that can be gotten with optical telescopes. But VLBI techniques are difficult and time-consuming. Also, we sample only a small area of sky at any time when we work at high resolutions, so it takes longer to study a region at high resolution than it does at low resolution. Therefore, VLBI techniques can be applied only to very small areas of sky. But for those few areas, chosen for their special interest, our knowledge of the structure of radio sources has been fantastically improved.

We have already mentioned (in Section 7.11) that VLBI techniques have provided an accurate measurement for the deflection of electromagnetic waves by the mass of the sun, providing a confirmation of Einstein's general theory of relativity. The discovery of extremely strong sources at the nuclei of our galaxy and of the galaxy M87 are also VLBI results.

The next stage in VLBI work is to set up permanent networks of radio telescopes devoted to this purpose. Britain's MERLIN (**M**ulti-**E**lement **R**adio-**L**inked **I**nterferometer **N**etwork) with a 64-km-maximum baseline is already in operation. Other plans are under way in the U.S., in Canada, in Australia, and elsewhere.

14.6c Aperture-Synthesis Techniques

By suitably arranging a set of radio telescopes across a landscape, one can simultaneously make measurements over a variety of baselines, because each pair of telescopes in the set has a different baseline from each other pair (Fig. 14–25). With such an arrangement one can more rapidly map a radio source than one can with two-dish interferometers. Also, several dishes instead of just two gives that much more collecting area.

Since the resolution of an interferometer depends on the baseline, at any one time the resolution is quite good along the line in which the telescopes lie but is only the resolution of a single dish in the perpendicular direction. Fortunately, we can take advantage of the fact that the earth's rotation changes the orientation of the telescopes with respect to the stars (Fig. 14–26). Thus by observing over a 12-hour period, one can improve the resolution in all directions.

In effect, we have synthesized a large telescope that covers an elliptical area whose longest diameter is the same as the maximum separation of the outermost telescopes. Alternatively, one can synthesize a large telescope by distributing smaller telescopes so that the baselines between them are in different directions. This interferometric technique is known as *aperture synthesis;* it was used to make the radio images of Cygnus A and Centaurus A earlier in this chapter.

The most fantastic aperture-synthesis radio telescope has been constructed in New Mexico by the National Radio Astronomy Observatory. It is composed of 27 dishes, each 26 m in diameter, arranged in the shape of a "Y" over a flat area 27 km in diameter (Fig. 14–27). Because the array is in the shape of a "Y" rather than a straight line, it does not need to wait for the earth to rotate in order to synthesize the aperture. The "Y" is delineated by railroad tracks, on which the telescopes can be transported to 72 possible observing sites. The control room at the center of the "Y" contains a large computer to analyze the signals. The system is prosaically called the *Very Large Array (VLA).*

The VLA can make pictures of a field of view a few minutes of arc across, with resolutions comparable to the 1 arc sec of optical observations from large telescopes, in about 10 hours. Though this resolution is lower than that obtainable for a limited

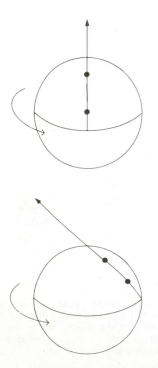

Figure 14–26 The earth's rotation carries around a straight line of dishes so that it maps out an ellipse in the sky. The figure shows the view of a line of dishes on the earth that we would see from the direction of the source being observed.

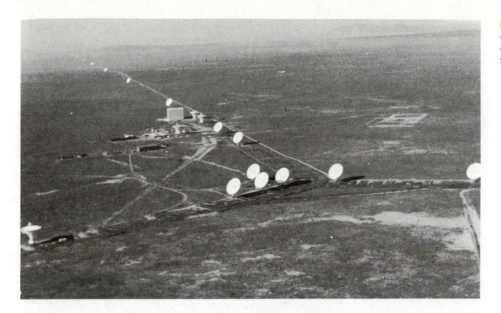

number of sources with VLBI techniques, the ability to make pictures in such a short time is invaluable. And we can learn a lot at a resolution of 1 arc sec (Color Plate 52).

14.6d Aperture-Synthesis Observations

At Westerbork in the Netherlands, twelve telescopes, each 25 m in diameter, are spaced over a 1.6-km baseline. Westerbork, an older aperture-synthesis system than the VLA, has discovered several giant double radio sources (Fig. 14–28), much larger than any of the double-lobed sources previously known. Some are hundreds of times larger than our own galaxy (Fig. 14–29). These are the largest single objects currently known in the universe.

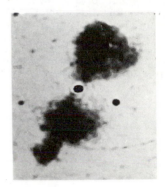

Figure 14–28 A Westerbork synthesis map of DA 240, a giant radio galaxy with two lobes of emission. It is 34 arc min across in our sky, larger than the full moon. Its central radio component coincides with a distant galaxy observed in visible light.

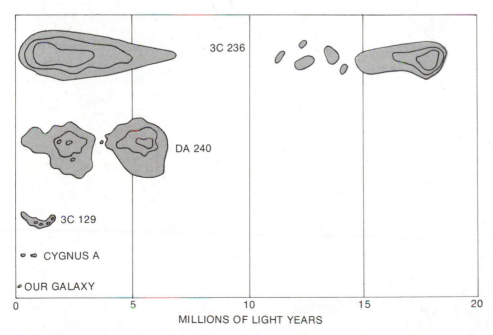

3C 236

DA 240

3C 129

CYGNUS A

OUR GALAXY

0   5   10   15   20

MILLIONS OF LIGHT YEARS

Figure 14–29 A comparison of the sizes of giant double-lobed radio galaxies with our own galaxy.

Figure 14–30 A Westerbork radio map of the head-tail radio source 3C 129, converted to a radio-photograph in which the brightness of the image corresponds to the intensity of radio emission at a wavelength of 21 cm (near but not at the hydrogen wavelength). The resolution is about 30 arc sec. The front end of the head of the radio galaxy corresponds to the position of a giant optical galaxy, a member of a rich cluster of galaxies.

Interferometer observations have revealed the existence of a class of galaxies with ''tails.'' They are called *head-tail galaxies,* and resemble tadpoles in appearance (Fig. 14–30). These galaxies expel the clouds of gas that we see as tails. The objects are double-lobed radio sources with the lobes bent back as the objects move through intergalactic space. High-resolution observations of one such galaxy with the VLA (Fig. 14–31) show that the source at the nucleus is less than 0.1 arc sec across, corresponding at the distance of this galaxy to a diameter of only 0.01 parsec. A narrow continuous stream of emission leads away from the nucleus and into the tail. Such observations are being used to understand both the galaxy itself and the intergalactic medium, and tie in with the x-ray observations of clusters of galaxies. Depending on the velocity of the galaxy and the density of the intergalactic medium, the lobes can be bent back by different amounts, so there is a range from lobes opposite each other to lobes slightly bent back to lobes bent back enough to make head-tail galaxies. Thus it makes sense that most head-tail galaxies are found in rich clusters of galaxies.

The discrete blobs that we can see in the tails indicate that the galaxies give off puffs of ionized gas every few million years as they chug through intergalactic space. Perhaps by studying these puffs, we can learn about the main galaxies themselves as they were at earlier stages in their lives. Head-tail galaxies seem to be a common although hitherto unknown type.

14.7 The Origin of Galaxies

From the time that it was realized that galaxies were separate units of space, it has generally been believed that galaxies originated through some sort of *gravitational instability*. In this theory, a fluctuation in density either developed or pre-existed in the gas from which the galaxy was to form. This fluctuation grew in mass, collapsed (controlled by the force of gravity), and then cooled until the galaxy was formed.

Figure 14–31 The head-tail galaxy NGC 1265 examined with the high resolution of the VLA at increasing resolution from left to right. The resolution of the close-up at the right is about 1 arc sec. The effective size of the telescope beam in each is shown by the ellipse in a lower corner. (Courtesy of F. N. Owen, J. O. Burns, and L. Rudnich, National Radio Astronomy Observatory)

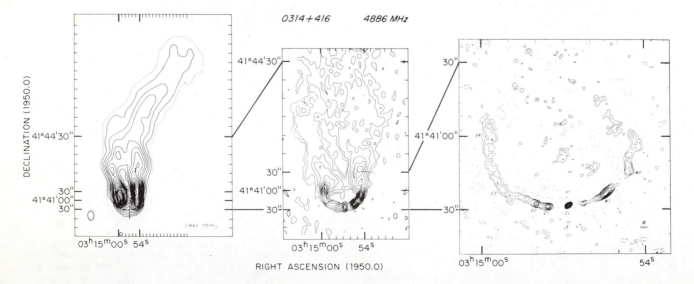

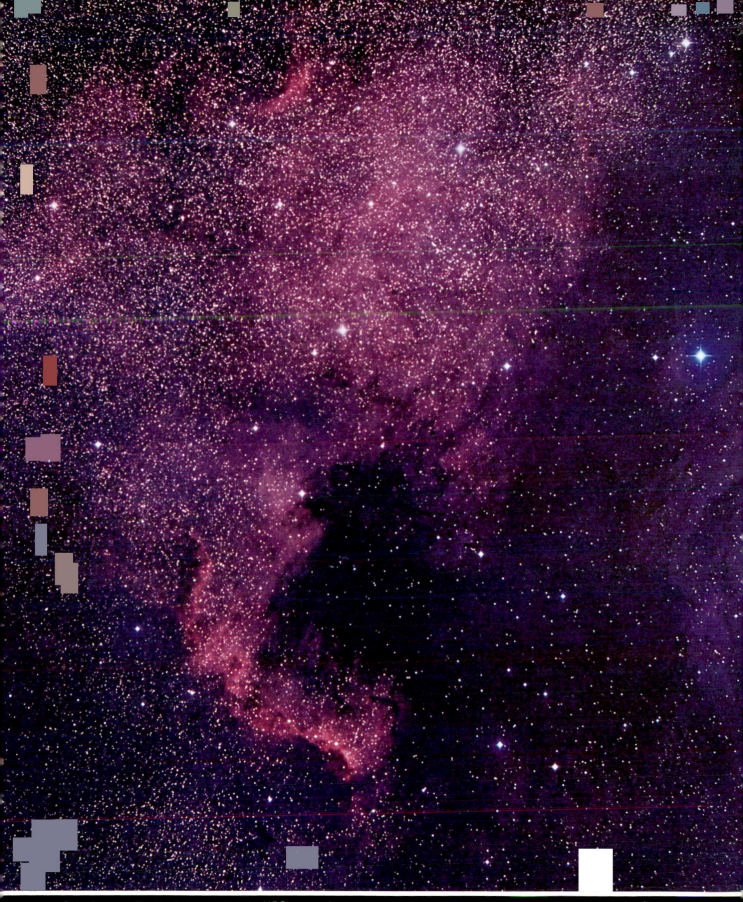

Color Plate 60: The North America Nebula, NGC 7000, in Cygnus. Absorbing dust is silhouetted against glowing gas to give the shape of the North American continent. (Palomar Observatory, California Institute of Technology photo with the 1.2-m Schmidt camera)

Color Plate 61 (top): M16, the Eagle Nebula in the constellation Serpens. Hydrogen radiation makes it appear red. The bright stars at the upper right are hot and young, and are part of a galactic cluster. The small, dark regions may be protostars. (Palomar Observatory, California Institute of Technology photo with the 5-m telescope)

Color Plate 62 (bottom): M17, the Omega Nebula in Sagittarius. (Palomar Observatory, California Institute of Technology photo with the 1.2-m Schmidt camera)

Color Plate 63: The Trifid Nebula, M20, in Sagittarius is glowing gas divided into three visible parts by absorbing lanes. The blue nebula at the top is unconnected to the Trifid. (Palomar Observatory, California Institute of Technology photo with the 5-m telescope)

Color Plate 64 (top): The Lagoon Nebula, M8, in Sagittarius. Red hydrogen light is emitted by gas that is excited by the radiation of very hot stars buried within the nebula; dark filaments of material within the cloud emit strong infrared radiation. Also, several peculiar variable stars in this nebula occasionally flare up. The Lagoon Nebula is about 60 light years across and is located about 6500 light years away from us. (Kitt Peak National Observatory photo with the 4-m telescope)

Color Plate 65 (bottom): The Large Magellanic Cloud, a satellite galaxy of our own Milky Way Galaxy. It is best visible from the southern hemisphere. (Photo by Hans Vehrenberg)

Color Plate 66 (top): The Whirlpool Galaxy, M51, in Canes Venatici, a type Sc spiral galaxy. At the end of one of its arms, a companion galaxy, NGC 5195, appears. (U.S. Naval Observatory photo)

Color Plate 67 (bottom): "The Mice," two interacting galaxies (NGC 4676 A and B). The original photograph was computer-enhanced with a special system at the Kitt Peak National Observatory. The computer system provided the colors in order to bring out the details. Tidal forces, caused by gravity, have distorted the shapes of the galaxies.

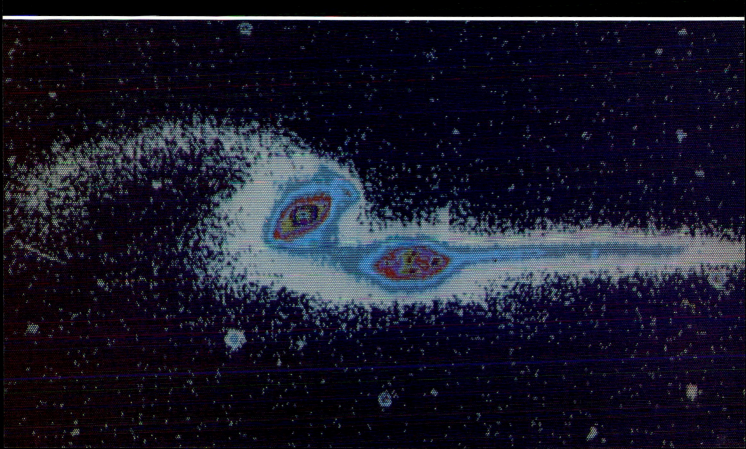

Color Plate 68: The Andromeda Galaxy, also known as M31 and NGC 224, the nearest spiral galaxy to the Milky Way. It is accompanied by two elliptical galaxies, NGC 205 (below) and M32 (above). M31 is Hubble type Sb. (Palomar

Color Plate 69: NGC 2997, a spiral galaxy (Sc) in Antlia. (© 1980 Anglo-Australian Telescope Board)

Color Plate 70 (bottom): Centaurus A, a powerful radio source whose optical image, NGC5128, shows an elliptical galaxy around which a heavy zone of dust appears to be wrapped. (Cerro Tololo Inter-American Observatory with the 4-m telescope)

Color Plage 71: The spiral galaxy NGC 253 (Sc) with its light-absorbing dust lanes. (© 1980 Anglo-Australian Telescope Board)

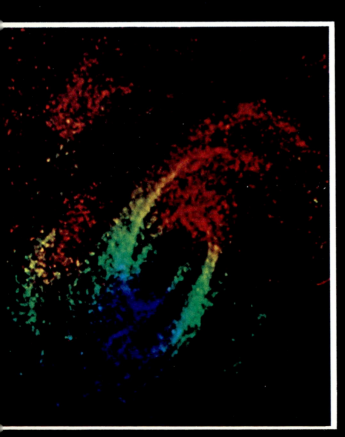 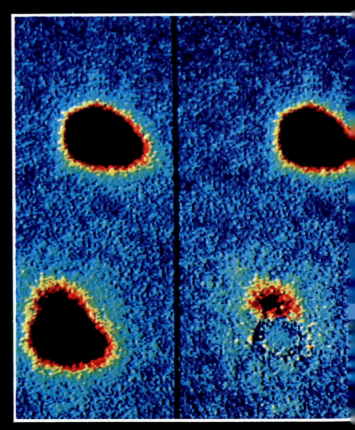

Color Plate 72: An image of 21-cm radiation from the Sb galaxy M81 in Ursa Major. Brightness represents intensity, color represents the Doppler shift; red=recession, violet=approach, green=no shift. (Westerbork data from A. H. Rots and W.W. Shane; imaged at NRAO)

Color Plate 73: The double quasar. At left we see a false-color view of the two images, which are separated by only 6 seconds of arc. At right bottom, the top image has been subtracted from the bottom image. The remainder is the galaxy that is acting as a gravitational lens. (Courtesy of Alan Stockton, Institute for Astronomy, University of Hawaii)

Current theories of galaxy formation have to take into account the fact that the universe is expanding, which we discussed in this chapter, and the fact that the universe is bathed in a glow called the background radiation, which we will discuss in Chapter 16. This background radiation was much stronger at the era when galaxies were formed than it is now.

There are still many unanswered questions with the theory that galaxies arose from gravitational instabilities. Theoreticians can't even agree whether individual galaxies formed first and later combined into clusters, or whether clusters of galaxies formed first, with individual galaxies condensing out of the larger bodies.

The latter possibility follows from ideas of Yakov Zeldovich of the Institute of Applied Mathematics in Moscow. He showed that early in the universe flat sheets of matter formed, each taking approximately the shape of a pancake. In this pancake model, the pancake may fragment later on into clusters of galaxies and, in turn, into galaxies. In between the pancakes and filaments of galaxies are giant *voids,* making a void the opposite of a supercluster. There is now observational evidence for such giant voids; new techniques are being used to measure redshifts in order to plot three-dimensional distributions of galaxies.

For the gravitational-instability theory, one idea gaining currency is that shock waves are important for starting the matter collapsing. Jeremiah Ostriker of Princeton has suggested that extreme shock waves from massive supernovae started the collapse. He proposes that the core of a galaxy is formed in a rapid chain reaction of very massive supernovae. The chain reaction would take place in only a million years or so, a very brief time on an astronomical scale. Where several shock waves met, the effect would be enhanced, which could account for the fact that galaxies are found in clusters.

In one sense, the gravitational instability theories hark back to the Aristotelian view that the universe was fundamentally simple, and that order cannot follow from basic disorder. The contrary view is that the universe was very complex at first and has evolved to its current stage of relative simplicity. The latest version of these alternative theories involves a fundamental *cosmic turbulence* that existed since the origin of the universe; we can visualize turbulence as a disturbed state with swirls and eddies in motion. At a certain stage in the expansion of the universe, amounts of mass suitable to become galaxies or clusters of galaxies would tend to separate out from the overall distribution of matter, carrying an intrinsic spin with them from the turbulence.

Such theories had fallen into disrepute until the last dozen years because theoretical calculations had indicated that pre-existing turbulence would have disappeared in the early stages of the universe. But Leonid Ozernoi and his collaborators at the Lebedev Physics Institute in Moscow have found reasons why this turbulence might not have disappeared, and have elaborated on cosmic turbulence theories. The idea of pre-existing turbulence makes it easier to understand why galaxies spin than do theories of gravitational instability. But one can object that we have merely changed the question from "where does the spin of the galaxies come from?" to "where does the turbulence come from?" without providing a fundamental answer.

The formation of galaxies is intimately connected with whatever else was going on in the early years of the universe, and thus is connected with the theories of cosmology we will discuss in Chapters 16 and 17. The new theory, continually gaining acceptance, that the universe inflated very rapidly in size in a fraction of a second of time implies that any preexisting structure or turbulence would have been smoothed out. Theoreticians are now working to explain how galaxies might have formed in this "inflationary universe" (Section 17.4) but haven't succeeded yet in predicting fluctuations of the proper size.

Summary and Outline

Observations and catalogues of non-stellar objects
  Lord Rosse's early observations of spiral forms
  Messier's catalogue, General Catalogues by the Herschels,
    New General Catalogue (NGC) and Index Catalogues
    (IC) by Dreyer
  Galaxies as ''island universes''
  When Hubble observed Cepheids in galaxies, he proved
    that galaxies were outside our own Milky Way galaxy
Hubble classification (Section 14.1)
  Elliptical galaxies (E0–E7)
  Spiral galaxies (Sa–Sc) and barred-spiral galaxies (SBa–
    SBc)
  Irregular galaxies (Irr); also SBm
  Peculiar galaxies E(pec), S(pec)
  Amount of gas, and of star formation, increases toward Sc
  IRAS infrared observations show star formation
Origin of galactic structure (Section 14.2)
  Density waves, chain reactions of supernovae, and gravi-
    tational interactions pulling out ''tails''
Clusters of galaxies (Section 14.3)
  Local Group includes our galaxy, 2 other spirals, and about
    2 dozen other galaxies
  Rich clusters, such as Virgo and Coma, are x-ray sources,
    containing hot intergalactic gas
  Local Supercluster exists

The universe is expanding (Section 14.4)
  Hubble's law, $v = H_0 d$, expresses how the velocity of ex-
    pansion increases with increasing distance
  Best current values for Hubble's constant are 50 or
    100 km/s/Mpc
  The expansion is universal, and has no center
Active galaxies (Section 14.5)
  Some objects are powerful sources of radiation in the radio,
    x-ray, and infrared spectral regions
  Double-lobed shape is typical of radio galaxies; sometimes
    a peculiar optical object is present at the center
Radio interferometry (Section 14.6)
  Single-dish radio telescopes have low resolution; interfer-
    ometers give resolution as high or higher than optical
    observations
  Aperture-synthesis arrays provide quicker maps, while still
    retaining high resolution
  VLBI (very-long-baseline interferometry) uses widely sep-
    arated dishes to provide the highest resolution now
    possible
  Current interferometers at Cambridge in England, Wester-
    bork in the Netherlands, and VLA (Very Large Array)
    in U.S.; MERLIN and other networks
  Giant radio galaxies and head-tail galaxies studied

Key Words

Messier numbers, Shapley-Curtis debates, spiral galaxies, Hubble type, elliptical galaxies,
giant ellipticals, dwarf ellipticals, bar, types E0, E7, Sa, Sb, Sc, SBa, SBb, SBc, Irr, peculiar
galaxy, tuning-fork diagram, S0, clusters of galaxies, Local Group, rich cluster, poor cluster,
regular cluster, irregular cluster, supercluster, Local Supercluster, Hubble's law, Hubble's
constant, primary (secondary, tertiary) distance indicators, galactic cannibalism, look-back
time, extragalactic radio sources, radio galaxy, active galaxy, lobes, double-lobed structure,
baseline*, very-long-baseline interferometry*, VLBI*, aperture synthesis*, Very Large Array,
VLA*, head-tail galaxies*, gravitational instability, voids, cosmic turbulence

---

*These words are found in an optional section.

Questions

1. What shape do most galaxies have?

2. Since we see only a two-dimensional outline of an elliptical galaxy's shape, what relation does this outline have to the galaxy's actual 3-dimensional shape?

3. Sketch and compare the shapes of types Sb and SBb.

4. Draw side views of types E7, S0, Sa, Sb, and Sc, showing the extent of the nuclear bulge.

5. The sense of rotation of galaxies is determined spectroscopically. How might this be done?

6. Which classes of galaxies are the most likely to have new stars forming? What evidence supports this?

7. Describe 3 pieces of evidence that galaxies collide.

8. How is it possible that galaxies could exist close to our own yet not have been discovered before?

9. To measure the Hubble constant, you must have a means (other than the redshift) to determine the distances to galaxies. What are three methods that are used?

10. Discuss IRAS observations of galaxies.

11. Discuss the distribution of x-ray emission from rich clusters of galaxies, and why different distributions may exist.

12. Does alpha Centauri, the nearest set of stars to us, show a redshift that follows Hubble's law? Explain.

‡13. At what velocity is a galaxy 3 million light years from us receding?

‡14. A galaxy is receding from us at a velocity of 1000 km/s. (a) If you could travel at this rate, how long would it take you to travel from New York to California? (b) How far away is the galaxy from us?

†15. (a) At what velocity in km/s is a galaxy 100,000 parsecs away from us receding? (b) Express this velocity in km/s, mi/s and mi/hr.

†16. (a) At what velocity in km/s is a galaxy 1 million light years away receding from us?

---

†This indicates a question requiring a numerical solution.
‡Answers: For H_0 = 50 km/s/Mpc **13.** 46 km/s
14. (a) 4 s (b) 20 Mpc

†17. How far away (in Mpc) is a galaxy with a redshift of 0.2? In km?

18. Compare the radio and x-ray emission of a radio galaxy with that of the Milky Way Galaxy.

19. What comment can generally be made about the optical appearance of active galaxies?

20. Contrast VLA and VLBI.

21. Why does interferometry allow you to get finer detail than does a single-dish telescope?

22. Briefly list 5 specific VLBI results, found by looking through the pictures, text, color plates, and index of this book.

23. Briefly list 5 specific VLA results, found by looking through the pictures, text, color plates, and index of this book.

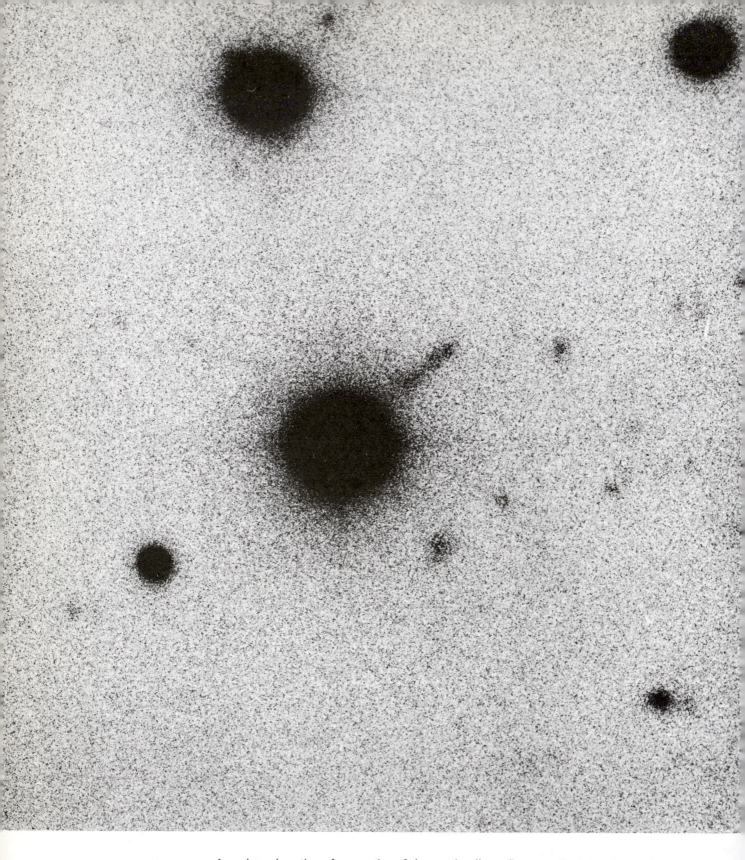

An enlarged portion of a negative of the quasi-stellar radio source 3C 273, taken with the 5-m Hale telescope of the Palomar Observatory. The object, the largest of the black disks, looks like any 13th-magnitude star, except for the faint jet that is visible out to about 20 seconds of arc, equivalent to 50,000 parsecs, from the quasi-stellar object.

Quasars   15

Aims: To describe the discovery of quasars, their significance as probes of the early stages of the universe, their connections with galaxies, and our new knowledge of the giant black holes powering them

Quasars are enigmatic objects that appear almost like stars: points of light in the sky. But unlike the stars that we see, quasars occur in the farthest reaches of the universe and thus may be a key to our understanding the history and structure of space. Although quasars are relatively faint in visible light, they are among the strongest radio sources in our sky, and therefore must be prodigious radiators of energy. Quasars turn out to be small (on a galactic scale), so ordinary methods of generating energy are not sufficient. Similarly, we would be surprised if we got a tremendous explosion out of a tiny firecracker. Consequently, the most exotic and efficient method of generating energy—matter being gobbled by a giant black hole—has been invoked to explain the energy of quasars.

The word *quasar* originated from QSR, a contraction of "quasi-stellar radio source" (*quasi:* as if, seemingly). However, the most important characteristic of quasars is not that they are emitting radio radiation but that they are travelling away from earth at tremendous speeds. After quasars were named, radio-quiet "quasi-stellar objects" (QSO's) were discovered and are also called quasars. So some quasars are radio-quiet, and others are radio-loud. In both cases, astronomers deduce distances by observing the quasar spectra, measuring the Doppler shifts, and applying Hubble's law (Section 14.4). Since some quasars have the largest redshifts known, Hubble's law tells us that they are the most distant objects that we see. We shall discuss the evidence on this point; only a few holdouts don't accept it these days.

After the initial discoveries of hundreds of quasars over two decades ago, advances in understanding quasars came slowly for a while. But in the 1980's, new technologies have enabled us to observe them better, which has led to an improved and revised understanding and a few surprises, as we shall discuss.

15.1 The Discovery of Quasars

Quasars are a discovery resulting from the interaction of optical astronomy and radio astronomy. When maps of the radio sky turned out to be very different in appearance from maps of the optical sky, many astronomers in the 1950's set out to correlate the radio objects with visible ones.

Single-dish radio telescopes did not give sufficiently accurate positions of objects in the sky to allow identifications to be made, so interferometers had to be used. Most of the radio objects catalogued were identified with optical objects, but a few had no clear identifications. At least one of the strong radio sources, 3C 48 (the 48th source in the 3rd Cambridge catalogue), seemed suspiciously near a faint (16th magnitude) bluish star (Fig. 15–1). At that time, 1960, no stars had been found to emit radio waves with the sole exception of the sun, whose radio radiation we can detect only because of the sun's proximity to earth.

Figure 15–1 The first quasar, 3C 48, photographed with the 5-m telescope of the Palomar Observatory.

In Australia, a large radio telescope was used to observe the passage of the moon across the position in the sky of another bright radio source, 3C 273. We know the position of the moon in the sky very accurately. When it occults—hides—a radio source, then we know that at the moment the signal strength decreases, the source must have passed behind the advancing limb of the moon. Later on, the instant the radio source emerges, the position of the lunar limb marks another set of possible positions. The source must be at one of the two points where these two curved lines meet. Three lunar occultations of 3C 273 occurred within a few months, and from the data, a very accurate position for the source and a map of its structure were derived. The optical object, which is 13th magnitude, is not completely starlike in appearance, for a luminous jet appears to be connected to the point nucleus (as shown in the photograph opening this chapter). It is thus "quasi-stellar."

Accurate positions can be measured for radio sources either by interferometry or by lunar occultation.

The radio emission from 3C 273 has two components. The discovery that one coincides with the jet, and the other coincides with the bluish stellar object, clinched the identification of the optical object with the radio object. Maarten Schmidt photographed the spectrum of this "quasi-stellar radio source" with the 5-m Hale telescope on Palomar Mountain. The spectra of 3C 273 and 3C 48, both bluish quasi-stellar objects, showed emission lines, but the lines did not agree in wavelength with the spectral lines of any of the elements. The lines had the general appearance of spectral lines emitted by a gas of medium temperature, though.

The breakthrough came in 1963. At that time, Schmidt noted that the spectral lines of 3C 273 (Fig. 15–2) seemed to have the same pattern as lines of hydrogen under normal terrestrial conditions. Schmidt then made a major scientific discovery: he asked himself whether he could simply be observing a hydrogen spectrum that had been greatly shifted in wavelength by the Doppler effect. The Doppler shift required would be huge: each wavelength would have to be shifted by 16 per cent toward the red to account for the spectrum of 3C 273. This would mean that 3C 273 is receding from us at approximately 16 per cent of the speed of light. Immediately, Schmidt's colleague Jesse Greenstein recognized that the spectrum of 3C 48 could be similarly explained. All the lines in the spectrum of 3C 48 were shifted by 37 per cent, a still more astounding redshift.

Later, absorption lines were discovered in quasars, in addition to the emission lines already known. Many quasars have several systems of emission and absorption lines of differing redshifts. Some of these lines seem to be formed in clouds of gas of differing velocities surrounding the quasars. Others are formed farther from the quasar as the light travels from the quasar to us. Recently, IUE spectra showed that our galaxy and the Large and Small Magellanic Clouds have extensive halos of gas that cause absorption lines detectable in the ultraviolet. This discovery indicates that

Figure 15–2 Spectrum of the quasar 3C 273. The lower spectrum consists of hydrogen and helium lines; it establishes the scale of wavelength. The upper part is the spectrum of the quasar, an object of 13th magnitude. The Balmer lines Hβ, Hγ, and Hδ in the quasar spectrum are at longer wavelengths than in the comparison spectrum. The redshift of 16 per cent corresponds, according to Hubble's law (H_0 = 50), to a distance of three billion light years. Note that the comparison spectrum represents hydrogen and helium sources on earth (more particularly, located inside the Palomar dome).

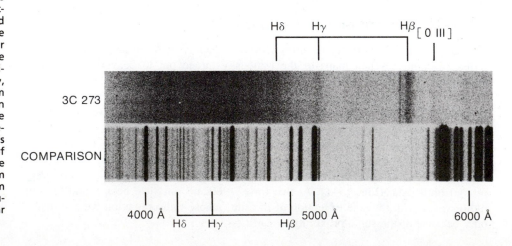

the absorption lines in quasars might be formed in the halos of otherwise unseen galaxies between the quasars and us. Since each galaxy has a different redshift, we can see the same line at different wavelengths.

The basic hydrogen line Lyman alpha often appears at many different redshifts for each galaxy. Thus, through the ultraviolet, we see a *Lyman-alpha forest* of lines. The gas clouds in which these lines are formed seem to be made almost entirely of hydrogen, which indicates that they are located out near the quasars where we are seeing to earlier times in the universe. The nearby quasar 3C 273 shows no such lines, while high-redshift quasars show many, indicating that the hydrogen clouds were more numerous early in the universe's history.

Over 1500 quasars have been discovered, with redshifts ranging up to over 350 per cent. Candidate objects that have a high probability of being quasars can be found by looking for star-like objects that seem unusually strong in the blue and ultraviolet (Fig. 15–3). Examining x-ray sources has also proved fruitful. But one must take spectra to prove that the objects are quasars, a time-consuming procedure on such faint objects. Astronomers estimate that there have been perhaps a million quasars in the universe. By now, however, most of them have probably lived out their lifetimes; that is, they have given off so much energy that they are no longer quasars. Perhaps about 35,000 quasars now exist.

Figure 15–3 This object appears stellar but is known to be not a star but a quasar.

15.2 The Redshift in Quasars

It seems most reasonable that the redshift arises from the Doppler effect. Most objects in the universe show some Doppler shift with respect to the earth. For velocities that are small compared to the velocity of light, the amount of shift in the spectrum is written simply, for rest wavelength λ, velocity v, and speed of light c, as

$$\frac{\Delta\lambda}{\lambda} = \frac{v}{c} \text{ (as discussed in Section 5.6).}$$

Astronomers often use the symbol z to stand for $\Delta\lambda/\lambda$, the amount of the redshift.

The Doppler shifts of quasars are much greater than those of most galaxies. Even for 3C 273, the brightest quasar, $z = 0.16$ (read "a redshift of 16 per cent"). Thus Hubble's law implies that 3C 273 is as far away from us as distant galaxies. Only a few galaxies are known to have greater redshifts. To find the velocity of recession for quasars with redshifts larger than about 0.4, we must use a formula based on Einstein's special theory of relativity (Section 15.2b). Some quasars are receding at over 90 per cent of the speed of light. Hubble's law then tells us they are over 15 billion light years away.

The star in our galaxy with the largest radial velocity with respect to earth has a Doppler shift of 0.2%; $z = 0.2\% = 0.002$. Thus the star is moving away from us at only 0.2% the speed of light. The only object in our galaxy with a Doppler shift in the range of those of quasars is SS 433 (Section 10.12).

*15.2a Working with Non-Relativistic Doppler Shifts

Let us consider a redshift of 0.2, to pick a round number, and calculate its effect on a spectrum (Fig. 15–4). $z = 0.2$ implies that $v/c = 0.2$, and therefore

$$v = 0.2c = 0.2 \times (3 \times 10^5 \text{ km/s}) = 6 \times 10^4 \text{ km/s}.$$

If a spectral line were emitted at 4000 Å in the quasar, at what wavelength would we record it on earth on our CCD (or, if we were old-fashioned, on our film)?

$$\Delta\lambda/\lambda = \Delta\lambda/4000 = 0.2.$$

Therefore

$$\Delta\lambda = 0.2 \times 4000 \text{ Å} = 800 \text{ Å}.$$

Figure 15–4 *(A)* The wavelengths of the visible part of the spectrum are shown at top, and the wavelengths at which spectral lines would appear after undergoing a redshift of 0.2 ($\Delta\lambda/\lambda = 0.2$) are at bottom. The lines are shifted by 0.2 times their original wavelengths, and wind up at 1.2 times their original wavelengths. The non-relativistic formula $\Delta\lambda/\lambda = v/c$ can be applied to give the approximate velocity; at still smaller redshifts this formula is even more accurate.

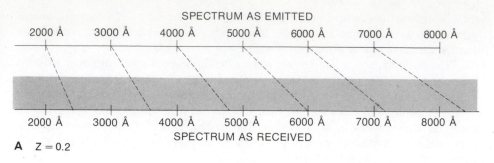

A $Z = 0.2$

Note that we have only calculated $\Delta\lambda$, the shift in wavelength. The new wavelength is equal to the old wavelength plus the shift in wavelength,

$$\lambda_{\text{new}} = \lambda_{\text{original}} + \Delta\lambda = 4000 \text{ Å} + 800 \text{ Å} = 4800 \text{ Å}.$$

Thus the line that was emitted at 4000 Å in the quasar would be recorded at 4800 Å on earth.

Similarly, a line that was emitted at 5000 Å is also shifted by 0.2, and 0.2 of 5000 Å is 1000 Å. Then $\lambda + \Delta\lambda = 5000 \text{ Å} + 1000 \text{ Å} = 6000 \text{ Å}$. The spectrum is thus not merely displaced by a constant number of angstroms, but is also stretched more and more toward the higher wavelengths.

15.2b Working with Relativistic Doppler Shifts

The simple Doppler formula above is valid only for velocities much less than c, the speed of light. For speeds closer to the speed of light, we must use a formula from the special theory of relativity,

$$\frac{\Delta\lambda}{\lambda} = \sqrt{\frac{1 + v/c}{1 - v/c}} - 1.$$

Positive values of v correspond to receding objects. When v is much less than c, then the relativistic formula approximates the non-relativistic formula. (Note that if you substitute $v = 0$ in the relativistic formula, the redshift derived is indeed 0.) But when v is close to c, $\Delta\lambda/\lambda$ is greater than 1 even though v is still less than c. We still use the letter z to stand for $\Delta\lambda/\lambda$.

For example, if $v = 90$ per cent of c, $v/c = 0.9$.

$$\frac{\Delta\lambda}{\lambda} = \sqrt{\frac{1 + 0.9}{1 - 0.9}} - 1 = \sqrt{\frac{1.9}{0.1}} - 1 = \sqrt{19} - 1 = 4.4 - 1 = 3.4.$$

Figure 15–4 *(B)* For a redshift of 1 ($\Delta\lambda/\lambda = 1$), as shown here, the lines are shifted *(bottom)* by an amount equal to their original wavelengths *(top)*, and wind up at twice their original wavelengths. The velocity is a significant fraction of the speed of light and the simple non-relativistic formula for the Doppler shift cannot be applied. The relativistic formula allows z to be greater than 1 without the velocity exceeding the speed of light.

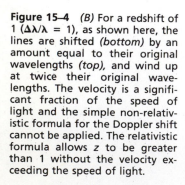
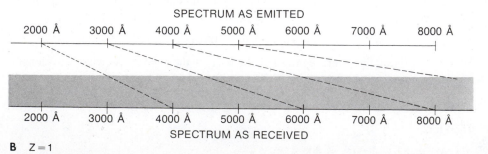

B $Z = 1$

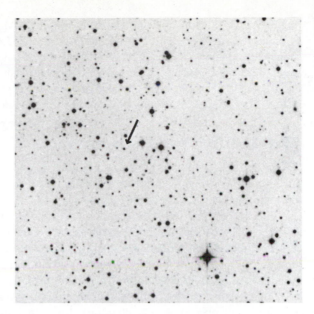

Figure 15–5 The farthest known quasar and thus the farthest known object in the universe, PKS 2000-330. Its redshift is 3.78. Its name comes from its position in the catalogue of radio sources compiled at the Australian National Radio Observatory at Parkes.

There is no physical significance for $z = 1$; we merely have the shift, $\Delta\lambda$, equalling the original wavelength, λ. The quasar with the largest known redshift (Fig. 15–5) has $z = 3.78$, which means that its wavelengths are shifted by 378 per cent. This makes the new wavelengths $3.78 + 1 = 4.78$ times the original wavelengths. But its velocity of recession is just over 90 per cent of the speed of light. The velocities of recession of all quasars are still less than the speed of light, as the special theory of relativity tells us they must always be.

15.3 The Importance of Quasars

If we accept that the redshifts of quasars are caused by the Doppler effect, and we apply Hubble's law, we realize that the quasars with the largest redshifts are the farthest known objects in the universe. At the moment, we have no other method than Hubble's law of measuring distances to objects so far away, and so we would like to use quasars to test Hubble's law since deviations from the law would no doubt show up in the farthest objects.

If the quasars were found not to satisfy Hubble's law, on the other hand, then doubt would be cast on all distances derived by Hubble's law. In that case, when we were observing an object, we would never know whether the object satisfied the law or not. We would never be able to trust a distance derived from Hubble's law. At present, the evidence seems overwhelming that quasars do follow Hubble's law, that is, that their redshifts and distances are proportional with the same Hubble constant that we find for other galaxies.

If we accept the quasars as the farthest objects, then they are billions of light years away, and their light has taken billions of years to reach us. Thus we are looking back in time when we observe the quasars, and we hope that they will help us understand the early phases of our universe. For example, a survey of the entire sky to look for quasars has turned up many new ones, and has shown that the number of quasars per volume of space increases as you go outward. Thus there were more quasars in the universe long ago. Among other things, this shows that the universe has evolved.

If quasars are at the distances determined by Hubble's law, as almost everyone agrees, then when we observe quasars with large redshifts, we are seeing way back in time. Patrick Osmer of the Cerro Tololo Inter-American Observatory searched especially hard for quasars beyond the farthest now known. Though his techniques should have shown new quasars out there if they existed, he didn't detect any. This strengthens the belief that quasars started glowing suddenly at a certain time billions of years ago. This exciting moment marked the beginning of a new stage in the history of our universe, as though giant birthday candles were lighted all at once.

15.4 Feeding the Monster

If the quasars are as far away as the orthodox view holds, then they must be intrinsically very luminous to appear to us at their observed intensities. They are more luminous than entire ordinary galaxies.

But at the same time, the quasars must be very small because of the following argument: The optical brightness of quasars was measured on old collections of photographic plates, and turned out to vary on a time scale of weeks or months. If something is, say, a tenth of a light year across, you would expect that it could not vary in brightness in less than a tenth of a year. After all, one side cannot signal to the other side, so to speak, to join in the variation more quickly than that (Fig. 15–6). Somehow the whole object has to be coordinated, and that ability is limited by the speed of light. So the rapid variations in intensity mean that the quasars are fairly small. Even VLBI techniques have not succeeded in resolving the nuclei of quasars. The Steward Observatory's CCD scanning of the sky, described as a supernova search in Section 10.2, will also give data about quasar variations.

So quasars had not only to give off a tremendous amount of energy but also to do so from an exceedingly small volume. The difficulty of giving off so much energy from such a small volume is known as *the energy problem*.

A model of many exploding supernovae explained the observed sporadic variations in the intensity as variations in the number going off from one moment to the next. But this model required too many supernovae, a dozen a day in some cases. Other unconvincing suggestions to explain the source of energy in quasars included the idea that matter and antimatter are annihilating each other there.

At present, the consensus is that quasars have giant black holes in their centers, larger scale versions of the maxi black holes that may well exist in the centers of our own and other galaxies. The black hole would contain millions or billions of times as much mass as the sun. As mass falls into the black hole, energy is given off. Black holes resulting from collapsed stars give off energy in a similar way, but on a

Figure 15–6 *(A)* The figure illustrates why a large object can't fluctuate in brightness as rapidly as a smaller object. Say that each object abruptly brightens at one instant. The wave emitted from the top of the object takes somewhat longer to reach us than the wave emitted from the side of the object nearest us, just because of the additional distance it has to travel. We don't see the full effect of the variation in brightness until we have the waves from all parts of the object. This simply takes longer for large objects than it does for smaller ones. *(B)* Both optical and radio radiation from the quasar OJ 287 vary in step in a matter of weeks or less. The radio radiation is shown here.

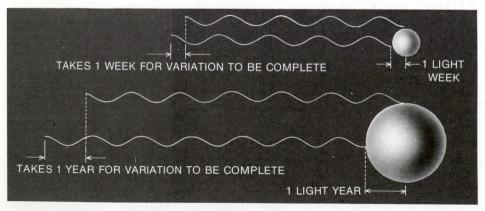

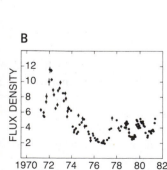

much smaller scale. The central black hole in a quasar is known as "the engine" that produces the energy. The gas and dust around the engine is the "fuel." Together the engine and the fuel are called "the machine."

Theoretical models show how matter falling into such a giant black hole would form an accretion disk. Data across the ultraviolet, visible, and infrared spectrum, including space observations with IUE, have supported this picture. The strength of ultraviolet radiation, for example, matches the predictions that the inner part of the accretion disk—where the friction of the gas is especially high—would reach 35,000 K. The matter in the accretion disk would eventually fall into the central black hole, so the process is called "feeding the monster." We will come back to the question of where quasars get their fuel.

15.5 The Origins of the Redshifts

15.5a Non-Doppler Methods

The most obvious explanation of quasar redshifts is the Doppler effect caused by the expansion of the universe, a "cosmological redshift." But the distances implied are so large and the energy problem at first seemed so difficult that other explanations have been investigated very thoroughly.

If the quasars are close, we must think of some other way of accounting for their great redshifts. There are at least three other ways. They are all unattractive to most astronomers because they challenge Hubble's law and thus cast doubt on our knowledge of the whole scale of the universe. If Hubble's law were to fail for the quasars, we would never know when we could trust it.

A first non-cosmological way also relies on velocity and assumes that quasars are close but are going away from us very rapidly. Quasars could be relatively close to us yet show high redshifts if they had been ejected from the center of a galaxy at very high velocities (Fig. 15–7). If such quasars occurred in another galaxy, we would expect to see some of them going away from us and some coming toward us. But all the quasars have redshifts; none of them has a blueshift, so this idea is unlikely.

If quasars exploded from the center of our galaxy or long enough ago from a nearby galaxy, all the quasars would have expanded past us and are now receding, so no blueshifted quasars would be expected. But so much energy would be required for such tremendous and perhaps repeated explosions that the energy problem would not be resolved.

A second non-cosmological possibility is the gravitational redshift that has been verified for the sun and for white dwarfs. But even before the time this method was ruled out by observations, scientists had not been able to work out such a system for quasars that would be free of other, unobserved consequences. And eventually, measurements were made of redshifts from the faint material that surrounds some quasars. The gravitational redshift theory predicts that the redshift would vary with distance from the quasar, but the observations contradict this. The theory has thus been discarded.

A third possibility is that quasars act on principles that we do not yet understand—some new kind of physics. But the basic philosophy of science forbids us from inventing new physical laws as long as we can satisfactorily use the existing ones to explain all our data.

Now that the energy problem has all but disappeared, since giant black holes form a reasonable "engine," there is little need to invoke these non-cosmological methods of redshifts.

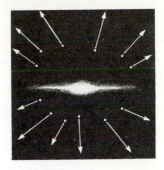

Figure 15–7 The idea that quasars were local, and were ejected from our galaxy, could explain why they all have high redshifts. But a tremendous amount of energy would be necessary to make the ejection. So we might not find ourselves any closer to a solution of the energy problem—where all this energy comes from.

To provide the observed luminosity, the black hole in the center of the quasar 3C 273 would have more than 2.5 billion solar masses, and would be accreting about 25 solar masses per year.

15.5b Evidence for the Doppler Effect

Some of the strongest evidence that the quasars are at their *cosmological distances* (their distances according to Hubble's law) instead of being *local* comes from a major survey of the number of quasars carried out over a dozen years by Richard Green, now at the University of Arizona, and Maarten Schmidt of Caltech. The quasars increase in number with distance from us in a manner that is difficult to account for on a local model but occurs naturally if quasars are at their distances according to Hubble's law.

Now there is even stronger evidence that the redshift is a valid estimator of distance for quasars. A strange object named BL Lacertae has been known for years. At first, it was thought to be merely a variable star (as its name shows), but in recent years it was suspected to be stranger than that, partly because of its rapidly varying radio emission. Its spectrum showed only a featureless continuum, with no absorption or emission lines, so little could be learned about it.

Two Caltech astronomers had the idea of blocking out the bright central part of BL Lacertae with a disk held up in the focus of their telescope, thereby obtaining the spectrum of the haze of gas that seems to surround the star. On one of the several nights that they observed, the central source was relatively faint, and the spectrum of the surrounding haze of gas could be observed with an electronic device. Its spectrum turned out to be a faint continuum with absorption lines typical of the stars in a galaxy. Thus BL Lacertae is apparently a galaxy with a bright central core. Further, the lines are redshifted by 7 per cent. Nobody doubts that a redshift of this amount arises from the expansion of the universe. Yet BL Lacertae, with its point-like appearance and its rapidly varying brightness, appears almost like an extremely nearby quasar! Other similar objects (now called Lacertids or BL Lacs) have also been found to have redshifts of this order.

The Lacertids appear to be a missing link between galaxies and quasars. The properties of quasars (assuming that they are at their Hubble's law distances)—such as the brightness of the core relative to that of the other gas that may be present—are more extreme than, but a continuation of, the properties of Lacertids. Since few doubt that the Lacertids are at their Hubble's law distances, the comparison with quasars strengthens the notion that quasars are also at their own Hubble's law distances.

Quasars with small redshifts have been found (Fig. 15–8). We had not noticed them before because they are located near the galactic plane and their light is thus severely dimmed. The quasar shown was discovered as the optical identification of an x-ray source, which may well be the best way to discover radio-quiet quasars. It is only 800 million light years away. This source is dimmed by 5 magnitudes—a factor of 100—by dust in the plane of our galaxy. If it had been located outside the galactic plane, it would appear brighter than 12th magnitude and would undoubtedly have been discovered long ago. The discovery of this and other objects intermediate in properties between ordinary galaxies and traditional quasars endorses the interpretation of quasars as distant objects.

15.6 Associations of Galaxies with Quasars

15.6a Statistics and Non-Hubble Redshifts

A principal attack on the theory that quasars are at great distances from us has come from Halton Arp of the Mount Wilson and Las Campanas Observatories. On the basis of his observations, he concludes that quasars might be physically linked to galaxies, both the peculiar and the ordinary types. Arp contends that he has found many examples in which two or more quasars lie on a straight line, with at least one

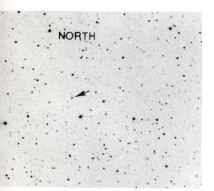

Figure 15–8 QSO 0241+622, studied by Bruce Margon, now of the University of Washington, and Karen Kwitter, now of Williams College, appears stellar and has a luminosity far exceeding that of normal galaxies, just as quasars do. Its spectrum resembles those of many quasars and Seyferts. Its redshift is only 4 per cent.

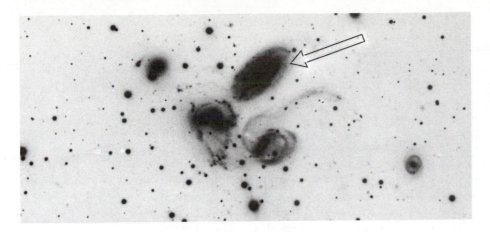

Figure 15–9 Stephan's Quintet, in a negative print of a photograph taken by Halton Arp with the 5-m telescope at Palomar Observatory. Less-exposed photographs show that the dark area to the lower left of the galaxy with the discordant redshift (arrow) has two nuclei and so is two interacting galaxies. Thus there are indeed five galaxies in this "quintet."

on exactly the opposite side of the galaxy from another. He argues that the quasars and the galaxy, which have different redshifts, must therefore be linked. We know the distances to the galaxies from Hubble's law; if the quasars and galaxies are physically linked, then the quasars must be at the same distance from us as the galaxies. These distances are much smaller than the quasars' redshifts indicate.

Any argument that Hubble distances cannot be trusted would show that we cannot rely on Hubble distances for quasars. One such set of observations applies to a group of galaxies rather than to a mixture of galaxies and quasars. The redshift of one of the galaxies in Stephan's Quintet (Fig. 15–9), five apparently linked galaxies, is different from the redshift of the others. Such cases can be explained without rejecting Hubble's law if the object with a discordant redshift only accidentally appears in almost the same line of sight as do the other objects. Further, associations of a quasar with a distant group of galaxies with redshifts that do agree with each other have been found (Fig. 15–10).

In some cases, it seems that a galaxy and a quasar with different redshifts are actually linked by a bridge of material. But this too can be a projection effect (Fig. 15–11). Even in the "best" examples of galaxy-quasar bridges, different sci-

Figure 15–10 Special printing techniques have been applied to this photographic plate of the low-redshift (z = 0.2) quasar 3C 206. 3C 206 is a quasar in an elliptical galaxy that is a member of a compact cluster of galaxies. All objects in the central 40 arc sec of the photograph are galaxies in the cluster, and have the same redshift as the quasar; images of other galaxies in the cluster merge with the sides of the quasar's image. A faint nebulosity around the quasar, between the two vertical lines, is barely visible among the noise from photographic grains on this negative print. Note that the quasar image is fuzzier than the images of the many field stars visible. The fuzziness has been verified by digital study of the plates. (Observations obtained by Susan Wyckoff, Peter Wehinger and Thomas Gehren with the 3.6-m telescope of the European Southern Observatory's station in Chile.)

Figure 15–11 One example of how even improbable things can happen. The galaxies shown at center, called 145-IG-03, appear to be interacting on this Cerro Tololo plate: But are two galaxies interacting or three? Surprisingly, the stellar-appearing object at the end of the luminous bridge—the round dot at the lower right end of the apparently connected objects—turns out (from its spectrum) to be a normal star of spectral type K with a redshift very close to 0. This star is superimposed on the more distant galaxies to an accuracy of less than 0.1 arc sec. The example shown here points up the difficulty of concluding on the basis of probability arguments that objects are physically associated.

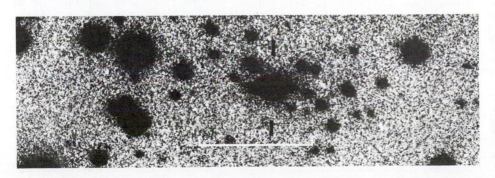

entists do not agree whether the connection is real. In some cases, the resolution of the controversy will have to await the Space Telescope.

Arp's data are subject to several objections, which are mainly statistical. Arp feels that the probability of finding quasars and galaxies so close together in the sky by chance is very small. But statistical methods are valid only before rather than after you know what is actually present. For example, if I flip a coin, show you that it has come up ''heads,'' and then ask you what is the chance that it is ''heads,'' you may answer ''50 per cent.'' But the correct answer is 100 per cent. Once the deed is done, the odds are determined; after all, I showed you that the coin had come up heads. At a racetrack, you can't place your bets after the race is over. Similarly, once we look at a quasar and a galaxy that are apparently linked together, the odds are now 100 per cent that they are apparently linked. We can no longer apply the simple statistical argument that the odds of finding two objects so close together by accident are so small that they must be physically linked.

The associations of galaxies with a quasar of a different redshift would be harder to explain as a projection effect if we found a cluster of galaxies in which not just one but two discrepant redshifts were found. No such cluster is known. So far, most astronomers do not accept Arp's arguments and feel that quasars are in fact very far away.

Alan Stockton of the University of Hawaii carried out an important statistical test of the association of quasars with galaxies. He selected 27 relatively nearby quasars that were so luminous that the energy problem would be particularly severe. He carefully examined the regions around the quasars, and for 17 of the quasars found one or more galaxies apparently associated with them, as shown by their small angular separations.

For 8 of the quasars, at least one of the galaxies nearby had a redshift identical with that of the quasar (Fig. 15–12). Since Stockton had chosen his sample of quasars before knowing what he would find, he was able to apply standard statistical tests to assess the probability that he could randomly find galaxies so apparently close in the sky to quasars. The probability of a projection effect making galaxies appear so close to a high fraction of quasars is less than one in a million. Since the quasar and galaxy redshifts agree, the results strongly endorse the idea that the distances to quasars are those we derive from Hubble's law. In 1984, a Maryland/Caltech/Washington/Leiden team found similar results from even closer companions, clinching the cosmological interpretation.

15.6b Quasars and Galaxy Cores

In recent years, evidence has been mounting that quasars are extreme cases of galaxies rather than being truly different phenomena. Some spiral galaxies have especially bright nuclei. A *Seyfert galaxy,* a type discovered by Carl Seyfert of the Mount Wilson Observatory in 1943 and named after him, has a very bright nucleus indeed (Figs. 15–13 and 15–14) compared with the spiral arms. Another type of galaxy, an *N galaxy,* also has an especially bright core. Can the quasars that we see be only the bright nuclei of galaxies? After all, although we can see both the nuclei

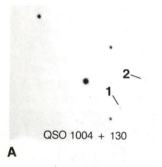

A   QSO 1004 + 130

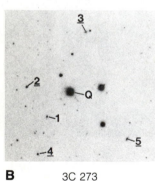

B   3C 273

Figure 15–12 In these photographs taken by Alan Stockton on Mauna Kea, the central objects are quasars, whose catalogue numbers are given below the photos, and the numbered objects are galaxies. The redshifts of the galaxies and the quasars agree. Note how the quasars are brighter than galaxies at the same distances. *(A)* From the original survey. *(B)* The subsequent discovery that the first known and brightest quasar, 3C 273, is associated with a group of galaxies of the same redshift (#2–5); one galaxy with a disparate redshift (#1, not underlined) is also present.

Table 15–1 Energies of Galaxies and Quasars

| | Relative Luminosity | | |
	X-Ray	Optical	Radio
Milky Way	1	1	1
Radio galaxy	100–5,000	2	2,000–2,000,000
Seyfert/N galaxy	300–70,000	2	20–2,000,000
Quasar: 3C 273	2,500,000	250	6,000,000

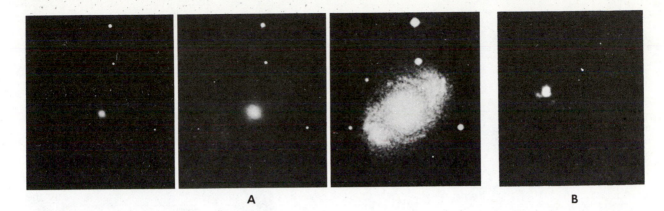

A

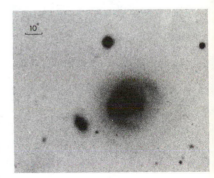

B

Figure 15–13 (A) Increasing exposure time reveals more and more galactic structure around the Seyfert galaxy NGC 4151 (redshift z = 0.001), which appears almost stellar on the least-exposed photograph. (B) The nucleus of the Seyfert galaxy photographed from an altitude of 25,000 m with the 90-cm telescope carried aloft in 1970 by the Stratoscope II balloon.

and the spiral arms of nearby Seyferts, only the nuclei would be visible if such galaxies were very far away.

Seyfert galaxies not only have relatively bright nuclei but also have broad emission lines in their spectra, a sign that hot gas is present. The exceptional breadth in wavelength of the lines could be caused by rapidly moving matter in the galaxies' cores, a sign of violent activity there. Two to five per cent of galaxies are Seyferts.

In the early 1970's, Jerome Kristian of the Hale Observatories studied a sample of about 24 quasars to see whether they could be the bright cores of distant galaxies. He attempted to predict which quasars would be close enough to reveal some structure around the core—nebulosity, which quasars would be so far away that they appear as points like stars, and which quasars would be in between. He then found on the best photographs of these objects that almost all objects in which one would expect to see structure or nebulosity did in fact show it, that almost all objects that would not show structure or nebulosity because they were too far away did not show it, and that the middle group was mixed.

In the last few years, this work has been carried further by several investigators taking advantage of the exceptionally good seeing available at certain observatories and of new, sensitive equipment like CCD's. For example, structure was found around 3C 48, one of the first quasars to be identified. Then Susan Wyckoff, Peter Wehinger, and colleagues were able to resolve structure in 13 of 15 low-redshift quasars they studied (Fig. 15–15). John Hutchings and colleagues in Canada similarly resolved other quasars (Fig. 15–16). The fuzzy structure detectable is technically known as *fuzz*. In many cases, the quasar fuzz could be gas the quasar has ejected. The diameter of the fuzz decreases with increasing redshift, just as it would do if the fuzz were from a galaxy surrounding a bright central region. This point further strengthened (as if it needed strengthening) the case that quasars are at their cosmological distances.

In only two cases—3C 273 and one other—is the material near quasars in the form of jets. In at least some other cases, the structure shows the characteristic spectrum of a galaxy. The fuzz around 3C 48 has been found to have spectral lines from hot stars. So star formation must have been going on in that source during the past billion years. Further, the observation strengthens the case that quasars are objects in spiral rather than elliptical galaxies.

The results indicate that quasars are somehow related to the cores of galaxies. Perhaps galaxies go through a quasar stage during which their nuclei are very bright, or quasars are an extreme case somewhat different from but similar to explosive events in galaxy cores. The surrounding material usually seems to resemble material in spiral galaxies, so the quasar phenomenon may usually or always take place in spirals.

Figure 15–14 A Seyfert galaxy, ESO 113-IG 45, with an exceptionally luminous nucleus. This is a case of a quasar (or almost quasar) in the center of a spiral galaxy.

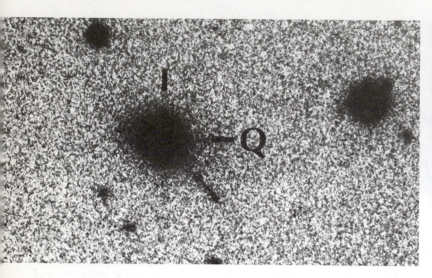

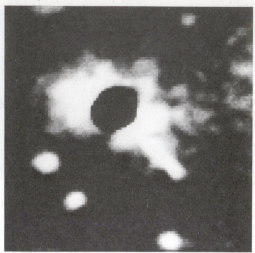

Figure 15–15 Fuzz (faint nebulosity), in addition to the jet, can be seen around the brightest quasar, 3C 273. The redshift of the nebulosity is the same as that of the quasar. The left image is photographic. In the right image, based on data taken with a CCD camera, the image of a star has been digitally subtracted from the quasar's image; a central disk had masked the brightest middle part of the quasar and of the star on their respective plates, but some spillover light had remained to be considered. The contrast of the remainder was then enhanced to bring out the surrounding galaxy. It is 400,000 times fainter than the quasar, comparable with a giant elliptical galaxy. (*Left:* Susan Wyckoff, Peter Wehinger, and Thomas Gehren; *right:* Anthony Tyson, Bell Labs)

Further, properties of quasars like the types of lines in their spectra are now known to be similar to those of certain types of galaxies. Seyfert nuclei and quasars, for example, have similar ultraviolet and optical emission lines. Though the quasars may be extreme cases, they are not as different from other types of known objects as had once been thought.

In sum, the relation between quasars and the cores of active or peculiar galaxies has been consistently strengthened as we have been able to observe finer detail to greater distances. This linkage makes the quasars seem somewhat less strange, but at the same time causes galaxies to seem more exotic. Now that structure can be detected around so many quasars, quasars have lost their original attribute of appearing point-like. Lacertids, Seyfert galaxies, and quasars are probably products of the same basic phenomenon, though are in different stages of evolution or activity. If we had known of BL Lacertae's redshift earlier on, quasars would not have looked as strange to us when they were discovered. And if collapsed objects like neutron stars and black holes had been in our minds when quasars were discovered, we would have had an obvious way to produce a lot of energy in a small space. In that case, we would not have said that "the energy problem" for quasars existed.

15.6c Quasars and Interacting Galaxies

Since almost all quasars are so far away, it seems that whatever provided the fuel to make them so bright must have soon been exhausted. So why can we see a few quasars that are so close? We now newly realize that these closest quasars may be rejuvenated, having received a new supply of fuel.

Some evidence that galaxies interact gravitationally has been known for a while. For example, we discussed (Section 14.2) the tails that have been explained as gravitational interactions. The occurrence of Seyfert galaxies is more common in galaxy pairs than in individual galaxies. And during the last half-dozen years, astronomers have discovered more signs of galactic cannibalism. There is also increased evidence

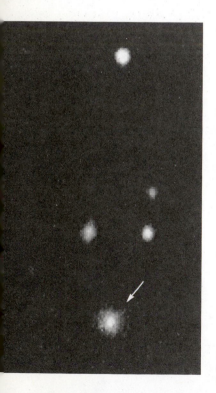

Figure 15–16 Note the fuzz around the quasar IE 2344+18, compared with the sharpness of the image around the star. (Palomar observations by Matthew Malkan)

that some active galaxies like Centaurus A have absorbed their companions.

New studies of quasars, carried out with high resolution, have revealed that some apparent bumps on the quasar images are actually independent objects. In other cases, independent objects were already known from the work of Arp and others. These objects have the same redshifts as the nearby quasars, and statistics now seem to show that the objects and the quasars are actually associated (Figs. 15–17 and 15–18). These could be the objects that have been stripped of their outer layers to provide fuel for the quasar, rejuvenating the engine. Close encounters of galaxies with quasars in a way to provide the fuel for the black holes are rare, which could explain why nearby quasars are few in number.

The quasars that are farthest away, those with the largest redshifts, are using up an original store of fuel. We see these quasars far enough back in time—the same time that rich clusters had just formed. The quasars may have consumed some of the gas present in these clusters. The closer quasars might be having a second youth, with new gas being introduced from encounters with other galaxies. Or, since it takes longer for the small clusters or groups of galaxies (where we see these quasars) to be stripped of their gas, these close quasars could be getting fuel for their black holes for the first time. Observationally, at least 30 per cent of quasars with redshifts up to 0.6—a fair way out into space—appear to be interacting, so the interaction model can be commonly applied.

According to this interaction model, Seyferts may be similar to quasars, except that Seyferts have smaller black holes or less fuel. Radio galaxies may be older quasars or objects that never received much fuel. (Quasars resemble radio galaxies by also being double-lobed radio sources.)

Certainly quasars will be among the first objects studied with the Hubble Space Telescope. Just imagine being able to see them seven times more clearly!

*15.7 Superluminal Velocities

Very-long-baseline interferometry (VLBI) observations with extremely high resolution (0.001 arc sec) have revealed the presence of a few small components in radio images of jets in a few of the quasars. Further, the observations have shown

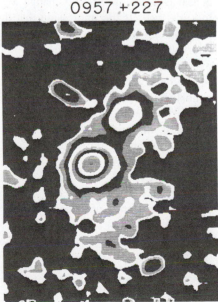

Figure 15–17 The larger circle is the image of the quasar 3C 323.1, and the smaller circle apparently touching it is a nearby galaxy. Stripping off the outer layers of this galaxy may have provided the fuel for the quasar. (Mauna Kea observations by Alan Stockton)

1747 + 684

0957 + 227

Figure 15–18 Interacting quasars and galaxies show in these computer scans. (Canada-France-Hawaii Telescope observations at Mauna Kea by John Hutchings and colleagues)

that in some cases these components are separating at angular velocities across the sky that seem to correspond (at the distances of these objects based on Hubble's law) to velocities greater than the speed of light. Since the special theory of relativity tells us that the apparent *superluminal velocities*—velocities greater than the speed of light—could not be real, other theoretical explanations of the data have been sought.

The jets are presumably forced out of the quasar centers above the quasars' poles. The compression that formed the accretion disks would have made such high magnetic fields in the disks that particles couldn't cross them, leaving the poles as the best way out. The material ejected in this way presumably forms the lobes detectable with radio telescopes. The jets may be a larger-scale version of the jets of SS 433 (Section 10.12), a source in our galaxy.

A "Christmas tree" model, in which components are flashing on and off, had seemed possible at first. In this model, no rapid velocities need be implied, since the sources we see this year were not necessarily the same ones we saw last year. But now, years of observations (Fig. 15–19) show that the components continue to separate from each other, rather than flashing on and off. The calculated velocity of this source is $10c$ (10 times the speed of light), while other sources show velocities ranging up to $45c$.

The current model involves the special theory of relativity, depending on the fact that light travels at a finite speed. Picture a jet of gas that is moving rapidly almost directly toward us. The jet almost (but not quite) keeps up with the light it emits (Fig. 15–20). Let us say that the jet is moving almost directly toward us at 99 per cent of the speed of light. If we look now, we see the jet at a certain position in the sky. But when we looked a year ago for us, the jet was a lot farther away, about 99 per cent of a light year farther back, making it about 1.99 light years away. Its radiation from that time has had only a year to travel, so hasn't reached us yet.

We saw, a year ago for us, the quasar jet's radiation from much longer ago than a year ago. In the interval between our observations, the quasar jet had several times as long to move as we would naively think it had. So it could, without exceeding the speed of light, move several times as far. When we derive its apparent velocity across the sky, by simply dividing the angle it appeared to move across the sky by the 1 year that elapsed for us, we thus derive an apparent angular velocity across the sky several times greater than its real velocity. And when we consider how far the quasar is away, in order to change its jet's angular velocity into an apparent velocity in km/s, we get the impression that the jet is moving faster than the speed of light, even though it really isn't. We have divided how far the **quasar jet** moved by the interval between **our** observations, not how far the quasar jet moved by the interval over which **it** moved that distance.

Where was the quasar jet one year ago? It has been chasing its own radiation, almost (but not quite) keeping up with it. So we received its radiation from one year ago very recently, less than 1 year ago. Again, we could be fooled into thinking that the jet is moving faster than the speed of light if we divide how far the quasar appeared to move from side to side across the sky in that short time interval by the length of the time interval, since the quasar really had a full year to move that distance.

Under these circumstances, the apparent separation of the objects is not the actual separation at any one instant of time. For certain angles of view, the apparent

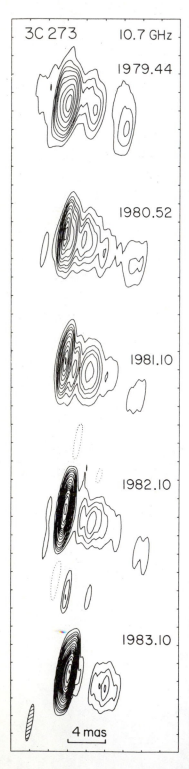

3C 273   10.7 GHz

1979.44

1980.52

1981.10

1982.10

1983.10

4 mas

Figure 15–19 A series of photographs of the extreme center of 3C 273, made with radio telescopes spread over intercontinental distances (between California and Germany). In the earliest map, the two components are separated by less than $\frac{6}{1000}$ arc sec. The series shows first one and then another knot of emission separating from the core at an apparently superluminal velocity of $10c$. They go off in the direction of 3C 273's optical jet, and may become part of the jet when they move farther out.

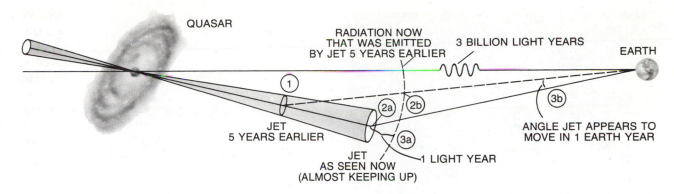

QUASAR

RADIATION NOW
THAT WAS EMITTED
BY JET 5 YEARS EARLIER   3 BILLION LIGHT YEARS

EARTH

①

②a ②b

③b

JET
5 YEARS EARLIER

ANGLE JET APPEARS TO
MOVE IN 1 EARTH YEAR

③a

JET
AS SEEN NOW
(ALMOST KEEPING UP)

1 LIGHT YEAR

Figure 15–20 The leading model for explaining how a jet of gas emitted from a quasar, shown at left with its accretion disk surrounding a central black hole, can seem to be travelling at greater than the speed of light when seen from earth (shown at right). (1) marks the position of the jet 5 years earlier than the jet we are seeing now. (It then takes another 3 billion years for the radiation from 3C 273 to reach us, an extra duration that we can ignore for the purposes of this example.) After an interval of 5 years, the light emitted from point (1) has reached point (2b), while the jet itself has reached point (2a). The jet has been moving so fast that points 2a and 2b are separated by only 1 light-year (3a) in this example. So with a 1-year interval we receive light from two positions that the jet has taken 5 years to go between. We thus think that the jet is moving faster than it actually is, since it appears to move over angle (3b) in 1 year while it really did so in 5 years.

rate of separation can look much greater than the real rate of separation. The special theory of relativity provides many such apparent paradoxes, which are resolved when we take careful account of exactly when light was emitted.

The ''superluminal motions'' might just give an important key to the understanding of quasars. The model we have just discussed requires that the knot of gas is moving almost right at us (within about 10°), which may seem statistically unlikely. But theory predicts that the radiation from the rapidly moving components may be concentrated in beams. We might detect only those that are beaming radiation toward us (which would also explain why we sometimes see only one jet). If quasars concentrate energy in beams this way, then their total energy emission would not be as high as we had calculated on the assumption that they were emitting radiation in all directions at an equal rate. This would make quasar luminosities similar to luminosities of galaxies. But current thinking is that only a few of the quasars are beaming radiation at us, and thus enhancing our measurements of their power.

The apparent superluminal expansion has been detected in several quasars and in at least one galaxy, and so may be a fairly common phenomenon. There presumably would be a second jet of gas oriented away from us, to make the situation symmetric, but this jet would be moving away so fast it would be redshifted too much for us to detect it.

*15.8 Quasars Observed from Space

The earth's atmosphere prevents the ultraviolet and x-ray region of quasar spectra from reaching us, but our ability to launch telescopes in spacecraft has eliminated this handicap. Since many of the spectral lines we observe in the visible spectra of quasars were originally emitted by the quasar in the ultraviolet and have since been redshifted into the visible, it is particularly important to study the ultraviolet spectra of relatively bright nearby quasars.

The first ultraviolet quasar spectrum was made by a Johns Hopkins University group, who launched a 40-cm telescope with a high-resolution spectrograph on a rocket. More recently, the International Ultraviolet Explorer is allowing the obser-

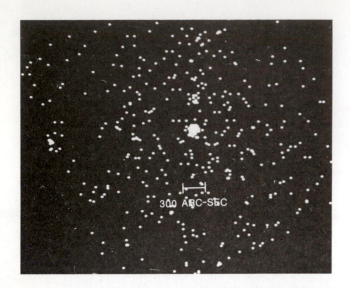

300 ARC-SEC

Figure 15–21 The quasar B2 1225+31 at a redshift of 2.2 is visible in this 100-minute Einstein Observatory exposure. Quasar numbers like this refer to their celestial coordinates. (Observations by Harvey Tananbaum and colleagues at the Harvard-Smithsonian Center for Astrophysics)

vation of many more quasars. Many images of the Lyman-alpha line of hydrogen appear (the Lyman-alpha forest).

The breakthrough in observing quasars in x-rays came with the Einstein Observatory. Until then, only a few quasars had been observed in this region of the spectrum. The sensitivity of Einstein enabled astronomers to observe dozens of x-ray emitting quasars. Some of the quasars observed are among the farthest known (Fig. 15–21). Even an observation of 3C 273 turned up a new distant quasar in the background (Fig. 15–22). The objects are confirmed as quasars only when optical spectra are taken and large redshifts found.

Surprisingly, more of the quasars being discovered with the Einstein Observatory have low redshifts rather than high redshifts. Combined x-ray and optical observations of these quasars may change our ideas of how quasars are distributed in space and how they evolved with time.

*15.9 Double Quasars

The astronomical world was agog in 1979 at the discovery of a pair of quasars so close to each other that they might be two images of a single object. The story began when inspection of the Palomar Sky Survey charts (which were taken years

Figure 15–22 The bright object at lower right is the prominent quasar 3C 273, but a newly discovered quasar is visible at upper left as well in this x-ray image taken with the Einstein Observatory. The objects are too small to be resolved, though the image of 3C 273 has spread considerably because of overexposure. (Observations by Harvey Tananbaum and colleagues at the Harvard-Smithsonian Center for Astrophysics)

QUASAR

INTERVENING GALAXY

EARTH

Figure 15–23 The gravity of a massive object, perhaps a galaxy, in the line of sight can form multiple images of an object. If the alignment of the earth, the intermediate object, and the distant quasar is perfect, we would see a ring. A slight misalignment would make crescents or individual images.

ago with the larger of the Palomar Schmidt telescopes) revealed a close pair of 17th-magnitude objects at the position corresponding to a radio source.

When the spectra of the objects were taken at Kitt Peak, both objects turned out to be quasars, and their spectra looked identical. Even stranger, their redshifts were essentially identical. Some of the first spectra taken with the then-new Multiple Mirror Telescope were of these objects. The MMT redshift measurements showed that the quasars were even closer to each other than the previous observations had shown.

It seemed improbable that two independent quasars should be so similar in both spectral lines and redshifts, so the scientists who took the first spectra suggested that both images showed the same object! They suggested that a gravitational bending of the quasar's radiation was taking place. We have discussed such gravitational bending as a consequence of Einstein's general theory of relativity and have seen that it has been verified for the sun (Section 7.11). For the quasars, a massive object between the quasar and us is acting as a gravitational lens, bending the radiation from the quasar one way on one side and the other way on the other side (Fig. 15–23). As a result, we see the image of the quasar in at least two places.

Detailed radio maps of the region have been made with several interferometers, most notably with the VLA. The VLA's map (Fig. 15–24) shows the two sources and others as well.

One test of whether a gravitational lens is present is to see if the brightness fluctuations of the two optical images are similar. Since the light from the two images travels along different paths through space, though, the fluctuations could be displaced in time from each other by several years or more. Only time will tell if the

The two images are separated by only 6 seconds of arc. Their redshifts are identical to the third decimal place: $z = 1.4136 \pm 0.0015$.

Dennis Walsh of the University of Manchester, Robert F. Carswell of Cambridge University, and Ray J. Weymann of the University of Arizona first identified the double quasar and suggested that it resulted from a gravitational lens.

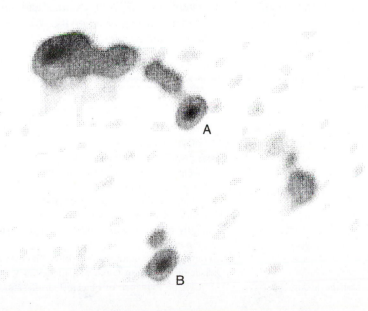

A

B

Figure 15–24 A radio map of the double quasar, Q 0957+561 A and B, made with the VLA. The sources marked A and B correspond to the optical objects. Additional images are also seen, which could result from an off-center gravitational lens. (Courtesy of B. F. Burke and P. E. Greenfield, MIT, and D. H. Roberts, Brandeis)

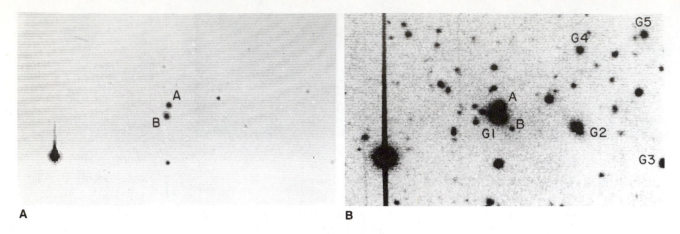

A   B

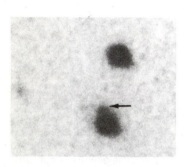

Figure 15–26 This optical photograph of the double quasar, 0957 + 561A and B, shows the intervening galaxy. The 17th-magnitude quasar images are bluish and are separated by only 5.7 seconds of arc. Their redshifts are identical to the third decimal place: $z = 1.4136 \pm 0.0015$. This photograph by Alan Stockton of the University of Hawaii shows the intervening galaxy only 1 second of arc from the "B" quasar image. Its redshift is about 0.37. The objects are in the constellation Ursa Major.

Figure 15–25 Two views of a photograph of the double quasar taken with a CCD camera at the 5-m Hale telescope. *(A)* The frame is reproduced so that the twin images, marked A and B, stand out. *(B)* The frame is reproduced so that the fainter parts of the image show. We then see a cluster of galaxies, some of which are marked (G1–G5). 90 per cent of the objects on the frame are galaxies. A and B are now overexposed and blur out. G1 provides most of the gravity for the lens effect. Though its image does not show clearly, its redshift can be measured spectroscopically. (Courtesy of Jerome Kristian, James Westphal, and Peter Young, Palomar Observatory)

pattern of fluctuation of one object over a period of time matches that of the other. Theoretical work indicates that ten years might be long enough to wait.

Our knowledge of the double quasar was greatly increased by a picture taken with an electronic camera (using CCD's, Section 2.11a). The observation shows (Fig. 15–25) that a cluster of galaxies lies between us and the quasar. Quasar image B is seen through the brightest member of the cluster. The gravitational effect of the cluster adds to that of this galaxy in making the gravitational lens.

The gravitational lens picture was strongly endorsed by a photograph that seems actually to show the intervening galaxy (Fig. 15–26 and Color Plate 73). It seems to be a giant elliptical galaxy with a redshift about one-fourth that of the quasar.

The double quasar was the first known example of a gravitational lens. Additional multiple quasars have since been discovered, and searches are under way for still more.

The best understood case, the double quasar, is an exciting and valuable verification of a prediction of Einstein's general theory of relativity. We expect the additional multiple quasars to provide further confirmation.

Summary and Outline

Discovery of quasars (Section 15.1)
    QSR—quasi-stellar radio source
    Huge Doppler shifts in spectra
    Radio-quiet objects also included as quasars
Doppler shift in quasars (Section 15.2)
    Large redshifts mean large velocities of recession and, by Hubble's law, great distances.
    Relativistic formula used for redshifts close to or greater than 1
Importance of quasars (Section 15.3)
    Test of Hubble's law, which is thus a test of the accuracy

    of the distance scale
    Great distance, which means we view most of them as they were in an early stage of the universe
Energy problem (Section 15.4)
    Too much energy required from a small volume to be accounted for by ordinary processes
    Generally accepted explanation: giant black hole is the engine
Non-Doppler explanations of redshifts (Section 15.5)
    Quasar ejected from cores of galaxies at very high velocities; no evidence of blueshifted quasars, however

Gravitational redshifts; conflict with observations

Quasars operating under new physical laws?

All these alternative explanations challenge Hubble's law, but are not now needed since quasars can be understood as being powered by black holes

BL Lacertae and nearby quasars show that quasars are indeed at their Hubble-law distances

Galaxies and quasars (Section 15.6)

Question of physical link between galaxies and quasars

Stephan's Quintet and other sources with discrepant redshifts might be explained as chance alignments.

Other observations and their statistical analysis show that quasars are indeed at their cosmological distances.

Seyfert and N galaxies also have bright nuclei; BL Lacertae and other nearby objects known to have distances corresponding to Hubble's law have properties in common with quasars; are quasars truly a new phenomenon?

Quasars seem to be events in cores of galaxies at some

evolutionary stage; many quasars have surrounding fuzz that may be the underlying galaxies; the fuel may come from interactions with other galaxies.

Superluminal velocities of quasars (Section 15.7)

Components detected in several quasars seem to be moving apart at speeds greater than the speed of light.

Current explanation involves special theory of relativity and jets of gas pointing almost at us moving at velocities close to but less than the speed of light.

Space observations of quasars (Section 15.8)

Rocket and IUE studies in ultraviolet

Einstein Observatory observations of quasars with large redshifts and of numerous faint quasars

The double quasar (Section 15.9)

Two quasars with identical properties located very close together

Apparently two images of one quasar formed by a gravitational lens

Key Words

quasar, Lyman-alpha forest, the energy problem, cosmological distances, local, Seyfert galaxy, N galaxy, fuzz, superluminal velocity*

---

*This word is found in a optional section.

Questions

1. Why is it useful to find the optical objects that correspond in position with radio sources?

†2. (a) You are heading toward a red traffic light so fast that it appears green. How fast are you going? (b) At \$1 per mph over the speed limit of 55 mph, what would your fine be in court?

3. A quasar is receding at 1/10 the speed of light. (a) If its distance is given by the Hubble relation, how far away is it? (b) At what wavelength would the 21-cm line appear?

†4. A quasar has $z = 0.3$. What is its velocity of recession in km/s, using the non-relativistic formula?

†5. We observe a quasar with a spectral line whose rest wavelength is 3000 Å, but which is observed at 4000 Å. (a) How fast is the quasar receding, using the non-relativistic formula? (b) How far away is it if its distance is given by the Hubble relation? Specify the value you are using for Hubble's constant.

†6. The farthest known quasar, PKS 2000-330, has $z = 3.78$. At what wavelength does the Lyman-α line, whose wavelength is 1216 Å from a source at rest, appear?

‡7. A quasar is receding with a velocity 85 per cent of the speed of light. At what wavelength would a spectral line appear if it appears at 5000 Å when we observe it emitted from a gas in a laboratory on earth? What part of the spectrum is it

in when it is emitted, and in what part of the spectrum do we observe it from the quasar?

†8. A quasar is receding with a velocity 95 per cent of the speed of light. At what wavelength would the Lyman-alpha spectral line appear? (It is emitted at 1216 Å.) What part of the spectrum is it in when it is emitted, and in what part of the spectrum do we observe it from the quasar?

†9. A quasar has $z = 2$. At what velocity is it receding from us, and what is its distance from us?

10. What do quasi-stellar radio sources and radio-quiet quasars have in common?

11. Why does the rapid time variation in some quasars make the "energy problem" even more difficult to solve?

12. What are three differences between quasars and pulsars?

13. Briefly list the objections to each of the following "local" explanations of quasars: (a) They are local objects flying around at large velocities. (b) The redshift is gravitational.

14. If quasars were proved to be local objects, would this help solve the "energy problem"? Explain.

15. Explain how parts of a quasar could appear to be moving at greater than the speed of light, without violating the special theory of relativity.

16. Describe the implications of the observations of quasars in the x-ray spectrum. What new ability allowed these observations to be made?

17. What features of some quasars suggest that quasars may be closely related to galaxies?

---

†This indicates a question requiring a numerical solution.

‡Answer: **7.** 17,500 Å; emitted in visible (blue); observed in infrared.

Albert Einstein in the Swiss Patent Office in Berne in 1905, at the time his special theory of relativity was published. Cosmological theories are based on his general theory of relativity (1916).

Cosmology 16

Aims: To study the origin of the universe and the earliest stage from which we receive radiation

In Armagh, Ireland, in the mid-seventeenth century, Bishop Ussher declared that the universe was created the evening preceding Sunday, October 23rd, in the year 4004 B.C. Nowadays we are less certain of the date of our origin (though we do set out a split-second agenda for the first few minutes of time).

We study the origin of the universe as part of the study of the universe as a whole, *cosmology*. Even more than in other parts of astronomy, in order to study cosmological problems we use simultaneously both theoretical calculations and all our abilities to observe a wide variety of celestial objects.

The study of where we have come from and of what the universe is like leads us to consider where we are going. Is the universe now in its infancy, in its prime of life, or in its old age? Will it die? It is difficult for us who, after all, spend a lot of time thinking of topics like "what shall I watch on TV tonight?" or "what's for dinner?" to realize that we can think seriously about the structure of space around us. It is awesome to realize that we can conclude what the future of the universe will be. One must take a little time every day, as Alice was told when she was in Wonderland, to think of "impossible things." By and by we become accustomed to concepts that may seem overwhelming at first. You must sit back and ponder when studying cosmology; only in time will many of the ideas that we shall discuss take shape and form in your mind.

*16.1 Olbers' Paradox

Many of the deepest questions of cosmology can be very simply phrased. Why is the sky dark at night? Analysis of this simple observational question leads to profound conclusions about the universe.

We can easily see that the night sky is basically dark, with light from stars and planets scattered about on a dark background. But a bit of analysis shows that if stars are distributed uniformly in space, then the sky shouldn't be dark anywhere. If we look in any direction at all we will eventually see a star (Fig. 16–1), so the sky

Figure 16–1 *(A)* If we look far enough in any direction in an infinite universe, our line of sight will hit the surface of a star. This leads to Olbers' paradox. *(B)* This painting by the Belgian surrealist René Magritte shows a situation that is the equivalent to the opposite of Olbers' paradox.

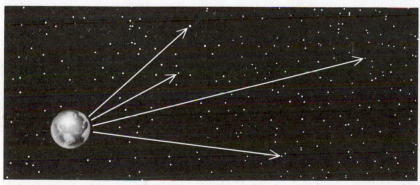

A

B

Figure 16-2 The lower halves of trees are often painted in Mexican parks, and provide an example of a similar phenomenon to seeing a uniform expanse of starlight.

"Olbers" has an "s" on its end, so we write of "Olbers' paradox" (or, alternatively, "Olbers's paradox"); "Olber's paradox," with the apostrophe before a sole "s," is incorrect, since his name wasn't "Olber."

The energy, E, of a quantum at a given wavelength, λ, corresponding to a certain frequency, ν, is $E = h\nu = hc/\lambda$. As its wavelength gets longer because of the Doppler effect and Hubble's law, the formula shows that the quantum's energy diminishes. This energy is actually lost from the quantum.

should appear uniformly bright. We can make the analogy to our standing in a forest. There, we would see some trees that are closer to us and some trees that are farther away. But if the forest is big enough, our line of vision will always eventually stop at the surface of a tree. If all the trees were painted white, we would see a white expanse all around us (Fig. 16-2). The white on the trees that are farther away is of the same brightness as the white on the trees that are closer. Similarly, when looking up at the night sky we would expect the sky to have the uniform brightness of the surface of a star.

The fact that this argument implies that the sky is uniformly bright, while observation shows that the sky is dark at night, is called *Olbers' paradox*. (A paradox occurs when you reach two contradictory conclusions, both apparently correct.) Wilhelm Olbers phrased it in 1823, although the question had been discussed at least a hundred years earlier. Solving Olbers' paradox leads us into considering the basic structure of the universe. The solutions could not have been advanced in Olbers' time because they depend on more recent astronomical discoveries.

We have phrased Olbers' paradox in terms of stars, though we know that the stars are actually grouped into galaxies. But we can carry on the same argument with galaxies, and deduce that we must see the average surface brightness of galaxies everywhere. This is patently not what we see.

One might think that the easiest way out of this paradox, as was realized in Olbers' time, is simply to say that interstellar dust is absorbing the light from the distant stars and galaxies. But this doesn't solve the problem: because the sky is generally dark in all directions, the dust would have to be everywhere. This widely distributed dust would soon absorb so much energy that it would heat up and begin glowing. Given a long enough time, all the matter in the universe would begin glowing with the same brightness, and we would have our paradox all over again.

One solution to Olbers' paradox lies in part in the existence of the redshift, and thus in the expansion of the universe (Section 14.4). This does not mean that the answer to the paradox is simply that visible light from distant galaxies is redshifted out of the visible, for at the same time ultraviolet light is continually being redshifted into the visible. The point is rather that each quantum of light, each photon, undergoes a real diminution of energy as it is redshifted. The energy emitted at the surface of a faraway star or galaxy is diminished by this redshift effect before it reaches us.

But the redshift is not the whole solution of Olbers' paradox. As we look out into space, we are looking back in time, because the light we see has taken a finite amount of time to travel to us. If we could see out far enough, we would possibly see back to a time before the stars were formed. E. R. Harrison of the University of Massachusetts has pointed out that this is another way out of the dilemma. In fact, Harrison calculates that this explanation is a more important contribution to the solution of the paradox than is the existence of the redshift. In most directions, we would not expect to see the surface of a star for 10^{24} light years. We would thus have to be able to see stars 10^{24} years back in time for the sky to appear uniformly bright. However, the stars don't burn that long, and the universe is simply not that old. On the basis of Hubble's law, an age of somewhat over 10^{10} years seems to be the maximum.

The fact that we have to know about the expansion and the age of the universe to answer Olbers' question—why is the sky dark at night?— shows how the most straightforward questions in astronomy can lead to important conclusions. In this case, we find out about the expansion of the universe or about the lifetimes of the universe and the stars in it.

16.2 The Big-Bang Theory

Astronomers looking out into space have noticed that on a sufficiently large scale the universe looks about the same in all directions. That is, ignoring the presence of local effects such as our being in the plane of a particular galaxy and thus seeing a Milky Way across the sky, the universe has no direction that is special. Further, it seems that there is no change with distance either, except insofar as time and distance are linked.

These notions have been codified as the *cosmological principle*: **the universe is homogeneous and isotropic throughout space.** The assumption of *homogeneity* says that the distribution of matter doesn't vary with position (that is, with distance from the sun), and the assumption of *isotropy* says that the universe looks about the same no matter in which direction we look. Actually, of course, we know that we have to look in certain directions to see out of our galaxy without having our vision ended by the interstellar dust, but remember that we are ignoring inhomogeneities or lack of isotropy on this small scale.

The explanations of the universe that astronomers now accept are known as *big-bang* theories. Basically, these cosmologies say that once upon a time there was a great big bang that began the universe. From that instant on, the universe expanded, and as the galaxies formed they shared in the expansion. The big-bang theories satisfy the cosmological principle.

Many students ask whether the fact that there was a big bang means that there was a center of the universe from which everything expanded. The answer is no; first of all, the big bang may have been the creation of space itself. Furthermore, the matter of this primordial cosmic egg was everywhere at once. There may be an infinite amount of matter in the universe, so it is possible that at the big bang an infinite amount of matter was compressed to an infinite density while taking up all space. We would have to imagine our expanding raisin cake extending infinitely in all directions, with no edge.

Consider a two-dimensional analogy to a universal expansion: the surface of a rubber balloon covered with polka dots. (Though the balloon is three-dimensional, its surface has only two dimensions and we consider only its surface.) Let us consider the view if we are sitting on one dot. As the balloon is blown up, all the other dots seem to recede from us. No matter which dot we are on, all the other dots seem to

"Big bang" is the technical as well as the popular name for these cosmologies; professionals write about the big bang in the scientific journals. The theories are based on Einstein's general theory of relativity.

Remember that the galaxies themselves are not expanding; the stars in a galaxy don't tend to move away from each other.

Figure 16–3 To trace back the growth of our universe, we would like to know the rate at which its rate of expansion is changing. Big-bang models of the universe are shown; the vertical axis represents a scale factor, *R*, that represents some measure of distances and how they change as a function of time, *t*. The universe could be open and expand forever, or be closed and begin to contract again. If it is closed, we do not know whether it would oscillate or whether we are in the only cycle of expansion or contraction it will ever undergo. We will see later that current evidence favors the open model.

Astronomers use a quantity called the *deceleration parameter*, which is given the symbol q_0, to describe how fast the expansion is slowing down. $q_0 = \frac{1}{2}$ marks the dividing line between an open universe (q_0 less than $\frac{1}{2}$, including 0) and a closed universe (q_0 greater than $\frac{1}{2}$, including 1).

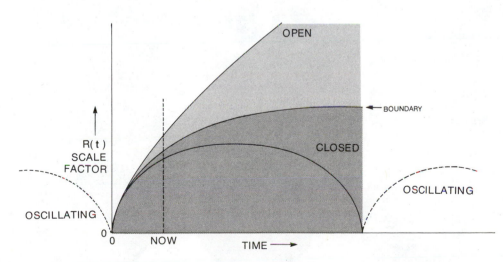

recede. The **surface** of the balloon is two-dimensional yet has no edge. Even if we go infinitely far in any direction, we never reach a boundary. (Remember, Columbus did not fall off the edge of the earth.) There is no center to the surface from which all dots are actually expanding. It is much more difficult for most of us to visualize a three-dimensional situation like an infinite raisin cake as in Section 14.4a (or, including time as a fourth dimension, a four-dimensional space-time). Yet the above analogy is valid, because our universe can expand uniformly yet have no center to the expansion.

What will happen in the future? One possibility is that the universe will continue to expand forever. This case is called an *open universe*. It corresponds to the case where the universe is infinite. The other possibility is that at some time in the future the universe will stop expanding and will begin to contract. This case is called a *closed universe* (Fig. 16–3), and corresponds to the case where the universe is finite (though it may still have no boundary, just as we can continue straight ahead forever on the surface of a balloon). In a closed universe, we would eventually reach a situation that we might call a "big crunch."

The open and closed universe can be considered in an analogy from geometry

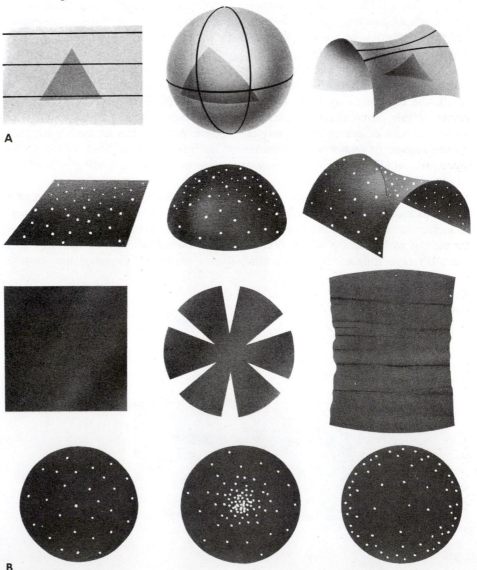

Figure 16–4 Two-dimensional analogues to three-dimensional space. *(A)* A closed universe is like the surface of a sphere in several ways, including the fact that great circles (full-diameters) always intersect and cannot be parallel. An open universe is like the surface of a saddle in several ways, including the fact that an infinite number of parallel lines can be drawn through the same point. A flat universe is in between. These concepts of geometry follow to the geometrical calculation of the volume of a shell. *(B)* Consider radio sources uniformly distributed throughout space *(top row),* and count how the number changes with distance from us. If we use the formula for volume from a flat space, we would essentially be flattening out the curvature *(middle row).* The result for the different curvatures is shown at bottom. Though such counts were made for a long time, it now seems that we are seeing so far out into space that such arguments tell us more about how radio sources evolve over time than they do about cosmology.

(Fig. 16–4). Einstein's general theory of relativity predicts that space itself could show properties of being curved, just as the surface of a saddle or the surface of a sphere is fundamentally curved. On none of these surfaces, for example, can we draw straight lines that remain parallel out to infinity, that is, never crossing and always remaining the same distance from each other. Is space in the universe curved positively like a sphere, curved negatively like a saddle, or flat? We do not know. The closed universe would be like the surface of a sphere, which bends inward, while the open universe would be like the surface of a saddle, which bends outward.

In principle, one could decide whether the universe is open or closed by counting the number of radio sources or of other galaxies in shells of constant thickness going outward from our galaxy, since the different geometries imply different changes in the number of sources. But the fact that radio sources and other galaxies may have been brighter or fainter far back in time complicates matters too much. Such studies really tell us mostly about the evolution of radio sources and of other galaxies.

What was present before the big bang? There is no real way to answer this question. For one thing, we can say that time began at the big bang, and that it is meaningless to talk about "before" the big bang because time didn't exist. We do not now think that the universe can remain in a static condition, so it seems unlikely that the universe was always just sitting there in an infinitely compressed state, whatever "always" means.

Of course, these possibilities don't answer the question of why the big bang happened. It is possible that there had been a prior big bang and then a recollapse, and that our current big bang was one in an infinite series. This version of a closed-universe theory is called the *oscillating universe*. But the oscillating-universe theory doesn't really tell us anything about the origin of the universe because in this case there was no origin. The concepts discussed in these paragraphs will not be easy to digest. They may take hours, years, or a lifetime to come to terms with.

If we consider Hubble's law with the current value for Hubble's constant—and there is good evidence that Hubble's constant is the same for relatively nearby galaxies as it is for very distant galaxies—we can extrapolate backward in time (Fig. 16–5). We simply calculate when the big bang would have had to take place for the universe to have reached its current state at its current rate of expansion. This calculation indicates that, for a Hubble constant of 50 km/sec/Mpc, the universe is somewhere between 13 billion and 20 billion years old. The range of values indicates the uncertainty in the calculation of how much the effect of gravity would have by now slowed down the expansion of the universe.

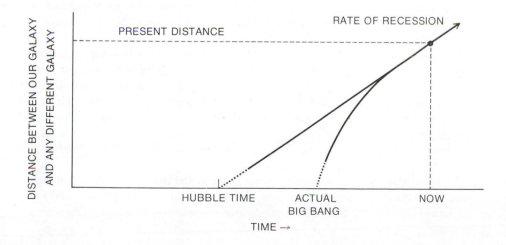

Figure 16–5 If we could ignore the effect of gravity, then we could trace back in time very simply; the Hubble time corresponds to the inverse of the Hubble constant ($1/H_0$). Actually, gravity has been slowing down the expansion. The vertical axis again represents some scale factor.

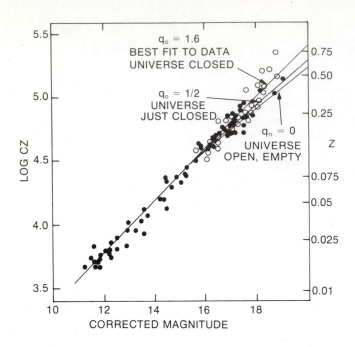

Figure 16–6 This Hubble diagram, plotted in terms of redshift and magnitude, shows faint clusters of galaxies. Measurements of J. Kristian, A. Sandage, and J. Gunn are shown as open circles, added to the dots of prior observations. The horizontal axis shows the magnitude, which is a measure of distance. The magnitude has been corrected for various subtle effects, such as interstellar absorption and the appearance of a galaxy to observing equipment.

In principle, we can determine the future of the universe from the slight deviations at upper right of the curves from a straight line. If the curve appears to the upper left of a straight line, the expansion is slowing down. If the curve falls far enough to the left, then the universe is closed, finite, and will eventually begin to contract. If it is not so far to the left, then the universe is open, infinite, and will expand forever. Even though the data curve slightly upward from a straight line, the data are currently not sufficient to definitely decide between these possibilities. A special difficulty that prevents us from doing so is the effect of the galaxies' evolution, which is very uncertain.

A larger Hubble constant would correspond to a younger age, given the assumption that the universe is expanding at a constant or at a decreasing rate. (Some astronomers consider a ''cosmological constant'' in Einstein's equations—a constant Einstein himself first included and then dropped. Such a cosmological constant would have the universe expand at an increasing rate. Such an expansion might be necessary if Hubble's constant is 100, since we know that the globular clusters would otherwise be older than the universe, which is obviously impossible.)

Open or closed, the universe will last at least another forty billion years, so we have nothing to worry about for the immediate future. Nevertheless, the study of the future of the universe is an exceptionally interesting investigation. Some of the methods of tackling this question involve measuring the amount of mass in the universe and thus the amount of gravity (as we will discuss in Section 17.2). Other methods of determining our destiny involve looking at the most distant detectable objects to see if any deviation from Hubble's law can be determined. This investigation of distant bodies has been carried out many times, but a deviation from the straight-line relation between velocity and distance known as Hubble's law has not been found conclusively. The deviations have always been within the uncertainty of the measurement. One major uncertainty comes from the fact that the best independent measure of the distances to the farthest galaxies comes from their brightnesses (Fig. 16–6).

We now realize that the farthest galaxies, from which we had hoped to best determine a deviation from Hubble's law, may well have been very different long ago when they emitted the light we are now receiving. Thus we cannot assume that even such basic properties as their size and brightness were similar then to their size and brightness now. Since determining the distance to these distant galaxies depends on understanding their properties, it now seems that such considerations of galaxy evolution are too uncertain to allow us to determine in this way whether the universe is open or closed.

The assumption that the universe is homogeneous and isotropic is always subject to question. Although inhomogeneities on a scale up to clusters of galaxies and superclusters are known, it is usually assumed that the universe is homogeneous on a larger scale.

As Beatrice Tinsley's research pointed out, how galaxies evolve with time is not known. Were they brighter or fainter long ago? One possibility is that galaxies grow fainter with age. There is also evidence that some giant galaxies in clusters may be consuming other galaxies in those clusters. This galactic cannibalism (Section 14.4) would make the remaining galaxy become brighter.

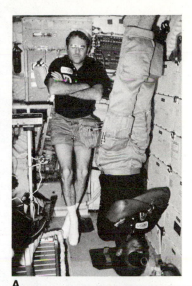

A

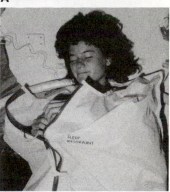

B

Big-bang theories actually arise as solutions to a set of equations that Einstein advanced as part of his general theory of relativity. Early solutions by Einstein himself and others (Fig. 16–9) were not valid. We now use solutions worked out in the early 1920's by the Soviet mathematician Alexander Friedmann. The Belgian abbé Georges Lemaître independently discovered similar solutions.

*16.3 The Steady-State Theory

The cosmological principle, that the universe is homogeneous and isotropic, is very general in scope, but starting in the late 1940's three British scientists began investigating a principle that is even more general. Hermann Bondi, Thomas Gold, and Fred Hoyle considered what they called the *perfect cosmological principle:* the universe is not only homogeneous and isotropic in space but also **unchanging in time.** A criterion for science is that we must accept the simplest theory that agrees with all the observations, but it is a matter of personal preference whether the cosmological principle or the perfect cosmological principle is simpler.

The theory that follows from the perfect cosmological principle is called the *steady-state theory* (Fig. 16–10); it has certain philosophical differences from the big-bang cosmologies. For one thing, according to the steady-state theory, the universe never had a beginning and will never have an end. It always looked just about the way it does now and always will look that way.

The steady-state theory must be squared with the fact that the universe is expanding. How can the universe expand continually but not change in its overall ap-

Figure 16–7 In "free fall," the spacecraft is accelerating toward earth with the same acceleration that the astronauts have; they thus do not sense any gravity, since they are not pressed against anything. *(A)* Space-shuttle astronauts Richard H. Truly *(left)* and Guy Bluford *(right)* resting. *(B)* Space-shuttle astronaut Sally Ride in a sleep restraint, which keeps her from floating around the cabin.

Figure 16–8 A warped golf green is a two-dimensional analogy to curved space. Though the golf ball is rolling "straight," it appears to curve as Tom Watson putts here because the surface of the green is curved. Similarly, light travelling through curved space appears to curve even though it is travelling "straight." This test of the principle of equivalence has been verified for light at eclipses and for radio waves by observing quasars apparently passing near the sun.

Figure 16–9 Einstein and Willem de Sitter at the Mount Wilson Observatory in 1931. Einstein's and de Sitter's solutions to the equations of general relativity are not now used. Big-bang theories are based on solutions by Friedmann and Lemaître.

The rate at which new matter would have to be created works out to be only one hydrogen atom per cubic centimeter of space every 10^{15} years, equivalent to one thousand atoms of hydrogen per year in a volume the size of the Astrodome in Houston. This is far too small for us to be able to measure. The "law of conservation of mass-energy" would thus not be valid at this level, in the steady-state theory.

Figure 16–10 In the steady-state theory, as the dotted box at left expands to fill the full box at right, new matter is created to keep the density constant. In the picture, the four galaxies shown at left can all still be seen at right, but new galaxies have been added so that the number of galaxies inside the dotted box is about the same as it was before.

pearance? For the density of matter to remain constant, new matter must be created at the same rate that the expansion would decrease the density. Only in this way can the density remain the same.

The matter created in the steady-state theory is not simply matter that is being converted from energy by $E = mc^2$. No, this is **matter that is appearing out of nothing,** and is thus equivalent to energy appearing out of nothing.

For many years a debate raged between proponents of the big-bang theories and proponents of the steady-state theory. The evidence that came in—usually seeming to indicate that distant objects were somehow different from closer ones, which would show that the universe was evolving—seemed to favor the big-bang cosmologies over the steady-state theory. But none of this evidence was conclusive because alternative explanations for the data could be proposed or the steady-state theory itself could be modified (sometimes extensively) to be consistent with the discoveries.

The discovery of quasars provided some of the strongest evidence against the steady-state theory. The quasars are for the most part located far away in space, and so were more numerous at an earlier time. Thus something has been changing in the universe, and change is not acceptable in the steady-state theory. As the evidence that the quasars were indeed at these distances grew, the status of the steady-state theory diminished.

In the next section we shall discuss still stronger evidence that provided the crushing blow against the steady-state theory.

16.4 The Primordial Background Radiation

In 1965, a discovery of the greatest importance was made: radiation was detected that is most readily explained as a remnant of the big bang itself.

That much is easy to state, and if we have in fact discovered radiation from the big bang itself, then clearly the steady-state theory is discredited. But the explanation of just how the observed radiation is interpreted to come from the big bang is fairly technical and requires an understanding of abstract concepts. However, the sheer importance of the discovery warrants our taking the time to consider this matter thoroughly.

The discovery was made by Arno A. Penzias and Robert W. Wilson of the Bell Telephone Laboratories in New Jersey (Fig. 16–11). They were testing a radio telescope and receiver system (Fig. 16–12) to try to track down all possible sources of static. The discovery, in this way, parallels Jansky's discovery of radio emission from space, which marked the beginning of radio astronomy.

Penzias and Wilson were observing at a wavelength in the radio spectrum. After they had subtracted, from the static they observed, the contributions of all known

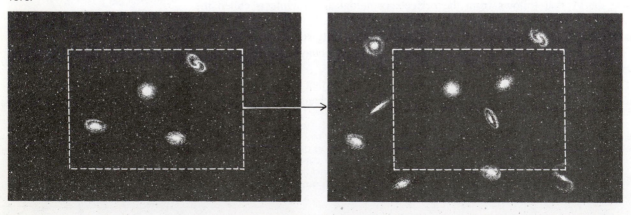

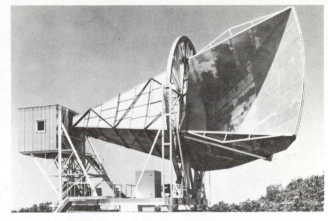

Figure 16–11

Figure 16–12

Figure 16–11 Arno Penzias (*left*) and Robert W. Wilson (*right*) with their antenna in the background. Penzias and Wilson won the 1978 Nobel Prize in Physics for their discovery of the cosmic background radiation.

Figure 16–12 The large horn-shaped antenna at the Bell Laboratories' space communication center in Holmdel, New Jersey. Penzias and Wilson found more radio noise than they expected at the wavelength of 7 cm at which they were observing. After they removed all possible sources of noise (by fixing faulty connections and loose antenna joints, and by removing "sticky white contributions" from nesting pigeons), a certain amount of radiation remained. It was the 3° background radiation.

sources, they were left with a residual signal that they could not explain. The signal was independent of the direction they looked, and did not vary with time of day or season of the year. The remaining signal corresponded to the very small amount of radiation that would be put out at that frequency by a black body at a temperature of only 3 K, 3° above absolute zero.

At the same time, Robert Dicke, P. J. E. Peebles, David Roll, and David Wilkinson at Princeton University had, coincidentally, predicted that radiation from the big bang should be detectable by radio telescopes. First, they concluded that radiation from the big bang permeated the entire universe, so that its present-day remnant should be coming equally from all directions. Second, they concluded that this radiation would have the spectrum of a black body, which just means that the amount of energy coming out at different wavelengths can be described by giving a temperature. Third, they predicted that though the temperature of the radiation was high at some time in the distant past, the radiation would now correspond to a black body at a particular very low temperature, only a few degrees above absolute zero (Fig. 16–13). If radiation could be found that came equally from all directions and had the low-temperature black-body spectrum that matched the prediction, then the theory would be confirmed.

The Princeton group continued the process of building their own receiver to observe at a different radio wavelength. They were soon able to measure the intensity at this wavelength, and found that it also corresponded to that of a black body at 3°. This tended to confirm the idea that the radiation was indeed from a black body, and thus that it resulted from the big bang.

Because the big bang took place simultaneously everywhere in the universe, radiation from the big bang filled the whole universe. The radiation thus has the property of being isotropic to a very high degree, that is, it is the same in any direction that we observe. It was generated all through the universe at the same time, so its remnant must seem to come from all around us now. The fact that the observed

We have discussed black bodies in Section 4.2. Basically, the emission from a black body follows Planck's law of radiation (Fig. 4–1) in that for a given temperature there is an equation that tells us the intensity of radiation at each wavelength. The key fact to remember about Planck's law is that specifying just one number—the temperature—is enough to define the whole Planck curve.

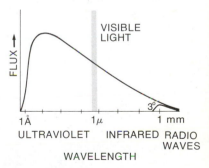

Figure 16–13 Planck curves for black bodies at different temperatures. Radiation from a 3° black body (the inner curve at lower right) peaks at very long wavelengths. On the other hand, radiation from a very hot black body (the upper curve) peaks at very short wavelengths.

It is tempting to think of black bodies as physical rather than conceptualized objects; we must fight that temptation in order to understand the radiation from the big bang. It is quite normal for us to try to visualize abstract concepts in concrete terms. Horatio, for example, speaks to Hamlet of "the morn, in russet mantle clad. . . ." Of course, the morn wasn't wearing any such thing; it has no particular shape to be clad.

radiation was highly isotropic was thus strong evidence that it came from the big bang. It also shows that the early universe was very homogeneous.

Most other mechanisms to account for the black-body radiation predict that the intensity of the radiation would vary slightly from direction to direction in the sky. For example, if the radiation originated in discrete sources, then we would expect the radiation to be most intense in the direction of the strongest sources.

16.4a The Origin of the Background Radiation

The leading models of the big bang consider a hot big bang. The temperature was billions upon billions of degrees in the fractions of a second following the beginning of time.

In the millennia right after the big bang, the universe was opaque. Photons did not travel very far before they were absorbed by hydrogen ions. After the photons were absorbed, they were soon reradiated. This recycling process results in black-body radiation corresponding to the temperature of the matter.

Gradually the universe cooled. After about a million years, when the temperature of the universe reached 3000 K, the temperature and density were sufficiently low for the hydrogen ions to combine with electrons to become hydrogen atoms. This *recombination* took place suddenly. Since hydrogen has mainly a spectrum of lines rather than a continuous spectrum and no electrons remained free, the gas suddenly began to absorb photons at only a few wavelengths instead of at all wavelengths. Thus from this time on, most photons could travel all across the universe without being absorbed by matter. The universe had become transparent.

Since matter rarely interacted with the radiation from that time to the present, it no longer continually recycled the radiation. We were thus left with the radiation at the temperature it had at the instant when the universe became transparent. This radiation is travelling through space forever. Observing it now is like studying a fossil. As the universe continues to expand, the radiation retains the shape of a Planck curve but the curve corresponds to cooler and cooler temperatures.

Actually, the Princeton group was not the first to predict that such radiation might be present. Many years earlier Ralph Alpher and Robert Herman (in 1948) and George Gamow (in 1953) had made similar predictions, but this earlier work was at first overlooked.

Since the radiation is present in all directions, no matter what we observe in the foreground, it is known as the *cosmic background radiation (CBR)*. In reference to its origin, it is also called the *primordial background radiation*.

In the above scenario, the universe has changed from a hot opaque place to its current cold, transparent state. Such a change is completely inconsistent with the steady-state theory.

16.4b The Temperature of the Background Radiation

Planck curves corresponding to cooler temperatures have lower intensities of radiation and have the peaks of their radiation shifted toward longer wavelengths. We have seen that radiation from the sun, which is 6000 K, peaks in the yellow-green and that radiation from a cool star of 3000 K peaks in the infrared. The universe, much cooler yet, peaks at still longer wavelengths: the peak of the black-body spectrum is at the dividing wavelength between infrared and radio waves. In a moment, we shall see just how cool the universe is.

Now, we recall that Penzias and Wilson measured a particular flux at the wavelength measured by their equipment. This flux corresponds to what a black body at a temperature of only 3 K would emit. Thus we speak of the universe's *3° background radiation**. Following the original measurement at Bell Labs and at Princeton, other groups soon measured values at other radio wavelengths.

---

*In the Système International, this would be called 3 K radiation, but we shall join the astronomical community in continuing to call it 3° radiation.

Unfortunately, though, for many years we were able to observe only at wavelengths longer than that of the peak, so only the right-hand half of the curve was defined. All the points corresponded to a temperature of approximately 3°, but still, a black-body curve does have a peak. It would have been more satisfying if some of the points measured lay on the left side of the peak. Points lying on the left side of the peak would prove conclusively that the radiation followed a black-body curve and was not caused by some other mechanism that could produce a straight line, or some other form that happened to mimic a 3° black-body curve in the limited region of the spectrum that had been studied.

Though it would have been desirable to measure a few points on the short-wavelength side of the black-body curve, astronomers were faced with a formidable opponent that frustrated their attempts to measure these points: the earth's atmosphere. Our atmosphere absorbs most radiation from the long infrared wavelengths.

In 1975, infrared observations were made from balloons that seem to have proved unequivocally that the radiation follows a black-body curve. More recent measurements appear in Figure 16–14.

So at present, astronomers consider it settled that radiation has been detected that could only have been produced in a big bang. Accepting this clearly rules out the steady-state theory. It means that some version of the big-bang theories must hold, though it doesn't settle the question of whether the universe will expand forever, or will eventually contract.

16.4c The Background Radiation as a Tool

Now that the spectrum has been measured so precisely on both sides of the peak of intensity, we have moved beyond the point of confirming that the radiation is black-body. We can now study its deviations from the general black-body curve and from isotropy.

A very slight difference in temperature—about one part in three thousand—has been measured from one particular direction in space to the opposite direction (Fig. 16–15). This *anisotropy* (deviation from isotropy) is what would result from the Doppler effect if our sun was moving at 350 km/s with respect to the background radiation. Since we know the velocity with which our sun is moving as it orbits the center of our galaxy, we can remove the effect of our sun's orbital motion. We deduce that our galaxy is moving 520 km/s relative to the background. Our galaxy is moving in roughly the direction of the constellation Hydra. Since some measure-

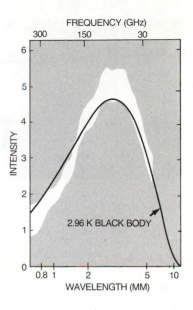

Figure 16–14 The balloon observations of D. Woody and P. L. Richards showing the agreement of the data *(white, unshaded region)* with the theoretical curve for a black body *(solid line)*. The radio observations we have discussed previously are all located in the lower right-hand portion of the graph.

The importance of the discovery of the background radiation cannot be overstressed. Dennis Sciama, the British cosmologist, put it succinctly by saying that up to 1965 we carried out all our calculations knowing just one fact: that the universe expanded according to Hubble's law. After 1965, he said, we had a second fact: the existence of the background radiation. That may be a bit oversimplified, but it is essentially true.

A

B

Figure 16–15 *(A)* The U-2 jet used to carry out observations of the anisotropy of the background radiation from an altitude of 20 km. *(B)* Radiometers to measure anisotropy at two different frequencies in the modified upper hatch of the U-2. The larger pair of horn antennas measures radiation at a wavelength of 0.9 mm; the smaller pair of horns measures radiation at a wavelength of 0.5 mm. Two horns are used for each wavelength because comparative measurements can be made more accurately than absolute measurements of the intensity.

Figure 16–16 An artist's conception of the COBE spacecraft, scheduled for launch in October 1987. The success of IRAS, in the same class of satellite, has been good for the Explorer program and thus for the prospects of funding COBE.

ments indicate that the Virgo Cluster of galaxies is moving at a much smaller velocity with respect to us, it may mean that the whole Virgo Cluster is moving in the direction that we derive from the measurements of anisotropy. We do not know any particular reason why our galaxy and the Virgo Cluster should have such a large velocity.

Observations that relate infrared brightnesses of galaxies with distances measured from radio studies give a result similar to that deduced from the anisotropy in the background radiation.

16.4d Future Studies of the Background Radiation

COBE (Fig. 16–16), **CO**smic **B**ackground **E**xplorer, is a NASA spacecraft now scheduled for launch in 1987 to study the background radiation. It should be able to study the spectrum over a wide wavelength range and measure how isotropic the radiation is, at four wavelengths. COBE should yield measurements 10 times more precise than the current measurements.

Summary and Outline

Olbers' paradox and its solutions (Section 16.1)
  The expansion of the universe diminishes energy in quanta
  We are looking back in time to before the formation of the stars
Big bang theory (Section 16.2)
  Cosmological principle: universe is homogeneous and isotropic
  Hubble's constant and the "age" of the universe—13 to 20 billion years
Steady-state theory (Section 16.3)
  Perfect cosmological principle: universe is homogeneous,

  isotropic, and unchanging in time
  Continuous creation of matter is implied
  Demise of the steady-state theory with discovery of quasars and 3° black-body radiation
Primordial background radiation (Section 16.4)
  Provides strong support for the big-bang theory
  Direct radio measurements were made but only on one side of the peak of the black-body curve.
  Infrared observations confirm peaked nature of the radiation
  Anisotropy indicates motion of the sun and the Virgo cluster

Key Words

cosmology, Olbers' paradox, cosmological principle, homogeneity, isotropy, big-bang theories, open universe, closed universe, oscillating universe, deceleration parameter, perfect cosmological principle*, steady state theory*, recombination, cosmic background radiation, primordial background radiation, 3° background radiation, anisotropy

*These words are found in an optional section.

Questions

1. Hindsight has allowed solutions to be found to Olbers' paradox. For example, knowing that the universe is expanding, we can come up with a solution. However, based on the reasoning in this chapter, do you think it is possible that scientists might have used Olbers' paradox to reach the conclusion that the universe is expanding before it was determined observationally? Explain.

†2. Our universe is about 10^{10} years old. If we were riding

on a light beam emitted now, how long would we have to wait on the average to arrive on a star, given that the average line of sight ends 10^{24} light years away? (Assume for the calculation that the stars are still there.) The question demonstrates the length of this time span. The current 10^{10}-year age of the universe is what percentage of this time?

†3. Using the formula for the energy of a photon, how many times less energy does a photon corresponding to a wavelength of 1 mm (the peak of the cosmic background curve) have compared with a photon corresponding to 1 Å in

†This indicates a question requiring a numerical solution.

wavelength (which would have been emitted by the background in the early seconds of the universe)?

4. We say in the chapter that the universe is homogeneous on a large scale. Referring to the discussions in Chapter 14, what is the largest scale on which the universe does not seem to appear homogeneous?

5. List observational evidence in favor of and against each of the following: (a) the big-bang theory, and (b) the steady-state theory.

6. What is the relation of Einstein's general theory of relativity to the big bang?

7. What is "perfect" about the perfect cosmological principle?

8. If galaxies and radio sources increase in luminosity as they age, then when we look to great distances, we are seeing them when they were less intense than they are now. We thus cannot compare them directly to similar nearby galaxies or radio sources. If this increase in luminosity with time is a valid assumption, does it tend to make the universe seem to decelerate at a greater or lesser rate than the actual rate of deceleration? Explain.

†9. For a Hubble constant of 50 km/s/Mpc, show how you calculate the Hubble time, the age of the universe ignoring the effect of gravity. (Hint: Take $1/H_0$, and simplify units so that only units of time are left.)

†10. What is the Hubble time if the Hubble constant is 100 km/s/Mpc? Compare with the answer from question 9. Comment on the additional effect that gravity would have.

11. In actuality, if current interpretations are correct, the 3° background radiation is only indirectly the remnant of the big bang, but is directly the remnant of an "event" in the early universe. What event was that?

†12. Apply Wien's displacement law to the solar photospheric temperature and spectral peak in order to show where the spectrum of a 3° black body peaks.

13. What does the anisotropy of the background radiation tell us?

†14. By how much would the H-alpha line (whose rest wavelength is 6563 Å) be shifted by the sun's velocity of 350 km/s with respect to the background radiation?

Topics for Discussion

1. Which is more appealing to you: the cosmological principle or the perfect cosmological principle? Discuss why we should adopt one or the other as the basis of our cosmological theory.

2. Who do you think deserved the Nobel Prize: the scientists who first observed the background radiation observationally without having a theory, the scientists who had made rough predictions decades earlier, or the scientists who made theoretical models but hadn't yet carried out their planned observations? Note that the Nobel Prize rules allow the award to go to no more than three individuals.

The central part of the cluster of galaxies that we see lying far beyond the constellation Hercules. Studies of the motions of the galaxies show that over 10 times more mass is present than is visible.

The Past and Future of the Universe   17

Aims: **To study the first second of time in the universe, how the elements were formed, and what the future of the universe will be**

How did the tremendous explosion we call the big bang result in the universe we now know, with galaxies and stars and planets and people and flowers? Obviously, many complex stages of formation have taken place, and what was torn asunder at the beginning of time has now taken the form of an organized system.

We can trace the expansion backward in time and calculate how long ago all the matter we see would have been so collapsed to a point. The result of this calculation is that the big bang took place some 13 to 20 billion years ago. It is difficult to comprehend that we can meaningfully talk about the first few **seconds** of that time so long ago. But we can indeed set up sets of equations that satisfy the physical laws we have derived, and make computer simulations and calculations that we think tell us a lot about what happened right after the origin of the universe. In this chapter we will go back in time beyond the point where the background radiation was set free to travel through the universe. This phase of time is known as the *early universe*.

Our knowledge of the structure of the universe and of the nuclear and other particles in it seems to be able to take us back to 10^{-43} second. Before 10^{-43} second, the universe was so compressed that not only the laws of general relativity but also those of quantum mechanics have to be taken into account, and we are not at present able to do so simultaneously. So we can't even say 10^{-43} second after what, since we don't really know that there was a zero of time. But we measure time from the instant at which everything would have been together, extrapolating the measurable part of the expansion backwards.

In this chapter, we will first discuss how the chemical elements were formed. Then we will go on to discuss how our studies of the elements enable us to make predictions using the big-bang theories that have been current for the last decades. Next, we will describe our knowledge of the basic physical forces that govern the universe, and what they tell us about the universe's first second of time. Finally, we will see how our understanding of this early time has given us a new picture of the universe's evolution, and leads to a different understanding of our fate.

17.1 The Creation of the Elements

Modern cosmology is merging with *particle physics*—the studies of the particles inside atoms, which are often carried out with giant atom smashers. As we push farther back toward the beginning of time in our understanding of the universe, it has become necessary to understand what particles were around and how they interacted with each other.

One striking discovery has been that for every subatomic particle, there is a corresponding *antiparticle*. This antiparticle has the same mass as its particle, but is opposite in all other properties. For example, an antiproton has the same mass as a proton, but has negative charge instead of positive charge. Some particles have a property called spin; their antiparticles spin in the opposite direction. Antiparticles together make up *antimatter*. If a particle and its antiparticle meet each other, they annihilate each other; their total mass is 100 per cent transformed into energy in the amount $E = mc^2$.

We know that our universe, up to at least the scale of a cluster of galaxies, is made of matter rather than antimatter. If there were a substantial amount of antimatter anywhere, it would meet and annihilate some matter, and we would see the resulting energy as gamma rays. We do not detect enough gamma rays for this annihilation to be taking place commonly. And there is enough matter between the planets, between the stars, and between the galaxies to provide a continuous chain of matter between us and the Local Group of galaxies, showing that the Local Group is made of matter rather than antimatter. The chain probably extends to the Local Supercluster as well.

Baryons are a kind of subatomic particle that make up nuclei; protons and neutrons are the most familiar examples of baryons. Up to about 10^{-35} second after the big bang, there was a balance between baryons and their antiparticles on the one hand and photons (which can be thought of as particles of light) on the other. Up to this time, the number of particles and antiparticles were the same, and there was a continual annihilation of the two; each time a particle and its antiparticle met, their mass was transformed into energy. Similarly, particle/antiparticle pairs (including protons and antiprotons) kept forming out of the energy that was available.

Since matter and antimatter should have been formed in equal quantities, we must explain why our universe is now made almost entirely out of matter. Only recently have scientists found a reasonable (though speculative) explanation. Though few nuclear reactions produce different amounts of matter and antimatter, enough of these imbalanced reactions could have taken place at a sufficient rate in the early universe to provide a slight imbalance: 100,000,001 particles for each 100,000,000 antiparticles. Then all the antimatter annihilated an equal amount of matter. The relatively small residuum of matter left over is the matter that we find in our universe today! The annihilations created 100,000,000 photons for each baryon, a ratio that—as we can measure from the background radiation—holds true today. The 1980 Nobel Prize in Physics went to the scientists who first detected experimentally (Fig. 17–1) that particles can occasionally decay asymmetrically into matter and antimatter.

Following the annihilation of most of the protons and essentially all of the antiprotons, the universe had expanded enough so that less energy was available in any

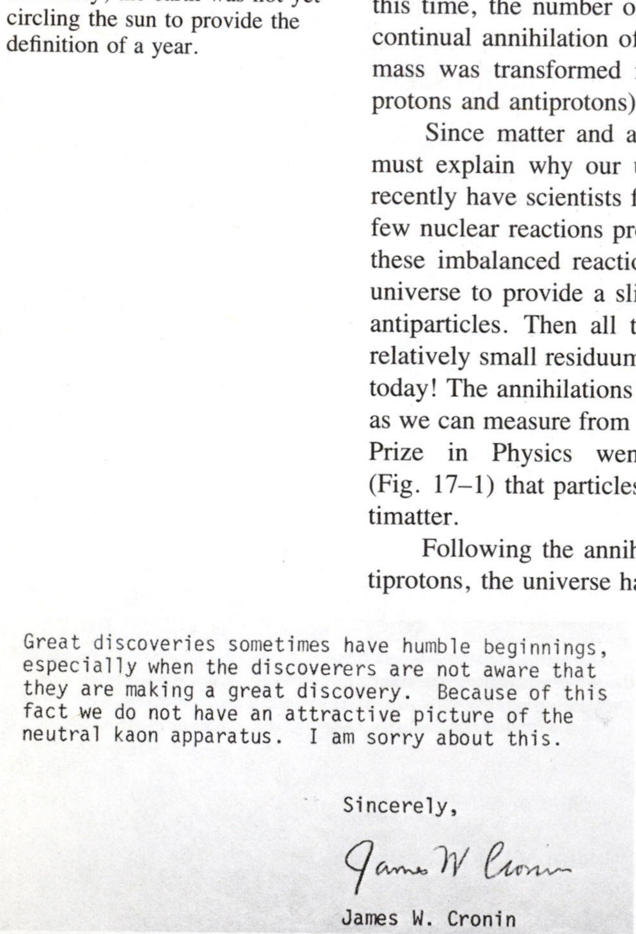

We have to define time in terms of the properties of atoms; obviously, the earth was not yet circling the sun to provide the definition of a year.

Great discoveries sometimes have humble beginnings, especially when the discoverers are not aware that they are making a great discovery. Because of this fact we do not have an attractive picture of the neutral kaon apparatus. I am sorry about this.

Sincerely,

James W Cronin

James W. Cronin

Figure 17–1 James Cronin and Val Fitch received the 1980 Nobel Prize in Physics for their experiment that showed an asymmetry in the decay of a certain kind of elementary particle. This result may point the way to the explanation of why there is now more matter than antimatter in the universe.

given place. Since $E = mc^2$, no longer was enough energy available to form protons and antiprotons. But until about a second has passed, there was still enough energy to form lighter particles like electrons and positrons (anti-electrons). Then the electrons and positrons annihilated each other. We were left with a sea of hot radiation, which dominated the universe, and which we now detect as the background radiation. At that time, unlike the present, the photons each had so much energy that most of the universe's energy was in the form of photons. Since neutrinos interact so rarely, they may not have been annihilated and many presumably remain from this era.

The results of computer calculations of what happened after the first few seconds of the universe are less speculative than the reason why we wound up with a universe made of matter. According to these calculations, the universe's fantastically high density and temperature continued to diminish. After about 5 seconds, the universe had cooled to a few billion degrees. Only simple kinds of matter—protons, neutrons, electrons, neutrinos, and photons—were present at this time. The number of protons and neutrons was relatively small, but the number grew so that these particles had a relatively larger share of the universe's energy as the temperature continued to drop.

After about a hundred seconds, the temperature dropped to a billion degrees (which is low enough for a deuterium nucleus to hold together). The protons and neutrons began to combine into heavier assemblages—the nuclei of the heavier isotopes of hydrogen and elements like helium and lithium. The study of the formation of the elements is called *nucleosynthesis*. The standard model of how the elements formed in the big bang (Fig. 17–2) was worked out by William A. Fowler of Caltech and Robert V. Wagoner, now of Stanford.

The first nuclear amalgam to form was simply a proton and neutron together. We call this a *deuteron*; it is the nucleus of deuterium, an isotope of hydrogen (see Fig. 8–6). Then two protons and a neutron could combine to form the nucleus of a helium isotope, and then another neutron could join to form ordinary helium. Within minutes, the temperature dropped to 100 million degrees, too low for nuclear reactions to continue. Nucleosynthesis stopped, with about 25 per cent of the mass of the universe in the form of helium. Nearly all the rest was and is hydrogen.

Some of the earliest quantitative work on nucleosynthesis in the big bang was described in an article published under the names of Ralph Alpher, Hans Bethe, and George Gamow in 1948. Actually, Alpher and Gamow did the work, and just for fun included Bethe's name in the list of authors so that the names would sound like the first three letters of the Greek alphabet: alpha, beta, gamma. These letters seemed particularly appropriate for an article about the beginning of the universe.

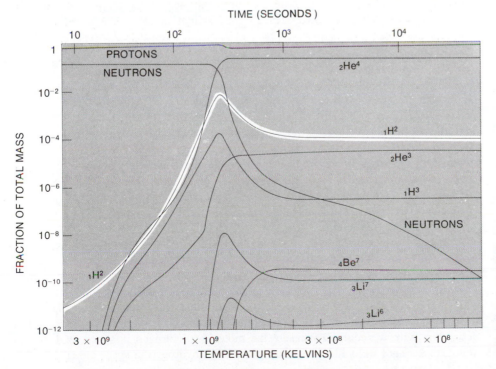

Figure 17–2 A theoretical model, worked out by Robert V. Wagoner of Stanford and William A. Fowler of Caltech, that shows the changing relative abundances in the first minutes after the big bang. Time is shown on the top axis and the corresponding temperature is shown on the bottom axis.

To test this theory, astronomers have studied the distribution of helium around the universe, to see if it tends to be approximately this percentage everywhere. But helium is very difficult to observe when it is at a relatively low temperature. Furthermore, even when we can observe helium in stars we are observing only the surface layers, which do not necessarily have the same abundances of the elements as the interiors of the stars. "The helium problem"—whether the helium abundance is constant throughout the universe—has been extensively studied. Though the investigation is difficult, the helium is apparently uniformly distributed.

Only the lightest elements were formed in the early stages of the universe. The nuclear state of mass 5 (the lithium isotope with three protons and two neutrons, or else the helium isotope with two protons and three neutrons), which can be reached by adding a proton to a helium-4 nucleus, is not stable; it doesn't hold together long enough to be built upon by the addition of another proton to form still heavier nuclei. If this gap is bridged, perhaps by fusing two helium-3 nuclei, another unstable nuclear state is mass 8. As a result, heavier nuclei are not formed in the early stages of the universe. Only hydrogen (and deuterium) and helium (and perhaps traces of low-mass isotopes such as those of lithium with nuclear mass of 6 or 7) would be formed in the big bang.

The heavier elements must, then, have been formed at times after the big bang. E. Margaret Burbidge, Geoffrey Burbidge, William A. Fowler, and Fred Hoyle showed in 1957 how both heavier elements and additional amounts of lighter elements can be synthesized in stars. In a stellar interior, processes such as the triple-alpha process (Section 8.4) can get past these mass gaps. Elements are also synthesized in supernovae (Section 10.2).

At present, as a result of their work, we believe that element formation took place in two stages. First, light elements were formed soon after the big bang. Later, the heavier elements and additional amounts of most of the lighter elements were formed in stars or in stellar explosions.

The theoretical calculations of the formation of the elements use values measured in the laboratory for the rate at which particles and nuclei interact with each other. For his role in making these measurements and in the theoretical work on nucleosynthesis, William Fowler (Fig. 17–3) shared in the 1983 Nobel Prize in Physics.

17.2 The Future of the Universe

How can we predict how the universe will evolve in the distant future? We know that for the present it is expanding, but will that always continue? We have seen that the steady-state theory now seems discredited, so let us discuss the alternatives that are predicted by different versions of standard big-bang cosmologies.

The basic question is whether there is enough gravity in the universe to overcome the expansion. If gravity is strong enough, then the expansion will gradually stop, and a contraction will begin. If gravity is not strong enough, then the rate of expansion might slow, but the universe would continue to expand forever, just as a rocket sent up from Cape Canaveral will never fall back to earth if it is launched with a high enough velocity.

To assess the amount of gravity, we must determine the average mass in a given volume of space, that is, the density. It might seem that to find the mass in a given volume we need only count up the objects in that volume: one hundred billion stars plus umpteen billion atoms of hydrogen plus so much interstellar dust, etc. But there are severe limitations to this method, for many kinds of mass are invisible to us. How much matter is in black holes or in the form of neutrinos, for example? Until a few years ago, we couldn't measure the molecular hydrogen in space, and until a few decades ago, we couldn't measure the atomic hydrogen in space either.

Figure 17–3 William A. Fowler, shortly after receiving the 1983 Nobel Prize in Physics for his theoretical and experimental work on nucleosynthesis.

Still another place mass could be hidden is connected to the fact that an intergalactic medium would be hard to detect if it were hot enough. Above a certain temperature, the matter would be almost entirely ionized, and no 21-cm radiation would be emitted. A hot gas, however, would radiate x-rays, so through x-ray observations we can set limits on the amount of hot intergalactic gas that can be present.

We must turn to methods that assess the amount of mass by effects that don't depend on the visibility of the mass. All mass has gravity, for example, so we are led to study the gravitational attraction on a large scale.

17.2a The Missing-Mass Problem

One place to investigate large-scale gravitational attractions is in clusters of galaxies. Clusters of galaxies appear to be large-scale stable configurations that have lasted a long time. But since galaxies have random velocities in various directions with respect to the center of the mass of the cluster, why don't the galaxies escape?

If we assume that the clusters of galaxies are stable in that the individual galaxies don't disperse, we can calculate the amount of gravity that must be present to keep the galaxies bound. Knowing the amount of gravity in turn allows us to calculate the mass that is causing this gravity. When this supposedly simple calculation is carried out for the Virgo Cluster, it turns out that there should be fifty times more mass present than is observed; 98 per cent of the mass expected is not found. This deficiency is called the *missing-mass problem*.

17.2b Deuterium and Cosmology

Perhaps the major method now being used to assess the density of the universe concerns itself with the abundance of the light elements. The light elements incude hydrogen, helium, lithium, beryllium, and boron. Since these light elements were formed soon after the big bang (as we saw in Section 17.1), they tell us about conditions at the time of their formation. If we can find what the density was then, we can use our knowledge of the rate that the universe has been expanding to determine what the density is now.

But somehow we must distinguish between the amount of these elements that was formed in the big bang, and the amount subsequently formed in stars. This consideration complicates the calculations for helium, for example, because helium formed in stellar interiors and then spewed out in supernovae has been added to the primordial helium.

Fortunately, the deuterium isotope of hydrogen is free of this complication. We do not think that any is formed in stars, so all the deuterium now in existence was formed at the time of the big bang.

Deuterium has a second property that makes it an important probe of the conditions that existed in the first fifteen minutes after the big bang, when the deuterium was formed. The amount of deuterium that is formed is particularly sensitive to the density of matter at the time of formation. A slight variation in the primordial density makes a larger change in the deuterium abundance than it does in the abundance of other isotopes. We measure the amount of deuterium relative to the amount of hydrogen.

Why is the ratio of deuterium to (ordinary) hydrogen sensitive to density? Deuterium very easily combines with an additional neutron. This combination of one proton and two neutrons is another hydrogen isotope (tritium). The second neutron quickly decays into a proton, leaving us with a combination of two protons and one neutron, which is an isotope of helium. If the universe was very dense in its first few

Physicists and astronomers were excited in 1980 by a report of an experiment that indicated indirectly that neutrinos may have mass. The universe contains so many neutrinos that if neutrinos have mass, they could "close" the universe. But experimental verification that neutrinos have mass has not yet been forthcoming. If neutrinos have mass, they could be the invisible matter that makes up the galactic corona (Section 12.2).

The *missing-mass problem* is the discrepancy found when the mass derived from consideration of the motions of galaxies in clusters is compared with the mass that we can observe. It is really more a "missing-light problem"; the mass must be there.

Deuterium, "heavy hydrogen," contains one proton and one neutron in its nucleus instead of just the single proton of ordinary hydrogen. We are most familiar with deuterium as a constituent of "heavy water," HDO, which is used in atomic reactors. Deuterium is potentially an important constituent of the fusion process we are trying to harness to provide energy on earth.

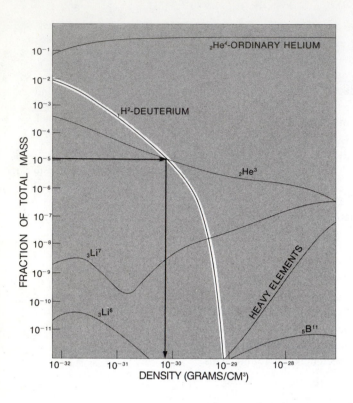

Figure 17–4 The horizontal axis shows the current cosmic density of matter. From our knowledge of the approximate rate of expansion of the universe, we can deduce what the density was long ago. The abundance of deuterium, outlined in white, is particularly sensitive to the time when the deuterium was formed. Thus present-day observations of the deuterium abundance tell us what the cosmic density is, by following the arrow on the graph.

minutes, then it was easy for the deuterium to meet up with neutrons, and almost all the deuterium "cooked" into helium. If, on the other hand, the density of the universe was low, then most of the deuterium that was formed still survives.

The result of theoretical calculations showing the relation of the amount of surviving deuterium and the density of the universe is shown in Figure 17–4. This graph can be used to find the density that would have been present soon after the big bang, if the ratio of deuterium to hydrogen can be determined observationally.

Unfortunately, deuterium is very difficult to observe. It has no spectral lines accessible to optical observation. Deuterium makes up one part in 6600 of the hydrogen in ordinary seawater, but it was not known how this related to the cosmic abundance of deuterium.

It was not until 1972 that anybody succeeded in detecting deuterium in interstellar space. Diego A. Cesarsky, Alan T. Moffet, and I observed a very faint absorption feature that, if real, is the absorption line caused by the spin-flip transition of deuterium (Section 13.4). At the same time, Penzias and Wilson were discovering deuterium as one of the constituents of a "deuterated" molecule, DCN, in the cloud of gas and dust in the constellation Orion. It is difficult to find a deuterium abundance from such molecular observations because the processes of interstellar molecular formation are not well understood. Penzias and Wilson have since mapped the strength of this DCN emission line in several molecular clouds in our galaxy. The amount of DCN seems to be less in the direction of the center of the galaxy than it is in other directions, probably because there are more stars in the direction consuming the deuterium.

Later on, the Lyman lines of deuterium were detected in absorption by the Princeton experiment aboard the Copernicus Orbiting Astronomical Observatory. They looked in the direction of a nearby star to try to detect an absorption line caused by deuterium between the star and the earth. The Copernicus and IUE satellites can study, however, only the nearest few hundred parsecs to us, 5 per cent of the distance to the galactic center. The Copernicus results (Fig. 17–5) seemed to have established

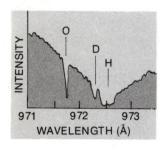

Figure 17–5 Observations from the Copernicus satellite of the Lyman (gamma) lines of interstellar deuterium and normal hydrogen. The broad absorption that takes up the whole graph is from hydrogen; its center is marked with an *H*. The corresponding deuterium line is the narrow dip marked with a *D*. It is narrower than the overall hydrogen line because the abundance of deuterium is less than that of hydrogen. In some observations, Doppler-shifted radiation from a small cloud of *H* has been confused with *D*. A narrow oxygen line (marked with an *O*) also appears.

that the ratio of deuterium to hydrogen is 14 parts per million in this region, but it now seems that at least some of the apparent deuterium lines were, in fact, Doppler-shifted hydrogen. A restudy is going on and my group is planning new radio observations.

Studies of deuterium thus far agree that the amount of deuterium is such that there is not, and was not, enough mass to reverse the expansion of the universe. The universe is apparently open. The results indicate, thus, that the universe will **not** fall back on itself in a big crunch.

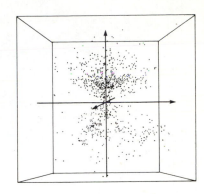

17.2c Galaxies, Quasars, and Cosmology

Even though over the years optical searches for deviations from Hubble's law have tended to favor a closed universe, new optical studies by the same group of people give a different result. Allan Sandage of the Mt. Wilson and Las Campanas Observatories and Gustav Tammann of the University of Basel in Switzerland have long studied the distances to galaxies. Joined by Amos Yahil of SUNY at Stony Brook, they have now analyzed the motion of nearby galaxies (Fig. 17–6). They tried to see if there is enough mass in concentrations like the Virgo cluster to distort the uniform outward flow of galaxies that Hubble's law predicts. The results indicate that there is not enough mass present, and so that the universe is open.

One of the latest contributions comes from space observations of x-rays. A diffuse background of x-ray emission had been discovered, and HEAO-1 had measured its spectrum. The shape of the spectrum had suggested that a lot of hot previously undiscovered intergalactic material was present. There might have been enough to "close the universe," that is, to provide enough gravity to make the universe closed.

But when the Einstein Observatory was launched, and was able to examine the x-ray background in detail, this idea changed. Riccardo Giacconi and his colleagues showed that much of this background really resulted from discrete quasars that could not be resolved by the earlier satellites' instruments. Long exposures on arbitrarily chosen fields that were apparently blank on shorter exposures revealed that faint quasars are present (Fig. 17–7). Thus, much of this background really resulted from discrete quasars that could not be resolved by the instruments on earlier satellites.

Figure 17–6 The group of nearby galaxies under study. The box is at 75 Mpc. Note the large complex of galaxies in the upper part of the diagram. They surround the Virgo Cluster, and extend all the way down to the Local Group, which is plotted at the center. Yahil, Sandage, and Tammann analyzed the rate at which the nearby galaxies are slowing down because of the gravity from the Virgo complex of galaxies, independent of whether the mass was in the galaxies themselves or in intergalactic matter.

Figure 17–7 A 12-hour exposure with the Einstein Observatory revealed new sources where none had been known before. The x-ray image *(left)* can be compared with an optical plate obtained with the Palomar Schmidt telescope *(right)*. Three sources are apparent. The upper source corresponds to a previously unknown quasar of magnitude 19.8 and $z = 0.5$. The source to the right corresponds to a previously unknown quasar of magnitude 17.8 and $z = 1.96$. The source at the left corresponds to an ordinary star of spectral type G0 and magnitude 13. The optical image includes objects down to 22nd magnitude. (X-ray image from Riccardo Giacconi and colleagues at the Harvard-Smithsonian Center for Astrophysics. Optical image from Wallace Sargent and Charles Kowal at the Palomar Observatory of Caltech.)

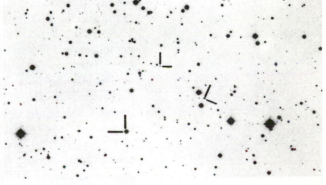

Emission from these faint, distant quasars seemed to be enough to provide much or all of the total x-ray background. So the need for otherwise invisible matter to provide the x-ray background seemed to have been reduced. Without this matter the universe again seemed open. But Bruce Margon of the University of Washington with colleagues at Columbia and UCLA more recently discovered that quasars found in this way were not typical ones. Their analysis showed that the contribution of quasars to the x-ray background, when they considered the full actual range of types of quasars, may be small. So the question of what causes the x-ray background is still open.

Studying the ages of galaxies and objects in them can also help foretell the fate of the universe. Globular clusters are the oldest objects in our galaxy; we have seen in Section 6.4 how we tell their ages from the position of turnoff stars in an H-R diagram. Our knowledge is limited by uncertainties in the abundances of heavy elements and in the amount of interstellar reddening. A new study, which included analyzing the abundances and the reddening, detected a cluster at least 16 billion years old. The universe can be that old only if Hubble's constant is less than about 60 and if the density is low (and thus the universe is open). The result is not compatible for any density with a higher H_0; the density could be high enough to close the universe only for H_0 unrealistically low (40 or less).

17.2d Neutrinos

Neutrinos (Section 8.6) have the very interesting property of travelling very rapidly; indeed, we have long thought that they always travel at the speed of light. Now, most matter cannot travel at the speed of light according to Einstein's special theory of relativity, because its mass gets larger and larger and approaches infinity as its speed approaches the speed of light. Relativity theory shows that the mass of an object is equal to a constant quantity called the *rest mass* divided by a quantity that gets smaller and smaller (approaching zero) as the object goes faster and faster in approaching the speed of light. So the mass of an object gets larger and larger as the object goes faster and faster, and at speeds close to the speed of light is so large that it takes a tremendous amount of energy to accelerate it a little more. The mass would become infinite at the speed of light if an object's rest mass is any number other than zero, because we would have that number divided by zero, which is infinite. And we would be unable to provide enough energy to accelerate the mass to that speed.

The only way that an object can travel at the speed of light is if its rest mass is zero, because then the mass is equal to zero divided by a number that gets smaller and smaller. The quotient is therefore always zero. In the limit of increasing velocity, the quotient remains zero. Thus photons, which are particles of light and thus travel at the speed of light, must have zero rest mass. We can also deduce that if neutrinos travel at the speed of light, they have no rest mass. That is, the neutrino, like the photon, would have no mass if it were at rest (which it never is), so there would be no rest mass to grow larger.

The evidence in terrestrial laboratories that neutrinos actually have some rest mass is weak, and requires more testing. But if neutrinos did have even a little bit of rest mass, there are so many neutrinos that they would make up the major part of the mass of the universe. They would provide an answer to the missing-mass problem, and could be the unseen matter in the coronas of galaxies. They could provide enough mass to close the universe! You will be reading more about the question of whether neutrinos have mass in years to come, in newspapers and magazines.

17.2e The Long Run

Most of the above measurements indicate that the universe will not stop its expansion, and that the galaxies will continue to fly away from each other. But the issue is not settled; there are still reasons to think that the universe may be closed. Of course, new theoretical ideas, as we shall discuss in the following sections, or new observations may yet change this picture. In any case, it is interesting to note that these results were obtained from large telescopes on earth, radio observations using a variety of techniques, and ultraviolet and x-ray observations using telescopes in space. We are making use of all the methods we know to tell us the future of the universe.

17.3 Forces in the Universe

There are four known types of forces in the universe:

1. The *strong force,* also known as the *nuclear force,* is the strongest. It is the force that binds particles together into atomic nuclei. Although it is very strong at close range, it grows weaker rapidly with distance.

Only some particles "feel" the strong force. These particles, which include protons and neutrons, are composed of 6 kinds of particles called *quarks* (Fig. 17–8). (James Joyce used the word "quark" in *Finnegan's Wake,* and scientists have appropriated it.) There are six kinds (*flavors*) of quarks called "up," "down," "strange," "charmed," "truth," and "beauty." Each kind of quark can have one of three properties called *colors,* often called red, blue, and green. (These names are whimsical and do not have the same meaning that the words have in general speech, though the "strange" quark got its name because it had a property called "strangeness" that made it different from more ordinary particles.) Each of the six flavors in three colors makes 18 subtypes of quarks in all; since an antiquark corresponds to each quark, there are really 36. This number is so large that even quarks may not be truly basic.

The strong force is carried between quarks by a particle; since this particle provides the "glue" that holds nuclear particles together, it is known (believe it or not) as a "gluon." Recent studies with atomic accelerators have led to the discovery of evidence for the existence of all six of the quarks and of the gluon.

2. The *electromagnetic force,* 1/137 the strength of the strong force, leads to electromagnetic radiation in the form of photons. So most of the evidence we have discussed in this book, since it was carried by light, x-rays, etc., was carried by the electromagnetic force. It is also the force involved in chemical reactions.

3. The *weak force,* important only in the decay of certain elementary particles, is currently being carefully studied. It is very weak, only 10^{-13} the strength of the strong force, and also has a very short range.

4. The *gravitational force* is the weakest of all over short distances, only 10^{-39} the strength of the strong force. But the effect from the masses of all the particles is cumulative—it adds up—so that on the scale of the universe gravity dominates the other forces.

Theoretical physicists studying the theory of elementary particles have in the past years made progress in unifying the theory of electromagnetism and the theory of the weak force into a single theory. Thus just as the forces known separately as electricity and magnetism were unified a hundred years ago into the electromagnetic force, the force of electromagnetism and the weak force have now been unified into the *electroweak force.* (The 1979 Nobel Prize in Physics was awarded for this work.) The forces would be indistinguishable from each other at the extremely high temperatures, above 10^{28} K, that may have existed in the earliest moments of the universe.

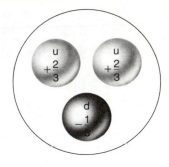

PROLON   +1 CHARGE UNIT

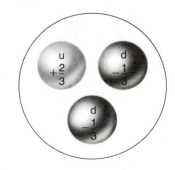

NEUTRON   0 CHARGE UNIT

Figure 17–8 The most common quarks are *up (u)* and *down (d);* ordinary matter in our world is made of them. The *up* quark has an electric charge of +⅔, and the *down* quark has a charge of −⅓. This fractional charge is one of the unusual things about quarks; prior to their invention (discovery?), it had been thought that all electric charges came in whole numbers. Note how the charge of the proton and of the neutron is the sum of the charges of their respective quarks.

Figure 17–9 Carlo Rubbia of Harvard, the head of the group that discovered the W and Z particles at the European accelerator known as CERN, in front of some of his measuring apparatus.

The early universe, to physicists studying elementary particles, is known as "that great big accelerator in the sky." It provides a source of higher energy than can be reached in any accelerator on earth.

The electroweak theory predicted that certain new particles could be discovered if a particle accelerator (informally called an "atom smasher") that was powerful enough could be built. The new particles, known as W and Z, were predicted to be much more massive than the proton, so they could be created out of energy only if a lot of energy were available. A European atom smasher, upgraded for the purpose, reached such high energies in 1982 and 1983 (Fig. 17–9) that the W and the Z particles were indeed created and detected (Fig. 17–10), a great triumph for the electroweak theory.

Progress is even being made in a family of *grand unified theories (GUTs)* that unifies the electroweak and the strong force. Such a theory would be necessary to provide a thorough explanation of the early universe. (In spite of the overblown name, gravity is still not included in the "grand unification.") The grand unified theories being considered imply, surprisingly, that protons are not stable—they should decay with a half-life greater than 10^{30} seconds. So GUTs imply that protons decay. Though this is a much longer time than the lifetime of the universe (10^{10} years equals only 3×10^{17} seconds), given a sufficiently large quantity of protons, such a decay should be detectable. After all, 1000 tons of matter contain about 10^{32} protons (Fig. 17–11), so perhaps 10 or 100 could decay each year in that volume. It is just such a change in the total number of baryons in the universe that is necessary to account for the current excess of matter over antimatter. Many scientists had expected to detect signs of proton decay by now, but the evidence has not been forthcoming. Here is another case where you will have to watch newspapers and magazines for the latest developments.

Many theoreticians are working on theories to incorporate gravity with the other forces; *supergravity* is one of the examples. But such theories are not yet satisfactory, and are not as advanced as GUTs.

At early times in the universe, the exotic particles now being observed by physicists in giant accelerators had been present as well. Even isolated quarks may have been present, though they would have combined to form baryons (protons, neutrons, etc.) by the time 1 microsecond went by. The mini black holes described in Section 11.6 may have been formed in this era.

17.4 The Inflationary Universe

The application to cosmology in the last few years of the exciting new theoretical and experimental results of particle physics has led to some big surprises. The results from particle physics have allowed scientists to push back our understanding of the universe much farther back in time than previously. In particular, now we can go back beyond the era of element formation.

The new GUTs theories got scientists thinking about the early universe. A major new theory holds that starting at the first 10^{-35} second of time and lasting 10^{-32}

Figure 17–10 A charged particle leaves a trail of bubbles as it passes through a "bubble chamber," a large volume of helium on the verge of boiling. By studying the trails, scientists concluded that the first W particle to be discovered *(arrow)* had passed.

Figure 17–11 In this search for decaying protons, this cavity— deep underground in a salt mine near Cleveland—has been filled with 10,000 tons of water. Water, H_2O, is mostly protons, since an H nucleus is a proton and an oxygen nucleus contains 8 protons. If any of the 2.5×10^{33} neutrons or protons bound in nuclei decay, the resulting particles will give off flashes of light. Twenty-four hundred photomultipliers have been installed to detect these flashes.

second, the existing universe grew rapidly larger. The rate of this "inflation" of the universe was astounding. During this fraction of a second, the volume of the universe grew 10^{100} times more than it would have grown according to the "standard" big-bang theory. The new theory, whose first form was advanced by Alan Guth (Fig. 17–12), now of M.I.T., is called the *inflationary universe*. A subsequent version, developed by Andrei Linde of the Lebedev Institute in Moscow, Andreas Albrecht, now of the University of Texas, and Paul Steinhardt of the University of Pennsylvania, was based on an even newer approach and resolved some major flaws in the first version. As we shall see below, the inflationary-universe theory solves a number of outstanding problems of cosmology.

*17.4a Explanation of the Inflationary Universe

The crucial assumption of the inflationary-universe model is that the universe underwent a special kind of "phase transition" during its early history. A phase transition is a change in the way elementary particles behave in the universe, and is analogous to the change in the way water molecules behave as they undergo a phase transition from liquid (water) to solid (ice).

In the case of water molecules, what decides whether they form water or ice is the temperature. If one calculates the energy of a group of molecules that have formed a water droplet and compares it to the energy of the same group of molecules that have formed an ice crystal, one finds that the relation between the energies of the water and ice are different at different temperatures. At temperatures above the freezing temperature, the energy of the water droplet is less than that of the ice crystal. Since matter apparently prefers to be in a state with the lowest possible energy, water molecules prefer to form liquid water at such temperatures. At temperatures below the melting temperature, the energy of the ice crystal is lower than that of the water droplet, so water molecules prefer to form ice crystals at such temperatures.

However, as one cools water below the freezing temperature, it may not immediately be transformed into ice, even though the ice has the lower energy. There will generally be an energy barrier that must be crossed before the water can crystallize. In this way, even as one continues to lower the temperature, water molecules can be trapped in the form of liquid water even though water continues to have a greater energy than ice. This phenomenon is known as "supercooling," and is found in nature. (Water can be supercooled several degrees below freezing, a situation that is commonly found in clouds aloft.)

Almost all theories that unify elementary-particle forces imply that the universe

Figure 17–12 Alan Guth, originator of the first version of the inflationary-universe theory.

as a whole underwent phase transitions in its early history. Let us consider grand-unified theories, for example. The basic notion of grand-unified theories is that the strong, weak, and electromagnetic forces are different parts of a single unified force and are related to one another through a symmetry. (A cube could be flipped from left to right or from top to bottom without changing its appearance, for example, because of its symmetries. A rectangular bar of soap has some of the cube's symmetries but not others.) Today the symmetries among the forces of nature aren't apparent—the forces have drastically different strengths and characteristics. But at high temperatures, above 10^{28} K, the forces change and obtain equal strengths and characteristics, making the symmetry apparent.

Thus in the first second of time, as the universe cooled below 10^{28} K, grand-unified theories predict that the universe underwent a phase transition from a state where the forces were symmetric to one in which the symmetry was broken. (Again, this is analogous to the transformation from a liquid, which is rotationally symmetric, to an ice crystal, which has preferred axes and so does not have rotational symmetry. We say that the symmetry is broken.)

The inflationary-universe model assumes that as the universe expanded and the temperature decreased, the universe underwent such a phase transition in which it supercooled—remained trapped in a state where the forces are symmetric even though that state had a greater energy than the state in which the symmetry among the forces is broken. Because it was trapped in that higher-energy state, the universe had excess energy.

As the universe continued to cool, the excess energy became the dominant contribution to the total energy in the universe. (It was much greater, for example, than the total kinetic and mass energies of the particles contained within the universe.) However, unlike other kinds of energy, the amount of such energy in a unit volume—the energy density—does not change as the universe continues to expand. (This energy is associated with the state in which the universe was and not with the individual energies of the particles contained within it. As the universe continued to expand, the particles got spread out, so the kinetic and mass energy densities decreased, but the state and therefore the state energy density remained the same everywhere in the universe.)

This constant energy density, just as any form of energy, led to a gravitational force, but the gravitational force of a universe with a constant energy density is very peculiar—it causes the universe to expand so rapidly, the so-called ''inflation.'' (This is not intuitively obvious, but follows from calculations.) In a period of 10^{-32} second, the universe expanded in volume by a factor of 10^{100} or more than in the hot-big-bang picture.

17.4b The New Inflationary Universe

Of course, to be a useful cosmological picture, the expansion eventually had to stop, because our universe is not inflating today; it is expanding at a much more modest rate. Guth's original model was flawed in that the phase transition would have occurred violently, making the resulting universe unacceptably nonuniform.

In the current ''new inflationary universe'' model, the inflation stops smoothly after a small fraction of a second. At this time, the universe is transformed from the state in which the forces are symmetric into the lower energy state, in which the symmetry among the forces is broken. All the state energy is rapidly converted into ordinary matter and radiation. The universe continues to expand, but at the slower rate we observe today, just as in the standard big-bang theories (Fig. 17–13). The elements form, the background radiation is set free, and so on, on the same time scale.

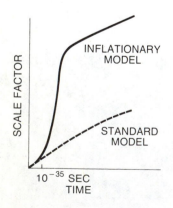

Figure 17–13 The rate at which the universe expands, for both inflationary and non-inflationary models.

But the theory does imply that the universe is very much larger than we had thought, in that the parts with which we could have been in touch have expanded far beyond our sight. The boundary of this volume, our "past light cone," is defined by signals travelling at the speed of light (Fig. 17–14), since relativity theory tells us that it is impossible for signals to travel faster. Since we are out of contact with some regions, those outside our past light cone, those regions could even be made of antimatter.

One major question that had been open in cosmology is how the universe became so homogeneous. According to the hot-big-bang model, our universe was supposed to have always been so large that neither light, nor any information about conditions at any given location, had had time since the big bang to travel across it. Thus matter in one part of the observed universe has always been out of contact with matter on the other side, as we see it. It has been a mystery how such widely different regions could have reached identical temperatures and densities. After all, we know that the temperature and density were nearly identical everywhere at the time the cosmic background radiation was emitted, since this background radiation is so isotropic.

In the inflationary-universe picture, our observed universe was much tinier before the phase transition than in the hot-big-bang picture, small enough that signals travelling at the speed of light had time to travel across it. The universe could thus be smoothed out. The whole part of the universe that we can potentially observe—the visible universe—was, before the inflation, smaller than a single proton. There might have been as many as 10^{15} pointlight particles (such as quarks, electrons, etc.) crammed into this tiny region, but essentially all of the approximately 10^{86} particles in the observed universe today were produced from the energy released in the phase transition. The inflation then accounts for how such a tiny region could grow to be the size of our observed universe. It thus explains why the universe is so homogeneous.

Another previously unresolved question is why the density of matter in the universe is so close to the critical amount that marks the difference between the universe's being open or closed. Although we do not yet know the precise value, we do know that the density of matter is within one or two factors of 10 of the critical density. Being even this close to the critical value is rather surprising, since it can be shown that such a situation is highly unstable, like balancing a pencil on its point! Why are we in such an unstable situation? The inflationary-universe theory shows that during the inflation, space-time became very flat (Section 16.2), just as a small bit of the surface of a balloon flattens out as the balloon grows larger. It follows that the universe is ever so slightly either open or closed, depending on what it was before the inflation.

We may thus have been asking the wrong question all these years when we asked whether the universe is open or closed, wondering what kinds of studies would tell us the answer. The inflationary-universe theory implies that the universe is incredibly close to the borderline case in which it would expand forever at an ever-decreasing rate; the longer it expands, the less its rate of expansion will be. If the theory is right, then the universe is so close to this borderline that it would be impossible for any observations to tell us on which side it lies.

The inflationary universe also provides satisfactory answers for another question plaguing some theoreticians: why don't we see magnetic monopoles in the universe? A *magnetic monopole* is a point source of magnetic force, analogous to an electron, which is a single particle that has an electric force. But bar magnets always have two opposite poles, and a single-pole magnet—a monopole—has never been discovered. Although scientists have toyed with the notion of monopoles for over a hundred years, it turns out that GUTs predict that there would be monopoles. Further, these

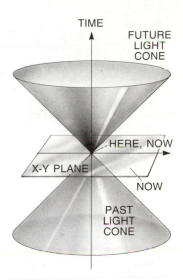

Figure 17–14 If we plot a three-dimensional graph with two spatial dimensions (omitting the third dimension) and a time axis, our location can receive signals only from regions close enough for signals to reach us at the speed of light or less, our "past light cone." We can never be in touch with regions outside our light cone; we say that we are "not causally connected" with those regions. In the inflationary universe, the volume of space with which we are causally connected is much greater than the observable universe. It is easy to explain how a volume that is causally connected can be homogeneous—collisions, for example, can smooth things out.

The situation is even worse than might seem; GUTs imply that so many monopoles were produced that their mass would dominate the density of mass in the universe, totally altering the history of the universe and destroying all successes of the hot-big-bang model.

monopoles should be very massive and should have been produced in enormous numbers in the early universe. The monopoles never decay and rarely meet up with antimonopoles (in which case they would annihilate each other), so they should be around today. Yet, none are seen. It is crucial to find some explanation of how the universe might have gotten rid of the high density of monopoles.

Since the monopoles are produced before inflation, the inflationary universe can account for the fact that we don't now see any by spreading the magnetic monopoles out. The inflation spreads them out so much that there is at most only a handful in the part of the universe observable by us.

A startling feature of the inflationary-universe theory is that it accounts for the existence of all the ordinary matter and radiation in the universe. In the theory, the energy from the peculiar state of the early universe is transformed, after the period of tremendous expansion, into the matter and radiation that we detect today. Almost all the energy we observe in the universe today was produced by this transformation of state energy that occurred at the end of the inflation. The model also shows why there is so much matter in the universe.

In Guth's words, the creation of energy in this way was "the ultimate free lunch."

Many scientists are now studying aspects of the inflationary-universe theory (Fig. 17–15), and the pace of discovery is fast. There are still some problems with the theory. For example, though the theory can explain the formation of galaxies in principle, the explanation depends on the numerical details of the unified-field theory and more progress must be made in particle physics before we can tell if predictions can match the observations with sufficient accuracy.

We cannot yet know whether these problems will be overcome through future refinement. But we can predict that the relation of physicists studying subnuclear particles and scientists studying cosmology will become closer and closer.

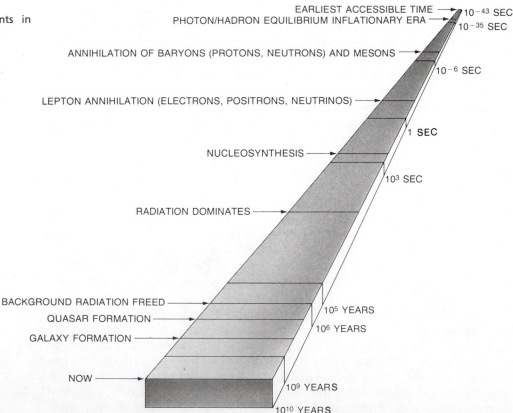

Figure 17–15 Major events in the history of the universe.

EARLIEST ACCESSIBLE TIME — 10^{-43} SEC
PHOTON/HADRON EQUILIBRIUM INFLATIONARY ERA — 10^{-35} SEC

ANNIHILATION OF BARYONS (PROTONS, NEUTRONS) AND MESONS — 10^{-6} SEC

LEPTON ANNIHILATION (ELECTRONS, POSITRONS, NEUTRINOS) — 1 SEC

NUCLEOSYNTHESIS — 10^3 SEC

RADIATION DOMINATES

BACKGROUND RADIATION FREED — 10^5 YEARS
QUASAR FORMATION — 10^6 YEARS
GALAXY FORMATION

NOW — 10^9 YEARS

10^{10} YEARS

Summary and Outline

Creation of the elements (Section 17.1)

Pairs of particles and antiparticles created out of energy in the early universe

Light elements formed in the first minutes after the big bang; heavier elements formed in interiors of stars and in supernovae

Future of the universe (Section 17.2)

The missing-mass problem: discrepancy between mass derived by studying motions of galaxies in clusters of galaxies and the visible mass

Deuterium-to-hydrogen ratio depends on cosmic density: all deuterium was formed in first minutes after the big bang; current evidence from deuterium is that universe is open.

If neutrinos have mass, the universe must be closed

Optical evidence from motions of nearby galaxies agrees that universe is open

Discovery from Einstein Observatory that some of the x-ray background is from faint quasars is being further interpreted

Forces in the universe (Section 17.3)

Four forces of nature, in declining order of strength: strong, electromagnetic, weak, gravitational

Particles like protons and neutrons are made of quarks

Theory of electroweak force unifies electromagnetic and weak forces; recently led to discovery of W and Z particles

Grand unified theories (GUTs) unify electroweak and strong forces; predict proton decay

The inflationary universe (Section 17.4)

Based on GUTs

Universe grew rapidly larger in first 10^{-32} second, and later began expanding as it would from standard big-bang theory

Explains homogeneity of background radiation, implies that universe is on boundary between being open and closed, and explains why we don't detect magnetic monopoles

Key Words

early universe, particle physics, antiparticle, antimatter, baryons, nucleosynthesis, deuteron, missing-mass problem, rest mass, strong force, nuclear force, quarks, colors, flavors, electromagnetic force, weak force, gravitational force, electroweak force, grand unified theories (GUTs), supergravity, inflationary universe, magnetic monopole, principal of equivalence

Questions

1. An electron has a negative charge. What is the charge of a positron, which is an antielectron? What is its mass?

†2. The mass of an electron is given in Appendix 2. What is the energy released in the annihilation of an electron and a positron? (Use the erg for the unit of energy, with 1 erg = 1 g · cm²/s².) Compare this with the energy being given off by the sun each second, also given in Appendix 2.

†3. Referring to the graph given, what are the relative abundances of several light elements or isotopes 1 hour after the big bang? Compare these abundances to the abundance of protons.

4. What is the advantage of studying deuterium over

studying helium for assessing the density of the universe?

5. In your own words, explain why the abundance of deuterium is linked with the density of the universe.

6. What are three pieces of evidence that the universe is open, and two that the universe may be closed?

7. Why must we resort to indirect methods to find out if the universe is open or closed?

8. How much energy would you have to put in to accelerate a proton until it was travelling at the speed of light? Explain.

9. Why do we feel gravity, given that the gravitational force is so weak relative to the other fundamental forces?

10. What are the advantages of the inflationary-universe theory?

†This indicates a question requiring a numerical solution.

Topics for Discussion

1. What effect would finding out whether the universe will expand forever, will begin to contract into a final black hole, or will contract and then oscillate with cycles extending into infinite time, have on the conduct of your life or on your thoughts?

2. Discuss whether astronomy or physics is more fundamental than the other, given the increasing overlap.

Part VI The Solar System

The Earth and the rest of the solar system may be important to us, but they are only minor companions to the stars. In *Captain Stormfield's Visit to Heaven,* by Mark Twain, the Captain races with a comet and gets off course. He comes into heaven by a wrong gate, and finds that nobody there has heard of "the world." ("The world, there's billions of them!" says a gatekeeper.) Finally, the gatekeepers send a man up in a balloon to try to detect "the world" on a huge map. The balloonist has to travel so far that he rises into clouds. After a day or two of searching he comes back to report that he has found it: an unimportant planet named "the Wart."

We too must learn humility as we ponder the other objects in space. And while it is no doubt the case that the heavens are filled with a vast assortment of suns and planets more spectacular than our own, still, the solar system is our own local environment. We would like to understand it and come to terms with it as best we can. Besides, in understanding our own solar system, we may even find some keys to understanding the rest of the Universe.

To get an idea of its scale, imagine that the solar system is scaled down and placed on a map of the United States. Let us say that the Sun is a hot ball of gas taking up all of Rockefeller Center, about a kilometer across, in the center of New York City.

Mercury would then be a ball 4 meters across at the distance of mid-Long Island, and Venus would be a 10-meter ball one and a half times farther away. The Earth, only slightly bigger, is located at the distance of Trenton, New Jersey. Mars, half that diameter, 5 meters across, is located past Philadelphia.

Only for the planets beyond Mars would the planets be much different in size from the Earth, and the separations become much greater. Jupiter is 100 meters across, the size of a baseball stadium, past Pittsburgh at the Ohio line. Saturn without its rings is a little smaller than Jupiter (including the rings it is a little larger), and is past Cincinnati toward the Indiana line. Uranus and Neptune are each about 30 meters across, about the size of a baseball infield, and are at the distance of Topeka and Santa Fe, respectively. And Pluto, a 4-meter ball like Mercury, travels on an elliptical track that extends as far away as Los Angeles, 40 times farther away from the Sun than the Earth is. (At present, Pluto is in Montana, closer to us than Neptune.) Occasionally a comet sweeps in from Alaska, or some other random direction, passes around the Sun, and returns in the general direction from which it came.

The planets fall naturally into two groups. The first group, the *terrestrial planets,* consists of Mercury, Venus, Earth, and Mars. All are

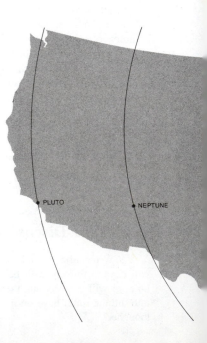

rocky in nature. The terrestrial planets are not very large, and have densities about five times that of water. (When the metric system was set up, the gram was defined so that water would have a convenient density of exactly 1 gram/cubic centimeter. Thus the terrestrial planets have densities of about 5 g/cm³.)

The second group, the *giant planets,* consists of Jupiter, Saturn, Uranus, and Neptune. All these planets are much larger than the terrestrial planets, and also much less dense, ranging down to slightly below the density of water. Jupiter and Saturn are largely gaseous in nature, similar to the Sun in composition. Uranus and Neptune have large gaseous atmospheres but have rocky or icy surfaces and interiors. Pluto, planet number nine, is anomalous in several of its properties, and so may have had a very different history from the other planets. Between the orbits of the terrestrial planets and the orbits of the giant planets are the orbits of thousands of chunks of small ''minor planets.'' These *asteroids* range up to 1000 kilometers across. Sometimes much smaller chunks of interplanetary rock penetrate the Earth's atmosphere and hit the ground. We shall discuss these *meteorites* and where they came from together with asteroids and comets.

Many people are interested in the planets in order to study their history—how they formed, how they have evolved since, and how they will change in the future. Others are more interested in what the planets are like today. Still others are interested in the planets mainly to consider whether they are harboring intelligent life.

Because the Moon and some other objects in the solar system have undergone less erosion on their surfaces than has the Earth, we now see them as they appeared eons ago. In this way the study of the planets and other objects in the solar system as they are now provides information about the solar system's early stages and its origin. Other objects, like Jupiter's moon Io, change even more rapidly in their surface appearance than does the Earth.

Now that spacecraft have explored so many planets close up, we can discuss many general properties of planets or of groups of planets. This study of *comparative planetology* gives us important insights into our own planet. But studies too restricted to comparative planetology can make us lose sight of the interesting individual nature of each planet and of the stages by which our knowledge has jumped during the past decades. In the following chapters, we thus treat each planet individually and in a somewhat chronological fashion, while also stressing common features. We can then better appreciate not only the planets themselves but also the projects and individual researchers who have been uncovering their mysteries.

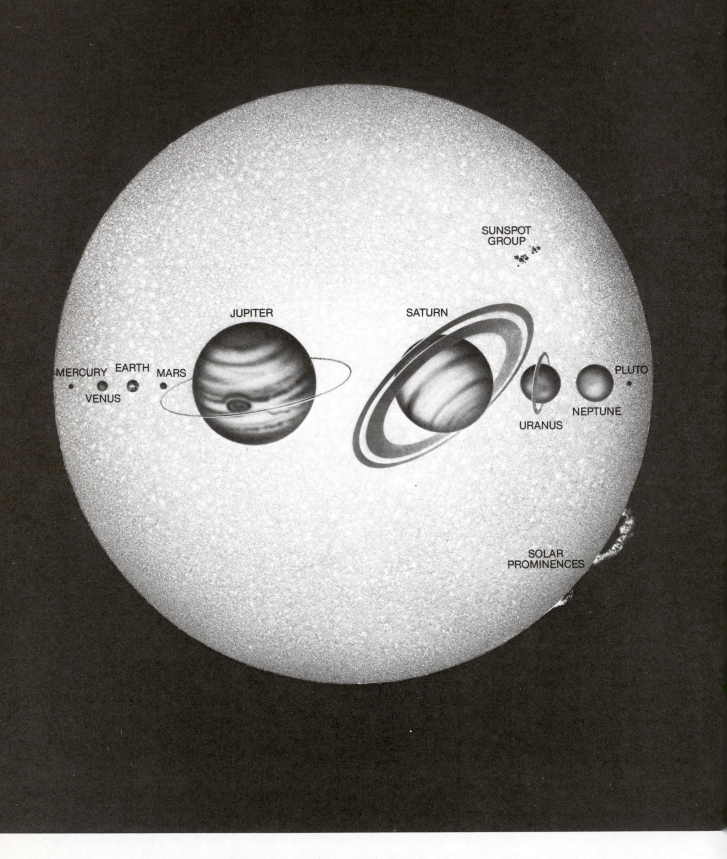

The relative sizes of the planets and the Sun. The sizes of the planets are exaggerated with respect to the Sun by a factor of 2 in this drawing. Some of the moons of Jupiter, Saturn, and Uranus are as large as the smaller planets.

The Early History of Astronomy

18

Aims: To discuss the scale and structure of the solar system, the history of our understanding of the geocentric and the heliocentric theories, Kepler's laws about the motions of the planets, Newton's law of gravity, and the origin of the solar system

In this chapter we describe some general ideas about the solar system and its structure. First we describe such commonly observed phenomena as phases of the Moon and eclipses. Then we discuss the origins of astronomy, which began with such practical things as keeping time and marking the arrival of the seasons. We trace the development of early astronomy through the work of some of the principal scientists who contributed to the field.

In this chapter we consider the period up to 1687, when Isaac Newton published his major work. His work marks the origin of what we might call the modern era of astronomy. We also consider the historical development of our ideas of the structure and origin of the solar system. Historical developments since the time of Newton are treated throughout this book in the context of the various separate topics.

18.1 The Phases of the Moon and Planets

From the simple observation that the apparent shapes of the Moon and planets change, we can draw conclusions that are important for our understanding of the mechanics of the solar system. The fact that the Moon goes through a set of *phases* approximately once every month is perhaps the most familiar everyday astronomical observation (Fig. 18–1). In fact, the name "month" comes from the word "moon." The actual period of the phases, the interval between a particular phase of the Moon and its next repetition, is approximately $29\frac{1}{2}$ Earth days (Fig. 18–2). This period can vary by as much as 13 hours.

The explanation of the phases is quite simple: The Moon is a sphere, and at all times the side that faces the Sun is lighted and the side that faces away from the Sun is dark. The phase of the Moon that we see from the Earth, as the Moon revolves around us, depends on the relative orientation of the three bodies: Sun, Moon, and Earth. The situation is simplified by the fact that the plane of the Moon's revolution around the Earth is nearly, although not quite, the plane of the Earth's revolution around the Sun.

Basically, when the Moon is almost exactly between the Earth and the Sun, the dark side of the Moon faces us. We call this a "new moon." A few days earlier or

The *phases* of moons or planets are the shapes of the sunlighted areas as seen from a given vantage point.

"The Moon" is often capitalized to distinguish it from moons of other planets; in general writing, it is usually written with a small "m." In Part VI, we shall capitalize "Earth" to put it on a par with the other planets and the Moon. For consistency, we shall also capitalize "Sun" and "Universe."

325

Figure 18–1 The phases of the Moon depend on the Moon's position in its orbit around the Earth. Here we visualize the situation as if we could be high above the Earth's orbit, looking down.

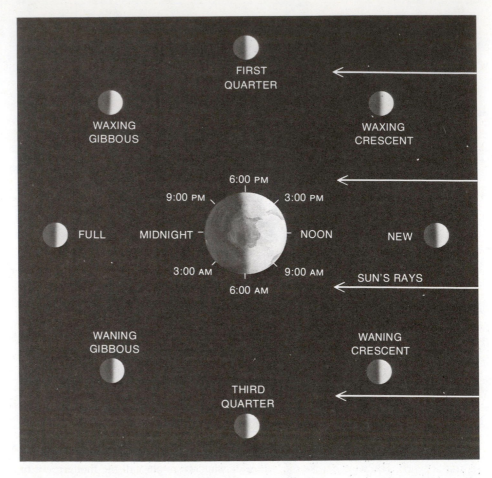

later we see a sliver of the lighted side of the Moon, and call this a "crescent." As the month wears on, the crescent gets bigger, and about 7 days after new moon, half the face of the Moon that is visible to us is lighted. We sometimes call this a "half moon." Since this occurs one-fourth of the way through the phases, the situation is also called a "first-quarter moon." (Instead of apologizing for the fact that astronomers call the same phase both "quarter" and "half," I'll just continue with a straight face and try to pretend that there is nothing strange about it.)

When over half the Moon's disk is visible, we have a "gibbous" moon. One week after the first-quarter moon, the Moon is on the opposite side of the Earth from

Figure 18–2 The phases of the Moon.

4 days
waxing crescent

7 days
1st quarter

10 days
waxing gibbous

14 days
full

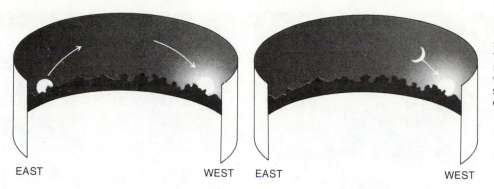

Figure 18–3 Because the phase of the Moon depends on its position in the sky with respect to the Sun, we can see why a full moon is always rising at sunset while a crescent moon is either setting at sunset, as shown here, or rising at sunrise.

the Sun, and the entire face visible to us is lighted. This is called a "full moon." One week later, when we see a half moon again, we have a "third-quarter moon." Then we go back to "new moon" again and repeat the cycle of phases.

Note that since the phase of the Moon is related to the position of the Moon with respect to the Sun, if you know the phase, you can tell when the Moon will rise. For example, since the Moon is 180° across the sky from the Sun when it is full, a full moon is always rising just as the Sun sets (Fig. 18–3). Each day thereafter, the Moon rises about 50 minutes later (24 hours divided by the period of revolution of the Moon around the Earth). The third-quarter moon, then, rises near midnight, and is high in the sky at sunrise. The new moon rises with the Sun in the east at dawn. The first-quarter moon rises near noon and is high in the sky at sunset. It is often visible in the late afternoon.

The Moon is not the only object in the solar system that is seen to go through phases. Mercury and Venus both orbit inside the Earth's orbit, and so sometimes we see the side that faces away from the Sun and sometimes we see the side that faces toward the Sun. Thus at times Mercury and Venus are seen as crescents, though it takes a telescope to observe their shapes. Spacecraft to the outer planets have looked back and seen the Earth as a crescent (Fig. 18–4).

18.2 Eclipses

Because the Moon's orbit around the Earth and the Earth's orbit around the Sun are not precisely in the same plane (Fig. 18–5), the Moon usually passes slightly above or below the Earth's shadow at full moon, and the Earth usually passes slightly above or below the Moon's shadow at new moon. But every once in a while, up to seven times a year, the Moon is at the part of its orbit that crosses the Earth's orbital

Figure 18–4 The crescent of Earth seen from a distance of 450,000 km by the Pioneer Venus spacecraft, whose trajectory took it briefly outside the Earth's orbit.

20 days
waning gibbous

22 days
third quarter

24 days
waning crescent

26 days
waning crescent

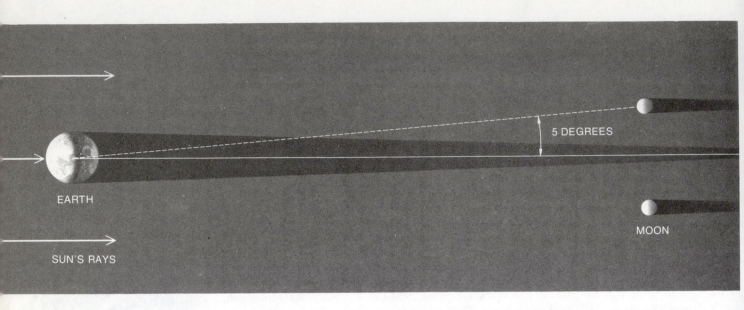

EARTH

5 DEGREES

MOON

SUN'S RAYS

Figure 18–5 The plane of the Moon's orbit is tipped with respect to the plane of the Earth's orbit, so the Moon usually passes above or below the Earth's shadow. Therefore, we don't have lunar eclipses most months.

plane at full moon or new moon. When that happens, we have a lunar or solar eclipse. Most of the eclipses are only partial: the Moon is only partly in the Earth's shadow or the Sun is only partly blocked by the Moon. The total eclipses, which are rarer, are much more interesting to watch.

We have already discussed eclipses of the Sun, when the Moon comes directly between the Earth and the Sun. Many more people see a lunar eclipse than a solar eclipse when one occurs. At a total lunar eclipse, the Moon lies entirely in the Earth's shadow and sunlight is entirely cut off from it (Fig. 18–6). So anywhere on the Earth that the Moon has risen, the eclipse is visible. In a total solar eclipse, on the other hand, the alignment of the Moon between the Sun and the Earth must be precise, and only those people in a narrow band on the surface of the Earth see the eclipse.

The total phase of a lunar eclipse—when the Moon is entirely within the Earth's shadow—is a much more leisurely event to watch than the total phase of a solar eclipse. During this time, the sunlight is not entirely shut off from the Moon (Fig. 18–7). A small amount is refracted around the edge of the Earth by our atmosphere. Most of the blue light is taken out during the sunlight's passage through our atmosphere; this explains how blue skies are made for the people part way around the globe from the point at which the Sun is overhead. The remaining light is reddish, and this is the light that falls on the Moon. Thus, the eclipsed Moon appears reddish.

The next two total lunar eclipses visible from the United States and Canada will be on April 24, 1986 (western regions only), and August 17, 1989.

Figure 18–6 When the Moon is between the Earth and the Sun, we observe an eclipse of the Sun. When the Moon is on the far side of the Earth from the Sun, we see a lunar eclipse. The part of the Earth's shadow that is only partially shielded from the Sun's view is called the *penumbra;* the part of the Earth's shadow that is entirely shielded from the Sun is called the *umbra.* The distances in this diagram are not to scale.

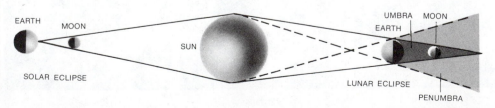

EARTH   MOON   UMBRA   MOON
                                                      EARTH
                        SUN
SOLAR ECLIPSE   LUNAR ECLIPSE
                                                         PENUMBRA

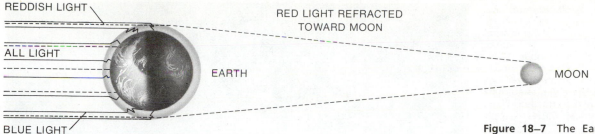

REDDISH LIGHT

ALL LIGHT

EARTH

RED LIGHT REFRACTED TOWARD MOON

MOON

BLUE LIGHT SCATTERED MORE BY ATMOSPHERE

Figure 18–7 The Earth's atmosphere scatters shorter wavelengths much more than it scatters longer wavelengths, so red light is scattered much less than blue light. Some of the red light survives its passage through the Earth's atmosphere and is bent, or refracted, toward the Moon. Thus the Moon often appears reddish during the total phase of a lunar eclipse, unless there is so much volcanic dust in the Earth's atmosphere that the moon appears gray.

18.3 Ancient Astronomy—The Main Line

18.3a Egyptian Astronomy

In these days of digital watches, electronic calculators, and accurate time given regularly on radio and television, it is hard to imagine what life was like in 1000 B.C. or earlier. No mechanical clocks existed and even the idea of a calendar was vague. Yet, each spring, the Nile River flooded the neighboring farmlands, and Egyptian farmers wanted to know how to predict when this would happen. Farmers everywhere, even those not dependent on annual flooding wanted to know when to expect the spring and the spring rains so that they could plant their crops in ample time.

Eventually, some people noticed that they could base a calendar on Sirius, the brightest star in the sky. It became visible in the eastern sky before sunrise just at a time of year when the crops should be planted and when the Nile was about to flood. The first visibility of Sirius in the morning sky—the *heliacal rising* of Sirius—served as a marker for Egyptian astronomers and priests and enabled them to let everybody know that the springtime had come. After all, the Egyptian climate does not show the changing of the seasons as obviously as more northern climates do.

Figures in the sky—constellations—had been recorded by the Sumerians as far back as 2000 B.C., and maybe earlier. The bull, the lion, and the scorpion are constellations that date from this time. Other constellations still recognized today were recorded by a Greek, Thales of Miletus, in about 600 B.C., at the dawn of Greek astronomy. Certain constellations were common to several civilizations. The Chinese also had a lion, a scorpion, a hunter, and a dipper, for example.

It has long been known that Egyptian pyramids and temples erected after 1500 B.C. have astronomical alignments. The Great Pyramid of Giza (from 2600 B.C.) has its four faces accurately aligned north, east, south, and west. The main axis of the temple of Karnak is aligned with mid-winter sunrise. The main temple at Abu Simbel originally faced the rising sun of a jubilee day, and has a side chapel that faces sunrise at the winter solstice. Also, the Sun God, Ra, was pictured widely. Astronomy was used for religious as well as practical purposes.

18.3b Greek Astronomy: The Earth at the Center

When we observe the planets in the sky, we notice that their positions vary from night to night with respect to the stars. The stars, on the other hand, are so far away that their positions are relatively fixed with respect to each other. The fact that the planets appear as ''wandering stars'' was known to the ancients; our word ''planet'' comes from the Greek word for ''wanderer.'' In ancient times, five planets were known—Mercury, Venus, Mars, Jupiter, and Saturn.

Of course, both stars and planets revolve together across our sky essentially once every 24 hours; they all rise approximately in the east and set approximately in the west. By the wandering of the planets we mean that the planets appear to move at a rate slightly different from that of the stars, so that over a period of weeks or months they change position with respect to the fixed stars. The planets and Moon appear to move near the *ecliptic,* the path of the Sun across the sky.

Figure 18–8 A planetarium simulation of the path of Mars through the region of Sagittarius from approximately December 1985 through November 1986, showing the retrograde loop. (Photograph by Allen Seltzer, using the Zeiss Mark VI planetarium projector at the Hayden Planetarium-American Museum, New York)

We give dates in our conventional B.C. and A.D. notation, but obviously the ancient Greeks couldn't do that. They actually dated events from Olympiads, just as the Romans mostly dated events from the accession of consuls, their rulers. By the late Roman period, a traditional epoch of the foundation of Rome was used, just as we use A.D. 1 as the origin of our system of dates. But A.D. 1 was assigned with hindsight in the sixth century. (Don't buy any coins dated 43 B.C.)

The motion of the planets in the sky is not always in the same direction with respect to the stars. Most of the time the planets appear to drift eastward with respect to the background stars. But sometimes they drift backwards, that is, westward. We call the backward motion *retrograde motion* (Fig. 18–8).

The ancient Greeks began to explain the motions of the planets in terms of the geometry of the solar system. For example, by determining which of the known planets had the longest periods of retrograde motion, they were able to discover the order of distance of the planets.

One of the earliest and greatest philosophers, Aristotle, lived in Greece about 350 B.C. He summarized the astronomical knowledge of his day into a qualitative cosmology that remained dominant for 1800 years. On the basis of what seemed to be very good evidence—what he saw—Aristotle thought, and actually believed that he knew, that the Earth was at the center of the Universe and that the planets, the Sun, and the stars revolved around it (Fig. 18–9). The Universe was made up of a set of 55 celestial spheres that fit around each other and that had rotation as their natural motion. Each of the heavenly bodies was carried around the heavens by a sphere; the motion of each sphere was determined by the motion of other spheres. These motions combined to account for the various observed motions, including revolution around the Earth, retrograde motion, and motion above and below the ecliptic. Each planet had several spheres to account for its various motions. The outermost sphere was that of the fixed stars, beyond which lay the prime mover, *primum mobile,* the sphere that caused the general rotation of the stars overhead.

Aristotle's theories ranged through much of science. He held that below the sphere of the Moon everything was made of four basic "elements": earth, air, fire, and water. The fifth "essence"—the quintessence—was a perfect, unchanging, transparent element of which the celestial spheres are formed.

Aristotle's theories dominated scientific thinking for almost two millennia, until the Renaissance. Unfortunately, most of his theories were far from what we now consider to be correct, so we tend to think that the widespread acceptance of Aristotelian physics impeded the development of science.

Figure 18–9 Aristotle's earth-centered theory. The Earth is at the center, orbited in larger and larger circles by the Moon, Mercury, Venus, the Sun, Mars, Jupiter, Saturn, and the stars.

In about A.D. 140, almost 500 years after Aristotle, the Greek astronomer Claudius Ptolemy (Fig. 18–10), in Alexandria, presented a detailed theory of the universe that explained the retrograde motion. Ptolemy's model, like Aristotle's, was earth-centered. To account for the retrograde motion of the planets, the planets had to be moving not simply on large circles around the earth but rather on smaller circles, called *epicycles*, whose centers moved around the earth on larger circles, called *deferents* (Fig. 18–11). (The notion of epicycles and deferents had been advanced earlier by such astronomers as Hipparchus, whose work on star catalogues we men-

tioned in Section 5.1.) It seemed natural that the planets should follow circles in their motion, since circles were thought to be "perfect" figures.

Sometimes the center of the deferent was not centered at the earth (and thus the circle was *eccentric*). The epicycles moved at a constant rate of angular motion (that is, the angle through which they moved was the same for each identical period of time). However, another complication was that the point around which the epicycle's angular motion moved uniformly was neither at the center of the Earth nor at the center of the deferent. The epicycle moved at a uniform angular rate about still another point, the *equant*. The equant and the Earth were equally spaced on opposite sides of the center of the deferent, as the figure shows.

Ptolemy's views were very influential in the study of astronomy, because versions of his ideas and of the tables of planetary motions that he computed were accepted for nearly 15 centuries. His major work, the *Almagest*, contained both his ideas and a summary of the ideas of his predecessors (especially those of Hipparchus), and is the major source of our knowledge of Greek astronomy.

*18.3c Other Ancient Astronomy

At the same time that Greek astronomers were developing their ideas, Babylonian priests were computing positions of the Moon and planets. The major work was carried out between about 700 B.C. and A.D. 50.

The Babylonian tablets that have survived show lists and tables of planetary positions and eclipses. The tablets also show predictions, for example, of the times when planets would be closest to (in *conjunction* with) and farthest from (in *opposi-*

Figure 18–10 Ptolemy. (Burndy Library, photograph by Owen Gingerich)

The equant was invented based on the knowledge that the seasons are not equal in length. The present-day values for the northern hemisphere are

spring	92^{d}19^h
summer	93^{d}15^h
autumn	89^{d}20^h
winter	89^{d}00^h

If the Sun goes in a circular orbit at uniform speed, then the orbit cannot be centered at the Earth.

This work is sometimes referred to as Chaldean—pronounced something like "kal-dē'-an" in English. Chaldea was the southern part of Babylon. Babylonian methods were communicated to the Greeks, and it is through the Greeks that Babylonian astronomy influenced Western thought.

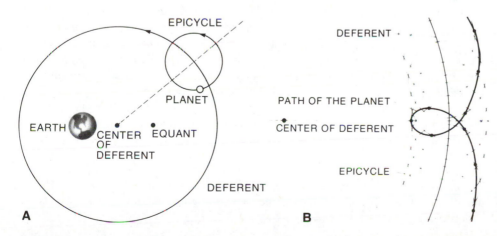

A
B

Figure 18–11 (A) In the Ptolemaic system, a planet would move around on an epicycle which, in turn, moved on a deferent. Variations in the apparent speed of the planet's motion in the sky could be accounted for by having the epicycle move uniformly around a point called the equant, instead of moving uniformly around the Earth or the center of the deferent. The Earth and equant were on opposite sides of the center of the deferent and equally spaced from it. When the planet was in the position shown, it would be in retrograde motion. (B) In the Ptolemaic system, the projected path in the sky of a planet in retrograde motion is shown.

Figure 18–12 Ulugh Begh, a grandson of the Tartar chief Tamerlane, compiled a star catalogue in the early 15th century. It was accurate to 1/10 the Moon's diameter, showing the high quality of his pre-telescopic instruments. His observatory is still standing at Samarkand. Soon thereafter in Europe, Regiomontanus (as John Muller became known) compared observations of planets and of eclipses with tables of predictions, and started a printing press at which he published both theoretical works and tables of positions. Regiomontanus, as had Arab astronomers in the 8th–10th centuries, translated and interpreted Ptolemy's principal work, the *Almagest*. The observational predictions in use in the 13th–16th centuries were those computed for King Alfonso X in Spain and known as the *Alphonsine Tables.*

tion to) the Sun in the sky and when objects would be visible for the first or last time in a year.

*18.3d Astronomy of the Middle Ages

The Middle Ages, which we might loosely define as the thousand years up to about 1500, were not marked by epochal astronomical discoveries or significant models, such as those that were advanced by Ptolemy and Aristotle. Nonetheless, the period was not as dead as commonly presented. Though we shall not go into details here, Aristotelian ideas were studied, developed, and criticized, and some observations were made (Fig. 18–12).

Other civilizations were also developing astronomical ideas. Chinese astronomers, for example, recorded the occurrences not only of eclipses but also of novae, and discussed the possible motion of the Earth. And they had known basic ideas, like that of the calendar, for centuries. India had observatories thousands of years ago. In Polynesia, the stars were long used for accurate navigation at sea. Stonehenge and other sites in Great Britain are evidence of prehistoric observations.

Helios was the sun god in Greek mythology.

18.4 Nicolaus Copernicus: The Sun at the Center

The credit for the breakthrough in our understanding of the solar system belongs to Nicolaus Copernicus (Fig. 18–13), a Polish astronomer whose 500th birthday was celebrated by the astronomical community in 1973. Copernicus advanced a *heliocentric* —Sun-centered—theory (Figs. 18–14 and 18–15). He suggested that the retrograde motion of the planets could be readily explained if the Sun, rather than the Earth, was at the center of the Universe; that the Earth is a planet; and that the planets move around the Sun in circles.

Aristarchus of Samos, a Greek scientist, had suggested a heliocentric theory 18 centuries earlier, though we do not know how detailed a picture of planetary motions he presented. His heliocentric suggestion required the then apparently ridiculous notion that the Earth itself moved, in contradiction to our sense and to the theories of Aristotle. If the Earth is rotating, for example, why aren't birds and clouds left behind the moving Earth? Only the 18th-century discovery by Isaac Newton of laws of motion substantially different from Aristotle's solved this dilemma.

Aristarchus's heliocentric idea had long been overwhelmed by the *geocentric*— Earth-centered—theories of Aristotle and Ptolemy.

Copernicus's theory, although it put the Sun instead of the Earth at the center of the solar system (and, for then, the Universe), still assumed that the orbits of celestial objects were circles. As a result, in order to improve the agreement between theory and observation, Copernicus still invoked the presence of some epicycles, though he eliminated the equant. Further, the detailed predictions that Copernicus himself computed on the basis of his theory were not in much better agreement with the existing observations than tables based on Ptolemy's model, because in many cases Copernicus still used Ptolemy's observations. The heliocentric theory appealed to Copernicus and to many of his contemporaries on philosophical grounds, rather than because direct comparison of observations with theory showed the new theory to be better.

Copernicus' heliocentric theory was published in 1543 in the book he called *De Revolutionibus*, which can be translated as *Concerning the Revolutions* (Fig. 18–17 and Color Plate 22). In it, he explained the retrograde motion of the planets as follows (Fig. 18–18):

Figure 18–13 Copernicus, in a 16th-century woodcut.

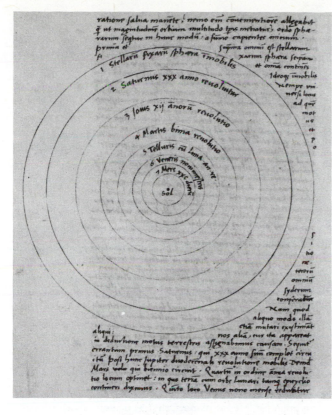

Figure 18–14 The page of Copernicus's original manuscript in which he drew his heliocentric system. The Sun (Sol) is at the center surrounded by Mercury (Merc), Venus (Veneris), Earth (Telluris), Mars (Martis), Jupiter (Jovis), Saturn (Saturnus), and the fixed stars. The manuscript is in the University library in Cracow.

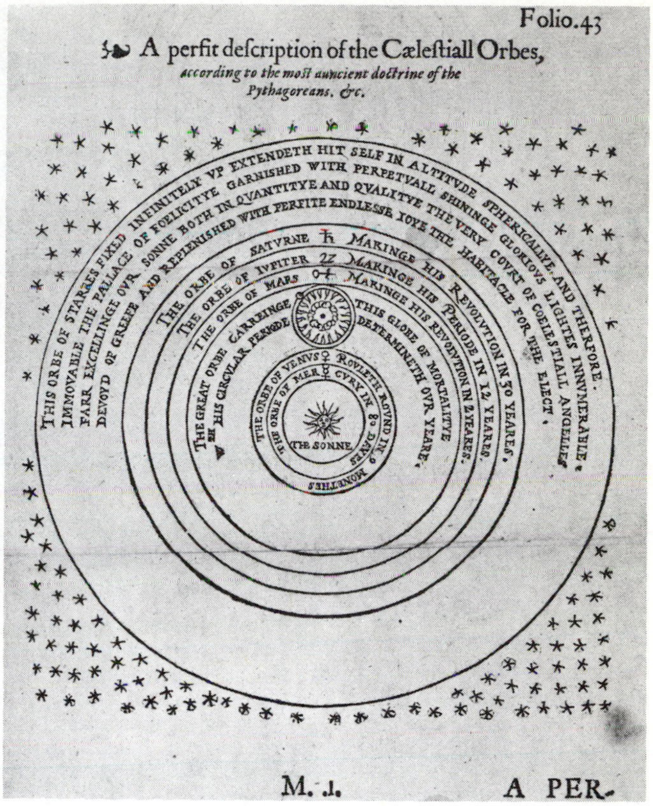

Figure 18–15 The first English diagram of the Copernican system, which appeared in an appendix by Thomas Digges to a book by his father in 1568.

Let us consider, first, an outer planet like Mars as seen from the Earth. Mars orbits the Sun more slowly than does the Earth. As the Earth approaches the part of its orbit that is closest to Mars, the projection of the Earth-Mars line outward to the stars (which are essentially infinitely far away compared to the planets) moves slightly against the stellar background. As the Earth comes to the point in its orbit closest to Mars, and then passes it, the projection of the line joining the two planets can actually seem to go backward, since the Earth is going at a greater speed than Mars. Then, as the Earth continues around its orbit, Mars appears to go forward again. A similar explanation can be demonstrated for the retrograde loops of the inner planets.

Figure 18–16 Copernicus' signature from the Uppsala University Library in Sweden, photographed by Charles Eames and reproduced courtesy of Owen Gingerich.

NICOLAI CO-
PERNICI TORINENSIS
DE REVOLVTIONIBVS ORBI-
um cœlestium, Libri VI.

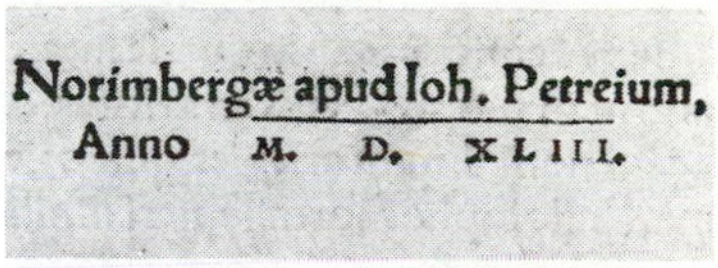

Norimbergæ apud Ioh. Petreium,
Anno M. D. XLIII.

Figure 18–17 From the title page of Copernicus's *De Revolutionibus*. Osiander added the words "Orbium Celestium." About 200 copies of this work are currently known to be extant.

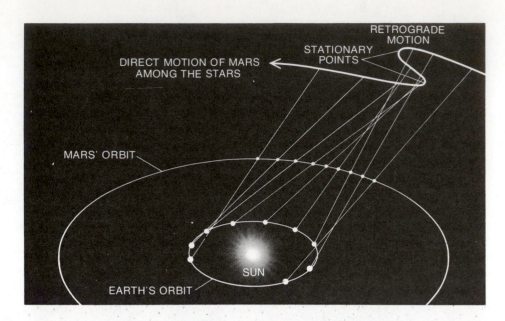

Figure 18–18 The Copernican theory explains retrograde motion as an effect of projection. For each of the nine positions of Mars shown from right to left, follow the line from Earth's position through Mars' position to see the projection of Mars against the sky. Mars' forward motion appears to slow down as the Earth overtakes Mars. Between the two stationary points, Mars appears in retrograde motion; that is, it appears to move backwards with respect to the stars. The drawing shows the explanation of retrograde motion for Mars or for other superior planets, that is, planets whose orbits lie outside that of the Earth. Similar drawings can explain retrograde motion for inferior planets, that is, planets whose orbits lie inside that of the Earth (namely Mercury and Venus).

18.5 Galileo Galilei

Leonardo da Vinci, for example, deduced in about A.D. 1500 why we are sometimes able to see the dark part of the Moon faintly lighted. We see the reflection of sunlight off the Earth and over to the Moon, light known as *earthshine.*

As art, music, and architecture began to flourish after the Middle Ages, astronomy also developed. Copernicus's work can be considered to be the beginning of the astronomical renaissance, and from that time on there was continual development.

The Italian scientist Galileo Galilei (Fig. 18–19) began to believe in the Copernican system in the 1590's, and later provided important observational confirmation of the theory. In late 1609 or early 1610, simultaneously with the first settlements in the American colonies, Galileo was the first to use a telescope for astronomical observation and to report his findings (Section 2.3).

In his book *Sidereus Nuncius (The Starry Messenger),* published in 1610, he reported that with this telescope he could see many more stars than he could with the naked eye, and could see that the Milky Way and certain other hazy-appearing regions of the sky actually contained individual stars. He described views of the Moon (Fig. 18–20), including the discovery of mountains, craters, and the relatively dark regions he called (and that we still call) *maria* (pronounced mar'-ē-a), seas. And, perhaps most important, he discovered that small bodies revolved around Jupiter (Fig. 18–21). This discovery proved that all bodies did not revolve around the Earth, and also, by displaying something that Aristotle and Ptolemy obviously had not known about, showed that Aristotle and Ptolemy had not been omniscient.

Subsequently, Galileo found that Saturn had a more complex shape than that of a sphere (though it took better telescopes to actually show the rings). He was one of several people who almost simultaneously discovered sunspots, and he studied their motion on the surface of the Sun. Galileo also discovered that Venus went through an entire series of phases (Fig. 18–22). This could not be explained on the basis of

Figure 18–19 Galileo Galilei, 1564–1642.

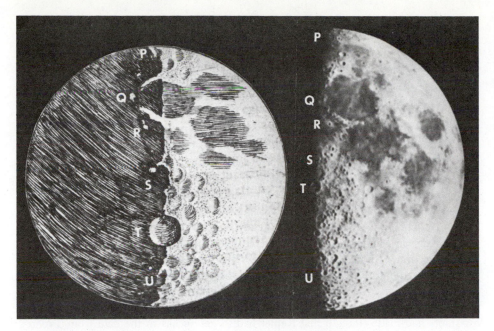

Figure 18–20 One of Galileo's original drawings of the moon, made with his small, unmounted (and therefore shaky) telescope, compared with a modern photograph. Notice how well Galileo did.

Thomas Harriot in England observed the Moon with a telescope a few months before Galileo, but did not report the features. Perhaps Galileo was sensitized to interpreting surfaces by the Italian Renaissance.

Ptolemaic theory, because if Venus travelled in an epicycle located between the Earth and the Sun, Venus should always appear as a crescent (Fig. 18–23). Also, more generally, Galileo's observations showed that Venus was a body similar to the Earth and the Moon in that it received light from the Sun rather than generating its own.

But although the Roman Catholic Church was not concerned about models made to explain observations, it was concerned about assertions that those models represented physical truth. It is difficult to say how much the Church feared that Galileo would generalize his ideas to a philosophical and religious level. In his old age, Galileo was forced by the Inquisition to recant his belief in the Copernican theory (Fig. 18–24). The controversy begun then has continued to the present day. Even now, the Vatican has a group studying how Galileo was treated. Pope John Paul II set up the group in 1979 "in loyal recognition of wrongs from whatever side they come . . . dispel the mistrust that still opposes, in many minds, a fruitful concord between science and faith." In 1983, he reported to a group celebrating the 350th anniversary of the publication of Galileo's *Dialogue* that it was "progressing very encouragingly" and that the Church upheld the freedom of scientific research.

Figure 18–21 A translation *(left)* of Galileo's original notes *(right)* summarizing his first observations of Jupiter's moons in January 1610. The shaded areas were probably added later. It had not yet occurred to Galileo that the objects were moons in revolution around Jupiter.

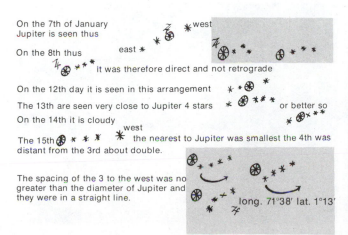

On the 7th of January Jupiter is seen thus

On the 8th thus   east

It was therefore direct and not retrograde

On the 12th day it is seen in this arrangement

The 13th are seen very close to Jupiter 4 stars   or better so

On the 14th it is cloudy

The 15th   west   the nearest to Jupiter was smallest the 4th was distant from the 3rd about double.

The spacing of the 3 to the west was no greater than the diameter of Jupiter and they were in a straight line.

long. 71°38' lat. 1°13'

Figure 18–22 The phases of Venus. Note that Venus is a crescent only when it is in a part of its orbit that is relatively close to the Earth, and so it looks larger at those times.

18.6 Tycho Brahe

In the last part of the 16th century, not long after Copernicus's death and while Galileo was a child, Tycho Brahe began a series of observations of Mars and other planets. Tycho, a Danish nobleman, set up an observatory on an island off the mainland of Denmark (Fig. 18–25). The building was called Uraniborg (after Urania, the muse of astronomy). The telescope had not yet been invented, but Tycho used giant sighting and angle-measuring instruments to make observations of unprecedented positional accuracy. In 1597, Tycho lost his financial support in Denmark and moved to Prague, arriving two years later. A young assistant, Johannes Kepler, came there to work with him. At Tycho's death, in 1601, Kepler—who had worked with Tycho for only 10 months—was left to analyze all the observations that Tycho and his assistants had made (though first Kepler had to get access to the data from Tycho's family, which proved troublesome).

18.7 Johannes Kepler

Tycho's observational data showed that the tables then in use did not adequately predict planetary positions. Kepler started to study them in 1600, and carried out detailed numerical calculations. (Nowadays we could use a computer to calculate in minutes results that took Kepler weeks to work out.) Kepler was eventually able to

Figure 18–23 In the Ptolemaic theory *(A)*, Venus and the Sun both orbited the Earth, but because it is known that Venus never gets far from the sun in the sky, the center of Venus' epicycle was restricted to always fall on the line joining the center of the deferent and the Sun. In this diagram Venus could never get farther from the Sun than the region restricted by the dotted lines. Thus, Venus would always appear as a crescent, though before the telescope was invented, this could not be verified. In the heliocentric theory *(B)* Venus is sometimes on the near side of the Sun, where it appears as a crescent, and it is sometimes on the far side, where we can see half or more of Venus illuminated. This agrees with Galileo's observations, a modern version of which appears as Figure 18–22. The actual observations can also be described using Tycho's theory (Fig. 18–25).

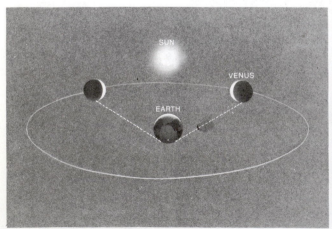

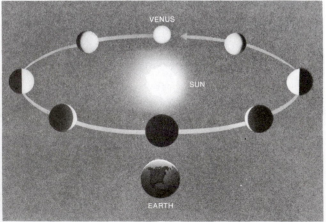

A   **B**

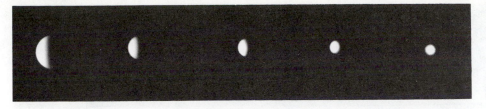

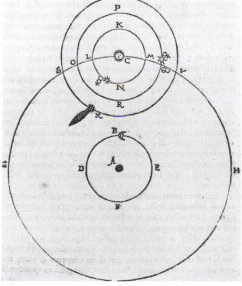

Figure **18–24** The frontispiece
to Galileo's *Dialogue Concerning
the Two World Systems,* 1632,
which led to his being taken before the Inquisition since he had
been warned in 1615 not to
teach the Copernican theory. According to the labels, Copernicus
is at the right, with Aristotle and
Ptolemy at the left; Copernicus
was drawn with Galileo's face,
however.

> ### Box 18.2 *Kepler's Laws of Planetary Motion*
>
> 1. The planets orbit the sun in ellipses, with the Sun at one focus.
> 2. The line joining the Sun and a planet sweeps through equal areas in equal times.
> 3. The squares of the periods of revolution are proportional to the cubes of the planets' distances from the Sun. (More accurately: the square of the period of a planet is proportional to the cube of the semimajor axis of its orbit—half the longest dimension of the ellipse.)

make sense out of the observations of Mars that Tycho had made, and thus clear up the discrepancies between the predictions and the observations of Mars' position in the sky. Kepler published three laws based on his empirical analyses, that is, laws based on experience and observation rather than theory, the first two appearing in 1609 (shortly before Galileo first turned a telescope on the sky) and the third—following years of additional work—in 1618. Let us first state the laws, and then discuss them in detail.

18.7a Kepler's First Law (1609)

Until Kepler worked out these laws, even the heliocentric calculations assumed that the planets followed "perfect" orbits, namely, circles. The discovery by Kepler that the orbits were in fact ellipses greatly improved the accuracy of the calculations.

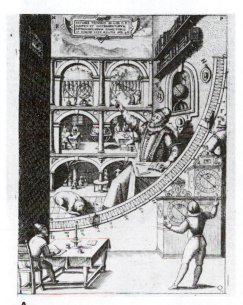

A   **B**

Figure **18–25** (*A*) Tycho's observatory at Uraniborg, Denmark.
Here Tycho is seen showing the
mural quadrant (that is, a big
quarter-circle on a wall) that he
used to measure the altitudes at
which stars and planets crossed
the meridian. (*B*) Tycho, known
throughout Europe, advanced his
own, earth-centered cosmology,
in which the Sun and Moon revolved around the Earth. Though
his cosmology was shortly forgotten, Tycho's current reputation
results from his having provided
the observational base for Kepler's research. The Comet of 1577
also appears on this plate reproduced from Tycho's book about
that comet, published in 1588.

Figure 18–26   It is easy to draw an ellipse. Put two nails or thumbtacks in a piece of paper, and link them with a piece of string that has some slack in it. If you pull a pen around while the pen keeps the string taut, the pen will necessarily trace out an ellipse. The string doesn't change in length, so the sum of the lengths of the lines from the pen to the foci equals the length of the string, which is constant. The shape of the ellipse will change if you change the length of the string, or if you change the distance between the foci.

An ellipse is a curve defined in the following way: First choose any two points on a plane; these points are called the *foci* (each is a *focus*). From any point on the ellipse, we can draw two lines, one to each focus. The sum of the lengths of these two lines is the same for each point on the ellipse (Fig. 18–26).

The *major axis* of an ellipse is the line within the ellipse that passes through the two foci, or the length of that line (Fig. 18–27). We often speak of the *semimajor axis*, which is just half the length of the major axis (semi- is a prefix from the Greek, and means "half"). The *minor axis* is the part of the line lying within the ellipse that is drawn perpendicular to the major axis and bisects it, or the length of that line. When one of the foci lies on top of the other, the major and minor axes are the same length, and the ellipse is the special case that we call a *circle*.

We often describe how far an ellipse is "out of round" by giving its *eccentricity* (Fig. 18–28), the distance between the foci divided by the length of the major axis. For a circle, the distance between the foci is zero, so its eccentricity is zero.

Ellipses (and therefore circles), parabolas, and hyperbolas are *conic sections,* that is, sections of a cone (Fig. 18–29). If you slice off the top of a cone, the shape of the slice is an ellipse. If your cut is parallel to the side of the cone, a parabola results. If your cut is tipped still further over, a hyperbola results.

For a planet orbiting the Sun, the Sun is at one focus of the elliptical orbit; nothing special marks the other focus. (We say that it is "empty.")

Figure 18–27   (A) The parts of an ellipse. (B) The ellipse shown has the same *perihelion* distance (closest approach to the Sun) as does the circle. Its *eccentricity,* the distance between its foci divided by its major axis, is 0.5. If the perihelion distance is kept constant but the eccentricity is allowed to reach 1, then we have a parabola. For eccentricities greater than 1, we have hyperbolas.

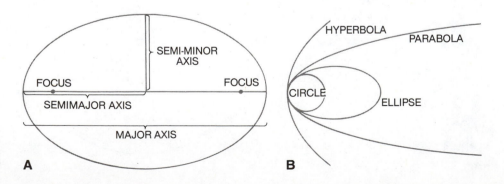

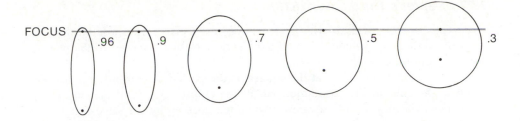

Figure 18–28 A series of ellipses of the same major axis but different eccentricities. The foci are marked; these are the two points inside with the property that the sum of the distances from any point on the circumference to the foci is constant. As the eccentricity—distance between the foci divided by the major (longer) axis—approaches 1, the ellipse approaches a straight line. As the eccentricity approaches zero, the foci come closer and closer together. A circle is an ellipse of zero eccentricity.

18.7b Kepler's Second Law (1609)

The second law, also known as the *law of equal areas,* governs the speed with which the planets travel in their orbits. Kepler's second law says that **the line joining the Sun and a planet sweeps through equal areas in equal times.** When a planet is at its greatest distance from the Sun in its elliptical orbit, the line joining it with the Sun sweeps out a long, skinny sector. (A *sector* is the area bounded by two straight lines from a focus of an ellipse and the part of the ellipse joining their outer ends.) The planet travels relatively slowly in this part of its orbit. By Kepler's second law, the long skinny sector must have the same area as the short, fat sector formed in the same period of time when the planet is closer to the Sun (Fig. 18–30). The planet, therefore, travels faster in its orbit then.

Kepler's second law is especially noticeable for objects with very elliptical orbits such as comets. It explains why Halley's Comet sweeps so quickly, within months, through the inner part of the solar system where the Earth is, and takes the rest of its 76-year period moving slowly through the outer parts.

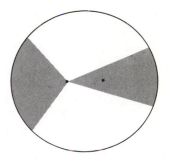

Figure 18–30 Kepler's second law states that the two shaded sectors, which represent the areas covered by a line drawn from a focus of the ellipse to an orbiting planet in a given length of time, are equal in area. The Sun is at this focus; nothing is at the other focus.

Figure 18–29 An ellipse is a "conic section" (a slice of a cone), in that the intersection of a cone and a plane that passes through the sides of a cone (and not the bottom) is an ellipse. If the plane is parallel to an edge of the cone, a parabola results. If the plane is tipped further over so that it is neither parallel to the edge nor intersects the side, then a hyperbola results. So parabolas and hyperbolas are conic sections as well.

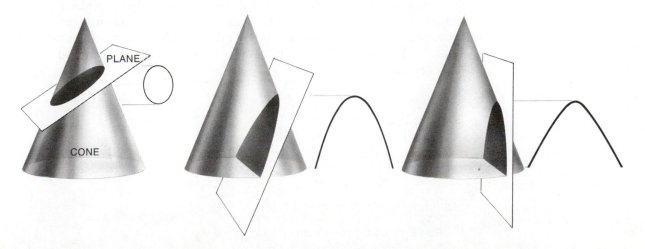

18.7c Kepler's Third Law (1618)

Kepler's third law deals with the length of time a planet takes to orbit the Sun, called its *period* of revolution. It relates the period to some measure of the planet's distance from the Sun (Fig. 18–31).

Kepler's third law says that **the square of the period of revolution is proportional to the cube of the semimajor axis of the ellipse.** That is, if the cube of the semimajor axis goes up, the square of the period goes up by the same factor. It is often easiest to express Kepler's third law by relating values for a planet to values for the earth. If P is the period of revolution of a planet and R is the semimajor axis,

$$\frac{P_{\text{Planet}}^{2}}{P_{\text{Earth}}^{2}} = \frac{R_{\text{Planet}}^{3}}{R_{\text{Earth}}^{3}}.$$

We can choose to work in units that are convenient for us on the Earth. We call the average distance from the Sun to the Earth 1 *Astronomical Unit* (1 *A.U.*). Similarly, the unit of time that the Earth takes to revolve around the Sun is defined as 1 *year*. Using these values, Kepler's third law appears in a simple form, since the numbers on the bottom of the equation are just 1.

EXAMPLE: If we know from observation that Jupiter takes 11.86 years to revolve around the sun, we can find Jupiter's distance from the sun in the following way:

$$\frac{(11.86)^{2}}{P_{\text{Earth}}^{2}} = \frac{R_{\text{Jupiter}}^{3}}{R_{\text{Earth}}^{3}}, \text{ and}$$

$$\frac{(11.86)^{2}}{1} = \frac{R_{\text{Jupiter}}^{3}}{1}.$$

Nowadays, we can calculate R easily with a pocket calculator. But astronomers are often content with approximate values that can be calculated in your head. For example, $(11.86)^{2}$ can be rounded off to $12^{2} = 144$. We can then easily find the cube root of 144 by trial-and-error. It is not hard to calculate that $4^{3} = 64$ (which is too small), $5^{3} = 125$ (which is too small), and $6^{3} = 216$ (which is too large). So the radius of Jupiter's orbit around the Sun is between 5 and 6, and is a little over 5.5 A.U. (The actual value is 5.2 A.U., so our rough calculation was fine.)

We have used Kepler's third law to determine how the period of a planet revolving around the Sun is related to the size of its orbit. But how do we find the constant of proportionality between P^{2} and R^{3}, that is, the number by which R^{3} must be multiplied to get P^{2} if we don't do the simplification of comparing with the values for the Earth? In the next section, we shall see that Isaac Newton derived mathematically that the constant of proportionality depends on the mass of the central body, which is the sun in this case.

Figure 18–31 Kepler's third law relates the period of an orbiting body to the size of its orbit. The outer planets orbit at much slower velocities than the inner planets and also have a longer path to follow in order to complete one orbit. The distances the planets travel in their orbits in 1 year is shown here.

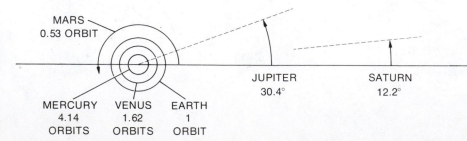

MARS
0.53 ORBIT

JUPITER
30.4°

SATURN
12.2°

MERCURY   VENUS   EARTH
4.14   1.62   1
ORBITS   ORBITS   ORBIT

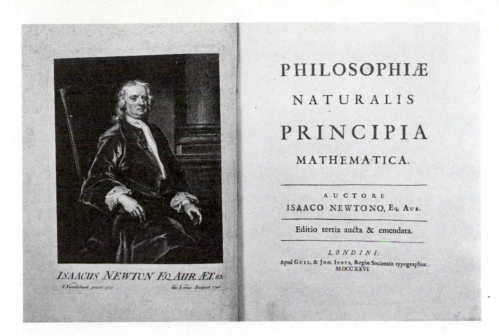

Figure 18–32 The title page and frontispiece of the third edition of Isaac Newton's *Principia Mathematica,* published in 1726. The first edition had appeared in 1687.

The period of revolution of satellites around planetary bodies follows Kepler's laws as well. This applies to artificial satellites in orbit around the Earth, and to moons of other planets.

18.8 Isaac Newton

Isaac Newton was born in England in 1642, the year of Galileo's death. He was to become the greatest scientist of his time and perhaps of all time. Previously in this book we have described his discovery that visible light can be broken down into a spectrum and his invention of the reflecting telescope. Here we shall speak only of his ideas about motion and gravitation.

For many years Newton developed his ideas about the nature of motion and about gravitation. In order to derive them mathematically, he invented calculus. Newton long withheld publishing his results, possibly out of shyness. Finally, his friend Edmond Halley, whose name we associate with the famous comet, persuaded him to publish his work. The *Philosophiae Naturalis Principia Mathematica (Mathematical Principles of Natural Philosphy),* known as *The Principia,* appeared in 1687 (Fig. 18–32). In it, Newton showed that the motions of the planets and comets could all be explained by the same law of gravitation that governs bodies on Earth. In fact, he derived Kepler's laws on theoretical grounds.

URANUS
4.29°

NEPTUNE
2.18°

PLUTO
1.5°

*Box 18.3 Kepler's Third Law, Newton, and Planetary Masses

Astronomers nowadays often make rough calculations to test whether physical processes under consideration could conceivably be valid. Astronomy has also had a long tradition of exceedingly accurate calculations. Pushing accuracy to one more decimal place sometimes leads to important results.

For example, Kepler's third law, in its original form—the period of a planet squared is proportional to its distance from the Sun cubed (P^2 = constant × R^3)—holds to a reasonably high degree of accuracy and seemed completely accurate when Kepler did his work. But now we have more accurate observations. If we consider each of the planets in turn, and if a term involving the sum of the masses of the Sun and the planet under consideration is included in the equation

$$P^2 = \frac{\text{constant}}{m_{\text{Sun}} + m_{\text{Planet}}} \times R^3,$$

the agreement with observation is improved in the fourth decimal place for the planet Jupiter and to a lesser extent for Saturn. The masses of the other planets are too small to have a detectable effect on the relation between P^2 and R^3.

Isaac Newton derived the equation in its general form:

$$P^2 = \frac{4\pi^2}{G(m_1 + m_2)} a^3$$

for a body with mass m_1 revolving in an elliptical orbit with semimajor axis a around a body with mass m_2. The constant G is the *universal gravitational constant* (Appendix 2). Newton's formula shows that the planet's mass contributes to the value of the proportionality constant, but its effect is very small. (The effect can be detected even today only for the most massive planets—Jupiter and Saturn.)

Kepler's third law, and its subsequent generalization by Newton, applies not only to planets orbiting the Sun but also to any bodies orbiting other bodies under the control of gravity. Thus it also applies to satellites orbiting planets. We determine the mass of the earth by studying the orbit of our Moon, and determine the mass of Jupiter by studying the orbits of its moons. Until recently, we were unable to reliably determine the mass of Pluto because we could not observe a moon in orbit around it. The discovery of a moon of Pluto in 1978 finally allowed us to determine Pluto's mass; we discovered that our previously best estimates (based on what we thought were Pluto's gravitational effects on Uranus) were way off. The same formula can be applied to binary stars to find their masses.

In the derivations, Newton used the law of gravitation that he had discovered; Newton—whether or not you believe that an apple fell on his head—was the first to realize the universality of gravity. He formulated the law that the force of gravity *(F)* between two bodies varies directly with the product of the masses (m_1 and m_2) of the two bodies and inversely with the square of the distance between them:

force of gravity $\propto m_1 m_2/d^2$.

$\propto$ stands for "is proportional to"; it means that the left-hand side is equal to a constant times the right hand side. We can write $F = -Gm_1 m_2/d^2$, the law of universal gravitation. (The minus sign means that the force tends to pull things together instead of push them apart.) It is a universal law in that it works all over the Universe rather than being limited to local applicability (on Earth or even in the solar system). G (Appendix 2) is called the *universal gravitational constant*.

Figure 18–33 It is all too easy to "misspeak" when discussing rotation, which refers to a body turning on its axis, and revolution, which refers to a body orbiting around another body.

18.9 The Revolution and Rotation of the Planets

The motion of the planets around the Sun in their orbits is called *revolution*. The spinning of a planet is called *rotation* (Fig. 18–33). The Earth, for example, revolves around the Sun in 1 year and rotates on its axis in 1 day (thus defining these terms).

The orbits of all the planets lie in approximately the same plane. They thus take up only a disk whose center is at the Sun, rather than a full sphere. Little is known of the parts of the solar system away from this disk, although the comets may originate in a spherical cloud around the Sun.

The plane of the Earth's orbit around the Sun is called the *ecliptic plane* (Fig. 18–34). Of course, since we are on the Earth rather than outside the solar system looking in, the position of the Sun appears to move across the sky with respect to the stars. The path the Sun takes among the stars, as seen from the Earth, is called the *ecliptic* (see also Section 3.3).

The *inclination* of the orbit of a planet is the angle that the plane in which its orbit lies makes with the plane in which the Earth's orbit lies. The inclinations of the orbits of the other planets with respect to the ecliptic are small, with the exception of Pluto's 17° (Fig 18–35). Of the other planets, Mercury's inclination is 7°; the remainder have inclinations of less than 4°. Pluto's much larger inclination is discrepant and is just one of the pieces of evidence suggesting that Pluto may not have formed under the same circumstances as the other planets in the solar system.

The fact that the planets all orbit the Sun in essentially the same plane is one of the most important facts that we know about the solar system. Its explanation is at the base of most models of the formation of the solar system. Central to that explanation is a property that astronomers and physicists use in analyzing spinning or revolving objects: *angular momentum*. The amount of angular momentum of a small body revolving around a large central body is (distance from the center) × (velocity)

Figure 18–34 The ecliptic plane is the plane of the Earth's orbit around the Sun.

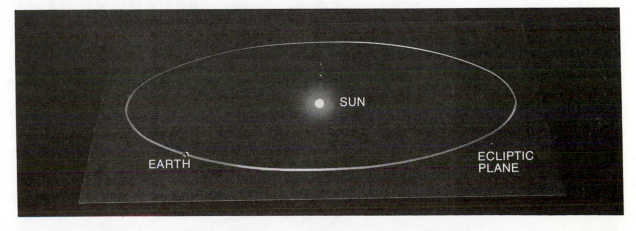

SUN

EARTH

ECLIPTIC PLANE

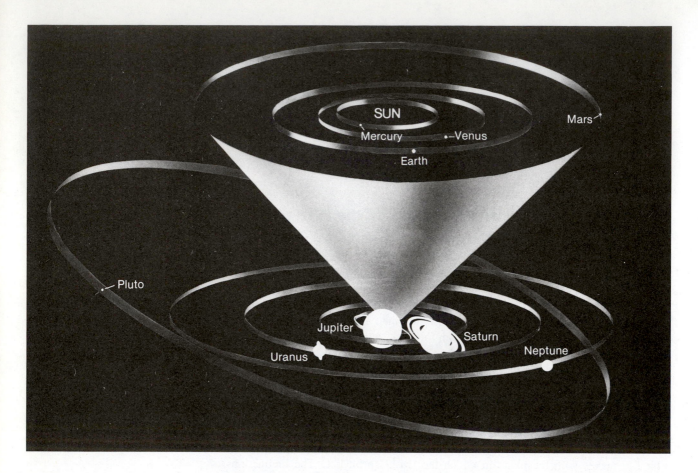

Figure 18–35 The orbits of the planets, with the exception of Pluto, have only small inclinations to the ecliptic plane. An enlarged view of the inner solar system is included at the top of the cone.

× (mass). The importance of angular momentum lies in the fact that it is "conserved"; that is, the total angular momentum of the system (the sum of the angular momenta of the different parts of the system) doesn't change even though the distribution of angular momentum among the parts may change. The total angular momentum of the solar system should thus be the same as it was in the past, unless some mechanism is carrying angular momentum away.

The most familiar example of angular momentum is an ice skater (Fig. 18–36). She may start herself spinning by exerting force on the ice with her skates, thus giving herself a certain angular momentum. When she wants to spin faster, she draws her arms in closer to her. This changes the distribution of her mass so that it is effectively closer to her center (the axis around which she is spinning), and to compensate for this she starts to rotate more quickly so that her angular momentum remains the same.

Thus from the fact that every planet is revolving in the same plane and in the same direction, we deduce that the solar system was formed of primordial material that was rotating in that direction. We would be very surprised to find a planet revolving around the Sun in a different direction—and we do not—or to a lesser extent, to find a planet rotating in a direction opposite to that of its fellows. There are, however, two examples of such backward rotation, which must each be carefully considered.

Such rotation in the opposite sense is called *retrograde rotation*. Do not confuse it with *retrograde motion*, which has to do with the apparent motion of the position of a planet in the sky as seen from the Earth.

18.10 Theories of Cosmogony

The solar system exhibits many regularities, and theories of its formation must account for them. The orbits of the planets are almost, but not quite, circular, and all lie in essentially the same plane. All the planets revolve around the Sun in the same direction, which is the same direction in which the Sun rotates. Moreover, almost all the planets and planetary satellites rotate in that same direction. Some planets have families of satellites that revolve around them in a manner similar to the way that the planets revolve around the Sun. And cosmogonical theories must explain the spacing of the planetary orbits and the distribution of planetary sizes and compositions.

René Descartes, the French philosopher, was one of the first to consider the origin of the solar system in what we would call a scientific manner. In his theory, proposed in 1644, circular eddies—called vortices—of all sizes were formed at the beginning of the solar system in a primordial gas and eventually settled down to become the various celestial bodies.

After Newton proved that Descartes's vortex theory was invalid, it was 60 years before the next major developments. The Comte de Buffon suggested in France in 1845 that the planets were formed by material ejected from the Sun when what he called a "comet" hit it. (At that time, the composition of comets was unknown, and it was thought that comets were objects as massive as the Sun itself.) Later versions of Buffon's theory, called *catastrophe theories,* followed a similar line of reasoning, although they spoke explicitly of collision with another star. A variation postulated that the material for the planets was drawn out of the Sun by the gravitational attraction of a passing star. This latter possibility is called a *tidal theory*.

Catastrophe and tidal theories are currently out of fashion, for they predict that only very few planetary systems would exist. This prediction is based on calculations that show that only very few stellar collisions or near-collisions would have taken place in the lifetime of the galaxy. There is some observational evidence that many stars may have planets (Section 28.2), which would require a method of formation that can account for more planetary systems. Also, theoretical calculations show that gas drawn out of a star in collision or by a tidal force would not condense into planets, but would rather disperse. Hence, catastrophe and tidal theories are currently out of favor.

The theories of cosmogony that astronomers now tend to accept stem from another 18th-century idea. Immanuel Kant, the noted German philosopher, suggested in 1855 that the Sun and the planets were formed by the same type of process. In 1896, the Marquis Pierre Simon de Laplace (pronounced La Plahce), the French mathematician, independently advanced a similar kind of theory to Kant's when he postulated that the Sun and the planets all formed from a spinning cloud of gas called a *nebula*. Laplace called this the *nebular hypothesis,* using the word "hypothesis" because he had no proof that it was correct. The spinning gas supposedly threw off rings that eventually condensed to become the planets. But though these beginnings of modern cosmogony were laid down in the time of Benjamin Franklin and George Washington, the theory still has not been completely understood or quantified, for not all the stages that the primordial gas would have had to follow are understood.

Further, some of the details of Laplace's theory were later thought for a time to be impossible. For example, it was calculated in the last century, by the great British physicist James Clerk Maxwell, that the planets would not have been able to condense out of the rings of gas, and also that they could not have been set to rotate as fast as they do.

So for a time, the nebular hypothesis was not accepted. But the current theories of cosmogony again follow Kant and Laplace (Fig. 18–37). In these *nebular theories*

Figure 18–36 An ice skater spins faster when she draws her arms in, as Katarina Witt (Gold Medal winner at the 1984 Winter Olympics) has done here. Since angular momentum is conserved, concentrating her mass closer to the spin axis leads to a compensating increase in spin.

The study of the origin of the solar system is called *cosmogony* (pronounced with a hard "g").

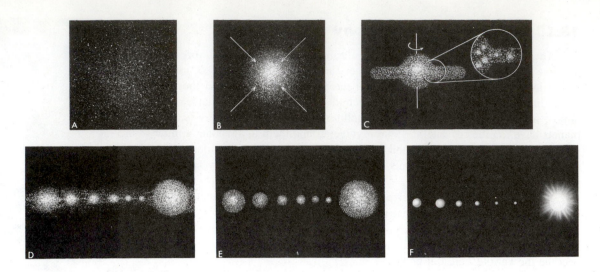

Figure 18–37 The leading model for the formation of the solar system has the protosolar nebula condensing and, between stages C and D, contracting to form the protosun and a large number of small bodies called planetesimals. The planetesimals clumped together to form protoplanets (D), which in turn contracted to become planets. Some of the planetesimals may have become moons or asteroids (which are discussed in Section 27.3).

the Sun and the planets condensed out of what is called a *primeval solar nebula*. Some five billion years ago, billions of years after the galaxies began to form, smaller clouds of gas and dust began to contract out of interstellar space. Similar interstellar clouds of gas and dust can now be detected at many locations in our galaxy.

There is increasing evidence that the collapse of gas to form the solar system was set off by shock waves (Fig. 18–38) from a nearby supernova. The evidence concerns the unusual abundances in meteorites (Section 27.2) of certain isotopes; as far as we know, the isotopes could only have been formed in such supernova explosions. The fact that they have such short lifetimes indicates that they could not have been formed too long before our solar system's collapse.

Because of random fluctuations in the gas and dust from which it formed, the primeval solar nebula probably would have had a small net spin from the beginning. As it contracted, it would have begun to spin faster because of the conservation of angular momentum (the same reason that the ice skater spins faster and that pulsars and black holes spin so rapidly). Gravity would have contracted the spinning nebula into a disk: in the plane of the nebula's rotation, the spin had the effect of a force that counteracts gravity. In the directions perpendicular to this plane, though, there was no force to oppose gravity's pull, and the nebula collapsed in that direction.

Figure 18–38 This photograph shows a shock wave in air made by a bullet travelling faster than the speed of sound. The sharp curved line at the right is the shock wave; it represents a sharp change in pressure. (Courtesy of Harold E. Edgerton, MIT)

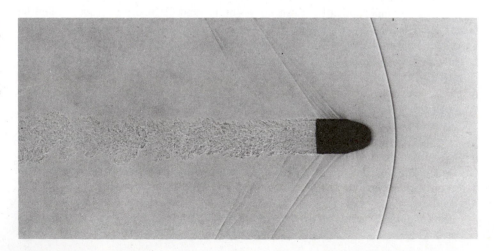

Perhaps rings of material were left behind by the solar nebula as it contracted towards its center. Perhaps there were additional agglomerations beside the *protosun,* the part of the solar nebula that collapsed to become the sun itself. These additional agglomerations, which may have been formed from interstellar dust, would have grown larger and larger. Millimeter-size particles would have clumped into larger bodies and then larger still; the size of the clumps would have increased until the bodies were kilometers and then thousands of kilometers across. The intermediate bodies, hundreds of kilometers across, are called *planetesimals.*

The planetesimals combined under the force of gravity to form *protoplanets.* These protoplanets may have been larger than the planets that resulted from them because they had not yet contracted, though gravity would ultimately cause them to do so. (We discussed a possible protoplanet around T Tauri in Section 8.1.) Some of the larger planetesimals may themselves have become moons.

As the Sun condensed, some of the energy it gained from its contraction heated it until its center reached the temperature at which nuclear-fusion reactions began taking place. The planets, on the other hand, were simply not massive enough to heat up sufficiently to have nuclear reactions start.

Several modifications of this basic theory have been worked out to accommodate particular observational facts. For example, we have to explain why the inner planets are small, rocky, and dense, while the next group of planets out are large and primarily made of light elements. Since the protosun would have been made primarily out of hydrogen and helium, with just traces of the heavier elements formed in earlier cyclings of the material through stars and supernovae, we need a mechanism for ridding the inner regions of the solar system of this lighter material. One possible way is to say the Sun flared fiercely andor often in its younger days (similar to a T Tauri star, as described in Section 8.1), and that the lighter elements were blown out of the inner part of the solar system. But proto-Jupiter and the other outer protoplanets, which were much farther away from the Sun, would have retained thick atmospheres of hydrogen and helium because of their high gravity.

A major modern method of calculation considers how the temperature decreases with distance from the center of the solar nebula, ranging from almost 2000 K close to the protosun, down to only about 20 K at the distance of Pluto. Various elements are able to condense—solidify—at different locations because of their differing temperatures. For example, in the positions occupied by the terrestrial planets it was too hot for icy substances to form, though such ices are present in the giant planets. Such calculations are perhaps the dominant consideration in much of the current cosmogonic research. The models of John Lewis of M.I.T. and A.G.W. Cameron of Harvard are noteworthy.

The nebular theories have received backing from the IRAS (1983's Infrared Astronomy Satellite) discovery that some stars, including the bright stars Vega and Fomalhaut (Appendix 5), shine over 10 times more brightly in the infrared beyond 60 microns than had been expected on the basis of their visible radiation or from IRAS observations of similar stars. Vega's radiation is coming from a region extending 80 A.U. all around the star, and there is no sign that Vega is losing mass. The most reasonable explanation is that we are seeing a ring of material around the star (Fig. 18–39) at a temperature of 90 K. We are likely to be seeing small material from millimeter size up to planetesimals, rather than a few bits of larger material (planets). An estimate of the total mass of the particles came out to roughly the same as the total mass of the planets and asteroids in our solar system. We could be seeing a new solar system in formation.

Infrared astronomers working with ground-based telescopes to do speckle interferometry (Box 6.1) have since found even more debris around two other stars. Steven Beckwith of Cornell and Ben Zuckerman of UCLA detected solar-system–sized

There are still more possibilities that bedevil cosmogonical research. Some of the planets and moons may not have been formed in their present configuration. Perhaps the Earth captured another protoplanet, which became the Moon. Perhaps Pluto was ejected from an orbit around Neptune into an orbit as an independent planet. Perhaps gravitational encounters put some of the planets and moons into retrograde rotation.

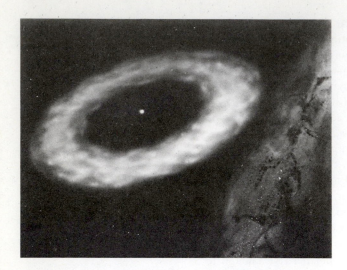

Figure 18–39 An artist's conception of the ring of material around Vega, discovered in the infrared by IRAS. An actual IRAS view of the Milky Way is in the corner of the picture.

clouds of dust around the stars HL Tauri and R Monocerotis. The clouds are elongated, as though they are disk-shaped. From the way in which the particles scatter infrared radiation, the astronomers deduced that an earth's mass of tiny dust particles, and our solar system's mass of dust and gas together, is present around HL Tau; R Mon has perhaps five times as much dust. It seems encouraging, though it is not proven, that planetary systems are forming in these young stars, only 100,000 years of age.

Summary and Outline

The scale of the solar system and the planets themselves
The phases of the Moon and the planets (Section 18.1)
Lunar and solar eclipses (Section 18.2)
Ancient astronomy (Section 18.3)
  Egyptian astronomy (Section 18.3a)
    Astronomy served the practical purposes of the farmers
  Greek astronomy (Section 18.3b)
    The apparent motion of the planets, including retrograde
    Geocentric theory of Aristotle and Ptolemy
    Epicycles, deferents, equants needed to explain the planetary motions
Heliocentric theory (Section 18.4)
  Copernicus, *De Revolutionibus* published in 1543
  Retrograde motion explained as a projection effect
  Copernican system still used circular orbits and epicycles
Galileo (Section 18.5)
  First to use telescope for astronomical observations
  His observations supported the heliocentric theory and showed that Aristotle and Ptolemy had not known everything

Tycho Brahe (Section 18.6)
  Amassed the best set of observations then ever obtained
Kepler (Section 18.7)
  Studied Tycho's observations and discovered three laws of planetary motion: (1) orbits are ellipses with sun at one focus; (2) equal areas are swept out in equal times; (3) period squared is proportional to distance cubed
Newton (Section 18.8)
  Newton worked out laws of gravity and of motion; derived Kepler's laws and so found the masses of the planets
The ecliptic; inclinations of the orbits of the planets; conservation of angular momentum (Section 18.9)
All planets revolve and most rotate in the same direction
Theories of cosmogony (Section 18.10)
  Catastrophe theories
  Nebular theories of Kant and Laplace have newer versions: protosun, planetesimals, and protoplanets
  A nearby supernova may have begun our solar system; infrared observation may show solar systems in formation.

Key Words

terrestrial planets, giant planets, asteroids, meteorites, comparative planetology, phases, helical rising, ecliptic, retrograde motion, primum mobile, epicycles, deferents, eccentric, equant, conjunction, opposition, heliocentric, geocentric, earthshine, maria, focus (foci), major axis, semimajor axis, minor axis, circle, eccentricity, conic sections, law of equal areas, sector, period, Astronomical Unit (A.U.), one year, universal gravitational constant, revolution, rotation, ecliptic plane, inclination, angular momentum, retrograde rotation, catastrophe theories, tidal theory, cosmogony, nebula, nebular hypothesis, nebular theories, primeval solar nebula, protosun, planetesimals, protoplanets

Questions

1. What are the features that distinguish the terrestrial from the giant planets? What features does Pluto have in common with either group?

2. Suppose that you live on the Moon. Sketch the phases of the Earth that you would observe for various times during the Earth's month.

3. If you lived on the Moon, would the motion of the planets appear any different than from Earth?

4. If you lived on the Moon, how would the position of the Earth change in your sky over time?

5. If you lived on the Moon, what would you observe during an eclipse of the Moon? How would an eclipse of the Sun by the Earth differ from an eclipse of the Sun by the Moon that we observe from Earth?

6. Sketch what you would see if you were on Mars and the Earth passed between you and the Sun. Would you see an eclipse? Why?

7. (a) If you lived on Saturn, describe the phases that the Earth would seem to go through on the Ptolemaic system. (b) Now describe the Copernican prediction for the phases that the Earth would seem to go through.

8. (a) If you lived on Saturn, describe the phases that Uranus would seem to go through on the Ptolemaic system. (b) Now describe the Copernican prediction for the phases that Uranus would seem to go through.

9. Compare the velocity that a planet must have around its epicycle in the Ptolemaic theory with the velocity that the epicycle must have around the deferent for us to observe retrograde motion.

10. Discuss the following statement: "With the addition of epicycles, the geocentric theory of the solar system could be made to agree with observations. Since it was around first, and therefore better known, it should have been kept."

11. How do the predictions of the Ptolemaic and Copernican systems differ for Venus' variation in apparent size?

12. What did the evidence of parallax indicate about models of the universe, and how and why did this change with time?

13. Discuss four new observations made by Galileo, and show how they supported, opposed, or were irrelevant to the Copernican theory.

14. Do Kepler's laws permit circular orbits? Explain.

15. Illustrate with a sketch the point in its orbit at which the Earth moves fastest.

‡16. Use Kepler's third law and the fact that Mercury's orbit has a semimajor axis of 0.4 A.U. to deduce the period of Mercury. Show your work.

†17. Use Kepler's third law and the fact that Jupiter's orbit has a semimajor axis of 5.2 A.U. to deduce the period of Jupiter. Show your work.

‡18. Use Kepler's third law and Saturn's period (Appendix 3) to derive its semimajor axis. Show your work.

†19. Halley's Comet returns every 76 years. Derive the semimajor axis of its orbit, and plot it on a sketch of the planetary orbits.

†20. What is the period of an object orbiting 1 A.U. from a 9-solar-mass star?

†21. Given that a satellite 100 km above the Earth's surface (Appendix 2) orbits in 90 minutes, how high would a satellite have to be to orbit once every 24 hours (and thus seem to hover overhead, a "synchronous satellite" of the type used for communications)?

22. What was the difference in the approaches of Kepler and Newton to the discovery of laws that controlled planetary orbits?

†23. (a) Use the data in Appendix 3 to show for which planets the square of the period does not equal the cube of the semimajor axis to the accuracy given. (b) Use the formula in Box 18.3 to show that including the effects of planetary masses removes the discrepancy.

†24. Astronomers on Planet X note that they are 1 I.U. from their sun, Zimga (where they use an Interplanetary Unit, I.U.), and they orbit every 9 X-months. They observe that the green planet, Planet Y, orbits in 72 X-months. How many I.U. is Planet Y from Zimga?

25. Explain how Pluto can have a longer period than Neptune, even though Pluto is and will be closer to the Sun until the year 2000?

26. Explain how conservation of angular momentum applies to a diver doing somersaults or twists. What can divers do to make sure they are vertical when they hit the water?

27. If two planets are of the same mass but different distances from the Sun, which will have the higher angular momentum around the Sun? (Hint: Use Kepler's third law.)

28. If the planets condensed out of the same primeval nebula as the Sun, why didn't they become stars?

29. Which planets are likely to have their original atmospheres? Explain.

30. Why might the ring around Vega be detectable only in the infrared and not in the visible? Explain in terms of its temperature and the radiation laws.

---

†This indicates a question requiring a numerical answer.
‡Answers: **16.** $P_{Mercury}^2/(1 \text{ year})^2 = (0.4 \text{ A.U.})^3/(1 \text{ A.U.})^3$;
$P_{Mercury} = 0.25$ year **18.** $(30 \text{ years})^2/(1 \text{ year})^2 = a_{Saturn}^3/(1 \text{ A.U.})^3$; $a_{Saturn} = \sqrt[3]{900} = 9.5$ A.U.

Earth photographed from a satellite in synchronous orbit. A grid of latitude and longitude and outlines of states and continents have been added to make it easier to follow weather patterns. Computer calculations using the data gathered by satellites and equations similar to those that govern stars are making it possible to improve weather predictions.

Our Earth   19

Aims: To describe and understand the Earth on a planetary scale

We know the Earth intimately; a mere wrinkle on its surface is a mountain to us. Weather satellites now give us a constant global view, and show us large atmospheric systems linking all parts of the Earth. More rarely, earthquakes or volcanoes bring us messages from below the Earth's surface. But no longer do we have to treat the Earth as one of a kind; by studying the other planets we can make comparisons that allow us to understand both the other planets and our own. In this chapter, we will summarize some of our knowledge of the Earth's structure and history. In later chapters, we will compare the Earth with the other planets.

19.1 The Earth's Interior

The study of the Earth's interior and surface is called *geology*. Geologists study how the Earth vibrates as a result of large shocks, such as earthquakes. Much of our knowledge of the structure of the Earth's interior comes from *seismology*, the study of these vibrations. The vibrations travel through different types of material at different speeds. Also, when the *seismic waves* strike the boundary between different types of material, they are reflected and refracted, just as light is when it strikes a glass lens. From piecing together seismological and other geological evidence, geologists have been able to develop a picture of the Earth's interior (Fig. 19–1).

The innermost region is called the *core*. It consists primarily of iron and nickel. The central part of the core may be solid, but the outer part is probably a very dense liquid. Outside the core is the *mantle*, and on top of the mantle is the thin outer layer called the *crust*. The upper mantle and crust together are called the *lithosphere*. The lithosphere is a rigid layer, and surrounds a zone that is partially melted.

Figure 19–1 The structure of the Earth and stages in its evolution. (*A*) Tens of millions of years after its formation, radioactive elements along with gravitational compression and impact of debris produced melting and differentiation. Heavy materials sank inward and light materials floated outward. During this time, the original atmosphere (consisting mostly of hydrogen) was blown away by the solar wind. The atmosphere that replaced it contained methane, ammonia, and water. (*B*) The heaviest materials form the core, and the lightest materials form the crust. (*C*) The crust has broken into rigid plates that carry the continents and move very slowly away from areas of sea-floor spreading.

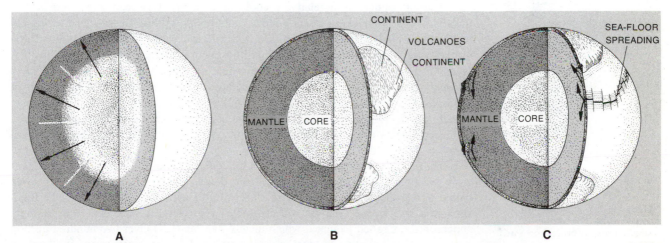

CONTINENT

VOLCANOES

CONTINENT

SEA-FLOOR SPREADING

MANTLE   CORE

MANTLE   CORE

A   B   C

How did such a layered structure develop? The Earth formed surrounded by a lot of debris in the form of dust and rocks. The young Earth was probably subject to constant bombardment from this debris. This bombardment heated the surface to the point where it began to melt, producing lava. However, this process could not have been responsible for heating the interior of the Earth, because the rock out of which the Earth was made conducts heat very poorly.

Much of the original heat for the Earth's interior came from gravitational energy released as particles accreted to form the Earth. But the major source of heat for the interior is the natural radioactivity within the Earth. Certain isotopes are unstable, even though they are formed naturally. That is, they spontaneously undergo nuclear reactions and eventually form more stable isotopes. In these reactions, high energy particles such as electrons or alpha particles (helium nuclei) are given off. These energetic particles collide with the atoms in the rock and give some of their energy to these atoms. The rock becomes hotter. When the Earth was forming, there was a sufficient amount of radioactive material in its interior to cause a great deal of heating. And since the material conducted energy poorly, the heat generated was trapped. As a result, the interior got hotter and hotter.

After about a billion years, the Earth's interior had become so hot that the iron melted and sank to the center, forming the core. Eventually other materials also melted. As the Earth cooled, various materials, because of their different densities and freezing points (the temperature at which they change from liquid to solid), solidified at different distances from the center. This process is called *differentiation;* it is responsible for the present layered structure of the Earth.

19.2 Continental Drift

In the process of differentiation, most of the radioactive elements ended up in the outer layers of the Earth. Thus these elements provide a heat source not far below the ground. This leads to a general *heat flow* outward through the upper mantle and crust (the lithosphere). The power flowing outward through each square centimeter of the lithosphere is small—10 million times less than that necessary to light an average light bulb and thousands of times less than that reaching us from the Sun. However, the terrestrial heat flow does have important geological consequences.

In some geologically active areas (Fig. 19–2), the heat-flow rate is much higher than average, which indicates that the source of heat is close to the surface. In a few places, the outflowing *geothermal energy* in these regions is being tapped as an energy source.

The lithosphere is a relatively cool, rigid layer; it is segmented into *plates,* thousands of kilometers in extent but only about 50 km thick. Because of the internal heating, the lithosphere sits on top of a hot layer where the rock is soft, though it is not hot enough to melt completely. The lithosphere actually floats on top of this soft layer. The hot material beneath the lithosphere's rigid plates churns very slowly in convective motions (an up-and-down circulation of material of which boiling is an example), carrying the plates around over the surface of the Earth. This theory, called *plate tectonics,* explains the observed *continental drift*—the drifting over eons of the continents from their original positions. ("Tectonics" is from the Greek word meaning "to build.")

That continental drift took place, which once seemed unreasonable, is now generally accepted. The continents were once connected as two supercontinents, one called Gondwanaland (after a province of India of geologic interest) and a northern supercontinent called Laurasia. (These may have, in turn, separated from a single supercontinent called Pangaea, which means "all lands.") Over two hundred million years or so, the continents have moved apart as plates have separated. We can see

Figure 19–2 Thermal activity beneath the Earth's surface results in geysers. The thermal area shown here is in Rotorua, New Zealand. Geothermal steam can be used to generate electricity; The Geysers, a geothermal area in California, provides much of San Francisco's electricity.

Figure 19–3 The San Andreas fault marks the boundary between the California and Pacific plates.

Figure 19–4 The NASA satellite LAGEOS (Laser Geodynamic Satellite), launched into a circular Earth orbit in 1976, provides information about the Earth's rotation and the crust's movements. It is expected to survive in orbit for 8 million years. In case the satellite is discovered in the distant future, it bears a series of views of the continents in their locations 270 million years in the past *(top)*, at present *(middle)*, and 270 million years in the future *(bottom)*, according to our knowledge of continental drift. (The dates are given in the binary system.) Data on continental drift are gathered by reflecting laser beams from telescopes on earth off the 426 retroreflectors that cover its surface. (Each retroreflector is a cube whose interior reflects incident light back in the direction from which it came.) Measuring the time until the beams return can be interpreted to show the accurate position of the ground station.

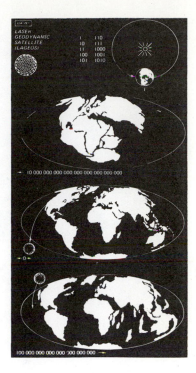

from their shapes how they originally fit together (Fig. 19–4). Finding similar fossils and rock types along two opposite coastlines, which were once adjacent but are now widely separated, provides proof. In the future, we expect California to separate from the rest of the United States, Australia to be linked to Asia, and the Italian "boot" to disappear.

The boundaries between the plates (Fig. 19–3) are geologically active areas. Therefore, these boundaries are traced out by the regions where earthquakes (Fig. 19–5) and most of the volcanoes (Fig 19–6) occur. The boundaries where two plates are moving apart mark regions where molten material is being pushed up from the hotter interior to the surface, such as the *mid-Atlantic ridge* (Fig. 19–7). Molten material is being forced up through the center of the ridge and is being deposited as lava flows on either side, producing new sea floor. The motion of the plates is also responsible for the formation of the great mountain ranges. When two plates come

Figure 19–5 This plot of earthquakes greater than 4.5 on the Richter scale from 1963 through 1973 shows that earthquakes occur preferentially at plate boundaries. The principal tectonic plates are labelled, and arrows show whether the plates are converging or diverging.

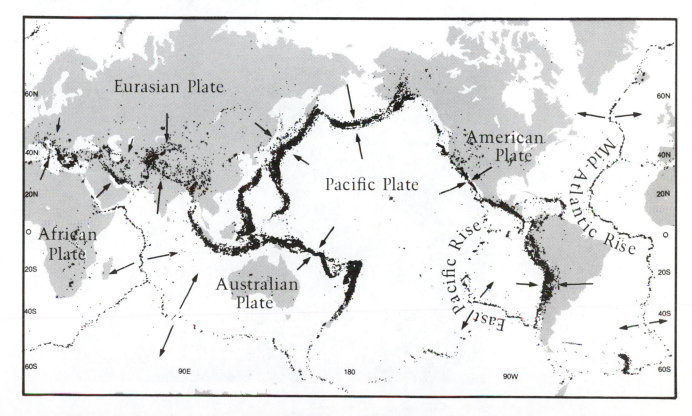

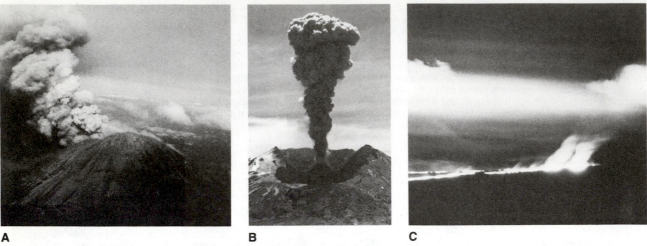

A   **B**   **C**

Figure 19–6 *(A)* Mt. St. Helens, shown here during its May 18, 1980, eruption, lies on a plate boundary. *(B)* After its devastating eruption. *(C)* Mauna Loa on the island of Hawaii during its 1984 eruption.

Figure 19–7 This map of the Atlantic ocean floor shows how the continents sit on the plates. North and South America are at the left; Europe and Africa are at the right. The feature running from north to south in the middle of the ocean is the *mid-Atlantic ridge*. It marks the boundary between plates that are moving apart. (Courtesy National Geographic Society, 1973)

© 1973 National Geographic Society

together, one may be forced under the other, and the other may be raised. The "ring of fire" volcanoes around the Pacific Ocean (including Mt. St. Helens in Washington) were formed in this way. Another example is the Himalayas, which were formed when the Indian subcontinent collided with the Asian continent.

19.3 Tides

It has long been accepted that tides are most directly associated with the Moon. This is because the tides—like the Moon—occur about an hour later each day. Tides result from the fact that the force of gravity exerted by the Moon (or any other body) gets weaker as you get farther away from it. Tides depend on the **difference** between the gravitational attraction of a massive body at different points on a less massive one, so the forces that cause tides are often called *differential forces*.

To explain the tides in Earth's oceans, consider, for simplicity, that the Earth is completely covered with water. We might first say that the water closest to the Moon is attracted toward the Moon with the greatest force and so is the location of high tide as the Earth rotates. If this were all there were to the case, high tides would occur about once a day. However, two high tides occur daily, separated by roughly 13 hours (Fig. 19–8).

To see how we get two high tides a day, consider three points, A, B, and C, where B represents the solid Earth, A is the ocean nearest the Moon, and C is the ocean farthest from the Moon (Fig. 19–9). Since the Moon's gravity weakens with distance, it is greater at A than at B, and greater at point B than at C. If the Earth and Moon were not in orbit about each other, all these points would fall toward the Moon, moving apart as they fell.

Thus the high tide on the side of the Earth that is near the Moon is a result of the water being pulled away from the Earth. The high tide on the opposite side of the Earth results from the Earth being pulled away from the water. In between the locations of the high tides, the water has rushed elsewhere so we have low tides. As the Earth rotates once a day, a point on its surface moves through two cycles of tides: high-low-high-low. Since the Moon is moving in its orbit around the Earth, a point on the Earth's surface has to rotate longer than 12 hours to return to a spot

Figure 19–8 Low and high tides in Nova Scotia on the Bay of Fundy, site of the world's highest tides. Though the explanation of tides as differential forces is valid, details of tides at individual locations on the shore also depend on such factors as the depth of the ocean floor or the shape of the channel; the latter is especially important here.

Figure 19–9 A schematic representation of the tidal effects caused by the Moon. The arrows represent the acceleration of each point that results from the gravitational pull of the Moon (exaggerated in the drawing). The water at point A has greater acceleration toward the Moon than does point B; since the Earth is solid, the whole Earth moves with point B. (Tides in the solid Earth exist, but are much smaller than tides in the oceans.) Similarly, the solid earth is pulled away from the water at point C. Tidal forces are important in many astrophysical situations, including accretion disks around neutron stars and black holes, and rings around planets.

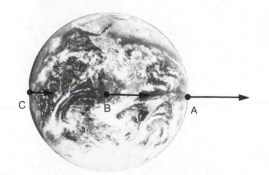

nearest to the Moon, and so to begin a new cycle. Thus the tides repeat about every 13 hours.

The Sun's effect on the Earth's tides is only about half as much as the Moon's. Though the Sun exerts a greater gravitational force on the Earth than does the Moon, the Sun is so far away that its force does not change very much from one side of the Earth to the other. And it is only the change in force that accounts for tides.

At a time of new or full moon, the tidal effects of the Sun and Moon work in the same direction, and we have especially high tides, called *spring tides* (from the German word "springen": to rise up; the word has nothing to do with the season "spring"). At the time of the first and last quarter moon, the effects of the Sun and Moon do not add, and we have *neap tides,* when high and low tides differ least.

19.4 The Earth's Atmosphere

The Earth's atmosphere presses down on us all the time, though we are used to it. The force pressing down per unit of area is called *pressure*. We define the pressure at the surface of the Earth to be *one atmosphere*. When we express the pressure in atmospheres, we are really taking the ratio of the pressure at any level to that at the surface. The pressure in the atmosphere (Fig. 19–10) falls off sharply with increasing height. The temperature (Fig. 19–11) varies less regularly with altitude.

It is convenient to divide the atmosphere into layers (Fig. 19–12), according to the composition and the physical processes that determine the temperature.

The Earth's weather is confined to the very thin *troposphere*. At the top of the troposphere, the pressure is only about 10 per cent of its value at the ground. A major source of heat for the troposphere is infrared radiation from the ground, so the temperature of the troposphere decreases with altitude.

Above the troposphere are the *stratosphere* and the *mesosphere*. The upper stratosphere and lower mesosphere contain the *ozone layer*. (Ozone is a molecule, consisting of three oxygen atoms—O_3—that absorbs ultraviolet radiation from the

Figure 19–11 Temperature in the atmosphere as a function of altitude. In the troposphere, the energy source is the ground, so the temperature falls off with altitude. Higher temperatures in other layers result from direct absorption of solar ultraviolet and x-rays.

Figure 19–10 Pressure in the Earth's atmosphere as a function of altitude. Note how rapidly it decreases as the altitude increases.

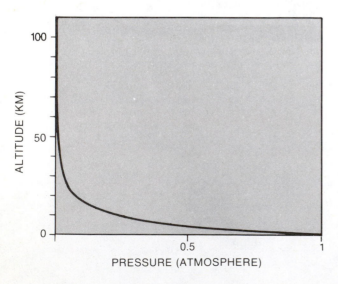

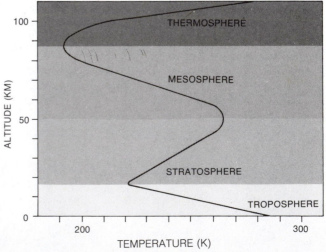

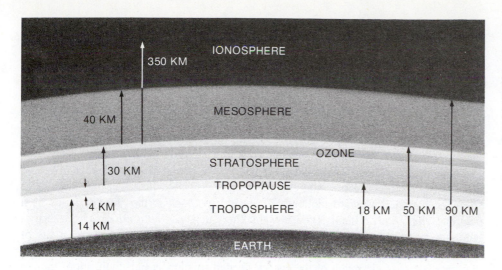

IONOSPHERE

350 KM

MESOSPHERE

40 KM

OZONE

STRATOSPHERE

30 KM

TROPOPAUSE

4 KM TROPOSPHERE 18 KM 50 KM 90 KM

14 KM

EARTH

Figure 19–12 The layers of the Earth's atmosphere showing the location of the part of the stratosphere where ozone is formed. Once formed, the ozone is transported to lower stratospheric layers.

Sun. The oxygen our bodies use is O_2.) The absorption of ultraviolet radiation by the ozone means that these layers get much of their energy directly from the Sun. The temperature is thus higher there than it is at the top of the troposphere.

Above the mesosphere is the *ionosphere,* where many of the atoms are ionized. The most energetic photons from the Sun (such as x-rays) are absorbed here. Thus the temperature gets quite high. Because of this rising temperature, this layer is also called the *thermosphere.* Since many of the atoms in this layer are ionized, the ionosphere contains many free electrons. When the conditions are right, radio waves bounce off the ionosphere, which allows us to tune in distant radio stations. When solar activity is high, there may be a lot of this "skip."

Our knowledge of our atmosphere has been greatly enhanced by observations from satellites above it. (See the figure opening this chapter.) Scientists carry out calculations using the most powerful supercomputers to interpret the global data and to predict how the atmosphere will behave. The equations are essentially the same as those for the internal temperature and structure of stars, except that the sources of energy are different.

The rotation of the Earth also has a very important effect in determining how the winds blow. Comparison of the circulation of winds on the Earth (which rotates in 1 Earth day), on slowly rotating Venus (which rotates in 243 Earth days), and on rapidly rotating Jupiter and Saturn (each of which rotates in about 10 Earth hours), helps us understand the weather on Earth.

19.5 The Van Allen Belts

In January 1958, the first American space satellite carried aloft, among other things, a device to search for charged particles that might be orbiting the Earth. This device, under the direction of James A. Van Allen of the University of Iowa, detected a region filled with charged particles of high energies. We now know that there are actually two such regions—the *Van Allen belts*—surrounding the Earth, like a small and a large doughnut (Fig. 19–13).

The particles in the Van Allen belts are trapped by the Earth's magnetic field. Charged particles can only move in the direction of magnetic-field lines, and not across the field lines. Usually a charged particle will follow a path that spirals around a magnetic-field line. As a particle moves from above the equator toward one of the

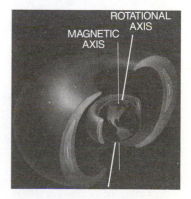

ROTATIONAL AXIS

MAGNETIC AXIS

Figure 19–13 The doughnut-shaped Van Allen belts. Though often called "radiation belts," they are actually regions of charged particles trapped by the Earth's magnetic field.

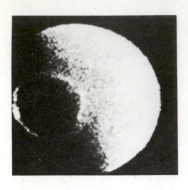

Figure 19–14   The aurora forms an oval around the north pole in this view from the Dynamics Explorer spacecraft, launched in 1981. The observations were made in ultraviolet light emitted by glowing oxygen; glowing oxygen also causes the typical green auroral light. The auroral oval traces the "feet" of the outer Van Allen belt and extends across the terminator between the sunlighted and night sides of the Earth. The aurora is caused by an interaction of the solar wind with the Earth's magnetosphere. Charged particles hit atoms in the earth's atmosphere, giving them energy that they then radiate.

magnetic poles, the field gets stronger, so the field lines get closer together. Eventually as the field lines converge on the poles, they make considerable angles with each other (though they never cross). When this happens the particle can travel no farther, since otherwise it would cross a neighboring field line. The magnetic force stops the particle's forward motion and makes it go back in the other direction as it continues its spiral. Scientists are now making such *magnetic mirrors* in terrestrial laboratories as part of one of the leading methods to keep a hot plasma (a gas of charged particles) trapped in one place long enough to start fusion. Controlled fusion may one day provide much of the energy needed to support our civilization, though major technical problems still remain to be solved.

Charged particles, often from solar storms, guided by the Earth's magnetic field toward the Earth's poles cause our atmosphere to glow, which we see as the beautiful northern and southern lights—the *aurora borealis* or *aurora australis,* respectively (Fig. 19–14 and Color Plate 36).

Summary and Outline

Structure of Earth's interior (Section 19.1)
    Core, mantle, and crust; natural radioactivity is source of
        heat for interior
    Differentiation: layered structure
Continental drift (Section 19.2)
    Geothermal energy from heat flow at active areas
    Continents are on moving plates; plate tectonics
Tides (Section 19.3)
    From differential forces; spring and neap tides

Earth's atmosphere (Section 19.4)
    All weather confined to thin troposphere
    Ozone layer between stratosphere and mesosphere
    Ionosphere (also called thermosphere) reflects radio waves.
Radiation belts (Section 19.5)
    Discovered by James Van Allen from data obtained from
        first American satellite; composed of charged particles
        trapped by Earth's magnetic field.

Key Words

geology, seismology, seismic waves, core, mantle, crust, lithosphere, differentiation, heat flow, geothermal energy, plates, plate tectonics, continental drift, mid-Atlantic ridge, differential forces, spring tides, neap tides, pressure, one atmosphere, troposphere, stratosphere, mesosphere, ozone layer, ionosphere, thermosphere, Van Allen belts, magnetic mirror, aurora borealis, aurora australis

Questions

†1. (a) Imagine that you are observing the Earth from an altitude of 100 km. Estimate the angular size of a house, a football field, and a city. What features might point to the existence of intelligent life? (b) What about from 1000 km

up? (c) Compare with the resolution of the eye and of a telescope.

2. Consult an atlas and compare the sizes of the Grand Canyon in Arizona and the Rift Valley in Africa. How do they compare in size with the giant canyon on Mars, which is about the diameter of the United States?

---
†This indicates a question requiring a numerical solution.

3. Plan a set of experiments or observations that you, as a Martian scientist, would have an unmanned spacecraft carry out on Earth. What data would your spacecraft radio back if it landed in a corn field? In the Sahara? In the Antarctic? In Times Square?

4. Of the following in the Earth, which is the densest? (a) crust; (b) mantle; (c) lithosphere; (d) core.

5. (a) Explain the origins of tides. (b) If the Moon were twice as far away from the Earth as it actually is, how would tides be affected?

†6. Using the inverse-square law of gravity, calculate the differential effect of the Moon's pull given the radius of the Earth (Appendix 3) and the Earth-Moon distance (Appendix 4), compared with the Moon's gravitational pull on Earth.

7. Draw a diagram showing the positions of the Earth, Moon, and Sun at a time when there is the least difference between high and low tides.

8. What is the source of most of the radiation that heats the troposphere?

9. Look at a globe and make a list or sketches of which pieces of the various continents probably lined up with each other before the continents drifted apart.

†10. At what height above the Earth's surface is the pressure 1 per cent of its value at the ground?

11. How does the temperature vary with height in the troposphere? In the stratosphere?

12. Space-shuttle astronauts orbit 175 km above the Earth. Communications satellites are $5\frac{1}{2}$ Earth radii above the surface. Compare their locations with the Van Allen belts, which can cause false readings on instruments.

Scientist-astronaut Harrison Schmitt collecting small rocks and rock chips with a lunar rake during the Apollo 17 mission to the Taurus-Littrow region of the Moon.

The Moon   20

Aims: To see how direct exploration of the Moon has increased our knowledge manyfold, although such fundamental questions as how the Moon was formed remain unanswered

The Earth's nearest celestial neighbor—the Moon—is only 380,000 km (238,000 miles) away from us on the average, close enough that it appears sufficiently large and bright to dominate our nighttime sky. The Moon's stark beauty has called our attention since the beginning of history, and studies of the Moon's position and motion led to the earliest consideration of the solar system, to the prediction of tides, and to the establishment of the calendar.

20.1 The Appearance of the Moon

The fact that the Moon's surface has different kinds of areas on it is obvious to the naked eye. Even a small telescope—Galileo's, for example—reveals a surface pockmarked with craters. The *highlands* are heavily cratered. Other areas, called *maria* (pronounced mar'-ē-a; singular, *mare*, pronounced mar'-ā), are relatively smooth, and indeed the name comes from the Latin word for sea (Fig. 20–1). But there are no ships sailing on the lunar seas and no water in them; the Moon is a dry, airless, barren place. The Moon's mass is only $\frac{1}{81}$ that of the Earth, and the gravity at its surface is only $\frac{1}{6}$ that of the Earth. Any atmosphere and any water that may once have been present would long since have escaped into space. The Moon is about one-fourth the diameter of the Earth, a very large fraction for a moon.

Other types of structures visible on the Moon besides the smooth maria and the cratered highlands include *mountain ranges* and *valleys*. The mountains are formed by debris, though, unlike mountains on Earth. Lunar *rilles* are clefts that can extend for hundreds of kilometers along the surface (Fig. 20–2). Some are relatively straight while others are sinuous. Raised *ridges* also occur. The craters themselves come in all sizes, ranging from as much as 295 km across for Bailly down to tiny fractions of a millimeter. Crater *rims* can be as much as several kilometers high, much higher than the Grand Canyon's rim stands above the Colorado River.

The Moon rotates on its axis at the same rate as it revolves around the Earth, always keeping the same face in our direction. The Earth's gravity has locked the Moon in this pattern, interacting with a bulge in the distribution of the lunar mass to prevent the Moon from rotating freely. As a result of this interlock, we always see essentially the same side of the Moon from our vantage point on Earth. Because of *librations* of the Moon—the apparent slight turning of the Moon—we see, at one time or another, $\frac{3}{5}$ of the whole surface. Librations arise from several causes, including the varying speed of the Moon in its elliptical orbit.

When the Moon is full, it is bright enough to cast shadows or even to read by. But full moon is a bad time to try to observe lunar surface structure, for any shadows we see are short. When the Moon is a crescent or even a half moon, however, the part of the Moon facing us is covered with long shadows. The lunar features then stand out in bold relief.

Figure 20–1 This Earth-based photograph shows Mare Imbrium at the upper left, the Apennine Mountains at the lower right, and many craters, some of which have central peaks (which arise from a rebound as the crater forms).

Figure 20–2 The Aridaeus Rille.

Shadows are longest near the *terminator,* the line separating day from night. The Moon makes a complete orbit of the Earth in 29½ days; the interval between successive new moons is the *synodic revolution period* of the Moon. Because the same side of the Moon always faces the Earth, different regions face the Sun as the Moon orbits. As a result, the terminator moves completely around the Moon with this 29½-day synodic period. Most locations on the Moon are thus in sunlight for about 15 days, during which time they become very hot—130°C (265°F)—and then in darkness for about 15 days, during which time their temperature drops to as low as −110°C (−170°F). The cycle of phases that we see from Earth also repeats with this 29½-day period—a *synodic month.*

In some sense, before the period of exploration by the Apollo program, we knew more about almost any star than we did about the Moon. As a solid body, the Moon reflects the solar spectrum rather than emitting one of its own, so we were

Figure 20–3 The full moon. Note the dark maria and the lighter, heavily cratered highlands. The positions of the 6 American Apollo (A) and 3 Soviet Luna (L) missions from which material was returned to Earth for analysis (Table 20–1) are marked. This photograph is oriented with North up, as we see the Moon with our naked eyes or through binoculars; a telescope normally inverts the image.

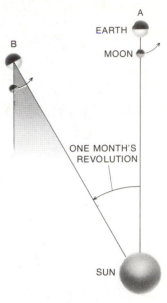

hard pressed to determine even the composition or the physical properties of the Moon's surface.

20.2 Lunar Exploration

The space age began on October 4, 1957, when the U.S.S.R. launched its first Sputnik (the Russian word for *travelling companion*) into orbit. The shock of this event galvanized the American space program, and within months American spacecraft were also in Earth orbit.

The ability to travel into space not only freed us from the obscuration of the Earth's atmosphere but also allowed us to explore the solar system directly. The Moon, as the closest celestial body, was obviously the place to begin.

In 1959, the Soviet Union sent its Luna 3 spacecraft around the Moon; Luna 3 radioed back the first murky photographs of the Moon's far side. Now that we have high-resolution maps, it is easy to forget how big an advance that was.

In 1961, President John F. Kennedy proclaimed that it would be a U.S. national goal to put a man on the Moon, and bring him safely back to Earth, by 1970. This grandiose goal led to the largest coordinated program of any kind in world history.

The American lunar program, under the direction of the National Aeronautics and Space Administration (NASA), proceeded in gentle stages. The ability to carry out manned space flight was developed first with single-astronaut suborbital and orbital capsules, called Project Mercury, and then with two-astronaut orbital spacecraft, called Project Gemini. Simultaneously, a series of unmanned spacecraft was sent to the Moon.

The manned and unmanned trains of development merged with Apollo 8, in which three astronauts circled the Moon on Christmas Eve, 1968, and returned to Earth. The next year, Apollo 11 brought humans to land on the Moon for the first time. It went into orbit around the Moon after a three-day journey from Earth, and a small spacecraft called the Lunar Module separated from the larger Command Module. On July 20, 1969—a date that from the long-range standard of history may be the most significant of the last millennium—Neil Armstrong and Buzz Aldrin left Michael Collins orbiting in the Command Module and landed on the Moon (Fig. 20–5). In the preceding days there had been much discussion of what Armstrong's historic first words should be, and millions listened as he said "One small

Figure 20–4 After the Moon has completed one revolution around the Earth with respect to the stars, which it does in 27⅓ days, it has moved from A to B. The Moon has still not swung far enough around to again be in the same position with respect to the Sun because the Earth has revolved one month's worth around the Sun. It takes about an extra two days for the Moon to complete its revolution with respect to the Sun, which gives a synodic revolution period of about 29½ days. The extra angle it must cover is shaded. In the extra two days, both Earth and Moon have continued to move around the Sun (toward the left in the diagram).

Figure 20–5 Neil Armstrong, the first person to set foot on the Moon, took this photograph of his fellow astronaut Buzz Aldrin climbing down from the Lunar Module of Apollo 11 on July 20, 1969. The site is called Tranquility Base, as it is in the Sea of Tranquility (Mare Tranquillitatis).

Table 20–1 Missions to the Lunar Surface and Back to Earth

Apollo 11	U.S.A.	1969	manned
Apollo 12	U.S.A.	1969	manned
Luna 16	U.S.S.R.	1970	unmanned
Apollo 14*	U.S.A.	1971	manned
Apollo 15	U.S.A.	1971	manned
Luna 20	U.S.S.R.	1972	unmanned
Apollo 16	U.S.A.	1972	manned
Apollo 17	U.S.A.	1972	manned
Luna 24	U.S.S.R.	1976	unmanned

*An explosion on Apollo 13 while en route to the Moon led to an emergency return of the astronauts to Earth, though they had to circle the Moon to do so.

step for man, one giant leap for mankind.'' (He meant to say ''for a man.'' When, in 1984, a space shuttle astronaut became the first human to fly freely in orbit, untethered, he joked, ''That may have been one small step for Neil, but it's a heck of a big leap for me.'')

The Lunar Module carried many experiments, including devices to test the soil, a camera to take stereo photos of lunar soil, a sheet of aluminum with which to capture particles from the solar wind, and a seismometer. Later Lunar Modules carried additional experiments, some even including a vehicle (Fig. 20–6). Six Apollo missions in all, ending with Apollo 17 in 1972, carried people to the Moon (Color Plates 24–27). Unfortunately, the missions became confused in the popular mind with just one of the experiments—the collection of rocks. These rocks and dust, returned to Earth for detailed analysis, were indeed important, but were only a fraction of each mission.

The Soviet Union sent three unmanned spacecraft to land on the lunar surface, collect lunar soil, and return it to Earth. The first two each collected a few grams of lunar soil. The third, Luna 24, drilled to $\frac{1}{2}$ meter below the lunar surface and brought a long, thin cylinder of material back to Earth. In 1970 and 1973, two other Soviet spacecraft had carried remote-controlled rovers, Lunokhods 1 and 2, that travelled across 10 km of the lunar surface over a period of months.

Figure 20–6 *(A)* Eugene Cernan riding on the Lunar Rover during the Apollo 17 mission. The mountain in the background is the east end of the South Massif. *(B)* Drawing by Alan Dunn; © 1971 The New Yorker Magazine, Inc.

A

B

20.3 The Results from Apollo

The kilometers of film exposed by the astronauts, the 382 kg of rock brought back to Earth (Fig. 20–7), the lunar seismograph data recorded on tape, and other data have been studied by hundreds of scientists from all over the Earth. The data have led to new views of several basic questions, and have raised many new questions about the Moon and the solar system.

20.3a The Composition of the Lunar Surface

The rocks that were encountered on the Moon are types that are familiar to terrestrial geologists. All the rocks are *igneous*, which means that they were formed by the cooling of lava. The moon has no *sedimentary* rocks, which are formed by deposits in water. (Earth's sedimentary rocks include limestone and shale.)

In the maria, the rocks are mainly *basalts* (Fig. 20–8A). The highland rocks are *anorthosites*. (Anorthosites—defined by their particular combination of minerals— are rare on Earth, though the Adirondack Mountains are made of them.) Anorthosites, though they have also cooled from molten material, have done so under different conditions than basalts and have taken longer to cool.

In both the maria and the highlands, some of the rocks are *breccias* (Fig. 20–8B), mixtures of fragments of several different types of rock that have been compacted and welded together. The ratio of breccias to crystalline rocks is much higher in the highlands than in the maria because there have been many more impacts in the visible highland surface, as shown by the greater number of craters.

The astronauts also collected some *lunar soils,* bits of dust and larger fragments from the Moon's surface. This *regolith* was built up by bombardment of the lunar surface by meteorites of all sizes over billions of years. Microscopes showed that some of these soils contain small glassy globules (Fig. 20–9), which are not common on Earth. Some of these glassy globules and glassy coatings undoubtedly resulted from the melting of rock during its ejection from the site of a meteorite impact and the subsequent cooling of the molten material.

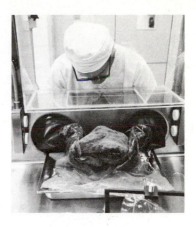

Figure 20–7 A moon rock from Apollo 14 being handled in the Lunar Receiving Laboratory at the Johnson Space Center in Houston.

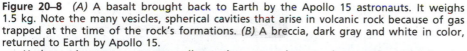

Figure 20–8 *(A)* A basalt brought back to Earth by the Apollo 15 astronauts. It weighs 1.5 kg. Note the many vesicles, spherical cavities that arise in volcanic rock because of gas trapped at the time of the rock's formations. *(B)* A breccia, dark gray and white in color, returned to Earth by Apollo 15.
 Under a microscope, one can easily see the contrast between lunar and terrestrial rocks. The lunar rocks contain no water, so we know that they never underwent the reactions that terrestrial rocks undergo. The lunar rocks also show that no oxygen was present when they were formed.

A   B

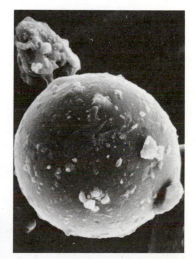

Figure 20–9 An enlargement of a glassy spherule from the lunar dust collected by the Apollo 11 mission. The shape and composition of the spherule indicate that it was created when a meteorite crashed into the Moon, melting lunar material and splashing it long distances. The glassy bead is enlarged 3200 times in this photograph taken with a scanning electron microscope.

Almost all of these rocks and soils have lower proportions of elements with low melting points (*volatile* elements) than do rocks and soils on the Earth. On the other hand, they contain relatively high proportions of elements with high melting points (*refractory* elements) like calcium, aluminum, and titanium. Elements that are even rarer on Earth, such as uranium, thorium, and the rare-earth elements, are also found in greater abundances. (Will we be mining on the Moon one day?) The Moon and the Earth seem to be similar chemically, though significant differences in overall composition do exist. Since none of the lunar rocks contain any trace of water bound inside their minerals, clearly water never existed on the Moon. This eliminates the possibility that life evolved there.

20.3b Lunar Chronology

One way of dating the surface of a moon or planet is to count the number of craters in a given area, a method that was used before Apollo. Even from the Earth we can count the larger lunar craters. If we assume that the events that cause craters—whether they are impacts of meteors or the eruptions of volcanoes—continue over a long period of time and that no strong erosion occurs, surely those locations with the greatest number of craters must be the oldest. Relatively smooth areas—like maria—must have been covered over with volcanic material at some relatively recent time (which is still billions of years ago). When one crater is superimposed on another (Fig. 20–10), we can be certain that the superimposed crater is the younger one.

A few craters on the moon, notably Copernicus (Figs. 20–3 and 20–11), have thrown out obvious rays of lighter-colored matter. The rays are material ejected when

Figure 20–10 This view from the orbiting Apollo 15 Command Module shows a smaller crater, Krieger B, superimposed on a larger crater, Krieger. Obviously, the smaller crater is younger than the larger one. Several rilles (clefts along the lunar surface that can be hundreds of kilometers in length) and ridges are also visible. Sometimes the areas in the centers of craters, the floors, are smooth, but sometimes craters have central peaks.

Figure 20–11 The crater Copernicus, seen in this ground-based photograph, has rays of light material emanating from it. This light material was thrown out radially when the meteorite impacted and formed the crater.

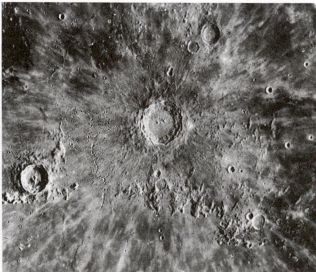

the crater was formed. Since these rays extend over other craters, the craters with rays must have formed later. The youngest rayed craters may be very young indeed—perhaps only a few hundred million years. The rays darken with time, so rays that may have once existed near other craters are now indistinguishable from the rest of the surface. Rays are visible for about 3 billion years.

Crater counts and the superposition of one crater on another give only relative ages. We could find the absolute ages only when rocks were physically returned to Earth. Scientists worked out the dates by comparing the current ratio of radioactive isotopes to nonradioactive isotopes present in the rocks with the ratio that they would have had when they were formed. From the known rate of radioactive decay, they can work out the ages. The oldest rocks that were found at the locations sampled on the Moon were formed 4.42 billion years ago. The youngest rocks were formed 3.1 billion years ago. (Ages are given to the accuracy to which they can be measured; the 4.42-billion-year age was especially precise and resulted from a catastrophic event on the moon.)

The ages of highland and maria rocks are significantly different. The highland rocks were formed between about 3.9 and 4.4 billion years ago, and the maria between 3.1 and 3.8 billion years ago. Several highland rocks are exactly the same age, all 4.42 billion years old. So 4.42 billion years ago may therefore have been the origin of the lunar surface material, that is, the time when it last cooled.

All the observations can be explained on the basis of the following general picture (Fig. 20–12): The Moon formed 4.6 billion years ago. We know that the top 100 km or so of the surface was molten after about 200 million years. The surface could have been entirely melted by the original heat or by an intense bombardment of meteorites (or debris from the period of formation of the Moon and the Earth). The decay of radioactive elements also provided heat. Then the surface cooled. From 4.2 to 3.9 billion years ago, bombardment (perhaps by planetesimals) caused most of the craters we see today. About 3.8 billion years ago, the interior of the Moon heated up sufficiently (from radioactive elements inside) that vulcanism began; lava flowed on the lunar surface and filled the largest basins that resulted from the earlier bombardment, thus forming the maria (Fig. 20–13). By 3.1 billion years ago, the era of vulcanism was over. The Moon has been geologically pretty quiet since then.

Up to this time, the Earth and the Moon shared similar histories. But active lunar history stops about 3 billion years ago, while the Earth continued to be geologically active. Because the Earth's interior continued to send gas into the atmosphere

Radioactive isotopes are those that decay spontaneously; that is, they change into other isotopes even when left alone. *Stable* isotopes remain unchanged. For certain pairs of isotopes—one radioactive and one stable—we know the proportion of the two when the rock was formed. Since we know the rate at which the radioactive one is decaying, we can calculate how long it has been decaying from a measurement of what fraction is left.

Meteorites hit the Moon with such high velocities that huge amounts of energy are released at the impact. The effect is that of an explosion, as though it had been TNT or an H-bomb exploding.

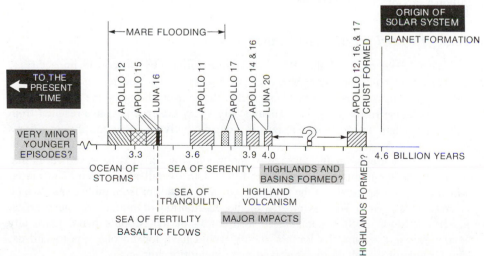

Figure 20–12 The chronology of the lunar surface, based on work carried out at the Lunatic Asylum, as the Caltech laboratory of Gerald Wasserburg is called. The ages of rocks found in eight missions are shown in boxes. The names and descriptions below the line took place at the times indicated by the positions of the words. Note how many missions it took to get a sampling of many different ages on the lunar surface.

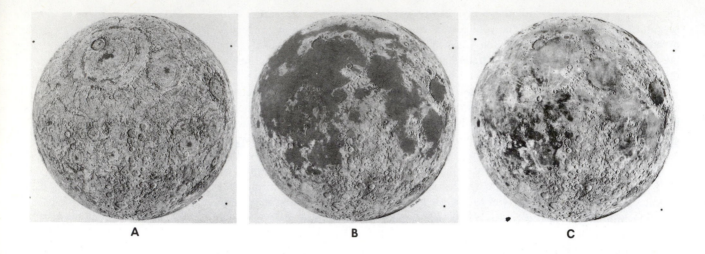

A   B   C

Figure 20–13 Artist's views of the formation of the lunar surface compared with a recent photographic map. *(A)* The Moon before the formation of the present mare surface material about 4 billion years ago. The concentric rings of the Imbrium basin, now Mare Imbrium, show prominently at upper left. From current lunar features, the artist has removed most mare material and late craters, and freshened certain early craters. *(B)* The Moon soon after the formation of most of the mare material approximately 3.3 billion years ago. Young craters such as Tycho and Copernicus are absent. The Moon looks much like it does now, though surface details differ. *(C)* A modern lunar map from the U. S. Geological Survey. (Drawings by Donald E. Davis under the guidance of Don E. Wilhelms of the U. S. Geological Survey)

Figure 20–14 *(A)* Mare Orientale, shown here, is at the edge of the surface of the Moon we see from Earth. Because it appears so foreshortened from Earth, it was little known before space exploration of the Moon. One of the unmanned pre-Apollo spacecraft radioed back this picture, which shows concentric rings over 450 km in radius extending outward. They were probably caused by a meteor impact. The origin of most other craters not surrounded by such rings is more ambiguous. The horizontal stripes are artifacts introduced while the data were being transmitted.

and because the Earth's higher gravity retained that atmosphere, the Earth developed conditions in which life evolved. The Moon, because it is smaller than the Earth, presumably lost its heat more quickly and also generated a thicker crust.

Almost all the rocks on the Earth are younger than 3 billion years of age; erosion and the remolding of the continents as they move slowly over the Earth's surface, according to the theory of plate tectonics (Section 19.2), have taken their toll. The oldest single rock ever discovered on Earth has an age of 4.1 or 4.2 billion years and few rocks are older than 3 billion years. So we must look to extraterrestrial bodies— the Moon or meteorites—that have not suffered the effects of plate tectonics or erosion (which occurs in the presence of water or an atmosphere) to study the first billion years of the solar system.

20.3c The Origin of the Craters

The debate over whether the craters were formed by meteoritic impact (Fig. 20–14) or by volcanic action began in pre-Apollo times. The results from Apollo indicate that most craters resulted from meteoritic impact, though some small fraction may have come from vulcanism. Though only a few craters may have resulted from vulcanism, there are many other signs of volcanic activity, including the lava flows that filled the maria.

The photographs of the far side of the moon (Fig. 20–15) have shown us that the near and far hemispheres are quite different in overall appearance. The maria, which are so conspicuous on the near side, are almost absent from the far side, which is cratered all over. The asymmetry in the distribution of maria may have arisen inside the Moon itself by an uneven distribution of the Moon's mass. Once any asymmetry was set up, the Earth's gravity would have locked one side toward us. The far side of the Moon thus received more meteoritic impacts.

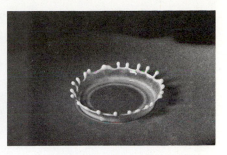

Figure 20–14 *(B) (above)* This series, made with a very short-exposure "strobe" light, shows the result of a falling milk drop. The formation of a lunar crater by a meteorite is similar, because the energy of the impact makes the surface material flow like a liquid. (Courtesy of Harold E. Edgerton, MIT)

In later chapters, we shall meet other cratered objects in our solar system. Careful study has apparently shown that the Moon and Mercury received their craters at the same general time. The family of small objects that made the craters on Mars came later, as did other families that made craters on the moons of Jupiter and Saturn.

20.3d The Lunar Interior

Before the Moon landings, it was widely thought that the Moon was a simple body, with the same composition throughout. But we now know it to be differentiated (Fig. 20–16), like the planets. It has a *crust* of relatively light material at its surface and a silica-rich *mantle* making up most or all of the interior. It may also have a metallic (iron-rich) *core*, though any such core would take up a much smaller fraction of the lunar interior than the Earth's core does of Earth's.

The lunar crust is perhaps 65 km thick on the near side and twice as thick on the far side. This asymmetry may explain the different appearances of the sides, because lava would be less likely to flow through the far side's thicker crust.

We cannot observe the interior of a moon or planet directly, though, so we must use indirect methods. In one of the best methods, scientists use seismographs to detect the lunar version of earthquakes. The speeds at which different types of waves are transmitted through the lunar interior tell us about the condition of the interior.

Figure 20–15 The far side of the Moon looks very different from the near side in that there are few maria (compare with Fig. 20–3). This photograph was taken from Apollo 16, and shows some of the near-side maria at the upper left.

Figure 20–16 *(above)* The Moon's interior. The depth of basalts is greater under maria, which are largely on the side of the Moon nearest the Earth. Almost all the 10,000 moonquakes observed originated in a zone halfway down toward the center of the Moon, a distance ten times deeper than most terrestrial earthquakes. This fact can be used to interpret conditions in the lunar interior. If too much of the interior of the Moon were molten, the zone of moonquakes probably would have sunk instead of remaining suspended there. The deep moonquakes came from about 80 locations, and were triggered at each location twice a month by tidal forces resulting from the variation in the Earth-Moon distance.

Working seismometers were left by Apollo astronauts at four widely spaced locations on the Moon. They enabled us to locate the origins of the thousands of weak moonquakes that occur each year—almost all of which are magnitudes 1 to 2 on the Richter scale. Perhaps three moonquakes per year reach Richter 4, a strength which on Earth could be felt but which would not cause damage. On one occasion (July 21, 1972) a meteorite hit the far side of the Moon and generated seismic waves strong enough that we would have expected them to travel through to the near side, where the seismometers are all located. From the fact that one type of seismic wave—so-called *shear waves*—did not travel through the core while the other types did, most researchers deduce that the core is molten or at least plastic in consistency. Shear waves involve a twisting, and if the material in the moon's center is molten, it would not act to restore the twisted material and the wave would not be transmitted.

The Apollo data are still being studied. A team of scientists reported in 1981 that earlier estimates of the amount of energy released by moonquakes were 100 times too low. The fact that seismic waves lose energy as they jump back and forth between the boulders in the Moon's surface layer had not been properly taken into account. The instrumental sensitivity of the seismometers was also newly taken into account. Still, the Moon seems seismically quiet compared with the Earth. Our lunar stations worked for only 8 years, however, so we do not know if giant moonquakes occur more rarely than that. Unfortunately, the seismometers and other instruments on the Moon were shut off by NASA in 1977 as an economy measure.

We think that magnetic fields are formed in molten planetary interiors, for then the material inside the crust can circulate. If the interior had always been solid, no magnetic field would have resulted.

If the interior of the Moon were molten, as is the Earth's interior, we would have expected to find a more intense magnetic field than the one we have detected. Yet the Moon has no general magnetic field. We do detect a weak magnetic field frozen into lunar rocks. It might be left over from a core that was molten long ago but has since cooled. We do not have a good understanding of the magnetic fields of other planets either.

Apollo astronauts made direct measurements of the rate at which heat flows upward through the top of the lunar crust (Fig. 20–17). The rate is one-third that of the heat flow on Earth. The value is important for checking theories of the lunar interior.

Tracking the orbits of the Apollo Command Modules and other satellites that orbited the Moon also told us about the lunar interior. If the Moon were a perfect, uniform sphere, the spacecraft orbits would have been perfect ellipses. We interpret the deviation of the orbits from an ellipse as an effect of an asymmetric distribution of lunar mass.

One of the major surprises of the lunar missions was the discovery in this way of *mascons*, regions of mass concentrations near and under most maria. (A few large mascons are known on Earth.) The mascons led to anomalies in the gravitational field, that is, deviations of the gravitational field from being spherical. The mascons may be lava that is denser than the surrounding matter. The existence of the mascons is evidence that the whole lunar interior is not molten, for if it were, then these mascons could not remain near the surface. However, the mascons could be supported by a crust of sufficient thickness.

In sum, we have contradictory evidence as to whether the lunar interior is hot or cold, molten or solid. Still, as a result of the meteorite impact detected with the seismic experiment, most scientists believe that the Moon's core is molten. If even such a small body as the Moon formed with a hot interior, then we know that the heat did not have time to be radiated away as it was forming. From this observation, we can conclude that the planetesimals (Section 18.10) would have had to coalesce very rapidly into planets. Thus the study of the Moon gives us insight into the formation of the whole solar system.

Figure 20–17 Astronaut Buzz Aldrin with some of the experiments the Apollo 11 astronauts deployed on the lunar surface.

20.3e The Origin of the Moon

Among the models that have been considered in recent years for the origin of the Moon are the following:

1. *Fission:* the Moon was separated from the material that formed the Earth;

2. *Capture:* the Moon was formed far from the Earth in another part of the solar system, and was later captured by the Earth's gravity; and

3. *Condensation:* the Moon was formed near to and simultaneously with the Earth in the solar system.

Comparing the chemical composition of the lunar surface with the composition of the terrestrial surface has been important in narrowing down the possibilities. The mean lunar density of 3.3 grams/cm^3 is close to the average density of the Earth's major upper region (the mantle), which had led some to believe in the fission hypothesis. However, detailed examination of the lunar rocks and soils indicates that the abundances of elements on Moon and Earth are sufficiently different to indicate that the Moon did not form directly from the Earth. (Though some minerals (Fig. 20–18) that do not exist on Earth have been discovered on the Moon, this results from different conditions of formation rather than from abundance differences.) Also, the theory of continental drift (Section 19.2) now explains the formation of the Pacific Ocean basin. Before the theory was accepted, it seemed more likely that the Pacific Ocean could be the hole left behind when the Moon was ripped from the Earth.

Still, the fission hypothesis, at least in the case that the Moon separated from the Earth very early on, cannot be excluded. The interval between the Earth's and Moon's formations in which fission could have happened, however—given the extreme ages of lunar soil and some of the rocks—was only a short one. But new evidence that the ratios of three oxygen isotopes are the same on the Moon as they are on the Earth but different from the ratio for meteorites indicates that the fission hypothesis is still to be reckoned with.

The capture model also has evidence against it. The conditions necessary for the Earth to capture such a massive body by gravity seem too unlikely for this to have occurred. However, the evidence against the model is not conclusive—one can't apply statistical methods to one example—and several possible ways have been suggested in which the Moon could have been captured by the Earth. Perhaps a larger body was broken up by tidal forces and only part was captured.

Although there are chemical differences between the Moon and the Earth, they are not so overwhelming as to exclude the condensation possibility. The probability that the Moon has an iron core backs this theory. Thus many astronomers think it probable that the Earth and the Moon formed near each other as a double planet, probably by accretion of planetesimals. The condensation model thus closely connects the question of the origin of the Moon to the larger question of the origin of the solar system.

It had been hoped that landing on the Moon would enable us to clear up this problem of the lunar origin. But though the lunar programs have led to modifications and updating of the models described above, none of these models has been entirely ruled out. Nobody can say definitely which, if any, of the three is correct.

20.4 Future Lunar Studies

For those of us who remember the shock of Sputnik, and followed each small step farther and farther off the Earth's surface and into space, it is hard to believe that the era of manned lunar exploration not only has begun but also has already

We are obtaining additional evidence about the capture model by studying the moons of Jupiter and Saturn. The outermost moon of Saturn, for example, is apparently a captured asteroid.

Figure 20–18 A crystal of armalcolite (**Arm**strong-**Al**drin-**Col**lins, the crew of Apollo 11), examined under a polarization microscope. This mineral has been found only on the Moon.

372

ended. At present we have no plans to send more people to the Moon. The United States does not even have any unmanned missions definitely planned, though some are in a prospective group of spacecraft suggested to NASA.

One surprise came in 1982, when it was realized that a meteorite found in Antarctica probably came from the Moon (Fig. 20–19). The rock may have been ejected from the Moon when a crater was formed. This rock, and all the lunar samples that have not been distributed or destroyed in analysis, are kept free of contamination in a fairly new Lunar Curatorial Facility in Houston for future study. Another lunar rock, also found in Antarctica, was reported in 1984 by Japanese scientists.

The orbiting manned space stations (Fig. 20–20) now planned for the 1990's by the U.S. and the U.S.S.R. may provide the capability of again sending people out beyond earth orbit. (We should stress, though, that the U.S. National Academy of Sciences found that the U.S. space station is not justified on scientific grounds in the next couple of decades.) Perhaps by the turn of the 21st century, manned lunar exploration will resume. Twenty or thirty years from now, we may each be able to visit the Moon as researchers or even as tourists.

Figure 20–19 This rock is one of many meteorites found in the Earth's continent of Antarctica. Under a microscope and in mineralogical and isotopic analyses, it seems like a sample from the lunar highlands (including anorthosites, for example) and quite unlike any terrestrial rock or any other meteorite. The cube is 1 cm on each side.

Figure 20–20 Space shuttle astronauts flew freely in space near the shuttle in 1984, practicing capabilities for future work that will be vital for space stations.

The Moon Treaty, "Agreement Governing Activities of States on the Moon and Other Celestial Bodies," was approved and made available in 1980 at the United Nations for countries to sign formally. It is an attempt to provide international law in advance of when it is needed to control competition for the Moon's wealth. The treaty is controversial because its phrase that the Moon and its resources are "the common heritage of mankind" leaves ambiguous whether the Moon (and other celestial bodies such as asteroids) would be common property or open to private enterprise (mining, etc.).

Summary and Outline

Lunar features (Section 20.1)
Maria, highlands, craters, mountains, valleys, rilles, ridges
Visibility dependent on sun angle; terminator
Revolution and rotation (Box 20.1)
Sidereal and synodic
Lunar exploration (Section 20.2)
Manned and unmanned lines of development merge in Apollo
Six Apollo landings: 1969 to 1972
Three Soviet unmanned spacecraft brought samples back
Two Soviet Lunokhods roved many kilometers
Composition of the lunar surface (Section 20.3a)
More basalts and highland anorthosites from cooled lava
More breccias—broken up and re-formed—in highlands
Chronology (Section 20.3b)
Relative dating by crater counting
Radioactive dating gives absolute ages
Highland rocks (3.9–4.4 billion years) older than maria (3.1–3.8 billion years)
Probable model: Moon formed 4.6 b.y. At 4.4 b.y., the

surface, which had been molten, cooled. Meteorite bombardment from 4.2 to 3.9 b.y. made most of the craters. Interior heated up from radioactive elements; the resulting vulcanism from 3.8 to 3.1 b.y. caused lava flows that formed the maria. Then lunar activity stopped.
Craters (Section 20.3c)
Almost all formed in meteoritic impact
Signs of vulcanism also present on surface
Far side has no maria, quite different from near side
Interior (Section 20.3d)
Moon is differentiated into a core, a mantle, and a crust
Seismographs revealed that interior is molten, though lack of strong lunar magnetic field indicates the opposite
Weak moonquakes occur regularly
Mascons discovered from their gravitational effects
Origin (Section 20.3e)
Condensation, fission, and capture theories still viable; most astronomers believe in condensation

Key Words

highlands, maria, mare, mountain ranges, valleys, rilles, ridges, rims, librations, terminator, sidereal, synodic, igneous, sedimentary, basalts, anorthosites, breccias, lunar soils, regolith, volatile, refractory, radioactive, stable, crust, mantle, core, shear waves, mascons, fission, capture, condensation

Questions

1. Compare the lengths of sidereal and synodic months. Which is longer? Why?
†2. About how many Moons would fit inside one Earth?
3. If the Moon's mass is $\frac{1}{81}$ Earth's, why is gravity at the Moon's surface as great as $\frac{1}{6}$ that at the Earth's surface?
4. To what location on Earth does the terminator on the Moon correspond?
5. Why is the heat flow rate related to the radioactive material content in the lunar surface?
6. What does cratering tell you about the age of the surface of the Moon, compared to that of the Earth's surface?

7. Why is it not surprising that the rocks in the lunar highlands are older than those in the maria?
8. Why are we more likely to learn about the early history of the Earth by studying the rocks from the Moon than those on the Earth?
9. Using the lunar-chronology chart, describe the relative ages of the lava flows in the seas of Fertility and Serenity.
10. What do mascons tell us about the lunar interior?
11. Choose one of the proposed theories to describe the origin of the Moon, and discuss the evidence pro and con.
12. (a) Describe the American and Soviet lunar explorations. (b) What can you say about future plans?

Topic for Discussion

Discuss the scientific, political, and financial arguments for resuming (a) unmanned and (b) manned exploration of the Moon.

Mercury, photographed at a distance of 200,000 km from Mariner 10, revealed a cratered surface. The symbol for Mercury appears at the top of the facing page.

Mercury

Aims: To discuss the difficulties in studying Mercury from the Earth, and the results of a space mission that flew nearby

Mercury is the innermost planet, and is one of the least understood. Except for distant Pluto, its orbit around the Sun is the most elliptical. (The difference between the maximum and minimum distances of Mercury from the Sun is as much as 40 per cent of the average distance, compared with less than 4 per cent for the Earth.) Its average distance from the Sun is $\frac{4}{10}$ of the Earth's average distance. Thus Mercury is 0.4 A.U. from the Sun.

Since we on the Earth are outside Mercury's orbit looking in at it, Mercury always appears close to the Sun in the sky (Fig. 21–1). At times it rises just before sunrise, and at times it sets just after sunset, but it is never up when the sky is really dark. The Sun always rises or sets within an hour or so of Mercury's rising or setting. Consequently, whenever Mercury is visible, its light has to pass obliquely through the Earth's atmosphere. This long path through turbulent air leads to blurred images. Thus astronomers have never gotten a really good view of Mercury from the Earth, even with the largest telescopes. Many people have never seen it at all. Even the best photographs taken from the Earth show Mercury as only a fuzzy ball with faint, indistinct markings (Fig. 21–2).

21.1 The Rotation of Mercury

From studies of ground-based drawings and photographs, astronomers did as well as they could to describe Mercury's surface. A few features could barely be distinguished, and the astronomers watched to see how long those features took to

In order to know when Mercury and other planets will be favorably located in the sky for you to observe them, you should read the sky descriptions or charts published in monthly magazines like *Sky and Telescope* or *Astronomy* and in some local newspapers, or use *A Field Guide to the Stars and Planets*.

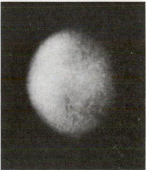

Figure 21–2 From the Earth, we cannot see much surface detail on Mercury. These views, among the best ever taken on Earth, were made by the New Mexico State University Observatory. Mercury was only 7.1 *(top)* and 5.1 *(bottom)* seconds of arc across, respectively.

Figure 21–1 Since Mercury's orbit is inside that of the Earth *(left)*, Mercury is never seen against a really dark sky. A view from the Earth appears at right, showing Mercury and Venus at their greatest respective distances from the Sun. The open view of their orbits is an exaggeration; we actually see the orbits nearly edge-on.

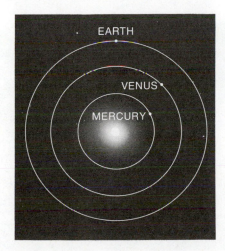

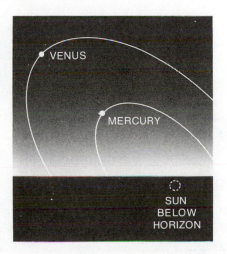

rotate around the planet. From these observations they decided that Mercury rotated in the same amount of time that it took to revolve around the Sun. Thus they thought that one side always faced the Sun and the other side always faced away from the Sun. This led to the fascinating conclusion that Mercury could be both the hottest planet and the coldest planet in the solar system.

But early radio-astronomy studies of Mercury indicated that the dark side of Mercury was too hot for a surface that was always in the shade. (Simply, the strength of the radio signal an object gives off depends on its temperature.)

Later, we became able not only to receive radio signals emitted by Mercury but also to transmit radio signals from Earth to Mercury and detect the echo. This technique is called *radar*, the acronym for **ra**dio **d**etection **a**nd **r**anging. Since Mercury is rotating, one side of the planet is always receding relative to the other. Such motion can be measured with the Doppler effect. The results were a surprise: scientists had been wrong about the period of Mercury's rotation. It actually rotates in 59 days.

Mercury's 59-day period of rotation is exactly $\frac{2}{3}$ of the 88-day period of its revolution, so the planet rotates three times for each two times it revolves around the Sun. Although its tendency to continue rotating is too strong to be overcome by the gravitational grip of the Sun, the Sun's steadying pull is strongest every $1\frac{1}{2}$ rotations. At those times, Mercury is in its perihelion position, so the gravitational bulge on Mercury is as near to the Sun as it can be. Mercury's spin was probably once much faster, but was slowed down because the Sun's gravity attracted the bulge more than it attracted the rest of the planet. A *gravitational interlock* is at work. But because Mercury's orbit is elliptical, the relation of rotation to revolution is 2:3 instead of the 1:1 relation of our Moon.

The rotation period is measured with respect to the stars; that is, the period is one Mercurian sidereal day, the interval between successive returns of the stars to the same position in the sky. Mercury's rotation and revolution combine to give a value for the rotation of Mercury relative to the Sun (that is, a Mercurian solar day) that is neither the 59-day *sidereal rotation period* nor the 88-day period of revolution. As we can see by careful analysis of Figure 21–3, if we lived on Mercury we would measure each day and each night to be 88 Earth days long. We would alternately be fried and frozen for 88 Earth days at a time. Mercury's *solar rotation period* is thus 176 days long, twice the period of Mercury's revolution.

Since we now know that different sides of Mercury face the Sun at different times, the temperature at the point where the Sun is overhead doesn't get as hot as it would otherwise. The temperature at this point is about 700 K (800°F).

It seemed reasonable that there could be a bulge in the distribution of Mercury's mass. The side that was bulging would be attracted to the Sun by gravity, locking the rotation to the revolution, just as the Moon is locked to the Earth. This is called *synchronous rotation.* Synchronous rotation implies that the periods of rotation and revolution are equal, and so the less massive body would always keep the same face toward the more massive body. Our Moon and all Jupiter's largest moons are in synchronous rotation.

Figure 21–3 Follow the arrow that starts facing rightward toward the Sun in the image of Mercury at the left of the figure, as Mercury revolves along the dotted line. Mercury, and thus the arrow, rotates once with respect to the stars in 59 days, when Mercury has moved only ⅔ of the way around the Sun. (Our view is as though we were watching from a distant star.) Note that after one full revolution of Mercury around the Sun, the arrow faces away from the Sun. It takes another full revolution, a second 88 days, for the arrow to again face the Sun. Thus the rotation period with respect to the Sun is twice 88, or 176, days.

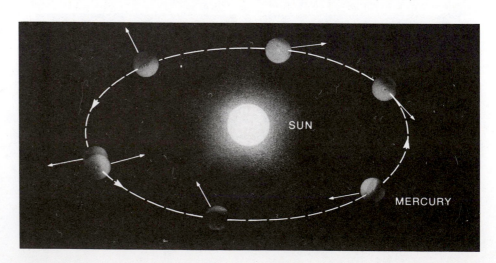

We know from Kepler's second law that Mercury travels around the Sun at different speeds at different times in its elliptical orbit. This effect, coupled with Mercury's slow rotation on its axis, would lead to an interesting effect if we could stand on its surface. From some locations we would see the Sun rise for an Earth day or two, and then retreat below the horizon from which it had just come, when the speed of Mercury's revolution around the Sun dropped below the speed of Mercury's rotation on its own axis. Later, the Sun would rise again, and then continue across the sky.

No harm was done by the scientists' original misconception of Mercury's rotational period, but the story teaches all of us a lesson: we should not be too sure of so-called facts. Don't believe everything you read here, either.

21.2 Mercury from the Ground

Even though the details of the surface of Mercury can't be studied very well from the Earth, other properties of the planet can be better studied. For example, we can measure Mercury's *albedo*, the fraction of the sunlight hitting Mercury that is reflected (Fig. 21–4). We can measure the albedo because we know how much sunlight hits Mercury (we know the brightness of the Sun and the distance of Mercury from the Sun). Then we can easily calculate at any given time how much light Mercury reflects, from both (1) how bright Mercury looks to us and (2) its distance from the Earth. Once we have a measure of the albedo, we can compare it with the albedo of materials on the Earth and on the Moon and thus learn something of what the surface of Mercury is like.

Let us consider some examples of albedo. An ideal mirror reflects all the light that hits it; its albedo is thus 100 per cent. (The very best real mirrors have albedos of as much as 96 per cent.) A black cloth reflects essentially none of the light that hits it; its albedo (in the visible part of the spectrum, anyway) is almost 0 per cent. Mercury's overall albedo is only about 6 per cent. Its surface, therefore, must be made of a dark—that is, poorly reflecting—material. The albedo of the Moon is similarly low. In fact, Mercury (or the Moon) appears bright to us only because it is contrasted against a relatively dark sky; if it were silhouetted against a bedsheet, it would look relatively dark, as if it had been washed in Brand X instead of Tide.

From Mercury's apparent angular size and its distance from the Earth—which can be determined from knowledge of its orbit—we have determined that Mercury is less than half the diameter of the Earth. Since Mercury has no moon, we can determine its mass only from its gravitational effects on bodies that pass near it, such as occasional asteroids and comets. The most accurate value we now have for Mercury's mass comes from tracking the Mariner 10 spacecraft flyby. We find that Mercury's mass is five times greater than that of our Moon and 5½ per cent that of Earth.

Mercury's density (its mass divided by its volume) can thus be calculated, and is roughly the same as the Earth's (Appendix 3). So Mercury's core, like those of

The *albedo* (from the Latin for whiteness) is the ratio of light reflected from a body to light received by it. The concept of albedo is widely used in the study of solid bodies in our solar system.

Figure 21–4 *Albedo* is the fraction of radiation reflected. A surface of low albedo looks dark.

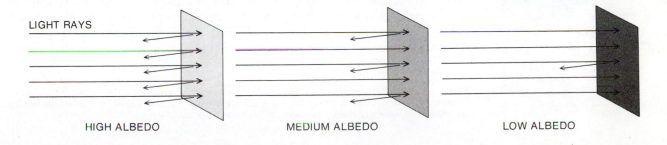

LIGHT RAYS

HIGH ALBEDO   MEDIUM ALBEDO   LOW ALBEDO

Figure 21–5 Mercury, photographed from a distance of 35,000 km as Mariner 10 approached the planet for the first time, shows a heavily cratered surface with many low hills. The valley at the bottom is 7 km wide and over 100 km long. The large flat-floored crater is about 80 km in diameter.

Venus and the Earth, must be heavy; it too is made of iron. Since Mercury has less mass than the Earth and is therefore less compressed, Mercury's core must contain even more iron than Earth's for the planets to have the same density.

21.3 Mariner 10

So astronomers, typically, had deduced a lot from limited data. In 1974, we learned much more about Mercury in a brief time. We flew right by. The tenth in the series of Mariner spacecraft launched by the United States went to Mercury. First it passed by Venus and then had its orbit changed by Venus' gravity to direct it to Mercury. Tracking its orbit improved our measurements of the gravity of these planets and thus of their masses. Further, the 475-kg spacecraft had a variety of instruments on board. One was a device to measure the magnetic fields in space and near the two planets. Another measured infrared emission of the planets and thus their temperatures. Two others of these instruments—a pair of television cameras—provided not only the greatest popular interest but also many important data.

21.3a Photographic Results

When Mariner 10 flew by Mercury the first time (yes, it went back again), it took 1800 photographs and transmitted them to Earth. It came as close as 750 km to Mercury's surface.

The most striking overall impression is that Mercury is heavily cratered (see the photograph that opened this chapter). At first glance, it looks like the Moon! But there are several basic differences between the features on the surface of Mercury and those on the lunar surface. We can compare how the mass and location in the solar system of these two bodies affected the evolution of their surfaces.

Mercury's craters seem flatter than those on the Moon, and have thinner rims (Fig. 21–5). Mercury's higher surface gravity may have caused the rims to slump more. Also Mercury's surface may have been more plastic when most of the cratering occurred. The craters may have been eroded by any of a number of methods, such as the impacts of meteorites or micrometeorites (large or small bits of interplanetary rock). Alternatively, erosion may have occurred during a much earlier period when Mercury may have had an atmosphere, or internal activity, or been flooded by lava.

Most of the craters seem to have been formed by impacts of meteorites. The secondary craters, caused by material ejected as primary craters were formed, are closer to the primaries than on the Moon, presumably because of Mercury's higher surface gravity. In many areas the craters appear superimposed on relatively smooth plains. The plains are so extensive that they are probably volcanic. A class of smaller, brighter craters are sometimes, in turn, superimposed on the larger craters and thus must have been made afterwards (Fig. 21–6). The rate at which objects hit and formed craters is probably about the same for all the terrestrial planets.

Figure 21–6 A fresh new crater, about 12 km across, in the center of an older crater basin. Because Mercury's gravitational field is higher than that of the Moon, material ejected by an impact on Mercury does not travel as far across the surface.

A

B

Figure 21–7 *(A)* Many rayed craters can be seen on this photograph of Mercury, taken six hours after Mariner 10's closest approach on its first pass. The north pole is at the top and the equator extends from left to right about ⅔ of the way down from the top. *(B)* A field of rays radiating from a crater off to the top left, photographed on the second pass of Mariner 10. The crater at top is 100 km in diameter.

Figure 21–8 A scarp *(arrow)* more than 300 km long extends from top to bottom in this second-pass picture.

Some craters have rays of higher albedo emanating from them (Fig. 21–7), just as some lunar craters do. The ray material represents relatively recent crater formation (that is, within the last hundred million years). The ray material must have been tossed out in the impact that formed the crater.

One interesting kind of feature that is visible on Mercury is lines of cliffs hundreds of miles long; on Mercury, as on Earth, such lines of cliffs are called *scarps*. The scarps are particularly apparent in the region of Mercury's south pole (Figs. 21–8 and 21–9). Unlike fault lines on the Earth, such as the San Andreas fault in California, on Mercury there are no signs of geologic tensions like rifts or fissures nearby. These scarps are global in scale, not just isolated.

Figure 21–9 This first-pass view of Mercury's northern limb shows a prominent scarp *(arrow)* extending from the limb near the middle of the photograph. The photograph shows an area 580 km from side to side.

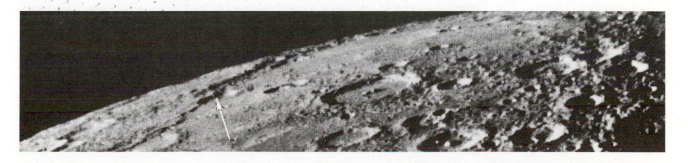

Figure 21–10 This mosaic of pictures from the first pass of Mariner 10 shows a "weird" terrain—hills and ridges cutting across many of the craters and the inter-crater areas.

Box 21.1 Naming the Features of Mercury

Seeing features on the surface of Mercury led to a need for names. The scarps were named for historical ships of discovery and exploration, such as Endeavour (Captain Cook's ship), Santa Maria (Columbus's ship), and Victoria (the first ship to sail around the world, which it did in 1519–1522 under Magellan and his successors). Some plains were given the name of Mercury in different languages, such as Tir (in ancient Persian), Odin (an ancient Norse god), and Suisei (Japanese). Craters were named for non-scientific authors, composers, and artists, in order to complement the lunar naming system, which honors scientists.

The scarps may actually be wrinkles in Mercury's crust. Mercury's core, judging by the fact that Mercury's average density is about the same as the Earth's, is probably iron and takes up perhaps 50 per cent of the volume or 70 per cent of the mass. Perhaps the core was once molten, and shrank by 1 or 2 km as it cooled. This shrinking would have caused the crust to buckle, creating the scarps in the quantity that we now observe. Another possibility is that Mercury formerly rotated more rapidly. As the rotation slowed down, the shape of the planet would have become less oblate (that is, closer to round instead of bulging at the equator). The crust might have cracked and wrinkled as it tried to match the new shape of the interior.

One part of the Mercurian landscape seems particularly different from the rest (Fig. 21–10). It seems to be grooved, with relatively smooth areas between the grooves. It is called the "weird terrain." No other areas like this are known on Mercury and only a couple have been found on the Moon. The weird terrain is 180° around Mercury from the Caloris Basin, the site of a major meteorite impact. Shock waves from that impact may have been focused halfway around the planet.

The Mariner 10 mission was a navigational coup not only because it used the gravity of Venus to get the spacecraft to Mercury, but also because scientists and engineers were able to find an orbit around the Sun that brought the spacecraft back to Mercury several times over. Every six months Mariner 10 and Mercury returned to the same place at the same time. As long as the gas jets for adjusting and positioning Mariner functioned, it was able to make additional measurements and to send back additional pictures in order to increase the photographic coverage. On its second visit, for example, in September 1974, Mariner 10 studied the region around Mercury's south pole for the first time. This pass was devoted to photographic studies. The spacecraft came within 48,000 kilometers of Mercury, much farther away than the 750 kilometer minimum of the first pass, but the data were still very valuable. On its third visit, in March 1975, it had the closest encounter ever—only 300 km above the surface (Fig. 21–11). Thus it was able to photograph part of the surface with such high resolution that it could see detail only 50 meters in size. Then the spacecraft ran out of gas for the small jets that control its pointing, so even though it still passes close to Mercury every few months, it can no longer take clear photographs or send them back to Earth.

21.3b Infrared Results

Mariner's infrared radiometer gave data that indicate that the surface of Mercury is covered with fine dust, as is the surface of the Moon, to a depth of at least several centimeters. Astronauts sent to Mercury, whenever they go, will leave footprints behind them.

Figure 21–11 A high-resolution view of the fractured and ridged plains of the Caloris Basin, photographed from a distance of 21,000 km on Mariner 10's third pass, 34 minutes after the spacecraft's closest approach.

The Mariner 10 mission gave us more accurate measurements of the temperature changes across Mercury than we had determined from the Earth. Within a few hundred kilometers of the terminator, the line between Mercury's day and night, the temperature falls from about 700 K to about 425 K (800°F to 300°F); it drops even lower farther across into the dark side of the planet. The minimum temperature is about 100 K ($-173°C = -280°F$).

21.3c Results from Other Types of Observations

Close-up spectral measurements showed that Mercury even has an atmosphere, although an all-but-negligible one. It is only a few billionths as dense as the Earth's, so slight that even someone standing on Mercury would need special instruments to detect it. Traces of helium, oxygen, carbon, argon, nitrogen, and xenon were detected with a spectrometer that operated in the ultraviolet. The presence of helium was a particular surprise, because helium is a light element and had been expected to escape from Mercury's weak gravity within a few hours. So it must be constantly replaced. The helium must come from the Sun in the solar wind, or, more likely, from radioactive decay of uranium and thorium on Mercury.

One more surprise—perhaps the biggest of the mission—was the detection of a magnetic field in space near Mercury. The field is weak; extrapolated down to the surface it is about 1 per cent of the Earth's. It had been thought that magnetic fields were generated by the rapid rotation of molten iron cores in planets, but Mercury is so small that its core would have quickly solidified. So the magnetic field is not now being generated. Perhaps the magnetic field has been frozen into Mercury since the time when its core was molten.

There are reasonable arguments against both the leading explanations for Mercury's magnetic field. If it were now being generated, we might expect to see some evidence on Mercury's surface of internal heat or of vulcanism during the past 3 billion years. But the alternative is that the magnetic field had been generated in the past. If so, the high temperatures that we think are required to generate a field would have destroyed the field once the generation stopped. The matter is not settled.

Mariner 10 detected lots of electrons near Mercury. Perhaps they are trapped in some sort of belt by the magnetic field, similar to the Van Allen belts around the Earth (Section 19.5). But perhaps they are bound to Mercury for shorter times than electrons trapped by the Earth's magnetic field.

No moons of Mercury have ever been detected, and Mariner 10 did an especially careful job of searching. The amount of light that any moon reflects depends on both its size and its albedo. After all, a moon is a certain brightness because it reflects a certain amount of sunlight. A relatively small size could be compensated for with a relatively high albedo. If we assume that a moon had the same low albedo as Mercury, we can calculate that Mariner 10 would have seen it unless it was smaller than 5 kilometers across. Since a moon with a higher albedo would have to be even smaller to have evaded detection, 5 km is a realistic upper limit.

Spacecraft encounters like those of Mariner 10 have revolutionized our knowledge of Mercury. The kinds of information that we can bring to bear on the basic questions of the formation of the solar system and of the evolution of the planets are much more varied now than they were just a short time ago.

Summary and Outline

Mercury is difficult to observe from the Earth; it is never far
    from the Sun in the sky
Radio astronomy (Section 21.1)
    Surface temperatures
    Rotation period linked to orbit: 1 day is 2 years
Albedo (Section 21.2)
    Low albedo means a poor reflector
    Mercury has a low albedo, only 6 per cent
Mass and density (Section 21.2)
    Density is similar to Earth's

Mariner 10 observations (Section 21.3)
    Photographic results
        Craters, maria, scarps, "weird terrain"
        Mechanisms: impacts, vulcanism, shrinkage of the crust
    Results from infrared observations
        Dust on the surface; temperature measurements
    Results from other types of observations
        A small atmosphere: where does the helium come from?
        A big surprise: a magnetic field, which tells us about the
          histories of Mercury and of the Earth

Key Words

synchronous rotation, radar, gravitational interlock, sidereal rotation period, solar rotation period, albedo, scarps

Questions

1. Assume that on a given day, Mercury sets after the Sun. Draw a diagram, or a few diagrams, to show that the height of Mercury above the horizon depends on the angle that the Sun's path in the sky makes with the horizon as the Sun sets. Discuss how this depends on the latitude or longitude of the observer.

2. If Mercury did always keep the same side towards the Sun, does that mean that the night side would always face the

same stars? Draw a diagram to illustrate your answer.

3. Explain why a day on Mercury is 176 Earth days long.

4. What did radar tell us about Mercury? How did it do so?

†5. Given Mercury's measured albedo, if about 1 erg hits a square meter of Mercury's surface per second, how much energy is reflected from that area? (An erg is a unit of energy.)

6. If ice has an albedo of 70–80 per cent, and basalt has an albedo of 5–20 per cent, what can you say about the sur-

---

†This indicates a question requiring a numerical answer.

face of Mercury based on its measured albedo?

7. If you increased the albedo of Mercury, would its temperature increase or decrease? Explain.

8. List those properties of Mercury that could best be measured by spacecraft observations.

9. Think of how you would design a system that transmits still photographs taken by a spacecraft back to Earth and then produces the pictures. The system can be simpler than your TV since still pictures are involved.

10. How would you distinguish an old crater from a new one?

11. What evidence is there for erosion on Mercury? Does this mean there must have been water on the surface?

12. List three major findings of Mariner 10.

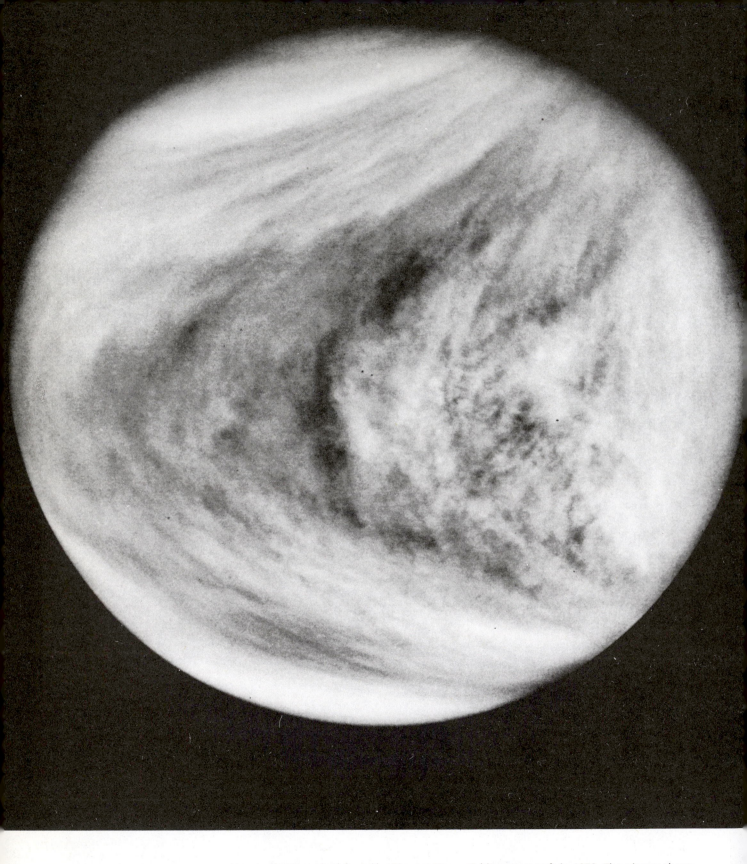

Venus, photographed from the Pioneer Venus Orbiter spacecraft in 1979. The picture shows the cloud tops; we see a turbulent atmosphere. The small mottled features near the center appear to be cells of convection caused by the heating of the atmosphere by solar radiation. The symbol for Venus appears at the top of the facing page.

Venus

Aims: To discuss Venus' clouds, the composition of its atmosphere, and the structure of its surface, and to use this knowledge to improve our understanding of the Earth's structure and atmosphere

Venus and the Earth are sister planets: their sizes, masses and densities are about the same. But they are as different from each other as the wicked sisters were from Cinderella. The Earth is lush; has oceans and rainstorms of water, an atmosphere containing oxygen, and creatures swimming in the sea, flying in the air, and walking on the ground. On the other hand, Venus is a hot, foreboding planet with temperatures constantly over 750 K (900°F), a planet on which life seems unlikely to develop. Why is Venus like that? How did these harsh conditions come about? Can it happen to us here on Earth?

Venus orbits the Sun at a distance of 0.7 A.U. Although it comes closer to us than any other planet—as close as 45 million kilometers—we did not know much about it until recently because it is always shrouded in heavy clouds (Fig. 22–1). Observers in the past saw faint hints of structure in the clouds, which seemed to indicate that these clouds might circle the planet in about 4 days, rotating in the opposite sense from Venus' orbital revolution. The clouds, though, never parted sufficiently to allow us to see the surface.

22.1 The Atmosphere of Venus

Studies from the Earth show that the clouds on Venus are primarily composed of droplets of sulfuric acid, H_2SO_4, with water droplets mixed in. Sulfuric acid may sound strange as a cloud constituent, but the Earth too has a significant layer of sulfuric acid droplets in its stratosphere, a higher layer of the atmosphere (Section 19.4). However, the water in the lower layers of the Earth's atmosphere, circulating because of weather, washes the sulfur compounds out of these layers, whereas Venus has sulfur compounds in the lower layers of its atmosphere in addition to those in its clouds.

Sulfuric acid takes up water very efficiently, so there is little water vapor above Venus' clouds. Painstaking work conducted at high-altitude sites on the Earth revealed the presence of the small amount of cytherean water vapor above Venus' clouds. This observation was difficult, because the spectral lines of water vapor from Venus were masked by the spectral water vapor lines that arise from the Earth's own atmosphere.

Spectra taken on Earth (Fig. 22–2) show a high concentration of carbon dioxide in the atmosphere of Venus. In fact, carbon dioxide makes up over 90 per cent of Venus' atmosphere (Fig. 22–3). The Earth's atmosphere, for comparison, is mainly nitrogen, with a fair amount of oxygen as well. Carbon dioxide makes up less than 1 per cent of the terrestrial atmosphere.

The surface pressure of Venus' atmosphere is 90 times higher than the pressure of Earth's atmosphere, as a result of the large amount of carbon dioxide in the for-

Figure 22–1 A crescent Venus, observed with the 5-m Hale telescope in blue light. We see only a layer of clouds.

Astronomers often use the adjective *cytherean* to describe Venus. Cythera was the island home of Aphrodite, the Greek goddess who corresponded to the Roman goddess Venus. The adjectives ''venusian'' and ''venerean'' are now in increasing use.

Figure 22–2 An infrared spectrum of Venus, showing the region from 8600 Å to 8822 Å, which includes spectral bands from carbon dioxide.

mer. Carbon dioxide on Earth dissolved in sea water and eventually formed our terrestrial rocks, often with the help of life forms. (Limestone, for example, has formed from marine life under the Earth's oceans.) If this carbon dioxide were released from the Earth's rocks, along with other carbon dioxide trapped in sea water, our atmosphere would become as dense and have as high a pressure as that of Venus. Venus, slightly closer to the Sun than Earth and thus hotter, had no oceans in which the carbon dioxide could dissolve nor life to help take up the carbon. Thus the carbon dioxide remains in Venus' atmosphere.

22.2 The Rotation of Venus

In 1961, radar astronomy penetrated Venus' clouds, allowing us to determine an accurate rotation period for the planet's surface. Venus, because it comes closer to Earth, is an easier target for radar than Mercury. Venus rotates in 243 days with respect to the stars in the direction opposite from the other planets; such backward motion is called retrograde rotation, to distinguish it from forward (direct) rotation. Venus revolves around the Sun in 225 Earth days. Venus' periods of rotation and revolution combine, in a way similar to that in which Mercury's sidereal day and year combine (Section 21.1), so that a solar day on Venus corresponds to 117 Earth days; that is, the planet's rotation brings the Sun back to the same position in the sky every 117 days.

The notion that Venus is in retrograde rotation seems very strange to astronomers, since the other known planets revolve around the Sun in the same direction, and all the planets (except Uranus) and most satellites also rotate in that same direction. Because of the conservation of angular momentum (Section 18.9), and since the original material from which the planets coalesced was undoubtedly rotating, we expect all the planets to revolve and rotate in the same sense.

Nobody knows definitely why Venus rotates "the wrong way." One possibility is that when Venus was forming, the planetesimals formed clumps of different sizes. Perhaps the second largest clump struck the largest clump at an angle that caused the result to rotate backwards. Scientists do not like *ad hoc* ("for this special purpose") explanations like this one, since ad hoc explanations often do not apply in general.

Another possible explanation of Venus' retrograde rotation deals with the effects of the circulation of the thick venerean atmosphere and the gravitational pull of the Sun on the solid surface of Venus and on the atmosphere. (It is the difference of the force of gravity on the near side and the far side of the planet—a tidal force—that counts.) Theories have been constructed to show that these effects could have gradually slowed up and reversed Venus' rotation.

The slow rotation of the solid surface contrasts with the rapid rotation of the clouds. The tops of the clouds rotate in the same sense as the surface of Venus rotates but about 60 times more rapidly, once every 4 days. This rapid rate may come from the law of conservation of angular momentum—the upper layers, which are of low density, must rotate rapidly to match the angular momentum of the slowly rotating dense lower layers.

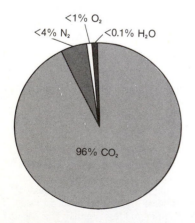

Figure 22–3 The composition of Venus' atmosphere.

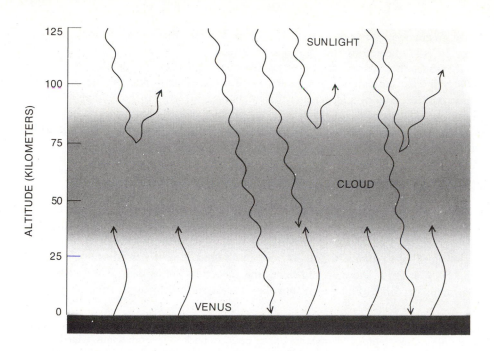

Figure 22–4 Sunlight can penetrate Venus' clouds, so the surface is illuminated with radiation in the visible part of the spectrum. Venus' own radiation is mostly in the infrared. This infrared radiation is trapped, a phenomenon called the greenhouse effect.

22.3 The Temperature of Venus

The surface of Venus can be detected from Earth with radio telescopes, since radio waves emitted by the surface penetrate the clouds. We can also determine the temperature of the cytherean surface by studying Venus' radio emission. The surface is very hot, about 750 K (900°F).

In addition to measuring the temperature on Venus, scientists theoretically calculate approximately what it would be if the cytherean atmosphere allowed all radiation hitting it to pass through it. This value—less than 375 K (215°F)—is much lower than the measured values. The high temperatures derived from radio measurements indicate that Venus traps much of the solar energy that hits it.

The process by which this happens on Venus is similar to the process that is generally—though incorrectly—thought to occur in greenhouses here on the Earth. The process is thus called the *greenhouse effect* (Fig. 22–4); it was suggested for Venus by Carl Sagan of Cornell and James B. Pollack of NASA/Ames.

Sunlight passes through the cytherean atmosphere in the form of radiation in the visible part of the spectrum. The sunlight is absorbed by, and so heats up, the surface of Venus. At the temperatures that result, the radiation that the surface gives off is mostly in the infrared. But the carbon dioxide and other constituents of Venus' atmosphere are together opaque to infrared radiation, so the energy is trapped. Thus Venus heats up above the temperature it would have if the atmosphere were transparent; the surface radiates more and more energy as the planet heats up until a balance is struck between the rate at which energy flows in from the sun and the rate at which it trickles out (as infrared) through the atmosphere. The situation is so extreme on Venus that we say a "runaway greenhouse effect" is taking place there. Understanding such processes involving the transfer of energy is but one of the practical results of the study of astronomy.

Greenhouses on Earth don't work quite this way. The closed glass of greenhouses on Earth prevents convection and the mixing in of cold outside air. The trapping of solar energy by the "greenhouse effect" is a less important process in an

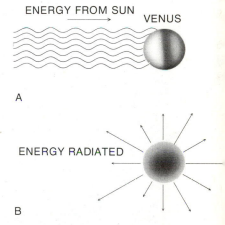

Figure 22–5 Venus receives energy from only one direction, and heats up and radiates energy (mostly in the infrared) in all directions. From balancing the energy input and output, astronomers can calculate what Venus' temperature would be if Venus had no atmosphere. This type of calculation—balancing quantities theoretically to make a prediction to be compared with observations—is typical of those made by astronomers.

actual greenhouse. Try not to be bothered by the fact that the greenhouse in your backyard is not heated by the "greenhouse effect."

If the Earth's atmosphere were to gain enough more carbon dioxide, the oceans could boil, and the carbon in rocks could be released and enter the atmosphere as carbon dioxide. This would increase the greenhouse effect, which could also become a runaway. We are presently increasing the amount of carbon dioxide in the Earth's atmosphere by burning fossil fuels; Venus has shown us how important it is to be careful about the consequences of our energy use.

22.4 Radar Mapping of the Surface of Venus

We can study the surface of Venus by using radar to penetrate Venus' clouds. Radars using huge Earth-based radio telescopes, such as the giant 1000-ft (304-m) dish at Arecibo, Puerto Rico, have mapped a small amount of Venus' surface with a resolution of up to about 20 km. Regions that reflect a large percentage of the radar beam back at Earth show up as bright, and other regions as dark. The Earth-based radar maps of Venus (Fig. 22–6) show large-scale surface features, some very rough

Figure 22–6 *(A)* A montage of radar maps showing one-fourth of the surface of Venus. Whiter areas show regions where more power bounces back to us from Venus, which usually corresponds to rougher terrain than darker areas. *(B)* The low reflectivity region Planum Lakshmi (a plain) and the high reflectivity region Montes Maxwell (mountains) make up most of a continental-sized region, Terra Ishtar. This image has a resolution of 10 km. *(C)* The craters are 30 to 65 km in diameter. They are relatively smooth and are surrounded by rough areas of ejected material. The craters are probably the result of impacts on the surface, but a volcanic origin cannot be ruled out as yet. This image, released in 1983, has a resolution of 3 km. (Courtesy of D. B. Campbell, Arecibo Observatory)

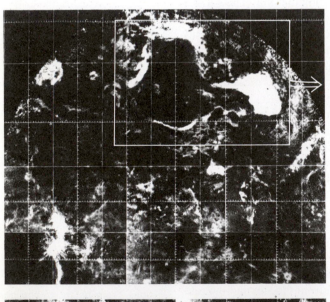
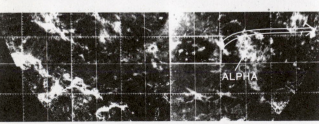

A

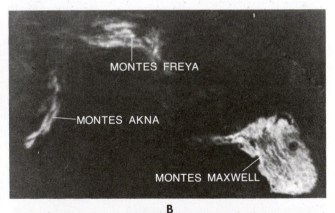

B

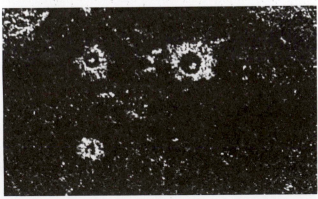

C

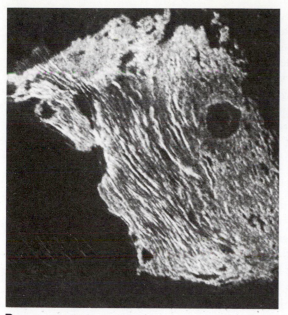

Figure 22–6 *(D)* The Arecibo Observatory's high-resolution radar image of Montes Maxwell, with a resolution of about 2 km. *(E)* An Arecibo image of Beta Regio.

D   **E**

and others relatively smooth, similar to the variation we find of surfaces on the Moon. Many round areas that are probably craters have been found, ranging up to 1000 km across. Most have probably been formed by meteoritic impact.

From Venus' size and from the fact that its mean density is similar to that of the Earth, we concluded that its interior is also probably similar to that of the Earth. This meant that we expected to find volcanoes and mountains on Venus, and that venusquakes probably occur too. A bright area near a large plain has been called Maxwell; we will see later on that Maxwell turns out to be a huge mountain.

The region Alpha, long known for its very high radar reflectivity, is circular and 1100 km across. It appears to have no counterpart on Earth. A central dark region may be a volcano.

A huge peak, Beta, whose base is over 750 km in radius and which has a depression 40 km in radius at its summit has also been detected by radar. It too is probably a giant volcano. It has long tongues of rough material extending as far as 480 km from it in an irregular fashion. A cluster of some 22 smaller peaks resembles the configuration of clusters of volcanoes on Earth.

A long trough at Venus' equator, 1500 km in length, seems to resemble the Rift Valley in East Africa, the Earth's largest canyon.

In Section 22.6, we will see how spacecraft radar observations have mapped Venus' surface more completely.

22.5 Early Spacecraft Observations

Venus was an early target of both American and Soviet space missions. During the 1960's, American spacecraft flew by Venus, and Soviet spacecraft dropped through its atmosphere.

In 1970, the Soviet Venera 7 spacecraft radioed 23 minutes of data back from the surface of Venus before it succumbed to the high temperature and pressure. Two years later, the lander from Venera 8 survived on the surface of Venus for 50 minutes. They confirmed the Earth-based results of high temperatures, pressures, and high carbon dioxide content.

The United States spacecraft Mariner 10 took thousands of photographs of Venus in 1974 as it passed by en route to Mercury. It went sufficiently close and its imaging optics were good enough that it was able to study the structure in the clouds with resolution of up to 220 meters (Fig. 22–7). The structure shows only when viewed in ultraviolet light. We could observe much finer details than from Earth.

The clouds appear as long, delicate streaks, like terrestrial cirrus clouds. In the cytherean tropics the clouds also show a mottling, which suggests that convection, the boiling phenomenon, is going on. A big "eye" of convection is visible just downwind from the point at which the Sun is overhead and, therefore, at which its heating actions are at the maximum. Having an "eye" downwind from the subsolar point is a perfect example of what atmospheric scientists would expect, since hot air rises and starts the process of convection. Strong winds blow the clouds at these upper levels around the planet at 300 km/hr, as rapidly as the jet stream blows on Earth but much more widespread.

No magnetic field was detected. This may indicate that either Venus does not have a liquid core or that the core does not have a high conductivity.

Such studies of Venus have great practical value. The better we understand the interaction of solar heating, planetary rotation, and chemical composition in setting up an atmospheric circulation, the better we will understand our Earth's atmosphere. We then may be better able to predict the weather and discover jet routes that would aid air travel, for example. The potential financial return from this knowledge is enormous: it would be many times the investment we have made in planetary exploration.

Soviet spacecraft that landed on Venus in 1975 survived long enough to send back photographs. The single photograph that the Venera 9 lander took in the 53 minutes before it succumbed to the tremendous temperature and pressure was the first photograph ever taken on the surface of another planet. It showed a clear image of sharp-edged, angular rocks (Fig. 22–8). This came as a surprise to some scientists, who had thought erosion would be rapid in the dense cytherean atmosphere, and that the rocks should therefore have become smooth or disintegrated into sand.

Three days later, the Venera 10 lander reached the surface and sent data for 65 minutes, including a photograph. The rocks at its site were not sharp; they resembled huge pancakes, and between them were sections of cooled lava or debris of weathered rock.

The Veneras measured the surface wind velocity to be only 1 to 4 km/hr. The low wind speed makes it seem likely that erosion is caused not by sand blasting but by melting, temperature changes, chemical changes (which can be very efficient at the high temperature of Venus), and other mechanisms.

The Venera landers also studied the soil, finding that it resembles basalt in chemical composition and density, in common with the Earth, the Moon, and Mars. They measured temperatures of 760 K and 740 K at the two sites, and a pressure over 90 times that of the Earth's atmosphere, confirming the earlier measurements. The Venera 8 lander three years earlier had reported a surface composition more like

Figure 22–7 A series of photographs of the circulation of the clouds of Venus, photographed at 7-hour intervals from Mariner 10 in the ultraviolet near 3550 Å. The contrast has been electronically enhanced so that small actual differences in contrast are made apparent. The size of the feature indicated by the arrows is about 1000 km.

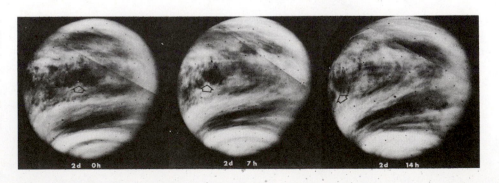

granite. But the Venera 8, 9, and 10 spacecraft could do only crude analyses of the soil, since they made measurements with gamma-ray measuring devices, which can detect only radioactive elements.

22.6 Current Views of Venus

The United States and the Soviet Union both sent spacecraft to Venus in 1978. NASA's Pioneer Venus 1 was the first spacecraft to go into orbit around Venus, allowing it to make observations over a lengthy time period. Its elliptical orbit ranges between 400 km and 65,000 km in altitude. Its dozen experiments included cameras to study Venus' weather by photographing the planet regularly in ultraviolet light (Fig. 22–9 and the figure opening this chapter).

The spacecraft also carried a small radar to study the topography of Venus' surface. The resolution was not quite as good as that of the best Earth-based radar studies of Venus, but the orbiting radar mapped a much wider area. It mapped over 90 per cent of the surface with a resolution typically better than 100 km. It was an early version of a new method of radar, called *synthetic-aperture radar*. In this method, the radar images are recorded over a period of time while the spacecraft moves. The views from these different aspects are assembled during the data reduction, giving resolution equivalent to a radar of much larger aperture; the larger aperture has been ''synthesized.''

From the radar map (Color Plate 35 and Fig. 22–10), we now know that 60 per cent of Venus' surface is covered by a rolling plain, flat to within plus or minus 1 km. Only about 16 per cent of Venus' surface lies below this plain, a much smaller fraction than the two-thirds of the Earth covered by ocean floor.

Two large features, the size of small Earth continents, extend several km above the mean elevation. A northern continent, Terra Ishtar, is about the size of the con-

Figure 22–8 The surface of Venus photographed from Venera 9. The photograph is reproduced in a flat strip, but was actually a scan from upper left, down toward the center, and up toward the right. Think of looking to your left at the horizon, then down at your feet, and then up to the horizon at the right. That is why the horizon is visible as tilted lines at upper left and at upper right.

Figure 22–9 Four photographs of Venus taken over a one-month span. They show relatively dark equatorial bands with many small features, probably from convection, superimposed. Polar bands of clouds are relatively bright.

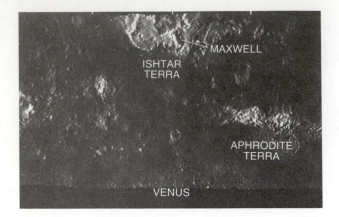

Figure 22–10 Venus' surface, based on the Pioneer Venus Orbiter's radar map, compared with Earth's surface at the same resolution (50-100 km). Two continents exist on Venus: Terra Aphrodite, which is comparable in scale with Africa, and Terra Ishtar, which is comparable in scale with the continental United States or Australia. The volcano Maxwell, 10.8 km above venusian "sea level" (the average radius of Venus), is on the right of Ishtar Terra and is the highest point on Venus. Sixty per cent of Venus' surface is covered with a huge rolling plain. Only about 16 per cent of Venus' surface is covered with lowlands, compared with over two-thirds of Earth's surface covered with oceans. Mid-ocean ridges, a sign of spreading plates, are not visible on Venus though this comparison shows that they would be detectable on Earth at this resolution.

Aphrodite, the Greek goddess of love, was the equivalent of the Roman goddess Venus. Ishtar was the Babylonian goddess of love and war, daughter of the moon and sister of the sun. Major features on Venus are being named after mythical goddesses, minor features after other mythical female figures, and still smaller circular features after famous women.

Figure 22–11 Pioneer Venus 1, the orbiter, is in the background. Pioneer Venus 2, which contains the basic cylinder called the "bus," is in the foreground. A large probe and three smaller probes (of which one is hidden) are mounted on the bus.

tinental United States. A giant mountain on it known as Maxwell is 11 km high, 2 km taller than Earth's Mt. Everest stands above terrestrial sea level. It had formerly been known only as a bright spot on Earth-based radar images. Ishtar's western part is a broad plateau, about as high as the highest plateau on Earth (the Tibetan plateau) but twice as large. A southern continent, Aphrodite Terra, is about twice as large as the northern continent and is much rougher.

A smaller elevated feature, Beta Regio, on the basis of Pioneer Venus and terrestrial radar results and on the results of the Soviet landers, appears to be a pair of volcanoes. The fourth highland feature, Alpha Regio, has rough terrain that may resemble the western United States. A giant canyon, 1500 km long, is 5 km deep and 400 km wide. It is the largest in the solar system and may be left over from an earlier stage of Venus' geological evolution.

Several circular features on the plains may be craters. This would indicate that the plains are geologically old, though there are many fewer craters than there are on the Moon, Mercury, or Mars. As we have seen, the Earth's current surface is geologically young, formed by moving "plates" as part of continental drift or "plate tectonics."

From studying the strength of gravity from place to place, by analyzing the orbit of the spacecraft, we learn what Venus' crust is like under the surface. Variations in Venus' gravity match the shape of the surface features observed with the radar. On Earth, most of the changes in gravity result from material hidden under the surface, connected with the processes that move plates, rather than with the visible continental boundaries.

The radar and gravity results show that Venus, unlike Earth, is now apparently made of only one continental plate. We observe nothing on Venus equivalent to Earth's mid-ocean ridges, at which new crust is carried up. Venus may well have such a thick crust that any plate tectonics that existed in the distant past was choked off.

The orbiter has provided evidence that volcanoes may be active on Venus. The abundance of sulfur dioxide it found on arrival was far higher than the upper limits of previous observations, and then dropped by a factor of 10 in the next five years. Over the same five-year period, the haze dropped drastically, as it had about 20 years earlier. The effect could come from eruptions at least ten times greater than Mexico's El Chichon in 1982.

Venus' wind patterns undergo long-term patterns of change. The jet-stream pattern seen at mid-latitudes when Mariner 10 flew by in 1974 had given way to a pattern of cloud and wind acting like a solid body by the time the Pioneer Venus Orbiter arrived. The upper atmosphere moves westward around the planet at tremendous speed, circling the planet every 4 days. The high-altitude haze layer (over the main cloud layer) that appears and disappears over a period of years contains tiny droplets of sulfuric acid.

Pioneer Venus 2 arrived at Venus at about the same time as Pioneer Venus 1, and carried several spacecraft together on a basic "bus" (Fig. 22–11). The bus and its four probes all penetrated Venus' atmosphere and descended to the surface. The Pioneer probes found that high-speed winds at the upper levels are coupled to other high-speed winds at lower altitudes. The lowest part of Venus' atmosphere, however, is relatively stagnant. Winds in this lower half of the atmosphere do not exceed about 18 km/hr. This lower half is much denser than the upper half, however, so the amount of momentum in the lower and upper-level winds may be the same.

The probes found that carbon dioxide makes up 96 per cent of the atmosphere. They found less water vapor but more forms of sulfur than had been expected in Venus' lower atmosphere. This is particularly important since the presence of carbon dioxide alone is not sufficient to cause the greenhouse effect. The spectrum of carbon dioxide has gaps that would let out too much infrared radiation. Thus Venus' carbon dioxide absorbs only 55 per cent of the solar heat entering Venus' atmosphere. But comparison of computer models with the Pioneer Venus results tells us that the gaps are plugged by water vapor (25 per cent), sulfuric acid droplets and other types of cloud and haze particles (15 per cent), and sulfur dioxide (5 per cent).

The Pioneer probes also discovered winds that blow somewhat toward the poles, at speeds of 25 km/hr. They thus carry equatorial heat poleward, setting up a planet-wide circulation of the atmosphere and making the temperature relatively constant all around Venus. The infrared sensing equipment aboard the Orbiter has revealed that the cloud tops are over 22°C warmer at the poles than at the equator (Fig. 22–12). Below 60 km in altitude, the polar atmosphere is cooler. Thus the lower atmosphere circulates from equator to pole: the warm atmosphere at the equator rises, cools as it expands, moves to the poles, and sinks. Above it, the upper atmosphere seems to circulate from pole to equator. On Earth, our planet's faster rotation stretches out the weather patterns, and instabilities make three circulating patterns between equator and pole in each hemisphere.

Venus' clouds, largely made of sulfuric acid droplets, start 48 km above its surface, and extend upward about 30 km. This is much higher than terrestrial clouds, which are made of water droplets and which rarely go above 10 km. (The upper layers of the Earth's atmosphere, above the clouds, also contain sulfuric acid droplets but are thin enough to be transparent, unlike Venus' layers of sulfuric acid droplets.) The Pioneer probes detected three distinct layers of venerean clouds, separated from each other by regions of relatively low density. The lowest layer, relatively dense, is only 2 km thick. Over it is the second layer, about 7 km thick. The uppermost layer does not have a sharp top. Below the clouds, the atmosphere is fairly free of dust or cloud particles.

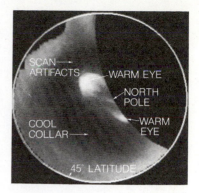

Figure 22–12 An infrared polar view from the Pioneer Venus Orbiter. The north pole is at the center, and the outer circle represents 45° latitude. The subsolar point is at the bottom. Brighter regions correspond to warmer cloud-top temperatures in these measurements at a wavelength of 11 microns. Two bright "eyes" straddle the pole. A dark cloud "collar" surrounds the pole. (NASA photo, courtesy of David J. Diner)

Now that the Pioneer Venus results have enabled us to understand the greenhouse effect on Venus, we can much better understand the effect of adding carbon dioxide from burning fossil fuels to Earth's climate. We have already increased the amount of carbon dioxide in Earth's atmosphere by 15 per cent; some scientists have predicted that it may even double in the next 50 years, which could result in a world-wide temperature rise of a few °C and eventually even in a runaway greenhouse effect. But such long-range predictions are very uncertain.

One of the probes measured that only about 2 per cent of the sunlight reaching Venus filters down to the surface, making it like a dim terrestrial twilight. Thus most of the Sun's heat is absorbed in the clouds, unlike the situation with Earth, and Venus' clouds are more important than Earth's for controlling weather.

When scientists studied the form of the gas argon (argon-36) that is thought to be left over from the original formation of Venus, they encountered an interesting mystery: The abundance of this ''primordial argon,'' compared to the amount of the form of argon (argon-40) that has been accumulating since the planets' formation, is 70 times larger on Venus than on Earth. If the planets' atmospheres had been accreted from comets or asteroids they encountered, the abundances should have been the same. Further, the relative amounts of the rare gases neon, argon, and krypton, compared with each other, are the same for Venus, the Earth, and Mars, and are different from the solar wind and the solar corona. This rules out the possibility that the planets' atmospheres came from the sun. The theory that the planets and their atmospheres resulted from the accretion of planetesimals fits all the data, since the planetesimals could have evolved differently at different locations in the solar nebula. So we are learning how the present atmospheres of the planets were formed. We are also learning how the material from the solar nebula was incorporated in the planets. No longer do we believe that the compositions of the inner planets differed because of temperature differences in the solar nebula. Increased pressure at Venus now appears to be the culprit; these higher pressures caused more of the rare gases to stick to the solid particles from which Venus formed. To interpret the observations, we are revising our understanding of the composition, temperature, and density of the solar nebula from which the solar system formed.

The spacecraft to Venus did not find any intrinsic magnetic field. Some magnetic field is present to hold the solar wind away from Venus, but it may result only from the solar wind's interaction with the ionized part of Venus' upper atmosphere. Though Venus rotates more slowly than Earth, and magnetic fields are thought to result from the rotation of a planet's interior, the difference in rotational period is not great enough to account for the difference in magnetic field strength. Perhaps the chemical differences between Venus and Earth, of which the argon abundance is an example, led to a different temperature structure in Venus' interior and thus a different magnetic field. Perhaps the lack of a magnetic field allowed the solar wind to scour Venus' atmosphere, and make it evolve differently from Earth's protected atmosphere.

The two 1978 Soviet missions, Venera 11 and 12, each consisted of a flyby, which acted as a communications relay, and a lander. They reached Venus shortly after the American missions, and the landers parachuted softly to the surface. On the way, they measured the chemical composition of the atmosphere. The high value for argon they found, for example, is in agreement with the high value found by the American probes. One of the Soviet landers sent back data from the surface for almost two hours and the other for an hour and a half, but did not transmit any pictures.

Both American and Soviet missions, from the radio signals they detected, reported lightning at a frequent rate. The Soviet landers detected up to 25 pulses of energy per second between 5 and 11 km above the surface, later confirmed by the Pioneer Venus Orbiter. The lightning is concentrated over Beta Regio and the eastern part of Aphrodite Terra. This is strong evidence that those regions are sites of active volcanoes. On Earth, it is common for the dust and ash ejected by volcanoes to rub together, generating static electricity that produces lightning. But other models are possible for producing the lightning.

Two Soviet spacecraft, Venera 13 and 14, arrived at Venus in March 1982. They carried cameras that showed a variety of sizes and shapes of rocks (Figs. 22–13 and 22–14). The color photographs showed that Venus' sky is orange. The landing sites were chosen in consultation with American scientists on the basis of Pioneer Venus data, the first example of such Soviet-American cooperation on planetary exploration.

Figure 22–13 The view from Venera 13, showing a variety of sizes and textures of rocks. The spacecraft landed in the foothills of a mountainous region south of Venus' equator and below the region Beta. The spacecraft survived for 127 minutes.
    The camera first looked off to the side, then scanned downward as though looking at its feet, and then scanned up to the other side. As a result, opposite horizons are visible as slanted boundaries at upper left and right.

The landers carried devices that measured the composition of the surface of the planet. Samples were drilled and brought into a low-pressure container at an Earth-like temperature. X-rays emitted when the samples were irradiated revealed which elements are present. (A ''fluorescent'' process occurs when radiation at one wavelength is given off in response to radiation at another, so the devices are known as x-ray fluorescence spectrometers.) The gamma-ray spectrometers on earlier Venera had been able to measure only naturally radioactive isotopes of uranium, thorium, and potassium. The chemical makeup of the regions of Venus where the new spacecraft landed is similar to that of basalt—volcanic rock. Since they had landed in regions near Beta Regio that had the overall topography of a volcanic region, the result indicates that the same chemical processes are at work on both Venus and the Earth.

The Soviet Veneras 15 and 16 have been orbiting Venus since October 1983. They are carrying infrared equipment to study the atmosphere, and a radar (Fig. 22–15).

Our recent space results, coupled with our ground-based knowledge, show us that Venus is even more different from the Earth than had previously been imagined. Among the differences are Venus' slow rotation, its one-plate surface, the absence of a satellite, the extreme weakness or absence of a magnetic field, the lack of water

Figure 22–14 The view from Venera 14, showing flat structures without the smaller rocks of the other site. The flat structures may be large flat rocks or may be fine material held together. The site is 1000 km east of the Venera 13 site and at a lower elevation. The chemical composition resembles that of basalts found on the Earth's ocean floor and the Moon's maria. Venera 13, on the other hand, found basalts more typical of the Earth's continental crust and the Moon's highlands.

Figure 22–15 A radar image apparently showing a young volcanic crater about 40 km wide, transmitted from the Venera 15 or 16 spacecraft in 1983. The region is Metis Regio, a highland area about 500 km across located just west of Ishtar. The linear features at left may be scarps and faults, perhaps formed or modified by vulcanism. The resolution is 1 or 2 km.

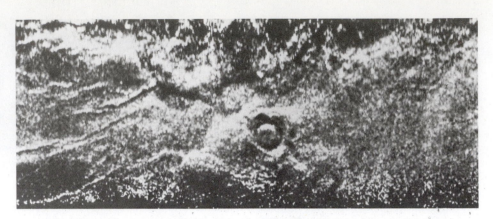

Figure 22–16 The Soviet-French spacecraft to Venus and Halley's Comet.

in its atmosphere, the high abundance of primordial argon, and its high surface temperature.

The next mission to Venus will be a joint Soviet-French probe carried in 1985 by a Soviet rocket en route to Halley's Comet. Plans to put a French balloon in Venus' atmosphere were dropped in favor of more comet experiments. The mission is called ''Vega'': ''Ve'' for Venus and ''ga'' for the Russian-language equivalent of Halley (there is no ''h'' in the Russian alphabet).

NASA's plans for an elaborate Venus Orbiting Imaging Radar spacecraft were cut for budgetary reasons, but a smaller spacecraft called Venus Radar Mapper is now being planned for launch from the Space Shuttle in 1988. Its synthetic-aperture radar should map most of Venus with $\frac{1}{4}$- to $\frac{1}{2}$-km resolution, compared with the 20-km best resolution of Pioneer Venus and the 1-km best resolution of Venera 15 and 16. It will cover Venus' equatorial regions while Venera 15 and 16 covered the north polar region. Maybe, with resolution on the order of 1 km, we can detect volcanic changes on Venus' surface during the 1-Venus-year lifetime of the Venera or Venus Radar Mapper missions.

Figure 22–17 The Venus Radar Mapper, which will use a synthetic aperture radar to obtain images of the surface that will appear like optical photographs, showing enough detail to be useful for geological studies. A 1988 launch is hoped for.

Summary and Outline

Venus' atmosphere (Section 22.1)
    Mainly composed of carbon dioxide
    Clouds primarily composed of sulfuric acid droplets
A slow rotation period, in retrograde (Section 22.2)
The temperature of Venus (Section 22.3)
    A high temperature, 750 K, caused by the greenhouse effect
    Greenhouse effect: visible light in, turns to infrared, infrared can't escape
The surface of Venus (Section 22.4)
    Radar maps show volcanoes and mountains
Early spacecraft observations (Section 22.5)
    U.S. Mariner 10 observations of clouds and atmosphere
    Soviet Venera 9 and 10 landers: temperature measurements and photographs of the surface
Later space exploration (Section 22.6)
    U.S. Pioneer Venus 1 and Pioneer Venus 2

Orbiter studied Venus for many months
    Followed evolution of clouds
    Mapped surface with radar; further evidence of volcanoes
    Temperature maps show circulation of atmosphere
Multiprobe contained bus and 4 probes
    Measured composition of atmosphere and temperatures
    Observed lightning
    Confirmed that greenhouse effect causes the high temperature
U.S.S.R. Venera
    11 and 12: radioed data from surface; observed temperature, lightning, composition of atmosphere. 13 and 14: landers, photographs: sky is orange; soil is basaltic in this region, which has volcanic topography. 15 and 16: signs of vulcanism.
U.S. radar mapper, U.S.S.R. flyby en route to Halley planned

Key Words

greenhouse effect, cytherean, synthetic-aperture radar

Questions

1. Make a table displaying the major similarities and differences between the Earth and Venus.

2. Why does Venus have more carbon dioxide in its atmosphere than does the Earth?

3. Why do we think that there have been significant external effects on the rotation of Venus?

4. If observers from another planet tried to gauge the rotation of the Earth by watching the clouds, what would they find?

5. Suppose a planet had an atmosphere that was opaque in the visible but transparent in the infrared. Describe how the effect of this type of atmosphere on the planet's temperature differs from the greenhouse effect.

6. Why do radar observations of Venus provide more data about the surface structure than a flyby with close-up cameras?

7. If one removed all the CO_2 from the atmosphere of Venus, the pressure of the remaining constituents would be how many times the pressure of the Earth's atmosphere?

8. The Earth's magnetic field protects us from the solar wind. What does this tell you about whether or not the solar-wind particles are charged?

9. Some scientists have argued that if the Earth had been slightly closer to the Sun, it would have turned out like Venus; outline the logic behind this conclusion.

10. Outline what you think Venus would be like if we sent an expedition to Venus that resulted in the loss of almost all of the CO_2 from its atmosphere.

11. Why do we say that Venus is the Earth's "sister planet"?

12. Compare and contrast the observations of the cytherean surface made from the Venera landers.

13. Compare the ground-based and Orbiter-based radar observations of Venus.

14. Do radar observations of Venus study the surface or the clouds? Explain.

15. Describe the major radar results from the Venus Pioneer Orbiter.

16. What is the advantage of synthetic-aperture radar over ordinary radar?

17. What signs of vulcanism are there on Venus?

18. Compare and contrast the circulation of the atmospheres of Venus and the Earth.

A view from Viking Orbiter 2 as it approached the dawn side of Mars, in early August 1976. At the top, with water-ice cloud plumes on its western flank, is Ascreaus Mons, one of the giant Martian volcanoes. In the middle is the great rift canyon called Valles Marineris, and near the bottom is the large, frosty crater basin called Argyre. The south pole is at the bottom. The symbol for Mars appears at the top of the facing page.

Mars

Aims: To discuss the atmosphere and surface features of Mars, to see the results of spacecraft in orbit and on the surface, and to assess the chances of finding life there

Mars has long been the planet of greatest interest to scientists and nonscientists alike. Its interesting appearance as a reddish object in the night sky and some past scientific studies have made Mars the prime object of speculation as to whether or not extraterrestrial life exists there.

In 1877, the Italian astronomer Giovanni Schiaparelli published the results of a long series of telescopic observations he had made of Mars. He reported that he had seen *canali* on the surface. When this Italian word for "channels" was improperly translated into "canals," which seemed to connote that they were dug by intelligent life, public interest in Mars increased.

In this country, Percival Lowell grew interested in the problem and in 1894 established an observatory in Flagstaff, Arizona, to study Mars. Over the next decades, there were endless debates over just what had been seen. We now know that the channels or canals Schiaparelli and other observers reported are not present on Mars—the positions of the *canali* do not even always overlap the spots and markings that are actually on the Martian surface (Fig. 23–1). But hope of finding life in the solar system springs eternal, and the latest studies have indicated the presence of considerable quantities of liquid water in Mars' past, a fact that leads many astronomers to hope that life could have formed during those periods. Still, as we shall describe, the Viking spacecraft that landed on Mars found no signs of life.

23.1 Characteristics of Mars

Mars is a small planet, 6800 km across, which is only about half the diameter and one-eighth the volume of Earth or Venus, although somewhat larger than Mercury. Mars' atmosphere is thin—at the surface its pressure is only 1 per cent of the surface pressure of Earth's atmosphere—but it might be sufficient for certain kinds of life.

From Earth, we have relatively little trouble in seeing through the Martian atmosphere to inspect that planet's surface, except when a Martian dust storm is raging. At other times, we are limited mainly by the turbulence in our own thicker atmosphere, which causes unsteadiness in the images and limits our resolution to 60 km at best on Mars. We can, however, follow features on the surface of Mars and measure the rotation period quite accurately. It turns out to be 24 hours 37 minutes 22.6 seconds from one Martian noontime to the next, so a day on Mars lasts nearly the same length of time as a day on Earth.

Unlike the orbits of Mercury or Venus, the orbit of Mars is outside the Earth's, so it is much easier to observe (Fig. 23–2) in the night sky (Fig. 23–3).

Mars revolves around the Sun in 23 Earth months. The axis of its rotation is tipped at a 25° angle from the plane of its orbit, nearly the same as the Earth's $23\frac{1}{2}°$ tilt. Because the tilt of the axis causes the seasons, we know that Mars goes through a year with four seasons just as the Earth does.

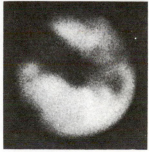

Figure 23–1 A drawing of Mars and a photograph, both made at the close approach of 1926. We now realize that the "canals" seen in the past do not usually correspond to real surface features.

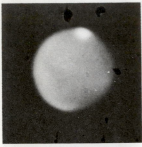

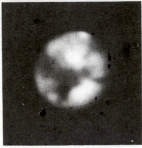

Figure 23–2 Photographs of Mars in violet *(top)* and in infrared *(bottom)* light. Mars has much less surface contrast in the violet.

When three celestial bodies are in a line, the alignment is called a *syzygy*, and the particular case when the planet is on the opposite side of the Earth from the Sun is called an *opposition*. The oppositions of Mars occur at intervals of 26 Earth months on the average.

Mars' orbit, about 1.5 A. U. in radius, has an eccentricity of 9 per cent, so at some of its oppositions it is closer to the Earth than at others. Obviously, these more favorable oppositions are better for observing, because at these times, the disk of Mars appears greater in angular size. It varies from 14 to 25 seconds of arc at opposition, from about $\frac{1}{30}$th to $\frac{1}{60}$th the size of the full Moon. At the opposition of September 1988, Mars will be especially close to Earth, only 59 million km away.

Figure 23–3 When Mars is at opposition, it is high in our nighttime sky and thus easy to observe. We can then simply observe in the direction opposite from that of the Sun. We are then looking at the lighted face.

We have watched the effect of the seasons on Mars over the last century. In the martian winter, in a given hemisphere, there is a polar cap. As the Martian spring comes to the northern hemisphere, the north polar cap shrinks and material at more temperate zones darkens. The surface of Mars is always mainly reddish, with darker gray areas that appear blue-green for physiological reasons of color contrast (see Color Plate 29). In the spring, the darker regions spread. Half a Martian year later, the same thing happens in the southern hemisphere.

One possible explanation that long ago seemed obvious for these changes is biological: Martian vegetation could be blooming or spreading in the spring. But there are other explanations, too. The theory that now seems most reasonable is that each year at the end of southern-hemisphere springtime, a global dust storm starts, covering Mars' entire surface with light dust. Then the winds reach velocities as high as hundreds of kilometers per hour and blow fine, light-colored dust off some of the slopes. This exposes the dark areas underneath. Gradually, as Mars passes through its seasons over the next year, the location where the dust is stripped away changes, mimicking the color change we would expect from vegetation. Finally, a global dust storm starts up again to renew the cycle. Astronomers still debate why the dust is reddish; the presence of iron oxide (rust) would explain the color in general, but might not satisfy the detailed measurements.

Mars has $\frac{1}{10}$ the mass of the Earth, and from the mass and radius one can easily calculate that it has an average density of slightly less than 4 grams/cm^3, not much higher than that of the Moon. This is substantially less than the density of $5\frac{1}{2}$ grams/cm^3 shared by the three innermost terrestrial planets, and indicates that Mars' overall composition must be fundamentally different from that of these other planets. Mars probably has a smaller core and a thicker crust than the Earth.

23.2 Space Observations from Mariner 9

Mars has been the target of series of spacecraft launched by both the United States and the Soviet Union. The earliest of these spacecraft were flybys, which attempted to photograph and otherwise study the Martian surface and atmosphere during the few hours they were in good position as they passed by (Table 23–1). The first flyby, Mariner 4, sent back photos showing only a cratered surface. This seemed to indicate that Mars resembles the Moon more than it does the Earth. Mariners 6 and 7 later confirmed the cratering, and also showed some signs of erosion.

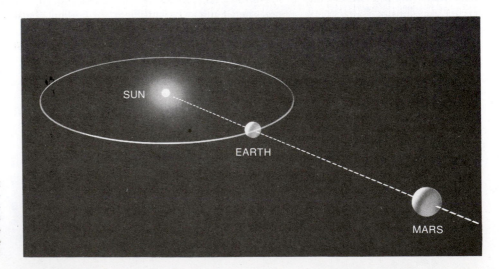

Table 23–1 Unmanned Probes Launched to Mars

Spacecraft	Of	Arrival Date	Comment
Mariner 4	USA	July 1965	Flyby
Mariner 6	USA	July 1969	Flyby
Mariner 7	USA	August 1969	Flyby
Mariner 9	USA	November 1971	Orbiter
Mars 2	USSR	November 1971	Orbiter; lander lost
Mars 3	USSR	December 1971	Orbiter; lander worked 20 s
Mars 4	USSR	February 1974	Flew by; no observations
Mars 5	USSR	February 1974	Orbiter
Mars 6	USSR	March 1974	Orbiter; lander failed
Mars 7	USSR	March 1974	Orbiter; lander failed
Viking 1	USA	June 1976	Orbiter and lander
Viking 2	USA	August 1976	Orbiter and lander

In 1971, the United States sent out a spacecraft not just to fly by Mars for only a few hours, but to actually orbit the planet and send back data for a year or more. This spacecraft, Mariner 9, went into orbit around Mars in November 1971, after a 5-month voyage from Earth. But when it reached Mars, our first reaction was as much disappointment as elation. While Mariner 9 was en route, a tremendous dust storm had come up, almost completely obscuring the entire surface of the planet. Only the south polar cap and 4 dark spots were visible. The storm began to settle after a few weeks, with the polar caps best visible through the thinning dust. Finally, three months after the spacecraft arrived, the surface of Mars was completely visible, and Mariner 9 could proceed with its mapping mission.

The data taken after the dust storm had ended showed four major ''geological'' areas on Mars: volcanic regions, canyon areas, expanses of craters, and terraced areas near the poles.

A chief surprise of the Mariner 9 mission was the discovery of extensive areas of vulcanism on Mars. The four dark spots on the Martian surface proved to be volcanoes. The largest of the volcanoes, which corresponds in position to the surface marking long known as Nix Olympica, ''the snow of Olympus,'' is named Olympus Mons, ''Mount Olympus.'' It is a huge volcano—600 kilometers at its base and about 25 kilometers high (Fig. 23–4). It is crowned with a crater 65 km wide; Manhattan island could be easily dropped inside. (The tallest volcano on Earth is Mauna Kea, in the Hawaiian islands, if we measure its height from its base deep below the ocean. Mauna Kea is taller than Everest, though still only 9 km high. And remember, it is on Earth, a much bigger planet than Mars.) The three other large volcanoes are on a 200-km-long ridge known as Tharsis.

Another surprise on Mars was the discovery of systems of canyons. One tremendous canyon—about 5000 kilometers long—is as big as the United States and comparable in size to the Rift Valley in Africa, the longest geological fault on Earth. Here again the size of the canyon with respect to Mars is proportionately larger than any geologic formations on the Earth. Venus, the Earth, and Mars each seem to have a large rift.

Perhaps the most amazing discovery on Mars was the presence of sinuous channels. These are on a smaller scale than the canali that Schiaparelli had seen, and are entirely different phenomena. Some of the channels show tributaries (Fig. 23–5), and the bottoms of some of the channels show the same characteristic features as stream beds on Earth. Even though water cannot exist on the surface of Mars under today's conditions, it is difficult to think of ways to explain the channels satisfactorily other than to say that they were cut by running water in the past.

Mariner 9 mapped the entire surface at a resolution of 1 km, and further photographed selected areas at a resolution 10 times better than this limit. The spacecraft lasted another year, and so more than completed its assigned task—in fact, we even got the scientific bonus of being able to see how the dust acted at the end of the storm. These observations gave us information both about the winds aloft and about the heights and shapes of Mars' surface features.

Figure 23–4 Olympus Mons, from Mariner 9. The image shows a region 800 km across.

Figure 23–5 Stream channels with tributaries indicate to most scientists that water flowed on Mars in the past.

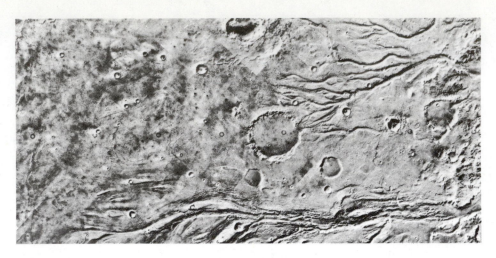

Figure 23–6 This mosaic of Mariner 9 views shows the north polar cap, which is shrinking as a result of Martian springtime. The volcano Olympus Mons is visible at the bottom.

The composition of a planet's atmosphere is measured by looking at its spectrum. This can even be done from the Earth, but the spectrum of the Earth's own atmosphere interferes, particularly at the wavelengths of the water-vapor lines.

This indication that water most likely flowed on Mars is particularly interesting because biologists feel that water is necessary for the formation and evolution of life. The presence of water on Mars, therefore, even in the past, may indicate that life could have formed and may even have survived.

If there had been water on Mars in the past, and in quantities great enough to cut river beds, where has it all gone? Most of the water would probably be in a permafrost layer beneath middle latitudes and polar regions.

Some of the water is bound in the polar caps (Fig. 23–6). Up to the time of Mariner 9, we thought that the polar caps were mostly frozen carbon dioxide—"dry ice"; the presence of water was controversial. We now know that the large polar caps that extend to latitude 50° during the winter are carbon dioxide. But when a cap shrinks during its hemisphere's summer, a residual polar cap of water ice remains.

If the water trapped in the polar caps or under the surface of Mars was released some time in the past, as liquid water on the surface or as water vapor in the atmosphere, it may have been sufficient to make Mars verdant. Perhaps the polar caps melted, releasing quantities of water, because dust that settled on the ice caused more sunlight to be absorbed or a widespread episode of vulcanism caused a general heating. Climatic change could also have resulted from a change in the orientation of Mars' axis of rotation, from a change in the solar luminosity, or from a variation in atmospheric composition (which would affect the greenhouse effect).

We found that the Martian atmosphere is composed of 90 per cent carbon dioxide with small amounts of carbon monoxide, oxygen, and water. We measured the density of the atmosphere by monitoring the changes of radio signals from spacecraft as they went behind Mars. As the spacecraft were occulted by the atmosphere of the planet, we could tell the rate at which the density drops. The surface pressure is less than 1 per cent of that near Earth's surface and decreases with altitude slightly more than it does on Earth. The atmosphere is too thin to affect the surface temperature, in contrast to the huge effects that the atmospheres of Venus and the Earth have on climate.

23.3 Viking!

In the summer of 1976, two U.S. spacecraft named Viking reached Mars after 10 month flights. Each contained two parts: an orbiter and a lander. The orbiter served two roles: it mapped and analyzed the Martian surface using its cameras and other instruments and relayed the lander's radio signals to Earth. The lander served

Figure 23–7 The Martian volcano Arsia Mons, the southernmost of the line of 3 volcanoes in Tharsis. The image was plotted based on Viking Orbiter data.

two purposes as well: it studied the rocks and weather near the surface of Mars and sampled the surface in order to decide whether there was life on Mars!

Viking 1 reached Mars in June of 1976 after a flight that led to spectacular large-scale views of the Martian surface (Color Plates 30 and 31). It orbited Mars for a month while it radioed back pictures to scientists on Earth who were trying to determine a relatively safe spot for Viking to land.

The orbiter observed the volcanoes (Figs. 23–7 and 23–8) and the huge valley (Fig. 23–9) at higher resolution than did Mariner 9.

These detailed views of Mars allow us to interpret better the similarities and differences that this planet—with its huge canyons and gigantic volcanoes—has with respect to the Earth. For example, Mars has exceedingly large, gently sloping volcanoes but no signs of the long mountain ranges or deep mid-ocean ridges that on Earth tell us that plate tectonics has been and is taking place. Many of the large volcanoes on Mars are ''shield volcanoes''—a type that has gently sloping sides formed by the rapid spread of lava (Fig. 23–10). On Earth, we also have steep-sided volcanoes, which occur where the continental plates are overlapping, as in Mt. St. Helens, the Aleutian Islands, or on Mount Fujiyama. No chains of such volcanoes exist on Mars.

Perhaps the volcanic features on Mars can get so huge because continental drift is absent there. If molten rock flowing upward causes volcanoes to form, then on Mars the features just get bigger and bigger for hundreds of millions of years, since the volcanoes stay over the sources and do not drift away.

On July 20, 1976, exactly seven years after the first manned landing on the Moon, Viking 1's lander descended safely onto a plain called Chryse. The views showed rocks of several kinds (Color Plates 32–34), covered with yellowish-brown material that is probably an iron oxide (rust) compound. Sand dunes were also visible (Fig. 23–11). The sky on Mars turns out to be yellowish-brown, almost pinkish

Figure 23–8 A high-resolution view of the caldera crater on top of Olympus Mons, photographed by the Viking 1 orbiter. We see a close-up of the upper half of the central crater shown in Fig. 23–4.

Figure 23–9 A mosaic of 102 photographs of Mars from Orbiter 1 in 1980, near the end of the spacecraft's lifetime. Valles Marineris, as long as the United States is wide, stretches across the center.

Figure 23–10 Shield volcanoes on Earth include Haleakala (shown here), Mauna Kea, and Mauna Loa in Hawaii, and Kilimanjaro in Africa. Martian volcanoes are all of this type, with gently sloping sides.

Figure 23–12 The first trench, 7 cm wide × 5 cm deep × 15 cm long, excavated by Viking 1 to collect Martian soil for the life-detection experiments. From the fact that the walls have not slumped, we can conclude that the soil is as cohesive as wet sand.

(Color Plates 32–34), from dust suspended in the air as a result of one of Mars' frequent dust storms. Craters up to 600 m across were visible near the horizon, and may have supplied the rocks in the view.

A series of experiments aboard the lander was designed to search for signs of life. A long arm was deployed (Color Plate 32), and a shovel at its end dug up a bit of the Martian surface (Fig. 23–12). The soil was dumped into three experiments that searched for such signs of life as respiration and metabolism (Box 23.1). The results were astonishing at first: the experiments sent back signals that seemed similar to those that would be caused on Earth by biological rather than by mere chemical processes. But later results were less spectacular, and non-biological explanations seem more likely. It is probable that some strange chemical process mimicked life in these experiments.

One important experiment gave much more negative results for the chance that there is life on Mars. It analyzed the soil and looked for traces of organic compounds. On Earth, many organic compounds left over from dead forms of life remain in the soil; the life forms themselves are only a tiny fraction of the organic material. Yet these experiments found no trace of organic material. On Mars, who knows? Perhaps life forms evolved that efficiently used up their predecessors. Still, the absence of organic material from the Martian soil is a strong argument against the presence of life on Mars.

The Viking 2 lander descended on September 3, 1976, on a site that was thought to be much more favorable to possible life forms than the Viking 1 site because it

Figure 23–11 This 100° view of the Martian surface was taken from the Viking 1 Lander, looking northeast at left and southeast at right. It shows a dune field with features similar to many seen in the deserts of Earth. From the shape of the peaks, it seems that the dunes move from upper left to lower right. The large boulder at the left is about 8 meters from the lander and is 1 × 3 meters in size. The boom that supports Viking's weather station cuts through the center of the picture. ("Chance of precipitation," the local newscaster would say, "0 per cent.")

was closer to the north pole with its prospective water supply. It too found yellowish-brown dust and a pinkish sky. Compared to Viking 1, Viking 2 found a smaller variety of types of rocks, with more large rocks and more pitted rocks (Fig. 23–13 and Color Plate 33). Such a distribution would result if the surface there had been ejected and flowed from a large crater 200 km away. The atmosphere at the second site, Utopia Planitia, contains three or four times more water vapor than had been observed near Chryse (the first landing site), 7500 km away. Viking 2's experiments in the search for life sent back data similar to Viking 1's.

Even if the life signs detected by Viking come from chemical rather than biological processes, as seems likely, we have still learned of fascinating new chemistry going on. When life arose on Earth, it probably took up chemical processes that had previously existed. Similarly, if life began on Mars in the past or will begin there in the future (assuming our visits didn't contaminate Mars and ruin the chances for indigenous life), we might expect the life forms to use chemical processes that already existed. So even if we haven't detected life itself, we may well have learned important things about its origin.

The presence of two orbiters circling Mars allowed a division of duties. One continued to relay data from the two Viking landers to Earth, while the orbit of the other was changed so that it passed over Mars' north pole. Summertime heat caused the main polar cap (thought to be made of dry ice—frozen carbon dioxide) to retreat, leaving only a residual ice cap. From the amount of water vapor detected with the orbiter's spectrograph, and from the fact that the temperature was too high for this residual ice cap to be frozen carbon dioxide, the conclusion has been reached that it is made of water ice. This important result indicates the probable presence of lots of water on the Martian surface in the past—in accordance with the existence of the stream beds—and seems to raise the possibility that life has existed or does exist on Mars. The layered look of the polar caps indicates that the general atmospheric conditions of Mars fluctuated over eons, as do Earth's.

Dating of the stream channels became possible late in the Viking mission when exceptionally high-resolution photos showed craters in the stream beds. The higher the density of craters, the older the underlying flow. So the earlier idea that all stream beds were formed at one stage early in Mars' history when the atmosphere was denser was not quite correct: volcanic flows and flows of water must have alternated over a long period of time.

Observations from the orbiters showed Mars' atmosphere (Fig. 23–14). The lengthy period of observation led to the discovery of weather patterns of Mars. With

The Vikings carried seismometers to search for marsquakes, but the effort was minor and the seismometers were mounted on the spacecraft, susceptible to wind shaking, instead of on the ground. Viking 1's failed, and Viking 2's reported only 1 event, though only 2 or 3 events would be detected in the same time on Earth with a device of this rudimentary sensitivity. One seismometer does not allow pinpointing a location, and we learned little from it.

The orbiters made a complete map of Mars with a resolution of 150–300 m, and even 8 m in some places. They took 52,000 pictures during their lifetimes; the landers have taken an additional 4,500 pictures.

Lasers are devices that amplify light very considerably by processes involving the atoms inside the laser. Many thousands of lasers have been built on Earth. A natural laser has been found in Mars' atmosphere, making several spectral lines in the infrared one billion times stronger than they would otherwise be.

Figure 23–13 The first photograph of the surface of Mars taken from Viking 2 on September 3, 1976. A wide variety of rocks are seen to lie on a surface of fine-grained material. Most of the boulders are 10–20 cm across. One of the Lander's footpads can be seen at the lower right.

Viking 1
  Orbiter:
    June 19, 1976–Aug. 7, 1980
  Lander:
    July 20, 1976–Nov. 13, 1982
Viking 2
  Orbiter:
    Aug. 7, 1976–July 25, 1978
  Lander:
    Sept. 3, 1976–April 12, 1980

*Box 23.1 Viking Experiments that Searched for Life

Each Viking lander carried three biological experiment: (1) a gas-exchange experiment; (2) a labelled-release experiment; and (3) a pyrolytic-release experiment. Each experiment examined Martian soil. The results were essentially the same at both landing sites.

1. The gas-exchange experiment looked for changes in the atmosphere caused by metabolism of organisms in the soil. Soil was mixed in the lander with a small amount of a nutrient medium. (The medium consisted of organic compounds dissolved in water, and was fondly known as "chicken soup.") Enough medium was added to humidify the chamber without wetting the soil. At first, carbon dioxide and oxygen—gases whose release on Earth is connected with life—were given off rapidly at both sites in amounts in excess of what would occur from terrestrial life forms. This seemed exciting, but then the release of gases stopped. Two and a half weeks later, enough medium was added to saturate the soil, and the mixture was incubated for several months. The production of carbon dioxide rose at first and then leveled off, while the production of oxygen continued to decline. All this action can be accounted for with chemical reactions; it seems that the original reactions, which seemed exciting at first, were the result of the chemical interaction of the soil with water vapor.

2. The labelled-release experiment used a medium containing simpler but cosmically common organic compounds that included some atoms of the radioactive isotope of carbon, carbon-14. The experiment was designed to reveal if Martian microbes breathed carbon dioxide into the atmosphere. When the nutrient was first added to the soil in the lander, carbon dioxide was released, as we could tell by counting the radioactive carbon-14 atoms that appeared in the atmosphere. But the CO_2 probably came from chemical decomposition of the soil, because the production of CO_2 stopped and no additional gas was released when more nutrient was added. Also, heating the mixture destroyed the reaction.

3. The pyrolytic-release experiment used a furnace to pyrolyze—cause a chemical change by heat—samples of Martian soil. The soil was sealed in a chamber along with Martian atmosphere and illuminated with simulated Martian sunlight from a xenon lamp. Small amounts of radioactive carbon monoxide and carbon dioxide were included, and after five days the soil was analyzed to see if any of the radioactive carbon had found its way into the organic material in the soil. Most of these tests indicated the presence of radioactive carbon dioxide, which was the expected indication that life was present in the soil. But when a sample of soil was exposed to high temperature for three hours before being tested, the production of carbon dioxide was reduced to 10 per cent of its previous value rather than being totally eliminated. It seems unlikely that any life would be able to survive this treatment, so a chemical explanation seems possible here too. The test was subsequently carried out in both dark and light and dry and humid surroundings, but none of the results matched the original high reading.

A fourth experiment relevant to the search for life was the chemical experiment: the mass spectrometer, which measures the mass of the molecules it samples and thus identifies them. It found no evidence for organic compounds in the soil, though it was able to make very sensitive measurements. Not even any of the organic materials expected to be left over from fallen meteorites were found. Neither did the cameras see large forms of life ("macrobes"), nor the water vapor sensors detect watering holes, nor the infrared sensors detect strange heat sources.

In sum, though the first results seemed to show that life might be present, the latter results did not back this idea. Though no terrestrial-type life was found, it is impossible to rule out conclusively all life forms.

Sterilization procedures have been worked out for spacecraft sent to Mars and the other planets so they do not bring along microorganisms that would contaminate the planet's atmosphere, leaving life forms from the contamination to be "discovered" by subsequent spacecraft. The spacecraft are baked under high temperatures for some time.

Figure 23–14 This oblique view from the Viking 1 Orbiter shows Argyre, the smooth plain at left center. It is surrounded by heavily cratered terrain. The Martian atmosphere was unusually clear when this photograph was taken, and craters can be seen nearly to the horizon. The horizon is bright mainly because of a thin haze. Detached layers of haze can be seen to extend from 25 to 40 km above the horizon and may be crystals of carbon dioxide.

its rotation period similar to that of Earth, many features of Mars' weather are similar to our own (Fig. 23–15). Surface temperatures measured from the landers ranged from a low of 150 K (190°F) at the northern site of Lander 1 to over 300 K (80°F) at Lander 2. Temperature varied each day by 35–50°C (60–90°F). Studies of Mars' weather have already helped us better understand windstorms in Africa that affect weather as far away as North America.

Both Viking landers and orbiters lasted long enough to show seasonal changes on Mars. The yearly temperature range spans the −123°C (−190°F) freezing point of carbon dioxide, which thus sometimes forms frost.

The orbiters showed dust storms (Fig. 23–16), which generally form in the early summer in the southern hemisphere and in winter in the northern hemisphere. The storms are seasonal partly because Mars is closest to the Sun at this time and the increased solar heating causes the atmosphere to circulate. The overall climates of the northern and southern hemispheres are different, probably because Mars' orbit is more elliptical than Earth's.

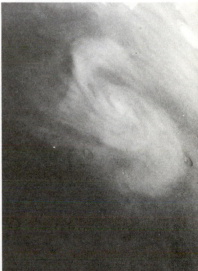

What's the next step? The Viking missions have been so spectacularly successful that further exploration of Mars would surely be valuable. The Soviets are planning a 1988 mission. NASA is preparing a Mars Geoscience-Climatology Orbiter for launch in 1990, to map surface chemistry and to study seasonal weather patterns. The European Space Agency is considering Kepler, an orbiter, among other space projects; the one chosen would be launched in about 1992.

Missions to penetrate the surface in order to implant seismometers would also be desirable. Many scientists would like to send a roving vehicle to travel over the Martian surface. A study of layers of Martian rock at a stream bed or at a canyon would reveal the ages of Mars, as the Grand Canyon does of Earth. But current budgetary limitations make it unlikely that we can hope before the mid-1990's for a Mars rover or a mission to bring Mars' soil back to Earth. Manned space flights to Mars would be much more expensive and so are even farther off.

Figure 23–15 This Martian storm on August 9, 1978, resembles satellite pictures of storms on Earth, showing the counter-clockwise circulation typical at northern latitudes. The temperature is too warm for the clouds to be carbon dioxide; they therefore must be made of water ice. The frost-filled crater at top right is 92 km in diameter. The white patches at top are outlying regions of the north polar remnant cap.

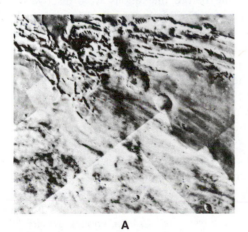

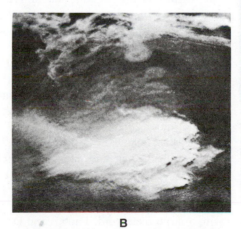

A

B

Figure 23–16 At left *(A)* we see a region of the Martian surface partly covered with clouds of water ice; at right *(B)* we see the same part of the surface 9 months later at an oblique angle. It includes a dust storm 600 km across, about the size of Colorado. The storm is moving eastward, from left to right; the leading edge is sharp and the trailing portion is diffuse.

Figure 23–17 *(left)* A Viking mosaic of Phobos. Its two largest craters have been named after Asaph Hall and after Angelina Stickney, his wife, who encouraged the search. *(right)* A close-up view of Phobos from Viking Orbiter 2 shows an area 3 × 3.5 km which, surprisingly, has grooves and chains of small craters. The grooves were probably associated with the formation of the crater Stickney. Similar chains of small craters on Earth's Moon, on Mars, and on Mercury were formed by secondary cratering from a larger impact.

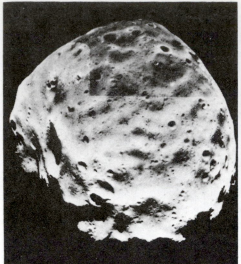
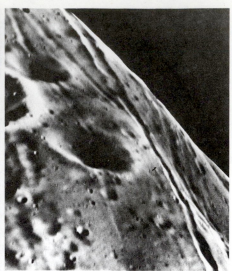

10.4 Satellites of Mars

In Greek mythology, Phobos (Fear) and Deimos (Terror) were companions of Ares, the equivalent of the Roman war god, Mars.

Mars has two moons—Phobos and Deimos. In a sense, these are the first moons we have met in our study of the solar system, for Mercury and Venus have no moons at all, and the Earth's Moon is so large relative to its parent that we may consider the Earth and Moon as a double planet system. The moons of Mars are mere chunks of rock, only 27 and 15 kilometers across, respectively.

Phobos and Deimos are very minor satellites. They revolve very rapidly; Phobos completes an orbit of Mars in only 7 hours and 40 minutes, more rapidly than a Martian day. Deimos orbits in about 30 hours, slower than Mars' rotation period, just as our Moon revolves more slowly than the Earth's rotation period. Because Phobos orbits more rapidly than Mars rotates, an observer on the surface of Mars would see Phobos moving conspicuously backwards compared to Deimos, the other planets, and the stars.

It is amusing to note that in 1727, long before the moons were discovered, Jonathan Swift invented two moons of Mars for *Gulliver's Travels*. This is widely considered to be a lucky guess, but it was a reasonable guess at that time, for numerological reasons. It may even have dated back to Kepler, who had considered in 1610 that the planets interior to the Earth—Mercury and Venus—had no known moons, the Earth had one, while Jupiter, the next planet out from Mars, had 4 known moons. He reasoned that Mars should have the intermediate value of 2. Both Swift's moons and the real ones are very small and revolve very quickly. The two nonfictional moons were discovered in 1877 by Asaph Hall of the U.S. Naval Observatory in Washington, D.C.

Mariner 9 and the Vikings made close-up photographs and studies of Phobos and Deimos (Figs. 23–17 and 23–18). Each turns out to be not at all like our Moon, which is a spherical, planet-like body. Phobos and Deimos are just cratered chunks of rock; they may be the remnants of a former single Martian moon that collided with a large meteoroid. Phobos and Deimos are too small to have enough gravity to have made them round.

Phobos, which is only about 27 km in its longest dimension and 19 km in its shortest, has a crater on it that is 8 kilometers across, a large fraction of Phobos' circumference. Deimos is even smaller than Phobos, only around 15 km in its longest

Figure 23–18 A Viking view of Deimos from a distance of 500 km. The lighted portion measures about 12 × 18 km. The surface is heavily cratered and thus presumably very old.

dimension and 11 km in its shortest. From their sizes, we can calculate their albedos; Phobos and Deimos turn out to be about as dark as the darkest maria on the Moon. From more detailed but similar comparisons of how the albedos vary with wavelength, we deduce that they may be made of dark carbon-rich (''carbonaceous'') rock, similar to some of the asteroids.

We can conclude that the moons are fairly old because they have been around long enough for their surfaces to be extensively cratered. The rotation of both moons is gravitationally linked to Mars with the same side always facing the planet, just as our Moon is linked to Earth.

The Soviet 1988 spacecraft will rendezvous with Phobos while orbiting Mars. Each will try to hover near Phobos, shoot a powerful laser beam at its surface, and analyze the spectrum of the debris.

Summary and Outline

Observations from the Earth (Section 23.1)
  Rotation period; surface markings; dust storms; seasonal changes
Earlier space observations (Section 23.2)
  Many space probes sent from the U.S. and the U.S.S.R.
  Mariner 9 orbiter mapped all of Mars
  Volcanic regions indicating geological activity; canyons larger than Earth counterparts; craters and terraced areas near the poles; signs of water including sinuous channels and the underlying layer of polar caps

Viking (Section 23.3)
  Orbiters and landers
  Close-up of photographs of surface rocks; yellowish-brown rocks, dust, and sky; analysis of soil
  Search for life: biological explanation seemed possible at first but chemical explanation now seems most likely
Satellites of Mars (Section 23.4)
  Phobos and Deimos are both irregular chunks of rock

Key Words

canali, syzygy, opposition

Questions

1. Outline the features of Mars that made scientists think that it was a good place to search for life.

2. If Mars were closer to the Sun, would you expect its atmosphere to be more or less dense than it is now? Explain.

†3. From the relative masses and radii (see Appendix 3), verify that the density of Mars is about 70 per cent that of Earth.

†4. Approximately how much more solar radiation strikes a square meter of the Earth than a square meter of Mars?

5. Compare the tallest volcanoes on Earth and Mars relative to the diameters of the planets.

6. What evidence is there that there is, or has been, water on Mars?

7. Why is Mars' sky pinkish?

8. Describe the composition of Mars' polar caps. Explain the evidence.

9. List the various techniques for determining the composition of the atmosphere of Mars.

10. Compare the temperature ranges on Mercury, Venus, Earth, and Mars.

11. List the evidence from Viking for and against the existence of life on Mars.

12. Discuss how (a) bacteria, and (b) humans would fare on the three biological experiments sent to Mars.

13. (a) Aside from the biology experiments, list three types of observations made from the Viking landers. (b) What are two types of observations made from the Viking orbiters?

14. Why aren't Phobos and Deimos regular, round objects like the Earth's moon?

†This indicates that the question requires a numerical solution.

Topics for Discussion

1. Discuss the problems of sterilizing spacecraft, and whether we should risk contaminating Mars by landing spacecraft.

2. Discuss the additional problems of contamination that manned exploration would bring. Analyze whether, in this context, we should proceed with manned exploration in the next decade or two.

A mosaic of Jupiter and its four Galilean satellites made from the Voyager 1 spacecraft. Callisto is at lower right, Ganymede is at lower left, relatively bright Europa is shown against Jupiter, and Io appears in the background at left. The symbol for Jupiter appears at the top of the facing page.

Jupiter **24**

Aims: To describe Jupiter, the dominant planet of the solar system; to discuss it as an example of a class of planets; and to meet a new major set of bodies—Jupiter's moons

The planets beyond the asteroid belt—Jupiter, Saturn, Uranus, and Neptune—are very different from the four "terrestrial" planets. These *giant planets*, or *Jovian planets*, not only are much bigger and more massive, but are also less dense. This suggests that the internal structure of these giant planets is entirely different from that of the four terrestrial planets.

Jupiter, also called Jove, was the chief Roman deity. Jovian is the adjectival form of Jupiter.

24.1 Fundamental Properties

The largest planet, Jupiter, dominates the Sun's planetary system. Jupiter is 5 A.U. from the Sun and revolves around it once every 12 years. It alone contains two-thirds of the mass in the solar system outside of the Sun, 318 times as much mass as the Earth. Jupiter has at least 16 moons of its own and so is a miniature planetary system in itself. It is often seen as a bright object in our night sky, and observations with even a small telescope reveal bands of clouds across its surface and show four of its moons.

Jupiter is more than 11 times greater in diameter than the Earth. From its volume and mass, we calculate its density to be 1.3 grams/cm^3, not much greater than the 1 gram/cm^3 density of water. This tells us that any core of heavy elements (such as iron) that Jupiter may have does not make up as substantial a fraction of Jupiter's mass as the cores of the inner planets make up of their planets' masses. Jupiter, rather, is mainly composed of the lighter elements hydrogen and helium. Jupiter's chemical composition is closer to that of the Sun and stars than it is to that of the Earth (Fig. 24–1). Jupiter isn't solid; it has no crustal surface at all. At deeper and deeper levels, its gas just gets denser and denser, eventually liquefying.

Jupiter rotates very rapidly, once every 10 hours. Undoubtedly, this rapid spin rate is a major reason for the colorful bands, which are clouds spread out parallel to the equator. The bands appear in subtle shades of orange, brown, gray, yellow, cream, and light blue, and are beautiful to see (Fig. 24–2). They are in constant turmoil; the shapes and distribution of bands change in a matter of days.

Interestingly, Jupiter's clouds do not rotate as the surface of a solid body would (Fig. 24–3): the clouds at the equator rotate slightly faster than the clouds nearer the poles, completing their rotation about five minutes earlier each time around. Thus every fifty days, the clouds at the equator have made an additional rotation. The deeper levels of clouds also probably rotate at differing velocities.

The most prominent feature of the visible cloud surface of Jupiter is a large reddish oval known as the *Great Red Spot* (see Figs. 24–2, 24–10, and Color Plates 37 and 38). It is about 14,000 km × 30,000 km, many times larger than the Earth, and drifts about slowly with respect to the clouds as the planet rotates. The Great Red Spot is a relatively stable feature, for it has been visible for at least 150 years, and, if an old report indeed refers to it, maybe 300 years. Sometimes it is relatively prominent, and at other times the color may even disappear for a few years. Other,

Jupiter's mass can be readily measured by applying Kepler's third law to the orbits of its satellites. Jupiter has only 1/1000 the mass of the Sun, too little for nuclear fusion to have begun in its interior.

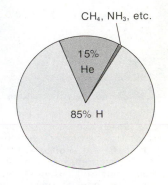

Figure 24–1 The composition of Jupiter.

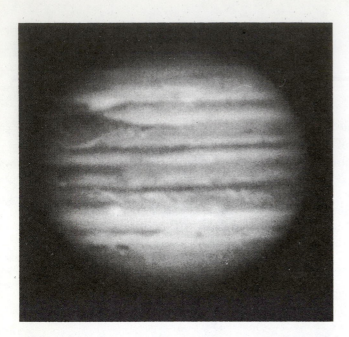

Figure 24–2 Jupiter, photographed from the Earth, shows belts and zones of different shades and colors. By convention (that is, the set of terms we arbitrarily choose to use), the bright horizontal bands are called *zones* and the dark horizontal bands are called *belts*. Adjacent belts and zones rotate at different speeds. Jupiter is shown with south up, since most telescopes invert images.

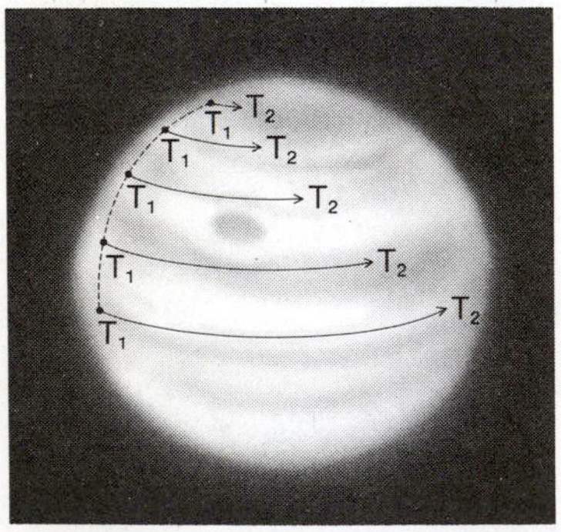

Figure 24–3 Jupiter's differential rotation, which causes the points that were aligned at time T_1 to be separated at time T_2, spreads out the clouds into bands. The rate at which this happens is exaggerated in this drawing.

smaller spots are also present. A pair of white ovals still present on Jupiter, for example, was first discovered in 1938.

Jupiter's rapid rotation makes the planet bulge at the equator. The distance from pole to pole is 7 per cent less than the equatorial diameter; we say that Jupiter is *oblate* (Fig. 24–4). Jupiter's axis of rotation is only 3° from the axis of Jupiter's orbit around the Sun, so that, unlike the Earth and Mars, Jupiter has no seasons.

In 1955, intense bursts of radio radiation were discovered to be coming from Jupiter. Though it was quite a surprise to detect any radio signals at all, several kinds of signals were detected. Some bursts came from the atmosphere and are correlated with passages of Jupiter's innermost moon, Io, above the regions from which the bursts are coming. At shorter radio wavelengths, Jupiter emits continuous radiation.

The fact that Jupiter emits radio waves indicated that Jupiter, even more so than the Earth, has a strong magnetic field and strong *radiation belts* (actually, belts filled with magnetic fields in which particles are trapped, large-scale versions of the Van Allen belts of Earth). We can account for the presence of this radio radiation only with mechanisms that involve such magnetic fields and belts. High-energy particles passing through space interact with Jupiter's magnetic field to produce the radio emission.

24.2 Jupiter's Moons: Early Views

Jupiter has at least 16 satellites. Four of the innermost satellites were discovered by Galileo in 1610 when he first looked at Jupiter with his telescope. These four moons are called the *Galilean satellites* (Fig. 24–5). One of these moons, Ganymede, at 5276 km in diameter, is the largest satellite in the solar system and is larger than the planet Mercury.

The Galilean satellites have played a very important role in the history of astronomy. The fact that these particular satellites were noticed to be going around another planet (Fig. 24–6), like a solar system in miniature, supported Copernicus' heliocentric model of our solar system. Not everything revolved around the Earth!

The dozen moons of Jupiter that were discovered before the last decade fall into three groups. The first group includes the five innermost satellites then known (Amal-

Box 24.1 Jupiter's Satellites in Mythology

All the moons except Amalthea are named after lovers of Zeus, the Greek equivalent of Jupiter. Amalthea, a goat-nymph, was Zeus' nurse, and out of gratitude he made her into the constellation Capricorn. Zeus changed Io into a heifer to hide her from Hera's jealousy; in honor of Io, the crescent moon has horns.

Ganymede was a Trojan youth carried off by an eagle to be Jupiter's cup bearer (the constellation Aquarius). Callisto was punished for her affair with Zeus by being changed into a bear. She was then slain by mistake, and rescued by Zeus by being transformed into the Great Bear in the sky. Jealous Hera persuaded the sea god to forbid Callisto to ever bathe in the sea, which is why Ursa Major never sinks below the horizon.

Europa was carried off to Crete on the back of Zeus, who took the form of a white bull. She became Minos' mother. Pasiphae (also the name of one of Jupiter's moons) was the wife of Minos and the mother of the Minotaur.

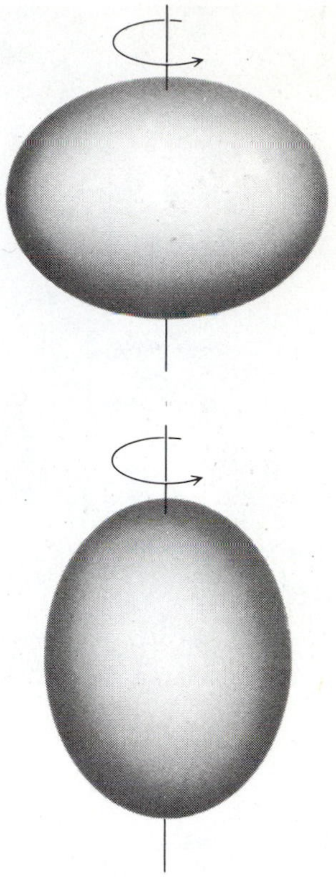

thea and the Galilean satellites), which are also the largest. Their orbits are all less than 2 million km in radius. The orbits of the four moons in the second group are all about 12 million km in radius. The orbits of the four outermost moons in the third group are 21 to 24 million km in radius. These outermost moons revolve in the direction opposite from that of Jupiter's rotation (that is, retrograde), while the other moons revolve prograde. Moons in the outer two groups may be bodies captured by Jupiter's gravity after their formation, for they have high inclinations and/or eccentricities.

In the last decade, the 1.2-m Palomar Schmidt telescope has been used to search for moons down to about 22nd magnitude, a level of intensity fainter than had been previously studied (Fig. 24–7). Spacecraft to Jupiter have discovered at least three other satellites.

24.3 Spacecraft Observations

Each of the planets we have discussed so far has recently been visited by spacecraft that have greatly advanced our knowledge about them. Jupiter is another spec-

Figure 24–4 The top ellipsoid is *oblate;* the bottom ellipsoid is *prolate.*

Figure 24–5 Jupiter with the four Galilean satellites. These satellites were named Io, Europa, Ganymede, and Callisto by Simon Marius, a German astronomer who independently discovered them in 1611.

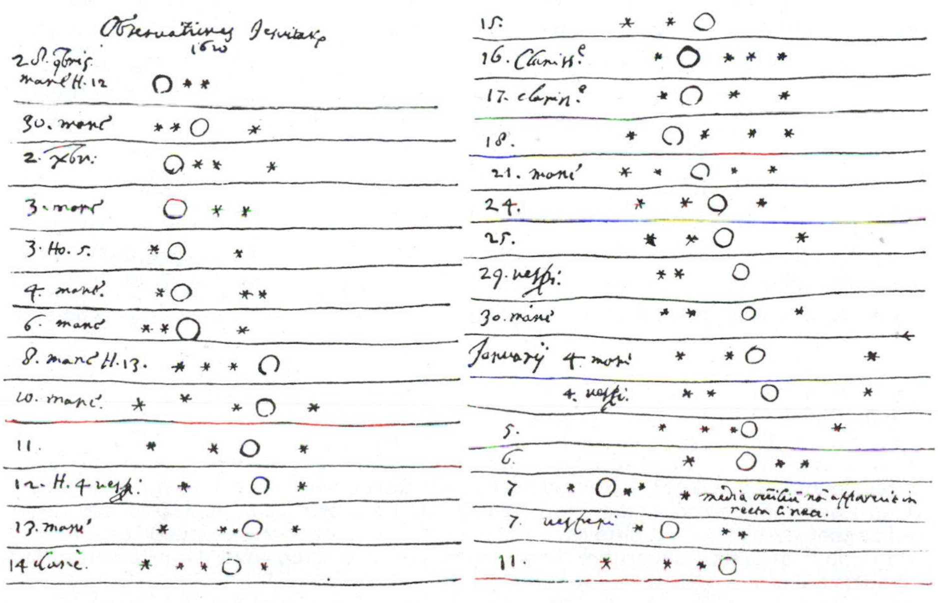

Figure 24–6 Galileo's sketches of the changes in the positions of the satellites.

Figure 24–7 The photograph on which the 13th moon of Jupiter, shown with an arrow, was discovered. The moon is named Leda. The telescope was set to track across the sky along with Jupiter; thus stars show as trails. The trails here correspond to anonymous—unnamed and unnumbered—faint stars.

tacular example. Two spacecraft, Pioneer 10 and Pioneer 11, gave us our first close-up views of the colossal planet in 1973 and 1974. A second revolution in our understanding of Jupiter occurred in 1979, when Voyager 1 and Voyager 2 also flew by Jupiter (Fig. 24–8).

Each spacecraft carried many types of instruments to measure properties of Jupiter, its satellites, and the space around them. The observations made with the imaging equipment were of the most popular interest. The resolution of the Voyager images was five times better than the best images we can obtain from Earth. The Voyagers provided a quantum leap in quality over even the Pioneers. In as little as 48 seconds, the Voyagers could send a full digitally-coded picture back to Earth, where detailed computer work improved the images. This was remarkable for a vehicle so far away that its signals travelled 52 minutes to reach us. The energy in the signal would have to be collected for billions of years to light a Christmas tree bulb for just one second!

The Pioneer and Voyager spacecraft are travelling fast enough to escape the solar system. As they depart, the spacecraft are radioing back valuable information about other planets and on interplanetary conditions. In about 80,000 years the spacecraft will be about one parsec (about three light years) away from the solar system. The Pioneers bear plaques that give information about their origin and about life on Earth, in case some interstellar traveller from another solar system happens to pick it up (Fig. 24–9). The Voyagers bear more elaborate mementos, as we shall describe in Section 28.4.

24.3a The Great Red Spot

The Great Red Spot shows very clearly in many of the images (Fig. 24–10 and Color Plate 38). It is a gaseous island many times larger across than the Earth. Its top is about 8 km above the neighboring cloud tops. The Spot is the vortex of a violent, long-lasting storm, similar to large storms on Earth.

Although we had not been able to see much structure in the Great Red Spot prior to 1979, we can now use time-lapse observations made from Voyager to study the Spot's rotation period. We also see how it interacts with surrounding clouds and smaller spots (Fig. 24–11). Many other similar storms (that is, spots) were observed on Jupiter from the Voyager spacecraft, but none was as large or as colorful as the Great Red Spot.

Why has the Great Red Spot lasted this long? Perhaps heat energy flows into the storm from below it, maintaining its energy supply. The storm also contains more mass than hurricanes or other cyclonic storms on Earth, which makes it more stable. Further, unlike Earth, Jupiter has no continents or other structure to break up the storm. Still another possible factor may be that Jupiter's clouds radiate less efficiently

Figure 24–8 This duplicate of Pioneers 10 and 11 hangs in the National Air and Space Museum in Washington, D.C.

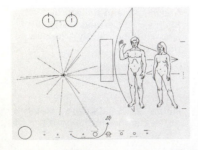

Figure 24–9 The plaques borne by Pioneers 10 and 11. A man and a woman are shown standing in front of an outline of the spacecraft, for scale. The spin-flip of the hydrogen atom (Section 13.4) is shown at top left. It also provides a scale, because even travellers from another solar system would know that the wavelength that results from the spin-flip is a certain length, which is 21 cm in our units. The Sun and planets (including distinctively ringed Saturn) and the spacecraft's trajectory from Earth are shown at bottom. (Will the visitors realize that we had not yet discovered the rings around Jupiter and Uranus?) The directions and periods of several pulsars are shown in the rays extending from the midpoint at left. Numbers are given in the binary system.

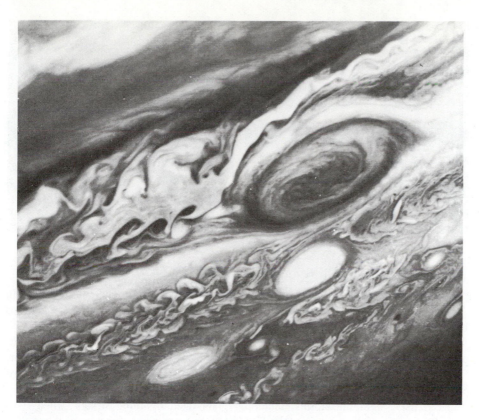

Figure 24–10 A mosaic showing the Great Red Spot, made from Voyager 2 at a range of 2 million km from Jupiter. The white oval south of the Great Red Spot is similar in structure. Both rotate in the anticyclonic—counterclockwise—sense.

Storms on Earth always rotate in the same direction in a given hemisphere (north or south). In the northern hemisphere, they rotate counterclockwise around low pressure regions; we call such storms "cyclonic." Around highs, we have "anticyclonic" storms, which rotate the other way. The directions are reversed in the southern hemisphere. The Great Red Spot is an anticyclonic storm, namely, counterclockwise in the southern hemisphere.

than Earth's. Also, we do not know how much energy the Spot gains from the circulation of Jupiter's upper atmosphere and eddies (rotating regions) in it. Until we can sample lower levels of Jupiter's atmosphere, we will not be able to decide definitively. Studying the eddies in Jupiter's atmosphere helps us interpret features on Earth. For example, one theory of Jupiter's spots holds that they are similar to circulating rings that break off from the Gulf Stream in the Atlantic Ocean.

24.3b Jupiter's Atmosphere

Heat emanating from the interior of Jupiter produces huge convection currents. The bright bands (Color Plate 39) on Jupiter, called *zones*, are rising currents of gas driven by this convection (Fig. 24–12). The dark regions, the *belts*, are falling gas. The tops of these dark belts are somewhat lower (about 20 km) than the tops of the

Figure 24–11 A time-lapse sequence of the Great Red Spot, showing the flow of gas. The pictures are taken every other rotation of Jupiter, making the interval 22 Earth hours. Note how a white cloud enters the Spot's circulation and begins to be swept around. The Great Red Spot rotates with a period of about 6 Earth hours.

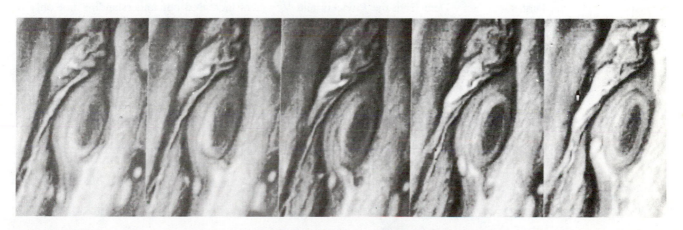

Figure 24–12 *(left)* Ground-based infrared observations with the 5-m Palomar telescope show bright regions where the temperature is the highest, presumably where we can see deepest into the clouds. The Great Red Spot appears on the left limb as a dark area encircled by a bright ring, indicating that the Spot is cooler than the region that surrounds it. *(right)* This view in the visible was taken one hour later from Voyager.

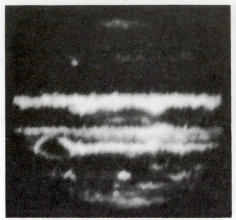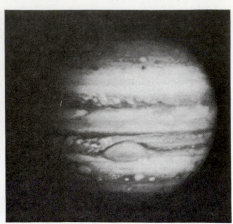

Comparative studies of the planets help us understand the circulation of clouds and winds, for example. The rapid rotation of Jupiter differs from the slow rotation of Venus, and from the rotation of Earth and Mars, which rotate in about 24 hours. Also, Jupiter receives heat from both below (internal) and above (the Sun), while the only major heat source of the Earth's atmosphere is the Sun. Further, our atmosphere is bounded below by a solid surface, while Jupiter's is not. Differences in composition are also extreme.

zones and so are about 10 K warmer. The cloud bands themselves are cyclonic patterns, pulled out by Jupiter's rotation to surround the planet. The colors we see in the belts may have to do with sulfur compounds in the middle clouds, or perhaps with organic molecules.

Voyager measurements of wind velocities showed that each hemisphere of Jupiter has a half-dozen currents blowing eastward or westward. The Earth, in contrast, has only one westward current at low latitudes (the trade winds) and one eastward current at middle latitudes (the jet stream).

Extensive lightning storms, including giant-sized lightning strikes called "superbolts," were discovered from the Voyagers, as were giant aurorae.

Pioneer 11 gave us the first opportunity to look at Jupiter from a different point of view than we have on Earth. The spacecraft came in over the polar region for two reasons: First, we wanted to avoid possible damage to the spacecraft from the intense Jovian radiation belts. Second, we can never see the polar regions from Earth, as Jupiter's axis is nearly perpendicular to our orbit. So scientists were anxious to get their first look at Jupiter's poles. The results were noteworthy; at high latitudes the circulation pattern of cloud bands that we see at the equator is destroyed and the bands break up into eddies. The Voyagers went a step further, and allowed detailed mapping of both poles (Fig. 24–13). With information like this, we can now study "comparative atmospheres": those of the Earth, Venus, Mars, and Jupiter.

24.3c Jupiter's Interior

Data from the Pioneers and Voyagers increased our understanding not only of the atmosphere but also of the interior of Jupiter (Fig. 24–14). Most of the interior is in liquid form. Jupiter's central temperature may be between 13,000 and 35,000 K. The central pressure is 100 million times the pressure of the Earth's atmosphere measured at our sea level because of Jupiter's great mass pressing in. Because of this high pressure, Jupiter's interior is composed of ultracompressed hydrogen surrounding a rocky core consisting of 20 Earth masses of iron and silicates. The heavy elements should have sunk to the center, but we have no direct evidence that they are there.

The lower levels of the liquid hydrogen are in the form of molecules that have lost their outer electrons. In this state, the hydrogen conducts heat and electricity. Since these properties are basic to our normal terrestrial definition of "metal," we say that the hydrogen is in a "metallic" state. This liquid metallic region makes up 75 per cent of Jupiter's mass. It is probably where Jupiter's magnetic field is gener-

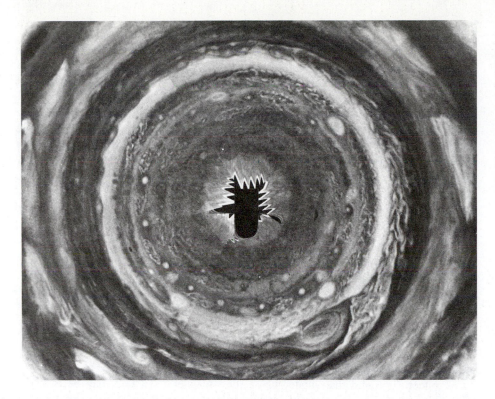

Figure 24–13 A computer reconstruction of Voyager images of Jupiter's south pole shows the view that would be seen from directly above the pole. An irregularly shaped region directly at the pole, for which no photographs were available at this 600-km resolution, appears black. The photograph shows the Great Red Spot and a disturbance trailing from it that extends half-way around the planet. The equal spacing of the smaller white spots suggests that some wave action may be taking place.

ated by the interaction of mass, motion, and an electric field. (This process also takes place in ''dynamos'' on Earth, which convert mechanical energy into electricity.)

Jupiter radiates 1.6 times as much heat as it receives from the Sun, leading to Jupiter's complex and beautiful cloud circulation pattern. There must be some internal energy source—perhaps the energy remaining from Jupiter's collapse from a primordial gas cloud 20 million km across to a protoplanet 700,000 km across, 5 times the present size of Jupiter. Jupiter is undoubtedly still contracting. It lacks the mass necessary by a factor of about 75, however, to have heated up enough for nuclear reactions to begin.

Figure 24–14 *(A)* We can use the infrared pictures of Jupiter to learn about the cloud structure in its atmosphere. We see into its atmosphere in the infrared until our view is blocked by clouds; only clouds and not general gas block our infrared view on both Jupiter and Earth. Since Jupiter's belts are brighter than the zones in the infrared, we must be seeing hotter gas there. Since the temperature of Jupiter increases downward toward its center, we must therefore be seeing lower levels. The zones are cloudy at higher levels. Thus the zones are rising atmosphere, while the belts are sinking atmosphere. *(B)* The current model of Jupiter's interior.

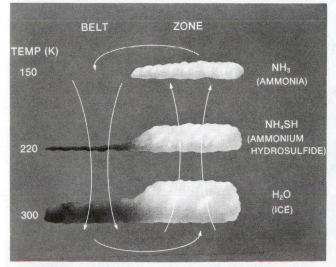

A

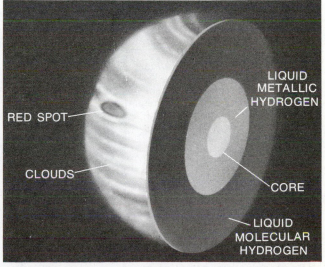

B

24.3d Jupiter's Magnetic Field

The Pioneer missions showed that Jupiter's tremendous magnetic field is even more intense than many scientists had expected, a result confirmed by the Voyagers (Fig. 24–15). At the height of Jupiter's clouds the magnetic field is 10 times that of the Earth, which itself has a strong field.

The inner field is toroidal, shaped like a doughnut, containing several shells like giant versions of the Earth's Van Allen belts. High-energy protons and electrons are trapped there. The satellites Amalthea, Io, Europa, and Ganymede travel through this region. The middle region, with charged particles being whirled around rapidly by the rotation of Jupiter's magnetic field, does not have a terrestrial counterpart. The outer region of Jupiter's magnetic field interacts with the solar wind as far as 7,000,000 km from the planet and forms a shock wave, as does the bow of a ship plowing through the ocean. When the solar wind is strong, Jupiter's outer magnetic field (shaped like a flattened pancake) is pushed in. Jupiter's magnetic field and radiation belts show up not only in the particle sensors but also by the x-rays discovered by Voyager.

Pioneer 11 observed a 10-hour fluctuation in the magnetic field. This fluctuation occurs because the magnetic axis is tilted 10 degrees from the axis of rotation, and the center of the field is not exactly at the center of the planet. The magnetic field thus resembles a wobbly disk. Also, Jupiter's magnetic poles are reversed compared with the Earth's: our north pole is on the same side of the plane of the planetary orbits as is Jupiter's south pole.

Pioneer 10 found that Jupiter has a long magnetic tail that may be a billion kilometers long. The magnetic tail results from the stretching of some of Jupiter's magnetic-field lines by the solar wind. These field lines prevent the solar wind from entering the tail region, which extends over 635 million km from Jupiter away from the Sun, 100 times longer than the Earth's magnetic tail.

24.4 Jupiter's Moons Close Up

Through Voyager close-ups, the satellites of Jupiter have become known to us as worlds with personalities of their own. The four Galilean satellites, in particular,

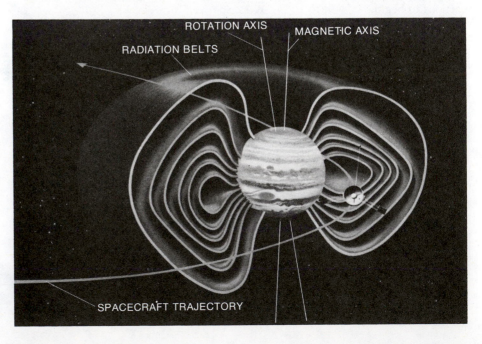

Figure 24–15 An artist's view of Jupiter's magnetosphere, the region of space occupied by the planet's magnetic field. It rotates at hundreds of thousands of kilometers per hour around the planet, like a big wheel with Jupiter at the hub. The inner magnetosphere is shaped like a doughnut with the planet in the hole. The highly unstable outer magnetosphere is shaped as though the outer part of the doughnut has been squashed. The outer magnetosphere is "spongy" in that it pulses in the solar wind like a huge jellyfish. It often shrinks to one-third of its largest size.

ROTATION AXIS   MAGNETIC AXIS

RADIATION BELTS

SPACECRAFT TRAJECTORY

were formerly known only as dots of light. Since these satellites range between 0.9 and 1.5 times the size of our own Moon, they are substantial enough to have interesting surfaces and histories (Color Plates 40–44 and the picture opening this Chapter).

Io provided the biggest surprises. Scientists knew that Io gave off particles as it went around Jupiter, but Voyager 1 discovered that these particles resulted from active volcanoes on the satellite (Fig. 24–16). Eight volcanoes were seen actually erupting, many more than erupt on the Earth at any one time. When Voyager 2 went by a few months later, most of the same volcanoes were still erupting.

Io's surface (Fig. 24–17 and 24–18 and Color Plates 39–41) has been transformed by the volcanoes, and overall is the youngest surface we have observed in our solar system. Gravitational forces from the other Galilean satellites pull Io slightly inward and outward from its orbit around Jupiter, flexing the satellite. Its surface moves in and out by as much as 100 meters. This squeezing and unsqueezing creates heat from friction, which presumably heats the interior and leads to the vulcanism. Io may have an underground ocean of liquid sulfur. The surface of Io is covered with sulfur and sulfur compounds, including frozen sulfur dioxide, and the atmosphere is full of sulfur dioxide. Io certainly wouldn't be a pleasant place to visit.

Io has mountains up to 10 km tall. Sulfur mountains that high couldn't support themselves, so perhaps these are silicate mountains with merely a coating of sulfur.

Sulfur, sodium, and other material given off by Io fill a doughnut-shaped region that surrounds Io's orbit. Further, as Io moves in its orbit through the Jovian magnetic field, a huge electric current is set up between Io and Jupiter itself in a tube of concentrated magnetic field called a *flux tube* (Fig. 24–19).

Europa (Color Plate 42), the brightest of Jupiter's Galilean satellites, has a very flat surface and is covered with narrow dark stripes (Fig. 24–20). This suggests that the surface we see is ice. The markings may be fracture systems in the ice, like

Figure 24–16 The volcanoes of Io can be seen erupting on this photograph. Linda Morabito, a JPL engineer, discovered the first volcano. At the time, she was studying images that had been purposely overexposed to bring out the stars so that they could be used for navigation. She saw the large volcanic cloud barely visible on the right limb. The plume extends upward for about 250 km and is scattering sunlight toward us. The bright spot just leftward of Io's terminator is a second volcanic plume projecting about the dark surface into the sunlight.

Figure 24–17 A mosaic of Io with a resolution of 8 km, showing volcanic features and no craters. The heart-shaped region at lower center surrounds an erupting volcano named for Pele, the Hawaiian volcano goddess. Its shape changed substantially between this Voyager 1 image and the Voyager 2 images.

Figure 24–18 A volcanic caldera 50 km in diameter on Io. Dark flows over 100 km long extend from it.
Io's surface, orange in color and covered with strange formations, led Bradford Smith of the University of Arizona, the head of the Voyager imaging team, to remark, "It's better looking than a lot of pizzas I've seen."

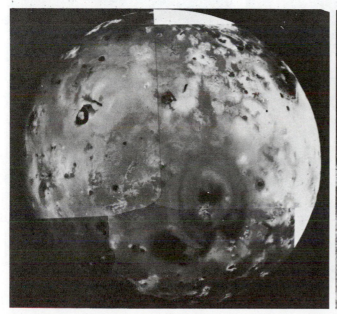

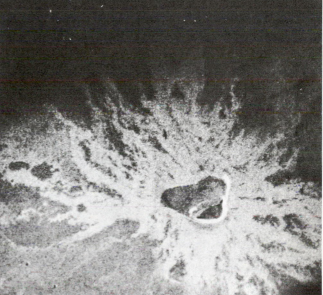

A

B

Figure 24–19 *(A)* A computer-processed image of one end of the Io torus in the light emitted by singly ionized sulfur, superimposed on a sketch showing Jupiter and a magnetic field line. The ionized sulfur is distributed in a ring around Jupiter by the planet's rotating magnetic field. (Courtesy of Carl B. Pilcher, University of Hawaii.) *(B)* As Io orbits Jupiter, a tube of magnetic field, a *flux tube*, extends up out of the plane of its orbit and carries a huge current.

fractures in the large fields of sea ice near the Earth's north pole. Few craters are visible, suggesting that the ice was soft enough below the crust to close in the craters. Either internal radioactivity or a gravitational heating like that inside Io may have provided the heat to soften the ice. Because it possibly has liquid water and some extra heating, some scientists consider it a worthy location to check for signs of life.

The largest satellite, Ganymede (Color Plate 43), shows many craters (Fig. 24–21) alongside weird grooved terrain (Fig. 24–22). Ganymede is larger than Mercury but less dense. It contains large amounts of water ice surrounding a rocky core. Ganymede's rotation is linked to its orbital period. But an icy surface is as hard as steel in the cold conditions that far from the Sun if it is not otherwise heated, so the cratered surface remains from perhaps 4 billion years ago. The grooved terrain is younger. The appearance of the side facing away from Jupiter is dominated by a dark cratered region. Few craters are visible near its center. The formation of the younger grooved terrain may have erased the evidence of the ancient impact there.

Ganymede also shows lateral displacements, where grooves have slid sideways, like those that occur in some places on Earth (for example, the San Andreas fault in California). Ganymede is the only place besides the Earth where such faults have been found. Thus further studies of Ganymede may help our understanding of terrestrial earthquakes. Other signs of modification of Ganymede's surface by tectonic processes in Ganymede's early years have also been seen.

Callisto (Color Plate 44) has so many craters (Fig. 24–23) that its surface is the oldest of Jupiter's Galilean satellites. Spectra indicate that Callisto is probably also covered with ice. A huge bull's-eye formation, Valhalla, contains about 10 concentric rings, no doubt resulting from a huge impact. Perhaps ripples spreading from the impact froze into the ice to make Valhalla. Despite its general appearance, Valhalla differs greatly in detail from ringed features on the Moon and Mercury; its rings are asymmetric, for example.

Voyager also observed Amalthea, formerly thought to be Jupiter's innermost satellite. This small chunk of rock is irregular and oblong in shape, looking like a dark red potato complete with potato "eyes." It is comparable in size to many aster-

Figure 24–20 The surface of Europa, which is very smooth, is covered by this complex array of streaks. Few impact craters are visible.

Figure 24–21 This mosaic of Ganymede, Jupiter's largest satellite, shows numerous impact craters, many with bright ray systems. The large crater at upper center is about 150 km across. The mountainous terrain at lower right is a younger region, as is the grooved terrain at bottom center.

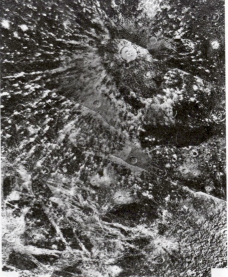

Figure 24–22 The terrain over large areas of Ganymede is covered with many grooves tens of kilometers wide, a few hundred meters high, and up to thousands of kilometers long. Younger grooves cover older grooves, and offsets sometimes occur along series of grooves.

Figure 24–23 Callisto is cratered all over, very uniformly. Several craters have rays or concentric rings. No craters large than 50 km are visible, from which we deduce that Callisto's crust cannot be very firm. The fact that the limb is so smooth indicates that there is no high relief. Because there are so many craters, the surface must be very old, probably over 4 billion years.

oids, ten times larger than the moons of Mars and ten times smaller than the Galilean satellites. The reddish color may be caused by material from Io.

Voyager 2 discovered three previously unknown satellites of Jupiter, making the total of known moons 16.

Studies of Jupiter's moons tell us about the formation of the Jupiter system, and help us better understand the early stages of the entire solar system.

24.5 Jupiter's Ring

Though Jupiter wasn't expected to have a ring, Voyager 1 was programmed to look for one just in case; Saturn's ring, of course, was well known, and Uranus's ring had been discovered only a few years earlier. The Voyager 1 photograph indeed showed a wispy ring of material around Jupiter at about 1.8 times Jupiter's radius, inside its innermost moon. As a result, Voyager 2 was targeted to take a series of photographs of the ring. From the far side looking back, the ring appeared unexpectedly bright (Fig. 24–24). This brightness probably results from small particles in the ring that scatter the light toward us. Within the main ring, fainter material appears to extend down to Jupiter's surface (Fig. 24–25).

The ring particles may come from Io, or else they may come from comet and meteor debris or from material knocked off the innermost moons by meteorites. The individual particles probably remain in the ring only temporarily. Perhaps Jupiter's newly discovered innermost moon causes the rings to end where they do.

In the next chapter, we shall discuss the fundamental ideas about how gravity leads to the formation of planetary rings, in the context of the most famous ring of all: the ring of Saturn.

24.6 Future Exploration of Jupiter

Voyager 1 and Voyager 2 were spectacular successes. The data they provided about the planet Jupiter, its satellites, and its ring, dazzled the eye and revolutionized our understanding.

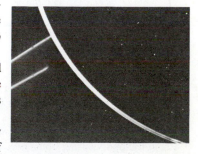

Figure 24–24 Jupiter's ring extends outward from Jupiter's bright limb in this photograph taken by Voyager 2 looking back from the dark side. The lower ring image is cut short by Jupiter's shadow.

Figure 24–25 A mosaic of Jupiter's ring from Voyager 2. It is slightly blurred by spacecraft motion, especially in the left-most image.

''And David put his hand in his bag, and took thence a stone, and slung it, and smote the Philistine [Goliath] in his forehead.''

I Samuel 17:49

Figure 24–26 An artist's drawing of the probe portion of Project Galileo, as it descends into the Jovian atmosphere. It is to be the first planetary mission launched from a space shuttle.

Jupiter's large mass makes it a handy source of energy to use to send probes to more distant planets. Just as a sling transfers energy to a stone, some of Jupiter's energy can be transferred to a spacecraft through a gravitational interaction. This *gravity assist* method has sent Pioneer 11 and the Voyagers to Saturn, and Voyager 2 to Uranus and Neptune.

NASA is planning Project Galileo, which will provide one spacecraft to orbit Jupiter and another to drop a probe into Jupiter's clouds. The Orbiter will orbit Jupiter a dozen times in a 20-month period, coming so close to several of Jupiter's moons that pictures will have a resolution 10 to 100 times greater even than those from the Voyagers. The Probe will transmit data for an hour as it falls through the Jovian atmosphere (Fig. 24–26). We expect to lose contact with it after it penetrates the clouds for 130 km or so. It should give us accurate measurements of Jupiter's composition. The comparison to the Sun's composition should help us understand Jupiter's origin.

Galileo had been scheduled for launch aboard a space shuttle in 1982. Unfortunately, the continued delays in the space shuttle program, budget cuts, and the fact that the shuttle's lifting capabilities are not as great as had been planned, have delayed the Galileo launches until 1986, postponing arrival at Jupiter until 1988.

The Hubble Space Telescope, in Earth orbit, should be able to take pictures of Jupiter with a resolution approximately equal to that of Voyager. So from 1986 on, we should get pictures of Jupiter's clouds every few days for a long time.

Summary and Outline

Fundamental properties (Section 24.1)
    Highest mass and largest diameter of any planet
    Low density: 1.3 grams/cm^3
    Composition primarily of hydrogen and helium
    Rapid rotation, causing oblateness of disk
    Colored bands on surface; Great Red Spot
    Intense bursts of radio radiation
    Strong magnetic field and radiation belt
Jupiter's moons (Section 24.2)
    Historic role in Copernican theory
Spacecraft observations (Section 24.3)
    Pioneer 10 and 11 flybys in 1973 and 1974
    Voyager 1 and 2 flybys in 1979, with greatly increased resolution

Many photographs taken of cloud structure
Great Red Spot is a giant anticyclonic storm
No solid surface: gaseous atmosphere and liquid interior
Jupiter radiates 2 to 3 times the heat it receives from the Sun (confirmed previous observations)
Colored bands result from huge convection currents and the rapid rotation
Temperatures of 400 to 450 K measured in middle layers of atmosphere
Pioneer 11 and Voyagers photographed the Jovian polar regions, a task impossible from Earth
Tremendous magnetic field detected
Spacecraft observations of moons (Section 24.4)
    Io is transformed by volcanoes

Europa has a flat surface, criss-crossed with dark markings
Ganymede has many craters and grooved terrain
Callisto is covered with craters, including a bull's-eye

Amalthea is irregular and oblong
Jupiter's ring (Section 24.5)
A narrow ring, composed of small particles

Key Words

giant planets, Jovian planets, Great Red Spot, radiation belts, Galilean satellites, zones, belts, flux tube, gravity assist

Questions

1. Why does Jupiter appear brighter than Mars despite its greater distance from the Earth?

2. Assume that the Jovians put a dye in their atmosphere such that a green line runs from the north pole to the south pole. Sketch the appearance of this line a few days later.

3. Even though Jupiter's atmosphere is very active, the Great Red Spot has persisted for a long time. How is this possible?

4. (a) How did we first know that Jupiter has a magnetic field? (b) What did the recent studies show?

†5. From Jupiter's distance from the Earth (Appendix 3), the size of the orbits of the Galilean satellites (Appendix 4), and the speed of light (Appendix 2), calculate how "late" the moons of Jupiter appear to arrive in their predicted positions when Jupiter is at its farthest point from the Earth compared to when Jupiter is closest to the Earth. Roemer used the opposite of this calculation in 1675 to measure the speed of light.

6. It has been said that Jupiter is more like a star than a planet. What facts support this statement?

7. What advantages over the 5-m Palomar telescope on Earth did Voyagers 1 and 2 have for making images of Jupiter?

8. What are two types of observations other than photography made from the Voyagers to Jupiter?

9. How does the interior of Jupiter differ from the interior of the Earth?

†10. From measurements made on Figure 24–11, calculate—in km/hr—the speed of rotation of the white blob caught in the Great Red Spot.

11. Which moons of Jupiter are icy? Why?

12. Contrast the volcanoes of Io with those of Earth.

13. Compare the surfaces of Callisto, Io, and the Earth's Moon. Show what this comparison tells us about the ages of features on the surfaces.

14. Using the information given in the text and in Appendices 3 and 4, sketch the Jupiter system, including all the moons and the ring. Mark the groups of moons.

15. (a) Describe the "gravity assist" method. (b) What does it help us do?

16. Explain how the Hubble Space Telescope might be able to equal Voyager's resolution even though it will be farther from Jupiter.

†17. Given the resolution of the Hubble Space Telescope, calculate the size of features visible on Jupiter. Provide sample distances on Earth for comparison.

†18. Repeat the calculation of Question 17 for Io, and comment on what structure will be seen.

---

†This indicates a question requiring a numerical solution.

Projects

1. Describe in some detail the origins of the names of each of Jupiter's moons in Greek mythology.

2. Over a four-hour interval one night, plot the positions of the Galilean satellites at half-hour intervals.

3. Over a one-week interval, plot the positions of the Galilean satellites from night to night. Using the observations in projects 2 and 3 together, deduce the projection on the sky of the orbits of the individual satellites.

Topics for Discussion

1. Although many of the most recent results about Jupiter came from spacecraft, prior ground-based studies had told us many things. Discuss the status of our pre-1973 knowledge of Jupiter, and specify both some things about which space research did not add appreciably to our knowledge and some things about which space research led to a major revision of our knowledge.

2. If we could somehow suspend a space station in Jupiter's clouds, the astronauts would find a huge gravitational force on them. What are some of the effects that this would have? (The novel *Slapstick* by Kurt Vonnegut deals with some of the problems that would arise if gravity were different in strength from what we are used to.)

Saturn and some of its moons, photographed from the Voyager 1 spacecraft. Dione is in the foreground, Tethys and Mimas are above it and to Saturn's right, Enceladus and Rhea are to the left, and Titan in its distant orbit is at the top. The symbol for Saturn appears at the top of the facing page.

Saturn

Aims: To describe Saturn, its rings, and its moons, and to see how the Voyager spacecraft multiplied our knowledge of them

Saturn is the most beautiful object in our solar system, and possibly even the most beautiful object we can see in the sky. The glory of its system of rings makes it stand out even in small telescopes.

Saturn, like Jupiter, Uranus, and Neptune, is a giant planet. Its diameter, without its rings, is 9 times greater than Earth's; its mass is also 9 times greater. Saturn is 9.5 A.U. from the Sun, and has a lengthy year, equivalent to 30 Earth years.

The giant planets have low densities. Saturn's is only 0.7 g/cm^3, 70 per cent the density of water (Fig. 25–1). The bulk of Saturn is hydrogen molecules and helium. Deep in the interior where the pressure is sufficiently high, the hydrogen molecules are converted into metallic hydrogen. Saturn could have a core of heavy elements, including rocky material, making up 20 per cent of its interior.

The rings extend far out in Saturn's equatorial plane, and are inclined to the planet's orbit. Over a 30-year period, we sometimes see them from above their northern side, sometimes from below their southern side, and at intermediate angles in between. When seen edge on, they are all but invisible (Fig. 25–2).

25.1 Saturn from the Earth

The rings of Saturn are either material that was torn apart by Saturn's gravity or material that failed to accrete into a moon at the time when the planet and its moons were forming. These bits of matter spread out in concentric rings around Saturn.

Every massive object has a sphere, called *Roche's limit* or the *Roche limit*, inside of which blobs of matter cannot be held together by their mutual gravity. The forces that tend to tear the blobs apart from each other are tidal forces (Section 19.3); they arise because some blobs are closer to the planet than others and are thus subject to higher gravity.

The radius of the Roche limit varies with the amount of mass in the parent body and the nature of the bodies but is usually about 2½ times the radius of the larger body. The Sun also has a Roche limit, but all the planets lie outside it. The natural moons of the various planets lie outside the respective Roche limits. Saturn's rings lie inside Saturn's Roche limit, so it is not surprising that the material in the rings is spread out rather than being collected into a single orbiting satellite.

Saturn has several concentric major rings visible from Earth. The brightest ring (the "B-ring") is separated from a fainter broad outer ring (the "A-ring") by an apparent gap called *Cassini's division*. Another ring (the "C-ring," or "crepe ring") is inside the brightest ring. It has long been thought that the gaps between the major rings result from gravitational effects from Saturn's satellites, though this is uncertain. Finer structure in the rings has been studied from spacecraft, as we shall see.

The rotation of Saturn's rings can be measured directly from Earth with a spectrograph, for different parts of the rings have different Doppler shifts. We know that

Figure 25–1 Since Saturn's density is lower than that of water, it would float, like Ivory Soap, if we could find a big enough bathtub.

Artificial satellites that we send up to orbit around the Earth are constructed of sufficiently rigid materials that they do not break up even though they are within the Earth's Roche limit.

Figure 25–2 The rings of Saturn as seen from Earth at various times over several years. They extend in the equatorial plane of Saturn from about 70,000 to over 135,000 km from the planet's center. They are inclined to Saturn's orbit by 27°. (Lowell Observatory photographs.)

*Box 25.1 The Roche Limit

E. A. Roche defined the limit in 1849 as the shortest distance within which a liquid body does not break up because the body's own gravitation pulling itself together is stronger than the tidal forces pulling it apart. Similar limits can be defined for solid bodies and non-spherical bodies, and for the case where the orbiting body is taking on individual molecules. The limits depend on both the mass of the planet and the density of the ring particles. Thus the existence of rings at a given radius can help distinguish whether the rings are made of ice or of stony material. Saturn's rings, to be within the Roche limit, must be made primarily of ice.

the rings are not solid objects, because they rotate at different speeds at different radii.

The rings are thin, for on at least one occasion, stars occulted by the rings could be seen shining dimly through. The rings are relatively much flatter than a phonograph record. Radar waves were first bounced off the rings in 1973. The result of the radar experiments showed that the particles in the ring are probably rough chunks of ice at least a few centimeters and possibly a meter across. Infrared spectra show that they are covered with ice.

Like Jupiter, Saturn rotates quickly on its axis, also in about 10 hours. As a result, it is oblate—bulges at the equator—by 10 per cent. Saturn's delicately colored bands of clouds rotate 10 per cent more slowly at high latitudes than at the poles. Methane, ammonia, and molecular hydrogen have been detected spectrographically.

Saturn gives off radio signals, as does Jupiter, an indication to earthbound astronomers that Saturn also has a magnetic field. And, like Jupiter, Saturn has a source of internal heating.

Until recent discoveries from the Earth and from space, we knew of 9 moons of Saturn. Additional moons of Saturn have been reported from time to time. The best time to search from the Earth is at the times when Saturn's rings are viewed edge on and thus fade from view, making it easier to see any moons. The discovery of at least two moons was reported from the 1966 passage of the Earth through the plane

Box 25.2 Cassini's Division

In 1610, Galileo discovered with his new invention, the telescope, that Saturn was not round; it seemed to have "ears." In 1656, the Dutch astronomer Christian Huygens published an anagram—a coded message made by scrambling letters (a procedure then often followed to establish priority for a discovery)—explaining that Saturn has a ring. Huygens' open publication of his result in a book followed three years later. Giovanni Cassini, who worked at the Paris Observatory some years later, observed the largest gap in the ring in 1675, and this gap is named after him.

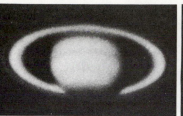   

of the rings, but the observations were difficult to confirm. The 1979–1980 passages revealed still additional moons (Fig. 25–3). (The Earth won't pass through the ring plane again until 1995–1996.) The spacecraft observations that we shall soon discuss also revealed new moons.

All the moons known from Earth-based observations, except Titan, range from about 130 km to 1600 km across. Planet-sized Titan, however, is a different kind of body. An atmosphere, including methane, had been detected on Titan by astronomers using terrestrial telescopes. Because an atmosphere implies a greenhouse effect, some scientists wondered whether Titan's surface may have been warmed enough for life to have evolved there.

25.2 Saturn from Space

Pioneer 11, which had passed Jupiter in 1974, reached Saturn in 1979. But Pioneer 11 had the handicap of having to travel across the diameter of the solar system from Jupiter to reach Saturn, while the Voyagers travelled a much shorter route. So Voyager 1 reached Saturn in 1980, only a year after Pioneer 11. Voyager 2 arrived in 1981. Between the time of the launch of Pioneer 11 and the launches of the Voyagers, experimental and electronic capabilities had improved so much that the Voyagers could transmit data at 100 times the rate of Pioneer 11.

25.2a Saturn's Rings

From Pioneer 11, the rings were visible for the first time from a vantage point different from the one we have on Earth. The backlit view obtained by Pioneer 11 (Fig. 25–4) showed that Cassini's division, visible as a dark (and thus apparently empty) band from Earth appeared bright, with its own dark line of material. The outermost major ring, the A-ring, showed structure as well. The brightest ring observed from Earth, the B-ring, appeared dark on the Pioneer 11 view, presumably because it is too opaque to allow light to pass through it. The rings appear to have low mass and therefore low density, endorsing the idea that they are made up largely

Saturn's moons (Appendix 4) are named after the Titans, in Greek mythology the children and grandchildren of Gaea, the goddess of the Earth, who had been fertilized by a drop of Uranus' blood.

All the moons except Phoebe and Iapetus are in orbits inclined no more than 1° or 2° to the plane of Saturn's equator. Iapetus' orbit is inclined 15°. The orbit of Phoebe, the outermost satellite, is inclined 150° and is very eccentric. And Phoebe is four times farther out than the next farthest satellite. So Phoebe may well be an asteroid captured at some time in the past, and we discuss it in Section 27.3 on asteroids.

Figure 25–3 When Saturn's rings were edge-on to the Earth in 1980, several new moons were discovered. An obstruction, necessary to block the image of Saturn's disk to prevent its relatively bright light from scattering about and hiding the moons' images, shows at the center of the photograph. Six of Saturn's previously known moons plus one newly discovered moon show to the sides of the rings on this April 9, 1980, exposure.

Figure 25–4 Though Pioneer 11, which passed Saturn in 1979, gave images far inferior to those from the Voyagers, it gave the first view of Saturn's rings when illuminated by the Sun from the other side. The rings appeared as inverses of the rings as seen from the front. Saturn's moon Rhea is below the planet. Compare the location of light and dark rings with those on the Voyager image opening this chapter. (NASA/Ames; U. of Arizona image processing.)

Figure 25–5 Saturn and its rings in a mosaic taken by Voyager 1 from a distance of 18 million km, 2 weeks before closest approach. (All Voyager photos courtesy of Jet Propulsion Laboratory/NASA.)

Figure 25–6 The complexity of Saturn's rings. About 100 can be seen on this enhanced image, taken by Voyager 1 when it was 8 million km from Saturn. Several narrow rings can even be seen inside the upper right part of the dark Cassini division. The Earth is shown to scale.

A moon discovered by Voyager 1, Saturn's 14th, appears at upper right, just inside the narrow F-ring. It apparently shepherds particles, like a sheep dog, keeping them from falling inward. Another newly discovered moon, just outside the ring, keeps the particles from escaping.

Figure 25–7 A close-up from Voyager 2 showing 6,000 km of a major ring, with the narrowest features 15 km wide.

Figure 25–8 A device on Voyager 2 followed the changes in brightness of the star delta Scorpii as it passed behind the rings. We see here a computer display of the data from a 50-km-wide section of rings with a resolution of 1 km. Saturn has thousands of such rings.

of ice. An icy composition would also explain why atomic hydrogen was found around the rings; it could result from dissociation of water ice.

The passage of Voyager 1 by Saturn was one of the most glorious events of the space program. The resolutions of the Voyager's cameras were much higher than the resolution of Pioneer 11's instrument, which had been designed for measuring polarization rather than for photography. Voyager 2's cameras turned out to be twice as sensitive as Voyager 1's and had higher resolution. Though during the Voyagers' approaches, the subtleness of colors in Saturn's clouds made photographs of the surface less spectacular than those of Jupiter, the structure and beauty of the rings dazzled everyone (Color Plate 46 and Fig. 25–5).

The closer the spacecraft got to the rings, the more rings became apparent. Before Voyager, scientists discussed whether there were three, four, five, or six rings. But as Voyager 1 approached, the photographs increasingly showed that each of the known rings was actually divided into many thinner rings. By the time Voyager 1 had passed Saturn, we knew of hundreds or a thousand rings (Fig. 25–6). Voyager 2 saw still more (Fig. 25–7). Further, a device on Voyager 2 was able to track the change in brightness of a star as it was seen through the rings, and found even finer divisions (Fig. 25–8). The number of these rings (sometimes called "ringlets") is in the hundreds of thousands.

Figure 25–6   Figure 25–7   Figure 25–8

Color Plate 22 (top, left): Copernicus' original manuscript of *De Revolutionibus*, the sixteenth-century work in which he set out his heliocentric theory. (Photo by Charles Eames)

Color Plate 23 (top right): Comet West, a bright comet visible in 1976. Broad bands are visible in the 50 million-kilometer-long dust trail. The faint blue gas tail extends straight up past the bright star Epsilon Pegasi. (Photo by Martin Grossmann)

Color Plate 24 (center): Apollo 17 view of the Taurus-Littrow Valley on the Moon. This huge, fragmented boulder had rolled a kilometer down the side of the North Massif to here. Scientist-astronaut Harrison Schmitt is at the left. The Lunar Rover is on the right. (NASA photo)

Color Plate 25 (bottom, left): The blast-off of Apollo 17, a night launch. (Photo by the author)

Color Plate 26 (bottom, center): An Apollo 17 astronaut with an instrument package. (NASA photo)

Color Plate 27 (bottom, right): The crescent Earth seen from Apollo 11 in orbit around the Moon. (NASA photo)

Color Plate 28 (top, left): Earth photographed from Apollo 11. Africa and the Middle East are clearly visible. (NASA photo)

Color Plate 29 (top, right): Mars, photographed from Earth as part of the International Planetary Patrol.

Color Plate 30 (bottom, left): Mars, photographed from the Viking 1 spacecraft as it approached in June 1976. Olympus Mons, the large volcano, is toward the top right of the picture. The Tharsis Mountains, a row of three other volcanoes, are also visible. To the left of these volcanoes, the irregular white area may be surface frost or ground fog. The large impact basin, Argyre, is the circular feature at the bottom of the disk. (NASA photo)

Color Plate 31 (bottom, right): The other side of Mars, photographed from Viking 1. The giant canyon in the upper hemisphere has a diameter larger than the coast-to-coast diameter of the United States.

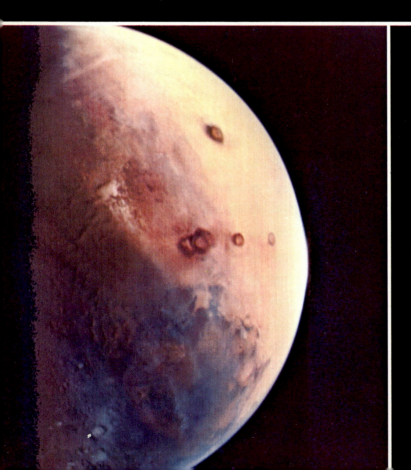

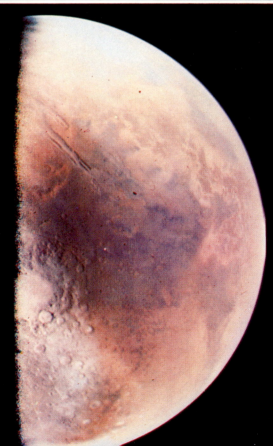

Color Plate 32 (top): Viking 1's sampler scoop is in the foreground of this view from the Martian surface. Angular rocks of various types can be seen. Large blocks one to two meters across can be seen on the horizon, which is about 100 meters from the spacecraft. The horizon may be the rim of a crater. (NASA photo)

Color Plate 33 (bottom, left): Mars' Utopia Planitia as seen from Viking 2, with the Lander's boom in the foreground. Many of the rocks are porous and sponge-like, similar to some of the Earth's volcanic rocks. (NASA photo; color-corrected version courtesy of Friedrich O. Huck)

Color Plate 34 (bottom, right): Mars' Utopia Planitia as seen from Viking 2. This winter scene reveals frost under some of the rocks (NASA photo; color-corrected version courtesy of Friedrich O. Huck)

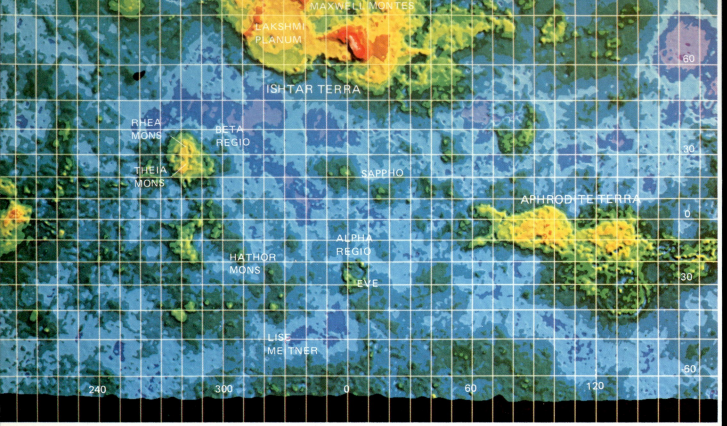

MAXWELL MONTES

LAKSHMI
PLANUM

60

ISHTAR TERRA

RHEA
MONS

BETA
REGIO

30

THEIA
MONS

SAPPHO

APHRODITE TERRA

0

ALPHA
REGIO

HATHOR
MONS

EVE

30

LISE
MEITNER

60

240   300   0   60   120

Color Plate 35 (top): A radar map of Venus from Pioneer Venus Orbiter. Most of Venus is covered by a rolling plain, shown in green and blue. The highlands (yellow and brown contours) sit atop the plain, like continents. Aphrodite Terra is half as big as Africa. Ishtar Terra is the size of Australia, but looks relatively large in this Mercator projection. The lowlands, though resembling Earth's ocean basin, cover only 16 percent of the planet. (Experiment and data—MIT; maps—U.S. Geological Survey; NASA/Ames spacecraft)

Color Plate 36 (bottom): An aurora borealis photographed in the Goldstream Valley, Alaska. (Gustav Lamprecht photo)

Color Plate 37 (top): Jupiter, its Great Red Spot, and 2 of its moons, in this Voyager 1 photograph. The satellite Io, can be seen against Jupiter's disk. The satellite Europa is visible off the limb at the right.

Color Plate 38 (bottom): Jupiter's Great Red Spot, photographed from Voyager 2. Gas in both the Spot and the adjacent white oval is circulating in the same direction.

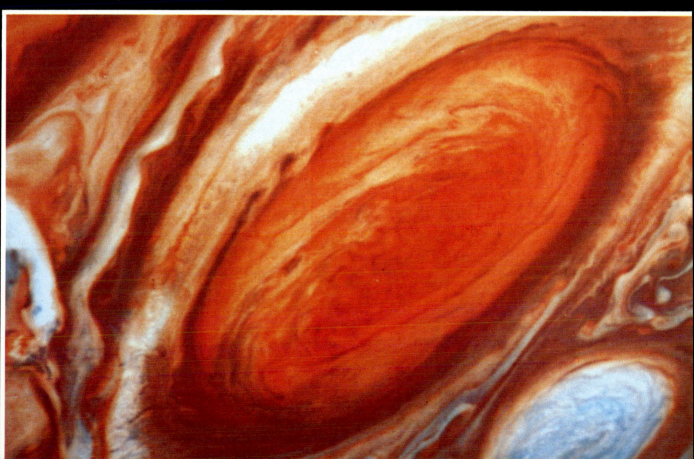

Color Plate 39 (top): Jupiter, photographed from Voyager 1. Io is visible above the Great Red Spot. Europa is also visible.

Color Plate 40 (bottom): (A) Io, photographed at a range of 826,000 km from Voyager 1. (B) Volcanoes erupting on Io, photographed from Voyager 2. Two volcanic eruption plumes that are about 100 km high and strongly scatter blue light appear on the limb.

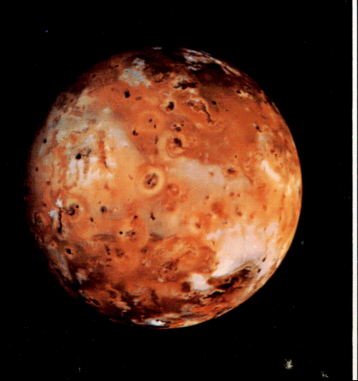

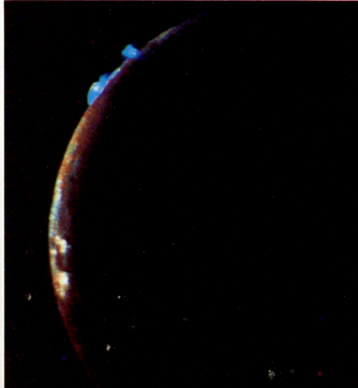

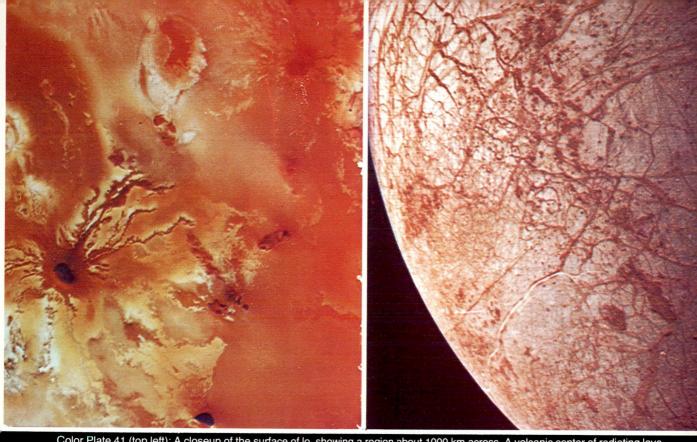

Color Plate 41 (top left): A closeup of the surface of Io, showing a region about 1000 km across. A volcanic center of radiating lava flows is visible at left center.

Color Plate 42 (top right): Europa, photographed from Voyager 2, showing its very smooth and fractured crust.

Color Plate 43 (bottom left): Ganymede, photographed from Voyager 2. The dark, cratered, circular feature is about 3200 km in diameter.

Color Plate 44 (bottom right): Callisto, photographed from Voyager 1. The bull's-eye, named Valhalla, is a large impact basin. The outer ring is about 2600 km across.

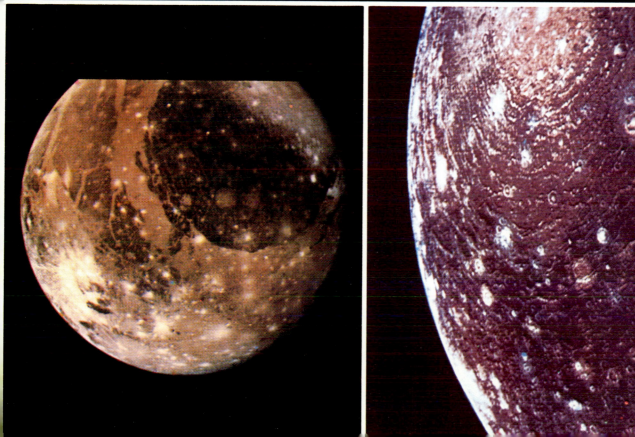

Color Plate 45 (top): Saturn, on March 11, 1974, showing bands on its disk and Cassini's division in its rings. (Lunar and Planetary Laboratory, University of Arizona photo)

Color Plate 46: A montage of Saturn and some of its moons, photographed from the Voyager 1 and 2 spacecraft. Clockwise from upper right, the moons are Titan (reddish), Iapetus (with one dark side, Tethys, Mimas (a small moon with a giant crater), Enceladus (lower left). Dione, and Rhea. (JPL/NASA)

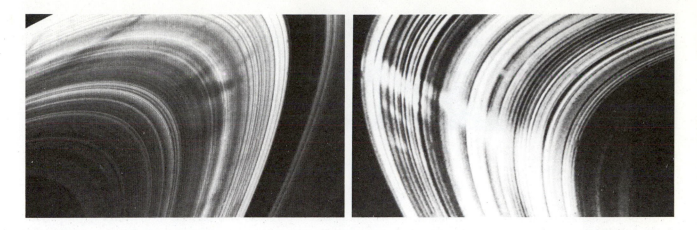

Figure 25–9 Dark radial spokes became visible in the rings as the Voyagers approached. They formed and dispersed within hours. This is a Voyager 2 image, part of a time-lapse series.

Figure 25–10 When sunlight scatters forward, the spokes appear bright. This is a Voyager 1 image.

Everyone had expected that collisions between particles in Saturn's rings would make the rings perfectly uniform. But there was a big surprise. As Voyager 1 approached Saturn, we saw that there was changing structure in the rings aligned in the radial direction. ''Spokes'' can be seen in the rings when they are seen at the proper angle (look back at Fig. 25–5). The spokes look dark from the side illuminated by the Sun (Fig. 25–9), but look bright from behind (Fig. 25–10). This information showed that the particles in the spokes were very small, about 1 micron in size, since only small particles—like terrestrial dust in a sunbeam—reflect light in this way. And it seems that the spoke material is elevated above the plane of Saturn's rings. Since gravity wouldn't cause this, electrostatic forces may be repelling the spoke particles.

It also came as a surprise that some of the rings are not round (Fig. 25–11). At least some of the rings have different brightness or are displaced from one location to another.

Studies of the changes in the radio signals from the spacecraft when it went behind the rings showed that the rings are only about 20 m thick, equivalent to the thinness of a phonograph record 30 km across—super long play.

A

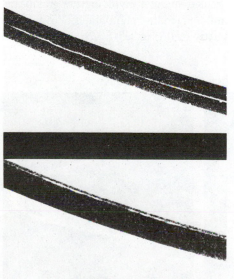

B

Figure 25–11 *(A)* A comparison of a small section of the rings seen on opposite sides of Saturn. The horizontal line marks the border between the two photos, each of which shows a different edge of the rings. In the dark gap on each picture we see a bright ring that is narrowed in the lower picture and slightly broadened and displaced in the upper in these Voyager 1 observations. *(B)* Voyager 2 discovered a new ''kinky'' ringlet inside the second darkest of the major ''gaps'' in the rings. We do not know if the kinky ring is off-center or if we are seeing different components in the upper and the lower frames. Resolution is about 15 km.

Figure 25–12 The outermost visible ring, the F-ring (which lies outside the A-ring), with its two "shepherding" satellites as seen from Voyager 1. The satellites are about 200 km and 220 km in diameter, respectively.

Before the flybys, it had been thought that the structure of the rings and the gaps betweeen them were determined by gravitational effects of Saturn's several known moons on the orbiting ring particles. If a particle in the rings has a period that is a simple fraction of a moon's period, then the particle and the moon would come back in the same configuration regularly and be relatively close together in their elliptical orbits. This pattern reinforces the gravitational pull of the moon and sweeps away particles at those locations. But this explanation had been worked out with fewer than a half-dozen rings known. The ring structure the Voyagers discovered is too complex to be explained in this way alone. Consideration of the sixteen or so satellites cannot explain thousands of rings. And the gaps turned out to be in the wrong place for this model to work.

The outer major ring turns out to be kept in place by a tiny satellite orbiting just outside it. And at least some of the rings are kept narrow by "shepherding" satellites (Fig. 25–12). By Kepler's laws, the outer satellite is moving slightly slower than the ring particles, and the inner one slightly faster. If a ring particle should move outward, it would be pulled back by the outer satellite, diminishing its angular momentum. This would make it sink back toward the ring. Conversely, if a ring particle should move inward, it would be pulled ahead by the inner satellite, increasing its angular momentum. This would push it back toward the ring. Thus material tends to stay in the narrow ring.

Scientists were astonished to find on the Voyager 1 images that the outer ring, the narrow F-ring discovered by Pioneer 11, seems to be made of three braids (Fig. 25–13A). But soon after scientists succeeded in finding explanations for the braided strands, involving the gravity of the pair of newly discovered moons shown in the preceding photograph, Voyager 2 images showed that the rings were no longer intertwined (Fig. 25–13B).

Figure 25–13 *(A)* Two narrow, braided, 10-km-wide rings are visible, as is a broader diffuse component about 35 km wide, on this Voyager 1 image of the F-ring. *(B)* The Voyager 2 images of the F-ring showed no signs of braiding. *(C)* Voyager 2 photographed the F-ring as the moons passed each other, but no kinkiness or braiding resulted.

A   B   C

A post–Voyager–1 theory said that many of the narrowest gaps may be swept clean by a variety of small moons embedded in them. These objects would be present in addition to the smaller icy snowballs that make up the bulk of the ring material. Unfortunately for the theoreticians, Voyager 2 did not find these larger objects, so the reason that such narrow rings and gaps exist is not yet understood.

25.2b Saturn Below the Rings

The structure in Saturn's clouds is of much lower contrast than that in Jupiter's clouds. After all, Saturn is colder so it has different chemical reactions. Even so, cloud structure was revealed to the Voyagers at their closest approach. A haze layer present for Voyager 1 cleared up by the time Voyager 2 arrived, so more details could then be seen (Fig. 25–14). Turbulence in the belts and zones showed up clearly. A few circulating ovals similar to Jupiter's Great Red Spot and ovals were detected (Fig. 25–15). Saturn's ovals are also storms in high-pressure regions, circulating oppositely to the low-pressure storms on Earth. Study of cloud features gave the first direct view of the rotation of Saturn's clouds and allowed detailed measurements of the velocities of rotation to be made.

Extremely high winds, 1800 km/hr and 4 times higher than the winds on Jupiter, were measured. On Saturn, the variations in wind speed do not seem to correlate with the positions of belts and zones, unlike the case with Jupiter. Also as on Jupiter but unlike the case for Earth, the winds seem to be driven by rotating eddies, which in turn get their energy from the planet's interior. Infrared scans showed that the temperature decreases 10 K from equator to pole.

Radio static detected at each Voyager encounter has proved to be from a massive thunderstorm spread 64,000 km around Saturn's equator, longer than the Earth's circumference. The storm was too deep in the haze to see directly. Though Earth, Venus, and Jupiter also have thunderstorms, Saturn's was by far the largest.

Analyses of gravity-field and temperature measurements from the spacecraft suggest that Saturn's core extends about 13,800 km from the center. Thus Saturn's core is about twice the size of the entire Earth. Saturn's interior is so compressed by the weight of the overlying gas that the core contains about 11 Earth masses of matter. The inner core is mostly iron and rock, and the outer core consists of ammonia, methane, and water. The next 21,000 km seem to consist of liquid metallic hydrogen, which is consistent with the discovery of Saturn's magnetic field.

Saturn radiates about 2.5 times more energy than it absorbs from the Sun. One interpretation is that only $\frac{1}{3}$ of Saturn's heat is energy remaining from its formation and from continuing gravitational contraction. The rest would be generated by the gravitational energy released by helium sinking through the liquid hydrogen in Saturn's interior. The helium that sinks has condensed because Saturn, unlike Jupiter, is cold enough.

25.2c Saturn's Magnetic Field

Pioneer 11 had revealed the magnetic field at Saturn's equator to be somewhat weaker than had been anticipated, only $\frac{2}{3}$ of the field present at the Earth's equator. Remember, though, that Saturn is much larger than the Earth and so its equator is much farther from its center. Saturn's magnetic axis appears to be aligned with its rotational axis, unlike Jupiter's, the Earth's, and Mercury's. This observation forces us to formulate new theories of how magnetic fields form inside planets. The total strength of Saturn's magnetic field is 1000 times stronger than Earth's and 20 times weaker than Jupiter's.

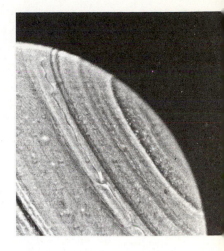

Figure 25–14 Waves and eddies in Saturn's clouds at northern latitudes from Voyager 2. The ribbonlike structure marks a high-speed westerly jet stream. Resolution is 100 km.

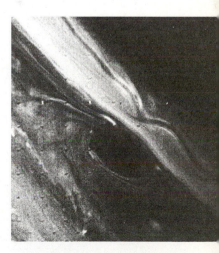

Figure 25–15 Time-lapse photos of this large brown spot in Saturn's atmosphere showed circulation in an anti-cyclonic sense, indicating that it is a high-pressure storm. Since such circulation often induces clearing, we may be seeing through an opening in higher clouds to darker underlying clouds. Resolution is 50 km.

Figure 25–16 Titan was disappointingly featureless even to Voyager 1's cameras because of its thick smoggy atmosphere. Its northern polar region was relatively dark.

Figure 25–17 Looking back at Titan's dark side, Voyager 2 recorded how extensive Titan's atmosphere is by photographing sunlight scattered by particles hundreds of kilometers high.

Figure 25–18 Haze layers can be seen at Titan's limb, with divisions at altitudes of 200, 375, and 500 km.

Saturn has belts of charged particles (Van Allen belts), which are larger than Earth's but smaller than Jupiter's. Though Saturn's magnetic field passes through the rings, the charged particles can't, so the Van Allen belts begin only beyond the rings. Even there, the number of charged particles is depleted by the satellites. Thus the belts are weaker than Jupiter's. The size of the belts fluctuates as the solar wind varies.

Since the rotational and magnetic axes are aligned, scientists were hard pressed to explain why a radio pulse occurred every rotation. By comparing Voyager 1 and 2 results, scientists pinpointed the area from which the pulse comes. It may be the result of an internal anomaly. The existence of the pulse allowed Saturn's rotation period to be accurately measured.

Titan is sometimes within the belts and sometimes outside. Many of the particles in Saturn's magnetosphere probably come from Titan's atmosphere. A huge torus—doughnut—of neutral hydrogen from this source fills the region from Titan inward to Rhea.

A variety of instruments aboard the Voyagers studied the ionized gas around Saturn and the magnetic field. The region between the moons Dione and Rhea is filled with a "plasma torus." (A "plasma" is a gas containing both ions and electrons. It is electrically neutral, but is affected by electric and magnetic fields.) A cloud of this plasma is at the tremendously high temperature of 600,000,000 K, making it the hottest known place in the solar system. The high temperature means that the velocities of individual particles are high. Since the density is extremely low, though, the total energy in the cloud is not large.

The three passes through the magnetic field and plasma around Saturn, from Pioneer 11 and the two Voyagers, will surely not allow us to understand this complex object as well as we would like.

25.2d Saturn's Family of Moons

Farther out than the rings, we find the satellites of Saturn. Like those of Jupiter, they now have the personalities of independent worlds and are no longer merely dots of light in a telescope.

The largest of Saturn's satellites, Titan, is larger than the planet Mercury and has an atmosphere (Figs. 25–16 and 25–17) with several layers of haze (Fig. 25–18). We had known something about Titan's atmosphere from ground-based studies. Spectra showed signs of methane, polarization measurements had found signs of haze, and infrared measurements had indicated that hydrocarbons might be present. But a better understanding of Titan's atmosphere had to wait for the Voyagers. Studies of how the radio signals faded when Voyager 1 went behind Titan showed that Titan's atmosphere is denser than Earth's. Surface pressure on Titan is $1\frac{1}{2}$ times that on Earth. Though we had thought that Titan was the largest moon in our solar system, Titan's atmosphere is so thick that Titan's surface is slightly smaller than the surface of Jupiter's Ganymede.

The Voyagers' ultraviolet spectrometers detected nitrogen, which makes up the bulk of Titan's atmosphere. The methane is only a minor constituent, perhaps 1 per cent. Titan's huge doughnut-shaped hydrogen cloud presumably results from the breakdown of its methane (CH_4).

The temperature near the surface, deduced from measurements made with Voyager's infrared radiometer, is only about $-180°C$, somewhat warmed by the greenhouse effect but still extremely cold. This temperature is near that of methane's "triple point," at which it can be in any of the physical states—solid, liquid, or gas. So methane may play the role on Titan that water does on Earth, though some evidence disagrees with this conclusion. Parts of Titan may be covered with methane lakes or oceans, and other parts may be covered with methane ice or snow. An ethane (C_2H_6) ocean, or a mixture of ethane and methane, has more recently been suggested to overcome some of the objections that the pure methane ocean did not satisfy certain predictions.

Titan's clouds are opaque, and probably contain both liquid nitrogen and liquid methane. It seems that Titan's atmosphere is opaque because of the action of sunlight on chemicals in it, forming a sort of "smog," and giving it its reddish tint. Smog on Earth forms in a similar way. Some of the color may result from the bombardment of Titan's atmosphere by protons and electrons trapped in Saturn's magnetic field. Other energy to run the chemical reactions may come from the solar wind, since Titan's orbit is large enough to go in part outside Saturn's protective magnetic field.

Some of the organic molecules formed in Titan's atmosphere may rain down on its surface. Thus the surface, hidden from our view, may be covered with an organic crust about a kilometer thick, perhaps partly dissolved in liquid methane. These chemicals are similar to those from which we think life evolved on the primitive Earth. The 1983 discovery (with the 4-m Kitt Peak telescope) of carbon monoxide provides further evidence for an atmosphere resembling that of the early Earth. But it is probably too cold on Titan for life to begin.

The surfaces of Saturn's other moons, all icy, are so cold that the ice acts as rigid as rock and can retain craters. The moons' mean densities are all 1.0 to 1.5, which means that they are probably mostly water ice throughout with some rocky material included. Unexpectedly, the densities do not decrease with distance from the planet, as do the densities of the planets with distance from the Sun and of Jupiter's moons with distance from that planet.

In addition to Titan, four of Saturn's moons are over 1000 km across, so we are dealing with major bodies (see Appendix 4), about $\frac{1}{3}$ the size of Earth's Moon. For Saturn as for Jupiter, most satellites perpetually have the same side facing their planet. Let us consider the major ones in order from the inside out.

Mimas boasts a huge impact structure that is $\frac{1}{4}$ the diameter of the entire moon (Fig. 25–19*A*). The crater has a raised rim and a central peak, typical of large impact craters on the Earth's Moon and on the terrestrial planets. The whole satellite is saturated with craters. The canyon visible half way around Mimas (Fig. 25–19*B*) may be a result of the impact. The energy of the impact may have shattered the satellite and been focused half way around. Many craters a few kilometers across are visible.

Enceladus' surface (Fig. 25–20) has both smooth regions and regions covered with impact craters. The existence of smooth regions suggests that Enceladus' surface

Moon	Diameter (km)
Mimas	392
Enceladus	510
Tethys	1,060
Dione	1,120
Rhea	1,530
Titan	5,150
Hyperion	410 × 220
Iapetus	1,460
Phoebe	220

Fig. 25–19 A, Mimas

Fig. 25–19 B, Mimas

Fig. 25–20 Enceladus

Figure 25–19 *(A)* The impact feature on Mimas is about 130 km in diameter. *(B)* The other side of Mimas, showing the trough crossing the center that may be the result of the impact.

Figure 25–20 A computer-enhanced image of Enceladus from Voyager 2. Enceladus resembles Jupiter's Ganymede in spite of being 10 times smaller. Resolution is 2 km on this Voyager 2 mosaic.

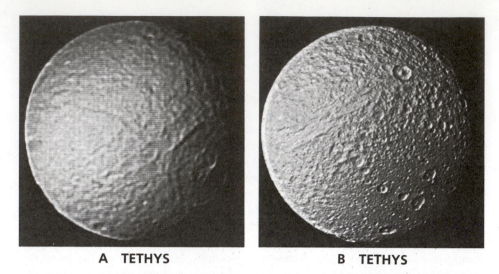

A TETHYS   **B TETHYS**

Figure 25–21 *(A)* A computer-enhanced view of Tethys, showing a large bright circular feature 180 km across. The crater has been flattened by the flow of softer ice, and probably formed early in Tethys' history when its interior was still relatively warm and soft. *(B)* Craters and a 750-km-long trench on Tethys. The large crater at the upper right sits on the trench. Both images are from Voyager 2, with resolution of 9 and 5 km, respectively.

was melted comparatively recently and so may be active today. The internal heating may be the result of a gravitational tug of war with Saturn and other moons, as for Jupiter's Io. Linear sets of grooves on Enceladus, tens of kilometers long, are probably geologic faults in the crust, resembling those on Jupiter's Ganymede.

Tethys also has a large circular feature (Fig. 25–21*A*). The difference between heavily and lightly cratered regions may indicate that internal activity took place early in Tethys' history. The side of Tethys that faces Saturn (Fig. 25–21*B*) shows not only many craters but also a large canyon that goes nearly ¾ of the way around the satellite. The canyon, or trench, several kilometers deep, could have resulted from the expansion of Tethys as its warm interior froze. Or it could be a fault through the entire planet, a sign of the brittleness of ice under these extremely cold conditions.

Figure 25–22 *(A)* Dione, showing impact craters, debris, ridges or valleys, and wispy rays. *(B)* The largest crater on Dione is about 100 km in diameter. The sinuous valleys were probably formed by geological faults. *(C)* Dione in transit in front of Saturn's clouds. Notice the difference between the trailing hemisphere *(left)* and the leading hemisphere.

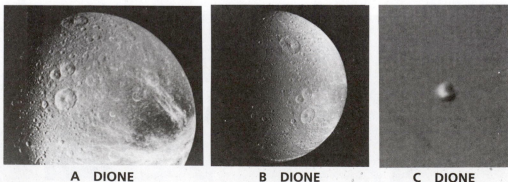

A DIONE   **B DIONE**   **C DIONE**

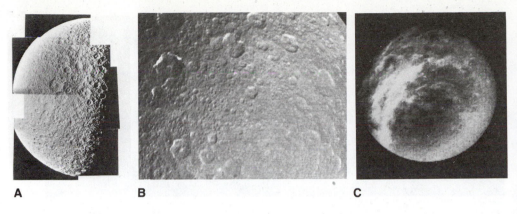

A   B   C

Figure 25–23 *(A)* Rhea, with craters as large as 300 km across, many with central peaks. *(B)* A Voyager 1 close-up of Rhea with resolution of 2.5 km. The area shows an ancient, heavily cratered region. White areas on the edges of several craters in the upper right are probably fresh ice either visible on steep slopes or deposited by leaks from inside. *(C)* This hemisphere of Rhea shows wispy streaks.

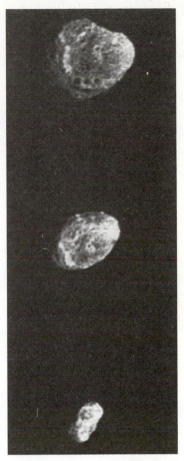

Figure 25–24 Three views of Hyperion, an irregularly shaped satellite roughly 410 × 220 km.

Dione, too, shows many impact craters, some of them with rays of debris (Fig. 25–22*A*). Many valleys are visible in Dione's icy crust (Fig. 25–22*B*). Scientists wonder why Dione's two sides are so different (Fig. 25–22*C*). The wisps of material shown may have erupted from inside.

Two associated satellites share Tethys' orbit, always slightly ahead of Tethys itself. A small moon also shares Dione's orbit. No moons were known to share orbits before we studied the Saturn system.

Rhea has impact craters (Fig. 25–23). Some craters, with sharp rims, must be fresh; others, with subdued rims, must be ancient. Rhea's other side has wispy light markings (Fig. 25–23*C*). Slush and mud could have flowed out of Rhea's interior to cover all the large craters in some areas. The debris in the Saturn system continued to make small craters.

Next out is Titan, which we have already discussed.

Hyperion, when photographed by Voyager 2, turned out to look like a hamburger or a hockey puck (Fig. 25–24). Its orientation does not seem stable, so it may have been jarred into this position by one of the impacts that show up as the many craters on its surface or possibly by an impact that broke apart a larger object. Hyperion and Mimas are essentially the same size but vastly different in appearance. Surely they had different histories.

Strangely, the side of Iapetus (Fig. 25–25) that precedes in its orbit is 5 times darker than the side that trails. The large circular feature is probably an impact feature outlined by dark material. Iapetus' density indicates that it may be the sole moon to contain an interior partly made of methane ice in addition to water ice. Water ice covers the bright surface. Perhaps the dark material is a hydrocarbon formed by sunlight reacting on some of the methane that wells up.

Saturn's outermost satellite, Phoebe, is probably a captured asteroid and will be discussed in that section, Section 27.3.

In addition to these 9 moons long known, several others have been detected from the ground in the last decades, and several have been confirmed or discovered by the Pioneer and Voyager spacecraft. Voyager even provided close-ups of one of the recent ground-based discoveries (Fig. 25–26). This satellite is too small (only 135 km × 70 km) for gravity to have pulled it into a spherical shape. Oddly, it

Figure 25–25 Iapetus, whose trailing side is 5 times brighter than its leading one. The dark side may have accumulated dust spiralling in toward Saturn or may be covered with hydrocarbons. It could be as black as pitch because it is pitch! A large circular feature is also visible.

A

B

Figure 25–26 *(A)* Saturn's tooth-shaped 11th moon, only 135 × 70 km. *(B)* The thin line that can be seen to move across this moon in the 13 minutes between views is the shadow of an otherwise unseen ring.

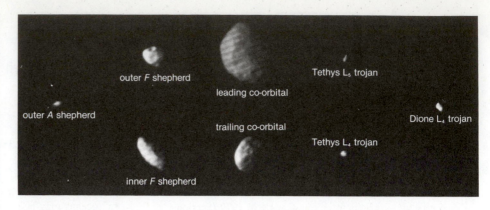

Figure 25–27 Several of Saturn's smaller satellites, viewed from Voyager 2. Two are in the "Trojan points" of Tethys and one in a "Trojan point" of Dione, gravitationally stable locations ⅙ of the way around the orbit from the major bodies.

apparently all but shares an orbit with another newly discovered satellite. The orbits are not absolutely identical, so the two approach each other every few years. Their gravitational fields are strong enough to affect the other only when they are very close to each other. When this happens, they twirl around each other, interchange orbits, and are off on their merry ways again, separating from each other. No other case of such a gravitational interaction is known in the solar system. They may be halves of a moon that broke apart.

It is difficult to say how many moons Saturn has. Indeed, there may be many orbiting in the rings that are 50 or 100 km across. A sampling appears in Fig. 25–27. They are probably fragments of larger bodies that broke up in the early years of the Saturn system. They may be mostly icy with a mixture of meteoritic rock.

Figure 25–28 The crescent of Saturn and the planet's rings, taken from 1,500,000 km from the far side of Saturn as Voyager 1 departed. The rings' shadows are visible cutting across the overexposed crescent.

25.3 Beyond Saturn

Yet another superlative for Voyager 1 was the view it gave us when it looked back on the crescent Saturn (Fig. 25–28). The spacecraft is now travelling up and out of the solar system.

The platform carrying the television cameras of Voyager 2 suffered a malfunction within the hour of passing through the ring plane behind Saturn. The problem may simply have been that the spacecraft was old, and a gear was wearing out. Some data were lost, though the main part of the mission was then over. JPL engineers were able to restore the spacecraft to working order in time to get some observations of the winds in the southern hemisphere to match the earlier northern-hemisphere results, a departing view of Saturn's rings, and images of Phoebe.

All parts of Voyager 2 are now working as the spacecraft speeds to the outer solar system. As Voyager 2 travels toward a 1986 rendezvous with Uranus and a 1989 rendevous with Neptune, it is recording data on interplanetary space.

The Voyagers' passages through the Saturn system exchanged a lot of unknowns for a lot of knowledge plus many specific new mysteries. Monitoring Saturn with the Hubble Space Telescope should give some new information. No spacecraft visits are now planned.

Summary and Outline

Giant planet with system of rings; low density: 0.7 g/cm^3
Ring system (Section 25.1)
   Inclined 27° to orbit; chunks of rock and ice orbiting inside Roche limit
   Cassini's division apparently dark
   Radar studies show the rings are very thin
   Rapid rotation makes planet oblate
   No solid surface
   Magnetic field present
   Pioneer 11 viewed rings from a new angle in 1979 (Section 25.2)
   Voyagers 1 and 2 made closeup studies in 1980 and 1981.
Rings (Section 25.2a)
   Divide into thousands of ringlets
   Radial spikes made out of small particles held up by electrostatic force
   Some rings not round
   Ring structure affected by shepherding satellites
Saturn's surface and interior (Section 25.2b)
   High-pressure storms found
   Winds 4 times higher than those on Jupiter
   Wind variations do not match belts and zones
Magnetic field (Section 25.2c)

Surface field only $\frac{2}{3}$ that of Earth but total field is 1000 times stronger
Saturn's rotational and magnetic axes are aligned
Hydrogen torus around Titan
Plasma torus of extremely high temperature present
Moons (Section 25.2d)
   Titan
      Atmosphere is mostly nitrogen; surface pressure is $1\frac{1}{2}$ Earth's; methane near its triple point; methane lakes or oceans may be present; reddish smog; surface may be covered by organic materials
   Other moons are icy
      Mimas: huge crater and giant canyon
      Enceladus: recent activity and internal heating
      Tethys: large crater and giant canyon
      Dione: craters, rays, wisps
      Rhea: craters and wisps; no large craters
      Hyperion: hamburger shaped but same size as Mimas
      Iapetus: one side very dark, the other bright
      Phoebe: captured asteroid
      Small moons: some with unusual orbits, several irregular in shape

Key Words

Roche's limit (Roche limit), Cassini's division

Questions

1. What are the similarities between Jupiter and Saturn?
†2. What is the angular size of the Sun as viewed from Saturn? How many times smaller is this than the apparent angular size of the Sun from Earth?
†3. When Jupiter and Saturn are closest to each other, what is the angular size of Jupiter as viewed from Saturn?
4. What is the Roche limit and how does it apply to Saturn's rings?
†5. Calculate the relative strength of the tidal force from Saturn itself on a moon the size of Titan at Saturn's B-ring and at Titan's orbit.
6. Sketch Saturn's major rings and the orbits of the largest moons, to scale.
7. Describe the major developments in our understanding of Saturn, from ground-based observations to Pioneer 11 to the Voyagers.
8. Why are the moons of the giant planets more appealing for direct exploration by humans than the planets themselves?
9. Why does Saturn have so many rings? What holds the material in a narrow ring?
10. Explain how part of the rings can look dark from one side but bright from the other.
11. What did the Voyagers reveal about Cassini's division?
12. What are "spokes" in Saturn's rings and how might they be caused?
13. What have we learned about Titan?
14. Explain how methane on Titan may act similarly to water on Earth.
15. Describe two of Saturn's moons other than Titan.

---

†This indicates a question requiring a numerical answer.

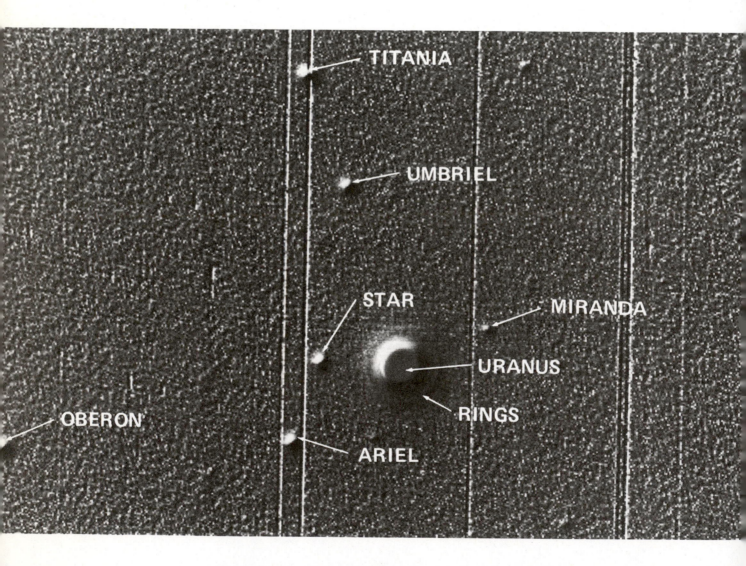

Uranus, its rings, and its moons. Special processing on this CCD image generated a false 3-D texture. Special instrumentation and exceptional observing conditions allowed this first clear picture of the rings to be obtained. The 1986 visit of Voyager 2 to Uranus will give us close-up views of the planet, its rings, and its moons. The symbols for Uranus, Neptune, and Pluto appear at the top of the facing page. (R. J. Terrile, JPL, and B. A. Smith, U. Arizona, at Carnegie Institution's Las Campanas Observatory.)

Uranus, Neptune, and Pluto

26

Aims: To study the two other giant planets—Uranus and Neptune—and to see how little we know about tiny Pluto

The two other giant planets beyond Saturn, Uranus (U'ranus) and Neptune, are each about 50,000 km across and about 15 times more massive than the Earth. Though Uranus and Neptune have no solid surfaces, like Jupiter and Saturn, they may have higher proportions of heavier elements mixed in with the hydrogen and helium of which they are almost entirely formed.

Current models indicate that Uranus and Neptune each has a hydrogen-helium atmosphere surrounding a liquid mantle of water, methane, and ammonia. Below that is a rocky core, containing mostly silicon and iron. The densities of Uranus and Neptune are low (1.2 and 1.7 grams/cm^3, respectively). Their albedos are high, which indicates that they are covered with clouds.

The outermost planet, Pluto, is a small, solid body. Its physical features and its orbit are quite different from those of the four giant planets.

26.1 Uranus

Uranus was the first planet to be discovered that had not been known to the ancients. The English astronomer and musician William Herschel reported the discovery in 1781. Actually, Uranus had been plotted as a star on several sky maps during the hundred years prior to Herschel's discovery, but had not been singled out.

Uranus revolves around the Sun in 84 years, at an average distance of more than 19 A.U. from the Sun. Uranus never appears larger than 4.1 seconds of arc, so studying its surface from the Earth is difficult (Fig. 26–1).

Even the photographs from the balloon Stratoscope, which in 1970 carried a 90-cm telescope up to an altitude of 24 km, above most of the Earth's atmosphere, showed no detail on Uranus' surface. Stratoscope's resolution was ⅙ arc sec, about 3 times better than the best ground-based observations. Since Stratoscope would have seen belts on Uranus like those of Jupiter and Saturn, we know that Uranus does not have them. We may be seeing sunlight scattered by molecules in Uranus' atmosphere instead of seeing belts; this scattering would also account for the high albedo.

The Stratoscope observations suggest that Uranus is surrounded by thick methane clouds, with a clear atmosphere of molecular hydrogen above them. The trace of methane mixed in with the hydrogen makes Uranus look greenish. Both methane and molecular hydrogen have been observed with ground-based spectrographs.

Uranus is so far from the Sun that its outer layers are very cold. Studies of its infrared radiation give a temperature of 58 K. There is no evidence for an internal

Uranus, in Greek mythology, was the personification of Heaven and ruler of the world, the son and husband of Gaea, the Earth. Neptune, in Roman mythology, was the god of the sea, and the planet Neptune's trident symbol reflects that origin. Pluto was, in Greek mythology, the god of the underworld; some say its symbol also incorporates the initials of Percival Lowell, who sponsored the search that led to its discovery.

Figure 26–1 Uranus, photographed from the ground in blue light, shows no surface markings.

heat source, unlike the case for Jupiter, Saturn, and Neptune. The discovery with IUE spacecraft observations of an aurora indicates the presence of a magnetic field.

Uranus has five moons (see the photograph opening this chapter), ranging from 320 to 1690 km across. Little is known about them, but spectra reveal water ice. Though we cannot yet see details on the surfaces of these moons, we can assume that they have many similarities with the moons of Jupiter and Saturn.

26.1a The Rotation of Uranus

The other planets rotate such that their axes of rotation are very roughly parallel to their axes of revolution around the Sun. Uranus is different, for its axis of rotation is roughly perpendicular to the other planetary axes, lying only 8° from the plane of its orbit (Fig. 26–2). Its rotation is, barely, retrograde. Sometimes one of Uranus' poles faces the Earth, 21 years later its equator crosses our field of view, and then another 21 years later the other pole faces the Earth. Polar regions remain alternately in sunlight and in darkness for decades. Thus there could be strange seasonal effects on Uranus. When we understand just how the seasonal changes in heating affect the clouds, we will be closer to understanding our own Earth's weather systems.

Over the last few years different studies of Uranus' rotation have given different values for the period. Values measured by various methods in recent years vary widely, from 12 to 24 hours. Each method seems to make sense, so we don't know why they give different results. Each method incorporates different assumptions, so perhaps some of these assumptions are wrong. The problem indicates how uncertain our knowledge is of the distant members of our solar system. It is important to know about the period of rotation because models for the structure of Uranus' interior depend on the rotation period. Only when we pin down the value of Uranus' rotation period can we settle on a model for the interior.

The north pole is now turning toward the Sun as part of Uranus' 84-year period of changing orientation. It will point closest to the Earth and Sun in 1987.

26.1b Uranus' Rings

Figure 26–2 Uranus' axis of rotation lies in the plane of its orbit. Notice how the planet's poles come within 8° of pointing toward the Sun, while $\frac{1}{4}$ of an orbit before or afterwards, the Sun is almost over the equator.

In 1977, Uranus occulted (passed in front of) a faint star. Predictions showed that the occultation would be visible only from the Indian Ocean southwest of Australia. James Elliot, then of Cornell and now of MIT, led a team that observed the

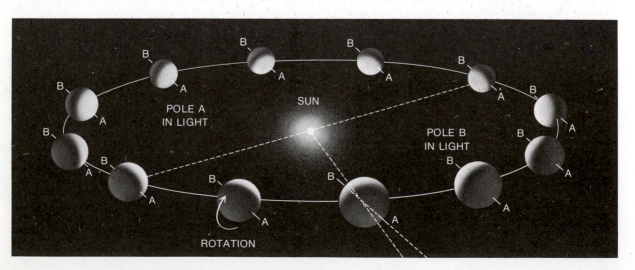

occultation from NASA's Kuiper Airborne Observatory, an instrumented airplane (Section 2.13). Surprisingly, about half an hour before the predicted time of occultation, they detected a few slight dips (Fig. 26–3) in the star's brightness. (The equipment had been turned on early because the exact time of the occultation was uncertain.) They recorded similar dips, in the reverse order, about half an hour after the occultation. Ground-based observers detected some of the dips, but none observed complete before-and-after sequences of all the rings, as was possible from the plane. The dips indicated that Uranus is surrounded by at least five rings (Fig. 26–4). More occultations confirmed the discovery of these rings and suggested the possibility of additional ones. Nine rings are now known (Fig. 26–5).

The rings have radii between 42,000 and 52,000 km, 1.7 to 2.1 times the radius of the planet. (Uranus' innermost moon is 130,000 km out.) Reexamination of the old photographs from the Stratoscope balloon revealed that the shadow of the rings could be detected on Uranus' disk.

The Uranian rings are very narrow from side to side; some are only a few km wide. No material has been detected between them. How can narrow rings exist, when the tendency of colliding particles would be to spread out? The discovery of Uranus' narrow rings led to the suggestion that a small satellite in each ring keeps the particles together. This model, set up for Uranus, turned out to be applicable to the narrow ringlets of Saturn discovered by the Voyagers.

The Caltech infrared group observed the rings from Earth in reflected sunlight at a wavelength of 2.2 microns. At that infrared wavelength, Uranus is dark because absorption from its atmospheric methane prevents sunlight from being reflected, while the reflectivity of the rings (always relatively poor) remains the same as at other wavelengths. The low reflectivity of the rings probably indicates that, unlike the case for Saturn's rings, the ring particles do not have icy surfaces.

Voyager 2 will reach Uranus on January 24, 1986. The spacecraft will approach as close as 107,000 km (a quarter of the distance from the Earth to the Moon.) If instrumental malfunctions don't silence its instruments, its close-up photographs will reveal details of Uranus, its rings, and its moons!

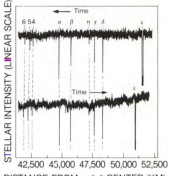

Figure 26–3 As Uranus drifted in front of a star (SAO 258687) on March 10, 1977, dips in the intensity of the star showed that rings around Uranus had occulted the starlight. On the bottom axis of the graphs, the times of observation have been converted into the distance of the star from the center of the rings. The top trace, from right to left, shows the changes in intensity in the ½ hour before the star went behind Uranus itself. The bottom trace, from left to right, shows what happened in the ½ hour after the star reappeared. Rings α, β, γ, and δ are round and have a common center. The ε (epsilon) ring appeared at different distances from the center on opposite sides of Uranus, indicating that it is elliptical.

Figure 26–4 The positions of the dips in the observations by which the rings of Uranus were discovered are marked. Some of the rings have been assigned Greek letters and others have been assigned numbers. The epsilon ring seems to be particularly out of round, and the orientation of the orbit's axis is precessing about 2° per day.

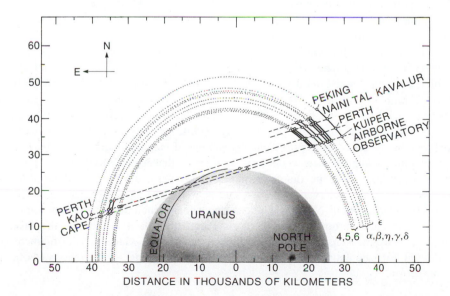

Figure 26–5 An artist's sketch of Uranus and its rings. (Drawn from data of J. L. Elliot, E. Dunham, L. H. Wasserman, R. L. Millis, and J. Churms.)

Figure 26–6 From Adams' diary, kept while he was in college: "1841. July 3. Formed a design in the beginning of this week, of investigating, as soon as possible after taking my degree, the irregularities in the motion of Uranus which are yet unaccounted for."

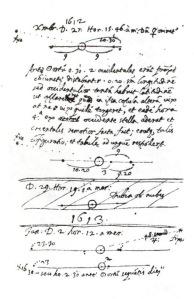

Figure 26–7 Galileo's notebook from late December 1612, showing a * marking an apparently fixed star to the side of Jupiter and its moons. The "star," at extreme left, was apparently Neptune. The horizontal line through Jupiter extends 24 Jupiter radii to each side. Jupiter's moons appear as dots on the line.

The episode shows the importance of keeping good lab notebooks, a lesson we should all take to heart.

26.2 Neptune

Neptune is even farther from the Sun than Uranus, 30 A.U. compared to about 19 A.U. Neptune takes 165 years to orbit the Sun. Its discovery was a triumph of the modern era of Newtonian astronomy. Mathematicians analyzed the deviations of Uranus (then the outermost known planet) from the orbit it would follow if gravity from only the Sun and the other known planets were acting on it. The small deviations could have been caused by gravitational interaction with another planet.

John C. Adams in England (Fig. 26–6) predicted positions for the new planet in 1845, but the astronomy professor at Cambridge did not bother to try to observe this prediction of a recent college graduate. Adams went to London to meet the Astronomer Royal (the chief British astronomer), but the A.R.'s butler did not permit dinner to be interrupted, and Adams went away. Though Adams left a copy of his calculations, when the Astronomer Royal requested further information, partly to test Adams' abilities, Adams did not take the request seriously and did not respond at first. The very "proper" Astronomer Royal took offense and did not choose to have further dealings with Adams. The story then continues in France, where a year later Urbain Leverrier was independently working on predicting the position of the undetected planet.

When the Astronomer Royal saw in the scientific journals that Leverrier's work was progressing well (Adams' work had not been made public), for nationalistic reasons he began to be more responsive to Adams' calculations. But the search for the new planet, though begun in Cambridge, was carried out half-heartedly. Neither did French observers take up the search. Leverrier sent his predictions to an acquaintance at Berlin, where a star atlas had recently been completed. The Berlin observer, Johann Galle, enthusiastically began observing and discovered Neptune within hours by comparing the sky against the new atlas.

Years of acrimonious nationalistic debate followed over who (and which country) should receive the credit for the prediction. (Adams and Leverrier themselves became friends.) We now credit both Adams and Leverrier. With hindsight, we see that they each assumed a radius for the new planet's orbit based on a numerological scheme (Bode's law) for the distances of planets. The value they assumed was incorrect, but luckily gave the right position anyway at that time.

Neptune has not yet made a full orbit since it was located in 1846. But it now seems that Galileo actually observed Neptune in 1613, which would more than double the period of time over which it has been observed. Charles Kowal, a Palomar astronomer, and Stillman Drake, a historian of science, tracked down Galileo's observing records from January 1613, when calculations showed that Neptune had passed close to Jupiter. Galileo, in fact, twice recorded in his notebooks stars that were very close to Jupiter (Fig. 26–7), stars that modern catalogues do not show. Galileo even once noted that one of the "stars" actually seemed to have moved from night to night, as a planet would. The objects that Galileo saw were very close to but not quite exactly where our calculations of Neptune's orbit show that Neptune would have been at that time. Presumably, Galileo saw Neptune, and we can use positions he measured to improve our knowledge of Neptune's orbit.

Neptune's angular size in our sky is so small that it is always very difficult to study. Even measuring its diameter accurately is hard, and is best done when it occults stars. From observations of the rate at which the stars dim, astronomers deduce information about Neptune's upper atmosphere, including its temperature and pressure structure. From the length of time that the star is hidden by the disk of Neptune, they measure Neptune's diameter. An accurate value for the diameter is important for calculating the planet's density, thus leading to deductions about its composition.

Figure 26–8 Neptune, photographed in 1983 with a CCD (Section 2.11) in the infrared light (8900 Å) that is strongly absorbed by methane gas in Neptune's atmosphere, making most of the planet appear dark. The 3 bright regions are clouds of methane ice crystals above the methane gas, reflecting sunlight. The CCD provided the infrared sensitivity necessary to take the picture with a short exposure. This photograph was one of the first to show Neptune's rotation period of 17 hours 50 minutes. (Image by Richard J. Terrile of JPL and Brad Smith of the University of Arizona with the 2.5-m du Pont Telescope in Chile)

Neptune, like Uranus, appears greenish in a telescope because of its atmospheric methane. Like Uranus, its rotational period is difficult to determine. Values from before Figure 26–8 ranged from 17 to 22 hours.

Structure on Neptune has been detected from the ground (Fig. 26–8) on images taken electronically. The images show bright regions in the northern and southern hemisphere that are probably discrete clouds separated by a dark equatorial band. Motion caused by Neptune's rotation can also be seen. Another sign that Neptune has weather is a set of drastic changes in its infrared brightness that was once observed to occur, even though its brightness in the visible remained fairly constant.

Neptune has two moons (Fig. 26–9). Triton (named after a sea god, son of Poseidon) is large, probably a little larger than our Moon. It is massive enough to have an atmosphere and a melted interior. Spectroscopy seems to show that Triton has a rocky rather than an icy surface, and a thin methane atmosphere. Since it apparently resembles Pluto, which has methane frost on its surface, Triton may have methane frost too. Dale Cruikshank and colleagues of the University of Hawaii have spectroscopically detected a lot of nitrogen, which could make up an ocean there. A second moon, Nereid (the Greek word meaning "sea nymph"), is much smaller, 600 kilometers across. It revolves in a very eccentric orbit with an average radius 15 times greater than Triton's.

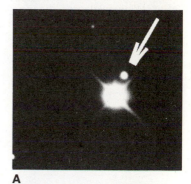

Figure 26–9 (A) Neptune (showing diffraction rings caused in the telescope) and its nearer satellite, Triton (shown with the arrow). (B) Nereid's orbit is much larger and more elliptical than that of Triton.

Does Neptune have rings, like the other giant planets? There is no obvious reason why it shouldn't. But some occultations have now been studied, and no rings were found (Fig. 26–10). The most recent occultation of a comparatively bright star (11th magnitude) was in 1983; my own site in Indonesia (where we were anyway to observe the solar eclipse 4 days previously) was clouded out, but good observations were obtained elsewhere. So Neptune does not have rings as extensive as those of Saturn or Uranus, though a ring as insubstantial as that of Jupiter would have escaped detection.

Voyager 2 will reach the Neptune system August 24, 1989, and will pass close to Neptune, Triton, and Nereid. If the instruments are still working, these points of light will then be transformed to images of objects easier to comprehend.

26.3 Pluto

Pluto, the outermost known planet, is a deviant. Its orbit is the most eccentric and has the greatest inclination with respect to the ecliptic plane, near which the other planets revolve.

Pluto will reach perihelion, its closest distance from the Sun, in 1989. Its orbit is so eccentric that part lies inside the orbit of Neptune. It is now on that part of its

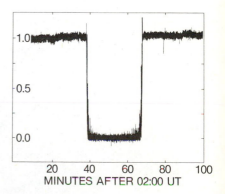

Figure 26–10 The 1981 occultation of a star by Neptune, observed at the 2.2-micron methane wavelength where Neptune is relatively dark compared with the star. The major dip is the occultation by the planet itself. No set of dips on either side appeared. The one dip on the right-hand side is too narrow to result from a ring.

Figure 26–11 Small sections of the plates on which Tombaugh discovered Pluto. On February 18, 1930, Tombaugh noticed that one dot among many had moved between January 23, 1930 *(left),* and January 29, 1930 *(right).*

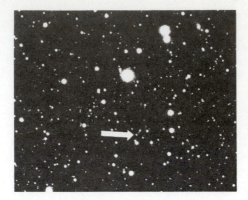 

Since Pluto is now within 10 years of perihelion out of its 248-year period, it is about as bright at opposition as it ever gets from Earth. It hasn't been as bright—about magnitude 13.5—for over 200 years. It should be barely visible through even a small telescope under dark-sky conditions.

orbit, and will remain there until 1999. Thus in a sense, Pluto is the eighth planet for a while, though the significant point is really that the semi-major axis of Pluto's orbit, 40 A.U., is greater than that of Neptune.

The discovery of Pluto was a result of a long search for an additional planet which, together with Neptune, was causing perturbations in the orbit of Uranus. Finally, in 1930, Clyde Tombaugh found the dot of light that is Pluto (Fig. 26–11) after a year of diligent study of photographic plates at the Lowell Observatory. From its slow motion with respect to the stars from night to night (Fig. 26–12), Pluto was identified as a new planet. Its period of revolution is almost 250 years.

26.3a Pluto's Mass and Size

Even such basics as the mass and diameter of Pluto are very difficult to determine. It has been hard to deduce the mass of Pluto because the procedure requires measuring Pluto's effect on Uranus, a more massive body. (The orbit of Neptune is too poorly known to be of much use.) Moreover, Pluto has made less than one revolution around the Sun since its discovery, thus providing little of its path for detailed study. As recently as 1968, it was concluded that Pluto had 91 per cent the mass of the Earth, though soon thereafter it was realized that the data were actually not reliable enough to make any conclusions about the value of Pluto's mass. Later studies of the orbit of Uranus indicated that Pluto's mass was 11 per cent that of Earth, but these observations were also very uncertain.

The situation changed drastically in 1978 with the surprise discovery that Pluto has a satellite. The presence of a satellite allows us to deduce the mass of the planet by applying Newton's form of Kepler's third law. In this age of space exploration, it is refreshing to see that important discoveries can be made with ground-based telescopes.

A U.S. Naval Observatory scientist, James W. Christy, was studying plates taken to refine our knowledge of Pluto's orbit. He noticed that some of the photo-

Figure 26–12 Pluto's motion can be seen in these photographs taken on successive nights.

graphs seemed to show a bump on the side of Pluto's image (Fig. 26–13). The bump was much too large to be a mountain, and turned out to be a moon orbiting with the period that we had previously measured for the variation of Pluto's brightness—6 days 9 hours 17 minutes. The elongated image of Pluto has since been seen both on older photographs of Pluto and on more recent ones in the positions that have been predicted on the basis of past observations. The moon has been named Charon, after the boatman who rowed passengers across the River Styx to Pluto's realm in Greek mythology (and pronounced "Sharon," similarly to the name of the discoverer's wife, Charlene). We will have to wait for the Space Telescope to see Pluto and Charon resolved from each other.

From the photograph we can get an approximate idea of the distance between Pluto and Charon. This measurement allows us to calculate the sum of the masses of Pluto and Charon. If we assume that Pluto and its moon have the same albedos and densities, we can compute the masses of each. Charon is 5 or 10 per cent of Pluto's mass, and Pluto is only 1/500 the mass of the Earth, ten times less than had been suspected even recently.

Until about the same time as the discovery of Charon, the best method for determining the radius of Pluto, as it is for Neptune, had been to observe a stellar occultation. Pluto passed near a 15th magnitude star in 1965, and this unique passage was observed very closely to see if the star would be occulted. But the star was never hidden from view. From this fact, astronomers knew that Pluto's radius is smaller in the sky than the smallest separation of the position of the center of Pluto and the position of the star, both of which were known accurately (Fig. 26–14). Since we know the distance to Pluto, simple trigonometry gave a limit to the radius of Pluto. This observation showed that Pluto had to be smaller than 6800 km across, which confirmed that Pluto was closer in size to the terrestrial than to the Jovian planets.

More recent infrared spectral studies have shown that Pluto is even smaller than this limit, for methane ice may be present on its surface. Since ice has a high albedo, Pluto could be smaller yet still reflect the amount of light that we measure. And some of the light comes from Pluto's moon, leaving less of the light to come from Pluto itself.

Only in 1979 was the diameter of Pluto measured directly. The technique of speckle interferometry, which involves studying a series of images taken rapidly enough to freeze out the effect of the Earth's turbulent atmosphere, was used with

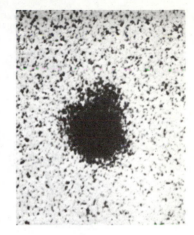

Figure 26–13 The discovery image of Charon and Pluto, taken on July 2, 1978. The bump at upper right was interpreted as the satellite. On the basis of photographs like this one, taken with the 1.5-m reflector of the U.S. Naval Observatory in Flagstaff, Arizona, the separation of Charon from Pluto has been estimated. From the separation and the period, a value of 0.2 per cent that of Earth has been derived for Pluto's mass.

Charon may be $\frac{1}{3}$ the size of Pluto, and is separated from Pluto by only about 8 Pluto diameters (compared to the 30 Earth diameters that separate the Earth and the Moon). So Pluto/Charon are almost a double-planet system.

The fact that no photographs taken since Charon's discovery show it any more clearly than the picture above illustrates why it had not been previously discovered.

Figure 26–14 Charon, revealed by computer reduction of observations made with a CCD on the University of Arizona's 1.5-m reflector. *(A)* An image of Pluto/Charon with contrast enhanced and contours drawn in. *(B)* A similar image of a star. Since we know the star is round, the apparent shape of this image shows how the telescope affects images. *(C)* The probable image of Pluto (calculated by scaling the star's image to Pluto's brightness) is subtracted out, leaving this residual image, which should be Charon alone. (Observations by H. J. Reitsema, F. Vilas, and B. A. Smith.)

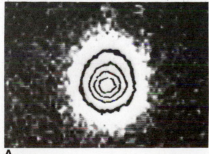

A

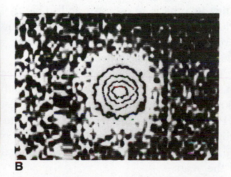

B

C

Figure 26–15 The fact that a 15th-magnitude star was not occulted by Pluto when the latter went nearby in 1968 gave us a limit for how large the diameter of Pluto could be. The result showed that Pluto was closer in size to the terrestrial than to the Jovian planets. Calculations based on the presence of ice on Pluto's surface, and speckle-interferometry measurements, have improved on this value.

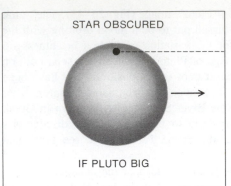 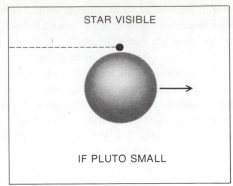

the 5-m Hale telescope. Pluto's diameter measured in this way is between 3000 and 3600 km. Pluto is much smaller than Mars (6800 km) or Mercury (4800 km), and is approximately the same size as our Moon (3500), adding to the evidence that we should perhaps not consider Pluto as a full-fledged planet. Speckle interferometry has also been used to measure the separation of Pluto and Charon; values of distance and mass agree with those given above.

26.3b What is Pluto?

The newer values of Pluto's mass and radius can be used to derive Pluto's density, which turns out to be very low, between 0.5 and 1 times the density of water. Since only ices have such low density, Pluto must be made of frozen materials. Its composition is thus more similar to that of the satellites of the giant planets than to that of the terrestrial planets. Spectral evidence indicates that non-icy materials (possibly silicates) are probably present in addition to methane ice.

Ironically, now that we know Pluto's mass, we can calculate that it is far too small to cause the perturbations in Uranus' orbit that originally led to Pluto's discovery. Thus the prediscovery prediction was actually wrong, and the discovery of Pluto was purely the reward of hard work in conducting a thorough search in a zone of the sky near the ecliptic.

Pluto is not massive enough to retain much of an atmosphere. But a tenuous atmosphere of methane has been detected from infrared spectral observations. The atmosphere could come from methane frost on its surface continually changing to methane gas. Since some astronomers think that even at Pluto's low temperature the methane gas would escape from Pluto's low gravity if it was pure, it may be mixed in with heavy gases like argon.

No longer does Pluto, with its moon and its atmosphere, seem so different from the other outer planets. Pluto remains strange in that it is so small next to the giants, and that its orbit is so eccentric and so highly inclined to the ecliptic. The new values of Pluto's mass and density revive the thinking that Pluto may be a former moon of one of the giant planets, probably Neptune, and escaped because of a gravitational encounter with another planet.

The orbits of Neptune and Pluto are affected at present by their mutual gravities in such a way that their orbital positions relative to each other repeat in a cycle every 20,000 years. The two planets can never come closer to each other than 18 A.U. But Pluto may still have broken away from Neptune when another planet passed nearby. The same event could have broken off a piece of Pluto to become Charon. In this model, their original joint orbit around the Sun has been modified by gravity over the years.

If we were standing on Pluto, the Sun would appear a thousand times fainter than it does to us on Earth. We would need a telescope to see the solar disk, which would be about the same size that Jupiter appears from Earth. No spacecraft to Pluto are planned, but we await observations with the Hubble Space Telescope.

Are there still further planets beyond Pluto? Tombaugh continued his search, and found none. But 1983's Infrared Astronomy Satellite, IRAS, recorded thousands of objects in the infrared. If a 10th planet exists, it would undoubtedly have been recorded, and may one day turn up in the data analysis.

Summary and Outline

Uranus (Section 26.1)

Large planet with low density and high albedo; discovered by Herschel in 1781; atmosphere of methane and molecular hydrogen; no surface structure can be seen; axis of rotation is near the plane of its orbit and retrograde; low or no internal heating; rings discovered at an occultation; 9 now known; 5 known moons; Voyager 2 arrives in January 1986

Neptune (Section 26.2)

Large planet with low density and high albedo; predictions by Adams and Leverrier—part science and part luck; atmosphere of methane and molecular hydrogen; 2 known moons; internal heating; no rings; Voyager 2 arrives in August 1989

Pluto (Section 26.3)

Most eccentric orbit in solar system; greatest inclination to ecliptic; rotation accurately determined due to varying albedo; discovered in 1930 after diligent search along the ecliptic; radius, mass (and thus density) difficult to determine; discovery of a moon (Charon) allows mass to be deduced; mass is only 0.2 per cent that of Earth; density thus is 0.7 that of water, similar to satellites of outer planets

Search for still further planets with IRAS

Questions

1. What is strange about the direction of rotation of Uranus?

2. Using Appendix 4, compare the sizes of the moons of Uranus and Neptune with other objects in the solar system.

3. Explain how the occultation of a star can help us learn the diameter of a planet.

4. Which planets are known to have internal heat sources?

5. What fraction of its orbit has Neptune traversed since it was discovered? Since it was first seen?

6. What fraction of its orbit has Pluto traversed since it was discovered?

7. What evidence suggests that Pluto is not a "normal" planet?

†8. At what wavelength would a cold body like a 10th planet give off the most radiation? What part of the spectrum is this and how does it affect the chance of discovering such a planet? Use a reasonable estimate for the temperature, and Wien's law.

9. Summarize the evidence that suggests that Pluto is not a giant planet.

10. From the separation of Pluto and Charon, show how to calculate the mass of Pluto.

---

†This indicates a question requiring a numerical solution.

Comet West, a bright comet that was visible to pre-dawn observers in the northern hemi-sphere in 1976.

Comets, Meteoroids, and Asteroids

27

Aims: To describe the non-planetary members of our solar system, and to see how their history may provide us with our best information about the origin of the solar system

Besides the planets and their moons, many other objects are in the family of the Sun. The most spectacular, as seen from Earth, are comets. Bright comets have been noted throughout history, instilling in observers great awe of the heavens.

It has been realized since the time of Tycho Brahe, who studied the comet of 1577, that comets are not merely in the Earth's atmosphere. From the fact that the comet did not show a parallax when observed from different locations on Earth, Tycho deduced that the comet was at least three times farther away from the Earth than the Moon.

Asteroids and meteoroids are other residents of our solar system. We shall see how they and the comets are storehouses of information about the solar system's origin.

27.1 Comets

Every few years, a bright comet fills our sky. From a small, bright area called the *head,* a *tail* may extend gracefully over one-sixth (30°) or more of the sky (Fig. 27-1). The tail of a comet is always directed away from the Sun.

Although the tail may give an impression of motion because it extends out only to one side, the comet does not move visibly across the sky as we watch. With binoculars or a telescope, however, an observer can accurately note the position of the head with respect to nearby stars, and detect that the comet is moving at a slightly different rate from the stars as comet and stars rise and set together. Within days or weeks a bright comet will have faded from below naked-eye brightness, though it can be followed for additional weeks with binoculars and then for additional months with telescopes.

Most comets are much fainter than the one we have just described. About a dozen new comets are discovered each year, and most become known only to astronomers. An additional few are ''rediscovered'' each year—that is, from the orbits derived from past occurrences, it can be predicted when and approximately where in the sky a comet will again become visible. Up to the present time, over 600 comets have been discovered.

Comets have long been seen as omens. ''When beggars die, there are no comets seen; The heavens themselves blaze forth the death of princes.''

Shakespeare, *Julius Caesar*

The ''long hair'' that is the tail led to the name *comet,* which comes from the Greek for ''long-haired star,'' *aster kometes.*

449

Figure 27–1 Comet Ikeya-Seki over Los Angeles in 1965, observed from the Mount Wilson Observatory. Photographs like this are taken with ordinary 35-mm cameras; this one was a 32-sec exposure on Tri-X film at f/1.6.

27.1a The Composition of Comets

At the center of a comet's head is its *nucleus,* which is at most a few kilometers across. It is composed of chunks of matter. The most widely accepted theory of the composition of comets, advanced in 1950 by Fred L. Whipple of the Harvard and Smithsonian Observatories, is that the nucleus is like a *dirty snowball*. The nucleus may be ices of such molecules as water (H_2O), carbon dioxide (CO_2), ammonia (NH_3), and methane (CH_4), with dust mixed in.

Figure 27–2 The head of Halley's Comet in 1910.

This model explains many observed features of comets, including why the orbits of comets do not appear to accurately follow the laws of gravity. When sunlight evaporates the ices, molecules are expelled from the nucleus. This action generates an equal and opposite reaction, the same force that runs jet planes. Since the comet nucleus is rotating, the force is not always directly away from the Sun, even though the evaporation is triggered on the sunny side. So comets show the effects of non-gravitational forces in addition to the effect of solar gravity.

The nucleus itself is so small that we cannot observe it directly from Earth. Radar observations have verified that it is a few km across. The rest of the head is the *coma* (pronounced cō′ma), which may grow to be as large as 100,000 km or so across (Fig. 27–2). The coma shines partly because its gas and dust are reflecting sunlight toward us and partly because gases liberated from the nucleus are excited enough by sunlight that they radiate.

With spacecraft, we became able to observe in comets the Lyman-alpha line of hydrogen in the ultraviolet. In 1970 we discovered that a huge hydrogen cloud a million km in diameter (Fig. 27–3) surrounds a comet's head. The hydrogen cloud probably results from the breakup of water molecules by ultraviolet light from the

A

B

Figure 27–3 These contours of intensity result from photographs of Comet Kohoutek taken by the Skylab astronauts outside the Earth's atmosphere. *(A)* The photograph was taken in the Lyman alpha line of hydrogen at 1216 Å in the ultraviolet. *(B)* These contours result from a photograph that was taken immediately afterward at exactly the same scale, but through a filter that did not pass the Lyman alpha line. The comparison shows that a huge hydrogen halo, 1° across or 2,500,000 km in diameter, was present. In *B*, we see the tail, 2° or 5,000,000 km in length. Hot stars from the background constellation, Sagittarius, also show.

Box 27.1 Discovering a Comet

A comet is generally named after its discoverer, the first person (or spacecraft) to detect it (or the first two or three if they independently find it within an interval of days). Comets are also temporarily assigned letters in their order of discovery in a given year. Then, a year or two later when all the comets that passed near the Sun in that given year are likely to be known, Roman numerals are assigned in order of their perihelion passage.

Many discoverers of comets are amateur astronomers, including some who examine the sky each night with large binoculars in hope of finding a comet. To do this, one must know the sky very well, so that one can tell if a faint, fuzzy object is a new comet or a well-known nebula. It is for that reason that Messier made his famous eighteenth-century list of nebulae (Appendix 8).

If you find a comet, telegraph or Telex the International Astronomical Union Central Bureau for Astronomical Telegrams, at the Smithsonian Astrophysical Observatory in Cambridge, Massachusetts (Telex II = TWX 710-320-6842 ASTROGRAM CAM) (Telegraph address: CENTRAL BUREAU FOR ASTRON, CAMBRIDGE MASS). Or you may telephone (617) 495-7244, though this is a less desirable method. In any case, you should identify the direction of the comet's motion, its position, and its brightness. Don't forget to identify yourself by giving your name, address, and telephone number. If you are the first (or maybe even the second or third) to find the comet, it will be named after you.

The record for discovering comets belongs to the former caretaker of the Marseilles Observatory, Jean Louis Pons, who discovered 37 between 1801 and 1827. Several Japanese amateur astronomers have found many comets: Minoru Honda has found a dozen. After work each night, he spends many hours scanning the sky. But a comet may become sufficiently bright for discovery only while it is daylight or cloudy in Japan, so there is hope for less dedicated observers.

Many comets, particularly the ones discovered by professional astronomers, are discovered not by eye but only by examination of photographs taken with telescopes. The photographs were usually taken for other purposes, so the discovery of a comet exemplifies serendipity—a fortuitous extra discovery. IRAS found half a dozen comets.

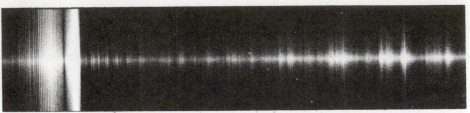

Figure 27–4 Part of the spectrum of Comet Tago-Sato-Kosaka, photographed in 1969 at the European Southern Observatory in Chile. Molecules give not just single lines but rather bands (groups) of lines; one of the bands of CN is visible at the left. Bands of CH and C_3 are seen in the rest of the photograph.

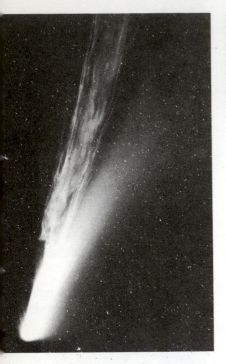

Figure 27–5 In Comet Mrkos in 1957, the straight *ion tail*, extending toward the top, and the *dust tail*, gently curving toward the right, were clearly distinguished.

Figure 27–6 These tiny extraterrestrial dust particles, magnified about 5000 times by a scanning electron microscope, were captured at an altitude of 20 km, as the particles drifted downward. Most of the particles are 0.01 mm across and weigh a billionth of a gram. Most of the particles were fluffy like this one. Each contains perhaps a million grains of different minerals.

Sun. In the visible, we have long been able to detect spectral lines from simple molecules (Fig. 27–4).

The tail can extend 1 A.U. (150,000,000 km), so comets can be the largest objects in the solar system. But the amount of matter in the tail is very small—the tail is a much better vacuum than we can make in laboratories on Earth.

Many comets actually have two tails (Fig. 27–5). Both extend generally in the direction opposite to that of the Sun, but are different in appearance. The *dust tail* is caused by dust particles released from the ices of the nucleus when they are vaporized. The dust particles are left behind in the comet's orbit, blown slightly away from the Sun by the pressure caused by photons of sunlight hitting the particles. As a result of the comet's orbital motion, the dust tail usually curves smoothly behind the comet.

The *gas tail* (also called the *ion tail*) is composed of ions blown out more or less straight behind the comet by the solar wind. As puffs of ionized gas are blown out and as the solar wind varies, the ion tail takes on a structured appearance. Each puff of matter can be seen. Magnetic fields in the solar wind carry only ionized matter along with them; the neutral atoms are left behind in the coma.

Dust that may be from comets (Fig. 27–6) has been captured by sending up sticky plates on a NASA U-2 aircraft sent high above terrestrial pollution. The abundances of the elements in the dust are similar to those of meteorites rather than to terrestrial material. The particles probably date back to the origin of the solar system. One reason to think so is the evidence of small deviations in the abundances of certain elements compared to the usual solar-system abundances. The extra amounts of certain elements would have been caused by a supernova exploding shortly before the dust's formation. This supernova may have triggered the formation of our solar system.

A comet—head and tail together—contains less than a billionth of the mass of the Earth. It has been said that comets are as close as something can come to being nothing.

27.1b The Origin and Evolution of Comets

It is now generally accepted that trillions of incipient comets surround the solar system in a sphere perhaps 50,000 A.U. (almost 1 light year) in radius. This sphere is known as the *Oort comet cloud* after Jan H. Oort, the Dutch astronomer who advanced the theory in 1950. The total mass of matter in the cloud is only 1 to 10 times the mass of the Earth. Occasionally one of the incipient comets leaves the Oort cloud, perhaps because gravity of a nearby star has tugged it out of place, and the comet approaches the Sun in a long ellipse. The comet's orbit may be altered if it passes near a Jovian planet. Because the Oort cloud is spherical, comets are not limited to the plane of the ecliptic and come in randomly from all angles.

As the comet gets closer to the Sun, the solar radiation begins to vaporize the molecules in the nucleus. We have not generally been able to detect the molecules in the nucleus directly—the *parent molecules*—though we would clearly love to do so. However, we have mostly detected in the head the simpler molecules into which the parent molecules break down—the *daughter molecules*. Examples of daughter molecules are H, OH, and O, which might be results of the breakdown of the parent H_2O. Similarly, daughters NH and NH_2 can result from parent NH_3.

The tail forms, and grows longer as more of the nucleus is vaporized. Even though the tail can be millions of km long, it is still so tenuous that only 1/500 of the mass of the nucleus may be lost. Thus a comet may last for many passages around the Sun. But some comets may hit the Sun and be destroyed (Fig. 27–7).

The comet is brightest and its tail is generally longest at perihelion. However, because of the angle at which we view the tail from the Earth, it may not appear the longest at this stage. Following perihelion, as the comet recedes from the Sun, its tail fades; the head and nucleus receive less solar energy and fade as well. The comet may be lost until its next return, which could be as short as 3.3 years (Encke's Comet) or as long as 80,000 years (Comet Kohoutek) or more. With each reappearance, a comet loses a little mass and eventually disappears. We shall see in Section 27.2 that some of the meteoroids are left in its orbit. Some of the asteroids, particularly those that cross the Earth's orbit, may be dead comet nuclei.

Comets are unpredictable. Comet Kohoutek, which was a popular bust and an astronomer's delight, bore out that idea. In 1974, it became bright enough to study from Skylab (Color Plate 17), and from the ground. But it didn't become the dazzling naked-eye object once predicted. Bright comets ordinarily appear with much less notice. In 1975, Richard West of the European Southern Observatory discovered a comet that within months became an object very easy to see with the naked eye (Color Plate 23).

In 1983, the IRAS satellite and then two amateur astronomers independently discovered a comet that passed exceptionally close to Earth, 4,600,000 km, making it the closest comet since 1770. On only a few days' notice, Comet IRAS-Araki-Alcock became visible to the naked eye, appearing for a couple of days like a hazy

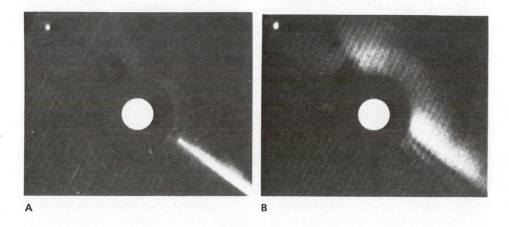

A **B**

Figure 27–7 *(A)* A coronagraph in orbit photographed this comet heading for the Sun in 1979. A dark disk made an artificial eclipse of the Sun; a white spot the size of the Sun has been added. The comet disappeared behind this disk at 1 million km/hr and did not re-emerge. *(B)* Eleven hours after the impact, this cometary material appeared. Although it looks like a splash, it may simply be a part of the comet's tail that had been blown into view by the pressure of the Sun's radiation. There may also be some contribution from a "coronal transient," a solar eruption. The white dot at upper left is Venus.

Figure 27–8 1983's Comet IRAS-Araki-Alcock came so close to the Earth that it moved 60° across the sky in a single day. It looked like a cloud in the sky three times the diameter of the moon.

cloud (Fig. 27–8). Scientists bounced radar off its nucleus, confirming the prediction of Whipple's model that solid material is present in cometary nuclei.

Because new comets come from the places in the solar system that are farthest from the Sun and thus coldest, they probably contain matter that is unchanged since the formation of the solar system. So the study of comets is important for understanding the birth of the solar system.

27.1c Halley's Comet

In 1705, the English astronomer Edmond Halley (Fig. 27–9) applied a new method developed by his friend Isaac Newton to determine the orbits of comets from observations of their positions in the sky. He reported that the orbits of the bright comets that had appeared in 1531, 1607, and 1682 were about the same. Because of this, and because the intervals between appearances were approximately equal, Halley suggested that we were observing a single comet orbiting the Sun, and predicted that it would again return in 1758. The reappearance of this bright comet on Christmas night of that year, 16 years after Halley's death, was the proof of Halley's hypothesis (and Newton's method); the comet has since been known as Halley's Comet. It seems probable that the bright comets reported every 74 to 79 years since 87 B.C. (and possibly even in 240 B.C.) were earlier appearances. The fact that Halley's Comet has been observed at least 27 times endorses the calculations that show that less than 1 per cent of a cometary nucleus' mass is lost at each perihelion passage.

Halley's Comet went especially close to the Earth during its 1910 return (Fig. 27–11), and the Earth actually passed through its tail. Many people had been frightened that the tail would somehow damage the Earth or its atmosphere, but the tail had no noticeable effect. Even then, most scientists knew that the gas and dust in the tail were too tenuous to harm our environment.

Figure 27–9 Edmond Halley.

Halley's Comet is in view, and we are awaiting the months around its February 9, 1986, perihelion (Figs. 27–10 and 27–12). NASA's plans to launch a spacecraft to fly nearby were cancelled for financial reasons. It would have taken magnificent close-up pictures, penetrated the nucleus, and sampled parent molecules directly.

The European Space Agency is sending a spacecraft, Giotto (after the 14th-century Italian artist who included Halley's Comet in a painting). The Japanese are also sending a pair of spacecraft. A Soviet pair of spacecraft, Vega ("Ve" from "Venus" and "ga" from "Halley," given that there is no "h" in Russian and the comet's name begins with a "g" sound in that language), will proceed from Venus to Halley's Comet.

Halley's Comet will not be as spectacular from the ground in 1986 as it was in 1910, for the Earth and Comet will be on opposite sides of the Sun when the comet is brightest. Thus the comet is not expected to appear spectacular to the unaided eye. To see it, you will need dark skies away from city lights; also, binoculars will help.

Though no U.S. spacecraft will go close to Halley's Comet, Pioneer Venus with its small ultraviolet telescope will be turned from Venus toward it. And an existing U.S. satellite (International Sun-Earth Explorer 3, now renamed International Cometary Explorer) has been sent to reach the tail of another comet—Comet Giacobini-Zinner—on September 11, 1985. We will get useful data, but not the fabulous pictures we have grown to expect. The launch of Space Telescope will come too late for Halley's Comet, though the University of Wisconsin and Johns Hopkins University are providing ultraviolet telescopes to view the comet from Spacelab.

Mark Twain was born during the appearance of Halley's Comet in 1835, and often said that he came in with the comet and would go out with it. And so he did, dying during the 1910 return of Halley's Comet.

Figure 27–10 *(A)* The appearance of Halley's Comet in 1985/6 from a latitude of 30°N. From 40°N, the Comet will be 10° lower in the sky on each date. Predictions for the comet's approximate total visual magnitude, including the tail, are given in parentheses. From southern latitudes, the Comet will appear much higher in the sky. *(B)* When Halley's Comet reaches perihelion, 0.6 A.U., it will be on the opposite side of the Sun from the Earth. When the comet is closest to the Earth, it will be much farther from the Sun. The part of the Comet's orbit south of the plane of the Earth's orbit is dotted.

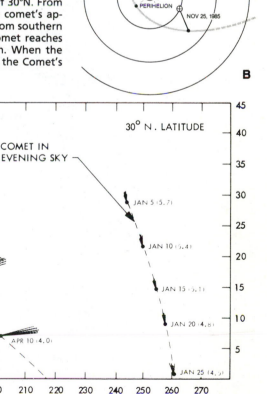

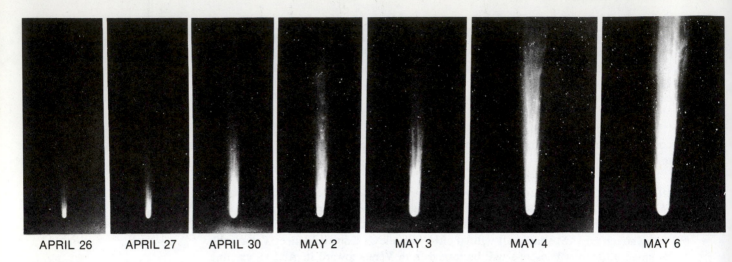

APRIL 26   APRIL 27   APRIL 30   MAY 2   MAY 3   MAY 4   MAY 6

Figure 27–11 Halley's Comet in 1910.

Figure 27–12 (A) For some time before its perihelion, Halley's Comet's path seems to circle in the sky as a result of the Earth's revolution around the Sun. At its closest (B), its apparent motion will grow more rapid.

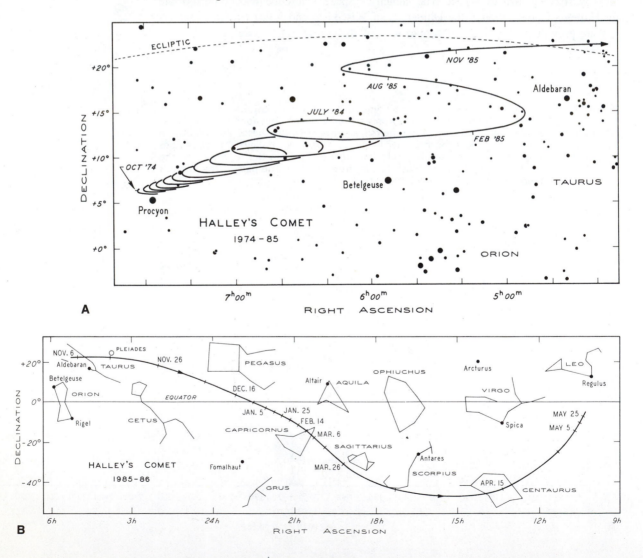

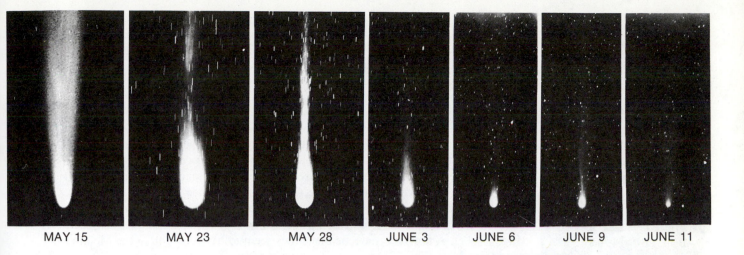

MAY 15 MAY 23 MAY 28 JUNE 3 JUNE 6 JUNE 9 JUNE 11

27.2 Meteoroids

There are many small chunks of matter in interplanetary space, ranging up to tens of meters across. When these chunks are in space, they are called *meteoroids*. When one hits the Earth's atmosphere, friction slows it down and heats it up—usually at a height of about 100 km—until all or most of it is vaporized. Such events result in streaks of light in the sky, which we call *meteors* (popularly known as *shooting stars*). The brightest meteors can reach magnitude −15 or −20, brighter than the full moon. We call such bright objects *fireballs* (Fig. 27–13). We can sometimes even hear the sounds of their passage and of their breaking up into smaller bits. When a fragment of a meteoroid survives its passage through the Earth's atmosphere, the remnant that we find on Earth is called a *meteorite*. We now also refer to objects that hit the surface of the Moon or of other planets as meteorites.

27.2a Types and Sizes of Meteorites

Tiny meteorites less than a millimeter across, *micrometeorites,* are the major cause of erosion on the Moon. Micrometeorites also hit the Earth's upper atmosphere all the time, and remnants can be collected for analysis from balloons or airplanes. The micrometeorites are thought to be debris from comet tails. They may have been only the size of a grain of sand, and are often sufficiently slowed down that they are not vaporized before they reach the ground. The resulting dust—100 tons a day of it—can be sampled by collecting ice from the Arctic or Antarctic, or from mountain tops.

Space is full of meteoroids of all sizes, with the smallest being most abundant. Most of the small particles, less than 1 mm across, come from comets. Most of the large particles, more than 1 cm across, come from collisions of asteroids in the asteroid belt. It is generally these larger meteoroids that become meteorites. Some meteorite types can even be assigned, from matching the spectrum of reflected sunlight, to certain asteroids.

There are several kinds of meteorites. Most of the meteorites that are found have a very high iron content—about 90 per cent; the rest is nickel. These *iron meteorites* (*irons,* for short) are thus very dense—that is, they weigh quite a lot for a given volume.

Most meteorites that hit the Earth are stony in nature, and are often referred to simply as *stones*. Most of these stony meteorites are of a type called *chondrites,*

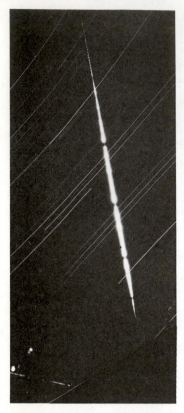

because they contain rounded particles called "chondrules." The stony meteorites without chondrules are called *achondrites;* they may have once been chondrites, but melted and cooled. Because stony meteorites resemble ordinary rocks and disintegrate with weathering, they are not usually discovered unless their fall is observed. That explains why most meteorites discovered at random are irons. But when a fall is observed, most meteorites recovered are stones. The stony meteorites have a high content of silicates; only about 10 per cent of their mass is nickel and iron.

A large terrestrial crater that is obviously meteoritic in origin is the Barringer Crater in Arizona (Fig. 27–14). It resulted from what was perhaps the most recent large meteor to hit the Earth, for it was formed only 25,000 years ago.

Every few years a meteorite is discovered on Earth immediately after its fall (Fig. 27–15). The chance of a meteorite's landing on someone's house is very small, but it has happened (Fig. 27–16)! Many meteorites have recently been found in the Antarctic, where they have been well-preserved as they accumulated over the years (Fig. 27–17). A few odd Antarctic meteorites seem to have come from the Moon or even from Mars (Fig. 27–18).

*27.2b Carbonaceous Chondrites

Objects that contain large quantities of carbon are *carbonaceous. Carbonaceous chondrites*—stony meteorites with a high carbon content are rare. One such carbonaceous chondrite—the Murchison Meteorite—contains simple amino acids, building blocks of life. It fell near Murchison, Victoria, Australia, in 1969. Analysis of these amino acids showed that they were truly extraterrestrial and had not simply contaminated the sample on Earth. The formation of such complicated organic chemicals in cold, isolated places like meteoroids is one of the several indications that the precursors of life develop naturally. We shall be developing more of this evidence in the following chapter. Some of the carbonaceous chondrites may be cometary debris. The largest carbonaceous chondrite available for scientific study is the Allende meteorite, which fell in Mexico in 1969.

We measure the ages of meteorites by studying the ratios of radioactive and non-radioactive isotopes in them, just as we date lunar or terrestrial rocks. The measurements show that the meteoroids were formed up to 4.6 billion years ago, the beginning of the solar system. The chondrites may date from the formation itself; some may be pieces of planetesimals. The abundances of the elements in meteorites thus tell us about the solar nebula from which the solar system formed. In fact, up

Figure 27–13 A fireball observed in 1970 by the "Prairie Network" of the Smithsonian Astrophysical Observatory, a network of wide-field cameras spaced around the mid-west of the United States in hope of pinpointing a meteor's path and thus permitting the meteorite to be found. A shutter in the camera rotates so that breaks occur in the image of the meteor trail at intervals of $\frac{1}{20}$ sec. (These short breaks are visible at the top, but are blurred out lower on the photo, where only periodic longer breaks show.) The trail persisted for about 8 seconds. The fireball pictured here led to the discovery of a meteorite at Lost City, Oklahoma.

Figure 27–14 The Barringer meteor crater in Arizona. It is 1.2 km in diameter. Dozens of other terrestrial craters are now known, many from aerial or space photographs. The largest may be a depression over 400 km across under the Antarctic ice pack, comparable with lunar craters. Another very large crater, in Hudson Bay, Canada, is filled with water. Most are either disguised in such ways or have eroded away.

Figure 27–15 A bright fireball was observed and photographed on February 5, 1977. Analysis indicated that it probably landed near Innisfree, Alberta, Canada. Scientists flew out to search the snow-covered wheat fields and found this 2-kg meteorite. It is only the third meteorite recovered whose previous orbit around the Sun is known.

Figure 27–16 A meteorite landed in 1982 in Wethersfield, Connecticut, coming through the roof and ceiling and bouncing off a bench.

Figure 27–17 Scientists picking up meteorites in the Antarctic. The meteorites have probably been long buried in ice, and are made visible when wind erodes the ice. Thousands of meteorites were found in this expedition.

Tektites, small, rounded glassy objects that are found at several locations on Earth, may have splashed out from meteorite impacts. The origin of tektites, including whether they came from the Earth or from the Moon, has long been controversial, though a terrestrial origin seems most probable.

to the time of the Moon landings, meteorites were the only extraterrestrial material we could get our hands on.

Most of our knowledge of the abundances of the elements in the solar system has come either from analysis of the solar spectrum or from analysis of meteorites. Until recently it was thought that all the analyses showed that abundances of each element were uniform throughout the solar nebula. But evidence is growing, including in particular the high quality data from the Allende meteorite, that abundances varied from place to place in the solar nebula. The abundances of isotopes of common elements like oxygen may vary by 5 per cent; the abundances of rare earths (elements 57–71) can vary by greater amounts.

27.2c Meteor Showers

Meteors often occur in *showers*, when meteors are seen at a rate far above average. On any clear night a naked-eye observer may see a few *sporadic* meteors an hour, that is, meteors that are not part of a shower. (Just try going out to a field in the country and watching the sky for an hour.) During a shower several meteors may be visible to the naked eye each minute, though this is rare. Meteor showers occur at the same time each year (Table 27–1), and represent the Earth's passing through the orbits of defunct comets and hitting the meteoroids left behind. Meteorites do not usually result from showers, so presumably the meteoroids that cause a shower are very small.

Figure 27–18 This meteorite found in Antarctica is made of material that makes us think it is a rock from Mars. It is 1.3 billion years old, one of 8 similar objects known, whereas most other meteorites are 4.5 billion years old. The separate discovery of a meteorite that probably came from the Moon (Fig. 20–19) makes it likely that other meteorites hitting the Moon or Mars can blast rocks off them without pulverizing the rocks.

Table 27–1 Meteor Showers*

Name	Date of Maximum	Duration Above 25% of Maximum	Approximate Limits	Number per Hour at Maximum
Quadrantids	Jan 4	1 day	Jan 1–6	110
Lyrids	April 22	2 days	April 19–24	12
Eta Aquarids	May 5	3 days	May 1–8	20
Delta Aquarids	July 27–28	7 days	July 15–Aug 15	35
Perseids	Aug 12	5 days	July 25–Aug 18	68
Orionids	Oct 21	2 days	Oct 16–26	30
Taurids	Nov 8	spread out	Oct 20–Nov 30	12
Leonids	Nov 17	spread out	Nov 15–19	10
Geminids	Dec 14	3 days	Dec 7–15	58

*The number of sporadic meteors per hour is 7 under perfect conditions. The visibility of showers depends mostly on how bright the Moon is on the date of the shower, which depends on its phase. Meteors are best seen with the naked eye; using a telescope or binoculars merely restricts your field of view.

Many theories have been advanced to explain an explosion near the Tunguska River in Siberia in 1908. Most scientists think it was a fragment of a meteoroid or comet entering our atmosphere. Since no remnant has been found, even though trees were blown outward for kilometers around, it may rather have been a small fragment of a comet, and vaporized completely, and/or exploded high above the Earth's surface. Signs of dust from the explosion have been found at widely separated regions of the Earth.

The rate at which meteors are usually seen increases after midnight on the night of a shower, because that side of the Earth is then turned so that it plows through the oncoming interplanetary debris. If the Moon is gibbous or full, then the sky is too bright to see the shower well. The meteors in a shower are seen in all parts of the sky, but their trajectories all seem to emanate from a single point in the sky, the *radiant.* A shower is usually named after the constellation that contains its radiant. The existence of a radiant is just an optical illusion; perspective makes the set of parallel paths approaching us appear to emanate from a point.

27.3 Asteroids

The nine known planets were not the only bodies to result from the agglomeration of planetesimals 4.6 billion years ago. Thousands of *minor planets,* called *asteroids,* also resulted. They are detected by their small motions in the sky relative to the stars (Fig. 27–19).

Most of the asteroids have elliptical orbits between the orbits of Mars and Jupiter, in a zone called the *asteroid belt* (Fig. 27–20). Indeed, it was not a surprise when the first asteroid was discovered, on January 1, 1801, because Bode's law, a numerological formula for the sizes of planetary orbits (still without known justification), predicted the existence of a new planet in approximately that orbit.

The first asteroid was discovered by a Sicilian clergyman/astronomer, Giuseppe Piazzi, and was named Ceres after the Roman goddess of harvests. Though Piazzi followed Ceres' motion in the sky for over a month, he became ill and the object moved into evening twilight. Ceres was lost. But the problem led to a great advance in mathematics: the young mathematician Carl Friedrich Gauss developed an important way of plotting the orbits of objects given only a few observations. The orbit Gauss worked out led to the rediscovery of Ceres. Modern versions of Gauss' methods are still in use today, though calculations are now done by computer rather than by hand.

Though the discovery of Ceres was a welcome surprise, even more surprising was the subsequent discovery in the next few years of three more asteroids. Bode's law hadn't called for them! The new asteroids were named Pallas, Juno, and Vesta,

Figure 27–19 Asteroids leave a trail on a photographic plate when the telescope is tracking at a sidereal rate, that is, with the stars.

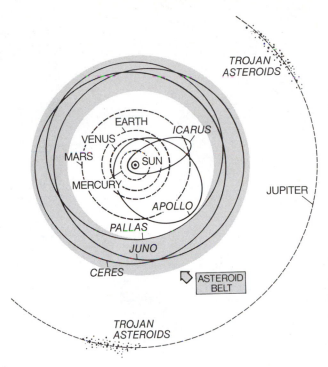

Figure 27–20 Some asteroid orbits. The asteroid belt, in which most asteroids lie, is shaded.

also after goddesses, which began the generally observed tradition of assigning female names to asteroids. The next asteroids weren't discovered until 1845, and over 2000 are now named and numbered. They are assigned numbers when their orbits become well determined. The number and name of an asteroid are often listed together: 1 Ceres, 16 Psyche, and 433 Eros, for example. Asteroids rarely come within a million km of each other, though occasionally collisions do occur, producing meteoroids.

Only half a dozen known asteroids are larger than 300 km across, and over 200 more are larger than 100 km across. The largest asteroids known (Fig. 27–21) are the size of some of the moons of the planets. Small asteroids may be only 1 km or less across. All the asteroids together contain less mass than the Moon. Perhaps 100,000 asteroids could be detected with Earth-based telescopes if we wanted to work on it.

Box 27.2 *Measuring the Sizes of Asteroids*

The most accurate way to measure the size of an asteroid is to follow its passage in front of a star. The stars are so far away that the shadow of the asteroid projected on the Earth in the light of this star is the full size of the asteroid. By timing when the star disappears and reappears, we can measure its diameter along one line across the asteroid. If several observers time the occultation and reappearance from different positions on the Earth, we get enough information to plot an accurate shape (Fig. 27–22).

Accurate measurements of the diameter of an asteroid are important for computing its albedo and its density. Sufficiently accurate knowledge of the albedo and the density allows us to distinguish among carbonaceous, stony and nickel-iron materials.

Figure 27–21 The sizes of the larger asteroids and their relative albedos.

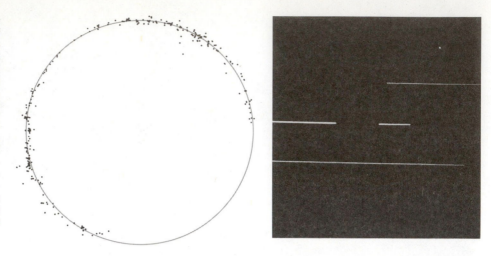

Figure 27–22 The points show the individual observations of dozens of amateur observers who timed the moments at which a star disappeared and reappeared when it was occulted by the asteroid Pallas in 1983. The result is best fit by an ellipse 520 by 516 km. The star is so far away that the shadow of the asteroid it casts on the Earth is essentially the same as the asteroid itself. (Paul D. Maley, International Occultation Timing Association) (Right) Star trails at the occultation of a star by Pallas. Pallas blocked the star for 43 seconds at this observing site; other stars were not blocked.

Box 27.3 Phoebe, a Captured Satellite

Voyager 2 revealed much about Saturn's outermost satellite, Phoebe (Fig. 27–23). Phoebe is 200 km across and very dark (albedo about 5 per cent). Ground-based observations had showed that Phoebe orbits Saturn in the ecliptic rather than in Saturn's equatorial plane, in which the other satellites orbit Saturn, and that Phoebe's orbit is retrograde, unlike the orbits of Saturn's other satellites. All this evidence indicates that Phoebe is actually a captured asteroid rather than a satellite formed in Saturn's original nebula along with the other satellites.

The Pioneers and Voyagers en route to Jupiter and beyond travelled through the asteroid belt for many months and showed that the amount of dust among the asteroids is not increased over the amount of interplanetary dust in the vicinity of the Earth. Particles the size of dust grains are only about three times more plentiful in the asteroid belt, and smaller particles are less plentiful. So the asteroid belt will not be a hazard for space travel to the outer parts of the solar system.

Figure 27–23 A Voyager 2 view of Saturn's moon Phoebe, taken on September 4, 1981, from a distance of 2.2 million km. Phoebe is 200 km in diameter, twice the size previously believed. Its albedo is only 5 per cent.

Box 27.4 The Extinction of the Dinosaurs

On Earth 65 million years ago, dinosaurs and many other species were extinguished. Evidence has recently been accumulating that these extinctions were sudden and were caused by an Apollo asteroid hitting the Earth. Among the signs is the worldwide distribution of the element iridium released in the impact. The impact would have raised so much dust into the atmosphere that sunlight could have been shut out for months.

Other evidence exists, on the other hand, that the species may have disappeared more gradually, and that an asteroid was therefore not the cause. Research is continuing.

A still newer idea is based on a possible periodicity of 28 million years between mass extinctions on Earth. It has been suggested that a faint, undiscovered companion to the sun comes to the inner part of its orbit with that period, and its gravity then sends a number of comets in toward the Earth where they crash. (Fortunately, the star isn't due back for 15 million years.) Obviously, a lot of verification for both the existence of periodic extinctions and for this theory would have to be found before it is generally accepted.

*27.3a Special Groups of Asteroids

Aten, *Apollo*, and about half of the *Amor asteroids* have orbits that cross the orbit of the earth. The Aten asteroids (Fig. 27–24), further, have orbits whose major axes are smaller than Earth's. (The perihelions of Amors are slightly farther from the sun than those of Apollos.) We have observed a few dozen asteroids of these types. We see them only when they come close to Earth and there may be 1000 in all. Two asteroids even go inside the orbit of Mercury. The first to be found was named Icarus, after the inventor in Greek mythology who flew too near the Sun. On June 14, 1968, Icarus came within 6 million km of the Earth; others have since come closer, one only 800,000 km of us. We may be able to send a spacecraft to such a closely approaching asteroid during the next decade.

Earth-crossing asteroids may well be the source of most meteorites, which could be debris of collisions when these asteroids visit the asteriod belt. Eventually, most earth-crossing asteroids will probably collide with the Earth. Luckily, we think that only a few dozen of them are greater than 1 km in diameter; statistics show that collisions of this tremendous magnitude should take place every few hundred million years, and could have drastic consequences for life on earth.

Another asteroid type is the Trojans, which always oscillate about the points 60° ahead of or behind Jupiter in their solar orbits (the ''Lagrangian points''). At the Lagrangian points, the gravitational pulls of the Sun and of Jupiter balance, so material does not move toward one or the other.

*27.3b The Compositions of Asteroids

Since 1970, new methods have been used to measure the sizes and albedos of asteroids. One method is based on the amount of infrared radiation emitted by asteroids at wavelengths of 10 or 20 microns. A second is based on the polarization of visible light, since the polarization of darker asteroids seems to change more as the asteroid's position changes than it does for lighter ones. Some asteroids are as dark as coal, with albedos of only 3 per cent. Indeed, the surface may be covered with carbon or carbon compounds. Other asteroids have albedos of 50 per cent.

Such considerations, and also asteroid spectra, lead us to the conclusion that asteroids are made of different materials from each other, and represent the chemical compositions of different regions of space. The asteroids at the inner edge of the asteroid belt are mostly stony in nature, while the ones at the outer edge are darker (as a result of being more carbonaceous). Most of the small asteroids that pass near the Earth—Icarus, Eros, Toro, Geographos, and Alinda—are among the stony group. Three of the largest asteroids—Ceres, Hygiea, and Davida—are among the carbonaceous group. A third group may be mostly composed of iron and nickel. The differences may be a direct result of the variation of the solar nebula with distance from the protosun. Differences in chemical composition also show that we must discard the old theory that the asteroids represent the breakup of a planet that once existed between Mars and Jupiter.

Asteroids have also had different types of histories since their formation. 4 Vesta, for example, seems to be covered with basalt, which would have resulted from lava. Since lava comes from internal melting, some asteroids are thus differentiated, that is, layered. However, we think that Vesta seems too small to have been heated by radioactive decay to the 1000 K necessary to cause the lava. And if radioactive materials were distributed uniformly in the asteroid belt, we do not understand why other big asteroids like Pallas and Ceres do not have basalt on their surfaces as well. Perhaps the distribution of short-lived radioactive isotopes—like one of aluminum—was not uniform. The asteroids and their meteoritic offspring may provide the data to show us the details of how the solar system was formed.

Figure 27–24 The streak shows the discovery of the Aten asteroid Ra-Shalom, which is in an Earth-crossing orbit. It was discovered in a search for high-speed objects with the small Schmidt camera at Palomar. Discovered at the time of the Israeli-Egyptian accords, Ra-Shalom was named as a symbol of hope for peace in the Middle East.

Figure 27–25 Chiron. This is the plate on which it was discovered. It was quite a surprise that a new type of object should be discovered within our solar system.

When the solar system was forming, proto-Jupiter probably became so massive that it disrupted a large region of space around it. The planetesimals in that region may have been sent into eccentric orbits inclined to the ecliptic plane to become the asteroids.

27.4 Chiron

Chiron (kī'ron), a small object in our outer solar system (Fig. 27–25), was discovered in 1977. (In Greek mythology, Chiron was the wisest of the centaurs and the teacher of Achilles.) Calculations of its orbit permitted observations to be found as far back as 1895, so its orbit is now well-determined. It has a 50-year period and an eccentricity of 0.38. Its distance from the Sun ranges from 8.5 A.U., within the orbit of Saturn, to almost as far out as the orbit of Neptune. Chiron is so faint that it must be small—100 to 300 km across. No other objects are known with similar orbits and brightness. Chiron's orbit is so elliptical that it will ultimately either collide with a planet or be ejected from the solar system.

Whether Chiron is an asteroid (perhaps the first discovered of a trans-Saturnian belt of asteroids; it is catalogued as number 2060) or an exceptionally large comet far from the Sun is unknown. As Chiron approaches the Earth during its 1995 perihelion, it will certainly be the subject of scrutiny.

Chiron may really be telling us that there is no sharp distinction between asteroids and comets. Traditionally, comets give off gases because they contain ices, and asteroids are rocky; but other evidence already indicates that asteroids can contain ices, and comets can contain rock. Most scientists think that comets originate far beyond Pluto, while asteroids were born close to their present location in the middle solar system. But other distinctions between comets and asteroids may be slight.

Summary and Outline

Comets (Section 27.1)
    The head (nucleus + coma together); the tail
    Dust tail: sunlight reflected off particles; ion tail (gas tail):
        sunlight re-emitted by ions blown back by solar wind
    Dirty snowball theory: nucleus of ice with dust mixed in
    Coma is gases vaporized from nucleus; molecules present
    Ultraviolet space observations revealed huge hydrogen
        cloud
    Origin of comets: Oort comet cloud; comets detached by
        gravity
    Halley's Comet followed for last 27 returns, but 1986 re-
        turn will not be as spectacular as some past passages

Meteoroids (Section 27.2)
    Meteoroids: in space; *meteors:* in the air; *meteorites:* on the
        ground
    Most meteorites that hit the Earth are chondrites, which
        resemble stones and are therefore often undetected;
        iron meteorites are more often found
    Meteorites may be chips of asteroids and thus give us ma-
        terial with which to investigate the origin of the solar
        system
    Some meteors arrive in showers, others are sporadic
Asteroids, also called minor planets (Section 27.3)
    Most in asteroid belt, 2.2 to 3.2 solar radii (between orbits
        of Mars and Jupiter)
    Apollo and Aten asteroids come within Earth's orbit
    Trojan asteroids in the Lagrangian points (Jupiter ± 60°)

Key Words

comet, head, tail, nucleus, dirty snowball, coma, dust tail, gas tail, ion tail, Oort comet cloud, parent molecules, daughter molecules, meteoroids, meteors, shooting stars, fireballs, meteorites, micrometeorites, iron meteorites, irons, stones, tektites, chondrites, achrondrites,* carbonaceous, showers, sporadic, radiant, minor planets, asteroids, asteroid belt, Apollo asteroids,* Aten asteroids,* Amor asteroids,* carbonaceous chondrites

_____

*These words are found in optional sections.

Questions

1. In what part of its orbit does a comet travel head first?

2. Why is the Messier catalogue important for comet hunters?

3. Would you expect comets to follow the ecliptic? Explain.

4. How far is the Oort comet cloud from the Sun, relative to the distance from the Sun to Pluto?

5. The energy that we see as light from a comet comes from what source or sources?

6. Many comets are brighter after they pass close to the Sun than they are on their approach. Why?

7. Which part of a comet has the most mass?

8. Explain why Comet West, seen in the photograph opening this chapter, showed delicate structure in its tail. Are we observing mainly the dust tail or the ion tail?

9. Why was the large cloud of hydrogen that surrounds the head of a comet not detected until 1970?

†10. Use Kepler's third law and the known period of Halley's Comet to deduce its semimajor axis. Sketch the orbit's shape on the solar system to scale.

†11. Use Kepler's third law and the 80,000-year period of Comet Kohoutek to deduce its semimajor axis. Relate the result to the size of planetary orbits.

12. What is the relation of meteorites and asteroids?

13. Why don't most meteoroids reach the Earth's surface?

14. Why do some meteor showers last only a day while others can last several weeks?

15. Why are meteorites important in our study of the solar system?

16. Compare asteroids with the moons of the planets.

17. How would the occultation of a star by an asteroid tell us whether the asteroid has an atmosphere?

_____

†This indicates a question requiring a numerical answer.

Observing Project

Observe the next meteor shower. Use your naked eyes. No instruments will be necessary.

Yoda. © Lucasfilm, Ltd. (LFL) 1980. All rights reserved. From the motion picture: *The Empire Strikes Back*, courtesy of Lucasfilm, Ltd.

Life in the Universe

28

Aims: **To discuss the possibility that intelligent life exists elsewhere in the Universe besides the Earth and to assess our chances of communicating with it**

We have discussed the nine planets and the dozens of moons in the solar system, and have found most of them to be places that seem hostile to terrestrial life forms. Yet some locations besides the Earth—Mars, with its signs of ancient running water, and perhaps Titan or Europa—have characteristics that allow us to convince ourselves that life may have existed there in the past or might even be present now or develop in the future. *Exobiology* is the study of life elsewhere than on Earth.

In our first real attempt to search for life on another planet, the Viking landers (Section 23.3) carried out biological and chemical experiments with Martian soil. The results showed that there is probably no life on Mars.

Since it seems reasonable that life, as we know it, anywhere in the Universe would be on planetary bodies, let us first discuss the chances of life arising elsewhere in our solar system. Then we will consider the chances that life has arisen in more distant parts of our galaxy or elsewhere in the Universe.

28.1 The Origin of Life

It would be very helpful if we could state a clear, concise definition of life, but unfortunately that is not possible. Biologists state several criteria that are ordinarily satisfied by life forms—reproduction, for example. Still, there exist forms on the fringes of life—viruses, for example, which need a host organism in order to reproduce—and scientists cannot always agree whether some of these things are ''alive'' or not.

In science fiction, authors sometimes conceive of beings that show such signs of life as the capability for intelligent thought, even though the being may share few of the other criteria that we ordinarily recognize. In Fred Hoyle's novel *The Black Cloud*, for example, an interstellar cloud of gas and dust is as alive as—and smarter than—you or I. But we can make no concrete deductions if we allow such wild possibilities, and exobiologists prefer to limit the definition of life to forms that are more like ''life as we know it.''

This rationale implies, for example, that extraterrestrial life is based on complicated chains of molecules that involve carbon atoms. Life on Earth is governed by deoxyribonucleic acid (DNA) and ribonucleic acid (RNA), two carbon-containing molecules that control the mechanisms of heredity. Chemically, carbon is able to form ''bonds'' with several other atoms simultaneously, which makes these long carbon-bearing chains possible. In fact, we speak of compounds that contain carbon atoms as *organic*. In addition, a few scientists consider silicon-based life, since sili-

Figure 28–1 Cyril Ponnamperuma is examining a "spark discharge" apparatus, used to simulate conditions on the primitive earth, or, at times, the outer planets of our solar system. The lower chamber is charged with water, and gases of the desired composition (methane, ammonia, nitrogen, etc.) are introduced into the upper chamber. Electric coils produce a spark discharge in the upper chamber, simulating lightning. By heating the outer arm of the apparatus, and cooling the inner core, a convection current is set up, simulating atmospheric flow between the "atmosphere" and the "ocean." After sparking for several days, the materials that have been synthesized are recovered from the liquid phase. These materials are commonly amino acids, sometimes sugars, and other complicated materials. Most recently, the laboratory has found evidence of nucleic acid bases in the products.

con is almost as abundant as carbon and can also form several simultaneous bonds (though its chemistry is less rich).

How hard is it to build up long organic chains? To the surprise of many, an experiment performed in the 1950's showed that it was much easier than had been supposed to make organic molecules. At the University of Chicago, at the suggestion of Harold Urey, Stanley Miller put several simple molecules in a glass jar. Water vapor (H_2O), methane (CH_4), and ammonia (NH_3) were included along with hydrogen gas. Miller exposed the mixture to electric sparks, simulating the lightning that may exist in the early stages of the formation of a planetary atmosphere. After a few days, long chains of atoms had formed in the jar; these organic molecules were even complex enough to include simple amino acids, the building blocks of life. Modern versions of these experiments have created even more complex organic molecules from a wide variety of simple actions on simple molecules.

Since the original experiment of Urey and Miller, most scientists have come to think that the Earth's primitive atmosphere was not made of methane and ammonia, which would have disappeared soon after the Earth's formation. Rather, the gases in today's oceans and atmosphere were released from the Earth's interior by volcanoes. When life formed, the atmosphere might have consisted of carbon dioxide, water, and carbon monoxide, but no methane and ammonia. The atmosphere, in short, resembled today's atmosphere but without oxygen. Cyril Ponnamperuma and colleagues at the University of Maryland have recently synthesized all five of the chemical bases of human genes in a single experiment that simulated the Earth's primitive atmosphere. They have also detected all five in the Murchison meteorite. Scientists have also found extraterrestrial amino acids in two meteorites that had been long frozen in Antarctic ice.

However, mere amino acids or even DNA molecules are not life itself. A jar containing a mixture of all the atoms that are in a human being is not the same as a human being. This is a vital gap in the chain; astronomers certainly are not qualified to say what supplies the "spark" of life.

Still, many astronomers think that since it is not difficult to form complex molecules, life may well have arisen not only on the Earth but also in other locations. Even if life is not found in our solar system, there are so many other stars in space that it would seem that some of them could have planets around them and that life could have arisen there independently of life on Earth.

28.2 Other Solar Systems?

We have seen how difficult it was to detect even the outermost planets in our own solar system. At the four-light-year distance of Proxima Centauri, the nearest star to the Sun, any planets around stars would be too faint for us to observe directly. There is hope, though, that the Hubble Space Telescope could see very large ones.

We now have a better chance of detecting such a planet by observing its gravitational effect on the star itself. One such method is the same that has been used to discover white dwarfs. We study the proper motion of a star—its apparent motion across the sky with respect to the other stars (as discussed in Section 5.7)—and see if the star follows a straight line or appears to wobble. Any wobble would have to result from an object too faint to see that is orbiting the visible star.

The easiest stars to observe in a search for planets are the stars with the largest proper motion, for these are likely to be the closest to us and thus any wobble would cover the greatest angular distance that we could hope for. The star with the largest proper motion, Barnard's star, is only 1.8 parsecs (6 light years) away from the Sun. It is the fourth nearest star to the Sun; only the members of the alpha Centauri triple system are nearer.

At the Sproul Observatory (Fig. 28–2), a major astrometric center, Peter van de Kamp studied observations of Barnard's star taken from 1937 on, and in 1962 reported that he had discovered a wobble in its proper motion (Fig. 28–3). He interpreted this wobble to indicate the presence of first one and then two giant planets, larger even than Jupiter. (An Earth-sized planet would not have enough mass to produce a perceptible wobble.) He also found evidence, though less confidently, for a planet around another of the nearest stars, eta Eridani.

However, George Gatewood of the Allegheny Observatory in Pittsburgh, and Heinrich Eichhorn of the University of South Florida, have since reported that their study of other long-term observations of Barnard's star does not show the wobble in its proper motion. A restudy of the Sproul astrometic plates by van de Kamp revealed a possible problem caused by a 1949 change in the telescope. A restudy of the more recent data significantly reduced the amount of wobble that may be present. The current indication of a wobble is so weak that scientists might not have reported it at all. And a new Naval Observatory study also did not show a wobble. In the meantime, a few other nearby stars show slight wobbles, though it is unclear whether they should be explained by large planets or mini-suns. The astrometric measurements from NASA's Space Telescope and ESA's Hipparcos should greatly improve the quality of the data in the next few years.

Donald McCarthy of the University of Arizona has been carrying out speckle interferometry in the infrared, which is able to discover faint companions that would not show up in the visible. Some large planets or mini-suns could be detected. Close stellar companions have been discovered for some stars, but no companion for Barnard's star has been detected. So, at least, Barnard's star's wobble—if real—is apparently not caused by a stellar companion.

A spectrographic survey carried out at Kitt Peak seems to endorse the idea that many stars have planets. One-tenth of the 123 solar-type stars and one-sixth of the hotter stars under study showed varying radial velocities. Such a varying motion would correspond to a star's going back and forth as a planet swings around it, in order to keep the gravitational center of the system moving straight. Some of the invisible companions have masses that may be too small to allow them to be stars. Some of these may be planets rather than black dwarfs; just how many, it is difficult to say. Methods of measuring the radial velocities of stars with greatly improved accuracy are under development, and new measurements are being carried out at Mauna Kea.

The indications from IRAS (Fig. 18–39) that solar systems are forming around stars, as deduced from the excess infrared measured, are not as satisfactory as a direct sighting of a planet or a detection of its gravitational effect. Nevertheless, the IRAS results surely encourage the point of view that solar systems are common.

28.3 *The Statistics of Extraterrestrial Life*

Instead of phrasing one all-or-nothing question about life in the Universe, we can break down the problem into a chain of simpler questions. This procedure was developed by Frank Drake, a Cornell astronomer, and extended by, among others, Carl Sagan of Cornell and Joseph Shklovskii, a Soviet astronomer.

Figure 28–2 The Sproul 61-cm visual reflector at Swarthmore College is used for photographic studies of our stellar neighborhood. A long-range search for very faint stellar and planet-like companions has been made since 1937 from the ever-increasing plate collection, the largest of its kind to come from one telescope. Several possible candidates for sub-stellar objects have been discovered with a number of unseen faint low-mass stellar companions to nearby stars.

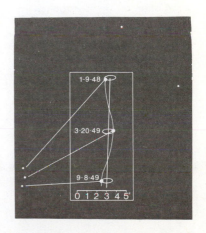

Figure 28–3 The image is a composite of three photos of Barnard's star, taken at the Sproul Observatory at intervals of approximately 6 months. The stars at the top have small proper motions and small parallaxes, and appear in the same position on each of the three negatives. Barnard's star, on the other hand, shows its proper motion of over 10 arc sec per year. It also shows a lateral displacement caused by parallax, the result of the Earth's orbit around the Sun. The displacement caused by the possible planets does not show at this scale.

Figure 28–4 As viewed from a distant star high above the sun's poles, the planets in our solar system would cause a slight deviation in the sun's proper motion. But the maximum deviation would be less than $\frac{1}{3}$ the sun's radius, and would be only $\frac{2}{100,000}$ arc sec at the distance of the nearest star.

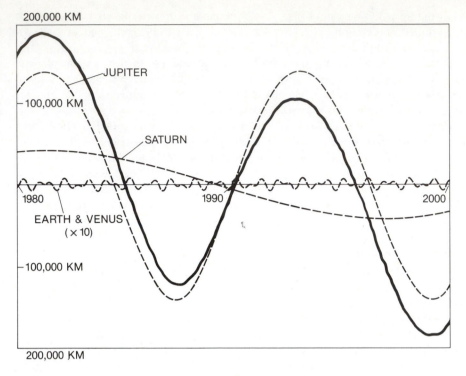

28.3a The Probability of Finding Life

First we consider the probability that stars at the centers of solar systems are suitable to allow intelligent life to evolve. For example, the most massive stars evolve relatively quickly, and probably stay in the stable state that characterized the bulk of their lifetimes for too short a time to allow intelligent life to evolve.

Second, we ask what the chances are that a suitable star has planets. With the proper motion studies (Barnard's star, etc.) no longer seeming to give clearly positive results, we have no accurate estimate of the probability, but most scientists think that the chances are probably pretty high.

Third, we need planets with suitable conditions for the origin of life. A planet like Jupiter might be ruled out, for example, because it lacks a solid surface and because its surface gravity is high. (Alternatively, though, one could consider a liquid region, if it were at a suitable temperature, to be as advantageous as were the oceans on Earth to the development of life here.) Also, planets probably must be in orbits in which the temperature does not fluctuate too much. If the planets are in multiple-star systems, as most seem to be, such orbits exist either far from a central pair of stars or, if the stars are far apart, close to the stars.

Fourth, we have to consider the fraction of the suitable planets on which life actually begins. This is the biggest uncertainty, for if this fraction is zero (with the Earth being a unique exception), then we get nowhere with this entire line of reasoning. Still, the discovery that amino acids can be formed in laboratory simulations of primitive atmospheres, and the discovery of complex molecules in interstellar space, indicate to many astronomers that it is not as difficult as had been thought to form

Figure 28–5 An archaebacterium through a microscope. (Courtesy of J. G. Zeikus, University of Wisconsin)

complicated molecules. Amino acids, much less complicated than DNA but also basic to life as we know it, have even been found in meteorites. The discovery from radio observations that molecules are associated with sites of star formation strengthens the idea that these molecules may eventually become part of planets that form. Many astronomers choose to think that the fraction of planetary systems on which life begins may be high.

Analysis of an odd kind of single-celled organism called *archaebacteria* (Fig. 28–5) indicates that life may have formed very early in the Earth's history. Some of these living creatures—found in airless places like Yellowstone's hot springs—take in carbon dioxide and hydrogen and give off methane. Their RNA (genetic material) is different from that of other bacteria or of plants or animals. The discovery supports the idea that life evolved before oxygen appeared in the Earth's primeval atmosphere. These and some other living bacteria do not survive in the presence of oxygen. Further, scientists have recently discovered organisms that thrive on sulfur from geothermal sources instead of solar energy. Some are bacteria that metabolize hydrogen sulfide to reproduce and grow, while others are animals that eat the bacteria. So environments on other planets may not be as hostile to life as we had thought. Even in the most apparently hostile places on Earth, we have now found living things (Fig. 28–6).

If we want to have meaningful conversations with aliens, we must have a situation where not just life but intelligent life has evolved (Fig. 28–7). We cannot converse with algae or paramecia. Furthermore, the life must have developed a technological civilization capable of interstellar communication. These considerations reduce the probabilities somewhat, but it has still been calculated that there are likely to be technologically advanced civilizations within a few hundred light years of the Sun.

Now one comes to the important question of the lifetime of the technological civilization itself. We now have the capability of destroying our civilization either dramatically in a flurry of hydrogen bombs or more slowly by, for example, altering our climate or increasing the level of atmospheric pollution. It is a sobering question to ask whether the lifetime of a technological civilization is measured in decades, or whether all the problems that we have—political, environmental, and otherwise—can be overcome, leaving our civilization to last for millions or billions of years.

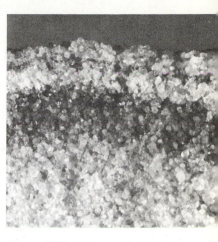

Figure 28–6 We see the inside of an Antarctic rock, with a lichen growing safely insulated from the external cold.

28.3b Evaluating the Probability of Life

At this point we can try to estimate (to guess, really) fractions for each of these simpler questions within our chain of reasoning. We can then multiply these fractions together to get an answer to the larger question of the probability of extraterrestrial life. Reasonable assumptions lead to the conclusion that there may be millions or billions of planets in this galaxy on which life may have evolved. The nearest one may be within dozens of light years of the Sun. On this basis, many if not most professional astronomers have come to feel that intelligent life probably exists in many places in the universe. Carl Sagan's latest estimate is that a million stars in the Milky Way Galaxy may be supporting technological civilizations.

Though this point of view has been gaining increasing favor over the last decade, a reaction has recently started. Evaluating Drake's equations with a more pessimistic set of numbers can lead to the conclusion that we earthlings are alone in our galaxy.

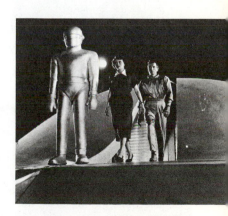

Figure 28–7 From *The Day the Earth Stood Still,* (© 1951 Twentieth Century-Fox Film Corporation. All rights reserved.) The movie described a mission to Earth sent in order to restrain our warlike nature.

The lack of agreement among astronomers is indicated by the fact that at an International Astronomical Union symposium, there were proponents of each of the following four ranges of possible values that may be assigned to N, the number of advanced civilizations in our galaxy: that N is very small, that N is very large, that N is neither very small nor very large, and that N is either very small or very large. An I.A.U. Commission on life in the universe was formed.

A

B

Figure 28–8 *(A)* Voyager 2 being prepared for launch. The workers are loading a gold-plated copper record bearing two hours' worth of Earth sounds and a coded form of photographs. The sounds include a car, a steamboat, a train, a rainstorm, a rocket blastoff, a baby crying, animals in the jungle, and greetings in various languages. Musical selections include Bach, Beethoven, rock, jazz, and folk music. *(B)* 116 slides are included on the record. This slide—which I took when in Australia for an eclipse—shows Heron Island on the Great Barrier Reef in Australia, in order to illustrate an island, an ocean, waves, beach, and signs of life.

Box 28.1 The Probability of Life in the Universe

Our discussion follows the lines of an equation written out to estimate the number of civilizations in our galaxy that would be able to contact each other. In 1961, Frank Drake of Cornell wrote

$$N = R_* f_p n_e f_\ell f_i f_c L$$

where R_* is the rate at which stars form in our galaxy, f_p is the fraction of these stars that have planets, n_e is the number of planets per solar system that are suitable for life to survive (for example, those that have Earth-like atmospheres), f_ℓ is the fraction of these planets on which life actually arises, f_i is the fraction of these life forms that develop intelligence, f_c is the fraction of the intelligent species that choose to communicate with other civilizations and develop adequate technology, and L is the lifetime of such a civilization. Now that we have seen that Jupiter's and Saturn's moons are worlds with personalities of their own, perhaps we should replace f_p with a new factor f_{mp} (moons and planets) that is f_p multiplied by the average number of suitable moons per planet in a solar system.

One of the largest uncertainties is f_ℓ, which could be essentially zero or could be close to 1. The result is also very sensitive to the value one chooses for L—is it 1 century or a billion years? Do civilizations destroy themselves? Or perhaps they just go off the airwaves. Depending on the alternatives one chooses, one can predict that there are dozens of communicating civilizations within 100 light years or that the Earth is unique in the galaxy in having one.

One reason for doubting that the universe is teeming with life is the fact that extraterrestrials have not established contact with us. Where are they all? To complicate things further, is it necessarily true that if intelligent life evolved, they would choose to explore space or to send out messages?

On the 20th anniversary of the original proposal that Philip Morrison and Giuseppe Cocconi had made to search for extraterrestrial radio signals from intelligent life, Morrison pointed out that we still knew three facts: (1) no radio signals from afar have been detected; (2) no extraterrestrials are known on Earth; and (3) we are here. He also noted that we now probably have a fourth fact: that there is no life on Mars (which would make the current direct evidence of the probability of life arising equal to $\frac{1}{2}$—Earth yes and Mars no).

What about multigenerational self-sustaining colonies voyaging through space? They need not travel close to the speed of light.

Michael Papagiannis of Boston University and others have pointed out that even if colonization of space took place at only 1 light year per century, that would still mean that the entire galaxy would have been colonized in less than 1 per cent of its lifetime, equivalent to 1 week of a human life. The Drake equation does not take colonization into account. The fact that we do not find extraterrestrial life here indicates that our solar system has not been colonized, which in turn indicates that life has not arisen elsewhere. Our descendants may turn out to be the colonizers of the galaxy!

28.4 Interstellar Communication

What are the chances of our visiting or being visited by representatives of these civilizations? Pioneers 10 and 11 and Voyagers 1 and 2 are even now carrying messages out of the solar system in case an alien interstellar traveller should happen to

Figure 28–9 The Arecibo telescope in Puerto Rico, used to send a message into space. The telescope is 305 meters across, the largest telescope on Earth.

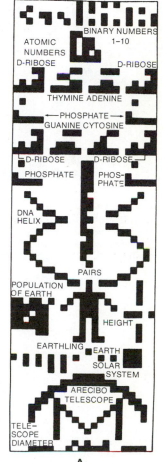

encounter these spacecraft (Fig. 28–8). But it seems unlikely that humans themselves can travel the great distances to the stars, unless some day we develop spaceships to carry whole families and cities on indefinitely long voyages into space.

Yet there is plenty of time in the future for interstellar space travel to develop. The Sun has another 5 billion years to go in its current stable state, while recorded history on Earth has existed for only about 5000 years, one millionth of the age of the solar system.

We can hope even now to communicate over interstellar distances by means of radio signals. We have known the basic principles of radio for only a hundred years, and powerful radio, television, and radar transmitters have existed for less than 50 years. Even now, though, we are sending out signals into space on the normal broadcast channels. A wave bearing the voice of Caruso is expanding into space, and at present is 50 light years from Earth. And once a week a new episode of *Dallas* is carried into the depths of the Universe. Military radars are even stronger.

From another star, the Sun would appear to be a variable radio source, because the Earth and Sun would be unresolved and the Earth's radio signals would therefore appear to be coming from the Sun. The signal would vary with a period of 24 hours since it would peak each day as specific concentrations of transmitters rotated to the side of the Earth facing our listener.

In 1974 the giant radio telescope at Arecibo, Puerto Rico (Fig. 28–9), was upgraded. At the rededication ceremony, a powerful signal was sent out into space at the fundamental radio frequency of hydrogen radiation, bearing a message from the people on Earth (Fig. 28–10). This is the only major signal that we have purposely sent out as our contribution to the possible interstellar dialogue.

The signal from Arecibo was directed at the globular cluster M13 in the constellation Hercules, on the theory that the presence of 300,000 closely packed stars

Figure 28–10 *(A)* The message sent to M13, plotted out and with a translation into English added. *(B)* The message was sent as a string of 1679 consecutive characters, in 73 groups of 23 characters each (two prime numbers). There were two kinds of characters, each represented by a frequency; the two kinds of characters are reproduced here as 0's and 1's. The basic binary-system count at upper right is provided with a position-marking square below each number.

in that location would increase the chances of our signal being received by a civilization on one of them. But the travel time of the message (at the speed of light) is 24,000 years to M13, so we certainly could not expect to have an answer before twice 24,000, or 48,000 years, have passed. If anybody (or any**thing**) is observing our Sun when the signal arrives, the radio brightness of the Sun will increase by 10 million times for a 3-minute period. A similar signal, if received from a distant star, could be the giveaway that there is intelligent life there.

We mustn't be too quick to close out other possibilities. For example, neutrinos travel at or close to the speed of light. Just because we find neutrinos hard to detect doesn't prove that other civilizations haven't chosen them to carry messages.

28.5 The Search for Life

For planets in our solar system, we can search for life by direct exploration (as we did with Viking on Mars), but electromagnetic waves, which travel at the speed of light, seem a much more sensible way to search for extraterrestrial life outside our solar system. How would you go about trying to find out if there was life on a distant planet? If we could observe the presence of abundant oxygen molecules or discover from the spectrum of a distant star that the molecules there are out of their normal balance, that would tell us that there must be life there. The Space Telescope, for example, should be able to detect the infrared band of radiation from molecular oxygen on a planet orbiting alpha Centauri.

The most promising way to detect the presence of life at great distances appears to be a search in the radio part of the spectrum. But it would be too overwhelming a task to listen for signals at all frequencies in all directions at all times. One must make some reasonable guesses on how to proceed.

A few frequencies in the radio spectrum seem especially fundamental, such as the spin-flip line of neutral hydrogen—the simplest element—at 21 cm. This corresponds to 1420 MHz, a frequency over ten times higher than stations at the high end of the normal FM band. We might conclude that creatures on a far-off planet would decide that we would be most likely to listen near this frequency because it is so fundamental.

On the other hand, perhaps the great abundance of radiation from hydrogen itself would clog this frequency, and one or more of the other frequencies that correspond to strong natural radiation should be preferred. The "water hole," the wavelength range between the radio lines of H and OH, has a minimum of radio noise from background celestial sources, the telescope's receiver, and the Earth's atmosphere, and so is another favored possibility. If we ever detect radiation at any frequency, we would immediately start to wonder why we had not realized that this frequency was the obvious choice.

In 1960, Frank Drake used a telescope at the National Radio Astronomy Observatory for a few months to listen for signals from two of the nearest stars—tau Ceti and epsilon Eridani. He was searching for any abnormal kind of signal, a sharp burst of energy, for example, such as pulsars emit. He called this investigation Project Ozma after the queen of the land of Oz in L. Frank Baum's later stories.

Figure 28–12 Dorothy and Ozma climb the magic stairway. (From "Glinda of Oz" by L. Frank Baum, illustrated by John R. Neill, copyright 1920)

The observations were later extended in a more methodical search called Ozma II. Starting in 1973, Ben Zuckerman of the University of Maryland (now at U.C.L.A.) and Patrick Palmer of the University of Chicago systematically monitored over 600 of the nearest stars of supposedly suitable spectral type. They used computers to search the data they recorded for any sign of special signals; Ozma II obtained data at 10 million times the rate of the original Ozma. Over the course of several days, they studied each star for a total of about ½ hour. Nothing turned up.

Over a dozen separate searches have been undertaken in the radio spectrum, some concentrating on all-sky coverage and others on coverage of individual stars over time. Paul Horowitz of Harvard tested nearly 12 million combinations of stars and radio frequencies at Arecibo. Radio telescopes at Berkeley's Hat Creek Observatory, Ohio State, and the Algonquin Observatory in Ontario have also been used. Drake and Sagan of Cornell searched five different galaxies with the Arecibo telescope. Their equipment wasn't very sensitive to any individual star's radiation, but had a hundred million stars in its beam at the same time. The telescope would have picked up a civilization if there was even a single one in those galaxies. Even a search for strong ultraviolet radiation from a laser-like source has been made from a telescope in space.

Many astronomers feel that carrying out these brief searches was worthwhile, since we had no idea of what we would find. But now many feel that the early promise has not been justified, and the chance that there is extraterrestrial life seems much lower.

More and more astronomers in both the United States and in the Soviet Union are interested in "communication with extraterrestrial intelligence" (CETI), and in the simpler and less expensive "search for extraterrestrial intelligence" (SETI). The programs include scanning a wide range of frequencies in all directions in the sky for suitable strange signals. Paul Horowitz of Harvard has developed a relatively inexpensive way of scanning over a million narrow-frequency bands simultaneously. He has set up his "suitcase SETI" on a 26-m Harvard radio telescope for long-term scanning of frequencies that seem most likely (21 cm, OH, $\frac{1}{2} \times 21$ cm, etc.), covering a wide range of sky. They will do a megaOzma per minute. Scientists at NASA/Ames Research Center are developing a scanner for first 74,000 and later

Scientific spinoffs of SETI include improved radio spectral-line surveys of stars leading to better knowledge of interstellar matter and the discovery of new radio sources. The information on the effects of terrestrial radiation and the atmosphere on radio signals will also help improve our deep-space tracking of spacecraft and communication links.

2,000,000 channels to be used on existing space-probe-tracking antennae. They will use computers to analyze the signals at 1-s intervals in frequencies including the water hole, starting on 773 nearby stars within 80 light-years. Very-narrow-frequency signals could be signs of life, since naturally arising signals cover wider ranges.

Our improved abilities to observe in the infrared might pick up evidence of advanced civilizations in another way. Freeman Dyson, at the Institute of Advanced Study at Princeton, has suggested that a civilization might redistribute some of its planet's mass around its star to soak up stellar energy that would otherwise escape. IRAS picked up the infrared glow caused by material that blocks even 1 per cent of the sunlight of a star like the Sun at a distance of 300 light years. This distance would include 100,000 stars, so we will see what further data reduction turns up.

The list of searches has grown too long to include completely. The next years, with radio searches under way and Space Telescope aloft, will bring us much more thorough studies and precise results than we have had in the quest for planets and life.

Summary and Outline

The origin of life (Section 28.1)
  Earth life based on complex chains of carbon-bearing (organic) molecules; carbon (and silicon) form such chains easily; organic molecules are easily formed under laboratory conditions that simulate primitive atmospheres
Evidence for other solar systems (Section 28.2)
  Search for wobbles in proper motions or radial velocities
  Evidence for a planet around Barnard's star is not as clear as it once was
  Data from Space Telescope and Hipparcos will be valuable.

Statistical chances for extraterrestrial life (Section 28.3)
  Problem broken down into stages; major uncertainties include what the chance is that life will form given the component parts, and what the lifetime of a civilization is likely to be
Signals we are sending from Earth (Section 28.4)
  Leakage of radio, television, and radar; signal beamed from Arecibo toward globular cluster M13
The search for life (Section 28.5)
  Experiments on Viking; Projects Ozma, Ozma II, Suitcase SETI

Key Words

exobiology, organic, archaebacteria

Questions

1. What is the significance of the Miller-Urey experiments?

2. How have the Miller-Urey experiments been improved upon?

3. Does a star with a large proper motion necessarily have planets? Explain.

4. Discuss the evidence for and against the idea that Barnard's star has planets around it.

5. Describe two ways that the Hubble Space Telescope will aid in the search for planets around distant stars.

6. What is one advantage of the radial-velocity method over the proper-motion method for detecting distant planets?

† 7. If $\frac{1}{10}$ of all stars are of suitable type for life to develop, 1 percent of all stars have planets, and 10 percent of planetary systems have a planet at a suitable distance from the star, what fraction of stars have a planet suitable for life? How many such stars would there be in our galaxy?

---

†This indicates a question requiring a numerical answer.

† 8. Supply your own numbers in the Drake equation, and calculate the result.

† 9. On the message sent to M13, work out the binary value given for the population of Earth and compare it with the actual value.

———————————

†This indicates a question requiring a numerical answer.

†10. On the message sent to M13, work out the binary value given for the Arecibo telescope, and compare it with the actual value.

11. Why do we think that intelligent life wouldn't evolve in planets around stars much more massive than the sun?

12. List three means by which we might detect extraterrestrial intelligence.

Topics for Discussion

1. Comment on the possibility of scientists in another solar system observing the Sun, suspecting it to be a likely star to have populated planets, and beaming a signal in our direction.

2. Using a grid 37 × 41 (two prime numbers), work out a message you might send. Show the string of 0's and 1's.

Figure 28–13 This first picture of what is apparently another solar system was taken with a ground-based telescope. An occulting disk blocked out the star β (beta) Pictoris; shielding its brightness revealed the material that surrounds it. About 200 times as much mass as the Earth is present in the form of solid material. β Pictoris is too far south in the sky to be seen from mid-northern latitudes. (R. J. Terrile, JPL, and B. A. Smith, U. Arizona, with Carnegie Institution's Las Campanas Observatory, Chile)

Epilogue

We have had our tour of the Universe. We have seen the stars and planets, the matter between the stars, and the distant objects in our Universe like quasars and clusters of galaxies. We have learned how our universe is expanding and that it is bathed in a glow of radio waves.

Further, we have seen the vitality of contemporary science in general and astronomy in particular. The individual scientists who call themselves astronomers are engaged in fascinating studies, often pushing new technologies to their limits. New telescopes on the ground and in space, new computer capabilities for studying data and carrying out calculations, and new theoretical ideas are linked in research about the Universe.

Our knowledge of the Universe changes so rapidly that within a few years much of what you read here will be revised. So your study of astronomy shouldn't end here; I hope that, over the years, you will keep up by following astronomical articles and stories in newspapers, magazines, books, and television, and by considering the role of scientific research as you vote. And I hope you will remember the methods of science—the mixture of logic and standards of proof by which scientists operate—that you have seen illustrated in this book.

Figure E–1 A space-shuttle astronaut, set free from the orbiter, becomes a human satellite in space and orbits the Earth at 27,000 km/hr. Our increasing capabilities in space join with ground-based and theoretical studies to improve our knowledge of the Universe.

HOPKINS OBSERVATORY
WILLIAMS COLLEGE
WILLIAMSTOWN, MASS. 01267
To: Universal Creation Foundation
REQUEST FOR SUPPLEMENT TO U.C.F. GRANT
#000-00-00000-001
"CREATION OF THE UNIVERSE"
This report is intended only for the
internal uses of the contractor.
Period: Present to Last Judgment
Principal Creator: Creator
Proposal Writers and Contract Monitors:
Jay M. Pasachoff and Spencer R. Weart

BACKGROUND

Under a previous grant (U.C.F. Grant #000-00-00000-001), the Universe was created. It was expected that this project would have lasting benefits and considerable spinoffs, and this has indeed been the case. Darkness and light, good and evil, and Swiss Army knives were only a few of the useful concepts developed in the course of the Creation. It was estimated that the project would be completed within four days (not including a mandated Day of Rest, with full pay), and the 50% overrun on this estimate is entirely reasonable, given the unusual difficulties encountered. Infinite funding for this project was requested from the Foundation and granted. Unfortunately, this has not proved sufficient. Certain faults in the original creation have become apparent, which it will be necessary to correct by means of miracles. Let it not be said, however, that we are merely correcting past errors; the final state of the Universe, if this supplemental request is granted, will have many useful features not included in the original proposal.

PROGRESS TO DATE

Interim progress reports have already been submitted ("The Bible," "The Koran," "The Handbook of Chemistry and Physics," etc.). The millennial report is currently in preparation, and a variety of publishers for the text (tentatively entitled, "Oh, Genesis!") will be created. The Gideon Society has applied for the distribution rights. Full credit will be given to the Foundation.

Materials for the Universe and for the Creation of Man were created out of the Void at no charge to the grant. A substantial savings was generated when it was found that materials for Woman could be created out of Man, since the establishment of Anti-Vivisection Societies was held until Phase Three. Given the limitations of current eschatological technology, it can scarcely be denied that the Contractor has done His work at a most reasonable price.

SUPPLEMENT

We cannot overlook a certain tone of dissatisfaction with the Creation which has been expressed by the Foundation, not to mention by certain of the Created. Let us state outright that this was to be expected, in view of the completely unprecedented nature of the project. The need for a supplement is to be ascribed solely to inflation (not to be confused with expansion of the Universe, which was anticipated). Union requests for the accrual of Days of Rest at the rate of one additional Day per week per millennium ($Dw^{-1}m^{-1}$) must also be met. Concerning the problem of Sin, we can assure you that extensive experimentation is under way. Considerable experience is being accumulated and we expect a breakthrough before long. When we are satiated with Sin, we shall go on to consider Universal Peace.

We cannot deny—in view of the cleverness of the Foundation's auditors—that the bulk of the supplemental funds will go to pay off old bills. Nevertheless we do not anticipate the need for future budget requests, barring unforeseen circumstances. If this project is continued successfully, additional Universe—anti-Universe pairs can be created without increasing the baryon number, and we would keep them out of the light cone of the Original. By the simple grant of an additional Infinity of funds (and note that this proposal is merely for Aleph Null), the officers of the Foundation will be able to present their Board of Directors with the accomplished Creation of one or more successful Universes, instead of the current incomplete one.

We will attempt to minimize the additional delays that may temporarily exist during the changeover from fossil fuels to fusion for some minor locations in the Universe. For the time being, elements with odd masses (hydrogen, lithium, etc.) will be created only on odd days of the month and those with even masses (helium, beryllium, etc.) on the even days, except, of course, for Sundays. The 31st of each month will be devoted to creation of trans-uranic elements. The Universe has been depleted of deuterium; new creation of this will take place only on February 29th in leap years.

PROSPECTIVE BUDGET

Remedial miracles on fish of the sea	∞
Remedial miracles on fowl of the air	∞
Creeping things that creep upon the earth, etc.	∞
Hydrogen	n/c (created)
Heavier elements	n/c (nucleosynthesis)
(Note: The carbon will be reclaimed and ecologically recycled)	
Mountains (Sinai, Ararat, Palomar)	∞
Extra quasars, neutron stars	∞
Black holes (no-return containers)	∞
Miscellaneous, secretarial, office supplies, etc.	∞
Telephone installation (Princess model, white, one-time charge, tax included)	$16.50

SALARIES

Creator (1/4 time)	at His own expense
Archangels	
Gabriel	1 trumpet (Phase 5)
Beelzebub	misc. extra brimstone (low sulfur)
Others	assorted halos
Prophets	
Moses	stone tablets (to replace breakage)
Geniuses	finite

N.B. Due to the Foundation's regulations and changes in Exchange Rates, we have not yet been able to reimburse Euclid (drachmae), Leonardo (lire), Newton (pounds sterling), Descartes (francs), or Reggie Jackson (dollars). Future geniuses will be remunerated indirectly via the Alfred Nobel Foundation.

Graduate students (2 at 2/5 time)	reflected glory

MONITORING EQUIPMENT & MISC.

1 5-meter telescope (maintenance)	finite
Misc. other instruments, particle accelerators, etc.	large but finite
Travel to meetings	∞
Pollution control equipment	+40%
Total	∞ + 40% = ∞
Overhead (51.97%)	∞
Total funds requested	∞

Starting date requested:
Immediate. Pending receipt of supplemental funds, layoffs are anticipated to reach the 19.5% level in the ranks of angels this quarter.

Figure E–2 A mock proposal similar to one that scientists submit for government funding.

Appendix 1
Measurement Systems

Metric units

SI units			SI abbrev.	Other abbrev.
	length	meter	m	
	volume	liter	L	ℓ
	mass	kilogram	kg	kgm
	time	second	s	sec
	temperature	kelvin	K	°K

Other metric units

	SI abbrev.	Other abbrev.
1 micron (μ) = 1 micrometer (μm) $= 10^{-6}$ meter	μm	μ
1 Ångstrom (Å or A); $= 10^{-10}$ meter $= 10^{-8}$ cm	0.1 nm	Å

Prefixes for use with basic units of metric system

Prefix	Symbol	Power			Equivalent
exa	E	10^{18}	=	1,000,000,000,000,000,000	
peta	P	10^{15}	=	1,000,000,000,000,000	
tera	T	10^{12}	=	1,000,000,000,000	Trillion
giga	G	10^{9}	=	1,000,000,000	Billion
mega	M	10^{6}	=	1,000,000	Million
kilo	k	10^{3}	=	1,000	Thousand
hecto	h	10^{2}	=	100	Hundred
deca	da	10^{1}	=	10	Ten
——	——	10^{0}	=	1	One
deci	d	10^{-1}	=	.1	Tenth
centi	c	10^{-2}	=	.01	Hundredth
milli	m	10^{-3}	=	.001	Thousandth
micro	μ	10^{-6}	=	.000001	Millionth
nano	n	10^{-9}	=	.000000001	Billionth
pico	p	10^{-12}	=	.000000000001	Trillionth
femto	f	10^{-15}	=	.000000000000001	
atto	a	10^{-18}	=	.000000000000000001	

Some other units used in astronomy

	SI	Other
Energy:	joule (J) = 1 $m^2 \cdot kg/s^2$	ergs, eV
Power:	watt (W) = 1 J/s	joule/sec
Frequency:	hertz (Hz)	cycles per sec

Conversion factors:

1 joule = 10^7 ergs	1 gm = 0.0353 oz
1 electron volt (eV) = 1.60207×10^{-12} ergs	1 in = 25.400 mm = 2.54 cm
1 watt = 1 joule per sec	1 ft = 0.3048 m
1 cm = 0.3937 in	1 yd = 0.9144 m
1 m = 1.0936 yd	1 mi = 1.6093 km = 8/5 km
1 km = 0.6214 mi = 5/8 mi	1 lb = 0.4536 kg
1 kg = 2.2046 lb	

Appendix 2
Basic Constants

Physical constants		
Speed of light*	c	$= 299\ 792\ 458$ m/s
Constant of gravitation†	G	$= 6.6726 \pm 0.0005 \times 10^{-11}$ m^3/kg/s^2
Planck's constant	h	$= 6.6262 \times 10^{-27}$ erg $\cdot$ s
Boltzmann's constant	k	$= 1.3806 \times 10^{-16}$ erg/kelvin
Stefan-Boltzmann constant	σ	$= 5.66956 \times 10^{-5}$ erg/cm$^2 \cdot$ deg$^4 \cdot$ s^1
Wien displacement constant	$\lambda_{max}T$	$= 0.289789$ cm·K $= 28.9789 \times 10^6$Å·K
Mass of hydrogen atom	m_H	$= 1.6735 \times 10^{-24}$ gm
Mass of neutron	m_n	$= 1.6749 \times 10^{-24}$ gm
Mass of proton	m_p	$= 1.6726 \times 10^{-24}$ gm
Mass of electron	m_e	$= 9.1096 \times 10^{-28}$ gm
Rydberg's constant	R	$= 1.09677 \times 10^5$/cm

Mathematical constants
$\pi = 3.1415926536$
$e = 2.7182818285$

Astronomical constants		
Astronomical unit*	1 A.U.	$= 1.495\ 978\ 70 \times 10^{11}$ m
Solar parallax*	$\pi_{\odot}$	$= 8.794148$ arc sec
Parsec	pc	$= 206\ 264.806$ A.U.
		$= 3.261633$ light years
		$= 3.085678 \times 10^{18}$ cm
Light year	ly	$= 9.460530 \times 10^{17}$ cm
		$= 6.324 \times 10^4$ A.U.
Tropical year (1900)*– equinox to equinox		$= 365.24219878$ ephemeris days
1 Julian century*		$= 36525$ days
1 day*		$= 86400$ s
Sidereal year		$= 365.256366$ ephemeris days
		$= 3.155815 \times 10^7$ s
Mass of sun*	$M_{\odot}$	$= 1.9891 \times 10^{33}$ gm
Radius of sun*	$R_{\odot}$	$= 696000$ km
Luminosity of sun	$L_{\odot}$	$= 3.827 \times 10^{33}$ erg/s
Mass of earth*	M_E	$= 5.9742 \times 10^{27}$ gm
Equatorial radius of earth*	R_E	$= 6\ 378.140$ km
Mean distance center of earth to center of moon		$= 384\ 403$ km
Radius of moon*	R_M	$= 1\ 738$ km
Mass of moon*	M_M	$= 7.35 \times 10^{25}$ gm
Solar constant	S	$= 135.3$ mW/cm^2
Direction of galactic center (precessed for 2000.0)	α	$= 17^h45.6^m$
	δ	$= -28° 56'$

Precession formulae
Change in dec. $= \Omega t \sin (23.5°)\cos(\text{r.a.})$
Change in r.a. $= \Omega t [\cos(23.5°) + \sin(23.5°)\sin(\text{r.a.})\tan(\text{dec.})]$

where Ω is the annual rate of precession. If Ω is used as 50 arc sec/year, declination will be given in arc sec. If Ω is used as 3.3^s/year, right ascension will be given in seconds of r.a.

*Adopted as "IAU (1976) system of astronomical constants" at the General Assembly of the International Astronomical Union. The meter was redefined in 1983 to be the distance travelled by light in a vacuum in 1/299,792,458 second.

†G. G. Luther and W. R. Towler, *Physical Review Letters 48,* 121, 1982.

Appendix 3
The Planets

Appendix 3a. Intrinsic and Rotational Properties

Name	Equatorial Radius km	Equatorial Radius ÷ Earth's	Mass ÷ Earth's	Mean Density (g/cm³)	Oblateness	Surface Gravity (Earth = 1)	Sidereal Rotation Period	Inclination of Equator to Orbit	Apparent Magnitude at 1984 Opposition
Mercury	2,439	0.3824	0.0553	5.43	0.0	0.378	58.646^d	0°	−1.8
Venus	6,052	0.9489	0.8150	5.24	0.0	0.894	243.01^{d}R	177.3	−4.3
Earth	6,378.140	1	1	5.515	0.0034	1	23^{h}56^{h}04.1^s	23.45	—
Mars	3,397.2	0.5326	0.1074	3.93	0.005	0.379	24^{h}37^{m}22.662^s	25.19	−1.9
Jupiter	71,398	11.194	317.89	1.36	0.061	2.54	9^{h}50^m to > 9^{h}55^m	3.12	−2.7
Saturn	60,000	9.41	95.17	0.71	0.109	1.07	10^{h}39.9^m	26.73	+0.1
Uranus	26,145	4.1	14.56	1.30	0.03	0.8	12^h to 24^h	97.86	+5.5
Neptune	24,300	3.8	17.24	1.8	0.03	1.2	18^h±10^m	29.56	+7.9
Pluto	1,500–1,800	0.3	0.02	0.5–0.8	?	~0.03	6^{d}9^{h}17^m	118?	+13.7

R signifies retrograde rotation.

The masses and diameters are the values recommended by the International Astronomical Union in 1976, except for new values for Jupiter, Saturn, Uranus, and Pluto. Surface gravities were calculated from these values. The length of the Martian day is from G. de Vaucouleurs (1979). Most Uranus values from Elliot, J. L., E. Dunham, D. J. Mink, and J. Churms (*Ap. J. 236*, 1026, 1980); Uranus radius from Elliot (private communication, 1981). Ranges are given for recent measurements of the rotation periods of Uranus and Neptune. Densities, rotational periods of inner planets, oblatenesses, inclinations, and magnitudes are from *The Astronomical Almanac 1984*.

Appendix 3b. Orbital Properties

Name	Semimajor Axis A.U.	Semimajor Axis 10⁶ km	Sidereal Period Years	Sidereal Period Days	Synodic Period (days)	Eccentricity	Inclination to Ecliptic
Mercury	0.3871	57.9	0.24084	87.96	115.9	0.2056	7°00′26″
Venus	0.7233	108.2	0.61515	224.68	584.0	0.0068	3°23′40″
Earth	1	149.6	1.00004	365.26	—	0.0167	0°00′14″
Mars	1.5237	227.9	1.8808	686.95	779.9	0.0934	1°51′09″
Jupiter	5.2028	778.3	11.862	4,337	398.9	0.0483	1°18′29″
Saturn	9.5388	1427.0	29.456	10,760	378.1	0.0560	2°29′17″
Uranus	19.1914	2871.0	84.07	30,700	369.7	0.0461	0°48′26″
Neptune	30.0611	4497.1	164.81	60,200	367.5	0.0100	1°46′27″
Pluto	39.5294	5913.5	248.53	90,780	366.7	0.2484	17°09′03″

Mean elements of planetary orbits for 1980, referred to the mean ecliptic and equinox of 1950 (Seidelmann, P. K., L. E. Doggett, and M. R. De Luccia, *Astronomical Journal 79*, 57, 1974). Periods are calculated from them.

"NOTES FOR APPENDIX 4"

Based on a table by Joseph Veverka in the *Observer's Handbook 1980* of the Royal Astronomical Society of Canada, updated for satellites of Mars, Saturn, Jupiter, and Pluto. Many of Veverka's values are from J. Burns. Masses of Phobos and Deimos from Veverka and Burns (*Ann. Rev. Earth Planet. Sci. 8*, 527–558, 1980). Names from David D. Morrison © 1983 Astronomical Society of the Pacific.

*Values recommended by the International Astronomical Union in 1976.

Diameters marked with a † from Cruikshank, D. (R. H. Brown, D. P. Cruikshank, D. Morrison, 1982, based on IRTF measurements). Values for the Galilean satellites are from Voyager (Smith, B. A. et al., *Science 204*, 951, 1979). Amalthea diameter from Voyager. Values for Phobos and Deimos from J. Veverka and J. A. Burns, *Annual Reviews of Earth and Planetary Science 8*, 527–558, 1980. Charon data from R. S. Harrington and J. W. Christy (*Astronomical Journal 85*, 168, 1980). Inclinations greater than 90° are retrograde rotation, and R signifies retrograde revolution. Apparent magnitudes at discovery given for Jupiter XIII and XIV. Apparent magnitude at discovery given for Pluto's satellite.

Speckle interferometry studies with the Canada-France-Hawaii telescope (I.A.U. Circular 3509, September 1980) give 4000 and 2000 km, respectively, for the diameters of Pluto and Charon, a total mass of ⅓₀₀ that of Earth, and densities of 0.5 g/cm³.

**Both observed in 1966 by Dollfus, by Fountain and Larson, and by others, and confused as "Janus."

Masses and opposition magnitudes for satellites of Saturn from *The Astronomical Almanac 1982*. Additional moons are still being discovered on the Voyager images. Only moons with determined orbits are listed above.

Appendix 4
Planetary Satellites

	Satellite	Semimajor Axis of Orbit (km)	Sidereal Revolution Period (d)	(h)	(m)	Orbital Eccentricity	Orbital Inclination (°)	Diameter (km)	Mass ÷ Mass of Planet	Mean Density (g/cm³)	Discoverer	Visible Magnitude at Mean Opposition Distance
SATELLITE OF THE EARTH												
	The Moon	384,500	27	07	43	0.055	18–29	3476	0.01230002*	3.34	—	−12.7
SATELLITES OF MARS												
	Phobos	9,378	0	07	39	0.015	1.1	$27 \times 21 \times 18$	1.5×10^{-8}	2	Hall (1877)	11.3
	Deimos	23,459	1	06	18	0.00052	0.9–2.7v	$15 \times 12 \times 10$	3×10^{-9}	2	Hall (1877)	12.4
SATELLITES OF JUPITER												
XIV	1979 J1 Adrastea	127,000	0	07	04			(40)			Synnott/Voyager 1 (1980)	
XVI	1979 J3 Metis	129,000	0	07	06			(20)			Jewett/Voyager 2 (1979)	
V	Amalthea	180,000	0	11	57	0.003	0.4	$270 \times 165 \times 153$	18×10^{-10}		Barnard (1892)	14.1
XV	Thebe	222,000	0	16	11			(80)			Synnott/Voyager 1 (1980)	
I	Io	422,000	1	18	28	0.000	0	3632 ± 60	4.70×10^{-5}*	3.5	Galileo (1610)	5.0
II	Europa	671,000	3	13	14	0.000	0.5	3126 ± 60	2.56×10^{-5}*	3.0	Galileo (1610)	5.3
III	Ganymede	1,070,000	7	03	43	0.001	0.2	5276 ± 60	7.84×10^{-5}*	1.9	Galileo (1610)	4.6
IV	Callisto	1,885,000	16	16	32	0.01	0.2	4820 ± 60	5.6×10^{-5}*	1.8	Galileo (1610)	5.6
XIII	Leda	11,110,000	240			0.147	26.7	(10)	5×10^{-13}		Kowal (1974)	20
VI	Himalia	11,470,000	251			0.158	27.6	170	8.5×10^{-10}		Perrine (1904)	14.7
X	Lysithea	11,710,000	260			0.130	29.0	(20)	0.010×10^{-10}		Nicholson (1938)	18.4
VII	Elara	11,740,000	260			0.207	24.8	80	0.35×10^{-10}		Perrine (1905)	16.4
XII	Ananke	20,700,000	617 R			0.169	147	(20)	0.007×10^{-10}		Nicholson (1951)	18.9
XI	Carme	22,350,000	692 R			0.207	164	(30)	0.020×10^{-10}		Nicholson (1938)	18.0
VIII	Pasiphae	23,330,000	735 R			0.378	145	(40)	0.077×10^{-10}		Melotte (1908)	17.7
IX	Sinope	23,700,000	758 R			0.275	153	(30)	0.015×10^{-10}		Nicholson (1914)	18.3
—											Kowal (1975)	20
SATELLITES OF SATURN												
	Atlas	137,670		14	27	0.002	0.3	$80 \times 60 \times 40$			Voyager 1	17
	Inner F ring shepherd	139,353		14	43	0.003	0.0	$140 \times 100 \times 80$			Voyager 1	16
	Outer F ring shepherd	141,700		15	05	0.004	0.05	$110 \times 90 \times 70$			Voyager 1	16
	Epimetheus**	151,422		16	40	0.007	0.1	$220 \times 200 \times 160$			Smith, Reitsema, Larson and Fountain	15
	Janus**	151,472		16	40	0.009	0.3	$14 \times 120 \times 100$			Cruikshank/Pioneer 11	
1	Mimas	185,600		22	37	0.020	1.5	390	6.6×10^{-8}	1.2	W. Herschel (1789)	12.9
2	Enceladus	238,100	1	08	32	0.005	0.0	510	1.3×10^{-7}	1.1	W. Herschel (1789)	11.7
3	Tethys	294,700	1	22	15	0.000	1.9	1,050	1.1×10^{-6}	1.0	Cassini (1684)	10.3
	Telesto	294,700	1	22	15			$34 \times 28 \times 26$				19
	Calypso	294,700	1	22	15			$34 \times 22 \times 22$				19
4	Dione	377,500	2	17	36	0.002	0.0	1,120	1.9×10^{-6}	1.4	Cassini (1684)	10.4
	Dione B	378,060	2	17	45			$36 \times 22 \times 30$			Laques and Lecacheux (1980)	19
5	Rhea	527,200	4	12	16	0.001	0.4	1,530	9×10^{-6}	1.3	Cassini (1672)	9.7
6	Titan	1,221,600	15	21	51	0.029	0.3	5,150	2.5×10^{-4}*	1.9	Huygens (1665)	8.3
7	Hyperion	1,483,000	21	06	45	0.104	0.4	$400 \times 250 \times 220$	2.0×10^{-7}	1.9	Bond (1848)	14.2
8	Iapetus	3,560,100	79	03	43	0.028	14.7	1,440	1.0×10^{-6}	1.2	Cassini (1671)	11.2
9	Phoebe	12,950,000	549	03	33	0.163	159	200	7×10^{-10}		W. Pickering (1898)	16.3
SATELLITES OF URANUS												
5	Miranda	130,000	1	09	56	0.01	3.4	(600)	1×10^{-5}		Kuiper (1948)	16.5
1	Ariel	191,000	2	12	29R	0.003	0	(1410 ± 105)	15×10^{-6}		Lassell (1851)	14.4
2	Umbriel	260,000	4	03	27R	0.004	0	(1160 ± 90)	6×10^{-6}		Lassell (1851)	15.3
3	Titania	436,000	8	16	56R	0.002	0	(1670 ± 90)	50×10^{-6}		W. Herschel (1787)	14.0
4	Oberon	583,000	13	11	07R	0.001	0	(1690 ± 90)	29×10^{-6}		W. Herschel (1787)	14.2
SATELLITES OF NEPTUNE												
3		75,000									Reitsema et al. (1981)	
	Triton	354,000	5	21	03R	0.000	160.0	3200 ± 400†	2×10^{-3}*		Lassell (1846)	13.6
	Nereid	5,570,000	365	5		0.76	27.4	(600)	10^{-6}		Kuiper (1949)	18.7
SATELLITE OF PLUTO												
	Charon	20,000?	6	9	17	0?	105?	(1000)	$0.05-0.1$?		Christy (1978)	16–17

Appendix 5
Brightest Stars

Star	Name	Position 1980.0 R.A.	Position 1980.0 Dec.	Apparent Magnitude (V)	Spectral Type		Absolute Magnitude	Distance (ly)	Proper Motion (s/yr)	Radial Velocity (km/s)
1. α CMa A	Sirius	06 44.2	− 16 42	− 1.46	A1	V	+ 1.42	8.7	1.324	− 7.6*
2. α Car	Canopus	06 23.5	− 52 41	− 0.72	F0	Ib-II	− 3.1	98	0.025	+ 20.5
3. α Boo	Arcturus	14 14.8	+ 19 17	− 0.06	K2	IIIp	− 0.3	36	2.284	− 5.2
4. α Cen A	Rigil Kentaurus	14 38.4	− 60 46	0.01	G2	V	+ 4.37	4.2	3.676	− 24.6
5. α Lyr	Vega	18 36.2	+ 38 46	0.04	A0	V	+ 0.5	26.5	0.345	− 13.9
6. α Aur	Capella	05 15.2	+ 45 59	0.05	G8 III? + F		− 0.6	45	0.435	+ 30.2*
7. β Ori A	Rigel	05 13.6	− 08 13	0.14v	B8	Ia	− 7.1	900	0.001	+ 20.7*
8. α CMi A	Procyon	07 38.2	+ 05 17	0.37	F5	IV-V	+ 2.6	11.4	1.250	− 3.2*
9. α Ori	Betelgeuse	05 54.0	+ 07 24	0.41v	M2	Iab	− 5.6	520	0.028	+ 21.0*
10. α Eri	Achernar	01 37.0	− 57 20	0.51	B3	Vp	− 2.3	118	0.098	+ 19
11. β Cen AB	Hadar	14 02.4	− 60 16	0.63v	B1	III	− 5.2	490	0.035	− 12*
12. α Aql	Altair	19 49.8	+ 08 49	0.76	A7	IV-V	+ 2.2	16.5	0.658	− 26.3
13. α Tau A	Aldebaran	04 34.8	+ 16 28	0.86v	K5	III	− 0.7	68	0.202	+ 54.1
14. α Vir	Spica	13 24.1	− 11 03	0.91v	B1	V	− 3.3	220	0.054	+ 1.0*
15. α Sco A	Antares	16 28.2	− 26 23	0.92v	M1 Ib + B		− 5.1	520	0.029	− 3.2
16. α PsA	Fomalhaut	22 56.5	− 29 44	1.15	A3	V	+ 2.0	22.6	0.367	+ 6.5
17. β Gem	Pollux	07 44.1	+ 28 05	1.16	K0	III	+ 1.0	35	0.625	+ 3.3
18. α Cyg	Deneb	20 40.7	+ 45 12	1.26	A2	Ia	− 7.1	1600	0.003	− 4.6*
19. β Cru	Beta Crucis	12 46.6	− 59 35	1.28v	B0.5	III	− 4.6	490	0.049	+ 20.0
20. α Leo A	Regulus	10 07.3	+ 12 04	1.36	B7	V	− 0.7	84	0.248	+ 3.5
21. α Cru A	Acrux	12 25.4	− 62 59	1.39	B0.5	IV	− 3.9	370	0.042	− 11.2*
22. ε CMa A	Adhara	06 57.8	− 28 57	1.48	B2	II	− 5.1	680	0.004	+ 27.4
23. λ Sco	Shaula	17 32.3	− 37 05	1.60v	B1	V	− 3.3	310	0.031	0
24. γ Ori	Bellatrix	05 24.0	+ 06 20	1.64	B2	III	− 4.2	470	0.015	+ 18.2
25. β Tau	Elnath	05 25.0	+ 28 36	1.65	B7	III	− 3.2	300	0.178	+ 8.0

Based on a table compiled by Donald A. MacRae in the *Observer's Handbook 1980* of the Royal Astronomical Society of Canada, amended with information from W. Gliese (private communication, 1979).

*Indicates an average value of a variable radial velocity.

Appendix 6
Greek Alphabet

Upper Case	Lower Case		Upper Case	Lower Case	
A	α	alpha	N	ν	nu
B	β	beta	Ξ	ξ	xi
Γ	γ	gamma	O	o	omicron
Δ	δ	delta	Π	π	pi
E	ε	epsilon	P	ρ	rho
Z	ζ	zeta	Σ	σ	sigma
H	η	eta	T	τ	tau
Θ	θ	theta	Υ	υ	upsilon
I	ι	iota	Φ	φ	phi
K	κ	kappa	X	χ	chi
Λ	λ	lambda	Ψ	ψ	psi
M	μ	mu	Ω	ω	omega

Appendix 7
The Nearest Stars

Name	R.A. (2000.0) h	m	Dec. (2000.0) °	'	Parallax π "	Distance pc	Proper Motion μ "/yr	θ °	Radial Vel km/s	Spectral Type	V	B-V	Mv	Luminosity (L☉ = 1)
1 Sun										G2 V	−26.72	0.65	4.85	1.0
2 Proxima Cen	14	30.0	−62	40	0.763	1.31	3.85	282	−16	M5.5 V	11.05	1.97	15.49	0.00006
α Cen A	14	39.6	−60	50	.750	1.33	3.68	281	−22	G2 V	−0.01	0.68	4.37	1.6
α Cen B										K1 V	1.33	0.88	5.71	0.45
3 Barnard's star	17	57.9	+04	41	.549	1.82	10.31	356	−108	M3.8 V	9.54	1.74	13.22	0.00045
4 Wolf 359 (CN Leo)	10	56.7	+07	00	.421	2.38	4.70	235	+13	M5.8 V	13.53	2.01	16.65	0.00002
5 BD +36°2147 = HD95735 (Lalande 21185)	11	03.4	+35	58	.400	2.50	4.78	187	−84	M2.1 V	7.50	1.51	10.50	0.0055
6 Sirius A	6	45.1	−16	43	.376	2.66	1.33	204	−8	A1 V	−1.46	0.00	1.42	23.5
Sirius B										DA	8.3	−0.12	11.2	0.003
7 L 726-8 = A	1	38.8	−17	57	.372	2.69	3.36	80	+29 }	M5.6 V	12.52		15.46	0.00006
UV Cet = B									+32 }		13.02	1.85	15.96	0.00004
8 Ross 154 (V1216 Sgr)	18	49.7	−23	49	.346	2.89	0.72	104	−4	M3.6 V	10.45	1.70	13.14	0.00048
9 Ross 248 (HH And)	23	41.9	+44	10	.313	3.19	3.18	176	−81	M4.9 V	12.29	1.91	14.78	0.00011
10 ε Eri	3	32.9	−09	28	.303	3.30	0.98	271	+16	K2 V	3.73	0.88	6.14	0.30
11 Ross 128 (FI Vir)	11	47.6	+00	48	.301	3.32	1.38	152	−13	M4.1 V	11.10	1.76	13.47	0.00036
12 61 Cyg A	21	06.9	+38	45	.295	3.39	5.22	52	−64	K3.5 V	5.22	1.17	7.56	0.082
61 Cyg B										K4.7 V	6.03	1.37	8.37	0.039
13 Procyon A	7	39.3	+05	14	.292	3.42	1.25	214	−3	F5 IV-V	0.37	0.42	2.64	7.65
Procyon B										DF	10.7		13.0	0.00055
14 L 789-6	22	38.4	−15	18	.291	3.44	3.26	46	−60	M5.5 V	12.18	1.96	14.49	0.00014
15 ε Ind	22	03.4	−56	47	.290	3.45	4.70	123	−40	K3 V	4.68	1.05	7.00	0.14
16 BD +43°44 A (GX And)	00	18.1	+44	00	.290	3.45	2.90	82	+13	M1.3 V	8.08	1.56	10.39	0.0061
+43°44 B (GQ And)	(Groombridge 34 AB)								+20	M3.8 V	11.06	1.80	13.37	0.00039
17 τ Ceti	1	44.1	−15	56	.288	3.47	1.92	297	−16	G8 V	3.50	0.72	5.72	0.45
18 BD +59°1915 A	18	42.9	+59	37	.287	3.48	2.29	325	0	M3.0 V	8.90	1.54	11.15	0.0030
+59°1915 B	(HD173739/40 = Struve 2398AB = ADS 11632AB)						2.27	323	+10	M3.5 V	9.69	1.59	11.94	0.0015
19 CD −36°15693 = HD217987 (Lacaille 9352)	23	05.9	−35	51	.283	3.53	6.90	79	+10	M1.3 V	7.35	1.48	9.58	0.013
20 G 51-15	8	29.9	+26	46	.276	3.62	1.27	242		M6.6 V	14.81	2.06	17.03	0.00001
21 L 725-32 (YZ Cet)	1	12.4	−16	59	.270	3.70	1.32	62	+28	M4.5 V	12.04	1.83	14.12	0.00020
22 BD + 5°1668	7	27.4	+05	13	.267	3.75	3.77	171	+26	M3.7 V	9.82	1.56	11.94	0.0015
23 CD −39°14192 = HD202560 (Lacaille 8760)	21	17.3	−38	52	.261	3.83	3.46	251	+21	K5.5 V	6.66	1.40	8.74	0.028
24 Kapteyn's star	5	11.2	−45	01	.259	3.86	8.72	131	+245	M0.0 V	8.84	1.56	10.88	0.0039
25 Krüger 60 A	22	28.1	+57	42	.253	3.95	0.86	246	−26	M3.3 V	9.85	1.62	11.87	0.0016
Krüger 60 B (DO Cep)										M5 V	11.3	1.8	13.3	0.0004
26 Ross 614 A	6	29.3	−02	48	.251	3.98	1.00	133	+24	M4.5 V	11.10	1.71	13.12	0.00049
Ross 614 B (V577 Mon)											14		16	0.00004
27 BD −12°4523	16	30.3	−12	39	.248	4.03	1.18	183	−13	M3.5 V	10.11	1.60	12.07	0.0013
28 van Maanen's star	0	49.0	+05	23	.232	4.31	2.99	155	+54	DB	12.37	0.56	14.20	0.00018
29 Wolf 424 A (FL Vir A)	12	33.4	+09	01	.229	4.37	1.76	279	−5 }	M5.3 V	13.16	1.80	14.97	0.00009
Wolf 424 B (FL Vir B)									}		13.4		15.2	0.00007
30 CD −37°15492 = HD225213	0	05.1	−37	21	.226	4.42	6.11	112	+23	M2.0 V	8.56	1.46	10.32	0.0065
31 L 1159-16 (TZ Ari)	2	00.2	+13	03	.224	4.46	2.09	149		M4.5 V	12.26	1.82	14.01	0.00022
32 BD +50°1725 = HD88230 (Groombridge 1618)	10	11.4	+49	27	.217	4.50	1.45	250	−26	K5.0 V	6.59	1.36	8.32	0.041
33 CD −46°11540	17	28.6	−46	54	.216	4.63	1.06	147		M2.7 V	9.37	1.53	11.04	0.0033
34 G 158-27	0	06.8	−07	32	.216	4.63	2.04	204		M5.5	13.74	1.95	15.39	0.00006
35 CD −49°13515 = HD204961	21	33.5	−49	00	.216	4.63	0.81	184	+8	M1.8 V	8.67	1.46	10.32	0.0065
36 BD +68°946	17	36.5	+68	20	.216	4.63	1.31	196	−22	M3.3 V	9.15	1.50	10.79	0.0042
37 CD −44°11909	17	37.1	−44	19	.212	4.72	1.16	217		M3.9 V	10.96	1.65	12.60	0.00079
38 G208−44 = A	19	53.9	+44	25	.212	4.72	0.74	143			13.41	1.90	15.03	0.00008
G208−45 = B											13.99	1.98	15.61	0.00005
39 L 145−141	11	45.4	−64	49	.210	4.76	2.68	97		DC	11.50	0.19	13.07	0.0052
40 BD −15°6290	22	53.2	−14	15	.209	4.78	1.14	124	+9	M3.9 V	10.17	1.60	11.77	0.0017
41 o²(40) Eri A	4	15.3	−07	39	.209	4.78	4.08	213	−42	K1 V	4.43	0.82	6.01	0.34
40 Eri B							4.07	212	−21	DA	9.52	0.03	11.10	0.0032
40 Eri C (DY Eri)									−45	M4.3 V	11.17	1.66	12.75	0.00069
42 BD +20°2465 (AD Leo)	10	19.6	+19	51	.206	4.85	0.49	264	+11	M3.3 V	9.43	1.54	11.00	0.0035
43 70 Oph A	18	05.5	+02	30	.200	5.00	1.12	167	−7	K0 V	4.22	0.86	5.76	0.43
70 Oph B										K5 V	6.00	0.86	7.54	0.084
44 BD +43°4305 (EV Lac)	22	46.9	+44	20	.200	5.00	0.83	236	−2	M3.8 V	10.2	1.6	11.7	0.0018
45 Altair	19	50.8	+08	52	.195	5.13	0.66	54	−26	A7 V	0.76	0.22	2.24	11.1
46 AC +79°3888	11	47.5	+78	41	.194	5.15	0.89	57	−119	M3.7 V	10.80	1.60	12.23	0.0011
47 G 9−38 = A	8	58.3	+19	45	.192	5.21	0.89	267		m	14.06	1.84	15.48	0.00006
LP426−40 = B										m	14.92	1.93	16.34	0.000025
48 BD +15°2620 = HD119850 (Lalande 25372)	13	45.6	+14	53	.189	5.29	2.30	129	+15	M1.7 V	8.49	1.44	9.91	0.0095

Courtesy of W. Gliese (1979), with new spectral types on a consistent scale by Robert F. Wing and Charles A. Dean (1983), new parallaxes from William van Altena (1983) and additional comments from Dorrit Hoffleit (1983). Transformed with precession and proper motion corrections to 2000.0 coordinates.

Appendix 8
Messier Catalogue

M	NGC	α h m 1980.0	δ ° '	m_v	Description
1	1952	5 33.3	+ 22 01	11.3	Crab Nebula in Taurus
2	7089	21 32.4	− 00 54	6.3	Globular cluster in Aquarius
3	5272	13 41.3	+ 28 29	6.2	Globular cluster in Canes Venatici
4	6121	16 22.4	− 26 27	6.1	Globular cluster in Scorpius
5	5904	15 17.5	+ 02 11	6	Globular cluster in Serpens
6	6402	17 38.9	− 32 11	6	Open cluster in Scorpius
7	6475	17 52.6	− 34 48	5	Open cluster in Scorpius
8	6523	18 02.4	− 24 23		Lagoon Nebula in Sagittarius
9	6333	17 18.1	− 18 30	7.6	Globular cluster in Ophiuchus
10	6254	16 56.0	− 04 05	6.4	Globular cluster in Ophiuchus
11	6705	18 50.0	− 06 18	7	Open cluster in Scutum
12	6218	16 46.1	− 01 55	6.7	Globular cluster in Ophiuchus
13	6205	16 41.0	+ 36 30	5.8	Globular cluster in Hercules
14	6402	17 36.5	− 03 14	7.8	Globular cluster in Ophiuchus
15	7078	21 29.1	+ 12 05	6.3	Globular cluster in Pegasus
16	6611	18 17.8	− 13 48	7	Open cluster and nebula in Serpens
17	6618	18 19.7	− 16 12	7	Omega Nebula in Sagittarius
18	6613	18 18.8	− 17 09	7	Open cluster in Sagittarius
19	6273	17 01.3	− 26 14	6.9	Globular cluster in Ophiuchus
20	6514	18 01.2	− 23 02		Trifid Nebula in Sagittarius
21	6531	18 03.4	− 22 30	7	Open cluster in Sagittarius
22	6656	18 35.2	− 23 55	5.2	Globular cluster in Sagittarius
23	6494	17 55.7	− 19 00	6	Open cluster in Sagittarius
24	6603	18 17.3	− 18 27	6	Open cluster in Sagittarius
25	IC4725	18 30.5	− 19 16	6	Open cluster in Sagittarius
26	6694	18 44.1	− 09 25	9	Open cluster in Scutum
27	6853	19 58.8	+ 22 40	8.2	Dumbbell Nebula; planetary nebula in Vulpecula
28	6626	18 23.2	− 24 52	7.1	Globular cluster in Sagittarius
29	6913	20 23.3	+ 38 27	8	Open cluster in Cygnus
30	7099	21 39.2	− 23 15	7.6	Globular cluster in Capricornus
31	224	0 41.6	+ 41 09	3.7	Andromeda Galaxy (Sb)
32	221	0 41.6	+ 40 45	8.5	Elliptical galaxy in Andromeda: companion to M31
33	598	1 32.8	+ 30 33	5.9	Spiral galaxy (Sc) in Triangulum
34	1039	2 40.7	+ 42 43	6	Open cluster in Perseus
35	2168	6 07.6	+ 24 21	6	Open cluster in Gemini
36	1960	5 35.0	+ 34 05	6	Open cluster in Auriga
37	2099	5 51.5	+ 32 33	6	Open cluster in Auriga
38	1912	5 27.3	+ 35 48	6	Open cluster in Auriga
39	7092	21 31.5	+ 48 21	6	Open cluster in Cygnus
40	—	—	—		Double star in Ursa Major
41	2287	6 46.2	− 20 43	6	Open cluster in Canis Major
42	1976	5 34.4	− 05 24		Orion Nebula
43	1982	5 34.6	− 05 18		Orion Nebula; smaller part
44	2632	8 38.8	+ 20 04	4	Praesepe; open cluster in Cancer
45	—	3 46.3	+ 24 03	2	The Pleiades; open cluster in Taurus

Appendix 8 continues on opposite page

M	NGC	α h m	1980.0	δ ° '	m_v	Description
46	2437	7 40.9		− 14 46	7	Open cluster in Puppis
47	2422	7 35.6		− 14 27	5	Open cluster in Puppis
48	2548	8 12.5		− 05 43	6	Open cluster in Hydra
49	4472	12 28.8		+ 08 07	8.9	Elliptical galaxy in Virgo
50	2323	7 02.0		− 08 19	7	Open cluster in Monoceros
51	5194	13 29.0		+ 47 18	8.4	Whirlpool Galaxy; spiral galaxy (Sc) in Canes Venatici
52	7654	23 23.3		+ 61 29	7	Open cluster in Cassiopeia
53	5024	13 12.0		+ 18 17	7.7	Globular cluster in Coma Berenices
54	6715	18 53.8		− 30 30	7.7	Globular cluster in Sagittarius
55	6809	19 38.7		− 31 00	6.1	Globular cluster in Sagittarius
56	6779	19 15.8		+ 30 08	8.3	Globular cluster in Lyra
57	6720	18 52.9		+ 33 01	9.0	Ring Nebula; planetary nebula in Lyra
58	4579	12 36.7		+ 11 56	9.9	Spiral galaxy (SBb) in Virgo
59	4621	12 41.0		+ 11 47	10.3	Elliptical galaxy in Virgo
60	4649	12 42.6		+ 11 41	9.3	Elliptical galaxy in Virgo
61	4303	12 20.8		+ 04 36	9.7	Spiral galaxy (Sc) in Virgo
62	6266	16 59.9		− 30 05	7.2	Globular cluster in Scorpius
63	5055	13 14.8		+ 42 08	8.8	Spiral galaxy (Sb) in Canes Venatici
64	4826	12 55.7		+ 21 48	8.7	Spiral galaxy (Sb) in Coma Berenices
65	3623	11 17.8		+ 13 13	9.6	Spiral galaxy (Sa) in Leo
66	3627	11 19.1		+ 13 07	9.2	Spiral galaxy (Sb) in Leo; companion to M65
67	2682	8 50.0		+ 11 54	7	Open cluster in Cancer
68	4590	12 38.3		− 26 38	8	Globular cluster in Hydra
69	6637	18 30.1		− 32 23	7.7	Globular cluster in Sagittarius
70	6681	18 42.0		− 32 18	8.2	Globular cluster in Sagittarius
71	6838	19 52.8		+ 18 44	6.9	Globular cluster in Sagittarius
72	6981	20 52.3		− 12 39	9.2	Globular cluster in Aquarius
73	6994	20 57.8		− 12 44		Open cluster in Aquarius
74	628	1 35.6		+ 15 41	9.5	Spiral galaxy (Sc) in Pisces
75	6864	20 04.9		− 21 59	8.3	Globular cluster in Sagittarius
76	650	1 40.9		+ 51 28	11.4	Planetary nebula in Perseus
77	1068	2 41.6		− 00 04	9.1	Spiral galaxy (Sb) in Cetus
78	2068	5 45.8		+ 00 02		Small emission nebula in Orion
79	1904	5 23.3		− 24 32	7.3	Globular cluster in Lepus
80	6093	16 15.8		− 22 56	7.2	Globular cluster in Scorpius
81	3031	9 54.2		+ 69 09	6.9	Spiral galaxy (Sb) in Ursa Major
82	3034	9 54.4		+ 69 47	8.7	Irregular galaxy (Irr) in Ursa Major
83	5236	13 35.9		− 29 46	7.5	Spiral galaxy (Sc) in Hydra
84	4374	12 24.1		+ 13 00	9.8	Elliptical galaxy in Virgo
85	4382	12 24.3		+ 18 18	9.5	Elliptical galaxy (S0) in Coma Berenices
86	4406	12 25.1		+ 13 03	9.8	Elliptical galaxy in Virgo
87	4486	12 29.7		+ 12 30	9.3	Elliptical galaxy (Ep) in Virgo
88	4501	12 30.9		+ 14 32	9.7	Spiral galaxy (Sb) in Coma Berenices
89	4552	12 34.6		+ 12 40	10.3	Elliptical galaxy in Virgo
90	4569	12 35.8		+ 13 16	9.7	Spiral galaxy (Sb) in Virgo
91	—	—		—		M58?
92	6341	17 16.5		+ 43 10	6.3	Globular cluster in Hercules
93	2447	7 43.6		− 23 49	6	Open cluster in Puppis
94	4736	12 50.1		+ 41 14	8.1	Spiral galaxy (Sb) in Canes Venatici
95	3351	10 42.8		+ 11 49	9.9	Barred spiral galaxy (SBb) in Leo

Appendix 8 continues on following page

M	NGC	α h m 1980.0	δ ° '	m_v	Description
96	3368	10 45.6	+11 56	9.4	Spiral galaxy (Sa) in Leo
97	3587	11 13.7	+55 08	11.1	Owl Nebula; planetary nebula in Ursa Major
98	4192	12 12.7	+15 01	10.4	Spiral galaxy (Sb) in Coma Berenices
99	4254	12 17.8	+14 32	9.9	Spiral galaxy (Sc) in Coma Berenices
100	4321	12 21.9	+15 56	9.6	Spiral galaxy (Sc) in Coma Berenices
101	5457	14 02.5	+54 27	8.1	Spiral galaxy (Sc) in Ursa Major
102	—	—	—		M101; duplication
103	581	1 31.9	+60 35	7	Open cluster in Cassiopeia
104	4594	12 39.0	−11 35	8	Sombrero Nebula; spiral galaxy (Sa) in Virgo
105	3379	10 46.8	+12 51	9.5	Elliptical galaxy in Leo
106	4258	12 18.0	+47 25	9	Spiral galaxy in (Sb) Canes Venatici
107	6171	16 31.8	−13 01	9	Globular cluster in Ophiuchus
108	3556	11 10.5	+55 47	10.5	Spiral galaxy (Sb) in Ursa Major
109	3992	11 56.6	+53 29	10.6	Barred spiral galaxy (SBc) in Ursa Major

Positions and magnitudes based on a table in the *Observer's Handbook 1980* of the Royal Astronomical Society of Canada.

Appendix 9
The Constellations

Latin Name	Genitive	Abbreviation	Translation
Andromeda	Andromedae	And	Andromeda*
Antlia	Antliae	Ant	Pump
Apus	Apodis	Aps	Bird of Paradise
Aquarius	Aquarii	Aqr	Water Bearer
Aquila	Aquilae	Aql	Eagle
Ara	Arae	Ara	Altar
Aries	Arietis	Ari	Ram
Auriga	Aurigae	Aur	Charioteer
Boötes	Boötis	Boo	Herdsman
Caelum	Caeli	Cae	Chisel
Camelopardalis	Camelopardalis	Cam	Giraffe
Cancer	Cancri	Cnc	Crab
Canes Venatici	Canum Venaticorum	CVn	Hunting Dogs
Canis Major	Canis Majoris	CMa	Big Dog
Canis Minor	Canis Minoris	CMi	Little Dog
Capricornus	Capricorni	Cap	Goat
Carina	Carinae	Car	Ship's Keel**
Cassiopeia	Cassiopeiae	Cas	Cassiopeia*
Centaurus	Centauri	Cen	Centaur*
Cepheus	Cephei	Cep	Cepheus*
Cetus	Ceti	Cet	Whale
Chamaeleon	Chamaeleonis	Cha	Chameleon
Circinus	Circini	Cir	Compass
Columba	Columbae	Col	Dove
Coma Berenices	Comae Berenices	Com	Berenice's Hair*
Corona Australis	Coronae Australis	CrA	Southern Crown
Corona Borealis	Coronae Borealis	CrB	Northern Crown
Corvus	Corvi	Crv	Crow
Crater	Crateris	Crt	Cup
Crux	Crucis	Cru	Southern Cross
Cygnus	Cygni	Cyg	Swan

Latin Name	Genitive	Abbreviation	Translation
Delphinus	Delphini	Del	Dolphin
Dorado	Doradus	Dor	Swordfish
Draco	Draconis	Dra	Dragon
Equuleus	Equulei	Equ	Little Horse
Eridanus	Eridani	Eri	River Eridanus*
Fornax	Fornacis	For	Furnace
Gemini	Geminorum	Gem	Twins
Grus	Gruis	Gru	Crane
Hercules	Herculis	Her	Hercules*
Horologium	Horologii	Hor	Clock
Hydra	Hydrae	Hya	Hydra* (water monster)
Hydrus	Hydri	Hyi	Sea serpent
Indus	Indi	Ind	Indian
Lacerta	Lacertae	Lac	Lizard
Leo	Leonis	Leo	Lion
Leo Minor	Leonis Minoris	LMi	Little Lion
Lepus	Leporis	Lep	Hare
Libra	Librae	Lib	Scales
Lupus	Lupi	Lup	Wolf
Lynx	Lyncis	Lyn	Lynx
Lyra	Lyrae	Lyr	Harp
Mensa	Mensae	Men	Table (mountain)
Microscopium	Microscopii	Mic	Microscope
Monoceros	Monocerotis	Mon	Unicorn
Musca	Muscae	Mus	Fly
Norma	Normae	Nor	Level (square)
Octans	Octantis	Oct	Octant
Ophiuchus	Ophiuchi	Oph	Ophiuchus* (serpent bearer)
Orion	Orionis	Ori	Orion*
Pavo	Pavonis	Pav	Peacock
Pegasus	Pegasi	Peg	Pegasus* (winged horse)
Perseus	Persei	Per	Perseus*
Phoenix	Phoenicis	Phe	Phoenix
Pictor	Pictoris	Pic	Easel
Pisces	Piscium	Psc	Fish
Piscis Austrinus	Piscis Austrini	PsA	Southern Fish
Puppis	Puppis	Pup	Ship's Stern**
Pyxis	Pyxidis	Pyx	Ship's Compass**
Reticulum	Reticuli	Ret	Net
Sagitta	Sagittae	Sge	Arrow
Sagittarius	Sagittarii	Sgr	Archer
Scorpius	Scorpii	Sco	Scorpion
Sculptor	Sculptoris	Scl	Sculptor
Scutum	Scuti	Sct	Shield
Serpens	Serpentis	Ser	Serpent
Sextans	Sextantis	Sex	Sextant
Taurus	Tauri	Tau	Bull
Telescopium	Telescopii	Tel	Telescope
Triangulum	Trianguli	Tri	Triangle
Triangulum Australe	Trianguli Australis	TrA	Southern Triangle
Tucana	Tucanae	Tuc	Toucan
Ursa Major	Ursae Majoris	UMa	Big Bear
Ursa Minor	Ursae Minoris	UMi	Little Bear
Vela	Velorum	Vel	Ship's Sails**
Virgo	Virginis	Vir	Virgin
Volans	Volantis	Vol	Flying Fish
Vulpecula	Vulpeculae	Vul	Little Fox

*Proper names
**Formerly formed the constellation Argo Navis, the Argonauts' ship.

Appendix 10
A Survey of the Sky

The Autumn Sky

As it grows dark on an autumn evening, the pointers in the Big Dipper will point upward toward Polaris. Almost an equal distance on the other side of Polaris as the distance from the pointers to Polaris is a W-shaped constellation named Cassiopeia. Cassiopeia, in Greek mythology, was married to Cepheus, the king of Ethiopia (and the subject of the constellation that neighbors Cassiopeia to the west). Cassiopeia appears sitting on a chair, as shown in the opening illustration of Chapter 4.

Continuing across the sky away from the pointers, we next come to the constellation Andromeda, who in Greek mythology was Cassiopeia's daughter. In Andromeda, you might see a faint hazy patch of light; this is actually the center of the galaxy nearest to our own and is known as the Great Galaxy in Andromeda. It takes a telescope or a long-exposure photograph to reveal that it is a galaxy. Though it is one of the nearest galaxies, it is much farther away than any of the individual stars that we see.

Southwest in the sky from Andromeda, but still high overhead, are four stars that appear to make a square known as the Great Square of Pegasus (Fig. 3–2).

If it is really dark out (which probably means that you are far from a city and also that the moon is not full or almost full), you will see the Milky Way crossing the sky high overhead. It will appear as a hazy band across the sky, with ragged edges and dark patches and rifts in it. The Milky Way passes right through Cassiopeia.

Moving northeast from Cassiopeia, along the Milky Way, we come to the constellation Perseus; he was the Greek hero who slew the Medusa. (He flew off on Pegasus, who is conveniently nearby in the sky and, from his winged mount, saw Andromeda, whom he saved.) On the edge of Perseus nearest to Cassiopeia, with a small telescope we can see two hazy patches of stars that are really clusters of hundreds of stars called *galactic clusters* (also called open clusters). This *double cluster in Perseus,* also known as h and χ (chi) Persei, provides two of the galactic clusters that are easiest to see with small telescopes.

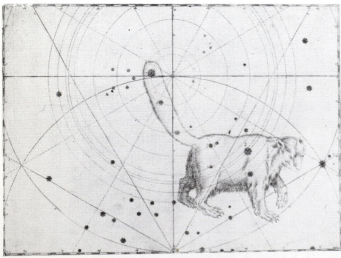

Figure A10–1 Polaris is the star at the tip of the tail of Ursa Minor, the Little Bear. The Little Dipper takes up most of the constellation. Because of precession, Polaris is somewhat closer to the north celestial pole than it was in about 1600, when this chart was drawn for Bayer's *Uranometria*. It was in this atlas of 1706 stars that Bayer began the system of assigning Greek letters to the brightest stars in each constellation. Polaris is α Ursae Minoris. (α, "alpha," is the first letter of the Greek alphabet.)

In the other direction from Cassiopeia (whose W is relatively easy to find), we find a cross of bright stars directly overhead. These are in the constellation Cygnus, the Swan. In the direction marked by Cygnus, in a location from which we receive a stream of wildly varying x-rays, we think a black hole is located, as we have discussed in Section 11.5. Also in Cygnus is a particularly dark region of the Milky Way, called the Northern Coal Sack. Dust in space in that direction prevents us from seeing as many stars as we do in other directions in the Milky Way. Cygnus contains a prominent grouping of stars known as the Northern Cross. Slightly to the west is another bright star, Vega, in the constellation Lyra (the Lyre). And farther westward we come to the constellation Hercules, named for the Greek hero who performed twelve great labors, of which the most famous was bringing back the golden apples. In Hercules is an older, larger type of star cluster called a globular cluster. It is known as M13, the *globular cluster in Hercules* (see the figure opening Chapter 6).

The Winter Sky

As the autumn proceeds and the winter approaches, the constellations we have discussed appear closer and closer to the western horizon for the same hour of the night. By early evening on January 1, Cygnus is setting in the western sky, while Cassiopeia and Perseus are overhead.

To the south of the Milky Way, near Perseus, we can now see a group of six stars close together in the sky. The grouping can catch your attention as you scan the sky. These are the Pleiades, traditionally the Seven Sisters of Greek mythology, the daughters of Atlas. This is another example of a galactic cluster. Binoculars or a small telescope reveal dozens of stars there; a large telescope will ordinarily show too small a region of sky for you to see the Pleiades well at all. So a bigger telescope isn't always better.

Further toward the east, rising earlier every evening, is the constellation Orion, the Hunter. Orion is perhaps the easiest constellation of all to pick out in the sky, for three bright stars close together in a line make up its belt. Orion is warding off Taurus, the Bull. A reddish star, Betelgeuse (beetle-juice would not be far wrong for pronunciation, though some say ''beh-tel-jooz''), marks Orion's shoulder, and symmetrically on the other side of his belt, the bright bluish star Rigel (rī′gel or ri′jel) marks his heel. Betelgeuse is an example of a red supergiant star; it is hundreds of

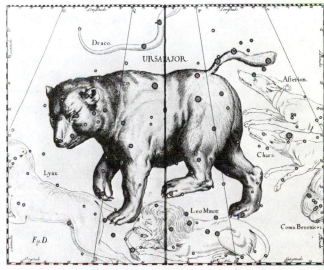

Figure A10–2 Ursa Major, the Big Bear, from Hevelius's atlas. It is a circumpolar constellation for most northern observers. The drawing is made from a point of view outside the celestial sphere, making the view reversed from what we see when we look up.

Figure A10–3 Orion, the Hunter, from Hevelius's atlas. The three stars in Orion's belt make it easy to find; Orion is prominent in the winter sky. The drawing is made from a point of view outside the celestial sphere, making the view reversed from what we see when we look up.

millions of kilometers across, bigger itself than the earth's orbit. Extending down from Orion's belt is his sword. A telescope, or a long exposure with a camera set to track the stars, reveals a beautiful region known as the Great Nebula in Orion, or the Orion Nebula (see Fig. 2–39). Its shape can be seen in even a smallish telescope; however, only photographs reveal its color (see Color Plate 56). It is the site of formation of new stars.

Rising after Orion is Sirius, the brightest star in the sky. Orion's belt points right to it. Sirius appears blue-white, which indicates that it is very hot. Sirius is sufficiently bright to stand out to the naked eye. It is part of the constellation Canis Major, the Great Dog. (You can remember its proximity to Orion by thinking of it as Orion's dog.)

Back toward the top of the sky, between the Pleiades and Orion's belt, is a V-shaped group of stars known as the Hyades. They mark the face of Taurus, the Bull, while the Pleiades are riding on Taurus's shoulder. The Hyades are an open cluster, as are the Pleiades. The origin of the bull and its riders comes from the Greek myth in which Jupiter turned himself into a bull to carry Europa over the sea.

The Spring Sky

We can tell that spring is approaching when the Hyades and Orion get closer and closer to the western horizon each night, and finally are no longer visible when the sun sets. Now the twins (Castor and Pollux), a pair of stars, are nicely placed for viewing in the western sky at sunset. Pollux is slightly reddish, while Castor is not, and they are about the same brightness. Castor and Pollux were the twins in the Greek pantheon of gods. The constellation is called Gemini, the Twins.

On spring evenings, the Big Bear is overhead, and anything in the dipper would spill out. Leo, the Lion, is just to the south of the zenith (follow the pointers backwards). Leo looks like a backward question mark, with the bright star Regulus at its base. Regulus marks the lion's heart. The rest of Leo, to the east of Regulus, is marked by a triangle of stars. Many people visualize a sickle-shaped head and a triangular tail.

If we follow an arc begun by the stars in the handle of the Big Dipper, we come to a bright reddish star, Arcturus, another supergiant. It is in the kite-shaped constellation Boötes, the Herdsman.

Figure A10–4   Argo Navis, the Argonauts' ship, is a southern constellation. This beautiful drawing from Bayer's atlas (1603) was made before the constellation was divided up into Puppis, the Stern; Pyxis, the Compass; Vela, the Sails; and Carina, the Keel.

Sirius sets right after sunset in the springtime; however, another bright star, Spica, is rising in the southeast in the constellation Virgo, the Virgin. It is farther along the arc from the Big Dipper through Arcturus. Vega, a star that is almost as bright, is rising in the northeast. And the constellation Hercules, with its globular cluster, is rising in the east in the evening at this time of year.

The Summer Sky

Summer, of course, is a comfortable time to watch the stars because of the warm weather. Spica is over toward the southwest in the evening. A bright reddish star, Antares, is in the constellation Scorpius, the Scorpion, to the south. ("Antares" means "compared with Ares," another name for Mars, because Antares is also reddish like the planet Mars.)

Hercules and Cygnus are high overhead, and the star Vega is prominent near the zenith. Cassiopeia is in the northeast. The center of our galaxy is in the direction of the dense part of the Milky Way that we see in the constellation Sagittarius, the Archer, low in the south.

Around August 12 every summer is a wonderful time to observe the sky, because that is when the Perseid meteor shower occurs. One bright meteor a minute may be visible at maximum. Just lie back on the grass and watch the sky in general— don't look in any direction specifically. (An outdoor concert is a good place to do this, if you can find a bit of grass away from spotlights.) Although the Perseids are the most observed meteor shower, partly because it occurs at a time of warm weather in the northern part of the country, many other meteor showers occur during the year. The most prominent are listed in Table 27–1.

The summer is a good time of year for observing a variable star, Delta Cephei; it appears in the constellation Cepheus, which is midway between Cassiopeia and Cygnus. Delta Cephei varies in brightness with a 4.3 day period (see Fig. 6–12). As we see in Chapter 6, studies of its variations have led to our being able to tell the distances to galaxies. And this takes us back to the real importance of studying the sky—not in just learning *where* things are but *what* they are and *how* they work. The study of the sky has led us to understand the universe, and this is the real importance and excitement of astronomy.

Appendix 11
Interstellar Molecules

Name of Molecule	Chemical Symbol	Year of Discovery	Part of Spectrum	First Wavelength Observed		Telescope Used for Discovery
methylidyne	CH	1937	visible	4300 Å	2.5-m	Mt. Wilson
cyanogen radical	CN	1940	visible	3875 Å	2.5-m	Mt. Wilson
methylidyne ion	CH^+	1941	visible	4232 Å	2.5-m	Mt. Wilson
hydroxyl radical	OH	1963	radio	18 cm	26-m	Lincoln Lab
ammonia	NH_3	1968	radio	1.3 cm	6-m	Hat Creek
water	H_2O	1968	radio	1.4 cm	6-m	Hat Creek
formaldehyde	H_2CO	1969	radio	6.2 cm	43-m	NRAO/Green Bank
carbon monoxide	CO	1970	radio	2.6 mm	11-m	NRAO/Kitt Peak
hydrogen cyanide	HCN	1970	radio	3.4 mm	11-m	NRAO/Kitt Peak
cyanoacetylene	HC_3N	1970	radio	3.3 cm	43-m	NRAO/Green Bank
hydrogen	H_2	1970	ultraviolet	1013–1108 Å		NRL rocket
methyl alcohol	CH_3OH	1970	radio	36 cm	43-m	NRAO/Green Bank
formic acid	HCOOH	1970	radio	18 cm	43-m	NRAO/Green Bank
"X-ogen"	HCO^+	1970	radio	3.4 mm	11-m	NRAO/Kitt Peak
formamide	$HCONH_2$	1971	radio	6.5 cm	43-m	NRAO/Green Bank
carbon monosulfide	CS	1971	radio	2.0 mm	11-m	NRAO/Kitt Peak
silicon monoxide	SiO	1971	radio	2.3 mm	11-m	NRAO/Kitt Peak
carbonyl sulfide	OCS	1971	radio	2.7 mm	11-m	NRAO/Kitt Peak
methyl cyanide, acetonitrile	CH_3CN	1971	radio	2.7 mm	11-m	NRAO/Kitt Peak
isocyanic acid	HNCO	1971	radio	3.4 mm	11-m	NRAO/Kitt Peak
methylacetylene	CH_3C_2H	1971	radio	3.5 mm	11-m	NRAO/Kitt Peak
acetaldehyde	CH_3CHO	1971	radio	28 cm	43-m	NRAO/Green Bank
thioformaldehyde	H_2CS	1971	radio	9.5 cm	64-m	Parkes
hydrogen isocyanide	HNC	1971	radio	3.3 mm	11-m	NRAO/Kitt Peak
hydrogen sulfide	H_2S	1972	radio	1.8 mm	11-m	NRAO/Kitt Peak
methanimine	H_2CNH	1972	radio	5.7 cm	64-m	Parkes
sulfur monoxide	SO	1973	radio	3.0 mm	11-m	NRAO/Kitt Peak
imidyl ion	N_2H^+	1974	radio	3.2 mm	11-m	NRAO/Kitt Peak
ethynyl radical	C_2H	1974	radio	3.4 mm	11-m	NRAO/Kitt Peak
methylamine	CH_3NH_2	1974	radio	3.5, 4.1 mm	11-m/6-m	NRAO and Tokyo
dimethyl ether	$(CH_3)_2O$	1974	radio	9.6 mm	11-m	NRAO/Kitt Peak
ethyl alcohol	CH_3CH_2OH	1974	radio	2.9–3.5 mm	11-m	NRAO/Kitt Peak
sulfur dioxide	SO_2	1975	radio	3.6 mm	11-m	NRAO/Kitt Peak
silicon sulfide	SiS	1975	radio	2.8, 3.3 mm	11-m	NRAO/Kitt Peak
acrylonitrile	H_2CCHCN	1975	radio	22 cm	64-m	Parkes
methyl formate	$HCOOCH_3$	1975	radio	18 cm	64-m	Parkes
nitrogen sulfide radical	NS	1975	radio	2.6 mm	5-m	Texas
cyanamide	NH_2CN	1975	radio	3.7 mm	11-m	NRAO/Kitt Peak
cyanodiacetylene	HC_5N	1976	radio	3.0 cm	46-m	Algonquin
formyl radical	HCO	1976	radio	3.5 mm	11-m	NRAO/Kitt Peak
acetylene	C_2H_2	1976	infrared	2.4 μ	4-m	KPNO
cyanohexatriyne	HC_7N	1977	radio	3 cm	46-m	Algonquin
cyanoethynyl radical	C_3N	1977	radio	3.4 mm	11-m	NRAO/Kitt Peak
ketene	H_2C_2O	1977	radio	3.0 mm	11-m	NRAO/Kitt Peak
nitroxyl	HNO	1977	radio	3.7 mm	11-m	NRAO/Kitt Peak
ethyl cyanide	CH_3CH_2CN	1977	radio	2.6–3.4 mm	11-m	NRAO/Kitt Peak
methane	CH_4	1977	radio	3.9 mm	11-m	NRAO/Kitt Peak
diatomic carbon	C_2	1977	infrared	1μ	1.5-m	Mt. Hopkins
butadinyl radical	C_4H	1978	radio	3 mm	11-m	NRAO/Kitt Peak
cyanoethylyne	C_3N	1978	radio	3 mm	11-m	NRAO/Kitt Peak
cyano-octatetra-yne	HC_9N	1978	radio	3 cm	46-m	Algonquin

Name of Molecule	Chemical Symbol	Year of Discovery	Part of Spectrum	First Wavelength Observed	Telescope	Used for Discovery
nitric oxide	NO	1978	radio	2 mm	11-m	NRAO/Kitt Peak
methyl mercaptan	CH_3SH	1979	radio	4 mm	7-m	Bell Labs
isothiocyanic acid	HNCS	1979	radio	3 mm	7-m	Bell Labs
ozone	O_3	1980	radio	1 mm	10.4-m	Owens Valley
cyano-tetracetylene	$HC_{11}N$	1981	radio	1.27 cm	46-m/37-m	Algonquin; Haystack
thioformyl radical	HCS^+	1981	radio	1.1–3.5 mm	7-m/11-m	Bell Labs; NRAO
methyl cyanoacetylene	CH_3C_3N	1983	radio	1.5–0.9 cm	37-m/43-m	Haystack; NRAO
methyl diacetylene	CH_3C_4H	1984	radio	1.5–1.2 cm	37-m	Haystack
silicon dicarbide	SiC_2	1984	radio	3.5–1.8 mm	7-m	Bell Labs; others
propynyl radical	C_3H	1984	radio	3.5–1.8 mm	7-m	Bell Labs; others
methyldiacetylene	CH_3C_4H	1984	radio	1.5–1.3 cm	37-m	Haystack

+ over 150 unidentified lines; only firm identifications are listed

Notes: Only the first wavelength or wavelengths observed are listed. Discoveries of forms including isotopes not included. Hat Creek is the site of the University of California's radio observatory. NRAO has telescopes at Green Bank, West Virginia, a millimeter-wave telescope on Kitt Peak in Arizona, and the VLA in New Mexico. The Naval Research Laboratory (NRL) is in Washington, D.C. The Australian National Radio Astronomy Observatory is at Parkes, N.S.W. The millimeter telescope at Fort Davis, Texas, is operated by the University of Texas. The Herzberg Institute's radio telescope is at Algonquin Park, Canada. The Kitt Peak National Observatory (KPNO) and Mt. Hopkins are in Arizona. The Lincoln Lab of MIT is in Lexington, Massachusetts. Bell Labs is in Holmdel, N. J. Caltech's Owens Valley Radio Observatory is in California. I thank Barry Turner of NRAO for information on the newest discoveries.

Appendix 12
Color Index

A measure of temperature often used for the horizontal axis in Hertzsprung-Russell diagrams is the difference between the apparent magnitudes measured in two spectral regions, for example B-V, blue minus visual. This standard system of filters was defined in Section 5.3a.

The difference B-V is called the *color index* (Fig. A12–1). The blue magnitude, B, measures bluer radiation than the visual magnitude, V. A very hot star is brighter in the blue than in the visible; thus V is fainter, that is, a higher number, than B. Therefore B-V is negative. Do not be confused by the fact that the magnitude of a brighter star is a lower number (more negative) than that of a fainter star.

Thus labeling the horizontal axis with the color index is equivalent to labeling it with temperature in kelvins. Since the color index can be measured easily with a telescope, and does not require theoretical interpretation or analysis of spectra before graphing, it is the measure of temperature ordinarily plotted. These plots, a type of H-R diagram, are often called *color-magnitude diagrams*.

The color index is zero for an A star of about 10,000 K. It falls in the range from -0.3 for the hottest stars to about $+2.0$ for the coolest. One can also compute a color index for the U and B (ultraviolet and blue) magnitudes, U-B, a color index for the m_{pg} and m_v (photographic and visual) magnitudes, or indeed a color index for magnitudes measured in any two spectral regions. These other color indices in the visual part of the spectrum have the same sense as the color index for B-V, that is, negative for hot stars and positive for cool ones. We must merely be certain that we are subtracting the longer wavelength measurement from the shorter wavelength measurement. Infrared astronomers use other standard systems of filters in their region of the spectrum.

Examples:
ζ Pup (blue-white) O5
$B = +1.96 \quad V = +2.25$
$C.I. = B\text{-}V = -0.29$
Betelgeuse (red) M2
$B = +2.55 \quad V = +0.69$
$C.I. = B\text{-}V = +1.86$

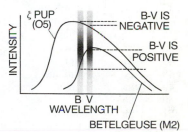

Figure A12–1 Hotter stars have negative color indices while cooler stars have positive color indices. (The star is brighter in the blue than in the visible, that is, B is a lower magnitude than V, so B-V is less than 0.)

Appendix 13
Elements and Solar Abundances

		Name	Atomic Weight	Abundance*			Name	Atomic Weight	Abundance*
1	H	hydrogen	1.01	10^{12}	55	Cs	cesium	123.91	$<7.9 \times 10^1$
2	He	helium	4.00	6.3×10^{10}	56	Ba	barium	137.34	1.2×10^2
3	Li	lithium	6.94	10×10^1	57	La	lanthanum	138.91	1.3×10^1
4	Be	beryllium	9.01	1.4×10^1	58	Ce	cerium	140.12	3.5×10^1
5	B	boron	10.81	1.3×10^2	59	Pr	praseodymium	140.91	4.6
6	C	carbon	12.01	4.2×10^8	60	Nd	neodymium	144.24	1.7×10^1
7	N	nitrogen	14.01	8.7×10^7	61	Pm	promethium	146	
8	O	oxygen	16.00	6.9×10^8	62	Sm	samarium	150.4	5.2
9	F	fluorine	19.00	3.6×10^4	63	Eu	europium	151.96	5.0
10	Ne	neon	20.18	3.7×10^7	64	Gd	gadolinium	157.25	1.3×10^1
11	Na	sodium	22.29	1.9×10^6	65	Tb	terbium	158.93	
12	Mg	magnesium	24.31	4.0×10^7	66	Dy	dysprosium	162.50	1.1×10^1
13	Al	aluminum	26.98	3.3×10^6	67	Ho	holmium	164.93	
14	Si	silicon	28.09	4.5×10^7	68	Er	erbium	167.26	5.8
15	P	phosphorus	30.97	3.2×10^5	69	Tm	thulium	168.93	1.8
16	S	sulphur	32.06	1.6×10^7	70	Yb	ytterbium	170.04	7.9
17	Cl	chlorine	35.45	3.2×10^5	71	Lu	lutetium	174.97	5.8
18	Ar	argon	39.95	1.0×10^6	72	Hf	hafnium	178.49	6.3
19	K	potassium	39.10	1.4×10^5	73	Ta	tantalum	180.95	
20	Ca	calcium	40.08	2.2×10^6	74	W	tungsten	183.85	5.0×10^1
21	Sc	scandium	44.96	1.1×10^3	75	Re	rhenium	186.2	≤ 0.5
22	Ti	titanium	47.90	1.1×10^5	76	Os	osmium	190.2	5.0
23	V	vanadium	50.94	1.0×10^4	77	Ir	iridium	192.2	7.1
24	Cr	chromium	52.00	5.1×10^5	78	Pt	platinum	195.09	5.6×10^1
25	Mn	manganese	54.94	2.6×10^5	79	Au	gold	196.97	5.6
26	Fe	iron	55.85	3.2×10^7	80	Hg	mercury	200.59	$<1.3 \times 10^2$
27	Co	cobalt	58.93	7.9×10^4	81	Tl	thallium	204.37	7.9
28	Ni	nickel	58.71	1.9×10^6	82	Pb	lead	207.19	8.5×10^1
29	Cu	copper	63.55	1.1×10^4	83	Bi	bismuth	208.98	$<7.9 \times 10^1$
30	Zn	zinc	65.37	2.8×10^4	84	Po	polonium	210	
31	Ga	gallium	69.72	6.3×10^2	85	At	astatine	210	
32	Ge	germanium	72.59	3.2×10^3	86	Rn	radon	222	
33	As	arsenic	74.92		87	Fr	francium	223	
34	Se	selenium	78.96		88	Ra	radium	226.03	
35	Br	bromine	79.90		89	Ac	actinium	227	
36	Kr	krypton	83.80		90	Th	thorium	232.04	1.6
37	Rb	rubidium	85.47	4.0×10^2	91	Pa	protactinium	230.04	
38	Sr	strontium	87.62	7.9×10^2	92	U	uranium	238.03	<4.0
39	Y	yttrium	88.91	1.3×10^2	93	Np	neptunium	237.05	
40	Zr	zirconium	91.22	3.6×10^2	94	Pu	plutonium	242	
41	Nb	niobium	92.91	7.9×10^1	95	Am	americium	242	
42	Mo	molybdenum	95.94	1.4×10^2	96	Cm	curium	245	
43	Tc	technetium	98.91		97	Bk	berkelium	248	
44	Ru	ruthenium	101.07	6.8×10^1	98	Cf	californium	252	
45	Rh	rhodium	102.91	2.5×10^1	99	Es	einsteinium	253	
46	Pd	palladium	106.4	3.2×10^1	100	Fm	fermium	257	
47	Ag	silver	107.87	7.1	101	Md	mendelevium	257	
48	Cd	cadmium	112.40	7.1×10^1	102	No	nobelium	255	
49	In	indium	114.82	4.5×10^1	103	Lr	lawrencium	256	
50	Sn	tin	118.69	1.0×10^2	104	Rf	rutherfordium	261	
51	Sb	antimony	121.75	1.0×10^1	105	Ha	hahnium	257–262	
52	Te	tellurium	127.60		106		not named; 1974	259–263	
53	I	iodine	126.90		107		not named; 1981	262	
54	Xe	xenon	131.30		108		not named; 1984	265	
					109		not named; 1982	266	

*Solar abundances are from John E. Ross and Lawrence H. Aller, *Science, 191,* 1223 (1976), relative to hydrogen, which is arbitrarily set to 10^{12}.

Atomic weights are averages for terrestrial abundances.

The year of discovery is given for unnamed elements.

Selected Readings

Monthly Non-Technical Magazines on Astronomy

Sky and Telescope, 49 Bay State Road, Cambridge, MA 02238.

Astronomy, 625 E. St. Paul Avenue, P.O. Box 92788, Milwaukee, WI 53202.

Mercury, Astronomical Society of the Pacific, 1290 24th Avenue, San Francisco, CA 94122.

The Griffith Observer, 2800 East Observatory Road, Los Angeles, CA 90027.

Magazines and Annuals Carrying Articles on Astronomy

Science 85, 86, 87, etc., P.O. Box 10790, Des Moines, IA 50340.

Science News, 1719 N Street, N.W., Washington, DC 20036. Published weekly.

Scientific American, 415 Madison Avenue, New York, NY 10017.

National Geographic, Washington, DC 20036.

Natural History, Membership Services, Box 4300, Bergenfield, N.J. 07621.

Physics Today, American Institute of Physics, 335 East 45 Street, New York, NY 10017.

Science Year (Chicago, IL: Field Enterprises Educational Corp.). The World Book Science Annual.

Smithsonian, 900 Jefferson Drive, Washington, DC 20560.

Yearbook of Science and the Future (Chicago, IL 60611: Encyclopaedia Britannica).

Science Digest, P.O. Box 10076, Des Moines, IA 50350.

Discover, 3435 Wilshire Blvd, Los Angeles, CA 90010.

Observing Reference Books

Donald H. Menzel and Jay M. Pasachoff, *A Field Guide to the Stars and Planets,* 2nd ed. (Boston: Houghton Mifflin Co., 1983). All kinds of observing information, including monthly maps and the 2000.0 sky atlas by Wil Tirion, and Graphic Timetables to locate planets and special objects like clusters.

Charles A. Whitney, *Whitney's Star Finder,* 3rd ed. (New York: Alfred A. Knopf, Inc., 1981).

Wil Tirion, *Star Atlas 2000.0* (Cambridge, MA: Sky Publishing Corp., 1981). Accurate and up-to-date. Available in black-on-white, white-on-black, and color versions.

Ben Mayer, *Starwatch* (New York: Perigee/Putnam, 1984). An inspirational introduction to observing.

Arthur P. Norton, *Norton's Star Atlas and Reference Handbook,* regularly revised by the successors of the late Mr. Norton (Cambridge, MA: Sky Publishing Corp.).

The Observer's Handbook (yearly), Royal Astronomical Society of Canada, 252 College Street, Toronto M5T IR7, Canada.

Guy Ottewell, *Astronomical Calendar* (yearly), Department of Physics, Furman University, Greenville, SC 29613.

The Astronomical Almanac (yearly), U.S. Government Printing Office, Washington, DC 20402.

Hans Vehrenberg, *Atlas of Deep Sky Splendors* (Cambridge, MA: Sky Publishing Corp., 4th ed., 1983). Photographs, descriptions, and finding charts for hundreds of beautiful objects.

General Reference Books

C. W. Allen, *Astrophysical Quantities,* 3rd ed. (London: The Athlone Press of the University of London, 1973). Tables and lists of almost every conceivable kind.

Kenneth R. Lang, *Astrophysical Formulae* (New York: Springer-Verlag New York, Inc., 1980). Formulas of all kinds, plus many tables.

Jeanne Hopkins, *Glossary of Astronomy and Astrophysics,* Revised ed. (Chicago: University of Chicago Press, 1980). A technical glossary compiled while editing the *Astrophysical Journal.*

National Geographic Society, *National Geographic Atlas of the World,* 5th ed. (Washington, DC 20036, 1981). Opens with 5 beautifully illustrated double-page spreads on astronomy.

For Information about Amateur Societies

American Association of Variable Star Observers (AAVSO), 187 Concord Avenue, Cambridge, MA 02138.

Association of Lunar and Planetary Observers (ALPO), 8930 Raven Drive, Waco, TX 76710.

Careers in Astronomy

Education Officer, American Astronomical Society, Sharp Laboratory, University of Delaware, Newark, DE 19711. A free booklet, *Careers in Astronomy,* is available on request.

Space for Women, derived from a symposium for women on careers. For free copies, write to: Center for Astrophysics, 60 Garden Street, Cambridge, MA 02138.

General Reading

Herbert Friedman, *The Amazing Universe* (Washington, DC: National Geographic Society, 1975). National Geographic's survey of modern astronomy, written by a pioneer in space science.

Kenneth R. Lang and Owen Gingerich, eds., *A Source Book in Astronomy and Astrophysics, 1900–1975* (Cambridge, MA: Harvard University Press, 1979). Reprints of fundamental articles.

Simon Mitton, ed., *The Cambridge Encyclopedia of Astronomy* (New York: Crown Publishers, 1977).

Harlow Shapley and H. E. Howarth, eds., *A Source Book in Astronomy* (New York: McGraw-Hill, 1929). Reprints.

Harlow Shapley, ed., *Source Book in Astronomy 1900–1950* (Cambridge, MA: Harvard University Press, 1960). Reprints.

Otto Struve and Velta Zebergs, *Astronomy of the Twentieth Century* (New York: Macmillan, 1962). A historical view.

George Seielstad, *Cosmic Ecology* (U. Cal. Press, 1983).

Collections of Articles

John C. Brandt and Stephen P. Maran, *The New Astronomy and Space Science Reader* (San Francisco: W. H. Freeman & Co., 1977). A varied collection of articles.

Owen Gingerich, ed., *Cosmology + 1* (San Francisco: W. H. Freeman & Co., 1977). Ten *Scientific American* reprints on cosmology, quasars, and black holes, plus one on the search for extraterrestrial intelligence.

Michael A. Seeds, ed., *Astronomy: Selected Readings* (Menlo Park, CA: Benjamin/Cummings, 1980). Articles from *Astronomy* magazine, including all their color illustrations.

Advanced Books

Frank H. Shu, *The Physical Universe* (Mill Valley, CA: University Science Books, 1982). An outstanding book; uses calculus.

Eugene H. Avrett, ed., *Frontiers in Astrophysics* (Cambridge, MA: Harvard University Press, 1976). A series of chapters, each written by an expert, on contemporary research.

Martin Harwit, *Astrophysical Concepts* (Ithaca: Cornell Univ. Press, 1984). For mathematical upper-division courses.

SOME ADDITIONAL BOOKS

Observatories and Observing

H. T. Kirby-Smith, *U.S. Observatories: A Directory and Travel Guide* (New York: Van Nostrand Reinhold, 1976).

David O. Woodbury, *The Glass Giant of Palomar* (New York: Dodd Mead, 1970). The story of the construction of the 5-m telescope.

Helen Wright, Joan N. Warnow, and Charles Weiner, eds., *The Legacy of George Ellery Hale* (Cambridge, MA: MIT Press, 1972). A beautifully illustrated historical treatment.

W. M. Smart, *Text-Book on Spherical Astronomy* (Cambridge: Cambridge University Press, 1977). The old standard, revised from the first (1931) edition and the later (1944) edition.

D. McNally, *Positional Astronomy* (New York: John Wiley & Sons, 1975). A more modern treatment than Smart.

Martin Cohen, *In Quest of Telescopes* (Cambridge, MA: Sky Publishing Corp., 1981). What it is like to be an astronomer.

Stars

Lawrence H. Aller, *Atoms, Stars, and Nebulae,* Revised ed. (Cambridge, MA: Harvard University Press, 1971). One of the Harvard Books on Astronomy series of popular works.

Bart J. Bok and Priscilla F. Bok, *The Milky Way,* 5th ed. (Cambridge, MA: Harvard University Press, 1981). A readable and well-illustrated survey from the Harvard Books on Astronomy series; includes good discussions of star clusters and H-R diagrams.

John A. Eddy, with Rein Ise, ed., *A New Sun: The Solar Results from Skylab* (NASA SP-402, 1979, GPO 033-000-00742-6).

Donald H. Menzel, *Our Sun,* Revised ed. (Cambridge, MA: Harvard University Press, 1959). A delightful general survey in the Harvard Books on Astronomy series.

Robert W. Noyes, *The Sun* (Cambridge, MA: Harvard University Press, 1982). A new member of the Harvard Books.

Robert Jastrow, *Red Giants and White Dwarfs,* 2nd ed. (New York: W. W. Norton Co., 1979).

Frederick Golden, *Quasars, Pulsars, and Black Holes* (New York: Scribner's, 1976).

Fire of Life, The Smithsonian Book of the Sun (Smithsonian Exposition Press, W. W. Norton Co., 1981).

Walter Sullivan, *Black Holes* (New York: Anchor Press/Doubleday, 1979).

Henry L. Shipman, *Black Holes, Quasars, and the Universe,* second edition (Boston: Houghton Mifflin, 1980). A careful discussion of several topics of great current interest.

Cecilia Payne-Gaposchkin, *Stars and Clusters* (Cambridge, MA: Harvard University Press, 1979).

Donald A. Cooke, *The Life and Death of Stars* (New York: Crown, 1984). A lavishly illustrated description of stellar evolution.

Solar System

Harold Masursky, G. W. Colton, and Farouk El-Baz, eds., *Apollo Over the Moon: A View from Orbit* (NASA SP 362, 1978, GPO033-000-00708-6).

Edgar M. Cortright, ed., *Apollo Expeditions to the Moon* (NASA SP 350, 1975, GPO033-000-00630-6).

Clark R. Chapman, Planets of Rock and Ice: from Mercury to the Moons of Saturn (New York: Charles Scribner's Sons, 1982). A nontechnical discussion of the planets and their moons.

Fred L. Whipple, *Orbiting the Sun: Planets and Satellites of the Solar System,* enlarged edition of *Earth, Moon, and Planets* (Cambridge, MA: Harvard University Press, 1981).

J. Kelly Beatty, Brian O'Leary, and Andrew Chaikin, *The New Solar System,* 2nd ed. Cambridge, MA: Sky Publishing Co., 1982). Each chapter written by a different expert.

David D. Morrison and Jane Samz, *Voyage to Jupiter* (NASA SP 439, 1980). The story of the Voyager missions to Jupiter and what we learned, masterfully told and profusely illustrated.

Bevan M. French, *The Moon Book* (New York: Penguin, 1977). A clear, authoritative, thorough, and interesting to read report on the lunar program and its results.

Bruce Murray, Michael C. Malin, and Ronald Greeley, *Earthlike Planets* (San Francisco: W. H. Freeman & Co., 1981).

Bruce Murray and Eric Burgess, *Flight to Mercury* (New York: Columbia Univ. Press, 1977).

James A. Dunne and Eric Burgess, *The Voyage of Mariner 10* (NASA SP 424, 1978).

Viking Lander Imaging Team, *The Martian Landscape* (NASA SP 425, 1978, GPO033-000-00716-7).

M. H. Carr and N. Evans, *Images of Mars—The Viking Extended Mission* (NASA SP-444, 1980).

Donald Goldsmith and Tobias Owen, *The Search for Life in the Universe* (Menlo Park, CA: Benjamin/Cummings, 1980).

David D. Morrison, *Voyages to Saturn* (NASA SP 451, 1982). Onward to Saturn, profusely illustrated.

Nigel Calder, *The Comet is Coming* (New York: Viking, 1981). About Halley's Comet.

Galaxies and the Outer Universe

Richard Berendzen, Richard Hart, and Daniel Seeley, *Man Discovers the Galaxies* (New York: Neale Watson Academic Publications, 1976). A historical review.

Charles A. Whitney, *The Discovery of Our Galaxy* (New York: Alfred A. Knopf, 1971). A historical discussion on a more popular level.

J. S. Hey, *The Evolution of Radio Astronomy* (New York: Neale Watson Academic Publications, 1973).

Gerrit L. Verschuur, *The Invisible Universe* (New York: Springer-Verlag, 1974). A non-technical treatment of radio astronomy.

David A. Allen, *Infrared, the New Astronomy* (New York: John Wiley & Sons, 1975). Includes a personal narrative.

Halton C. Arp, *Atlas of Peculiar Galaxies* (Pasadena, CA: California Institute of Technology, 1966). Worth poring over.

Allan Sandage, *The Hubble Atlas of Galaxies* (Washington, DC: Carnegie Institution of Washington, 1961), Publication No. 618. Beautiful photographs of galaxies and thorough descriptions. Everyone should examine this carefully.

Harlow Shapley, *Galaxies,* 3rd ed., revised by Paul W. Hodge (Cambridge, MA: Harvard University Press, 1972). A non-technical study of galaxies, written by the master. In the Harvard series.

Timothy Ferris, *The Red Limit* (New York: William Morrow & Co., 1977). Written for the general reader.

Nigel Calder, *Einstein's Universe* (New York: Viking, 1979). Mostly relativity.

William J. Kaufmann, *The Cosmic Frontiers of General Relativity* (Boston: Little, Brown, 1977).

William A. Fowler, *Nuclear Astrophysics* (Philadelphia: American Philosophical Society, 1967). A thin volume of popular lectures on the origin of the elements.

George Gamow, *One, Two, Three . . . Infinity* (New York: Bantam Books, 1971). A reprinting of a wonderful description of the structure of space that has introduced at an early age many a contemporary astronomer to his or her profession.

Steven Weinberg, *The First Three Minutes* (New York: Basic Books, 1977). A readable discussion of the first minutes after the big bang, including a discussion of the background radiation.

Timothy Ferris, *Galaxies* (New York: Stewart, Tabori, and Chang, 1982). Beautifully illustrated; paperback edition.

George O. Abell, consulting ed., and Sybil P. Parker, ed.-in-chief, McGraw-Hill Encyclopedia of Astronomy (New York: McGraw-Hill, 1982).

Rudolf Kippenhahn, *100 Billion Suns* (New York: Basic Books, 1983), a popular treatment of ''the birth, life, and death of the stars.''

John D. Barrow and Joseph Silk, *The Left Hand of Creation* (New York: Basic Books, 1983). Cosmology elucidated.

Glossary

absolute magnitude The magnitude that a star would appear to have if it were at a distance of ten parsecs from us.

absorption line Wavelengths at which the intensity of radiation is less than it is at neighboring wavelengths.

absorption nebula Gas and dust seen in silhouette.

accretion disk Matter that an object has taken up and which has formed a disk around the object.

achondrite A type of stony meteorite without chondrules. (See chondrite)

active galaxy A galaxy radiating much more than average in some part of the non-optical spectrum, revealing high-energy processes.

active regions Regions on the sun where sunspots, plages, flares, etc., are found.

active sun The group of solar phenomena that vary with time, such as active regions and their phenomena.

albedo The fraction of light reflected by a body.

allowed states The energy values that atoms can have by the laws of quantum mechanics.

alpha particle A helium nucleus; consists of two protons and two neutrons.

alt-azimuth A two-axis telescope mounting in which motion around one of the axes, which is vertical, provides motion in azimuth, and motion around the perpendicular axis provides up-and-down (altitude) motion.

altitude (a) Height above the surface of a planet, or (b) for a telescope mounting, elevation in angular measure above the horizon.

amino acid A type of molecule containing the group NH_2 (the amino group). Amino acids are fundamental building blocks of life.

Amor asteroids A group of asteroids with semimajor axes greater than earth's and between 1.017 A.U. and 1.3 A.U.; about half cross the earth's orbit.

angstrom A unit of length equal to 10^{-8} cm.

angular momentum An intrinsic property of a system corresponding to the amount of its revolution or spin. The amount of angular momentum of a body orbiting around a point is the mass of the orbiting body times its (linear) velocity of revolution times its distance from the point. The amount of angular momentum of a spinning sphere is the moment of inertia, an intrinsic property of the distribution of mass, times the angular velocity of spin.

angular velocity The rate at which a body rotates or revolves expressed as the angle covered in a given time (for example, in degrees per hour).

anisotropy Deviation from isotropy; changing with direction.

annular eclipse A type of solar eclipse in which a ring (annulus) of solar photosphere remains visible.

anorthosite A type of rock resulting from cooled lava, common in the lunar highlands though rare on earth.

antimatter A type of matter in which each particle (antiproton, antineutron, etc.) is opposite in charge and certain other properties to a corresponding particle (proton, neutron, etc.) of the same mass of the ordinary type of matter from which the solar system is made.

antiparticle The antimatter corresponding to a given particle.

aperture The diameter of the lens or mirror that defines the amount of light focused by an optical system.

aperture synthesis The use of several smaller telescopes together to give some of the properties, such as resolution, of a single larger aperture.

aphelion For an orbit around the sun, the farthest point from the sun.

Apollo asteroids A group of asteroids, with semimajor axes greater than earth's and less than 1.017 A.U., whose orbits overlap the earth's.

apparent magnitude The brightness of a star as seen by an observer, given in a specific system in which a difference of five magnitudes corresponds to a brightness ratio of one hundred times; the scale is fixed by correspondence with a historical background.

apsides (singular: apsis) The points at the ends of the major axis of an elliptical orbit. The *line of apsides* is the line that coincides with the major axis of the orbit.

archaebacteria A primitive type of organism, different from either plants or animals, perhaps surviving from billions of years ago.

association A physical grouping of stars; in particular, we talk of O and B associations or T associations.

asterism A special apparent grouping of stars, part of a constellation.

asteroid A "minor planet"; a non-luminous chunk of rock smaller than planet-size but larger than a meteoroid, in orbit around a star.

asteroid belt A region of the solar system, between the orbits of Mars and Jupiter, in which most of the asteroids orbit.

astrometric binary A system of two stars in which the existence of one star can be deduced by study of its gravitational effect on the proper motion of the other star.

astrometry The branch of astronomy that involves the detailed measurement of the positions and motions of stars and other celestial bodies.

Astronomical Unit The average distance from the earth to the sun.

astrophysics The science, now essentially identical with astronomy, applying the laws of physics to the universe.

Aten asteroids A group of asteroids with semimajor axes smaller than 1 A.U.

atom The smallest possible unit of a chemical element. When an atom is subdivided, the parts no longer have properties of any chemical element.

atomic clock A system that uses atomic properties to provide a measure of time.

atomic number The number of protons in an atom.

atomic weight The number of protons and neutrons in an atom, averaged over the abundances of the different isotopes.

aurora Glowing lights visible in the sky, resulting from processes in the earth's upper atmosphere and linked with the earth's magnetic field.

aurora australis The southern aurora.

aurora borealis The northern aurora.

autumnal equinox Of the two locations in the sky where the ecliptic crosses the celestial equator, the one that the sun passes each year when moving from northern to southern declinations.

A. U. Astronomical Unit.

azimuth The angular distance around the horizon from the northern direction, usually expressed in angular measure from 0° for an object in the northern direction, to 180° for an object in the southern direction, around to 360°.

background radiation See *primordial background radiation*.

Baily's beads Beads of light visible around the rim of the moon at the beginning and end of a total solar eclipse. They result from the solar photosphere shining through valleys at the edge of the moon.

Balmer series The set of spectral absorption or emission lines resulting from a transition down to or up from the second energy level (first excited level) of hydrogen.

bar The straight structure across the center of some spiral galaxies, from which the arms unwind.

baryons Nuclear particles (protons, neutrons, etc.) subject to the strong nuclear force; made of quarks.

basalt A type of rock resulting from the cooling of lava.

baseline The distance between points of observation when it determines the accuracy of some measurement.

beam The cone within which a radio telescope is sensitive to radiation.

Becklin-Neugebauer object An object visible only in the infrared in the Orion Molecular Cloud, apparently a very young star.

belts Dark bands around certain planets, notably Jupiter.

beta particle An electron or positron outside an atom.

big-bang theory A cosmological model, based on Einstein's general theory of relativity, in which the universe was once compressed to infinite density and has been expanding ever since.

binary pulsar A pulsar in a binary system.

binary star Two stars revolving around each other.

bipolar flow A phenomenon in young or forming stars in which streams of matter are ejected from the poles.

black body A hypothetical object that, if it existed, would absorb all radiation that hit it and would emit radiation that exactly followed Planck's law.

black dwarf A non-radiating ball of gas that results either when a white dwarf radiates all its energy or when gas contracts gravitationally but contains too little mass to begin nuclear fusion.

black hole A region of space from which, according to the general theory of relativity, neither radiation nor matter can escape.

blueshifted Wavelengths shifted to the blue; when the shift is caused by motion, from a velocity of approach.

B-N object See *Becklin-Neugebauer object.*

Bode's law A numerical scheme, known for 200 years, that gives the radii of the orbits of the seven innermost planets and the radius of the asteroid belt.

Bohr atom Niels Bohr's model of the hydrogen atom, in which the energy levels are depicted as concentric circles of radii that increase as (level number)^2.

Bok globule A type of round compact absorption nebula.

bolometer A device for measuring the total amount of radiation from an object.

bolometric magnitude The magnitude of a celestial object corrected to take account of the radiation in parts of the spectrum other than the visible.

bound-free transition An atomic transition in which an electron starts bound to the atom and winds up free from it.

breccia A type of rock made up of fragments of several types of rocks. Breccias are common on the moon.

burster One of the sources of bursts of x-rays.

calorie A measure of energy in the form of heat, originally corresponding to the amount of heat required to raise the temperature of 1 gram of water by 1°C.

canali Name given years ago to apparent lines on Mars.

capture A model in which a moon was formed elsewhere and then was captured gravitationally by its planet.

carbonaceous Containing a lot of carbon.

carbon cycle A chain of nuclear reactions, involving carbon as a catalyst at some of its intermediate stages, that transforms four hydrogen atoms into one helium atom with a resulting release in energy. The carbon cycle is only important in stars hotter than the sun.

carbon stars A spectral type of cool stars whose spectra show a lot of carbon; formerly types R and N.

carbon-nitrogen cycle The carbon cycle, acknowledging that nitrogen also plays an intermediary role.

carbon-nitrogen-oxygen tri-cycle A set of variations of the carbon cycle including nitrogen and oxygen isotopes as intermediaries.

Cassegrainian (Cassegrain) (telescope) A type of reflecting telescope in which the light focused by the primary mirror is intercepted short of its focal point and refocused and reflected by a secondary mirror through a hole in the center of the primary mirror.

Cassini's division The major division in the rings of Saturn.

catalyst A substance that participates in a reaction but that is left over in its original form at the end.

catastrophe theories Theories of solar-system formation involving a collision with another star.

CCD Charge-coupled device, a solid-state imaging device.

celestial equator The intersection of the celestial sphere with the plane that passes through the earth's equator.

celestial poles The intersection of the celestial sphere with the axis of rotation of the earth.

celestial sphere The hypothetical sphere centered at the center of the earth to which it appears that the stars are affixed.

center of mass The "average" location of mass; the point in a body or system of bodies at which we may consider all the mass to be located for the purpose of calculating the gravitational effect of that mass or its mean motion when a force is applied.

central star The hot object at the center of a planetary nebula, which is the remaining core of the original star.

Cepheid variable A type of supergiant star that oscillates in brightness in a manner similar to the star δ Cephei. The periods of Cepheid variables, which are between 1 and 100 days, are linked to the absolute magnitude of the stars by known relationships; this allows the distances to Cepheids to be found.

Chandrasekhar limit The limit in mass, about 1.4 solar masses, above which electron degeneracy cannot support a star, and so the limit above which white dwarfs cannot exist.

charm An arbitrary name that corresponds to a property that distinguishes certain elementary particles, including types of quarks, from each other.

chondrite A type of stony meteorite that contains small crystalline spherical particles called chondrules.

chromatic aberration A defect of lens systems in which different colors are focused at different points.

chromosphere The part of the atmosphere of the sun (or another star) between the photosphere and the corona. It is probably entirely composed of spicules and probably roughly corresponds to the region in which mechanical energy is deposited.

chromospheric network An apparent network of lines on the solar surface corresponding to the higher magnetic fields at the boundaries of supergranules; visible especially in the calcium H and K lines.

circle A conic section formed by cutting a cone perpendicularly to its axis.

circumpolar stars For a given observing location, stars that are close enough to the celestial pole that they never set.

classical When discussing atoms, not taking account of quantum mechanical effects.

closed universe A big-bang universe with positive curvature; it has finite volume and will eventually contract.

cluster (a) Of stars, a physical grouping of many stars; (b) of galaxies, a physical grouping of at least a few galaxies.

cluster variable A star that varies in brightness with a period of 0.1 to one day, similar to the star RR Lyrae. They are found in globular clusters. All cluster variables have approximately the same brightness, which allows their distance to be readily found.

CNO tri-cycle See *carbon-nitrogen-oxygen tri-cycle.*

color (a) Of an object, a visual property that depends on wavelength; (b) an arbitrary name assigned to a property that distinguishes three kinds of quarks.

color index The difference, expressed in magnitudes, of the brightness of a star or other celestial object measured at two different wavelengths. The color index is a measure of temperature.

coherent radiation Radiation in which the phases of waves at different locations in a cross-section of radiation have a definite relation to each other; in non-coherent radiation, the phases are random. Only coherent radiation shows interference.

color-magnitude diagram A Hertzsprung-Russell diagram in which the temperature on the horizontal axis is expressed in terms of color index and the vertical axis is in magnitudes.

coma   (a) Of a comet, the region surrounding the head; (b) of an optical system, an off-axis aberration in which the images of points appear with comet-like asymmetries.

comet   A type of object orbiting the sun, often in a very elongated orbit, that when relatively near to the sun shows a coma and may show a tail.

comparative planetology   Studying the properties of solar-system bodies by comparing them.

comparison spectrum   A spectrum of known elements on earth usually photographed on the same photographic plate as a stellar spectrum in order to provide a known set of wavelengths for zero Doppler shift.

composite spectrum   The spectrum that reveals a star is a binary system, since spectra of more than one object are apparent.

condensation   A region of unusually high mass or brightness.

conic sections   Geometric shapes obtained by slicing a cone.

conjunction   When two celestial objects reach the same celestial longitude; approximately corresponds to their closest apparent approach in the sky. When only one body is named, it is understood that the second body is the sun.

conservation law   A statement that the total amount of some property (angular momentum, energy, etc.) of a body or set of bodies does not change.

constellation   One of 88 areas into which the sky has been divided for convenience in referring to the stars or other objects therein.

contact binaries   Binaries in which both members fill their Roche lobes.

continental drift   The slow motion of the continents across the Earth's surface, explained in the theory of plate tectonics as a set of shifting regions called plates.

continuous spectrum   A spectrum with radiation at all wavelengths but with neither absorption nor emission lines.

continuum   (pl: continua) The continuous spectrum that we would measure from a body if no spectral lines were present.

convection   The method of energy transport in which the rising motion of masses from below carries energy upward in a gravitational field. Boiling is an example.

convection zone   The subsurface zone in certain types of stars in which convection dominates energy transfer.

convergent point   The point in the sky toward which the members of a star cluster appear to be converging or from which they appear to be diverging. Referred to in the moving cluster method of determining distances.

coordinate systems   Methods of assigning positions with respect to suitable axes.

core   The central region of a star or planet.

corona, galactic   The outermost region of our current model of the galaxy, containing most of the mass in some unknown way.

corona, solar or *stellar*   The outermost region of the sun (or of other stars), characterized by temperatures of millions of kelvins.

coronagraph   A type of telescope with which the corona can be seen in invisible light at times other than that of a total solar eclipse.

coronal holes   Relatively dark regions of the corona having low density; they result from open field lines.

correcting plate   A thin lens of complicated shape at the front of a Schmidt camera that compensates for spherical aberration.

cosmic abundances   The overall abundances of elements in the universe.

cosmic background radiation   The isotropic glow of 3° radiation.

cosmic rays   Nuclear particles or nuclei travelling through space at high velocity.

cosmic turbulence   A theory of galaxy formation in which turbulence originally existed.

cosmogony   The study of the origin of the universe, usually applied in particular to the origin of the solar system.

cosmological constant   A constant arbitrarily added by Einstein to an equation in his general theory of relativity in order to provide a solution in which the universe did not expand. It was only subsequently discovered that the universe did expand after all.

cosmological distances   Distances assigned by Hubble's law.

cosmological principle   The principle that on the whole the universe looks the same in all directions and in all regions.

cosmology   The study of the universe as a whole.

coudé focus   A focal point of large telescopes in which the light is reflected by a series of mirrors so that it comes to a point at the end of a polar axis. The image does not move even when the telescope moves, which permits the mounting of heavy equipment.

crust   The outermost solid layer of some objects, including neutron stars and some planets.

C-type stars   See *carbon stars*.

cytherean   Venusian.

dark nebula   Dust and gas seen in silhouette.

daughter molecules   Relatively simple molecules in comets resulting from the breakup of more complex molecules.

deceleration parameter (q_0)   A particular measure of the rate at which the expansion of the universe is slowing down. $q_0 < \frac{1}{2}$ corresponds to an open universe and $q_0 > \frac{1}{2}$ corresponds to a closed universe.

declination   Celestial latitude, measured in degrees north or south of the celestial equator.

deferent   In the Ptolemaic system of the universe, the larger circle, centered at the earth, on which the centers of the epicycles revolve.

degenerate matter   Matter whose properties are controlled, and which is prevented from further contraction, by quantum mechanical laws.

density   Mass divided by volume.

density wave   A circulating region of relatively high density, important, for example, in models of spiral arms.

density-wave theory   The explanation of spiral structure of galaxies as the effect of a wave of compression that rotates around the center of the galaxy and causes the formation of stars in the compressed region.

detached binaries   Binaries in which neither component fills its Roche lobe.

deuterium   An isotope of hydrogen that contains 1 proton and 1 neutron.

deuteron   A deuterium nucleus, containing 1 proton and 1 neutron.

diamond-ring effect   The last Baily's bead glowing brightly at the beginning of the total phase of a solar eclipse, or its counterpart at the end of totality.

differential effect   Different from location to location.

differential forces   A net force resulting from the difference of two other forces; a tidal force.

differential rotation   Rotation of a body in which different parts have different angular velocities (and thus different periods of rotation).

differentiation   For a planet, the existence of layers of different structure or composition.

diffraction   A phenomenon affecting light as it passes any obstacle, spreading it out in a complicated fashion.

diffraction grating   A very closely ruled series of lines that, through their diffraction of light, provide a spectrum of radiation that falls on it.

dirty snowball   A theory explaining comets as amalgams of ices, dust, and rocks.

discrete   Separated; isolated.

disk   (a) Of a galaxy, the disk-like flat portion, as opposed to the nucleus or the halo; (b) of a star or planet, the two-dimensional projection of its surface.

dispersion   (a) Of light, the effect that different colors are bent by different amounts when passing from one substance to another; (b) of the pulses of a pulsar, the effect that a given pulse, which leaves the pulsar at one instant, arrives at the earth at different times depending on the different wavelength or frequency at which it is observed. Both of these effects arise because light of different wavelengths travels at different speeds except in a vacuum.

D lines   A pair of lines from sodium that appear in the yellow part of the spectrum.

DNA   Deoxyribonucleic acid, a long chain of molecules that contains the genetic information of life.

Dobsonian   An inexpensive type of large-aperture amateur telescope characterized by a thin mirror, composition tube, and Teflon bearings on an alt-azimuth mount.

Doppler effect   A change in wavelength that results when a source of waves and the observer are moving relative to each other.

double star   A binary star; two or more stars orbiting each other.

double-lobed structure   An object in which radio emission comes from a pair of regions on opposite sides.

dust tail   The dust left behind a comet, reflecting sunlight.

dwarf ellipticals   Small, low-mass elliptical galaxies.

dwarf stars   Main-sequence stars.

dwarfs   Dwarf stars.

dynamo   A device that generates electricity through the effect of motion in the presence of a magnetic field.

dynamo theories   Explaining sunspots and the solar activity cycle through an interaction of motion and magnetic fields.

$E = mc^2$ Einstein's formula (special theory of relativity) for the equivalence of mass and energy.

early universe   The universe during its first minutes.

earthshine   Sunlight illuminating the moon after having been reflected by the earth.

eccentric   Deviating from a circle.

eccentricity   A measure of the flatness of an ellipse, defined as half the distance between the foci divided by the semimajor axis.

eclipse   The passage of all or part of one astronomical body into the shadow of another.

eclipsing binary   A binary star in which one member periodically hides the other.

ecliptic   The path followed by the sun across the celestial sphere in the course of a year.

ecliptic plane   The plane of the earth's orbit around the sun.

electric field   A force field set up by an electric charge.

electromagnetic force   One of the four fundamental forces of nature, giving rise to electromagnetic radiation.

electromagnetic radiation   Radiation resulting from changing electric and magnetic fields.

electromagnetic spectrum   Energy in the form of electromagnetic waves, in order of wavelength.

electromagnetic waves   Waves of changing electric and magnetic fields, travelling through space at the speed of light.

electron   A particle of one negative charge, 1/1830 the mass of a proton, that is not affected by the strong force. It is a lepton.

electron degeneracy   The state in which, following rules of quantum mechanics, the further compression of electrons generates a high pressure that balances gravity, as in white dwarfs.

electron volt (eV)   The energy necessary to raise an electron through a potential of one volt.

electroweak force   The unified electromagnetic and weak forces, according to a recent theory.

element   A kind of atom characterized by a certain number of protons in its nucleus. All atoms of a given element have similar chemical properties.

elementary particle   One of the constituents of an atom.

ellipse   A curve with the property that the sum of the distances from any point on the curve to two given points, called the foci, is constant.

elliptical galaxy   A type of galaxy characterized by elliptical appearance.

emission line   Wavelengths (or frequencies) at which the intensity of radiation is greater than it is at neighboring wavelengths (or frequencies).

emission nebula   A glowing cloud of interstellar gas.

emulsion   A coating whose sensitivity to light allows photographic recording of incident radiation.

energy   A fundamental quantity usually defined in terms of the ability of a system to do something that is technically called ''work,'' that is, the ability to move an object by application of force, where the work is the force times the displacement.

energy level   A state corresponding to an amount of energy that an atom is allowed to have by the laws of quantum mechanics.

energy problem   For quasars, how to produce so much energy in such a small emitting volume.

energy, law of conservation of   Energy is neither created nor destroyed, but may be changed in form.

ephemeris   A listing of astronomical positions and other data that change with time. From the same root as *ephemeral*.

epicycle   In the Ptolemaic theory, a small circle, riding on a larger circle called the deferent, on which a planet moves. The epicycle is used to account for retrograde motion.

equal areas, law of   Kepler's second law.

equant   In Ptolemaic theory, the point equally distant from the center of the deferent as the earth but on the opposite side, around which the epicycle moves at a uniform angular rate.

equator   (a) Of the earth, a great circle on the earth, midway between the poles; (b) celestial, the projection of the earth's equator onto the celestial sphere; (c) galactic, the plane of the disk as projected onto a map.

equatorial mount   A type of telescope mounting in which one axis, called the polar axis, points toward the celestial pole and the other axis is perpendicular. Motion around only the polar axis is sufficient to completely counterbalance the effect of the earth's rotation.

equinox   An intersection of the ecliptic and the celestial equator. The center of the sun is geometrically above and below the horizon for equal lengths of time on the two days of the year when the sun passes the equinoxes; if the sun were a point and atmospheric refraction were absent, then day and night would be of equal length on those days.

erg   A unit of energy in the metric system, corresponding to the work done by a force of one dyne (the force that is required to accelerate one gram by one cm/s^2) producing a displacement of one centimeter.

ergosphere   A region surrounding a rotating black hole (or other system satisfying Kerr's solution) from which work can be extracted.

escape velocity   The velocity that an object must have to escape the gravitational pull of a mass.

event horizon   The sphere around a black hole from within which nothing can escape; the place at which the exit cones close.

evolutionary track   The set of points on an H-R diagram showing the changes of a star's temperature and luminosity with time.

excitation   The raising of atoms to higher energy states than the lowest possible.

exclusion principle   The quantum mechanical rule that certain types of elementary particles cannot exist in completely identical states.

exit cone   The cone that, for each point within the photon sphere of a black hole, defines the directions of rays of radiation that escape.

exobiology   The study of life located elsewhere than earth.

exponent   The ''power'' representing the number of times a number is multiplied by itself.

exponential notation   The writing of numbers as a power of 10 times a number with one digit before the decimal point.

extended objects   Objects with detectable angular size.

extinction   The dimming of starlight by scattering and absorption as the light traverses interstellar space.

extragalactic   Exterior to the Milky Way Galaxy.

extragalactic radio sources   Radio sources outside our galaxy.

eyepiece   The small combination of lenses at the eye end of a telescope, used to examine the image formed by the objective.

featherweight stars   Stars of less than 0.07 solar masses.

field of view   The angular expanse viewable.

filament   A feature of the solar surface seen in Hα as a dark wavy line; a prominence projected on the solar disk.

fireball   An exceptionally bright meteor.

fission, nuclear   The splitting of an atomic nucleus.

flare   An extremely rapid brightening of a small area of the surface of the sun, usually observed in hydrogen-alpha and other strong spectral lines and accompanied by x-ray and radio emission.

flash spectrum   The solar chromospheric spectrum seen in the few seconds before or after totality at a solar eclipse.

flavors   A way of distinguishing quarks: up, down, strange, charmed, truth (top), beauty (bottom).

fluorescence   The transformation of photons of relatively high energy to photons of lower energy through interactions with atoms, and the resulting radiation.

flux   The amount of something (such as energy) passing through a surface per unit time.

flux tube   A torus through which particles circulate.

focal length   The distance from a lens or mirror to the point to which rays from an object at infinity are focused.

focus (pl: foci)   (a) A point to which radiation is made to converge; (b) of an ellipse, one of the two points the sum of the distances to which remains constant.

force   In physics, something that can or does cause change of momentum, measured by the rate of change of momentum with time.

Fraunhofer lines   The absorption lines of a solar or other stellar spectrum.

frequency   The rate at which waves pass a given point.

full moon   The phase of the moon when the side facing the earth is fully illuminated by sunlight.

fusion   The amalgamation of nuclei into heavier nuclei.

fuzz   The faint light detectable around nearby quasars.

galactic cannibalism   The incorporation of one galaxy into another.

galactic cluster   An asymmetric type of collection of stars that shared a common origin.

galactic corona   The outermost part of our galaxy.

galactic year   The length of time the sun takes to complete an orbit of our galactic center.

Galilean satellites   The four brightest satellites of Jupiter.

gamma rays   Electromagnetic radiation with wavelengths shorter than approximately 1 Å.

gas tail   The puffs of ionized gas trailing a comet.

general theory of relativity   Einstein's 1916 theory of gravity.

geocentric   Earth-centered.

geology   The study of the earth, or of other solid bodies.

geothermal energy   Energy from under the earth's surface.

giants   Stars that are larger and brighter than main-sequence stars of the same color.

giant ellipticals   Elliptical galaxies that are very large.

giant molecular cloud   A basic building block of our galaxy, containing dust, which shields the molecules present.

giant planets   Jupiter, Saturn, Uranus, and Neptune.

gibbous moon   The phases between half moon and full moon.

globular cluster   A spherically symmetric type of collection of stars that shared a common origin.

gluon   The particle that carries the color force (and thus the strong nuclear force).

grand unified theories (GUT's)   Theories unifying the electroweak force and the strong force.

granulation   Convection cells on the sun about 1 arc sec across.

grating   A surface ruled with closely spaced lines that, through diffraction, breaks up light into its spectrum.

gravitational force   One of the four fundamental forces of nature, the force by which two masses attract each other.

gravitational instability   A situation that tends to break up under the force of gravity.

gravitational interlock   One body controlling the orbit or rotation of another by gravitational attraction.

gravitational lens   In the gravitational-lens phenomenon, a massive body changes the path of electromagnetic radiation passing near it so as to make more than one image of an object. The double quasar was the first example to be discovered.

gravitationally   Controlled by the force of gravity.

gravitational radius   The radius that, according to Schwarzschild's solutions to Einstein's equations of the general theory of relativity, corresponds to the event horizon of a black hole.

gravitational redshift   A redshift of light caused by the presence of mass, according to the general theory of relativity.

gravitational waves   Waves that many scientists consider to be a consequence, according to the general theory of relativity, of changing distributions of mass.

gravity assist   Using the gravity of one celestial body to change a spacecraft's energy.

grazing incidence   Striking at a low angle.

great circle   The intersection of a plane that passes through the center of a sphere with the surface of that sphere; the largest possible circle that can be drawn on the surface of a sphere.

Great Red Spot   A giant circulating region on Jupiter.

greenhouse effect   The effect by which the atmosphere of a planet heats up above its equilibrium temperature because it is transparent to incoming visible radiation but opaque to the infrared radiation that is emitted by the surface of the planet.

Gregorian (telescope)   A type of reflecting telescope in which the light focused by the primary mirror passes its prime focus and is then reflected by a secondary mirror through a hole in the center of the primary mirror.

Gregorian calendar   The calendar in current use, with normal years that are 365 days long, with leap years every fourth year except for years that are divisible by 100 but not by 400.

ground level   An atom's lowest possible energy level.

ground state   See *ground level*.

GUT's   See *grand unified theories*.

H_0 The Hubble constant.

H I region   An interstellar region of neutral hydrogen.

H II region   An interstellar region of ionized hydrogen.

H line   The spectral line of ionized calcium at 3968 Å.

Hα   The first line of the Balmer series of hydrogen, at 6563 Å.

half-life   The length of time for half a set of particles to decay through radioactivity or instability.

halo   Of a galaxy, the region of the galaxy that extends far above and below the plane of the galaxy, containing globular clusters.

head   Of a comet, the nucleus and coma together.

head-tail galaxies   Double-lobed radio galaxies whose lobes are so bent that they look like a tadpole.

heat flow   The flow of energy from one location to another.

heavyweight stars   Stars of more than about 8 solar masses.

heliacal rising   The first time in a year that an astronomical body rises sufficiently far ahead of the sun that it can be seen in the morning sky.

heliocentric   Sun-centered; using the sun rather than the earth as the point to which we refer. A heliocentric measurement, for example, omits the effect of the Doppler shift caused by the earth's orbital motion.

helium flash   The rapid onset of fusion of helium into carbon through the triple-alpha process that takes place in most red giant stars.

Herbig-Haro objects   Blobs of gas ejected in star formation.

hertz   The measure of frequency, with units of /s (per second); formerly called cycles per second.

Hertzsprung gap   A region above the main sequence in a Hertzsprung-Russell diagram through which stars evolve rapidly and thus in which few stars are found.

Hertzsprung-Russell diagram   A graph of temperature (or equivalent) vs. luminosity (or equivalent) for a group of stars.

highlands   Regions on the moon or elsewhere that are above the level that may have been smoothed by flowing lava.

high-energy astrophysics   The study of x-rays, gamma rays, and cosmic rays, and of the processes that make them.

homogeneity   Uniformity throughout.

horizontal branch   A part of the Hertzsprung-Russell diagram of a globular cluster, corresponding to stars that are all at approximately zero absolute magnitude and have evolved past the red-giant stage and are moving leftward on the diagram.

hour angle   Of a celestial object as seen from a particular location, the difference between the local sidereal time and the right ascension (H. A. = L. S. T. − R. A.).

hour circle   The great circles passing through the celestial poles.

H-R diagram   Hertzsprung-Russell diagram.

Hubble constant (H_0)   The constant of proportionality in Hubble's law linking the velocity of recession of a distant object and its distance from us.

Hubble type   Hubble's galaxy classification scheme: E0, E7, Sa, SBa, etc.

Hubble's law   The linear relation between the velocity of recession of a distant object and its distance from us, $V = H_0 d$.

hyperfine level   A subdivision of an energy level caused by such relatively minor effects as changes resulting from the interactions among spinning particles in an atom or molecule.

IC   Index Catalogue, one of the supplements to Dreyer's *New General Catalogue*.

igneous   Rock cooled from lava.

image tube   An electronic device that receives incident radiation and intensifies it or converts it to a wavelength at which photographic plates are sensitive.

inclination   Of an orbit, the angle of the plane of the orbit with respect to the ecliptic plane.

Index Catalogue   See *IC*.

inferior conjunction   An inferior planet's reaching the same celestial longitude as the sun's.

inferior planet   A planet whose orbit around the sun is within the earth's, namely, Mercury and Venus.

inflationary universe   A model of the expanding universe involving a brief period of extremely rapid expansion.

infrared   Radiation beyond the red, about 7000 Å to 1 mm.

interference   The property of radiation, explainable by the wave theory, in which waves in phase can add (constructive interference) and waves out of phase can subtract (destructive interference); for light, this gives alternate light and dark bands.

interferometer   A device that uses the property of interference to measure such properties of objects as their positions or structure.

interferometry   Observations using an interferometer.

intergalactic medium   Material between galaxies in a cluster.

interior   The inside of an object.

International Date Line   A crooked imaginary line on the earth's surface, roughly corresponding to 180° longitude, at which, when crossed from east to west, the date jumps forward by one day.

interplanetary medium   Gas and dust between the planets.

interstellar medium   Gas and dust between the stars.

interstellar reddening   The relatively greater extinction of blue light by interstellar matter than of red light.

inverse-square law   Decreasing with the square of increasing distance.

ion   An atom that has lost one or more electrons. See also *negative hydrogen ion*.

ionized   Having lost one or more electrons.

ionosphere   The highest region of the earth's atmosphere.

ion tail   See *gas tail*.

IRAS   The Infrared Astronomical Satellite (1983).

iron meteorites   Meteorites with a high iron content (about 90%); most of the rest is nickel.

irons   Iron meteorites.

irregular cluster   A cluster of galaxies showing no symmetry.

irregular galaxy   A type of galaxy showing no shape or symmetry.

isotope   A form of chemical element with a specific number of neutrons.

isotropy   Being the same in all directions.

Jovian planet   Same as giant planet.

JPL   The Jet Propulsion Laboratory in Pasadena, California, funded by NASA and administered by Caltech; a major space contractor.

Julian calendar   The calendar with 365-day years and leap years every fourth year without exception; the predecessor to the Gregorian calendar.

Julian day   The number of days since noon on January 1, 4713 B. C. Used for keeping track of variable stars or other astronomical events. January 1, 2000, noon, will begin Julian day 2,451,545.

K line   The spectral line of ionized calcium at 3933 Å.

Keplerian   Following Kepler's law.

laser   An acronym for "light amplification by stimulated emission of radiation," a device by which certain energy levels are populated by more electrons than normal, resulting in an especially intense emission of light at a certain frequency when the electrons drop to a lower energy level.

latitude   Number of degrees north or south of the equator measured from the center of a coordinate system.

law of equal areas   Kepler's second law.

leap year   A year in which a 366th day is added.

lens   A device that focuses waves by refraction.

libration   The effect by which we can see slightly more than half the lunar surface even though the moon basically has one-half that always faces us; the back-and-forth motion of an object around one of the stable points in the three-body gravitational problem.

light   Electromagnetic radiation between about 3000 and 7000 Å.

lighthouse model   The explanation of a pulsar as a spinning neutron star whose beam we see as it comes around.

lightweight stars   Stars between about 0.07 and 4 solar masses.

light cone   The cone on a space-time diagram representing regions that can be in contact, given that nothing can travel faster than the speed of light.

light curve   The graph of the magnitude of an object vs. time.

light pollution   Excess light in the sky.

light year   The distance that light travels in a year.

limb   The edge of a star or planet.

limb darkening   The decreasing brightness of the disk of the sun or another star as one looks from the center of the disk closer and closer to the limb.

line profile   The graph of the intensity of radiation vs. wavelength for a spectral line.

lithosphere   The crust and upper mantle of a planet.

lobes   Of a radio source, the regions to the sides of the center from which high-energy particles are radiating.

local   In our region of the universe.

Local Group   The two dozen or so galaxies, including the Milky Way Galaxy, that form a subcluster.

local standard of rest   The system in which the average velocity of nearby stars is zero. We usually refer measurements of velocity of distant objects to the local standard of rest.

Local Supercluster   The supercluster of galaxies in which the Virgo Cluster, the Local Group, and other clusters reside.

logarithmic   A scale in which equal intervals stand for multiplying by ten or some other base, as opposed to linear, in which increases are additive.

longitude   The angular distance around a body measured along the equator from some particular point; for a point not on the equator, it is the angular distance along the equator to a great circle that passes through the poles and through the point.

long-period variables Mira variables.

look-back time The duration over which light from an object has been travelling to reach us.

luminosity The total amount of energy given off by an object per unit time.

luminosity class Different regions of the H-R diagram separating objects of the same spectral type: supergiants (I), bright giants (II), giants (III), subgiants (IV), dwarfs (V).

lunar eclipse The passage of the moon into the earth's shadow.

lunar occultation An occultation by the moon.

lunar soils Dust and other small fragments on the lunar surface.

Lyman alpha The spectral line (1216 Å) that corresponds to a transition between the two lowest major energy levels of a hydrogen atom.

Lyman-alpha forest The many Lyman-alpha lines, each differently Doppler-shifted, visible in the spectra of some quasars.

Lyman lines The spectral lines that correspond to transitions to or from the lowest major energy level of a hydrogen atom.

Magellanic Clouds Two small irregular galaxies, satellites of the Milky Way Galaxy, visible in the southern sky; the SMC may be split.

magnetic field lines Directions mapping out the direction of the force between magnetic poles; the packing of the lines shows the strength of the force.

magnetic lines of force See *magnetic field lines*.

magnetic mirror A situation in which magnetic lines of force meet in a way that they reflect charged particles.

magnetic monopole A single magnetic charge of only one polarity; may or may not exist.

magnetosphere A region of magnetic field around a planet.

magnification An apparent increase in angular size.

magnitude A factor of $\sqrt[5]{100} = 2.511886\ldots$ in brightness. See *absolute magnitude* and *apparent magnitude*. An *order of magnitude* is a power of ten.

main sequence A band on a Hertzsprung-Russell diagram in which stars fall during the main, hydrogen-burning phase of their lifetimes.

major axis The longest diameter of an ellipse; the line from one side of an ellipse to the other that passes through the foci. Also, the length of that line.

mantle The shell of rock separating the core of a differentiated planet from its thin surface crust.

mare (pl: maria) One of the smooth areas on the moon or on some of the other planets.

mascon A concentration of mass under the surface of the moon, discovered from its gravitational effect on spacecraft orbiting the moon.

maser An acronym for "**m**icrowave **a**mplification by **s**timulated **e**mission of **r**adiation," a device by which certain energy levels are more populated than normal, resulting in an especially intense emission of radio radiation at a certain frequency when the system drops to a lower energy level.

mass A measure of the inherent amount of matter in a body.

mass number The total number of protons and neutrons in a nucleus.

mass-luminosity relation A well-defined relation between the mass and luminosity for main-sequence stars.

Maunder minimum The period 1645–1715, when there were very few sunspots, and no periodicity, visible.

mean solar day A solar day for the "mean sun," which moves at a constant rate during the year.

meridian The great circle on the celestial sphere that passes through the celestial poles and the observer's zenith.

mesosphere A middle layer of the earth's atmosphere, where the temperature again rises above the stratosphere's decline.

Messier numbers Numbers of non-stellar objects in the 18th-century list of Charles Messier.

metal (a) For stellar abundances, any element higher in atomic number than 2, that is, heavier than helium. (b) In general, neutral matter that is a good conductor of electricity.

meteor A track of light in the sky from rock or dust burning up as it falls through the earth's atmosphere.

meteorite An interplanetary chunk of rock after it impacts on a planet or moon, especially on the earth.

meteoroid An interplanetary chunk of rock smaller than an asteroid.

micrometeorite A tiny meteorite. The micrometeorites that hit the earth's surface are sufficiently slowed down that they can reach the ground without being vaporized.

middleweight stars Stars between about 4 and 8 solar masses.

midnight sun The sun seen around the clock from locations sufficiently far north or south at the suitable season.

mid-Atlantic ridge The spreading of the sea floor in the middle of the Atlantic Ocean as upwelling material forces the plates to move apart.

Milky Way The band of light across the sky from the stars and gas in the plane of the Milky Way Galaxy.

minor axis The shortest diameter of an ellipse; the line from one side of an ellipse to the other that passes midway between the foci and is perpendicular to the major axis. Also, the length of that line.

minor planets Asteroids.

Mira variable A long-period variable star similar to Mira (omicron Ceti).

missing-mass problem The discrepancy between the mass visible and the mass derived from calculating the gravity acting on members of clusters of galaxies.

molecule Bound atoms that make the smallest collection that exhibits a certain set of chemical properties.

momentum A measure of the tendency that a moving body has to keep moving. The momentum in a given direction (the "linear momentum") is equal to the mass of the body times its component of velocity in that direction. See also *angular momentum*.

mountain ranges Sets of mountains on the earth, moon, etc.

naked singularity A singularity that is not surrounded by an event horizon and therefore kept from our view.

neap tides The tides when the gravitational pulls of sun and moon are perpendicular, making them relatively low.

nebular hypothesis The particular nebular theory for the formation of the solar system advanced by Laplace.

nebular theories The theories that the sun and the planets formed out of a cloud of gas and dust.

nebula (pl: nebulae) Interstellar regions of dust or gas.

negative hydrogen ion A hydrogen atom with an extra electron.

neutrino A spinning, neutral elementary particle with little or no rest mass, formed in certain radioactive decays.

neutron A massive, neutral elementary particle, one of the fundamental constituents of an atom.

neutron degeneracy A state in which, following rules of quantum mechanics, the further compression of neutrons generates a high pressure that balances gravity.

neutron star A star that has collapsed to the point where it is supported against gravity by neutron degeneracy.

New General Catalogue "A New General Catalogue of Nebulae and Clusters of Stars" by J. L. E. Dreyer, 1888.

new moon The phase when the side of the moon facing the earth is the side that is not illuminated by sunlight.

Newtonian (telescope) A reflecting telescope where the beam from the primary mirror is reflected by a flat secondary mirror to the side.

N galaxy A galaxy (probably elliptical) with a blue nucleus that dominates the galaxy's radiation. The emission lines are generally broader than those from Seyferts.

NGC New General Catalogue.

non-thermal radiation Radiation that cannot be characterized by a single number (the temperature). Normally, we derive this number from Planck's law, so that radiation that does not follow Planck's law is called non-thermal.

nova (pl: novae) A star that suddenly increases in brightness; an event in a binary system when matter from the giant component falls on the white dwarf component.

nuclear bulge   The central region of spiral galaxies.

nuclear burning   Nuclear fusion.

nuclear force   The strong force, one of the fundamental forces.

nuclear fusion   The amalgamation of lighter nuclei into heavier ones.

nucleosynthesis   The formation of the elements.

nucleus   (a) Of an atom, the core, which has a positive charge, contains most of the mass, and takes up only a small part of the volume; (b) of a comet, the chunks of matter, no more than a few km across, at the center of the head; (c) of a galaxy, the innermost region.

O and B association   A group of O and B stars close together.

objective   The principal lens or mirror of an optical system.

oblate   With equatorial greater than polar diameter.

occultation   The hiding of one astronomical body by another.

Olbers' paradox   The observation that the sky is dark at night contrasted to a simple argument that shows that the sky should be uniformly bright.

one atmosphere   The air pressure at the Earth's surface.

one year   The length of time the Earth takes to orbit the Sun.

Oort comet cloud   The trillions of incipient comets surrounding the solar system in a 50,000 A.U. sphere.

open cluster   A galactic cluster, a type of star cluster.

open universe   A big-bang cosmology in which the universe has infinite volume and will expand forever.

opposition   An object's having a celestial longitude 180° from that of the sun.

optical   In the visible part of the spectrum, 3900–6600 Å, or having to do with reflecting or refracting that radiation.

optical double   A pair of stars that appear extremely close together in the sky even though they are at different distances from us and are not physically linked.

organic   Containing carbon in its molecular structure.

Orion Molecular Cloud   The giant molecular cloud in Orion behind the Orion nebula, containing many young objects.

oscillating universe   The version of a closed universe in which our cycle is but one of many.

Ozma   One of two projects that searched nearby stars for radio signals from extraterrestrial civilizations.

ozone layer   A region in the earth's upper stratosphere and lower mesosphere where O_3 absorbs solar ultraviolet.

pancake model   A model of galaxy formation in which large flat structures exist and become clusters of galaxies.

paraboloid   A 3-dimensional surface formed by revolving a parabola around its axis.

parallax   (a) Trigonometric parallax, half the angle through which a star appears to be displaced when the earth moves from one side of the sun to the other (2 A.U.); it is inversely proportional to the distance. (b) Other ways of measuring distance, as in spectroscopic parallax.

parallel light   Light that is neither converging nor diverging.

parent molecules   Molecules in a comet that break up into daughter molecules.

parsec   The distance from which 1 A.U. subtends one second of arc. (Approximately 3.26 l-y).

particle physics   The study of elementary nuclear particles.

Pauli exclusion principle   The quantum-mechanical rule that certain types of elementary particles cannot exist in completely identical states.

peculiar velocity   The velocity of a star with respect to the local standard of rest.

penumbra   (a) For an eclipse, the part of the shadow from which the sun is only partially occulted; (b) of a sunspot, the outer region, not as dark as the umbra.

perfect cosmological principle   The assumption that on a large scale the universe is homogeneous and isotropic in space and unchanging in time.

periastron   The near point of the orbit of a body to the star around which it is orbiting.

perihelion   The near point to the sun of the orbit of a body orbiting the sun.

period   The interval over which something repeats.

phase   (a) Of a planet, the varying shape of the lighted part of a planet or moon as seen from some vantage point; (b) the relation of the variations of a set of waves.

photometry   The electronic measurement of the amount of light.

photomultiplier   An electronic device that through a series of internal stages multiplies a small current that is given off when light is incident on it; a large current results.

photon   A packet of energy that can be thought of as a particle travelling at the speed of light.

photon sphere   The sphere around a black hole, 3/2 the size of the event horizon, within which exit cones open and in which light can orbit.

photosphere   The region of a star from which most of its light is radiated.

plage   The part of a solar active region that appears bright when viewed in $H\alpha$.

Planck's constant   The constant of proportionality between the frequency of an electromagnetic wave and the energy of an equivalent photon. $E = h\nu = hc/\lambda$.

Planck's law   The formula that predicts, for gas at a certain temperature, how much radiation there is at every wavelength.

planet   A celestial body of substantial size (more than about 1000 km across), basically non-radiating and of insufficient mass for nuclear reactions ever to begin, ordinarily in orbit around a star.

planetary nebulae   Shells of matter ejected by low-mass stars after their main-sequence lifetime, ionized by ultraviolet radiation from the star's remaining core.

planetesimal   One of the small bodies into which the primeval solar nebula condensed and from which the planets formed.

plasma   An electrically neutral gas composed of approximately equal numbers of ions and electrons.

plates   Large flat structures making up a planet's crust.

plate tectonics   The theory of the earth's crust, explaining it as plates moving because of processes beneath.

plumes   Thin structures in the solar corona near the poles.

point objects   Objects in which no size is distinguishable.

polar axis   The axis of an equatorial telescope mounting that is parallel to the earth's axis of rotation.

pole star   A star approximately at a celestial pole; Polaris is now the pole star; there is no south pole star.

poor cluster   A cluster of stars or galaxies with few members.

Population I   The class of stars set up by Walter Baade to describe the younger stars typical of the spiral arms. These stars have relatively high abundances of metals.

Population II   The class of stars set up by Walter Baade to describe the older stars typical of the galactic halo. These stars have very low abundances of metals.

positive ion   An atom that has lost one or more electrons.

positron   An electron's antiparticle (charge of $+1$).

precession   The slowly changing position of stars in the sky resulting from variations in the orientation of the earth's axis.

precession of the equinoxes   The slow variation of the position of the equinoxes (intersections of the ecliptic and celestial equator) resulting from variations in the orientation of the earth's axis.

pressure   Force per unit area.

primary cosmic rays   The cosmic rays arriving at the top of the earth's atmosphere.

primary distance indicators   Ways of measuring distance directly, as in trigonometric parallax.

primeval solar nebula   The early stage of the solar system in nebular theories.

prime focus   The location at which the main lens or mirror of a telescope focuses an image without being reflected or refocused by another mirror or other optical element.

primordial background radiation   Isotropic millimeter and submillimeter radiation following a black-body curve for about 3 K; interpreted as a remnant of the big bang.

primum mobile   In Ptolemaic and Aristotelian theory, the outermost sphere around the earth, which gave its natural motion to inner spheres.

principal quantum number   The integer "*n*" that determines the main energy levels in an atom.

prolate   Having the diameter along the axis of rotation longer than the equatorial diameter.

prominence   Solar gas protruding over the limb, visible to the naked eye only at eclipses but also observed outside of eclipses by its emission-line spectrum.

proper motion   Angular motion across the sky with respect to a framework of galaxies or fixed stars.

proton   Elementary particle with positive charge 1, one of the fundamental constituents of an atom.

proton-proton chain   A set of nuclear reactions by which four hydrogen nuclei combine one after the other to form one helium nucleus, with a resulting release of energy.

protoplanets   The loose collections of particles from which the planets formed.

protosun   The sun in formation.

pulsar   A celestial object that gives off pulses of radio waves.

q_0 The deceleration parameter, a cosmological parameter that describes the rate at which the expansion of the universe is slowing up.

quantized   Divided into discrete parts.

quantum   A bundle of energy.

quantum mechanics   The branch of 20th-century physics that describes atoms and radiation.

quark   One of the subatomic particles of which modern theoreticians believe such elementary particles as protons and neutrons are composed. The various kinds of quarks have positive or negative charges of $\frac{1}{3}$ or $\frac{2}{3}$.

quasar   One of the very-large-redshift objects that are almost stellar (point-like) in appearance.

quiescent prominence   A long-lived and relatively stationary prominence.

quiet sun   The collection of solar phenomena that do not vary with the solar activity cycle.

radar   The acronym for **ra**dio **d**etection **a**nd **r**anging, an active rather than passive radio technique in which radio signals are transmitted and their reflections received and studied.

radial velocity   The velocity of an object along a line (the radius) joining the object and the observer; the component of velocity toward or away from the observer.

radiant   The point in the sky from which all the meteors in a meteor shower appear to be coming.

radiation   Electromagnetic radiation. Sometimes also particles such as alpha (helium nuclei) or beta (electrons).

radiation belts   Belts of charged particles surrounding planets.

radioactive   Having the property of spontaneously changing into another isotope or element.

radio galaxy   A galaxy that emits radio radiation orders of magnitude stronger than that from normal galaxies.

radio telescope   An antenna or set of antennas, often together with a focusing reflecting dish, that is used to detect radio radiation from space.

radio waves   Electromagnetic radiation with wavelengths longer than about one millimeter.

red giant   A post-main-sequence stage of the lifetime of a star; the star becomes relatively bright and cool.

reddened   See *reddening*.

reddening   The phenomenon by which the extinction of blue light by interstellar matter is greater than the extinction of red light so that the redder part of the continuous spectrum is relatively enhanced.

redshifted   When a spectrum is shifted to longer wavelengths.

red supergiant   Extremely bright, cool, and large stars; a post-main-sequence phase of evolution of stars of more than about 4 solar masses.

reflecting telescope   A type of telescope that uses a mirror or mirrors to form the primary image.

reflection nebula   Interstellar gas and dust that we see because it is reflecting light from a nearby star.

refracting telescope   A type of telescope in which the primary image is formed by a lens or lenses.

refraction   The bending of electromagnetic radiation as it passes from one medium to another or between parts of a medium that has varying properties.

refractory   Having a high melting point.

regolith   A planet's or moon's surface rock disintegrating into smaller particles.

regular cluster   A cluster of galaxies with spherical symmetry and a central concentration.

relativistic   Having a velocity that is such a large fraction of the speed of light that the special theory of relativity must be applied.

resolution   The ability of an optical system to distinguish detail.

rest mass   The mass an object would have if it were not moving with respect to the observer.

rest wavelength   The wavelength radiation would have if its emitter were not moving with respect to the observer.

retrograde motion   The apparent motion of the planets when they appear to move backwards (westward) with respect to the stars from the direction that they move ordinarily.

retrograde rotation   The rotation of a moon or planet opposite to the dominant direction in which the sun rotates and the planets orbit and rotate.

revolution   The orbiting of one body around another.

rich cluster   A cluster of many galaxies.

ridges   Raised surface features on the moon, apparently volcanic, perhaps having developed on the crust of lava lakes, as volcanic vents, or from faulting; Mercury also has ridges.

right ascension   Celestial longitude, measured eastward along the celestial equator in hours of time from the vernal equinox.

rilles   Sinuous depressions on the lunar surface, apparently volcanic, perhaps huge collapsed lava tubes or lava channels.

rims   The raised edges of craters.

Ritchey-Chrétien   A reflecting telescope design with the primary ground deeper than a paraboloid and a compensating secondary, to give a wide field.

Roche lobes   The figure-8 zone around binary stars in which the tidal forces lead to particles being able to flow freely without changing their total energy.

Roche's limit (Roche limit)   The sphere for each mass inside of which blobs of gas cannot agglomerate by gravitational interaction without being torn apart by tidal forces; normally about $2\frac{1}{2}$ times the radius of a planet.

rotation   Spin on an axis.

rotation curve   A graph of the speed of rotation vs. distance from the center of a rotating object like a galaxy.

RR Lyrae stars   A short-period "cluster" variable. All RR Lyrae stars have approximately equal absolute magnitude and so are used to determine distances.

S0   A transition type of galaxy between ellipticals and spirals; has a disk but no arms; all E7's are now known to be S0's.

scarps   Lines of cliffs; found on Mercury, earth, the moon, and Mars.

scattered   Light absorbed and then reemitted in all directions.

Schmidt camera   A telescope that uses a spherical mirror and a thin lens to provide photographs of a wide field.

Schwarzschild radius The radius that, according to Schwarzschild's solutions to Einstein's equations of the general theory of relativity, corresponds to the event horizon of a black hole.

scientific method No easy definition is possible, but it has to do with a way of testing and verifying hypotheses.

scientific notation Exponential notation.

scintillation A flickering of electromagnetic radiation caused by moving volumes of intermediary gas.

secondary cosmic rays High-energy particles generated in the earth's atmosphere by primary cosmic rays.

secondary distance indicators Ways of measuring distances that are calibrated by primary distance indicators.

sector Part of a circle bounded by an arc and two radii.

sedimentary Settled in a liquid, as for a type of rock.

seeing The steadiness of the earth's atmosphere as it affects the resolution that can be obtained in astronomical observations.

seismic waves Waves travelling through a solid planetary body from an earthquake or impact.

seismology The study of waves propagating through a body and the resulting deduction of the internal properties of the body. "Seismo-" comes from the Greek for earthquake.

semimajor axis Half the major axis, that is, for an ellipse, half the longest diameter.

Seyfert galaxy A type of spiral galaxy that has a bright nucleus and whose spectrum shows emission lines.

Shapley-Curtis debate The 1920 debate (and its written version) on the scale of our galaxy and of "spiral nebulae."

shear waves A type of twisting seismic waves.

shock wave A front marked by an abrupt change in pressure caused by an object moving faster than the speed of sound in the medium through which the object is travelling.

shooting stars Meteors.

showers A time of many meteors from a common cause.

sidereal With respect to the stars.

sidereal day A day with respect to the stars.

sidereal rotation period A rotation with respect to the stars.

sidereal time The hour angle of the vernal equinox; equal to the right ascension of objects on your meridian.

sidereal year A circuit of the sun with respect to the stars.

significant figure A digit in a number that is meaningful (within the accuracy of the data).

singularity A point in space where quantities become exactly zero or infinitely large; one is present in a black hole.

slit A long thin gap through which light is allowed to pass.

solar activity cycle The 11-year cycle with which solar activity like sunspots, flares, and prominences varies.

solar atmosphere The photosphere, chromosphere, and corona.

solar constant The total amount of energy that would hit each square centimeter of the top of the earth's atmosphere at the earth's average distance from the sun.

solar day A full rotation with respect to the sun.

solar dynamo The generation of sunspots by the interaction of convection, turbulence, differential rotation, and magnetic field.

solar flares An explosive release of energy on the sun.

solar rotation period The time for a complete rotation with respect to the sun.

solar time A system of time-keeping with respect to the sun such that the sun is overhead of a given location at noon.

solar wind An outflow of particles from the sun representing the expansion of the corona.

solar year (tropical year) An object's complete circuit of the sun; a tropical year is between vernal equinoxes.

solstice The point on the celestial sphere of northernmost or southernmost declination of the sun in the course of a year; colloquially, the time when the sun reaches that point.

space velocity The velocity of a star with respect to the sun.

spallation The break-up of heavy nuclei that undergo nuclear collisions.

special theory of relativity Einstein's 1905 theory of relative motion.

speckle interferometry A method obtaining higher resolution of an image by analysis of a rapid series of exposures that freeze atmospheric blurring.

spectral classes Spectral types.

spectral lines Wavelengths at which the intensity is abruptly different from intensity at neighboring wavelengths.

spectral type One of the categories O, B, A, F, G, K, M, C, S, into which stars can be classified from study of their spectral lines, or extensions of this system. The sequence of spectral types corresponds to a sequence of temperature.

spectrograph A device used to make and photograph a spectrum.

spectrometer A device to make and electronically measure a spectrum.

spectrophotometer A device to measure intensity at given wavelength bands.

spectroscope A device used to make and look at a spectrum.

spectroscopic binary A type of binary star that is known to have more than one component because of the changing Doppler shifts of the spectral lines that are observed.

spectroscopic parallax The distance to a star derived by comparing its apparent magnitude with its absolute magnitude deduced from study of its position on an H-R diagram (determined by observing its spectrum—spectral type and luminosity class).

spectroscopy The use of spectrum analysis.

spectrum A display of electromagnetic radiation spread out by wavelength or frequency.

speed of light By Einstein's special theory of relativity, the velocity at which all electromagnetic radiation travels, and the largest possible velocity of an object.

spicule A small jet of gas at the edge of the quiet sun, approximately 1000 km in diameter and 10,000 km high, with a lifetime of about 15 minutes.

spin-flip A change in the relative orientation of the spins of an electron and the nucleus it is orbiting.

spiral arms Bright regions looking like a pinwheel.

spiral galaxy A class of galaxy characterized by arms that appear as though they are unwinding like a pinwheel.

sporadic Not regularly.

sporadic meteor A meteor not associated with a shower.

spring tides The tides at their highest, when the earth, moon, and sun are in a line (from "to spring up").

stable Tending to remain in the same condition.

star A self-luminous ball of gas that shines or has shone because of nuclear reaction in its interior.

star clouds The regions of the Milky Way where the stars are so densely packed that they cannot be seen as separate.

star clusters Groupings of stars of common origin.

stationary limit In a rotating black hole, the location where space-time is flowing at the speed of light, making stationary particles that would be travelling at that speed.

steady-state theory The cosmological theory based on the perfect cosmological principle, in which the universe is unchanging over time.

Stefan-Boltzmann law The radiation law that states that the energy emitted by a black body varies with the fourth power of the temperature.

stellar atmosphere The outer layers of stars not completely hidden from our view.

stellar chromosphere The region above a photosphere that shows an increase in temperature.

stellar corona The outermost region of a star characterized by temperatures of 10^6 K and high ionization.

stellar evolution The changes of a star's properties with time.

stones A stony type of meteorite, including the chondrites.

stratosphere An upper layer of a planet's atmosphere, above the weather, where the temperature begins to increase. The earth's stratosphere is at 20–50 km.

streamers Coronal structures at low solar latitudes.

strong force The nuclear force, the strongest of the four fundamental forces of nature.

strong nuclear force The strong force.

S-type stars   A red giant showing strong ZrO instead of TiO.

subtend   The angle that an object appears to take up in your field of view; for example, the full moon subtends $\frac{1}{2}°$.

sunspot   A region of the solar surface that is dark and relatively cool; it has an extremely high magnetic field.

sunspot cycle   The 11-year cycle of variation of the number of sunspots visible on the sun.

superbolt   Giant lightning.

supercluster   A cluster of clusters of galaxies.

supergiant   A post-main-sequence phase of evolution of stars of more than about 4 solar masses. They fall in the upper right of the H-R diagram; luminosity class I.

supergranulation   Convection cells on the solar surface about 20,000 km across and vaguely polygonal in shape.

supergravity   A theory unifying the four fundamental forces.

superluminal velocity   An apparent velocity greater than that of light.

supermassive star   A stellar body of more than about 100 $M_{\odot}$.

supernova (pl: supernovae)   The explosion of a star with the resulting release of tremendous amounts of radiation.

supernova remnants   The gaseous remainder of the star destroyed in a supernova.

supernova-chain-reaction model   The explanation of spiral structure in terms of a chain of supernova explosions, each one leading to more than one more.

synchronous orbit   An orbit of the same period; a satellite in geosynchronous orbit has the same period as the earth's rotation and so appears to hover.

synchronous rotation   A rotation of the same period as an orbiting body.

synchrotron radiation   Nonthermal radiation emitted by electrons spiralling at relativistic velocities in a magnetic field.

synodic   Measured with respect to an alignment of 3 astronomical bodies.

synthetic-aperture radar   A radar mounted on a moving object, used with analysis taking the changing perspective into account to give the effect of a larger dish.

syzygy   An alignment of three celestial bodies.

T association   A grouping of several T Tauri stars, presumably formed out of the same interstellar cloud.

tail   Gas and dust left behind as a comet orbits sufficiently close to the sun, illuminated by sunlight.

tektites   Small glassy objects found scattered around the southern part of the southern hemisphere of the earth.

terminator   The line between night and day on a moon or planet; the edge of the part that is lighted by the sun.

terrestrial planets   Mercury, Venus, Earth, and Mars.

tertiary distance indicators   Ways to measure distance that are calibrated by secondary distance indicators.

thermal pressure   Pressure generated by the motion of particles that can be characterized by a temperature.

thermal radiation   Radiation whose distribution of intensity over wavelength can be characterized by a single number (the temperature). Black-body radiation, which follows Planck's law, is thermal radiation.

thermosphere   The uppermost layer of the atmosphere of the Earth and some other planets, the ionosphere, where absorption of high-energy radiation heats the gas.

three-color photometry   Measurements through U,B,V filters.

3° background radiation   The isotropic black-body radiation at 3 K, thought to be a remnant of the big bang.

tidal force   A force caused by the differential effect of the gravity from one body being greater on the near side of a second body than on the far side.

tidal theory   An explanation of solar-system formation in terms of matter being tidally drawn out of the sun by a passing star.

transit   The passage of one celestial body in front of another celestial body. When a planet is *in transit,* we understand that it is passing in front of the sun. Also, *transit* is the moment when a celestial

body crosses an observer's meridian, or the special type of telescope used to study such events.

transition zone   The thin region between a chromosphere and a corona.

transparency   Clarity of the sky.

transverse velocity   Velocity along the plane of the sky.

trigonometric parallax   See *parallax*.

triple-alpha process   A chain of fusion processes by which three helium nuclei (alpha particles) combine to form a carbon nucleus.

Trojan asteroids   A group of asteroids that precede or follow Jupiter in its orbit by 60°.

tropical year   The length of time between two successive vernal equinoxes.

troposphere   The lowest level of the atmosphere of the Earth and some other planets, in which all weather takes place.

T Tauri star   A type of irregularly varying star, like T Tauri, whose spectrum shows broad and very intense emission lines. T Tauri stars have presumably not yet reached the main sequence and are thus very young.

tuning-fork diagram   Hubble's arrangement of types of elliptical, spiral, and barred spiral galaxies.

21-cm line   The 1420-MHz line from neutral hydrogen's spin-flip.

Type I supernova   A supernova whose distribution in all types of galaxies, and the lack of hydrogen in its spectrum, make us think that it is an event in low-mass stars, probably resulting from the collapse and incineration of a white dwarf in a binary system.

Type II supernova   A supernova associated with spiral arms, and which has hydrogen in its spectrum, making us think that it is the explosion of a massive star.

UBV system   A system of photometry that uses three standard filters to define wavelength regions in the ultraviolet, blue, and green-yellow (visual) regions of the spectrum.

ultraviolet   The region of the spectrum 100–4000 Å, also used in the restricted sense of ultraviolet radiation that reaches the ground, namely, 3000–4000 Å.

umbra (pl: umbrae)   (a) Of a sunspot, the dark central region; (b) of an eclipse shadow, the part from which the sun cannot be seen at all.

uncertainty principle   Heisenberg's statement that the product of uncertainties of position and momentum is equal to Planck's constant. Consequently, both positon and momentum cannot be known to infinite accuracy.

universal gravitation constant   The constant G of Newton's law of gravity: force = Gm_1m_2/r^2.

uvby   A system of photometry that uses four standard filters to define wavelength regions in the ultraviolet, violet, blue, and yellow regions of the spectrum.

valleys   Depressions in the landscapes of solid objects.

Van Allen belts   Regions of high-energy particles trapped by the magnetic field of the earth.

variable star   A star whose brightness changes over time.

vernal equinox   The equinox crossed by the sun as it moves to northern declinations.

Very Large Array   The National Radio Astronomy Observatory's set of radio telescopes in New Mexico, used together for interferometry.

very-long-baseline interferometry   The technique using simultaneous measurements made with radio telescopes at widely separated locations to obtain extremely high resolution.

virial theorem   In the limited sense here, that half the gravitational energy of contraction goes into heating.

visible light   Light to which the eye is sensitive, 3900–6600 Å.

visual binary   A binary star that can be seen through a telescope to be double.

VLA   See *Very Large Array*.

VLBI   See *very-long-baseline interferometry*.

void   A giant region of the universe in which no galaxies are found.

volatile   Evaporating (changing to a gas) readily.

wavelength   The distance over which a wave goes through a complete oscillation.

wave front   A plane in which parallel waves are in step.

weak force   One of the four fundamental forces of nature, weaker than the strong force and the electromagnetic force. It is important only in the decay of certain elementary particles.

weight   The force of the gravitational pull on a mass.

white dwarf   The final stage of the evolution of a star of between 0.07 and 1.4 solar masses; a star supported by electron degeneracy. White dwarfs are found to the lower left of the main sequence of the H-R diagram.

white light   All the light of the visible spectrum together.

Wien's displacement law   The expression of the inverse relationship of the temperature of a black body and the wavelength of the peak of its emission.

winter solstice   For northern-hemisphere observers, the southernmost declination of the sun, and its date.

Wolf-Rayet star   A type of O star whose spectrum shows very broad emission lines.

W Virginis star   A Type II Cepheid, a fainter class of Cepheid variables characteristic of globular clusters.

x-rays   Electromagnetic radiation between 1–100 Å.

year   The period of revolution of a planet around its central star; more particularly, the earth's period of revolution around the sun.

Zeeman effect   The splitting of certain spectral lines in the presence of a magnetic field.

zenith   The point in the sky directly overhead an observer.

zero-age main sequence   The curve on an H-R diagram determined by the locations of stars at the time they begin nuclear fusion.

zodiac   The band of constellations through which the sun, moon, and planets move in the course of the year.

zodiacal light   A glow in the nighttime sky near the ecliptic from sunlight reflected by interplanetary dust.

zones   Bright bands in the clouds of a planet, notably Jupiter's.

Illustration Acknowledgments

STAR CHARTS—Wil Tirion

COLOR PLATES—**Cover and Plate 58** Photography by D. F. Malin of the Anglo-Australian Observatory. Original negative by U. K. Schmidt Telescope Unit, © 1981 Royal Observatory, Edinburgh; **Plates 1, 64, and 67** © National Optical Astronomy Observatory/Kitt Peak; **Plate 2** Courtesy of I. M. Kopylov, Special Astrophysical Observatory, U.S.S.R.; **Plates 3, 4, 10–12, and 25** Jay M. Pasachoff; **Plate 5** Deutsches Museum; **Plate 6** NASA and TRW Systems Group; **Plate 7** National Radio Astronomy Observatory; **Plate 8** Mario Grassi; **Plate 9** © 1972 Gary Ladd; **Plate 13** Dennis di Cicco photographs, Williams College Expedition; **Plates 14 and 15** Naval Research Laboratory/NASA. **Plate 16** NASA/JSC; **Plate 17** High Altitude Observatory/NASA; **Plates 18 and 19** Lewis House, Ernest Hildner, William Wagner, and Constance Sawyer/High Altitude Observatory, National Center for Atmospheric Research, National Science Foundation, and NASA; **Plates 20 and 21** Courtesy of Bruce E. Woodgate, Einar Tandberg-Hanssen, and colleagues at NASA's Marshall Space Flight Center; **Plate 22** Charles Eames; **Plate 23** Martin Grossmann; **Plates 24 and 26–28** NASA; **Plate 29** International Planetary Patrol; **Plates 30–32, 37–44, and 46** Jet Propulsion Laboratory/NASA, with the assistance of Jurrie van der Woude; **Plates 33 and 34** NASA; color-corrected version courtesy of Friedrich O. Huck; **Plate 35** Experiment and data—Massachusetts Institute of Technology; maps—U.S. Geological Survey; NASA/Ames spacecraft, courtesy of Gordon H. Pettengill; **Plate 36** Gustav Lamprecht; **Plate 45** Lunar and Planetary Laboratory, U. Arizona; **Plates 47, 57, and 70** © National Optical Astronomy Observatory/Cerro Tololo; **Plates 48–51, 54–56, 60–63, and 68** © The California Institute of Technology 1959, 1961, and 1965; **Plates 59, 69, and 71** Courtesy of the U.K. Schmidt Telescope Unit, Royal Observatory, Edinburgh, © 1980; **Plate 52** Courtesy of Philip E. Angerhofer, Richard A. Perley, Bruce Balick, and Douglas Milne with the VLA of NRAO; **Plate 53** Courtesy of S. S. Murray and colleagues at the Harvard-Smithsonian Center for Astrophysics; **Plate 65** Hans Vehrenberg; **Plate 66** U.S. Naval Observatory; **Plate 72** Westerbork data from Arnold H. Rots and William W. Shane; imaged at National Radio Astronomy Observatory; photograph © National Optical Astronomy Observatory/Kitt Peak; **Plate 73** Alan Stockton, Institute for Astronomy, U. Hawaii.

CHAPTER 1—**Opener** Lick Observatory Photograph; **Fig. 1–1** © David Scharf, 1977. All rights reserved; **Figs. 1–2 and 1–3** Jay M. Pasachoff; **Fig. 1–4** Skyviews Survey, Inc.; **Figs. 1–5, 1–6, 1–7, and 1–9** NASA; **Fig. 1–12** Harvard College Observatory; **Figs. 1–13, 1–14, and 1–15** Palomar Observatory photograph; **Fig. 1–16** © 1984 Mt. Wilson and Las Campanas Observatories, Carnegie Institution of Washington, Las Campanas du Pont reflector, observer: Sandage.

CHAPTER 2—**Opener** © 1973 NOAO/Kitt Peak; **Fig. 2–3** From "Ultraviolet Astronomy," Leo Goldberg, © 1969 by Scientific American, Inc. All rights reserved; **Fig. 2–6** Chris Jones, Union College; **Fig. 2–11** Yerkes Observatory; **Fig. 2–12** American Institute of Physics, Niels Bohr Library; **Figs. 2–14, 2–39A, 2–40, 2–41A, 2–51 and 2–54 to 2–58** Jay M. Pasachoff; **Fig. 2–19** © Royal Greenwich Observatory; **Figs. 2–21, 2–38, and 2–39C** Palomar Observatory Photograph; **Fig. 2–23** From the historical collection of the Mt. Wilson and Las Campanas Observatories, Carnegie Institution of Washington; **Fig. 2–24** Palomar Observatory—National Geographic Sky Survey. Reproduced by permission from the California Institute of Technology; **Fig. 2–25** California Institute of Technology, Pasadena, California; **Fig. 2–26** Courtesy of I. M. Kopylov, Special Astrophysical Observatory, U.S.S.R.; **Fig. 2–27** © 1972 NOAO/Kitt Peak; **Fig. 2–28** Arthur A. Hoag, NOAO/Cerro Tololo; **Fig. 2–29** Institute for Astronomy, U. Hawaii, photo by Duncan Chesley; **Fig. 2–30** Photo courtesy of the Fred Lawrence Whipple Observatory, a joint facility of U. Arizona and the Smithsonian Institution; **Fig. 2–31A** U. California, Lawrence Berkeley Laboratory; **Fig. 2–31B** U. Texas photo by Allan Mandel; **Fig. 2–31C** NOAO/ADP; **Fig. 2–32** NASA Marshall Space Flight Center; **Figs. 2–33, 2–44 and 2–45A** Perkin-Elmer; **Fig. 2–34** NASA; **Fig. 2–36** Produced from U.S. Air Force Defense Meteorological Satellite program (DMSP); **Fig. 2–39B** Harvard College Observatory; **Fig. 2–41B** © 1981 Anglo-Australian Telescope Board; **Fig. 2–43** © 1984 Anglo-Australian Telescope Board; **Fig. 2–45B** Gordon P. Garmire, Pennsylvania State U.; **Fig. 2–46** Giovanni Fazio, Harvard-Smithsonian Center for Astrophysics; **Fig. 2–47** *The Honolulu Advertiser,* photograph by Roy Ito; **Figs. 2–49 and 2–50** Courtesy of AT&T Bell Laboratories.

CHAPTER 3—**Opener** © 1977 Anglo-Australian Observatory; **Fig. 3–2** By permission of the Houghton Library, Harvard U.; **Fig. 3–3** Von del Chamberlain, *When Stars Came Down to Earth: Cosmology of the Skidi Pawnee Indians of North America.* Los Altos, CA, and College Park, MD, Ballena Press and Center for Archaeoastronomy 1983; **Figs. 3–4 and 3–16** Jay M. Pasachoff; **Figs. 3–11 and 3–12** © 1970 United Feature Syndicate, Inc.; **Fig. 3–13** Richard E. Hill from the site of the Case Western Reserve Universities Burrell Schmidt telescope on Kitt Peak; 3-hr exposure, f/5.6, Ektrachrome 400 film; **Fig. 3–14** Lick Observatory photograph; **Fig. 3–18** Emil Schulthess, Black Star; **Fig. 3–21** Mount Vernon Ladies Association of the Union.

PART II—**Opener** NOAO/Cerro Tololo

CHAPTER 4—**Opener** By permission of the Houghton Library, Harvard U.; **Fig. 4–1** From "The Little Prince" by Antoine de Saint-Exupéry, © 1943, by Harcourt Brace Jovanovich, Inc., in America and William Heinemann Ltd. in England; © 1971, by Consuelo de Saint-Exupéry. Reproduced by permission of the publishers; **Fig. 4–5** Palomar Observatory—National Geographic Society Sky Survey. Reproduced by permission from the California Institute of Technology; **Fig. 4–6** Bundesarchiv Koblenz; **Fig. 4–12** American Institute of Physics, Niels Bohr Library, Margrethe Bohr Collection; **Fig. 4–15** Harvard College Observatory; **Fig. 4–16** NOAO/Kitt Peak.

CHAPTER 5—Opener NOAO/Kitt Peak; **Fig. 5–2** Reproduced by special permission of *Playboy Magazine,* © 1971 by *Playboy;* **Fig. 5–8** Dorrit Hoffleit—Yale U. Observatory, courtesy of American Institute of Physics, Niels Bohr Library; **Fig. 5–9,** American Institute of Physics, Niels Bohr Library, Margaret Russell Edmondson Collection; **Fig. 5–10** After Bok and Bok, *The Milky Way,* courtesy Harvard University Press; **Fig. 5–11** Photography by Photolabs, Royal Observatory, Edinburgh. Original negative by U.K. Schmidt Telescope. © Royal Observatory, Edinburgh; **Fig. 5–16** Peter van de Kamp, Sproul Observatory.

CHAPTER 6—Opener Palomar Observatory Photograph; **Fig. 6–1** Mt. Wilson and Las Campanas Observatories, Carnegie Institution of Washington; **Figs. 6–2 and 6–20** Lick Observatory Photograph; **Fig. 6–3** Courtesy of Sproul Observatory; **Fig. 6–6** Peter van de Kamp; **Fig. 6–7** Based on data from D. L. Harris III, K. Aa. Strand, and C. E. Worley in *Basic Astronomical Data,* K. Aa. Strand, ed., University of Chicago Press, © 1963 by the University of Chicago; **Fig. 6–8** R. Hanbury Brown; **Fig. 6–9** Peter van de Kamp and Sarah Lee Lippincott, from *Vistas in Astronomy 19,* 231 (1975) courtesy Pergamon Press; **Fig. 6–10** Bernhard M. Haisch; **Fig. 6–11** American Association of Variable Star Observers/Janet Mattei; **Figs. 6–12, 6–16, and 6–19** After Bok and Bok, *The Milky Way,* courtesy Harvard University Press; **Fig. 6–14** Harvard College Observatory; **Fig. 6–15** Shigetsugu Fujinami; **Fig. 6–17** Williams College—Hopkins Observatory; **Fig. 6–18** Courtesy of the U. K. Schmidt Telescope Unit, Royal Observatory, Edinburgh; **Fig. 6–21** After Bok and Bok, *The Milky Way,* courtesy Harvard University Press, and a graph by Harold L. Johnson and Allan R. Sandage in the *Astrophysical Journal 124,* 379. Reprinted by permission of the University of Chicago Press, © 1956 by the American Astronomical Society.

CHAPTER 7—Opener and 7–21B William C. Livingston, NOAO/NSO; **Fig. 7–1** Anne Norcia; **Fig. 7–2** Courtesy of Instituto de Astrofisica de Canarias, Tenerife, and Kiepenheuer-Institute für Sonnenphysik, Freiburg; **Fig. 7–3A** Jorgen Christensen-Dalsgaard, NCAR, from: Proceedings from Conference on Pulsations in Classical and Cataclysmic Variables (J. P. Cox and C. J. Hansen, Ed., JILA, Boulder CO); **Fig. 7–3B** NOAO/NSO, courtesy of David H. Hathaway and Robin T. Stebbins; **Fig. 7–4** Mt. Wilson and Las Campanas Observatories, Carnegie Institution of Washington; **Figs. 7–5 and 7–20** NOAO/NSO; **Fig. 7–6** The Aerospace Corporation/David K. Lynch; **Fig. 7–7** M. Kanno, Hida Observatory, U. Kyoto; **Fig. 7–8** High Altitude Observatory/Richard R. Fisher, T. Baur, Lee Lacey, NCAR; **Figs. 7–9 and 7–28** Haleakala Observatory, Institute for Astronomy, U. Hawaii; **Fig. 7–10** Naval Research Laboratory *(interior)* and High Altitude Observatory *(exterior)*/NASA; **Fig. 7–11** American Science and Engineering, Inc./NASA; **Fig. 7–12** Naval Research Laboratory/NASA; **Fig. 7–15** Bryan Brewer, Earth View, Inc.; **Figs. 7–16, 7–17, and 7–19** Jay M. Pasachoff; **Fig. 7–18** Photograph by Daniel Fischer, Eclipse Staff, People's Observatory Bonn; **Fig. 7–21B** William C. Livingston, NOAO/Kitt Peak; **Fig. 7–22** A plot of Zurich sunspot numbers; updated from data provided by M. Waldmeier, Swiss Federal Observatory with values quoted in *Sky and Telescope;* **Figs. 7–23A and 7–23B** NASA; **Fig. 7–23C** Lewis House, Ernest Hildner, William Wagner, and Constance Sawyer, HAO/NCAR/NSF and NASA; **Figs. 7–26, 7–27, and 7–29** Courtesy of Harold Zirin, California Institute of Technology, Big Bear Solar Observatory photographs; **Fig. 7–30** Updated from Stephen H. Schneider and Clifford Mass, *Science 190,* 741, © 1975 by the American Association for the Advancement of Science; **Fig. 7–31** Palomar Observatory Photograph; **Fig. 7–32** Data from Richard Willson, JPL; photographs NOAO/NSO, courtesy of David H. Hathaway and Robin T. Stebbins; **Fig. 7–33** Courtesy of the Director of the Mt. Wilson and Las Campanas Observatories and of Otto Nathan; **Fig. 7–35** Lick Observatory Photograph; **Fig. 7–36** Courtesy of The Archives, California Institute of Technology; **Figs. 7–37 and 7–38** © 1919 by The New York Times Company. Reprinted by permission.

PART III—Opener Palomar Observatory Photograph

CHAPTER 8—Opener Bart Bok, labelling after R. D. Schwartz; **Fig. 8–2A** George H. Herbig, Lick Observatory; **Fig. 8–2B** From M. Cohen, J. H. Bieging, and P. R. Schwartz, *Astrophys. J. 253,* 707 (1982), courtesy of University of Chicago Press; **Fig. 8–2C** Courtesy of P. R. Schwartz, T. Simon, R. Campbell, and *Sky & Telescope;* **Fig. 8–3** George H. Herbig and B. F. Jones, *Astron. J. 86,* 1232 (1981); **Fig. 8–6** American Institute of Physics, Niels Bohr Library, Segrè collection; **Figs. 8–11, 8–12, and 8–13** Raymond Davis, Jr., Brookhaven National Laboratory.

CHAPTER 9—Opener © 1980 Anglo-Australian Telescope Board; **Fig. 9–1** After Richard L. Sears, *J. Royal Astron. Soc. Canada,* No. 1, Feb. 1974; originally from B. E. Paczyński, *Acta Astron. 20,* 47, 1970; **Fig. 9–3** Courtesy of R. R. Howell, C. Pilcher, and R. Hlivak, Institute for Astronomy, U. Hawaii; **Figs. 9–4 and 9–10** Palomar Observatory Photograph; **Fig. 9–7** Irving Lindenblad, U. S. Naval Observatory; **Fig. 9–8** Lick Observatory Photograph; **Fig. 9–9** © Ben Mayer, Los Angeles; **Fig. 9–11** Brian P. Flannery; **Figs. 9–14 and 9–15** NASA; **Figs. 9–16 and 9–17** Jay M. Pasachoff.

CHAPTER 10—Opener Lick Observatory Photograph; **Fig. 10–1** After Richard L. Sears, *J. Royal Astron. Soc. Canada,* No. 1, Feb. 1974; originally from B. E. Paczyński, *Acta Astron. 20,* 47, 1970; **Fig. 10–2** By permission of The Houghton Library, Harvard U.; **Fig. 10–3** Palomar Observatory Photograph; **Fig. 10–4** Paul Griboval, McDonald Observatory, U. Texas; **Fig. 10–5** William Miller and Museum of Northern Arizona; **Fig. 10–6A** Palomar Observatory Photograph, courtesy of Sidney van den Bergh; photograph by R. Minkowski; **Fig. 10–6B** Stuart Bowyer, U. California, Berkeley, and NASA; **Fig. 10–7A** Stephen S. Murray and colleagues, Harvard-Smithsonian Center for Astrophysics; **Fig. 10–7B** A. R. Thompson and colleagues, VLA/NRAO, courtesy of Robert M. Hjellming; **Fig. 10–8A** Jay M. Pasachoff; **Fig. 10–8B** L. N. Davhaev, V. M. Fedorov, Yu. A. Trubkin, Yu. N. Vavilov, Lebedev Physical Institute, courtesy of John Learned, U. Hawaii; **Fig. 10–10** Jocelyn Bell Burnell; **Fig. 10–11** Joseph H. Taylor, Jr., Marc Damashek, and Peter Backus, then U. Mass.—Amherst at the NRAO; **Fig. 10–12B** James Cordes and Arecibo Observatory (operated by Cornell U. under contract with the NSF); **Fig. 10–13** Joseph H. Taylor, Jr.; **Fig. 10–16** After Paul H. Serson; **Fig. 10–17** Lick Observatory Photograph; **Fig. 10–18A** H. Y. Chiu, R. Lynds, and S. P. Maran, photographed at NOAO/Kitt Peak; **Fig. 10–18B** F. R. Harnden, Jr., and colleagues, Harvard-Smithsonian Center for Astrophysics; **Marginal Note** Dan Stinebring, NRAO; **Fig. 10–20A** Joseph Weber, U. Maryland; **Fig. 10–20B** Peter Kramer; **Fig. 10–21** W. Priedhorsky, Los Alamos National Lab., and J. Petterson, New Mexico Institute of Mining and Technology; **Fig. 10–22** Bruce Margon, U. Washington; **Fig. 10–23** Courtesy of M. Watson, R. Willingale, Jonathan E. Grindlay, and Frederick D. Seward; *see Astrophys. J. 273,* 688 (1983), courtesy of University of Chicago Press; **Fig. 10–24** NRAO, operated by Associated Universities, Inc., under contract with the NSF.

CHAPTER 11—Opener and Fig. 11–9 Palomar Observatory Photograph/Jerome Kristian; **Fig. 11–5** Drawing by Chas. Addams; © 1974 The New Yorker Magazine, Inc.; **Fig. 11–6** After E. H. Harrison, U. Mass—Amherst; **Fig. 11–7** Chapin Library, Williams College; **Fig. 11–8A** Jean-Pierre Luminet, Groupe d'Astrophysique Relativiste, Observatoire de Paris; **Fig. 11–8B** John F. Hawley and Larry Smarr, U. Illinois at Urbana-Champaign; **Fig. 11–10** Lois Cohen—Griffith Observatory; **Fig. 11–11** Ricardo Giacconi and colleagues, then at Harvard-Smithsonian Center for Astrophysics.

PART V—Opener Reproduced by courtesy of The Trustees, The National Gallery, London

CHAPTER 12—Opener and Fig. 12–2 Palomar Observatory Photograph; **Fig. 12–1** Lick Observatory Photograph; **Fig. 12–4** Sgr A

West marked—William Liller, with the 4-m telescope of NOAO/Cerro Tololo; **Fig. 12–5** Dennis Downes, Max Planck Institut für Radioastronomie, Bonn, G.F.R.; **Figs. 12–6 and 12–7** NASA/JPL; **Fig. 12–8** Eric Becklin, Institute for Astronomy, U. Hawaii; **Fig. 12–9A** R. D. Ekers (NRAO) and U. J. Schwarz and W. M. Goss (Kapteyn Laboratories, the Netherlands) with the VLA of NRAO; **Fig. 12–9B** Mark Morris (U.C.L.A.) and Farhad Yusef-Zadeh and Don Chance (Columbia U.) with the VLA of NRAO; **Fig. 12–10** Lund Observatory, Sweden; **Fig. 12–11** Kent S. Wood, Naval Research Laboratory; **Fig. 12–12** Michael Watson, Paul Hertz, and colleagues, Harvard-Smithsonian Center for Astrophysics; **Fig. 12–13** Leon P. Van Speybroeck and colleagues, Harvard-Smithsonian Center for Astrophysics; **Fig. 12–14** H. A. Mayer-Hasselwander, K. Bennett, G. F. Bignami, R. Buccheri, N. D'Amico, W. Hermsen, G. Kanback, F. Lebrun, G. G. Lichti, J. L. Masnou, J. A. Paul, K. Pinkau, L. Scarsi, B. N. Swanenburg, and R. D. Wills; COS-B Observation of the Milky Way in High-Energy Gamma Rays, Ninth Texas Symposium on Relativistic Astrophysics, Eds. J. Ehlers, J. J. Perry, M. Walker, *Annals of the New York Academy of Sciences 336,* 211, 1980; courtesy of K. Pinkau; **Fig. 12–15** Data from W. Becker, Palomar Observatory Photograph; **Fig. 12–16** Agris Kalnajs, Mount Stromlo Observatory; **Fig. 12–17** Courtesy of Humberto Gerola and Philip E. Seiden, IBM Watson Research Center.

CHAPTER 13—Opener and Fig. 13–23 Jay M. Pasachoff; **Fig. 13–1B** David L. Talent, Abilene Christian U.; **Fig. 13–3** Palomar Observatory Photograph; **Fig. 13–7** Harvard U./E. M. Purcell; **Fig. 13–9A** Gart Westerhout, U. S. Naval Observatory; **Fig. 13–9B** Gerrit Verschuur; **Figs. 13–10 and 13–22** NRAO; **Figs. 13–12A and 13–13** Leo Blitz, U. Maryland; **Fig. 13–12B** B. J. Robinson, J. B. Whiteoak, R. N. Manchester, C.S.I.R.O., Australia, and W. H. McCutcheon, U. British Columbia; **Fig. 13–15** Bart J. Bok; **Fig. 13–16** Contours from Marc L. Kutner; Lick Observatory Photograph; **Fig. 13–18** Courtesy of Robert D. Gehrz, J. A. Hackwell, and Gary Grasdalen, Wyoming Infrared Observatory; **Figs. 13–19A and 13–20** NASA/JPL; **Fig. 13–19B** Meade Instruments Corp.; **Fig. 13–21** Graphic Films, Inc., Hollywood.

Part VI—Opener Lick Observatory Archives

CHAPTER 14—Opener and Figs. 14–1, 14–2, 14–5, 14–6B, 14–9, and 14–14 Palomar Observatory Photograph; **Fig. 14–3** Gerard de Vaucouleurs, U. Texas; **Fig. 14–4** Photography by Photolabs, Royal Observatory, Edinburgh, from original negatives by the U. K. 1.2-m Schmidt Telescope, © 1978 (14–4A), 1979 (14–4B) Royal Observatory, Edinburgh; **Fig. 14–6A** I. M. Kopylov, Special Astrophysical Observatory, U.S.S.R.; **Fig. 14–8** Alar Toomre and Juri Toomre; **Figs. 14–10 and 14–18** Hyron Spinrad, U. California at Berkeley; **Fig. 14–11** NOAO/Kitt Peak 4-m photograph by A. Oemler, courtesy of L. Thompson; **Fig. 14–12** Courtesy of Carlos S. Frenk and Simon White; **Fig. 14–13A** National Academy of Science; **Fig. 14–13B** From E. Hubble and M. L. Humason, *Astrophysical Journal 74,* 77, 1931, courtesy of University of Chicago Press; **Fig. 14–17** D. Schneider and J. Gunn; **Fig. 14–19** Radio image from the VLA of NRAO; optical images by Laird Thompson, U. Hawaii, *Astrophysical Journal 279,* L47, 1984, courtesy of University of Chicago Press; **Fig. 14–20A** 1974 NOAO/Kitt Peak; **Fig. 14–20B** Halton C. Arp, Palomar Observatory Photograph; **Fig. 14–21** Leftmost photograph from Palomar Observatory; **Fig. 14–27** NRAO, courtesy of John Lancaster and Robert M. Hjellming; **Fig. 14–28** Westerbork Synthesis Telescope, A. G. Willis, R. G. Strom, and A. S. Wilson, *Nature 250,* 625, © 1974 by Macmillan Journals Limited, London; **Fig. 14–29** Courtesy of Richard G. Strom, George K. Miley, Jan Oort, and *Scientific American;* **Fig. 14–30** Westerbork Synthesis Telescope, George K. Miley, G. C. Perola, P. C. van der Kruit, and H. van der Laan, *Nature 237,* 269, © 1972 by Macmillan Journals Limited, London; **Fig. 14–31** Courtesy of F. N. Owen, J. O. Burns, and L. Rudnick, NRAO.

CHAPTER 15—Opener and Fig. 15–2 Maarten Schmidt/Palomar Observatory photograph; **Figs. 15–1 and 15–3** Palomar Observatory photograph; **Fig. 15–5** Photos by Photolabs, Royal Observatory, Edinburgh. Original negative by UK Schmidt Telescope. © 1982 Royal Observatory, Edinburgh; **Fig. 15–6B** Thomas J. Balonek; **Fig. 15–8** Bruce Margon and E. A. Harlan, Lick Observatory photograph; **Fig. 15–9** Halton C. Arp/Palomar Observatory photograph; **Figs. 15–10 and 15–15A** Observations obtained by Susan Wyckoff, Peter Wehinger, and Thomas Gehren with the 3.6-m telescope of the European Southern Observatory; **Fig. 15–11** NOAO/Cerro Tololo; **Figs. 15–12, 15–17, and 15–26** Alan Stockton, Institute for Astronomy, U. Hawaii; **Fig. 15–13A** Palomar Observatory Photograph; montage by W. W. Morgan, Yerkes Observatory; **Fig. 15–13B** Princeton U. Observatory; **Fig. 15–14** R. M. West, A. C. Danks, and G. Alcaino, European Southern Observatory, *Astronomy and Astrophysics 62,* L113, 1978; **Fig. 15–15B** Anthony Tyson, courtesy of AT&T Bell Laboratories; **Fig. 15–16** Matthew Malkan, U. Arizona/Palomar Observatory photograph; **Fig. 15–18** John B. Hutchings and colleagues; Canada-France-Hawaii Telescope; **Fig. 15–19** Stephen Unwin, California Institute of Technology; **Fig. 15–21 and 15–22** Harvey Tanenbaum and colleagues, Harvard-Smithsonian Center for Astrophysics; **Fig. 15–24** B. F. Burke, P. E. Greenfield, D. H. Roberts, with the VLA of NRAO; **Fig. 15–25** Jerome Kristian, James A. Westphal, and Peter Young/Palomar Observatory photograph.

CHAPTER 16—Opener Lotte Jacobi; **Fig. 16–1B** From the collection of Mr. and Mrs. Paul Mellon; **Figs. 16–2 and 16–8** Jay M. Pasachoff; **Fig. 16–6** Jerome Kristian, Allan Sandage, and James Gunn; **Fig. 16–7** NASA/Johnson Space Center, courtesy of Mike Gentry; **Fig. 16–9** Wide World Photos; **Fig. 16–11** Courtesy of AT&T Bell Laboratories; **Fig. 16–14** D. Woody and P. L. Richards; **Fig. 16–15** Courtesy of George F. Smoot; **Fig. 16–16** NASA.

CHAPTER 17—Opener Palomar Observatory Photograph; **Fig. 17–1** James W. Cronin; **Figs. 17–2 and 17–4** Robert V. Wagoner, Stanford University; **Fig. 17–3** *Engineering and Science* magazine, California Institute of Technology; **Fig. 17–5** John B. Rogerson, Jr., and Donald H. York, Princeton U. Observatory; reprinted from the *Astrophysical Journal 186,* 195, 1973, with permission of the University of Chicago Press, © 1973 by the American Astronomical Society; **Fig. 17–6** Amos Yahil, Allan R. Sandage, and Gustav Tammann; **Fig. 17–7** X-ray image from Riccardo Giacconi and colleagues at the Harvard-Smithsonian Center for Astrophysics; optical image from Wallace Sargent and Charles Kowal, Palomar Observatory photograph; **Figs. 17–9 and 17–10** Photo CERN; **Fig. 17–11** Brookhaven National Laboratory; **Fig. 17–12** Alan Guth.

CHAPTER 18—Fig. 18–2 Lick Observatory Photograph; **Fig. 18–4** NASA; **Fig. 18–8** Allen Seltzer, American Museum—Hayden Planetarium, New York; **Figs. 18–9, 18–17, 18–25A&B, and 18–32** Chapin Library, Williams College; **Fig. 18–10** Burndy Library, photographed by Owen Gingerich; **Fig. 18–12** Joseph J. and Rose G. Mack; **Fig. 18–13** Courtesy of Owen Gingerich; **Fig. 18–14** Charles Eames; **Fig. 18–15** Henry E. Huntington Library and Art Gallery; **Fig. 18–16** Charles Eames, courtesy of Owen Gingerich; **Fig. 18–19** American Institute of Physics, Niels Bohr Library; **Fig. 18–20** Ewen Whitaker, Lunar and Planetary Laboratory, U. Arizona; **Fig. 18–21** U. Michigan Library, Dept. of Rare Books and Special Collections, translation by Stillman Drake, reprinted courtesy of *Scientific American;* **Fig. 18–22** New Mexico State U. Observatory; **Fig. 18–24** By permission of the Houghton Library, Harvard U.; **Fig. 18–26** Jay M. Pasachoff; **Fig. 18–33** By permission of Johnny Hart and News Group Chicago, Inc.; **Fig. 18–36** Photo courtesy of David Leonardi; **Fig. 18–38** Harold E. Edgerton, MIT; **Fig. 18–39** NASA.

CHAPTER 19—Opener Produced by Société Européenne de Propulsion (S.E.P., France), VISIR Laser Beam Recorder (Courtesy of Allied International Corporation, New York); **Fig. 19–1** from Raymond Siever, "The Earth," *Scientific American,* September 1975, and *The Solar System* (San Francisco: W. H. Freeman and Co., 1975), reprinted courtesy of *Scientific American;* **Fig. 19–2** Jay M. Pasachoff; **Fig. 19–4** U.S. Geological Survey, photograph by R. E. Wallace; **Fig. 19–3** NASA; **Fig. 19–5** Courtesy of Wilbur Rinehart, National Geophysical and Solar-Terrestrial Data Section, Environmental Data Service, National Oceanic and Atmospheric Administration; **Fig. 19–6***A* Courtesy of Stanley N. Williams, Dartmouth College, from *Science,* 13 June 1980 © the American Association for the Advancement of Science; **Fig. 19–6***B* Photograph by James Zollweg; **Fig. 19–6***C* Dale Cruikshank, U. Hawaii; **Fig. 19–7** © 1973 National Geographic Society; **Fig. 19–8** NFB Phototheque ONF (NFB-P-ONF) ©, photo by G. Blouin, 1949; **Fig. 19–14** L. A. Frank, U. Iowa.

CHAPTER 20—Opener NASA; **Figs. 20–1 to 20–3, 20–11** Lick Observatory photographs; **Figs. 20–5, 20–6***A,* **20–10, 20–14***A,* **20–15, and 20–17** NASA; **Fig. 20–6***B* Drawing by Alan Dunn, © 1971 The New Yorker Magazine, Inc.; **Figs. 20–7 and 20–8** Lunar Receiving Laboratory, Johnson Space Center, NASA; **Fig. 20–9** General Electric Research and Development Center; **Fig. 20–12** Data from Gerald J. Wasserburg and D. A. Papanastassiou, courtesy of Gerald J. Wasserburg, California Institute of Technology; **Fig. 20–13** Drawing by Donald E. Davis under the guidance of Don E. Wilhelms of the U. S. Geological Survey; **Fig. 20–14***B* Harold E. Edgerton, MIT; **Fig. 20–18** Judith W. Frondel, Harvard U.; **Fig. 20–19** NASA; **Fig. 20–20** NASA/JSC.

CHAPTER 21—Opener NASA; **Fig. 21–2** New Mexico State U. Observatory; **Figs. 21–5 to 21–11** NASA.

CHAPTER 22—Opener and Figs. 22–7, 22–9, and 22–11 NASA/Ames; **Fig. 22–1** Palomar Observatory photograph; **Fig. 22–2** Lick Observatory photograph; **Fig. 22–6** Courtesy of D. B. Campbell, Arecibo Observatory; **Fig. 22–8** Science Service; **Fig. 22–10** James W. Head, III, Brown U.; **Fig. 22–12** Courtesy of David J. Diner; **Figs 22–13 and 22–14** Courtesy of Valeriy Barsukov and Yuri Surkov; **Fig. 22–15** TASS; **Fig. 22–16** Institute for Space Research, U.S.S.R.; courtesy of the Planetary Society; **Fig. 22–17** NASA.

CHAPTER 23—Opener NASA/JPL; **Figs. 23–1 and 23–2** Lick Observatory Photographs; **Figs. 23–4 to 23–6, 23–8, 23–9, and 23–11 to 23–18** NASA/JPL; **Fig. 23–7** U.S. Geological Survey; **Fig. 23–10** Jay M. Pasachoff.

CHAPTER 24—Opener NASA/JPL; **Fig. 24–2** Lunar and Planetary Laboratory, U. Arizona; **Fig. 24–5** *Sky and Telescope;* **Fig. 24–6** Yerkes Observatory photograph; **Fig. 24–7** Palomar Observatory photograph/Charles T. Kowal; **Fig. 24–8** Jay M. Pasachoff; **Figs. 24–9 to 24–13, 24–15 to 24–18 and 20–19***B* to 20–26 NASA/JPL; **Fig. 24–19***A* Carl B. Pilcher, Jeffrey S. Morgan, and Jeanne H. Fertel, U. Hawaii, and Charles C. Avis, NASA/JPL.

CHAPTER 25—Opener NASA/JPL; **Fig. 25–2** Lowell Observatory; **Fig. 25–3** Stephen M. Larson, U. Arizona; **Fig. 25–4** NASA/Ames; **Figs. 25–5 to 25–28** NASA/JPL; obtained with the assistance of Jurrie van der Woude and Steven Edberg.

CHAPTER 26—Opener R. J. Terrile, JPL, and B. A. Smith, U. Arizona; **Fig. 26–1** Lunar and Planetary Laboratory, U. Arizona; **Fig. 26–3** Observations by J. L. Elliot, E. Dunham, and D. Mink; **Figs. 26–4 and 26–5** Data from J. L. Elliot, E. Dunham, L. H. Wasserman, R. L. Millis, and J. Churms; **Fig. 26–6** Master and Fellows of St. John's College of Cambridge; **Fig. 26–7** Courtesy of Charles T. Kowal; **Fig. 26–8** Images recorded by R. J. Terrile of JPL and B. A. Smith of U. Arizona at Carnegie Institution's Las Campanas Observatory; **Fig. 26–9** Lick Observatory photograph; **Fig. 26–10** James L. Elliot *et al.,* **Nature,** 10 Dec. 1981, © 1981 by Macmillan Journals Limited, London; **Fig. 26–11** Lowell Observatory photograph; **Fig. 26–12** Palomar Observatory photograph; **Fig. 26–13** U. S. Naval Observatory/James W. Christy, U. S. Navy photo; **Fig. 26–14** U. Arizona Observatories, courtesy of Harold J. Reitsema, Lunar and Planetary Laboratory.

CHAPTER 27—Opener Institute d'Astrophysique de Liège, Belgium with the Schmidt camera at the Observatoire de Haute Provence, France; **Fig. 27–1** William Liller; **Figs. 27–2, 27–5, and 27–11** Palomar Observatory photograph; **Figs. 27–3 and 27–6** NASA; **Fig. 27–4** European Southern Observatory/Jean Pierre Swings; **Fig. 27–7** Naval Research Laboratory, courtesy of Neil R. Sheeley, Jr.; **Fig. 27–8** Marian J. Warren, Hopkins Observatory, Williams College; **Fig. 27–9** National Portrait Gallery, London, painted by R. Phillips prior to 1721; **Fig. 27–12** *Sky and Telescope* by Roger Sinnott from orbital elements by Joseph Brady and Edna Carpenter, Lawrence Radiation Laboratory, U. California; **Fig. 27–10** Courtesy of Donald Yeomans, JPL; **Fig. 27–13** Smithsonian Astrophysical Observatory; **Fig. 27–14** Peter Bloomer, *Horizons West,* by permission of Meteor Crater Enterprises, Inc.; **Fig. 27–15** Arthur A. Griffin; **Fig. 27–16** Dan Haar, © 1982 Hartford Courant; **Fig. 27–17** Photo by Ursula B. Marvin; **Fig. 27–18** NASA; **Fig. 27–19** Harvard-Smithsonian Center for Astrophysics; **Fig. 27–22** *(left)* Paul D. Maley, International Occultation Timing Association; *(right)* Kevin Moody and Michael Verchota, U. Northern Colorado expedition, courtesy of Richard D. Dietz; **Fig. 27–23** NASA/JPL; **Fig. 27–24** Palomar Observatory photograph/E. Helin; **Fig. 27–25** Palomar Observatory photograph/Charles C. Kowal.

CHAPTER 28—Opener and Fig. 28–11 © Lucasfilm, Ltd. (LFL) 1980. All rights reserved. From the motion picture *The Empire Strikes Back,* courtesy of Lucasfilm, Ltd.; **Fig. 28–1** Cyril Ponnamperuma, U. Maryland; **Fig. 28–2** Sproul Observatory; **Fig. 28–3** Peter van de Kamp, Sproul Observatory; **Fig. 28–4** Calculations by W. N. Hubin; **Fig 28–5** J. G. Zeikus, U. Wisconsin, Madison; **Fig. 28–6** E. Imre Friedmann; **Fig. 28–7** From *The Day the Earth Stood Still.* © 1951 Twentieth Century Fox Film Corporation. All rights reserved; **Fig. 28–8***A* NASA; **Fig. 28–8***B* Jay M. Pasachoff; **Figs. 28–9 and 28–10** Cornell U. photographs; **Fig. 28–12** From ''Glinda of Oz,'' by L. Frank Baum, illustrated by John R. Neill, © 1920; **Fig. 28–13** R. J. Terrile, JPL, and B. A. Smith, U. Arizona, with Carnegie Institution's Las Campanas Observatory, Chile.

EPILOGUE—E–1 NASA; **E–2** © 1973 The Journal of Irreducible Results, Inc.

APPENDIX 10—Courtesy of The Houghton Library, Harvard U.

Index

References to illustrations, either photographs or drawings, are in italics. References to tables are followed by t. References to Color Plates are prefaced by CP. References to Appendices are prefaced by A.

Significant initial numbers are alphabetized as if they were spelled out in full. For example, 21 cm is alphabetized as *twenty-one cm*. Less important initial numbers and subscripts are ignored in alphabetizing. For example, 3C 273 appears at the beginning of the C's. M1 appears at the beginning of the M's. Greek letters are alphabetized under their English spellings.

A stars, A12, 80, *111*
A.U., *See* astronomical unit
Aaronson, Marc, 257
AAT, *See* Anglo–Australian Telescope
Abell 407, *260*
Abell, George, *194*
absolute magnitude, 92–94, 113
absolute zero, 70
absorption lines, 16, 73–74, *74*, 77, 126
  interstellar deuterium, 312
  quasars, 274
  solar, 126–127
  21–cm, 231–232
accelerator, *316*
accretion disk, *193*, 194–195, *194*, 204, *204*
achondrites, 458
active sun, *See* sun, active
ad hoc explanations, 386
Adams, John C., 442, *442*
Advanced X–Ray Astrophysics Facility (AXAF), 42, 221
albedo, *377*
  Deimos, 409
  Mercury, 377
Albireo, 105
Albrecht, Andreas, 317
Alcor, 108
Aldebaran, 81
Aldrin, Buzz, *363*, 370
Algol, 106, 173
Alice's Adventures in Wonderland, 204, *204*
Allende meteorite, 458
allowed state, 73
Almagest, 331
Alpha Centauri, 80
alpha particle, 158
Alpher, Ralph, 302, 309
alt–azimuth mount, 60–61, *60*
Amalthea, A4, 412–413, 418, 420
American Association of Variable Star Observers, 171
American Astronomical Society, 135, 244
amino acids, 468
ammonia, 235–236, 426, 431, 451, 468
Amor asteroids, 463
Andromeda Galaxy, Color essay, CP68, 180, 220, *220*, 242, *242*, 249, 254, 255, *264*; *See also* M31
Anglo–Australian Telescope (AAT), 29, 31t, *52*, *96*

angstrom, 14
Ångstrom, A.J., 14
angular measure, 19
angular momentum, 172, 343–344, *345*
anisotropy, 303, *303*
annular eclipse, *See* eclipses, annular
anorthosites, 365
Antarctic Circle, 61
Antares, CP58
Antennae, The, *255*
antimatter, 308
antiparticle, 308
aperture-synthesis techniques, 266–268, *267*, *268*
Apollo asteroids, 463
Apollo Program, 362–370, *362*, *364*, *370*
  Apollo 8, 363
  Apollo 11, CP27, CP28, *363*, *370*
  Apollo 14, *365*
  Apollo 15, *366*
  Apollo 16, *369*
  Apollo 17, CP24, CP25, CP26, *360*
  results from, 365–371
apparent magnitude, 87–89, *88*, 94, 113
archaebacteria, 470, *470*
Arctic Circle, 61
Arcturus, 81
Arecibo, 388, 473, *473*
argon, 394
Ariel, *438*
Aristarchus, 331, 332
Aristotle, 330, *330*, 331, *337*
Arizona, University of, *32*, 181, 189
armalcolite, *371*
Armstrong, Neil, *363*
Arp, Halton, 280–282
associations, *See* star clusters; O and B associations; T associations
asterisms, 54
asteroid belt, 460, 461
asteroids, 323, 460–464, *460*
  Amor, 463
  Apollo, 463
  Aten, 463, *463*
  composition, 463–464
  Trojan, 463
astrometric binaries, 108, *108*
astronomy, 90
Astron, 42, 218
Astronomer Royal, 442
Astronomical Almanac, 62

astronomical unit (A.U.), A2, 90, 340
Astronomy magazine, 375
Aten asteroids, 463, *463*
atmosphere, *See* Earth; individual planets; sun
atmosphere, one, 356
atomic number, 155
atoms, 72, 154–156
  allowed states, 76–78
  energy states, 73–74
  excited states, 74–75
  ground state, *73*, 75, 76
  ionized, 75
Augustus Caesar, 64
aurora, *358*
aurora australis, 139, 358
aurora borealis, CP36, 139, 358
Australian National Radio Observatory, CP4, *277*
AXAF, *See* Advanced X–Ray Astrophysics Facility

B stars, 80, 109, *111*, 116, 131, 152, 222, *239*, 253
B–V color index, A12
Baade, Walter, 180
background radiation, 269, 300–304, *303*
  origin of, 302
  temperature of, 302–303
Baily's beads, 132, *133*
Balmer, Johann, 77
Balmer series, *75*, 76, 77, 127
Barnard 5, Color essay
Barnard 335, *240*
Barnard, E.E., 100
Barnard's ring, *25*
Barnard's star, A7, 100–101, *100*, 468–469, *469*
Barringer meteor crater, 458, *458*
baryons, 308
basalts, 365, *365*
baseline, 90, 265
Bayer, Johann, 53, *68*, 178
beam, 46
beauty (quark), 315
Becklin, Eric, 241
Becklin–Neugebauer object, *240*, 241, *241*
Beckwith, Steven, 347
Begh, Ulugh, *332*
Bell Burnell, Jocelyn, 185
Bell Laboratories, 44
belts, *See* Saturn; Jupiter
β Cygni, 105

WINTER SKY

Facing North

Facing North

Facing South

Facing South

	TIME	D.S.T.
January 1	24 h	
January 15	23 h	
February 1	22 h	
February 15	21 h	
March 1	20 h	21 h
etc.		

Magnitudes: -1 0 1 2 3 4 (5)

◉ ○ Variable   ⊕ Globular Cluster   ⬭ Galaxy

○ Open Cluster   □ Nebula

MAP BY WIL TIRION
FOR JAY M. PASACHOFF

SPRING SKY

Facing North

Facing North

PUPPIS · HYDRA · SEXTANS · LEO · VIRGO · ZENITH 10°N · COMA BERENICES · ZENITH 20°N · LEO MINOR · ZENITH 30°N · CANES VENATICI · ZENITH 40°N · Arcturus · BOÖTES · SERPENS CAPUT · LIBRA · SCORPIUS · OPHIUCHUS · CORONA BOREALIS · M5 · Regulus · CANCER · ZENITH 50°N · URSA MAJOR · Big Dipper · M13 · HERCULES · SERPENS CAUDA · SCUTUM · Procyon · CANIS MINOR · Pollux · Castor · GEMINI · LYNX · Little Dipper · DRACO · URSA MINOR · Vega · LYRA · EAST · AQUILA · WEST · MONOCEROS · Polaris · HORIZON 10°N · HORIZON 30°N · VULPECULA · SAGITTA · ORION · M35 · AURIGA · Capella · CAMELOPARDALIS · CEPHEUS · HORIZON 30°N · CYGNUS · Deneb · TAURUS · Double · CASSIOPEIA · HORIZON 30°N · LACERTA · HORIZON 40°N · PERSEUS · Algol · HORIZON 50°N · ANDROMEDA · M31

NORTH

Facing South

Facing South

Vega · LYRA · Big Dipper · URSA MAJOR · ZENITH 50°N · LYNX · AURIGA · M35 · HERCULES · M13 · CORONA BOREALIS · BOÖTES · CANES VENATICI · ZENITH 40°N · LEO MINOR · Castor · Pollux · GEMINI · ORION · Arcturus · COMA BERENICES · ZENITH 30°N · ZENITH 20°N · ECLIPTIC · CANCER · AQUILA · SERPENS CAPUT · ZENITH 10°N · LEO · Regulus · Praesepe · CANIS MINOR · Procyon · OPHIUCHUS · M5 · VIRGO · SEXTANS · HYDRA · MONOCEROS · WEST · SERPENS CAUDA · EAST · SCUTUM · Spica · CRATER · LIBRA · CORVUS · HYDRA · PUPPIS · HORIZON 10°N · HORIZON 30°N · Antares · CENTAURUS · ANTLIA · HORIZON 30°N · CANIS MAJOR · SAGITTARIUS · SCORPIUS · LUPUS · Rigil Kent · ω · VELA · PYXIS · HORIZON 50°N · NORMA · CIRCINUS · CRUX · Mimosa · CARINA · VOLANS · ARA · MUSCA · TRIANGULUM AUSTRALE · CHA

SOUTH

	TIME	D.S.T.
April 1	24 h	01 h
April 15	23 h	24 h
May 1	22 h	23 h
May 15	21 h	22 h
June 1	20 h	21 h
etc.		

Magnitudes: -1 0 1 2 3 4 (5)

○ Open Cluster · □ Nebula · ⊙ Variable · ⊕ Globular Cluster · ◯ Galaxy

MAP BY WIL TIRION
FOR JAY M. PASACHOFF